Applied Calculus

Rodney D. Gentry
Department of Mathematics and Statistics
University of Guelph

HarperCollins*CustomBooks*

HarperCollins*CustomBooks* consists of products that are produced from camera-ready copy. Peer review, class testing, and accuracy are primarily the responsibility of the author(s).

You can order a copy at your local bookstore or call HarperCollins directly at 1-800-782-2665.

Manager of HarperCollins*CustomBooks*: Caralee Woods
Production Administrator: Nicole DuCharme
Cover Design: John Callahan

APPLIED CALCULUS

Copyright © 1995 by Rodney D. Gentry

All rights reserved. Printed in the United States of America. No part of this book may be used or reproduced in any manner whatsoever without written permission, except in the case of brief quotations embodied in critical articles and reviews. For information address HarperCollins Publishers, Inc., 10 East 53rd Street, New York, New York 10022.

HarperCollins ® and ® are registered trademarks of HarperCollins Publishers Inc.

ISBN: 0-673-99933-5

00 01 02 03 7 6 5 4

*Dedicated to Pat,
for
her perpetual support and encouragement.*

TABLE OF CONTENTS

Chapter 1   *Calculus Prerequisites*   1

1.1	Functions and Graphs	3
1.2	Polynomial functions	19
1.3	Elementary Functions	33
1.4	Exponential and Logarithm Functions	50
1.5	Trigonometric Functions	64
1.6	Form, Scale and Transformations	78
1.7	Limits and continuity	96

Chapter 2   *Discrete Models.*   105

2.1	An Introduction to Sequences and Series.	106
2.2	Discrete Dynamical Systems.	129
2.3	Graphing First-Order Dynamical Systems.	139
2.4	Linear First-Order Difference Equations.	151
2.5	Higher-Order Difference Equations.	164

Chapter 3   *Derivatives*   180

3.1	The Difference Quotient; the average rate of change over an interval	181
3.2	The Derivative.	189
3.3	Basic Derivative Rules	204
3.4	Derivatives of Elementary Functions	216
3.5	The Chain Rule	228
3.6	Higher Derivatives and Implicit Differentiation	240

Chapter 4   Applications of Derivatives.   254

- 4.1   Qualitative Aspects of a Function: Slope and Concavity.   255
- 4.2   Graphing with Calculus.   266
- 4.3   Local Extrema: Maximum and Minimum Values   274
- 4.4   Optimization with constraints   282
- 4.5   Optimal Resource Management   287
- 4.6   Optimization: Application in various disciplines.   296
- 4.7   Rolle's Theorem and the Mean Value Theorem   310
- 4.8   Differentials and Linear Approximations.   317
- 4.9   Taylor Polynomials and Taylor Series   324
- 4.10   Indeterminate limits; L'Hôpital's Rule   334

Chapter 5   Integration   343

- 5.1   Areas, averages and the definite integral.   345
- 5.2   The Definite Integral as a limit.   353
- 5.3   The Fundamental Theorem of Calculus and the Indefinite Integral.   371
- 5.4   Differentiating Integral Functions; The Second Fundamental Theorem of Calculus.   382
- 5.5   General Integral Rules.   390
- 5.6   The Area between curves.   398

Chapter 6   Methods of Integration   405

- 6.1   Integration by Substitution   406
- 6.2   Integrating Products: Integration by Parts   417
- 6.3   Integrating Rational Functions; The Partial Fractions Method.   425
- 6.4   Improper Integrals   434
- 6.5   Numerical Integration   444

Chapter 7 Applications Involving Integration — 455

7.1	Business and Economics	457
7.2	Volumes of Solids of Revolution	461
7.3	The Length of a Curve;	467
7.4	Probability and Statistics	474
7.5	Chemistry, Physics, and Hydrology	483
7.6	Physiology and Biology	495
7.7	Toxicology and Environmental Modeling	503

Chapter 8 Differential Equations — 509

8.1	First-order differential equations.	510
8.2	Stability and Equilibriums of Dynamical Systems.	520
8.3	Higher-order differential equations.	525

Chapter 9 Multiple Integration — 531

9.1	Iterated Integration.	531
9.2	The Double Integral.	537
9.3	Double Integrals Over General Regions: Determining the limits of integration.	544
9.4	Areas and Volumes as Multiple Integrals.	556
9.5	Applications Involving Multiple Integration:	564

Chapter 10 Partial Derivatives — 574

10.1	Introduction to Partial Derivatives.	575
10.2	Tangent Planes and Slopes of Surfaces.	583
10.3	Linear Approximations and the Differential of $F(x,y)$.	594
10.4	Linearization of Multivariable Functions; Stability Analysis of Dynamical Systems.	607

Chapter 11 Directional Derivatives — 621

11.1	Vectors and Vector valued-functions: Parametric Curves.	622
11.2	Derivatives of Vector Functions: Tangent Vectors. 630	
11.3	Directional Derivatives. 635	
11.4	Level Curves and Curves of Steepest Ascent or Descent	644

Chapter 12 Applications Using Higher-Order Partial Derivatives. — 652

12.1	Taylor Polynomials In Two-Dimensions: Nonlinear Approximations of F(x,y) 653	
12.2	Maximum and Minimum of Multivariable functions	662
12.3	Optimization with Constraints. 673	

Answers to Selected Exercises — 684

Chapter 1 *Calculus Prerequisites*

Section 1.1   Functions and Graphs 3
Function notation.
The algebra of functions.
Inverse functions.
Coordinate systems and graphs.
Shifted graphs.

Section 1.2   Polynomial functions 19
Power functions: $y = x^n$.
Polynomials

Section 1.3   Elementary Functions 33
The absolute value function.
Allometric functions: $y = x^r$
Discrete functions; $y = [\![x]\!]$.

Section 1.4   Exponential and Logarithm Functions ... 50
Exponential functions.
Logarithmic functions

Section 1.5   Trigonometric Functions 64
Periodic functions.
The circular functions: $\sin(t)$ and $\cos(t)$.
Trigonometric functions.

Section 1.6   Form, Scale and Transformations 78
Scaling of Coordinate Axes.
Log-log graphs.
Semi-log graphs.
Reciprocal scales.

Section 1.7   Limits and continuity 96
Limits of functions.
Limits and algebra.
Continuous functions.
The Extreme Value Theorem.
The Intermediate Value Theorem.

Introduction.

In this chapter a brief review is presented of the fundamental functions and algebraic properties that are used in Calculus. It is assumed that the reader is familiar with the elementary functions, polynomials, the exponential e^x, logarithms $\log(x)$ and $ln(x)$, and the basic trigonometric functions. When required, it is also assumed that these functions can be evaluated using a "scientific" calculator; no attempt has been made to incorporate problems for graphing calculators.

Some of the concepts may be introduced differently than the way you encountered them in high school. Perhaps the most obvious examples will be the emphasis placed on the point-slope equation of a line instead of the slope-intercept equation that many high school teachers insist on. Also, the $\sin(x)$ and $\cos(x)$ functions are introduced as circular functions.

The form of the graphs of specific elementary functions will be emphasized. Later, we will consider graph sketching using the tools of *calculus*. The objective now is for you to be able to know the shape of elementary graphs and to identify key points on a function's graph so that it may be roughly sketched without actually evaluating and plotting many points. Sketching the graphs of "shifted" or "scaled" functions will also be emphasized. This is a useful topic that is not always covered in high school courses.

In Section 1.6 the notion of scaling and transforming coordinate-axes is introduced. This section is important since most students now rely on computer programs to generate graphs for them. All computer graphing algorithms automatically scale graphs to fit on monitor screens or to have specific dimensions. Hence, it is important to understand the relationship between what you see and what the function actually is. In this section log-log and semi-log graphing is also introduced. Again, while computer programs now generate such graphs automatically, it is important that the user understand what the implications are of linear graphs on these systems. This material is essential for any scientist and is not usually covered in High School curriculums.

The last section of this Chapter provides a brief introduction to the notation of limits and properties of continuous functions. Again, it is expected that the reader has some familiarity with limits. The classical ε-δ definition of a limit is presented for completeness. However, proving limits exist has not been emphasized. Instead, we simply state that the elementary functions are continuous and that consequently most limits can be evaluated by simply plugging in the limiting x-value.

CAUTION   If you experience difficulties with any of these topics we encourage you to seek additional help immediately. The algebra and limit concepts introduced in this chapter will be routinely used in later chapters.

Section 1.1   Functions and Graphs

Function Notation

A real **function** is a rule that uniquely associates certain real numbers in a set called a **Domain** with numbers in a second set called a **Range**. Functions are referred to by letters and can be represented in various ways. The simplest functions can be indicated by sets of ordered pairs of numbers, such as the set

$$f = \{(3,1), (4,2), (5,3)\}$$

This describes a function with the name f whose **Domain** consists of the numbers occurring in the first component of the order pairs and whose **Range** consists of the numbers occurring in the second component:

$$\text{Domain}(f) = \{3, 4, 5\} \quad \text{and} \quad \text{Range}(f) = \{1, 2, 3\}$$

Sometimes the Domain and Range are abbreviated using simply D or R and a subscript to indicated the function name, e.g., D_f or R_f. This function f can be described using "function notation" by writing

$$f(3) = 1, \quad f(4) = 2, \quad \text{and} \quad f(5) = 3$$

What makes f a function is the fact that there is only one ordered pair for each number in its domain. If a set has two ordered pairs with the same first component and different second components then it is not a function. For instance, the set

$$g = \{(3,1), (3,2), (5,3)\}$$

is <u>not</u> a function because of the two pairs (3,1) and (3,2). In function notation this means that g(3) is not uniquely defined, g(3) = 1 and g(3) = 2. However, the second component can be repeated. The set

$$h = \{(3,6), (4,6), (5,3)\}$$

is a function, even though it has two pairs with 6 in the second component. Such functions are called *finite* since their Domains and Ranges consist of a finite sets of numbers. They are defined only for values in their Domains. If a number is not in the Domain of a function then the function can not be evaluated at the number. For instance, for the above function f, the value f(2) is not known as it simply has not been defined.

Functions are often indicated by explicit equations of the form y = f(x), which is read "y equals f of x". Such an equation indicates the name of the function is f, the **domain variable**, which is called the *independent variable*, is represented by x, and y represents the **range variable** and is called the *dependent variable*. An x-value is a number, called an *argument*, at which a function is evaluated to give a y-value. To indicate the paired association of x-values and y-values we shall use subscripted ordered pairs of the form (x_0,y_0) or (x_1,y_1) which indicate that $y_0 = f(x_0)$ and $y_1 = f(x_1)$, etc. The corresponding description of the function in set notation is

$$f = \{ (x,y) \mid y = f(x) \}$$

Explicit equations often do not indicate the name of the function and simply indicated the algebraic relationship between the independent variable appearing on the right side and the dependent variable which is solved for on the left side, e.g. $y = x^2 - 1$ or $A = \sin(\pi t)$. The most common symbols for the dependent variable are x and t, while y is the most common choice for the independent variable. Any letter can be used to represent a function, but the most commonly used letters are f and g. The function f(x) is repeatedly used to represent a generic function, changing in meaning each time a different algebraic form for f is specified, such as $f(x) = 3x$ or $f(x) = x^3$. In modeling applications, to help associate the mathematical concepts with the real phenomena that they represent, function names are often multi-letter acronyms that correspond to the names or descriptions of the events or physical quantities that they are describing numerically, e.g., FP for fluid pressure or Tmax for maximum temperature.

Example 1 **A function model.**

Problem The initial velocity of a ball dropped from a bridge is given as a function of time and a constant parameter g that represents the force of gravity. The function is the product of g and the square of the time. Express this function using the value $g = 9.8$ m s^{-2} and determine when the ball will be traveling at a speed of 50 meters per second.

Solution The independent variable is time, which is denoted by t. The dependent variables are the speed, denoted by s(t), and velocity, denoted by v(t). The speed is the absolute value of the velocity, it is always positive while velocity may be positive or negative and is associated with movement away from or towards a reference position. In this model they are the same since the velocity function

$$v(t) = 9.8t^2 \text{ m s}^{-1} \text{ (meters per second)}$$

is always non-negative. To find the time when s(t) = 50, we solve

$$50 = s(t) = |v(t)| = 9.8\, t^2$$

for t. The solution is[*] t = SQRT(50/9.8) ≈ 2.3 s. This model equation is only valid for a limited time range. It does not recognize that the ball may hit the surface below or that wind resistance will limit the velocity.

The **Natural Domain** of a function y = f(x) is the set of x-values for which the algebraic expression f(x) is defined.

$$D_f = \{\ x\ |\ f(x) \text{ exists}\ \}$$

Often the Natural Domain is simply referred to as **the Domain**.

Normally, the Natural Domain is found by determining the x-values for which the function is not defined and excluding these. The two obvious situations in which an expression is not defined are when its evaluation results in division by zero or the square root of a negative number. Other undefined expressions that are discussed in later sections of this chapter include logarithms of negative numbers and certain powers of negative numbers, like $(-2)^{0.25}$.

---

[*] The function name SQRT is commonly used to denote the square root function in computer programs like Lotus 1-2-3 or Quattro, and in symbolic mathematical programs like Maple, Mathematica, or DERIVE.

Alternative notations for SQRT($x^2 - 3x + 5$) include $(x^2 - 3x + 5)^{1/2}$, using a fractional exponent, or $\sqrt{x^2 - 3x + 5}$, which requires more sophisticated type setting. Sometimes the Root Symbol √ is used without the over bar extension as in √($x^2 - 3x + 5$).

Section 1.1   FUNCTIONS AND GRAPHS

Example 2 **Determining Natural Domains.**

Problem   Find the Natural Domain of the indicated function:

a) $f(x) = 0.5x^2/(1 - x)$ b) $g(x) = \text{SQRT}(100 - x^2)$

Solution   a) The function $f(x) = 0.5x^2/(1 - x)$ is not defined when its denominator is zero. Solving $1 - x = 0$ gives $x = 1$. Thus the

$$\text{Natural Domain of } 0.5x^2/(1 - x) = \{x \mid x \neq 1\} = (-\infty, 1) \cup (1, \infty)$$

This domain has been expressed both in set form and as intervals. (See the Appendix for a review of interval notation.) Sometimes, for simplicity, domains are indicated by equations or inequalities, without using the set notation. For instance, by writing

"The Natural Domain of f is $x \neq 1$."

b) To determine the Natural Domain of g we determine when the expression under the square root is positive or zero. The inequality $100 - x^2 \geq 0$ is rearranged to $100 \geq x^2$. Taking the square root gives the bounds for x:

$$-10 \leq x \leq 10$$

Thus, the Natural Domain of g is

$$D_g = \text{SQRT}(100 - x^2) = \{x \mid |x| \leq 10\} = [-10, 10]$$

☑

Often, when describing or modeling real phenomena with functions, it may be necessary to restrict or limit the domain values to an acceptable or realistic set that is smaller than the Natural Domain. This is done by specifying *constraints* on the independent variable in the form of a specific interval or by inequalities. For instance, if l denotes the length of a fish the variable l could not be negative. This would be indicated by the inequality $l \geq 0$. A formula relating the mass, M,[†] of a fish to its length, which might only be valid for small fish, say for l less than 0.2 m in length. To express the formula and the constraint we would write

$$M = M(l) = 1.89\, l^2 \quad ; \quad 0 < l \leq 0.2$$

The *Constrained Domain* of this mass function M is then the interval (0, 0.2]. (We intentionally omitted the value $l = 0$ since in this case the fish does not exist.)

> The **Range** of a function $y = f(x)$ consist of all the possible y-values corresponding to all x-values in the function's domain.
>
> $$R_f = \{y \mid y = f(x) \text{ for some } x \in D_f\}$$

[†] The common scientific way to designate a variable to represent a modeled quantity, such as mass, is to write the name of the quantity followed by the corresponding variable symbol enclosed in commas.

The domain used to establish a function's Range is its Natural Domain unless a Constrained Domain has been specified. Ranges can be more difficult to ascertain than domains. For some functions a direct algebraic approach can be used.

Example 3 **Algebraically determining a function's Range.**

Problem Determine the range of the fish mass function $M = M(l) = 1.89\, l^2$ for $l \in (0, 0.2]$.

Solution The process is to start with a variable in the Domain that is restricted by inequalities and to algebraically manipulate it (with the inequalities) to form the function, there by establishing inequalities for the range variable.

If l is in the Domain interval, then $\quad 0 < l \leq 0.2$

and squaring each term gives, $\quad 0 < l^2 \leq 0.04$

Multiplying by the coefficient 1.89 gives $\quad 0 < 1.89 l^2 \leq 1.89 \cdot 0.04 = 0.0756$

Thus, the range of the constrained function $M(l)$ is

$$R_M = (0, 0.0756]$$

In a modeling application l would have units of meters, the constant would have conversion units of grams per meter squared, $g\, m^{-2}$, and then M would have units of grams. Thus, the range of M is 0 to 0.0756 g.

Units are important, but are often not included in simple mathematical discussions that focus on the properties of a function. We will normally not indicate the units in calculations and intermediate algebra steps, but when appropriate they will be indicated in the final statement of a solution.

In the above example the range was determined by manipulating the inequalities that restrict the independent variable to the range to arrive at an inequality for the function value, i.e. the range variable. An algebraic method for determining a function's range is to solve the function equation $y = f(x)$ explicitly for x in terms of y. The range will be the set of y-values for which the solved equation is defined. This method works only when you can actually solve for the dependent variable, which may prove difficult or impossible for even relatively simple functions.

Example 4 **Determining a Range by solving for the independent variable.**

Problem Determine the Range of $y = f(x) = (x - 2)^2 + 3$.

Solution The Natural Domain of f is the entire real line $(-\infty, \infty)$. Solving for x gives the equation

$$x = (y - 3)^{1/2} + 2$$

The square root function is only defined for non-negative values, i.e., for $y - 3 \geq 0$. Thus:

$$\text{Range}(f) = [3, \infty)$$

Some further examples of functions and their Natural Domains and corresponding Ranges are indicated in Table 1.1.

Table 1.1

Equation	Dependent variable	Natural Domain	Independent variable	Range
$y = \text{SQRT}(3x - 2)$	x	$[2/3, \infty)$	y	$[0, \infty)$
$R = t^3$	t	$(-\infty, \infty)$	R	$(-\infty, \infty)$
$f(x) = x - x^2$	x	$(-\infty, \infty)$	not indicated	$(-\infty, 1/4]$
$u = 1/(2 + x)$	x	$x \neq 2$ or $(-\infty, -2) \cup (-2, \infty)$	u	$u \neq 0$ or $(-\infty, 0) \cup (0, \infty)$

The algebra of functions.

Functions can be combined algebraically using the normal arithmetic operations of +, -, ·, and / (which is used to denote division; the symbol ÷ is rarely used.) Two functions that are defined on the same domain can be added, subtracted, multiplied, or divided (in most cases). The resulting functions are defined *point-wise*, that is, by evaluating the individual functions at a particular x-value and then adding, subtracting, multiplying or dividing these numerical values. The **Natural Domain** of the algebraic combination of two functions is the intersection of the domains of the functions. The only *caveat* is that in the case of division, the denominator can not be zero and thus such occurrences must be excluded.

Example 5   **Algebraic Combinations of Functions.**

Problem   Determine the simplified form and Natural Domain of the algebraic combinations of

$$f(x) = 1/x^2 \quad \text{and} \quad g(x) = \text{SQRT}(4 - x^2)$$

Evaluate each combination at $x = 1$.

Solution   The function f is defined for $x \neq 0$ while $g(x)$ is only defined when $-2 \leq x \leq 2$. Except in the division case, the combined functions will be defined on the

$$\text{Natural Domain} = D_f \cap D_g = \{\, x \mid x \neq 0 \text{ and } -2 \leq x \leq 2 \,\} = [-2, 0) \cup (0, 2]$$

The domains of the two quotients are restricted by the condition that the denominator can not equal zero. Since $f(x) \neq 0$ for any x, the Natural Domain of g/f is also the two intervals $[-2, 0)$ and $(0, 2]$. However, as $g(x) = 0$ when $x = 2$ or -2, the

$$D_{f/g} = (-2, 0) \cup (0, 2)$$

As addition and multiplication of numbers are commutative operations, so is the addition and multiplication of functions,

$$f + g = g + f \quad \text{and} \quad f \cdot g = g \cdot f$$

The six different algebraic combinations of these functions that can be formed are as follows:

SUM: $(f + g)(x) = f(x) + g(x) = 1/x^2 + \text{SQRT}(4 - x^2)$

DIFFERENCES: $(f - g)(x) = f(x) - g(x) = 1/x^2 - \sqrt{4 - x^2}$

$(g - f)(x) = g(x) - f(x) = \sqrt{4 - x^2} - 1/x^2$

PRODUCT: $(f \cdot g)(x) = f(x) \cdot g(x) = 1/x^2 \cdot \sqrt{4 - x^2} = \sqrt{4 - x^2} / x^2$

QUOTIENTS: $(f/g)(x) = f(x) / g(x) = (1/x^2)/\sqrt{4 - x^2} = 1 / [x^2 \sqrt{4 - x^2}]$

$(g/f)(x) = g(x) / f(x) = \sqrt{4 - x^2} / (1/x^2) = x^2 \sqrt{4 - x^2}$

Notice that although the quotient g/f simplifies to a function defined at x = 0, the Natural Domain of g/f is still limited to the two intervals [-2,0) and (0,2] and does not include the value x = 0. This is because the quotient g/f is evaluated by first evaluating g and f and then performing the division; as f(0) is not defined, neither is g(0)/f(0) and hence the algebraic simplification is not legitimate for x = 0. At x = 1, these functions evaluate as:

$(f+g)(1) = 1 + \sqrt{3} \approx 2.73$ $(f \cdot g)(1) = \sqrt{3} \approx 1.73$ $(f-g)(1) = 1 - \sqrt{3} \approx -0.73$

$(g-f)(1) = \sqrt{3} - 1 \approx 0.73$ $(f/g)(1) = 1/\sqrt{3} \approx 0.58$ $(g/f)(1) = \sqrt{3} \approx 1.73$

☑

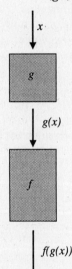

The composition of g with f is a sequential process.

Composition of Functions.

Another way to combine functions is sequentially, evaluating one function and then evaluating a second function using the first function's value as the argument of the second function. The mathematical name for this process is ***the composition*** of the functions. The symbol used to denote a composition is a small centered circle, ∘, which can sometimes look like a multiplication dot, ·, when it is poorly printed. For this reason compositions are frequently expressed as "nested" functions using multiple parentheses.

The composition of g and f, denoted by **f∘g** is defined by

$$f \circ g(x) \equiv f(g(x))$$

which is read as "f of g of x".

Composition of functions is <u>not</u> commutative. f∘g ≠ g∘f. The composition f∘g is defined on a subset of the domain of g:

Natural Domain of f∘g = { x ∈ Domain(g) | g(x) ∈ Domain(f) }

Example 6   **Evaluating compositions of functions.**

Problem   Evaluate the compositions f∘g and g∘f at x = 5 of the functions

$$f(x) = x^2 \text{ and } g(x) = 3x + 2$$

Solution   The two compositions are evaluated as:

$$f \circ g(x) = f(g(x)) = f(3x + 2) = (3x + 2)^2$$

$$g \circ f(x) = g(f(x)) = g(x^2) = 3x^2 + 2$$

When evaluated at x = 5,
$$f \circ g(5) = (3 \cdot 5 + 2)^2 = 17^2 = 289$$

while
$$g \circ f(5) = 3 \cdot 5^2 + 2 = 77$$

Example 7 The Domain of a compositions.

Problem Determine the domain of the compositions of $f(x) = x^{1/2}$ and $g(x) = 1/(x - 3)$.

Solution As $g(x)$ is defined for all $x \neq 3$ and $f(x)$ is only defined for non-negative values,

$$\text{Domain}(f \circ g) = \{\, x \neq 3 \mid g(x) = 1/(x-3) \geq 0 \,\} = (3, \infty)$$

$$\text{Domain}(g \circ f) = \{\, x \geq 0 \mid f(x) = x^{1/2} \neq 3 \,\} = [0, 9) \cup (9, \infty)$$

Inverse functions.

To solve equations you must perform a series of algebraic operations, these consist of the inverses of the operations used to algebraically de-construct the equation. Like undressing, to reverse the process of dressing, this is done in reverse order. You take off your hat, your coat then a sweater, etc. Mathematically, equations are formed by sequentially applying elementary functions, in composition, to form a function of the independent variable: $y = f(x)$. To solve this equation for x, each step involved in forming $f(x)$ must be reversed, starting with the last step and moving towards the first. These same reverse operations are simultaneously applied to the y-side of the equation $y = f(x)$ so that in the end, after each step in $f(x)$ is reversed, the resulting equation has the form $g(y) = x$. The resulting function g is called **the inverse of f**. It is more specifically the *inverse of f with respect to composition*. Because the inverse function g is directly related to f it is normally denoted using the letter f as f^{-1}. Unfortunately, not every function has an inverse function.

The inverse f^{-1} with respect to composition can only be defined when $f(x)$ is *one-to-one*. Remember, this means that there is only one x-value corresponding to each y-value. The function $f(x) = x^2$ is not one-to-one since $f(2) = 4$ and $f(-2) = 4$. If we wished to reverse the function, using the notation f^{-1} (which is read "f inverse"), say at the value 4, we would have two choices for $f^{-1}(4)$. Either $f^{-1}(4) = 2$ or $f^{-1}(4) = -2$. As a function must be unique, this means that the f^{-1} can not be a function. But we know what f^{-1} is in this case, the inverse of squaring, x^2, is taking the square root, denoted as SQRT(x) or $x^{1/2}$. The way to resolve this problem is to restrict the domain of the x^2 function to either non-negative values or to non-positive values. By default we will assume the domain is $D = [0, \infty)$ and that square root is positive.

DEFINITION OF AN INVERSE FUNCTION: f^{-1}.

> A function f that is one-to-one on a domain D has a composition inverse, denoted by f^{-1}. The **inverse function** f^{-1} is defined on the range of f corresponding to the given domain D,
>
> $$R = \{y \mid y = f(x) \text{ for some } x \in D\}$$
>
> by
>
> $$x = f^{-1}(y) \quad \text{if} \quad f(x) = y$$
>
> Thus, for all $y \in R$,
>
> $$f \circ f^{-1}(y) = f[f^{-1}(y)] = f(x) = y$$
>
> and for all $x \in D$,
>
> $$f^{-1} \circ f(x) = f^{-1}[f(x)] = f^{-1}(y) = x$$

Example 8 **Determining an inverse function.**

Problem Determine the inverse of the function:

(a) $f(x) = 3x + 7$ (b) $g(t) = t^3 - 4$ and (c) $h(x) = (3 + x)/(2x - 1)$

Solution (a) The method is to solve $y = f(x)$ by reversing each operation of f. If at any stage the operation can not be uniquely reversed, then the inverse function does not exist. This is done as follows:

$$y = 3x + 7$$

$$y - 7 = 3x$$

$$(y - 7)/3 = x$$

Thus we write $f^{-1}(y) = (y - 7)/3$. Note that the variable y can be replaced by any other symbol, such as

$$f^{-1}(t) = (t - 7)/3 \quad \text{or} \quad f^{-1}(x) = (x - 7)/3$$

(b) To determine g^{-1} we solve $y = g(t)$ for t as follows:

$$y = t^3 - 4$$

$$y + 4 = t^3$$

$$(y + 4)^{1/3} = t$$

Thus, $g^{-1}(y) = (y + 4)^{1/3}$ or, for a t-variable, $g^{-1}(t) = (t + 4)^{1/3}$.

(c) To find h^{-1} we solve $y = (3 + x)/(2x - 1)$ for x:

$$(2x - 1) \cdot y = 3 + x$$

$$2xy - y = 3 + x$$

$$2xy - x = 3 + y$$

$$x(2y - 1) = 3 + y$$

$$x = (3 + y)/(2y - 1)$$

Thus, the inverse of h is

$$h^{-1}(y) = (3 + y)/(2y - 1)$$

Notice that if we replace y by the variable x this becomes

$$h^{-1}(x) = (3 + x)/(2x - 1)$$

What? Did we make a mistake? This is the same algebraic form as h(x). No, this is correct. The function h(x) is one of the rare functions that is its own inverse.

Coordinate Systems and Graphs.

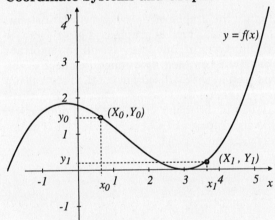

The point (x_0, y_0) on the graph of f has y-coordinate $y_0 = f(x_0)$.

The characteristics of a function can be illustrated by a graph. Visual images are perceived in relation to other nearby images that provide cues and references. For mathematical graphs the basic reference is provided by a coordinate system that combines a geometric reference with units of scale. We use the Cartesian coordinate system, consisting of two perpendicular number lines, called axes, drawn on a sheet of paper and intersecting at the origin. The horizontal line is associated with the independent variable and the vertical line represent the dependent variable or function. The horizontal axis is by default referred to as the "x-axis" unless another independent variable specified.

The graph of a function consists of all points whose coordinates satisfy the function's rule. If the function is $y = f(x)$ then the horizontal axis is the x-axis and the vertical axis is the y-axis. The graph of the function f then consists of the set of all points (x,y) such that $y = f(x)$. To refer to different specific points on a graph integer subscripts will be employed, such as (x_0, y_0) or (x_1, y_1) where $y_1 = f(x_1)$. The graph of a finite function will be a finite number of points.

Example 9 **The graph of a finite function.**

Problem   Sketch the graph of

$$f = \{ (2,3), (4,4), (5,-1), (6,2)\}$$

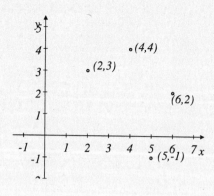

Solution   The points are plotted on a coordinate system as in the figure at teh right. The axis were labeled by default as x and y since the problem does not indicate the symbols to use for the variables.

☑

It is assumed that you have graphed points and simple equations in a prior course. You probably learned to graph an equation y = f(x) by plotting points and connecting them by lines or relatively smooth curves. Initially, you would determine a set of points on the graph by first evaluating f(x) for a given set of x-values, generating a table of point coordinates. These would be plotted and connected by lines. However, for simple nonlinear functions, like the quadratic $y = x^2$, curved lines connecting the points would provide a more accurate sketch. Another approach to making a sketched graph more accurate is to simply plot more points. Many computer programs that plot graphs do just this. They actually plot one point for each x-value that can possibly be distinguished on the computer screen.

In the later sections of this Chapter we will introduce the graphs of the elementary functions. Our objective is for you to associate a function with a particular type of graph, and *vis-a-versa*, to associate graphs having particular shapes with particular types of functions. It will be important that you can sketch the graphs of elementary functions and can identify a function whose graph has the characteristic of a given graph. We finish this section with a discussion of the relationship between equations of shifted graphs. To analytically determine the characteristics of a function's graph generally requires the use of calculus. Indeed, one of the major applications of differential calculus is to establish properties of functions and their graphs. In Chapter 3 we will consider curve sketching of more complicated functions using calculus as a tool.

Shifted Graphs.

Two curves that are identical in form, i.e. in their shape and orientation relative to the axes, will have related equations. One curve is said to be a *shifted* form of the another curve if its graph can be formed by rigidly moving the other curve horizontally and/or vertically, without any rotation.

When a graph is moved to the right or left we call this a *horizontal* shift. Algebraically, this is achieved by replacing the x-variable by a term x - *a* when the shift is to the right *a* units. For instance, if the graph of y = f(x) is shifted horizontally to the right 3 units we set *a* = 3 and the equation of the shifted curve will be y = f(x - 3). When the shift is to the left, x is replaced by a term x + *a*; for instance, the equation of a graph shifted to the left 2 units will have the form y = f(x + 2).

Section 1.1   FUNCTIONS AND GRAPHS

THE EQUATION OF A HORIZONTAL GRAPH SHIFT

> If for a positive number a, the graph of a function $y = f(x)$ is shifted horizontally a-units, the equation of the resulting curve is
>
> $y = f(x - a)$ if the graph of f is shifted is to the right a-units;
>
> $y = f(x + a)$ if the graph of f is shifted is to the left a-units.

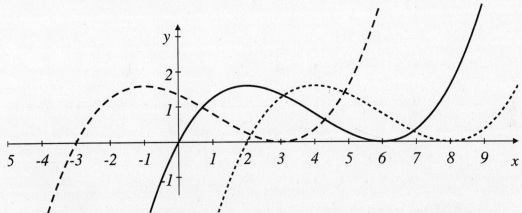

The graph of $y = f(x)$, solid curve, is shifted to the right 2 units to form the graph of $y = f(x - 2)$, the dotted curve, and is shifted to the left 3 units to form the graph of $y = f(x + 3)$, the dashed curve.

When a graph is moved upward or downward this is called a *vertical* shift. The corresponding equation of a vertically shifted curve is similarly formed by replacing the y-variable by a term of the form $y - b$ when the shift is upward b units or the term $y + b$ when the shift is downward b units. However, since we usual write the equations of functions in the explicit form $y = \ldots$ the corresponding equations for vertically shifted graphs are solved for y.

The equation of a graph shifted upward b units is $y - b = f(x)$ or $y = b + f(x)$.

The equation of a graph shifted downward b units is $y + b = f(x)$ or $y = -b + f(x)$.

For instance, $y = 5 + f(x)$ is the equation of the graph of $y = f(x)$ shifted upward 5 units and the equation of the graph of f shifted downward 7 units would be $y = -7 + f(x)$.

THE EQUATION OF A VERTICAL GRAPH SHIFT

> If b is a positive number and the graph of a function $y = f(x)$ is shifted vertically b-units, the equation of the resulting curve is
>
> $y = -b + f(x)$ if the graph of f is shifted is down b-units;
>
> $y = b + f(x)$ if the graph of f is shifted is upward b-units.

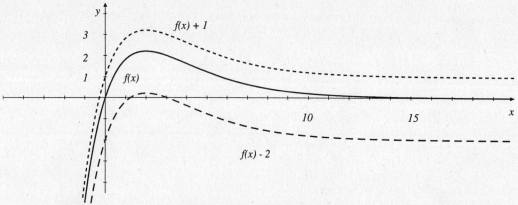

The graph of $y = f(x)$, the solid curve, is shifted upward 1 unit to form the graph of $y = f(x) + 1$, the dotted curve, and $y = f(x)$ is shifted downward 2 units to form the graph of $y = f(x) - 2$, the dashed curve.

To write the equations of graphs that are shifted both vertically and horizontally the above processes are combined to give an equation of the form $y = f(x \pm a) \pm b$. The $\pm$ sign can be confusing. To simplify the notation in a general formula we shall use the negative sign for the horizontal shift and the positive sign for the vertical shift. The constants a and b will then be allowed to be either positive or negative. The sign of the actual numerical values of a and b will indicate how the graph $y = f(x)$ is shifted. When the parameters are zero there is no shift.

GENERAL EQUATION OF A SHIFTED GRAPH

> The equation corresponding to the graph of $y = f(x)$ shifted horizontally and vertically has the form
>
> $$y = b + f(x - a)$$
>
> where the constants b and a can be either positive or negative.

The graph is shifted horizontal to the right when $a > 0$ and is shifted to the left, when $a < 0$.

The graph is shifted upward if $b > 0$ and is shifted downward if $b < 0$.

For instance, the graph of $y = 3 + (x - 2)^5$ is the graph of $y = x^5$ shifted horizontally to the right 2 units and vertically upward 3 units. It does not matter which shift is done first. The equation of the curve $y = x^2$ shifted to the left 7 units and downward 1.5 units is then $y = -1.5 + (x + 7)^2$.

Each point on a shifted graph is the image under the translation of a point on the original graph. If the image of the origin is the point (a,b) then the shifted graph is said to be **centered at the point (a,b)** and its equation is

$$y = b + f(x - a)$$

Each point on the shifted graph has the same relationship to the point (a,b) as its pre-shift image point on the original graph has to the origin, $(0,0)$.

For instance, the graph of $y = 0.6 + (x - 5)^3$ is said to be centered at $(5, 0.6)$ and is thought of as the graph of $y = x^3$ shifted horizontally 5 to the right and vertically 0.6 upward.

To determine the shift parameters a and b you need only to establish any two corresponding points on the original graph and on the shifted graph. If (x_0, y_0) is a point on the original graph of $y = f(x)$ and its image on the shifted graph is (x_0', y_0') then the shift constants are found as the difference in these coordinates:

$$a = x_0' - x_0 \quad \text{and} \quad b = y_0' - y_0$$

Example 10 **Equations of a shifted graph.**

Problem   Consider the graphs in the adjacent figure. Assuming the dashed graph is a shifted form of the solid graph, $y = f(x)$, what is its equation?

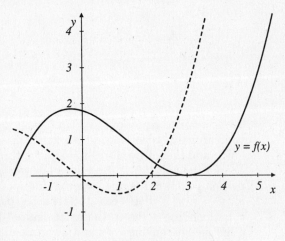

Solution   The dashed graph must be shifted both horizontally and vertically. The form of its equation is $y = b + f(x - a)$ and the problem is to establish the numerical value of the constants a and b. To do this we must identify a point (x_0, y_0) on the original graph of $y = f(x)$ and its image (x_0', y_0') on the shifted graph. The points must be easily identified on each graph. The most obvious point on the solid curve is perhaps the point $(3,0)$ where the graph of f touches the x-axis. It appears that this point is shifted to the point at the bottom of the dip on the dashed curve, with coordinates $(1, -0.5)$. Thus, setting $(x_0, y_0) = (3, 0)$ and $(x_0', y_0') = (1, -0.5)$, the shift constants are

$$a = 1 - 3 = -2 \quad \text{and} \quad b = -0.5 - 0 = -0.5$$

The equation of the shifted dashed graph is therefore $y = -0.5 + f(x + 2)$.

The graph of inverse functions.

The graphs of a function and its inverse are closely related. They are the "mirror" or symmetric image of each other through the line y = x. Algebraically, the points on one graph are obtained by interchanging the x and y coordinates of the points on the other graph. If (3,7) is on the graph of f then (7,3) is on the graph of f^{-1}. Similarly, if (-2,4) is on the graph of f^{-1} then (4, -2) is on the graph of f.

Example 11 Graphing an inverse function.

Problem Assume the graph of a function f(x) is as drawn at the left. Sketch the graph of its inverse function.

Solution Draw the line y = x on the graph and then form the mirror image of the given curve through this line. This can be done physically by rotating the paper on which the figure is sketched about the upper right and lower left corners of the graph. Then, holding the paper up to a bright light and tracing the curve from the opposite side of the paper. Using a ruler, the graph of f^{-1} can be sketched by drawing lines perpendicular to the line y = x, and where such lines intersect the graph of f plotting a point on the same line equidistant from the line y = x but on the opposite side from the point on the graph of f. The graph of f^{-1} is the curve through all such points. The graph of the corresponding inverse function is sketched below.

To sketch the graph of f^{-1} without using this process, you can simply plot several points on the graph as a guide and then draw a free-hand curve through these. To plot the points first draw several light "guide" lines perpendicular to the line y = x that intersect the graph of y = f(x). For each intersection point plot a corresponding point on the same "guide" line on the other side of the line y = x and the same distance from y = x as the intersection point. These will be the points on the graph y = f^{-1}(x).

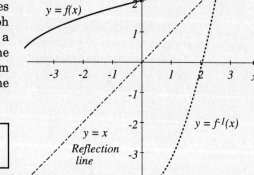

Exercise Set 1.1

1. Express the given set in interval notation.

   a) $\{x \mid 0 \leq x \leq 8\}$ b) $\{x \mid 0 < x \leq 8\}$ c) $\{x \mid |x - 3| \leq 1\}$ d) $\{x \mid |x + 2| > 3\}$

   e) $\{x \mid x > 2 \text{ and } x \leq 8\}$ f) $\{x \mid 0 < x \text{ or } x \leq -2\}$ g) $\{x \mid x^2 > 4\}$ h) $\{x \mid x^2 - 4 \leq 5\}$

2. Determine the Natural domain of the given function.

   a) $f(x) = x^2 - 1$ b) $g(x) = x/(x - 3)$ c) $h(x) = 2 - 1/x$

   d) $f(x) = \text{SQRT}(x - 4)$ e) $f(x) = 1/x^2$ f) $g(x) = 1/(x - 2) + 1/(x + 2)$

   g) $h(x) = \text{SQRT}(2 - 1/x)$ h) $f(x) = \text{SQRT}(x^2 - 4)$ i) $f(x) = \text{SQRT}(4 - x^2)$

3. Determine the ranges of the functions in Exercise 2.

4. Which of the following points lies on the graph of $y = 3x^2 - 1$?

   a) (1,1)   b) (1,2)   c) (2,12)   d) (0,-1)
   e) (-1,-4)   f) (-2,11)   g) ($\sqrt{2}$,5)   h) (-$\sqrt{2}$,-5)

5. Consider the finite functions $f = \{(2,3), (3,4), (5,2), (6,5)\}$ and $g = \{(2,3), (3,0), (4,2), (5,6)\}$.
   a) Indicate the Domain and Range of each function.
   b) Sketch the graph of both functions.
   c) Express each of the following combinations of f and g as a set and indicate the domain of the combined functions. i) $f + g$ ii) $f \cdot g$ iii) f/g iv) $f \circ g$ v) $g \circ f$
   d) Determine the inverse functions f^{-1} and g^{-1}.

6. For the given pair of functions evaluate f+g, g-f, f·g, f/g, f∘g and g∘f, and indicate their domains.

   a) $f(x) = 1/x$ $g(x) = x^2 - 3$ b) $f(x) = 2x - 1$ $g(x) = \sqrt{x}$ c) $f(x) = x^2 - 2x$ $g(x) = 1/(x + 1)$

7. What is the equation of the function whose graph is the same as the graph of f(x)

   a) shifted to the right 3 and up 2.   b) shifted down 3 and to the right 1.

   c) shifted down 1 and to the left 2.   d) shifted up 5 and to the left 4.

8. Indicate the relationship of the indicated function's graph to the graph of f(x).

   a) $y = 2 + f(x)$ b) $y = f(x - 3)$ c) $y = -2 + f(x - 5)$

   d) $y = 3 + f(x + 2)$ e) $y = f(x) - 2$ f) $y = -2 f(x - 2)$

9. Describe the graph of the given function relative to a simpler graph. You do not have to sketch the graphs.

   a) $y = 3 + x^2$ b) $y = (x + 3)^2$ c) $y = \text{SQRT}(x - 5)$

   d) $y = -2 + (x - 4)^2$ e) $y = x + 2 - (x + 2)^2$ f) $y = 3 + x - \text{SQRT}(x - 1)$

10. In the given graph the solid curve is y = f(x) and the dotted curve is y = g(x). Express the function g in terms of f.

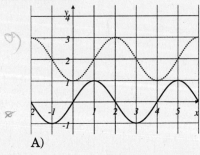
A)

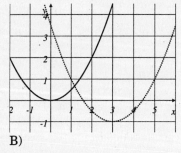
B)

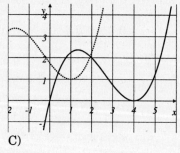
C)

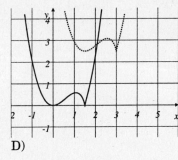
D)

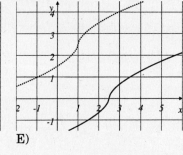
E)

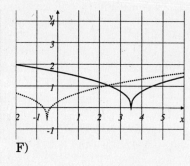
F)

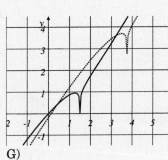
G)

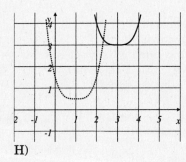
H)

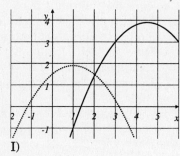
I)

11. For each of the graphs in the previous exercise, express the function f in terms of the function g.

12. Given the sketched graph of f(x) sketch the graph of its inverse function.

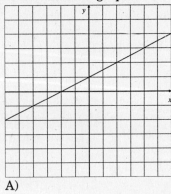
A)

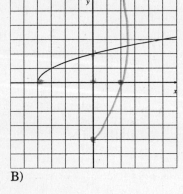
B)

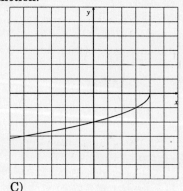
C)

Section 1.2 Polynomial Functions

The Power Functions.

The Basic Functions considered in this section are those that can be evaluated using ordinary arithmetic, by adding, subtracting, multiplying, or dividing. The building block of such functions are the **power functions** $y = x^n$ where the *exponent* n is a natural number (zero or a positive integer). Power functions can be defined recursively:

$$x^0 \equiv 1 \text{ and } x^n = x \cdot x^{n-1} \text{ for } n = 1, 2, 3, ...$$

for all real numbers x, except in the case when both x and n are zero. The expression 0^0 is not defined; it is discussed in Section 3.9. Without directly referring to this exception each time, we will simply say that the Domain of $y = x^n$ is all real numbers.

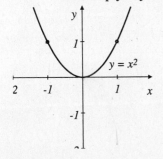

The character of the graph of $y = x^n$ depends on the exponent n. Each graph passes through the origin, (0,0), and the point (1,1).

For even n, n = 2, 4, ..., the graph passes through the point (-1,1), while for odd n, n = 3, 5, 7, ..., it passes through the point (-1,-1).

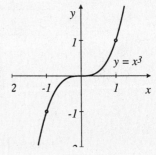

Adjectives describing graphs: Increasing, Decreasing, Even and Odd.

The adjectives *increasing* and *decreasing* are used to indicate how a function's values change as the independent variable increases. Algebraically, these terms are defined using inequalities.

INCREASING AND DECREASING FUNCTIONS

If, for **every** pair of numbers x_1 and x_2 in the interval (a,b),

$$f(x_1) < f(x_2) \text{ when } x_1 < x_2$$

then f(x) is **increasing on an interval (a,b)**;

if the function values satisfy the reverse inequality,

$$f(x_1) > f(x_2) \text{ when } x_1 < x_2$$

then f(x) is **decreasing on an interval (a,b)**.

Graphically, the terms *increasing* and *decreasing* indicate how the y-coordinates on a function's graph change as the x-coordinate increases. These concepts are associated with the order of the axis; like a number line, increasing means moving to the right along the horizontal x-axis. A curve or graph is increasing on an interval if, as it is traced for x-values starting at the left endpoint of the interval and proceeding to the right endpoint, the corresponding y-coordinates always moves upward, toward the top of the graph frame. If on the other hand, the y-coordinate on the graph being traced moves downward, toward the bottom of the graph frame, as the curve is traced from left to right over the interval, then the curve is decreasing on the interval.

When n is an odd number the power function $y = x^n$ and its graph are increasing on the entire real line, $(-\infty,\infty)$ as illustrated below for n = 1, 3 and 5. The left panel shows the graphs in a *window* with the x-range [-10,10] and the y-range of [-15,15]. Note that the size of the graphing window limits how much of a graph can be illustrated. For instance, the graph $y = x^3$ for $x > 15^{1/3} \approx 2.5$ can not be displayed in this window because the y-values are too great. In the right panel, the scale is increased so that the curves are sketched in a *window* with x in the interval [-1.5, 1.5].

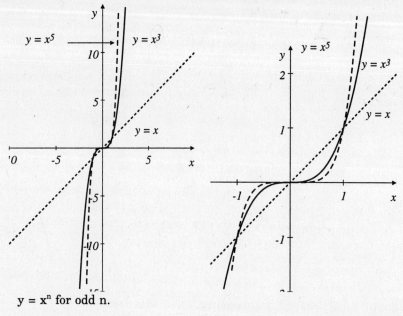

$y = x^n$ for odd n.

When n is even, and not zero, the graph of $y = x^n$ has the shape of the quadratic $y = x^2$. It is increasing when x is positive, on the half-line $(0,\infty)$, and decreasing when x is negative, on the half-line $(-\infty,0)$; this is seen in the following graphs, for n = 2, 4, and 6.

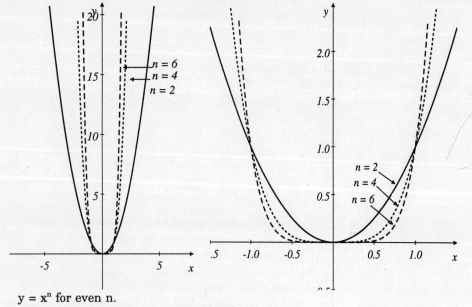

$y = x^n$ for even n.

For n = 0 the function $y = x^n$ reduces to the constant function y = 1, which is a degenerate power function whose graph is the horizontal line through (1,1). This graph is neither increasing nor decreasing.

The graphs of various power functions have an ordered relationship based on their relative exponents and the x-values.

$$\text{For } x > 1, \quad x < x^2 < x^3 < x^4 \ldots$$

Thus, for $x > 1$, as the exponent n increases, the value of x^n increases. This means the graphs of higher exponent power functions will be above the graphs corresponding to lower exponents. For instance, the graph of $y = x^5$ will be above the graph of $y = x^4$ for $x > 1$.

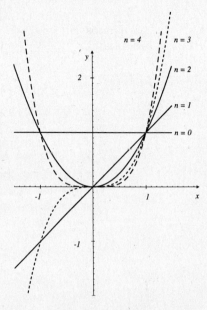

When x is between zero and one, the relationship of the power functions are reversed.

$$\text{For } 0 < x < 1, \quad x > x^2 > x^3 > x^4 \ldots$$

Consequently, the graphs of $y = x^n$ over the interval $(0,1)$ become lower as the exponent n becomes larger.

The graphs of the power functions can also be described in terms of their symmetry. The adjectives *even* and *odd* are used to describe the symmetry of functions and their graphs.

EVEN AND ODD FUNCTIONS

A function f is an **even function** if for all x

$$f(-x) = f(x) \qquad \text{EVEN}$$

The function f is an **odd function** if for all x

$$f(-x) = -f(x) \qquad \text{ODD}$$

The power functions $y = x^n$ is an even function when n is even
and is an odd function when n is an odd positive integer.

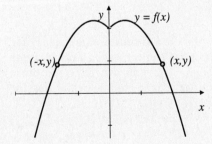

The graph of an **even function** is symmetric about the y-axis. This means that if a point (x,y) is on the function's graph then so is the point $(-x,y)$.

A horizontal line that intersect the graph of an *even function* will normally do so an even number of times, with paired points equidistant from the y-axis. The exception occurs if it intersects the graph at the y-axis.

The graph of an **odd function** is symmetric about the origin. This means that if a point (x,y) is on the graph then so is the point $(-x,-y)$.

If a line through the origin intersects the graph of an *odd function*, the intersection points (except at the origin) will be paired, the two paired points will be equidistant from the origin and the coordinates of one will be the negative of the other's coordinates.

Caution . . . .

Graphing $y = x^n$ for n large can be difficult because when x is not close to 1 the numerical values x^n are very large or very small. When a computer program generates a graph, the appearance will depend on the resolution of the out-put device, the screen or the printer and the window or graph frame that is employed. Sometimes the appearance can be deceiving.

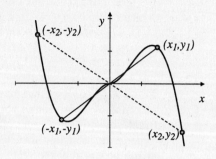

Consider the adjacent sketches of $y = x^{10}$, which is an even function, whose graph is said to be *u-shaped*. As illustrated, for x near the origin the graphs appear to coincide with the x-axis. This is not the case!

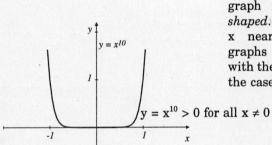

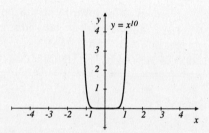

Because of the thickness of the lines, the resolution on the graphs is too coarse to distinguish between the actual function values and the x-axis.

The function $y = x^{11}$ is odd, and thus its graph is easily distinguishable from the graph of the even function $y = x^{10}$. The left segment of the graph $y = x^{11}$, for x < 0, is below the x-axis and thus appears as the left portion of the graph $y = x^{10}$ rotated downward. The graph of the right segment of $y = x^{11}$ appears to be very similar to that of $y = x^{10}$ for x > 0.

The curve $y = x^9$ would look very similar to the curve $y = x^{11}$. If x^9 were also plotted on this same graph frame, its graph would only seem to blur the graph of $y = x^{11}$.

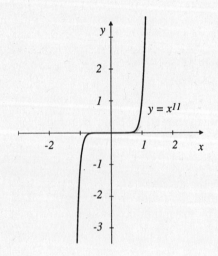

It is difficult to sketch such graphs on a uniform axis system for large x-values, such as x = 10, because the corresponding y-values are so great. The graphs of such functions can often be more usefully sketched utilizing nonlinear axis systems such as log-log graphs that will be discussed later.

Section 1.2   POLYNOMIAL FUNCTIONS   23

Scaled Power Functions

When a power term x^n is multiplied by a constant, c, such as cx^3, the constant c is called a *scalar*. Graphically, the effects of multiplying a power function by a constant is to stretch or compress the graph of the unscaled function in the y-direction. For c > 0, the graph of the

scaled power function $f(x) = c \cdot x^n$

will have a shape similar to the shape of the graph of $y = x^n$:

if **c > 1** it appears stretched away from the x-axis, and

if **0 < c < 1** it appears compressed towards the x-axis.

If the number c is negative, then the graph of $y = cx^n$ appears as symmetric image of the graph of $y = |c|x^n$ reflected through the x-axis, i.e., flipped so that positive y-values become negative and *vis-a-versa*.

A quick sketch of the scaled functions $y = cx^n$ can be made by first identifying the basic shape of the graph, plotting a few key points, and then drawing a characteristic curve through these. To make the sketch more accurate additional points can be plotted.

SHAPE: The graph of $y = cx^n$ will be u-shaped if n is even or shaped like $y = x^3$ if n is odd.

KEY POINTS: The curve $y = cx^n$ will pass through the origin (0,0), the point (1,c), and either through (-1,c) if n is even or through (-1,-c) when n is odd.

Example 1   **Graphs of $y = cx^n$.**

Problem   Sketch the graphs $y = cx^n$ for n = 1, 2, and 3 when c = 0.5, 1, 3, and -2.

Solution   By a sketch, at this stage, we mean a curve that has the general shape of the graph with only a few key points indicated. The illustrated graphs are very accurate, being generated by a computer program that in fact simply evaluates the functions at x-values so close that when the corresponding points are plotted they overlap to appear as a continuous line. Your sketches by hand will probably not be so accurate. The objective in this example is to emphasize the relative character of the graphs for various values of the constant c.

For each exponent n we plot the basic function, with c = 1, and the three scaled functions on the same coordinate system. The first step is to sketch the basic power function, the line $y = x$, the parabola $y = x^2$, or the cubic $y = x^3$. Then, to establish the relative scaling of each graph, the points (1,c) are plotted for c = 0.5, 3 and -2 and a curve similar to the basic power function is sketched with the same shaped passing through the plotted point and the origin. The graphs are sketched on the next page. Notice that the basic curve, the heavy solid line, passes through the point (1,1).

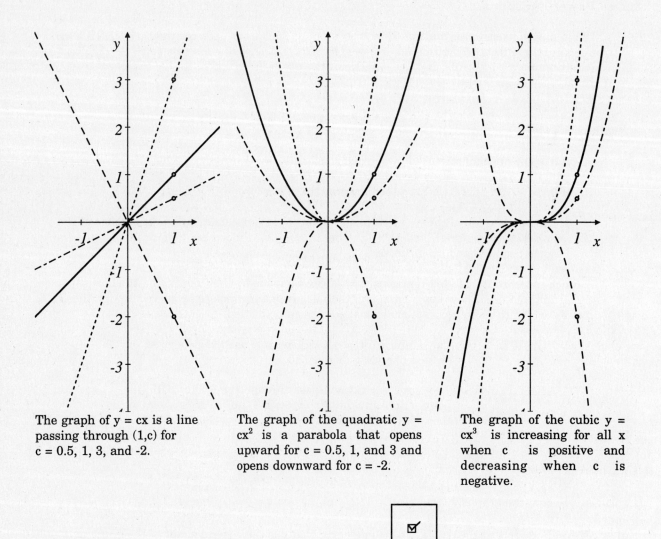

The graph of y = cx is a line passing through (1,c) for c = 0.5, 1, 3, and -2.

The graph of the quadratic y = cx² is a parabola that opens upward for c = 0.5, 1, and 3 and opens downward for c = -2.

The graph of the cubic y = cx³ is increasing for all x when c is positive and decreasing when c is negative.

☑

Shifted Power Functions.

The graphs of the functions y = xⁿ are "centered" at the origin. If these graphs are translated by shifting them horizontally a units and vertically b units they form the graphs of the **shifted power functions**

$$y = b + (x - a)^n$$

The graphs of these functions are "centered" at the point (a,b). Each such shifted power function passes through the point (a,b), which is the shifted image of the origin, and the point $(a + 1, b + 1)$, which is the shifted image of the point (1,1). When the graph of a scaled power function cxⁿ is shifted the resulting curve is the graph of the function

$$y = b + c(x - a)^n \quad \textbf{Scaled and shifted Power Function}$$

The graph of this function is the graph of y = xⁿ stretched or compressed in the y-direction by a factor of c and then shifted to be "centered" at the point (a,b). **It passes through the points (a,b) and $(a + 1, b + c)$.**

Identifying the graph of a Shifted Scaled Power function.

Assume that a curve is the sketch of a shifted scaled power function. Then its equation must have the form $y = b + c(x - a)^n$. To identify a graph with an equation the first step is to determine its basic shape, which will determine if the exponent n is an even or an odd number and the sign of the scaling constant.

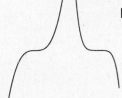

Is it always increasing or decreasing ?

If so then n must be an odd number, 1, 3, 5, etc.

The scaling constant c is positive if its increasing and negative if its decreasing.

Is it a *u-shape* or *an inverted u-shape*?

If so, then n must be an even number, 2, 4, 6, etc.

If it is *u-shaped* then c must be positive.

If it has an *inverted u-shape* then c must be negative.

To establish the shift constants a and b you must establish the point where the graph is centered.

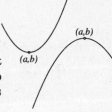

For n even, the graph of $y = b + c(x - a)^n$ is *centered* at the point (a,b) that is the bottom of a *u-shaped* graph, where it changes from decreasing to increasing, or the top point of an *inverted u-shaped* graph, where it changes from increasing to decreasing.

If n is odd, n > 1, then this *center* is the point where the graph seems to "level off" and then change its curvature.

The graph will be symmetric about this center point. Any line through the point (a,b) that intersects the graph will actually intersect at two points equidistant for the center in opposite directions along the line.

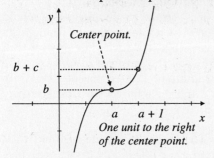

To establish the constant c you utilize the y-coordinate of the point on the graph one unit to the right of the center point.

Since this point on the graph will have y-coordinate $b + c$, set:

$$c = y(a + 1) - b$$

To establish the exponent n you can utilize the point on the graph two units to the right of the center, if its y-coordinate is y_2, then

$$y_2 = b + c\, 2^n \quad \text{or} \quad 2^n = (y_2 - b)/c$$

To solve this equation for n requires the use of logarithms, which will be discussed in Section 1.4. The basic solution has the form n

$$n = ln((y_2 - b)/c) / ln(2)$$

Example 2 Identifying the graphs of Shifted Power functions: a multiple choice.

Problem Identify which of the following functions describe the curves sketched below.

a) $y = -2 + x^2$
b) $y = 2 + (x - 4)^3$
c) $y = 2 - 0.5(x - 4)^2$
d) $y = -2 + 0.2(x + 2)$
e) $y = 2 - (x + 4)^5$
f) $y = 2 - 2(x + 4)^5$

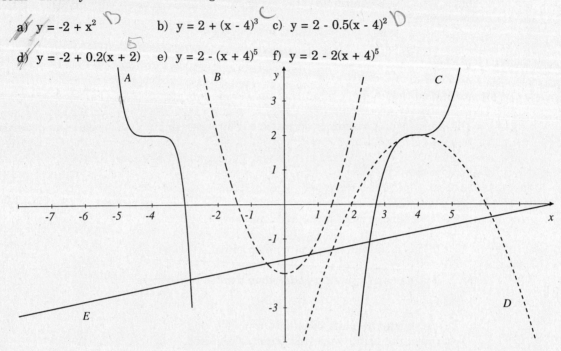

Solution Applying the above strategy we can establish the equations corresponding to such graphs. In the circumstance where a multiple choice is to be made (assuming that a correct answer is among the choices!) we can quickly eliminate some choices and pick the correct one.

Graph A is decreasing for all illustrated values and thus must correspond to an odd exponent with a negative coefficient. It is centered at (-4,2). Its equation must therefore have the form

$$y = 2 + c(x + 4)^n \quad \text{for some negative constant c and odd integer n.}$$

Two function choices have this form with n = 5, e) and f). To decide which is appropriate we must determine the constant c. This is done by identifying a second point on the graph. Any point would do, but it is normally easiest to use the fact that the graph of $y = b + c(x - a)^n$ passes through the point $(a + 1, b + c)$. Thus we determine the point on the graph A with x-coordinate one greater that the x-coordinate of the center point. At x = -4 + 1 = -3 the point on the graph of A appears to be (-3,0). The y-coordinate of this point must be the center y-coordinate plus the constant: 2 + c. To determine c we equate these two y-coordinates, setting 0 = 2 + c and solve for c = -2. Consequently curve A must be the graph of equation f): $y = 2 - 2(x + 4)^5$.

Graph B. This curve must correspond to an even exponent n with a positive scale coefficient. It is centered at (0,-2), i.e. is only shifted downward two units and a = 0. The only choice with this form is equation a) : $y = -2 + x^2$, whose graph will have the same form as curve B.

Graph C. This curve is centered at (4,2) and since it is increasing for all sketched x-values, it must correspond to an equation with an odd exponent and a positive coefficient. The only choice meeting these criteria is the cubic equation b): $y = 2 + (x - 4)^3$.

Curve D is the dashed curve also centered at (4,2) which has an *inverted u-shape*. Its function thus must have an even exponent and a negative scaling constant. Equation c): $y = 2 - 0.5(x - 4)^2$ is the only choice that has these features.

Curve E is a straight line. Of the function choices given, only d): $y = -2 + 0.2(x + 2)$ is linear with exponent n = 1. Equation d) is a line that passes through (-2,-2) and (-2 + 1, -2 + 0.2) = (-1,-1.8). As both of these points are on the line E it is the graph of equation d).

☑

Polynomials

Polynomials are linear combinations of scaled power functions, i.e., the sum or terms cx^n for different constants c and exponents n. These are the basic functions that you first experienced in algebra. The **degree of a polynomial** is the highest integer exponent of the power terms that form the polynomial. For instance, $f(x) = 2x + x^5 - 3x^2$ is a polynomial of degree 5. Notice that a polynomial does not need to include all powers of x less than its degree; there are no terms with x^0, x^3, or x^4 in the given polynomial. However, the most general polynomial of degree 5 could have such terms. The general polynomial function of degree 5 will be denoted by P with the subscript 5, P_5. It can be written using subscripted c's to represent the coefficients of the corresponding powers of x:

$$P_5(x) = c_0 + c_1 x + c_2 x^2 + c_3 x^3 + c_4 x^4 + c_5 x^5$$

The function f(x) given above has this form with the constants

$$c_0 = 0 \quad c_1 = 2 \quad c_2 = -3 \quad c_3 = 0 \quad c_4 = 0 \quad c_5 = 1$$

Normally, polynomials are written with the terms ordered, either from the lowest degree to the highest as was done for P_5 or in the reverse order. This is not a mathematical rule, simply a practice that makes the expression easier to read and diminishes the number of clerical errors when working with polynomials. The form of a general polynomial of degree N can be expressed, using the convention of three dots to indicate omitted terms, as

$$P_N(x) = c_0 + c_1 x + c_2 x^2 + c_3 x^3 + \ldots + c_{N-1} x^{N-1} + c_N x^N \qquad \textbf{N}^{\textbf{th}} \textbf{ Degree Polynomial}$$

or, using Sigma notation for the sum of terms with lower case n as an **index**, by

$$P_N(x) = \Sigma_{n=1,N} \; c_n x^n$$

This notation indicates that the *index variable* is n and the sum is taken of each general term $c_n x^n$ replacing n with the initial value 1, then incrementing by 1 unit, replacing n by 2, then 3, then 4, and so on, stopping when the index reaches the upper limit N.

Notation.... We will use this linear sigma notation with the index range indicated by a subscript following the Σ symbol as it is easily typed. Many algebra computer programs use the notation

$$\text{SUM}(\; c\_n * x\wedge n,\; n,\; 1,\; N)$$

where the bar notation c_n denotes a subscript c_n, and the caret ^ denotes exponents, x^n means x^n.

When writing the summation notation by hand, or when it is set by a mathematical type-setter, the index and limits of the index are often placed below and above the sigma sign as in

$$P_n(x) = \sum_{n=1}^{N} c_n x^n$$

Constant functions.

The simplest function is the polynomial of degree N = 0, the constant function y = c. The graph of a constant function is a horizontal line that intercepts the y-axis at c, i.e., it passes through the point (0,c).

Linear Functions.

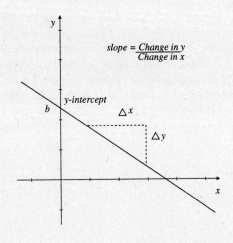

The polynomial of degree one is called a linear function since its graph is a straight line. You may remember that the equation of a line can be characterized by its slope and any point on the line. In high school algebra the most commonly used equation of a line is the

slope-intercept equation y = mx + b

This is the function y = x scaled by the slope coefficient m and shifted vertically b units, so that it intersects the y-axis at (0,b). Recall that the **slope** of a line is the amount by which the y-coordinate of a point on the line changes when the x-coordinate is increased by one unit. It is found by calculating the ratio of the change in y-coordinates between two points on the line to the change in the corresponding x-coordinates.

If (x_0, y_0) and (x_1, y_1) are any two points on a line,
then the **slope of the line** is given by

$$m = (y_1 - y_0) / (x_1 - x_0)$$

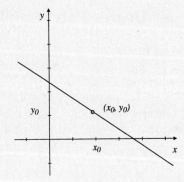

In calculus, our attention will be focused on lines passing through points that are generally not on the y-axis. For this reason we will most frequently utilize the

point-slope equation $y = y_0 + m(x - x_0)$

This is the equation of the generic line y = mx with slope m shifted to the point (x_0, y_0).

Example 3   Equations of lines.

Problem a) What is the equation of the line through the point (2,-3) with slope 0.5?
b) What is the equation of the line passing through the points (3,4) and (-1,2)?
c) What is the y-coordinate of the point on a line at x = 2.5 if the line passes through the point (1.5, 2.8) and has slope 0.7?

Solution a) The point-slope equation of the specified line is y = -3 + 0.5(x - 2).
b) First we compute the slope of the line:

$$m = (4 - 2)/(3 - (-1)) = 1/2$$

Two equivalent equations of the line are provided by the point slope equation using the calculated value for m and the two given points:

$$y = 4 + (1/2)(x - 3) \quad \text{and} \quad y = 2 + (1/2)(x + 1)$$

c) As the x-coordinate is the given point's x-value plus one, $2.5 = 1.5 + 1$, the y-coordinate is the sum of the given y-value plus the slope: $y = 2.8 + 0.7 = 3.5$.

Quadratic Functions.

Quadratic polynomials are called *parabolas*. The polynomial $y = A + Bx + Cx^2$ can be expressed as a scaled and shifted form of $y = x^2$. This is accomplished by a process known as "completing a square". This procedure gives a set of algebraic formulas relating the coefficients A, B, and C, to coefficients a, b, and c of a shifted quadratic:

$$A + Bx + Cx^2 = b + c(x - a)^2$$

$$a = -B/2C \qquad b = A - B^2/(4C) \qquad c = C$$

Shifted Quadratic Form

Thus the graph of a quadratic $y = A + Bx + Cx^2$ can be sketched by transforming it into a shifted form that is easily sketched. Its vertex is (a,b) and its scaling constant $c = C$ is simply the coefficient of the squared term in either the original quadratic or in the shifted form.

If c is positive the graph opens upward, like the graph of x^2. If c is negative, the graph opens downward, like the graph of $-x^2$. As $|c|$ becomes larger the graph becomes narrower, i.e., the horizontal distance between the two branches of the graph at any fixed y-value becomes smaller.

Example 4 **Graphing a quadratic.**
0
Problem Sketch the graph of $y = 3 + 5x - 2x^2$.

Solution The coefficients of the quadratic are $A = 3$, $B = 5$, and $C = -2$. We immediately know that the graph will be a parabola opening downward because C is negative. The coefficients of the equivalent shifted quadratic are

$$c = -2, \quad a = -5/(2 \cdot -2) = 1.25,$$

$$b = 3 - 5^2/(4 \cdot -2) = 3 + 25/8 = 6.125$$

The equivalent shifted quadratic form is thus

$$y = 6.125 - 2(x - 1.25)^2$$

The graph is centered at the point

$$(a,b) = (1.25 , 6.125)$$

The graph of
$y = 3 + 5x - 2x^2$

It is symmetric about the vertical line $x = 1.25$ and passes through the points $(a \pm 1, b + c)$, which are $(0.25 , 4.125)$ and $(2.25 , 4.125)$.

Notice that the quadratic sketched in the previous example crosses the x-axis at two points. The x-coordinates of these intersection points are the roots of the quadratic, the values of x for which $3 + 5x - 2x^2 = 0$. Using the Quadratic Formula these roots can be found as $x_1 = -0.5$ and $x_2 = 3$. They can be used to express the same quadratic in yet another equivalent form, called the *factored form*.

If x_1 and x_2 are roots of the equation $A + Bx + Cx^2 = 0$ then an equivalent **factored form** of the quadratic is given by

$$A + Bx + Cx^2 = C(x - x_1)(x - x_2)$$

Caution   Not all quadratics can be put into factored form. If the graph of a quadratic does not intersect the x-axis then it can not be factored as it has no real roots.

The graph of $y = C(x - x_1)(x - x_2)$ passes through the points $(x_1, 0)$ and $(x_2, 0)$ on the x-axis and opens upward if $C > 0$ and opens downward if $C < 0$.
The graph is symmetric about the vertical line that is midway between these points, i.e. about the line $x_m = (x_1 + x_2)/2$. Note x_m is the *mean* or average of x_1 and x_2.
The graph's vertex will be at (x_m, y_m) where

$$y_m = (-1/4)C(x_1 - x_2)^2$$

Example 5   **Graphing a quadratic in factored form.**

Problem Use the factored form of $y = -12 + 2x + 2x^2$ to sketch its graph.

Solution Factoring $C = 2$,

$$y = 2(x - 2)(x + 3)$$

The roots are $x_1 = 2$ and $x_2 = -3$ and the graph is symmetrical about the vertical line midway between these two roots:

$$x = (2 + (-3))/2 = -0.5$$

The parabola's vertex is on this line of symmetry. Its y-coordinate is found by substituting $x = -0.5$ into the equation:

$$y_m = (-1/4) \cdot 2 \cdot (2 - (-3))^2 = -12.5$$

As the x^2-coefficient $C = 2 > 0$ the graph opens upward.

The x-intercepts (2,0) and (-3,0) and the vertex (-0.5, -12.5) are plotted and then the graph is sketched as a smooth "parabolic" curve passing through these three points, as at the right.

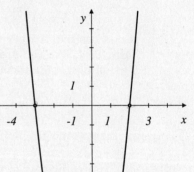
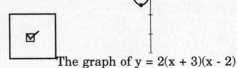
The graph of $y = 2(x + 3)(x - 2)$

A quick sketch by hand will probably not be as accurate as our computer generated drawings. You can improve your sketching perception by plotting additional points on the graph to confirm its shape.

Section 1.2 POLYNOMIAL FUNCTIONS 31

Exercise Set 1.2

1. Show algebraically that the function $f(x) = x^2$ is increasing on the interval $(0,\infty)$.

2. Show by example that $f(x) = x^2$ is neither increasing nor decreasing on the interval $(-5,5)$.

3. Show algebraically that: a) $f(x) = x^2$ is an even function. b) $f(x) = x^3$ is an odd function.

4. To examine how multiplication by a constant scales the graph of a function sketch on the same coordinate system the graphs of the functions $f(x)$, $2 \cdot f(x)$, $0.5\, f(x)$, and $-2\, f(x)$, for the given function.

   a) $f(x) = x$ b) $f(x) = x^2$ c) $f(x) = x^3$

5. Determine the degree of the polynomials and express it in the form with increasing powers of x.

   a) $y = 2(x+3)$ b) $y = (x+3)(x-2)$ c) $y = x - 1/2(x+2)$

   d) $y = x^5 - 2x^2 + 5$ e) $y = 5x + 1/2(1-10x)x$ f) $y = x^2 - 3x^3 + x$

6. i) Determine the point-slope equation of the line passing through the given pair of points. ii) Determine the y-intercept of the line, and iii) sketch the graph of the line.

   a) $(1,0), (2,5)$ b) $(3,-2), (-2,3)$ c) $(5,1), (6,1)$

   d) $(2,2), (4,1)$ e) $(2.1,-2), (0.1,3.4)$ f) $(-2,2), (4,1)$

7. Determine the point-slope equation of the indicated line, for the point with x-coordinate $x_0 = 1$.

   a) The line passing through $(3,3)$ with slope 6. b) The line with slope -3 and y-intercept 6.

   c) The line passing through $(3,3)$ with slope -1/6. d) The line with slope 3 and y-intercept 6.

   e) The line with slope 5 passing through $(-2,1.5)$. f) The line passing through $(-2.1,3.4)$ with slope 0.7.

   g) The line with slope 3/8 and x-intercept 4. h) The line with slope -0.05 and y-intercept 0.65.

8. If y is linearly related to x and z is linearly related to y, show that z is linearly related to x.

9. If the slopes of the lines l_1, l_2, l_3, and l_4 in the adjacent graph are m_1, m_2, m_3 and m_4, respectively, how are the numbers m_1, m_2, m_3 and m_4 related?

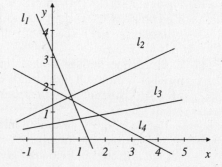

10. Transform the given quadratic to the shifted-scaled form and sketch its graph.

    a) $y = x^2 - 6x + 10$ b) $y = x^2 + 2x + 1$

    c) $y = (x - 3)(x - 1)$ d) $y = 3 - x^2$

    e) $y = -x(2x + 1)/4$ f) $5y = x + 2 - y^2$

    g) $x + y = x^2$ h) $y = (x - 2)(5 - x)$

11. If $f(x) = (x^2 - 1)^{1/3}$, evaluate a) $f(2)$ b) $f(x^{1/2})$ c) $f(x - 1)$ d) $f(x + h)$

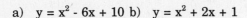

12. Sketch and give the equation of a parabola centered at the point (2,-3) that passes through (3,-5).

13. Given the sketch of the graph of a function g(x), sketch the graphs of the curves:
    a) $y = g(x - 1)$ b) $y = -2 + g(x + 1)$
    c) $y = 0.5g(x)$ d) $y = 1 + g(x + 2)$

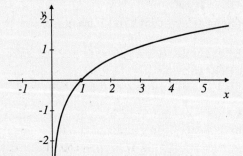

Fig. A $y = g(x)$

14. Complete the sentence.
    a) The graph of the equation $y = 3(x - 2)^3$ is obtained from the graph of $y = x^3$ by .....
    b) The graph of the equation $y = 1 - (x + 3)^4$ is obtained from the graph of $y = x^4$ by .....
    c) The graph of the equation $y = -2(x + 3)^3$ is obtained from the graph of $y = x^3$ by .....
    d) The graph of the equation $y = -2 + 0.5(x + 4)^2$ is obtained from the graph of $y = x^2$ by .....

15. Match each curve in the graph to one of these equations.
    a) $y = 2 + (x - 3)$
    b) $y = 2 - (1/3)(x - 3)$
    c) $y = 2 - (x - 2)^2$
    d) $y = 2 + (x - 2)^2$
    e) $y = 1 - (x - 1)^3$
    f) $y = 1 + (x - 1)^3$
    g) $y = -1 + (x + 0.5)^{2/3}$
    h) $y = -1 + (x + 0.5)^{1/3}$
    ** see the next section for more information about the graphs of these functions.

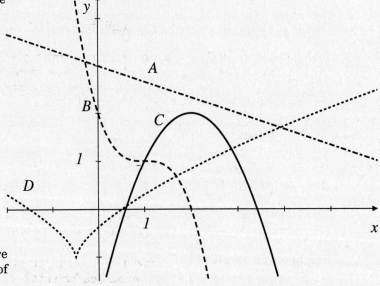

16. Assume that the growth rate of a plant, r(t), is a quadratic function of the number of days, t, since the seed was sown. If the plant is not growing at seeding and ceases to grow after 40 days, what is the equation for r(t) if the plant's maximum growth rate is $r_{max} = 4$ cm /d?

17. A population suffers from a contagious disease. Let the total size of the population be N and the number of infected individuals on day n be I(n).
    a) If the number of noninfected individuals on day n is S(n), give an equation relating S(n), I(n) and N.
    b) If there is a constant infection rate so that $I(n + 1) = I(n) + 5$, how many individuals are infected after 12 days, assuming initially there were 2 infected, I(0) = 2.?
    c) Assuming the number of new infected individuals is a constant proportion of the number of noninfected individuals, give an equation for the change $\Delta(n) = I(n + 1) - I(n)$, in the number of infected individuals.

Section 1.3 Basic Functions: $|x|$, x^r, and $[\![x]\!]$

The Absolute Value of a Function.

You know the absolute value of a number is positive!
By definition, $|x|$ equals x if $x \geq 0$ and -x if $x < 0$. An alternative way to define the absolute value is by

$$|x| \equiv \text{SQRT}(x^2)$$

You may think this definition complicated, but, as $\text{SQRT}(x^2)$ is by convention the positive root, this definition is consistent with our basic definition of $|x|$. This definition will prove useful when we consider "derivatives" of $|x|$ in a later chapter.

Given a function f(x), the function $y = |f(x)|$ is defined by

$$|f(x)| \equiv \begin{cases} f(x) & \text{if } f(x) \geq 0 \\ -f(x) & \text{if } f(x) < 0 \end{cases}$$

We could similarly define $|f(x)| = [f(x)]^{1/2}$.

The graph of $y = |f(x)|$ is identical with the graph of $y = f(x)$ when $f(x) > 0$ and is its mirror image, reflected about the x-axis, when $f(x) < 0$. Thus, to sketch $y = |f(x)|$ you must first identify the points where the graph of f(x) crosses the x-axis, if any. Then, to form the mirror image of the graph of f(x) over those x-intervals where $f(x) < 0$.

Example 1   Graphing $|f(x)|$.

Problem   Sketch the graph of $y = |x^2 - 4x + 3|$.

Solution   First, sketch the graph of

$$f(x) = x^2 - 4x + 3 = (x - 1)(x - 3)$$

This is a parabola that opens upward, intersecting the x-axis at $x = 1$ and $x = 3$. Since the shifted form of the function f is

$$f(x) = -1 + (x - 2)^2$$

its graph's vertex is at the point (2,-1).

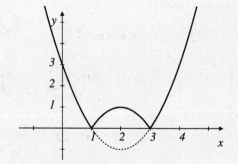

The graph of $|x^2 - 4x + 3|$.

$f(x) \geq 0$ on the intervals $(-\infty,1]$ and $[3,\infty)$

and hence on these intervals $|f(x)| = f(x)$. On the open interval (1,3) $f(x) \leq 0$ and hence

$$|f(x)| = -f(x) = -x^2 + 4x - 3 \quad \text{when } 1 < x < 3$$

The graph of $|f(x)|$ over the interval (1,3) coincides with the graph of $y = -f(x)$; it is the mirror image of the parabola $y = f(x)$ reflected through the x-axis; it has a vertex at (2,1). In the sketched graph the dotted curve is the portion of the graph of f that is flipped or rotated above the x-axis.

Allometric functions: $y = x^r$

Power functions with integer exponents, like $y = x^5$, can be defined by repeated multiplication. Functions with non-integer exponents, like $y = x^{5.2}$, are more difficult to define. The traditional name for a function $y = x^r$ where r is an arbitrary real number is an *allometric function*. Such functions can be defined in a systematic way:

First $y = x^n$ for n a natural number is defined as x multiplied by itself n-times.

Next, functions with negative integer exponents, $y = x^{-n}$, are defined as the multiplicative inverses of power functions.

Functions with fractional exponents or the form $r = 1/n$ are defined as the inverses of power functions with respect to function composition.

Rational exponent function, of the form $y = x^{m/n}$ where m and n are integers, are then defined as the function $x^{1/n}$ raised to the power m.

Finally, although we shall not explore the process in this text, $y = x^r$ for an irrational number r is defined as the limit of a sequence of functions with rational exponents, a limit as the rational exponents approach the irrational number r.

In the following we consider these functions and their graphs.

Reciprocal Functions: $y = x^{-n}$.

The **reciprocal of a natural number n** is denoted by 1/n and is by definition the *multiplicative inverse* of the number. i.e.,

$$1/n \text{ is the number that satisfies } n \cdot 1/n \equiv 1$$
$$\text{where n is an integer and } n \neq 0$$

Similarly, the **reciprocal of a function f(x)**, denoted by 1/f(x), is the multiplicative inverse of f(x).

DEFINITION OF THE RECIPROCAL FUNCTION 1/F(X).

> 1/f(x) is the function that satisfies
>
> $$f(x) \cdot 1/f(x) \equiv 1$$
>
> for all x such that $f(x) \neq 0$

The reciprocal of the power function x^n is $1/x^n$; by definition, this function satisfies

$$x^n \cdot 1/x^n = 1 \text{ for all } x \neq 0$$

An alternative notation for the reciprocal $1/x^n$ employs a negative sign in the exponent to indicate the reciprocal. This forms the basis for the definition of the allometric function x^r for $r = -n$, a negative integer exponent.

$$x^{-n} \equiv 1/x^n \qquad \text{for } x \neq 0 \qquad\qquad \textbf{Definition of } x^{-n}$$

Thus, by definition, x^{-n} satisfies the algebraic identity

$$x^n \cdot x^{-n} = 1 \text{ for } x \neq 0$$

CAUTION

By definition, an exponent -1 means the multiplicative reciprocal

$$x^{-1} \equiv 1/x$$

However, f^{-1} was used to denote the inverse of a function. To denote a reciprocal function, $1/f(x)$ the -1 exponent is applied to the right side of the argument, i.e., as

$$[f(x)]^{-1} = 1/f(x)$$

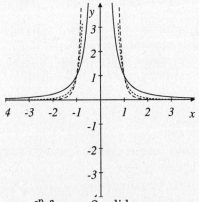

$y = x^{-n}$ for n = 2 solid curves, n = 4 dotted curves, n = 6 dashed curves.

The graph of the reciprocal function $y = 1/x^n$ or $y = x^{-n}$ consist of two segments, one defined for $x > 0$ and the other for $x < 0$.
The nature and relationship between these two graph segments depends on whether the number n is an even or odd number. The function x^{-n} is said to have a *singularity* at $x = 0$, where it is not defined.

When n is even, $y = x^{-n}$ is an even function and these segments are the mirror image of each other through the y-axis.

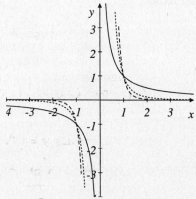

$y = x^{-n}$ for n = 1 solid curves, n = 3 dotted curves, n = 5 dashed curves.

When n is odd, $y = x^{-n}$ is an odd function and the two segments of $y = x^{-n}$ are the image of each other reflected through the origin, or, equivalently, under a rotation of 180° or π radians.

The right-graph segment of $y = x^{-n}$.

For all n, the right-segment of the graph $y = x^{-n}$, for $x > 0$, lies in the first quadrant and passes through the point (1,1). This curve is *decreasing*, and approaches the x-axis as the x increases. Mathematically, the segment is said to be *asymptotic* to the x-axis as x approaches ∞[†], i.e., graphically, as x approaches the positive end of the x-axis. Tracing this segment, moving from right to left, as x approaches zero from the right the segment approaches the upper portion of the y-axis. The curve is said to be *asymptotic* to the y-axis as x approaches zero from above.

For different exponents n, the positive segments of the curves $y = x^{-n}$ only intersect at the point (1,1). Otherwise, they never cross over each other. However, these graph segments satisfy a distinct graphical ordering. If $x > 1$ then the curves are closer to the x-axis for larger n, e.g., the graph of x^{-4} is lower than the graph of x^{-3} when $x > 1$. When x is between 0 and 1 this relationship is reversed, e.g., the graph of x^{-4} is above the curve x^{-3} when $0 < x < 1$.

[†] The symbol ∞ is not a number. It represents the concept of a numerical value greater than any real number; $r < \infty$ for any number r. Similarly, -∞ is used to represent the concept of a number smaller, i.e. more negative than, any real number; $-\infty < r$ for any number r. Graphically, ∞ and -∞ are the limits toward the right and left ends of an axis or number line.

The left-graph segment of $y = x^{-n}$.

The left-segment of the graph $y = x^{-n}$ is a continuous curve defined for negative x. This curve will lie in the second quadrant when n is even and in the third quadrant when n is odd. The left-segment will also be asymptotic to the x-axis, however, as x approaches $-\infty$. These left-segments are all asymptotic to the y-axis as x approaches zero through negative values, i.e., from below. However, the approach depends on whether n is even or odd; as x approaches zero from the left it is always negative and hence the sign of x^n and consequently its reciprocal x^{-n} will be negative if n is odd and positive if n is even.

Limits.

The verb *approaches* which we used to describe the graphs is very important in mathematics, especially in calculus, and corresponds to the mathematical concept of a *limit*. The phrases "as t approaches 5" indicates a limit, involving a variable t which is changing and becoming closer and closer to the limit number, 5. This limit process is denoted symbolically by

$$\lim_{t \to 5}$$

When the limit number is ∞, the phrase "as x approaches infinity" means that x increases, and becomes greater than any finite number. This is denote symbolically by

$$\lim_{x \to \infty}$$

When writing these limit phrases by hand or when they are set graphically the variable and its limit are often placed directly below the term "lim", which stands for limit, as in

$$\lim_{t \to 5} \quad or \quad \lim_{x \to \infty}$$

As x increases, the values of x^{-n} decrease. To indicate that as x becomes very large the values x^{-n} become very small and approach zero we write the limit equation

$$\lim_{x \to \infty} x^{-n} = 0$$

which is read "the limit as x goes to infinity of x to the minus n is equal to zero."

To indicate the limit of the values $y = x^{-n}$ on the right graph segment as x approaches zero we need to introduce a restricted type of limit, called a *one-sided limit*. This employs a sign following the limit number, a "+" sign to indicate that only values greater than (or to the right of) the limit number are considered and a "-" sign to indicate that only values less than (or to the left of) the limit number are considered:

$$\lim_{x \to a+} \text{ considers only } x > a, \text{ x-values to the right of } a.$$

$$\lim_{x \to a-} \text{ considers only } x < a, \text{ x-values to the left of } a.$$

The fact that the right-segment of the graph $y = x^{-n}$ approaches the y-axis is indicated algebraically by the one-sided limit

$$\lim_{x \to 0+} x^{-n} = \infty$$

This equation is read "the limit as x goes to zero from above (or from the right) of x to the negative n is equal to infinity." More mathematically, this means that for any given real number M there is some x-value x_0 such that if $x \in (0, x_0)$ then $y = x^{-n} > M$.

The fact that the left-segments of the graph of x^{-n} are asymptotic to the x-axis as x approaches $-\infty$, and consequently that the values of x^{-n} become closer to zero than any given non-zero number, is expressed by the limit equation:

$$\lim_{x \to -\infty} x^{-n} = 0$$

For $x < 0$, the values x^{-n} become very large positive numbers when n is even and very large negative numbers when n is odd as x approaches zero from the left, $x \to 0-$. If $x = -0.01$ then $x^{-4} = 10{,}000{,}000$ while $x^{-3} = -100{,}000$. This can be expressed by the limit statement

$$\lim_{x \to 0-} x^{-n} = \begin{cases} \infty, & \text{if n is even;} \\ -\infty, & \text{if n is odd.} \end{cases}$$

The limit part of this equation is read "the limit as x approaches zero from below (or from the left)..." The right side of this equation is a split statement with the various lines applying when the corresponding "if... " condition is valid; e.g., for n = 3 the limit is $-\infty$ and for n = 6 the limit is ∞.

Sketching the graph of $y = b + m(x - a)^{-n}$.

The standard approach introduced in Section 1.2 to sketch the shifted or scaled polynomial functions can be applied to sketch the graph of shifted or scaled reciprocal power functions. A basic reciprocal graph $y = x^{-n}$ can be scaled in the y-direction by multiplying by a constant, m, to form the graph of $y = mx^{-n}$. This scaled graph can then be translated by a horizontal shift of a units and a vertical shift of b units to give a graph that is *centered* at the point (a, b). The equation of the scaled, shifted graph is then

$$y = b + m(x - a)^{-n}$$

When sketching such shifted graphs, it is often helpful to sketch a *shifted axis* system, consisting of the horizontal line $y = b$ and the vertical line $x = a$. These lines intersect at the *center* point (a,b). The shifted graph will consist of two segments, sometimes called *branches*, that are asymptotic to the shifted axes.

KEY POINTS

The right branch   passes through the point $(a + 1, b + m)$.

It approaches the line $y = b$ as $x \to \infty$.

It approaches the line $x = a$ as $x \to a+$, i.e. x approaches a from the right and thus is always greater than a. If $m > 0$ it approaches the positive end of the line $x = a$ and if $m < 0$ it approaches the negative end of the line.

The left branch   passes through the point $(a - 1, b + m)$ if n is even and through $(a - 1, b - m)$ if n is odd.

It approaches the line $y = b$ as $x \to -\infty$.

It approaches the line $x = a$ as $x \to a-$, i.e. from the left (x is less than a but gets closer and closer to a).
For $m > 0$, when n is even it approaches the positive end of the line, toward the top of the graph, and when n is odd it approaches the negative end of the line. For $m < 0$ these are reversed.

When $m > 0$, the graph of $y = b + m(x - a)^{-n}$ will be similar to one of the two graphs illustrated in in the following sketches. For n even the graph will look like the graph on the left and for n odd it will resemble the graph on the right.

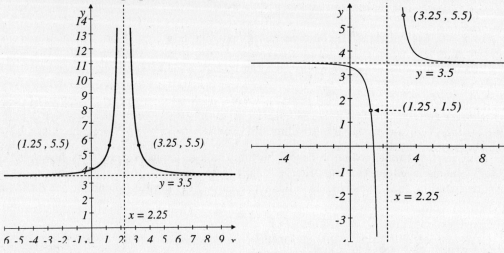

The graph of $y = 3.5 + 2(x - 2.25)^{-2}$.   The graph of $y = 3.5 + 2(x - 2.25)^{-3}$.

Example 2 **Graphing a shifted scaled reciprocal function.**

Problem Sketch the graph of the function

$$y = -2 + 5/(x - 4)$$

Solution This function is the basic reciprocal function $y = 1/x$, scaled by the factor 5 and shifted to be centered at the point (4,-2).

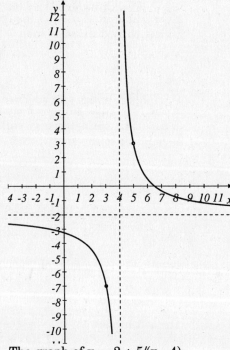

Its graph is a shifted hyperbola. It consist of two branches, the right branch is defined for $x > 4$ and the left branch for $x < 4$. These branches approach the same asymptotes, the horizontal line $y = -2$ and the vertical line $x = 4$. To sketch the graph we first draw these lines, the dotted lines in the sketch, as they form the "shifted axes" they are the image of the original x-y axes under the shift $(x, y) \to (x + a, y + b)$.

The right branch of the curve passes through the point $(4 + 1, -2 + 5) = (5,3)$.

The left branch of the curve passes through the point $(4 - 1, -2 - 5) = (3,-7)$.

The graph of $y = -2 + 5/(x - 4)$.

The nth root function: $y = x^{1/n}$.

As discussed above, an allometric function with a negative exponent, like x^{-3}, is defined as a reciprocal, as the multiplicative inverse of the function x^3. Allometric functions with fractional exponents of the form 1/n, like $x^{1/3}$, are defined as another type of inverse, the inverse with respect to composition.

The **inverse of the power function x^n**, which is referred to as the **n^{th} root** function, is denoted by **$x^{1/n}$** is defined by the property that

$$(x^n)^{1/n} = x \quad \text{and} \quad (x^{1/n})^n = x$$

for all $x \geq 0$ if n is even and for all x if n is odd.

As discussed in Section 1.1, the graphs of a function and its inverse are the symmetrical images of each other through the line $y = x$. Thus, the graph of $y = x^{1/n}$ and $y = x^n$ are reflections of each other through the line $y = x$. If you sketch the graph of $y = x^n$ on a coordinate system drawn on a square sheet of paper and then, holding the paper in the upper right and lower left corners, rotate the paper so that you are looking at the other side, you can see a graph through the paper when it is held up to a bright light. The graph you see through the paper is the graph of $y = x^{1/n}$. In this process the y-axis of the original x^n graph becomes the x-axis of the $x^{1/n}$ graph.

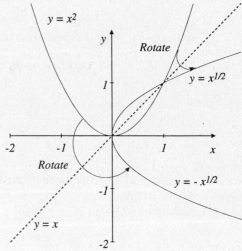

For even n, flipping the graph $y = x^n$ results in a graph that appears like a parabola opening to the right. This can not be the graph of a function. Since the image of the right branch of the graph $y = x^n$, for $x \geq 0$, is the upper portion of the flipped graph, by restricting the domain of x^n to $[0,\infty)$ gives just the portion of its graph to the right of the y-axis. When this portion is flipped it forms only the upper part of the parabola type curve and hence corresponds to the graph of a function, the positive value of the nth root function $x^{1/n}$. If n is odd, this restriction is not needed since the flipped image of the graph of x^n is the graph of a function defined for all x.

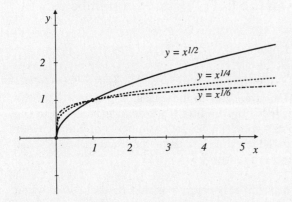

The graph of $x^{1/n}$ looks like the graph of the square root function $x^{1/2}$. It is a curve that originates at the origin, passes through the point (1,1) and is always *concave downward*, which means it is "turning" downward and is increasing at a slower rate as x increases. The curve $x^{1/n}$ appears to become flatter as x increases, although it never stops increasing. As x approaches ∞ the y-values of these functions approach ∞, however, much more slowly. The larger n is the more "level" the graph of $x^{1/n}$ appears, however it is never horizontal.

For odd n, the graph of $x^{1/n}$ is a curve that is symmetric about the origin. For n = 1 the graph is simply the straight line $y = x$. As seen in the sketch on the following page, for n = 3, 5, ... the curves for $x \geq 0$ looks like the graphs for even n. Rotating the graph for positive x about the origin by 180° gives the graph for $x \leq 0$.

The Domain of $x^{1/n}$

> The Natural Domain of $x^{1/n}$ is
>
> $[0,\infty)$ if n is an even integer,
>
> $(-\infty,\infty)$ if n is an odd integer.

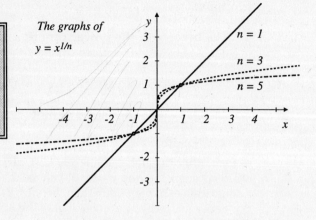

The graphs of $y = x^{1/n}$

Fractional Powers: $x^{m/n}$

A rational exponent is a fraction, a ratio of two integers. To evaluate x^r when r is a rational number we first, if necessary, reduce r so that

$$r = m/n, \text{ where m and n have no common divisors.}$$

Then, by definition, x^r is the nth root of x raised to the power m:

$$x^{m/n} \equiv (x^{1/n})^m$$

Thus $y = x^{m/n}$ is the composition of the power function $y = x^m$ and the nth root function $y = x^{1/n}$. Consequently, as x^m is defined for all x, the Natural Domain of $x^{m/n}$ is the same as the Natural Domain of $x^{1/n}$:

$$y = x^{m/n} \text{ is defined } \quad \text{for all x if n is odd and}$$
$$\text{only for } x \geq 0 \text{ if n is an even number.}$$

For an even denominator.

If n is even the graph of $x^{m/n}$ will be similar to the graph of the nth root function $x^{1/n}$ when the numerator m < n; the differences will be that as the numerator m increases, m = 2, m = 3, ... the corresponding graphs will become more linear, approaching the line y = x as m gets closer to the number n. As m gets larger, the graph of $x^{m/n}$ will rise from the origin less quickly and not appear to level off as quickly as x increases. This is illustrated, on the next page, for the function $x^{m/6}$ for m = 1, 2, ..., 6. When m = 6, $x^{m/6}$ = x and the graph is the line y = x.

When the numerator m > n the graph of $x^{m/n}$ are similarly related as m increases. But, when m > n, the graph of $x^{m/n}$ will resemble the right branch of a u-shape curve. $y = x^{m/n}$ is only defined only for $x \geq 0$, except when m is actually a multiple of n. The curves $y = x^{m/6}$ for m = 7, 8, .., 12 are sketched as dotted curves in the right panel of the following figure. For m = 12 the graph is just the graph of x^2 and hence it is actually defined for all x. These graphs are computer generated. You would have to plot many points to determine these shapes. One of the useful applications of CALCULUS will be to establish the shapes of such graphs by applying simple tests to the "derivatives" of the functions $y = x^{m/n}$.

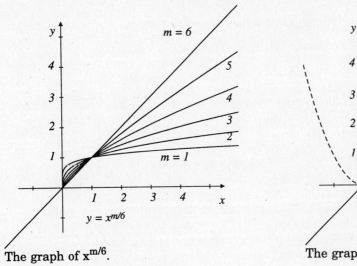

The graph of $x^{m/6}$.

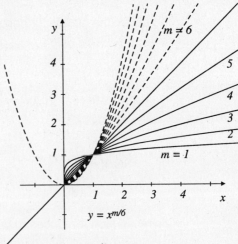

The graph of $x^{m/6}$ for $m = 1$ to 12.

For an odd denominator.

When the denominator n is odd, $x^{m/n}$ is defined for all x. If the numerator m is also an odd number then the graph of $x^{m/n}$ resembles the graph of $x^{1/n}$ when m < n, think of the graph of the cube root $y = x^{1/3}$; it has a flat s-shape, and is symmetric about the origin. When m > n the graph of $x^{m/n}$ more generally resembles the graph of a cubic $y = x^3$.

However, when the numerator m is even, the graph of $x^{m/n}$ is symmetric about the y-axis. When m < n the graph has the shape of a vee, with curved branches and a point at the origin. It is said to be *cusp shaped* at the origin. These shapes are illustrated below for the curves $y = x^{m/7}$ for odd m, 1, 3, 5, and 7, in the left panel and even m, 2, 4, and 6, in the right panel.

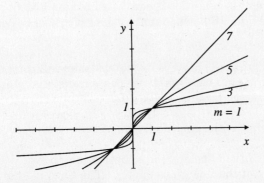

The graphs of $x^{m/7}$.
The left panel has m = 1, 3, 5, and 7.

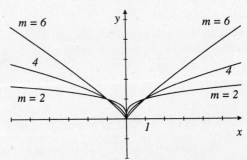

The right panel has m = 2, 4, and 6.
The shape of the graphs at the origin is a *cusp*.

A COMPUTER CAUTION . . . .

Computer programs are frequently not programmed to distinguish when a power x^r is defined for negative x-values. Consequently, in many programs, if you instruct a program to plot $x^{3/7}$ you will likely only get the plot for $x \geq 0$.

To obtain the graphs for negative x-values you must trick the computer. One way is to use the function

$$y = [SGN(x) \cdot |x|^{1/n}]^m$$

where the function SGN(x) (read "sign of x") is a standard function defined by

$$SGN(x) = \begin{cases} 1, & \text{if } x \geq 0 \\ -1, & \text{if } x < 0 \end{cases}$$

This ploy works and gives $x^{m/n}$ for all x-values except when m is a multiple of n, because then it reduces to powers of $|x|$ instead of powers of x.
The more mathematical name for this sign function is SIGNUM.

Shifted fractional powers.

The shifted and scaled function with a fractional power

$$y = b + k(x - a)^{m/n}$$

can be sketched by scaling the y-coordinates of the fractional power function $x^{m/n}$ by the factor k, and then shifting this graph to be centered at the point (a,b). If the scaling constant k is negative, the scaling will involve a reflection of the graph of $x^{m/n}$ about the x-axis, so that the *cusp* will point upward instead of downward. The shifted graph will be symmetric about the vertical line $x = a$.

Example 3 **Sketching a shifted fractional power.**

Problem Sketch the graph of $y = 2 + 0.25(x - 5)^{2/3}$.

Solution Because the denominator of the exponent is m/n = 2/3 is odd, this function is defined for all x. Because its numerator is even and less than the denominator, the graph will be *cusp shaped*.

The scaling coefficient is +0.25; being positive, this tells us that the *cusp* will point downward, being a number less than one means the graph will be compressed towards the x-axis.

The equation is in shifted form and *centered* at the point (5,2), which will be its *cusp point*.

To sketch the graph we plot at least two additional points; the easiest points to evaluate are those with x-coordinates one unit on either side of the line of symmetry. The term $(x - 5)^{2/3}$ is easily evaluated at these x-values, being $(\pm 1)^{2/3} = 1$, thus, at $x = 6$ and at $x = 4$ the y-coordinate will be the same, $y = 2 + 0.25$. If we wish to plot additional points to guide our sketch it is best to choose x-coordinates for which the term $(x - 5)^{2/3}$ can be readily evaluated. Since $8^{1/3} = 2$, if we let $x = 13$ or $x = -3$ the corresponding y-coordinate on the graph is

$$y = 2 + 0.25(8)^{2/3} = 3$$

Plotting first the points (5,2), (6,2.25), (4,2.25), (13,3) and (-3,3), the graph is sketched through these with a *cusp* at the vertex (5,2). This is a vee-shaped curve. Notice that this curve turns downward on both sides of the *cusp* point as it moves away from the vertex. It is said to be *concave downward* on the half-lines $(-\infty,5)$ and $(5,\infty)$.

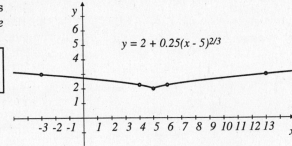

x^r when r is irrational.

A power function with an irrational number, like x^π, can be defined as the limit of power functions with rational exponents. Consider the function $y = x^r$ where r is an irrational real number. Irrational numbers can be approximated by rational numbers. For each number n there is a rational numbers r_n that is the decimal expansion or r with only n decimal places, i.e., the remaining infinite decimal expansion is set to zero. Then the difference

$$|r - r_n| < 10^{-n}$$

Hence, as n increases (we say n goes to infinity, $n \to \infty$) the difference $r - r_n$ approaches zero. The value of x^r is defined as the "limit" of the values of x raised to the powers r_n, using terminology that will be introduced in a later section:

$$x^r = \lim_{n \to \infty} x^{r_n}$$

Calculators use a different approach to calculate powers, they employ logarithms, which will be discussed in the next Section. For irrational r, x^r can only be defined when $x \geq 0$ as $x^{m/n}$ is only defined for $x \geq 0$ when n is even and the rational numbers r_n may have this form.

If r is irrational, the Natural Domain of x^r = $[0,\infty)$.

For practical purposes, if you wish to sketch the graph of a function like $y = x^\pi$ you can use a decimal approximation of π, say 3.1, and graph the function $y = x^{31/10}$. Using a graphing resolution of 0.001, i.e., plotting points with x-values increasing by 0.001, the graphs of x^π and $x^{31/10}$ over the interval [-2,3] will be indistinguishable when displayed on a graphing calculator or computer screen.

Discrete Functions

The function that rounds all numbers down is called the *greatest integer* function. A common representation for this function uses double square brackets as ⟦x⟧. This notation will be used in this text, but this function is denoted in computer applications using the function names like FLOOR(x), GINT(x), or @INT(x). In the computer literature it is often denoted by ⌊x⌋ and the graphic character ⌊ is called a "left floor" and ⌋ is a "right floor".

⟦x⟧ ≡ the greatest integer less than or equal to x

For positive numbers, ⟦x⟧ is simply the integer portion of the number when it is expressed as a decimal. Thus ⟦3.01⟧ = 3 and ⟦5.9999999⟧ = 5.

For negative numbers, rounding down a number between two integers results in the "larger" negative integer. For instance, ⟦-2.005⟧ = -3 and ⟦-8.99999⟧ = -9. The domain of the function ⟦x⟧ is $(-\infty,\infty)$ but

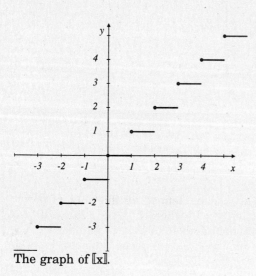

The graph of $[\![x]\!]$.

the Range of $[\![x]\!]$ is I, the set of integers. The graph of the function consists of horizontal line segments that include their left endpoints but do not include their right endpoints.

The function $[\![x]\!]$ is often called a *step* function. It can be combined with other functions to model phenomena that have abrupt changes, like the income tax rates for different income levels or a digital clock display. The simplest combination is with the scaling function that multiplies by a constant.

The function $[\![x]\!]$ can be scaled in two ways. Multiplying $[\![x]\!]$ by a constant c scales the y-values. The graph of

$$y = c \cdot [\![x]\!]$$

consists of horizontal line segments of length 1, including their left endpoints but not their right endpoints. The graph changes at each integer x-value and is constant on the intervals between integers. For instance,

for $3 \le x < 4$ $c \cdot [\![x]\!] = 3c$

for $4 \le x < 5$ $c \cdot [\![x]\!] = 4c$

and

for $-2 \le x < -1$ $c \cdot [\![x]\!] = -2c$

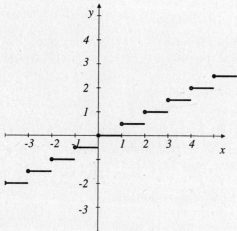

The graph of $y = 0.5 [\![x]\!]$.

The Range of $y = c [\![x]\!]$ consists of the values ...-3c, -2c, -c, 0, c, 2c, 3c, ... In the middle graph at the left, the constant c = 0.5 and hence the jump at each integer x-value is 0.5.

Alternatively, we can first scale the x-variable by multiplying x by a constant and then taking the greatest integer. Using a different constant, denoted by k, $[\![kx]\!]$ scales in the x-direction. It effectively stretches the graph of $[\![x]\!]$ away from the y-axis if k > 1 and compresses it towards the y-axis if 0 < k < 1. The graph of the function

$$y = [\![k \cdot x]\!]$$

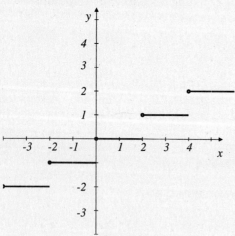

The graph of $y = [\![0.5x]\!]$.

consists of horizontal line segments at integer y-values of length $|1/k|$. If k > 0, the graph is similar to the graph of $[\![x]\!]$ except that it's jumps occur at integer multiples of 1/k. When k = 1/2, as in the figure ate the left, these intervals are 2 units long. For any integer n,

if $x \in [n/k, (n+1)/k)$ then $[\![k \cdot x]\!] = n$

since

$n/k \le x < (n+1)/k \implies n \le k \cdot x < n+1$

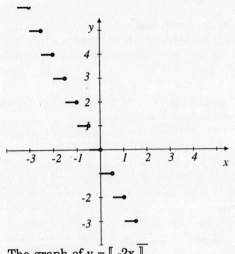

If the constant k is negative then the graph of y = ⟦k·x⟧ "steps down" instead of "up" as x increases. It is like the graph of ⟦x⟧ reflected about the y-axis, and then scaled in the x-direction by the factor 1/|k|. An important aspect in this case is that the line segments now include their right endpoints and not their left endpoints. This is because multiplying an inequality by a negative number reverses the inequality. If k < 0 then k/|k| = -1 and hence for any integer n

$$n/|k| < x \leq (n+1)/|k| \implies -n > kx \geq -(n+1)$$

Thus, if x is in the interval (n/|k| , (n + 1)/|k|],

$$⟦kx⟧ = -(n+1)$$

The graph of y = ⟦ -2x ⟧.

The graph of y = ⟦ f(x) ⟧

If f(x) is a given function then the composition of the greatest integer function with f results in a discrete-valued function, y = ⟦ f(x) ⟧:

$$⟦ f(x) ⟧ = n \text{ if } n \leq f(x) < n + 1$$

Example 4   The function ⟦ x^2 ⟧

Problem   Where does the graph of y = ⟦ x^2 ⟧ jump? What does its graph look like?

Solution   To determine the x-values at which the graph of y = f(x) = ⟦ x^2 ⟧ may "jump" we solve the algebraic equations f(x) = n for n an integer:

$$x^2 = n, \text{ for } n = 0, 1, 2, 3, ...$$

i.e., solve this for each n,

$$n = 0 \longrightarrow x = 0$$

$$n = 1 \longrightarrow x = 1 \text{ or } x = -1$$

$$n = 2 \longrightarrow x = \sqrt{2} \text{ or } x = -\sqrt{2}$$

$$n = 3 \longrightarrow x = \sqrt{3} \text{ or } x = -\sqrt{3}$$

$$n = 4 \longrightarrow x = 2 \text{ or } x = -2$$

Each of these x-values corresponds to a "jump" point, except for x = 0. The graph of the function ⟦x^2⟧ will remain constant on the intervals between these x-values. For positive x-values this constant is the value of ⟦x^2⟧ at the left endpoint of the interval:

$$\text{for } \sqrt{2} \leq x < \sqrt{3}, \ 2 \leq x^2 < 3 \text{ and hence } ⟦x^2⟧ = 2.$$

For negative x-values the constant is the value of ⟦x^2⟧ at the right endpoint of the interval:

$$\text{for } -2 < x \leq -\sqrt{3}, \ 3 \leq x^2 < 4 \text{ and hence } ⟦x^2⟧ = 3.$$

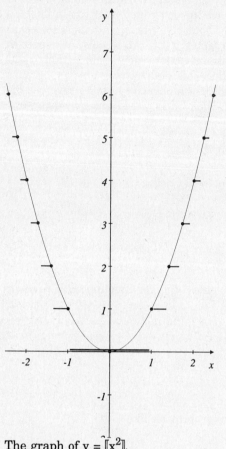

The reason the graph does not jump at x = 0 is because 0 is the left endpoint of the interval [0, 1) on which the function $[\![x^2]\!] = 0$ and the right endpoint of the interval (-1, 0] on which the function also is identically zero.

The graph of $y = [\![x^2]\!]$ is sketched by first lightly sketching the graph of $y = x^2$.

Then, a solid dot is sketched on this graph at each of the critical x-values, which are simply $\pm\sqrt{n}$ for n = 0, 1, 2, 3, ... For each positive integer n, this can be most easily done by placing a straight edge or ruler parallel to the x-axis and intersecting the y-axis at y = n. The points where this straight edge crosses the graph of $y = x^2$ will have x-coordinates $-\sqrt{n}$ and $\sqrt{n}$ and will be endpoints of horizontal line segments on the graph of $y = [\![x^2]\!]$.

Once these dots are sketched, the horizontal line segments are drawn.

Starting at a dot, draw a line proceeding horizontally toward an adjacent higher dot. The point on the line segment directly below an adjacent higher dot is the endpoint of the line segment but it <u>is not</u> included. The resulting sketch is illustrated at the left. Notice that the left endpoints are included when x^2 is increasing but the right endpoints are included when it is decreasing.

The graph of $y = [\![x^2]\!]$.

The graphical method employed in the previous example can be used to sketch the graph of $y = [\![f(x)]\!]$ for any function f(x) whose graph can be sketched first.

Start with the graph of y = f(x), like the curve in the graph at the right.

For each integer n <u>on the y-axis</u> lay a straight edge or ruler parallel with the x-axis forming the horizontal line y = n.

Place a dot on the graph of f(x) at each point where the rule crosses it. These dots indicate the points on the graph of $[\![f(x)]\!]$ where $[\![f(x)]\!] = f(x) = n$. The graph of $[\![f(x)]\!]$ may "jump" at these points.

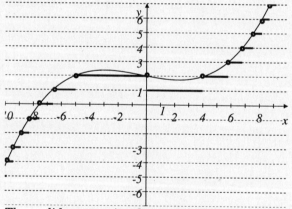

The solid curve is the graph of f(x). The horizontal line segments are the graph of $[\![f(x)]\!]$.

The graph of $[\![f(x)]\!]$ will be constant on the subintervals formed by the x-coordinates of these points.

To complete the sketch of $y = [\![f(x)]\!]$, draw horizontal line segments as follows. The x-coordinates of adjacent dots form the endpoints of an interval on which $y = [\![f(x)]\!]$ is constant. Starting at the lower of two adjacent dots, draw a horizontal line moving in the direction of the higher dot. The line segment includes the lower dot but not the endpoint below the higher dot.

Two special cases arise if two adjacent dots have the same height, i.e. the same y-coordinates. Consider the line segment connecting these dots. If the graph of f(x) lies above this segment then this segment forms part of the graph of ⟦ f(x) ⟧. On the other hand, if the graph of f(x) lies below this segment then the graph of ⟦ f(x) ⟧ includes not this segment but a parallel segment one unit lower.

This procedure can be applied if we know the graph of the function f(x) without even knowing its algebraic formula. This is illustrated in the previous figure. Notice that the line segment connecting the dots at (-5,2) and (0,2) since the graph of f lies above this. However, the segment from (0,1) to (4,1) lies below the graph of f and does not include its endpoints. You do not need to know the equation for f(x) to apply this graphical method.

Functions involving ⟦x⟧.

Sometimes a function will involve combinations of the greatest integer function and other functions, like

$$f(x) = x / ⟦x⟧ \quad \text{or} \quad f(x) = 3x - ⟦ 3x ⟧$$

To sketch the graph of such functions you must first establish the x-coordinates where the greatest integer function jumps. If the function involves the basic ⟦x⟧ then these jumps will occur at integer x-values. If the function involves ⟦ 3x ⟧ then, as discussed above, the jumps will occur at integer multiples of 1/3. The equation of the function f(x) will have a different form on each x-subinterval between two consecutive jump values.

Example 5   **Sketching a function involving ⟦x⟧.**

Problem   Sketch the graph of $f(x) = x / ⟦x⟧$.

Solution   Since ⟦x⟧ jumps at integer x-values, the graph of the function f(x) will change at points (n, f(n)). The nature of the graph must be consider on each interval [n, n + 1). For instance, for n = 0, 1, 2, and 3:

On the interval [0,1)   ⟦x⟧ = 0 and hence f(x) is not defined.

On the interval [1,2)   ⟦x⟧ = 1 and hence f(x) = x.

On the interval [2,3)   ⟦x⟧ = 2 and hence f(x) = x/2.

On the interval [3,4)   ⟦x⟧ = 3 and f(x) = x/3.

Continuing this way, we see that on the interval [n , n + 1)

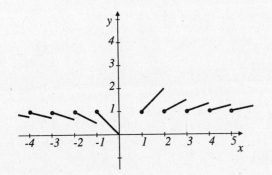

The graph of y = x / ⟦x⟧.

⟦x⟧ = n   and   f(x) = x/n

Notice that f(n) = 1 for each non-zero integer. The graph of f(x) consists of line segments with different slopes.

The line segment for x ∈ [n , n + 1) connects the point (n,1) and the point (n + 1 , 1 + 1/n) and has slope 1/n.

The left point (1, n) is included in the graph but the right point (n + 1, 1 + 1/n) is not on the graph. Notice that for negative intervals the slopes of these line segments are negative. For instance, on the interval [-3,-2) ⟦x⟧ = -3 and hence f(x) = x/-3 and the graph is a segment of the line y = -⅓x.

The slopes of these line segments approach zero as the intervals become father away from the origin. Thus, the line segments further from the origin are more horizontal.

Example 6   A "saw-tooth" function.

Problem   Sketch the graph of y = x - ⟦x⟧.

solution   On each interval between integer x-values the graph of this function looks the same. It is a segment of a line y = x shifted horizontally ⟦x⟧ units.

For $n \le x < n + 1$, ⟦x⟧ = n

and on this interval the function is

y = x - n

This is the line with slope 1 that passes throught (n,0).

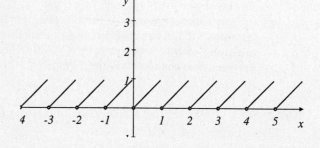

The graph of this function does not include the right end point of the line segments. This is a function that never attains its maximum value since y < 1 for all x.

Exercise Set 1.3

1   Use a scientific calculator to evaluate the given number, or indicate why it can not be evaluated.

   a) $2^{1.2}$ b) $-3^{1/3}$ c) $(-3)^{1/3}$ d) 2^π e) $(-2)^{3.25}$ f) $(-2)^{0.64}$ g) $(-4)^{0.8}$ h) $(-5)^{-1.1}$

2   Sketch the graph of |f(x)|.

   a) f(x) = x - 1   b) f(x) = -0.5x + 2   c) f(x) = 1.5 + 3(x - 6)   d) f(x) = x^2 - 1

   e) f(x) = $-0.5x^2 + 2$ f) f(x) = $-1 + (x - 4)^2$ g) f(x) = x^3 h) f(x) = $x^3 + 2$ i) f(x) = $(x - 4)^3$

3   Sketch the graph of the given function.

   a) $y = x^{-1}$ b) $y = (x + 1)^{-1}$ c) $y = 2 + (x - 1)^{-1}$ d) $y = x^{-2}$ e) $y = (x + 1)^{-2}$

   f) $y = 2 + (x - 1)^{-2}$ g) $y = x^{-3}$ h) $y = (x + 1)^{-3}$ i) $y = 2(x + 1)^{-3}$ j) $y = 1 + 2(x + 1)^{-3}$

Section 1.3 BASIC FUNCTIONS 49

4. Sketch the graph of the given function.

   a) $y = x^{1/2}$ b) $y = (x+1)^{1/2}$ c) $y = 2 + (x-1)^{1/2}$ d) $y = x^{1/3}$ e) $y = (x+1)^{1/3}$

   f) $y = 2 + (x-1)^{1/3}$ g) $y = x^{2/3}$ h) $y = (x+1)^{2/3}$ i) $y = 2 + (x-1)^{2/3}$

5. Identify the correspondence between the given functions and the curves in the graph.

   a) $y = 4 + (6-x)^{3/4}$ C

   b) $y = 3 + (x+1)^{1/3}$ D

   c) $y = 2 + (x-1)^{1/2}$ A

   d) $y = 2 + (x-4)^{3/5}$ B

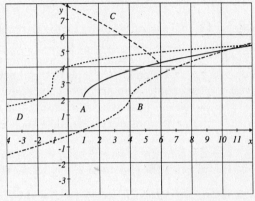

6. Sketch the graph of the function over the interval [-4,6].

   a) $[\![x]\!]$ b) $[\![2x]\!]$ c) $[\![x/4]\!]$ d) $[\![-x]\!]$ e) $2[\![x]\!]$ f) $0.2[\![x]\!]$

   g) $0.5[\![x/2]\!]$ h) $[\![x+2]\!]$ i) $[\![2x-1]\!]$ j) $2[\![x]\!] - 1$ k) $[\![2-x]\!]$ l) $[\![2-x/3]\!]$

7. Sketch the given function over the interval [-2,2].

   a) $[\![\tfrac{1}{4} x^2]\!]$ b) $[\![0.5x^3]\!]$ c) $[\![x^2 - x]\!]$ d) $[\![x]\!]^2$

   e) $[\![x]\!] - x$ f) $2x - [\![2x]\!]$ g) $x - 0.5[\![2x]\!]$ h) $[\![\sqrt{x}]\!]$

8. A taxi company charges $2.00 for the first 0.3 km and $0.50 per km or fraction of a km thereafter. Construct a function that describes the fare as a function of the distance x (km) traveled.

9. The 1994 Canadian Federal Income tax calculated on your Taxable income as 17% of the amount less than $29,590 plus 26% of Taxable income over this amount but less than $59,180, plus 29% of Taxable income over this amount. Then, the Basic Provincial Ontario income tax is 58% of the Federal tax. The Provincial sur-tax is 20% of the Basic Provincial tax exceeding $5,500 plus an additional 10% of the Basic Provincial tax exceeding $8,000.
   a) Calculate a formula that gives the total Federal Tax.
   b) Calculate a formula that gives the total Provincial Tax.
   c) Calculate a formula that gives the total Income Tax for an Ontario resident.

10. Sketch a graph of the combined Provincial Sales Tax and the Goods and Services Tax for cost P from $0.00 to $2.00.

Section 1.4 Exponential and Logarithm Functions

EXPONENTIAL FUNCTIONS

In this section we discuss **exponential** functions of the form $y = b^t$, where the independent variable t appears in the exponent and the positive number b is called the **base**. When b = 1 this function is simply the constant $y = 1^t = 1$. Consequently, we will normally consider $b \neq 1$. An alternative notation for an exponential employs the function name **exp**, with the base indicated by a subscript, and the exponent as an argument in parentheses:

$$y = b^t \quad \text{is equivalently to} \quad y = \exp_b(t)$$

which is read "y equals b to the power t" or "y equals the exponential, base b, of t."[†] The $\exp_b(--)$ notation will be useful when the exponent is a function of t or a subscripted variable. For instance,

$$2^{3t^2 - 4t + 5} \quad or \quad 10^{t_2}$$

can be written without using a graphic generator as

$$\exp_2(3t^2 - 4t + 5) \quad or \quad \exp_{10}(t_2)$$

These forms are easier to type on a word processor and may help avoid mistakes.

Algebraic Properties of Exponentials

E-P1 Since b is positive, $b^t > 0$ for all t.
 The Natural Domain of $y = b^t$ is $(-\infty, \infty)$.
 The Range of $y = b^t$ is $(0, +\infty)$.

E-P2 For any base b: $b^0 = 1$ and $b^1 = b$.

E-P3 $b^{t+k} = b^t \cdot b^k$ for any two numbers t and k

E-P4 $(b^t)^k = b^{t \cdot k}$

E-P5 If **b > 1**, then $y = b^t$ is an increasing function on $(-\infty, +\infty)$.

 If **0 < b < 1** then $y = b^t$ is a decreasing function on $(-\infty, +\infty)$.

The most common exponential functions are those with base 2, 10, or e; these are the functions $y = 2^x$, $y = 10^x$, and $y = e^x$. The number $e \approx 2.7182818182845904$ is a very important irrational number that arises naturally in many physical and biological models. As we shall see, all exponential functions are related and can actually be expressed in terms of the exponential function e^t. This will prove very useful when we study calculus as the function e^t is the only function that is equal to its calculus derivative and to its calculus integral.

The graphs of the exponential functions have one of two basic forms:

[†] Most computer programs use notation such as EXP(b,t) or @EXP(t*@ln(b)), which avoid the use of subscripts and exponents so that the function can be expressed on a single line of ordinary text.

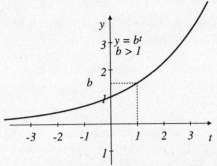 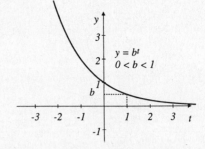

If **b > 1**, the graph of $y = b^t$ is increasing for all t-values. It passes through the points (0,1) and (1,b). As $t \to -\infty$ the graph asymptotically approaches the negative part of the t-axis.

For **0 < b < 1**, $y = b^t$ is a decreasing function that passes through the same y-intercept, (0,1), and the point (1,b). However, this graph is asymptotic to the positive x-axis as $t \to \infty$.

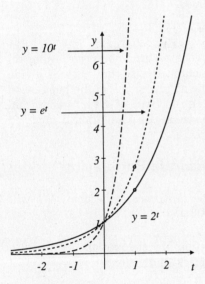

Graphs of $y = b^t$ for different bases will only intersect at the point (0,1). They do have an orderly relationship, as illustrated at the left for the base 2, e, and 10. Notice that, since these bases are greater than 1, as the base increase the values b^t become greater for t > 0, and smaller for t < 0, e.g.,

$$2^2 < e^2 \leq 10^2 \quad \text{and} \quad 10^{-1} < e^{-1} < 2^{-1}$$

However, these relationships vary in the opposite direction when the bases are less than one. This is illustrated at the right for the bases 1/5 = 0.2, 1/2 = 0.5, and 0.9. Notice that for b = 0.9 the graph appears almost horizontal in the limited *window* of the graph. If this curve were sketched over a larger range of t-values it would have the characteristic curved shape that is associated with exponential curves.

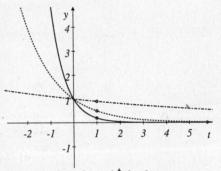

The graphs of $y = b^t$ for bases b = 0.2 solid curve, b = 0.5 dotted curve, and b = 0.9 dot-dashed curve.

The Exponential function: $y = e^t$.

The most common exponential function in calculus applications is **the exponential** with base *e*. The number *e* cannot be defined without utilizing the concept of a limit. In will later be defined as the limit of the term $(1+\varepsilon)^{1/\varepsilon}$ as ε becomes small and using the calculus concept of an *integral* in terms of the area under a curve $y = 1/t$. It is sufficient to think of *e* as a specific number that can be approximated as precisely as we wish. Often, for quick calculations only a few decimal places are required and for very rough estimates you may think of *e* as a number a little less than 3.

All the exponential curves have similar shapes and are algebraically equal to a scaled form of the basic exponential $y = e^t$. For any base b, there is a constant k such that

$$b^t = e^{kt} \quad \text{for all t.}$$

How the constant k is determined will be discussed later in this section. Because of this relationship we shall focus our discussion of exponential functions on ***the exponential*** function, e^t, and its scaled forms

$$y = c\, e^{kt} \quad \text{or} \quad y = c \exp(kt)$$

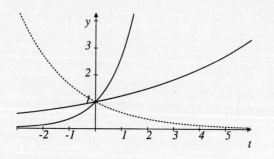

Graphs of $y = e^{kt}$ for $k = 1, 0.2$ and -0.5.

where c is a scaling constant in the y-direction and k is a scaling constant in the t-direction.

If the base is not indicated, it is assumed to be $b = e$,

$$\exp(t) \equiv e^t$$

Growth or decay of a population or substance that is proportional to the size of the population or substance leads to models in which the substance is described by exponential functions.

Example 1 A discrete model of a cell population: exponential growth.

Background Consider a population of bacteria cells in a homogeneous environment containing all nutrients necessary for growth. If the population is synchronized, so that all cells divide at the same time, the number N of cells present after the nth division (i.e. the population of the nth generation) is denoted by $N(n)$ for $n = 0,1,2,...$. As each cell of the nth generation divides to form two cells of the $(n + 1)^{st}$ generation, the size of the population is modeled by the following equation:

$$N(n + 1) = 2 \cdot N(n) \qquad \textbf{\textit{Discrete} Population Model}$$

for $n = 0,1,2,...$ Equations of this type are called *difference* equations. We will examine such equations in detail in Chapters 2 and 7.

Problem Determine the general solution of the given Discrete Population Model equation. Use this to predict the size of a cell culture that is initially 2,000 cells after 50 divisions.

Solution Assume that N_0 denotes the initial population $N(0)$, then

$$N(1) = 2 \cdot N(0) = N_0 \cdot 2$$

$$N(2) = 2 \cdot N(1) = 2 \cdot N_0 \cdot 2 = N_0 \cdot 2^2$$

$$N(3) = 2 \cdot N(2) = N_0 2^3$$

$$N(4) = 2 \cdot N(3) = N_0 2^4$$

$$\vdots$$

This sequence of evaluations establishes a pattern. It leads to a general equation by replacing 4 in the last equation by n:

$$N(n) = N_0 2^n$$

The function $N(n)$ is an exponential function. Its base is the number 2 and the independent variable is n. N_0 is a scaling coefficient. To verify that this function $N(n)$ is a solution of the general Model Equation, we substitute it, and the corresponding equation with n replaced by $n + 1$

$$N(n + 1) = N_0 2^{n+1}$$

into the Model Equation. Then, we algebraically rearrange the equation to verify that it is valid, i.e., the left side does equal the right side, for all values of n. For this example the substituted Model equation is

$$N_0 2^{n+1} = 2 \cdot N_0 2^n \; (= N_0 2^{n+1})$$ which is true for all n!

If $N_0 = 2{,}000$, substituting $n = 50$ we find

$$N(50) = 2{,}000 \cdot 2^{50} = 2{,}251{,}799{,}813{,}685{,}248{,}000$$

On your calculator, depending on its accuracy, you might evaluate this rather large number approximately, in scientific notation, as $N(50) \approx 2.2517998 \; 10^{18}$.

Exponential growth and decay.

Many processes and systems are characterized by the way they change with time. In Example 1 the population growth occurred at discrete times that were a constant generation or reproduction period apart. The model resulted in an exponential equation describing the cell population that did not depend on the length of the generations, simply that they were distinct, and that the change in the population at each discrete generation was directly proportional to the current size of the population at that generation. Many quantities change "instantaneously" over a period of time by a process in which the "rate of change" is proportional to the current size. The dynamics of how such quantities change continuously are modeled by *differential* equations, in which the rate of change is referred to as a "derivative". The first portion of our study of calculus will focus on "derivatives". As we will see later in the text, when the rate of change is proportional to the amount present at any time, the solution of such dynamical models express the amount present at time t as a scaled exponential function with base e.

More specifically, assume the amount of a substance at time t is $y(t)$ and the rate of change in $y(t)$ is proportional to $y(t)$. The dynamical model of how $y(t)$ changes is referred to as either a **growth model** if the amount is increasing or as a **decay model** if the amount is decreasing.

To describe both of these models we will use a positive constant k.

The Exponential Growth Model
If the quantity $y(t)$ increases or *grows* at a rate $k \cdot y(t)$ then $y(t) = y_0 e^{kt}$.

In this situation, the model parameter k is called a **growth rate** and the amount $y(t)$ is said to be increasing or **growing exponentially**. The initial amount of the quantity at time $t = 0$ is indicated by the coefficient y_0, as $y(0) = y_0 \, e^{k \cdot 0} = y_0 \, e^0 = y_0$, because $e^0 = 1$.

The Exponential Decay Model
If the quantity $y(t)$ decreases or *decays* at a rate $-k \cdot y(t)$ then $y(t) = y_0 e^{-kt}$.

If the amount $y(t)$ is decreasing at a constant proportional rate k ($k > 0$), in which case the actual rate of change is $-k$ times $y(t)$, it is said to be **decaying exponentially**.

In this text, we will always use a positive constant k and
employ the negative sign to indicate the decay model.

Although technically the rate of decay is $-k \cdot y(t)$, it is common practice is to simply refer to the parameter k as the *decay rate*. You should be aware that semantics are not consistent! You will encounter Professors and other text books and literature references that use $-k$ to denote the **decay rate**. This can be confusing but the context of a model should make it clear that a Decay Model is being considered. Again, the initial value of the solution $y(t)$ at $t = 0$ has been denoted by y_0.

Example 2 Radioactive decay.

Background Radioactive isotopes such as ^{14}C (carbon 14), ^{24}Na (sodium 24), ^{32}P (phosphorus 3), and ^{131}I (iodine 131) are widely used as tracers to study the movement or concentration of particular compounds both in vitro (i.e., in a laboratory) and in vivo (i.e., in their biological environment). The utility of these isotopes lies in the fact that they emit negatively charged particles and are thus easily monitored. However, these isotopes undergo transmutation in the process of emitting the particles and decay to nonradioactive forms. The amount of particles emitted is proportional to the amount of isotope present, which in turn indicates the amount of "labelled compound" present. To determine the amount of labelled compound present, it is necessary to account for the decay of the radioactive isotope. The basic assumption in radioactive assay is that the rate of decay is a constant k (which is indicated by -k to emphasize that the amount present is decreasing). The number of radioactive atoms present at time t is given by

$$n(t) = n_0 e^{-kt}$$

where n_0 is the initial number of radioactive atoms.

Problem The decay rate for ^{24}Na is 0.0462 hr^{-1}. If a sample contained 4 µg (this is read four micro grams) of this sodium isotope, how much isotope would be present after 5 hours?

Solution The model equation for the exponential decay of the sodium give the amount at time t after the initial determination as

$$n(t) = 4e^{-0.0462t}$$

At t = 5 hours, the amount is thus

$$n(5) = 4e^{-0.0462 \cdot 5} = 4e^{-0.2310} \approx 3.1748 \text{ µg}$$

Notice that although units are not indicated in the equations as they are being evaluated, they are indicated in the final answer.

Historically, prior to the advent of electronic calculators, evaluating exponentials was difficult, especially for base e. However, our decimal number system made it easier to evaluate 10^t and these values were put into tables and referred to as *anti-logarithms*. Many important models utilized the exponential base 10 instead of *e*. The following example introduces two models of this type.

Example 3 Extinction Coefficients in Chemical Laws.

Background Colorimetry and spectrophotometry are commonly used techniques in chemistry and biochemistry. They utilize the laws of Lambert and Beer concerning the passage of light through solutions.

Lambert's Law gives the intensity of light at a depth *l* in an absorbing medium, like a solution, when the intensity of light incident to the top surface is I_0. If the medium absorbs the light at a constant proportional rate, then an exponential decay model is appropriate and the equations is thus an exponential with a negative coefficient. Historically, the base 10 is used and the resulting equation is

$$I(l) = I_0 10^{-kl}$$ **Lambert's Law**

The constant k depends on the specific absorbing solution utilized and is called the *extinction coefficient*. The extinction coefficient k is the reciprocal of the depth l_{10} of the specific absorbing solution at which the intensity of the transmitted light is one-tenth of the intensity of the incident light, i.e. $k = 1/l_{10}$ where $I(l_{10}) = I_0 / 10$.

Beer's Law concerns the relationship between the Molar concentration, C (unit M = moles/liter), of a colored substance (being colored means it absorbs light of that color) in a solution and the transmittance of monochromatic light (the light only contains one color) through the solution. Again let I_0 denote the light intensity incident to a surface of the solution and let $I(l)$ denote the light intensity transmitted at a depth l. Based on the same exponential decay model, the function $I(l)$ is given by an exponential function, however, for Beer's Law the exponent is expressed in terms of two constants.

$$I(l) = I_0 10^{-\varepsilon C l} \qquad \textbf{Beer's Law}$$

The constant ε is the *molar extinction coefficient*; ε is a diffusion or scattering constant that depends on the wavelength (i.e., color) of the transmitted light and the solution. In practice one of the constants C or ε is known and the other is obtained using the above equation with experimentally determined values of I at different depths.

Problem The extinction coefficient of reduced nicotinamide-adenine dinucleotide phosphate (NADPH$_2$) at a wavelength of 340 μm is $\varepsilon = 6.22 \cdot 10^3$ cm^2/M. A 3-ml solution containing 0.2μ mole NADPH$_2$ is placed in a cuvette of length 1.05 cm. (A cuvette is a receptacle used to hold a liquid sample in a spectrophotometer, which measures light intensities.)
What percentage of the light is transmitted through the cuvette?

Solution The percentage of light transmitted is $100 \cdot I/I_0$. The equation for I is given by **Beer's Law**. The depth is $l = 1.05$. To evaluate $I(1.05)$ we first must calculate the Molar concentration of NADPH$_2$ by dividing the number of moles by the volume as follows:

$$C = \frac{0.2\mu \, moles}{3ml} = \frac{0.2 \times 10^{-6} moles}{3 \times 10^{-3} l} = 0.67 \times 10^{-4} M$$

The exponent in Beer's Law is then evaluated as

$$-\varepsilon C l = -(6.22 \cdot 10^3 \, cm^2/mole)(0.67 \cdot 10^{-4} \, moles/l)(1.05 \, cm) = -4.37577 \cdot 10^{-1} \, cm^3 / l$$

This term is dimensionless since a liter equals one thousand cubic centimeters, $l = 10^3 \, cm^3$:

$$-\varepsilon C l = -4.37577 \cdot 10^{-4}$$

Therefore the percentage of light transmitted through this solution at 340 mμ is

$$P = 100 \cdot [I(1.05) / I_0] = 100 \cdot 10^{-0.000437577} = 10^{-0.0437577} \approx 0.90\%$$

LOGARITHM FUNCTIONS.

To solve an equation for a variable in the exponent, for instance, to solve

$$3^t = 5$$

for t, we must reverse the exponential process that forms 3^t. This is possible because the exponentials are one-to-one functions. For $b > 1$, b^t is increasing on $(-\infty, +\infty)$ while for $0 < b < 1$ it is decreasing. Hence, for $b \neq 1$, the exponential function $f(t) = b^t$ has an inverse function (with respect to composition). Recall that inverse functions are characterized by the properties that

$$f[f^{-1}(t)] = t \quad \text{and} \quad f^{-1}[f(t)] = t$$

The inverse of $f(t) = \exp_b(t)$ is denoted by the function $f^{-1}(t) = \log_b(t)$.

DEFINITION: THE LOGARITHM WITH BASE b.

> For any positive number $b \neq 1$, the **logarithm function with base b** is the inverse of the exponential function with base b, and is denoted by
>
> $$\log_b(x)$$
>
> read "log base b of x". It satisfies the relation
>
> $$y = \log_b(x) \quad \text{if and only if} \quad b^y = x$$

By definition the function $\log_3(t)$ is the inverse of the function $\exp_3(t)$. Hence, $\log_3(3^t) = t$ and the equation $3^t = 5$ is solved for t by *taking the logarithm*, with base 3, of both sides of the equation:

$$t = \log_3(3^t) = \log_3(5)$$

What is the numerical value of $\log_3(5)$? The value of $\log_b(t)$ for particular numbers b and t is usually not obvious, unless t is recognized as a power of b. Of course the number $\log_3(5)$ can be evaluated on a scientific calculator, but what if the battery is dead? We may be able to estimate this value by evaluating $\log_3(x)$ for some x near 5. For instance, $\log_3(9) = 2$ since $9 = 3^2$ and as $3^1 = 3$, $\log_3(3) = 1$. Thus, since $3 < 5 < 9$, we might expect $\log_3(5)$ to be between 1 and 2. We discuss how to evaluate $\log_3(5)$ on a calculator later in this section.

You should be able to evaluate simple logarithms without a calculator! For example,

$\log_2(8) = 3$ since $2^3 = 8$ $\log_{10}(0.01) = -2$ since $10^{-2} = 0.01$

$\log_{\sqrt{2}}(16) = 8$ since $(\sqrt{2})^8 = 16$ $\log_7(1) = 0$ since $7^0 = 1$

$\log_{1/2}(0.25) = 2$ since $(1/2)^2 = 0.25$ $\log_5(125) = 3$ since $5^3 = 125$

When the base is 10 or *e* the logarithms are denoted as Log(t) or *ln*(t), respectively. On a scientific calculator there are usually two logarithm buttons: one typically is [Log] or [log] and the other is [ln]. The log or Log button refers to the $\log_{10}$ which is called the *common logarithm*. The function *ln*(x) is read "the *natural logarithm* of x", or simply as "el-en of x"; $ln(x) \equiv \log_e(x)$.

Algebraic Properties of Logarithms

L-P1 $\text{Domain}(\log_b) = (0,\infty)$, $\text{Range}(\log_b) = (-\infty,\infty)$.

L-P2 $\log_b(1) = 0$ for all b,   and $\log_b(b) = 1$.

L-P3 $\log_b(t_1 t_2) = \log_b(t_1) + \log_b(t_2)$.

L-P4 $\log_b(t^k) = k \cdot \log_b(t)$.

L-P5   If $b > 1$, $\log_b(t)$ is increasing on $(0,\infty)$.

       If $b < 1$, $\log_b(t)$ is decreasing on $(0,\infty)$.

Section 1.4 — Exponential and Logarithm Functions

The properties of logarithms are illustrated by the following.

L-P1 $\log_3(5)$ is defined but $\log_3(-5)$ is not defined. As the range of $\log_{0.1}$ is the set of all real numbers, there is a number x such that $\log_{0.1}(x) = -27$.

L-P2 $\log_6(1) = 0$ since $6^0 \equiv 1$, by definition of the zero exponent.
$\log_6(6) = 1$ since $6^1 = 6$.

L-P3 $\log_2(8) = \log_2(2^3) = 3 \log_2(2) = 3 \cdot 1 = 3$.

L-P5 $\log_2$ is increasing while $\log_{1/2}$ is decreasing.

The most useful properties of logarithms for the purpose of arithmetic calculations are L-P3 and L-P4. By utilizing these properties the operations of multiplication and exponentiation can be reduced to the simpler operations of adding and multiplying, respectively. We will utilize this property repeatedly in our study of calculus to transform functions so that they are easier to work with.

One particularly useful combination of these properties results in the formula.

$$\log_b\left(\frac{A}{B}\right) = \log_b(A) - \log_b(B)$$

This is accomplished by using the fact that when the exponent k = -1 in property L-P4,

$$-\log_b(B) = \log_b(B^{-1}) = \log_b(1/B)$$

The fact that all logarithm functions are multiples of each other is derived from the general "change-of-base" formula which relates logarithms of different bases:

$$\log_a(x) = \frac{\log_b(x)}{\log_b(a)} \qquad \textbf{Change of Base Formula}$$

This formula is obtained as follows. Assume that $y = \log_a(x)$. Then, by definition,

$$a^y = x$$

Taking the logarithm with base b of this equation gives

$$\log_b(a^y) = \log_b(x)$$

By the exponential rule, L-P4, the exponent y on the left side can be moved outside the logarithm as a multiplicative coefficient:

$$y \cdot \log_b(a) = \log_b(x) \quad \text{or} \quad y = \log_b(x) / \log_b(a)$$

Remembering that we began by assuming $y = \log_a(x)$, substituting $\log_a(x)$ for y in the last equation gives the change of base formula. The most common use of this change of base formula is to convert between $ln(x)$ and $\log(x)$. Setting

a = 10 and b = e: $\log(x) = ln(x)/ln(10) \approx 0.4343\, ln(x)$

and

a = e and b = 10: $ln(x) = \log(x)/\log(e) \approx 2.3026\, \log(x)$

Thus, to evaluate (approximate) $\log_3(5)$, you can use either the [Log] or [ln] calculator button:

$$\log_3(5) = \log(5)/\log(3) = 0.69897 / 0.47712 = 1.46497$$

or

$$\log_3(5) = ln(5)/ln(3) = 1.60943 / 1.09086 = 1.46497$$

Example 1 Solving equations with logarithms.

Problem Solve for x when $3^x = e^{x-2}$.

Solution This is done by taking the logarithm of both sides of the equation. Technically you could use any base, but it simplifies the calculation to use one of the bases present, either 3 or e.

$$ln(3^x) = ln(e^{x-2})$$

$$x\, ln(3) = x - 2 \quad \text{or} \quad x = 2/(1 - ln(3))$$

Alternatively, taking the $\log_3$ gives

$$\log_3(3^x) = \log_3(e^{x-2})$$

$$x = (x - 2)\log_3(e) \quad \text{or} \quad x = -2\log_3(e)/(1 - \log_3(e))$$

Graphing Logarithm Functions.

The graphs of logarithm functions are simply the graphs of the corresponding exponential functions reflected across the line $y = x$. The following graph illustrates this for the bases $b = 2$, e, and 10. Observe that the graphs of the logarithm functions have an ordered relationship that is just the inverse of how the exponential functions are related for different bases.

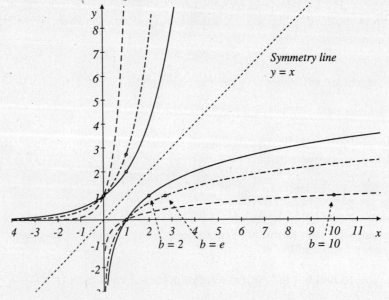

While people most commonly think of the graph of a log function as having the shape indicated in the figure above, when the base b is between zero and one, the graph of $\log_b$ has the decreasing form illustrated at the right. Notice that as the base number becomes closer to one the graphs become steeper. When b is close to one, log(b) is close to zero. From the Change of Base formula,

$$\log_b(x) = \log(x)/\log(b)$$

Then, $\log_b(x)$ is log(x) scaled by dividing by a small number, which is the same as multiplying by a large number. For instance,

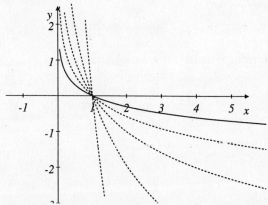

The graphs of $y = \log_b(x)$ for b = 0.1 (solid curve) increasing to 0.9 in steps of 0.2.

$\log_{0.9}(x) \approx -21.85 \log(x)$ and $\log_{0.99}(x) \approx -229.1 \log(x)$

Example 2 **Half-life of an exponentially decaying substance.**

Background The term *half-life* is used in many contexts and normally means a unit of time in which an exponentially decaying substance will decrease to one half its starting size. Traditionally the half-life is denoted as $t_{1/2}$. The important thing to remember is that the substance will decrease by 50% over an interval $[T, T + t_{1/2}]$ no matter what the starting time T, or the size of the substance at this starting time.

Problem Determine a formula for $t_{1/2}$ when the amount of material is described by the Exponential Decay Model with decay constant k.

Solution The amount of material is described by the exponential function $\quad y(t) = y_0 e^{-kt}$

$$y(t) = y_0 e^{-kt}$$

The number $t_{1/2}$ is determined by solving the equation $y(t_{1/2}) = y(0)/2$. The solution will express $t_{1/2}$ in terms of the decay constant k. To find $t_{1/2}$ we must solve the "implicit" equation

$$y_0 e^{-k \cdot t_{1/2}} = \frac{y_0}{2}$$

The solution does not depend on the initial value y_0 which cancels, leaving the equation

$$\exp(-kt_{1/2}) = 1/2$$

Taking the natural log of this equation gives $\quad t_{1/2} = \ln(2)/k.$

$$ln(\exp(-k\, t_{1/2}) = ln(1/2) \quad \text{or} \quad -k\, t_{1/2} = ln(1/2)$$

Dividing by -k, and using the fact that $-ln(1/2) = ln(2)$ gives

$$t_{1/2} = ln(2)\,/\,k \qquad \textbf{Half-life}$$

CAUTION.... remember this formula assumes a positive k and a negative sign in the exponential model $y = y_0 e^{-kt}$.

Example 3 Carbon-14 dating.

Background The radioactive isotope ^{14}C is present in all living matter. The main source of this isotope is carbon dioxide in the air, which contains ^{14}C isotopes as the result of cosmic radiation. In living material the level of ^{14}C is constant as the decay of the ^{14}C is counterbalanced by the assimilation of new ^{14}C. However, at death the assimilation stops and the ^{14}C present slowly disappears as it decays. The amount of ^{14}C in a sample t years after death is given by

$$A(t) = A_0 e^{-kt}$$

where -k is the decay rate and A_0 is the amount of ^{14}C present at the time of death, t = 0. The half-life of ^{14}C is $t_{1/2}$ = 5685 years.

Problem What is the decay constant for ^{14}C? What percentage of ^{14}C originally present in a sample will be present 10,000 years after it died? How long will it take for 60% of the ^{14}C to decay?

Solution The constant k is evaluated by solving the equation $A(t_{1/2}) = A_0/2$:

$$A_0 e^{-k t_{1/2}} = \frac{A_0}{2}$$

Cancelling A_0 gives

$$e^{-k t_{1/2}} = \frac{1}{2}$$

Taking the natural logarithm of both sides gives

$$-k \cdot t_{1/2} = ln(1/2) \quad \text{or} \quad k = -ln(1/2) / t_{1/2}$$

Since

$$-ln(1/2) = ln((1/2)^{-1}) = ln(2)$$

the constant k is given by

$$k = ln(2)/ t_{1/2} = 0.69314 / 5685 \approx 0.0001219 = 1.219 \; 10^{-4}$$

The percentage of the original ^{14}C left after T-years will be

$$A(T)/A_0 \cdot 100\%$$

Thus, the percentage left 10,000 years after it ceases to live is:

$$100 \cdot A(10,000)/A_0 = 100 \; e^{-k \cdot 10,000} = 100 \; e^{-1.21926} = 29.54\%$$

To establish the time required for 60 percent of the ^{14}C to decay, we determine when 40% will remain. Solving $A(t)/A_0 = 0.40$ for t as follows;

$$e^{-kt} = 0.4 \;\Rightarrow\; -kt = ln(0.4) \;\Rightarrow\; t = -ln(0.4) / k = 7515.16 \text{ years}$$

Example 4 Using "logs" to establish an allometric model of leaf area.

Background In forestry and horticulture the amount of surface area of a tree or plant is an important variable that is difficult to measure. The technique usually employed is to relate this surface area to a more easily measured quantity, namely the "dry weight" of the plant or branches. As a plant grows, its leaf area, abbreviated LA, and the dry weight, W, are usually found to increase at rates proportional to the current area or weight. Assume the proportionality constants are r for leaf area and R for dry weights. Thus LAI and W are modeled by exponential growth equations.

Problem Establish a relationship between the leaf area LA and the dry weight W.

Solution The exponential growth equations for LA and W as functions of time t are

$$LA = LA_0 \, e^{rt} \quad \text{and} \quad W = W_0 \, e^{Rt}$$

To express LA as a function of W the time variable t is eliminated by solving the W equation for t and substituting this into the LA equation. Proceeding, we take the natural log of the W equations and solve for t:

$$ln(W) = ln(W_0 \, e^{Rt}) = ln(W_0) + ln(e^{Rt}) = ln(W_0) + Rt$$

$$t = [ln(W) - ln(W_0)] / R = (1/R)ln(W / W_0)$$

Substituting this value for t in the leaf area equation gives

$$LA = LA_0 \, \exp\{r(1/R)[ln(W/W_0)]\}$$

While this looks like an exponential function, it is not. The exponent is the term inside the braces { }, which can be rearranged to as a ln term:

$$r(1/R)[ln(W/W_0)] = ln[(W/W_0)^{r/R}]$$

and since $\exp(ln(z)) = z$ for any expression z,

$$LA = LA_0 \, \exp\{r(1/R)[ln(W/W_0)]\} = LA_0 \, \exp\{ln[(W/W_0)^{r/R}]\} = LA_0 \, [(W/W_0)^{r/R}]$$

This is an allometric function of the relative dry weight, W/W_0. If the growth rates r and R are equal, then the equation

$$LA = LA_0 \, [(W/W_0)^{r/R}]$$

reduces to a linear relationship $LA = LA_0 \, W/W_0$.

Example 5 **Beer's Law: determining extinction coefficients.**

Problem Determine the molar extinction coefficient ε used in *Beer's Law* if the percentage of light transmitted at the depth l = 3 cm was I = 30% for a 2 Molar concentration of a solute.

Solution Recall (Example 3) that the equation describing the transmitted light is

$$I(l) = I_0 10^{-\varepsilon C l} \qquad \textbf{Beer's Law}$$

The percentage transmitted light is then $I = 100 \cdot I(l)/I_0 = 100 \cdot 10^{-\varepsilon \cdot 2 \cdot 3}$

Solving $30 = 100 \, 10^{-6\varepsilon}$ for ε by taking logs of both sides gives

$$\log(30/100) = \log(10^{-6\varepsilon}) = -6\varepsilon \quad \text{thus} \quad \varepsilon = (-1/6)\log(0.3) = 0.08714$$

EXERCISE SET 1.4

1. Determine which is the larger of the given pairs of numbers by using general properties of exponentials and logarithms. CONFIRM YOUR ANSWER WITH A CALCULATOR.

   a) $2^5, 2\sqrt{5}$ b) $2^a, (1/2)^a$ c) $e^x, \sqrt{10^x}$ d) $3^2, 2^3$ e) $3^5, 5^3$

   f) $3, \log(3)$ g) $3, ln(3)$ h) $\log(5), \log(2\sqrt{5})$ i) $\log(536), 4$

   j) $a^x, \log_a(x)$ if $x > 0$ and $a > 1$ k) $a^x, \log_a(x)$ if $x > 0$ and $0 < a < 1$

2. Assume that a bacteria cell culture grows by division and initially contained 3000 cells. Assume that the generation time for this culture is 20 minutes.
   a) How large will the culture be in two days (assuming there are no deaths?)
   b) If 20 percent of the cells die each generation, how many cells are in the nth generation?
   c) If 20 percent of the cells die each generation, what will be the size of the culture in two days? (HINT: To evaluate the number, use the fact that if $x = a^b$ then $x = \exp(b\, ln(a))$.)

3. Sketch the graph of the given function.

   a) $y = 2^x$ b) $y = 2^{x-2}$ c) $y = 2^t/4$ d) $y = 10^x$ e) $y = (0.1)^t$ f) $y = e^{-t}$

4. The half-life of ^{24}Na is approximately 15 hours. If the amount of ^{24}Na initially present is $n_0 = 400$, how much will be present after 10 hours; 15 hours; and 20 hours?

5. Assume a radioactive material is weighted at times $t = 2$ and $t = 5$ and the weights were 25 g and 10 g, respectively.
   a) What is the decay rate of this substance?
   b) What is the half-life $t_{1/2}$ of this substance?
   c) How much of the material was present at time $t = 0$?
   d) Graph the decay curve for this substance.

6. a) Compute the $t_{1/4}$ time of exponential decay curve where $t_{1/4}$ is the time in which one-fourth of the initial material will decay.
   b) Assuming that a substance has a $t_{1/4}$ time of 5 days, determine the initial amount of the substance given that 7 g remained after 12 days.

7. A body placed in water at 0°C will lose heat at a rate proportional to its temperature. The temperature t minutes after immersion will thus be given $T(t) = T_0 e^{-kt}$, where the constant k reflects the surface area of the body and any insulation. If a human with body temperature 37°C fell into arctic waters at 0°C, she or he would lose consciousness if her or his body temperature reached 25°C. Assuming the decay rate is $-k = -0.02$, how long would a person survive in the water? If $-k = -0.2$, what would be the survival time?

8. To test the functioning of the thyroid gland, radioactive iodine ^{131}I is introduced into the body and its accumulation and decay in the gland are monitored. The half-life of ^{131}I is $t_{1/2} = 8$ days.

   a) What fraction of a dose of ^{131}I will decay each day?
   b) If the maximum permissible amount of ^{131}I is 5 units, how long after the administration of a dose of 4 units can a second 4-unit dose be safely administered?
   c) If the maximum permissible amount of ^{131}I is 5 units, what is the maximum dose which can be administered every 24 hours when the initial dose is 5 units?

9. Simplify the given expression.

   a) $\log(x^3) + \log(y) - \log(z)$ b) $10^{\log(3x)}/2^{\log(x)}$ c) $ln(e^{3x-2})/x$ d) $10^{3\log(x-1)}$

10 Solve the given equation for x.

   a) $35 = 10e^{-2x}$ b) $\log(2x) = 0.1$ c) $e^{3x} = 7$ d) $\ln(2x-1) = 7$

   e) $5^x = 10$ f) $e^x = 7e^{4x}$ g) $\log_3(5x) = 2$ h) $e^{3(x-2)} = 1/2$

   i) $\log(3) - \log(x) = 7$ j) $\ln(x+1)^2 = 0$ k) $e^{-0.1x} = 2$ l) $\log_2(x/3) = 7$

11 Solve for t.   a) $R = A e^{-4(t-2)}$ b) $\log(t^3) = 6$ c) $10^{3t} = 2^t$ d) $\ln(3t-2) = \ln(10)\log(t)$

   e) $5 \cdot 10^{2t-1} + 2 = A$ f) $2^{3t} = 8 \cdot 2^t$ g) $\log_5(3t-2) = -2$ h) $2^3 - 2\log_2(2t-1) = \log_2(5)$

12 Graph the indicated function.   a) $y = \ln(x)$ b) $y = \log(x)$ c) $y = \log_2(x)$

   d) $y = \log_3(x)$ e) $y = \log_2(2x)$ f) $y = \log(2x)$

13 Using Beer's Law, determine the molar extinction coefficient of a substance if a 0.8 concentration registered a value $I_1 = 530$ at a 3-cm depth and a 0.4 concentration registered a value $I_2 = 1060$ at the same depth and wavelength.

14 A cuvette of 2-cm depth and hence containing 2 ml of solution was used to measure the colored light absorbency of a 0.18-mole solution of protein. At a given wavelength, μ, 16 percent of the light was transmitted. What is the molar extinction coefficient of the protein at wavelength μ?

15 Light transmitted through a solution registered 525 at a depth of 2 cm and 50 at a depth of 5 cm.
   a) What is the extinction coefficient of the solution?
   b) What was the incident intensity I_0?

16 The "loudness" of sound is measured on a scale of "bels" (after Alexander Graham Bell). The "loudness", L, of a sound is related to its intensity, I, by the exponential relation $I = k10^L$.

   The "loudness" is a relative or subjective scale of measuring sound. To see this, let I_0 be the intensity of a sound just below the faintest sound which a person can hear (such as the sound of a pin dropped on grass). Define the "loudness" corresponding to I_0 to be $L_0 = 0$.

   a) Show that the loudness L satisfies $L = \log(I/I_0)$ bels = $10 \log(I/I_0)$ db, where db is decibels.
   b) What is the "loudness" in decibels of a sound whose intensity is $3000 \times I_0$?
   c) What is the increase in relative intensity between the sound of a watch tick at 20 db and the sound of normal conversation at 70 db?
   d) What is the intensity of an automobile horn if it has a "loudness" of 100 db and $I_0 = 1$?

17 Earthquakes are measured on the Richter scale expressed in terms of magnitude variable R. The intensity I of an earthquake vibration is an exponential function with base b = 10 of the Richter scale magnitude R.
   a) Show that the magnitude R satisfies the equation $R = \log(I/I_0)$, where I_0 is the intensity of normal earth vibrations corresponding to $R_0 = 0$.
   b) How much greater is the intensity of an earthquake measuring 7.2 on the Richter scale than one measuring 4.6?

18 The 1979 population of Canada was 23,597,600, and in 1991 it was 26,853,500.
   a) Assuming an exponential grow, estimate the instantaneous growth rate of the Canadian population.
   b) Convert this to an *annual growth rate* R by expressing the exponential population model in the form $P(t) = c R^t$.

Section 1.5 Periodic and Trigonometric Functions.

PERIODIC FUNCTIONS

One of the most obvious and profound phenomena in the biological world is the cyclic repetition of particular events and processes. Examples of such repetition abound, from the annual seasons of the year to the constant replication of DNA in cells which then divide to form new cells. In this section we introduce basic terminology for describing analytically such repetition and then we define the trigonometric functions which are the most important periodic functions.

The notion of repetition or cycling is associated with a linear variable, usually representing time. If some measurable aspect of the process is described by a function f(t), then as t increases the function values f(t) will repeat. The shortest time between repetitions is called the "period" of the process. The cyclic aspect of the process is observed analytically as the function repeats its values over intervals of time corresponding to the complete cycles of the process. the length of these intervals is the period.

Example 1 **Cell division; a periodic process.**

Background Yeast cells, such as *Schizosaccharomyces pombe*, multiply by a process of mitosis in which each cell duplicates its DNA and then divides to form two sister cells. This is a cyclic process in which two stages can be measured: (1) the time of division of the cells and (2) the time it takes a cell to replicate its nucleus, the nuclear division time. The cell grows linearly in length over the period from its birth to the time of nuclear division. It then ceases to grow in size during the time it takes to form new cell walls and divide. Thus, if the cycle of a cell is associated with an interval [0,D], the graph illustrates a cell's length over one cycle.

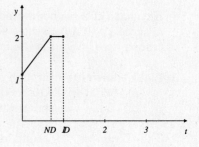

Problem Assume that the cell mitosis cycle is of length 1 time unit and the nuclear division occurs at 0.7 units of time. Assume a mature cell is of length 2 mm and a new daughter cell is 1.1 mm long. What is the equation f(t) expressing the length of a cell and its subsequent daughter cells? Sketch the graph of this periodic function for several periods.

Solution To sketch the graph we need only sketch it over one period, the interval {0,1) that corresponds to a single replication cycle. Then this sketch can be shifted horizontally to form the sketch over subsequent periods. The function f(t) is described by a split equation since the cell grows linearly over the interval [0, 0.7) and remains constant in length over the interval [0.7 , 1). As the line through the points (0, 1.1) and (0.7, 2) has slope m = 0.9/0.7 = 9/7, the equation for f is given by

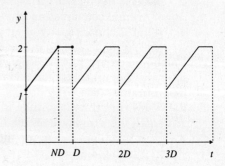

$$f(t) = \begin{cases} (9/7)t + 1.1 & \text{if } 0 \le t < 0.7 \\ 2.0 & \text{if } 0.7 \le t < 1.0 \end{cases}$$

Cell length over repeated division periods.

Over a longer interval of time the function f(t) describes the length of a cell as it continually divides, grows, and redivides, etc. The graph of f is formed by repeating the graph of f over the first interval [0,1) over each subsequent interval. In Fig. 1. the units on the t-axis were simply left as the length D of a cell division cycle. This is called the period of the process.

DEFINITION OF A PERIODIC FUNCTION

A function f(x) is **periodic** if there exist a number P such that

$$f(x + P) = f(x) \text{ for all } x.$$

The smallest positive number P for which this is true is called **the period** of the function f.

Observe that if a function is periodic with period P, then for any integer n,

$$f(x + nP) = f(x)$$

In particular, if f has period P then

$$f(x - P) = f(x)$$

This last identity implies that a horizontal shift of the graph of f(x) P-units to the right results in an identical graph.

Example 2   **A periodic function with "jumps".**

Problem   Show algebraically that the function $f(x) = x - [\![x]\!]$ is periodic with period 1. Sketch its graph.

Solution   To show algebraically that $f(x + 1) = f(x)$ we need to first observe that for any number x,

$$[\![x + 1]\!] = [\![x]\!] + 1$$

e.g., $4 = [\![4.2]\!] = [\![3.2 + 1]\!] = [\![3.2]\!] + 1 = 3 + 1$. Then,

$$f(x + 1) = (x + 1) - [\![x + 1]\!] = x + 1 - [\![x]\!] - 1 = x - [\![x]\!] = f(x)$$

To show that the period is 1 and not a positive number P < 1 we show that for any such number P the equation $f(x + P) = f(x)$ is not true for all x. In fact, it is not true for any x in the interval [0, 1 - P). To see this, assume that $0 < x < 1 - P$. Then $x + P < 1$ and hence

$$f(x) = x - [\![x]\!] = x - 0 = x$$

and

$$f(x + P) = x + P - [\![x + P]\!] = x + P - 0 \neq x$$

The graph of this function f(x) can be sketched by sketching it over the interval [0,1) and then shifting this segment of the graph right and left by integer amounts to achieve the periodic repetitive graph of f. On the interval [0,1) the function's graph is simply the line $y = x$ since $[\![x]\!] = 0$ on this interval. Thus the graph of f appears as the curve sketched in the following figure.

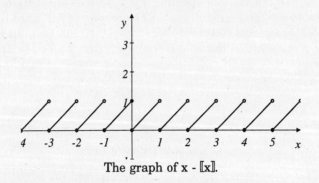

The graph of $x - [\![x]\!]$.

Circular periodic functions.

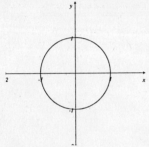

The unit circle.

Prehistoric people observed the periodic cycles of days and nights, the moon and even the movement of stars. A simple instance of a periodic phenomena is the repetition that occurs when an object traces around and around a closed curve. To describe such a phenomena mathematically we use the **unit circle** described as the graph of the relationship $x^2 + y^2 = 1$. This circle can be considered as the *path* or more mathematically as the *trajectory* of a dot, sometimes referred to as a *point*, that moves around the circle in a counter-clock-wise direction. If the dot begins at the coordinate (1,0) and in t-seconds moves exactly the distance t along an arc of the circle, its location at time t will be the point on the circle with coordinates that we can denote by x(t) and y(t).

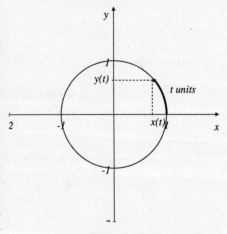

First, let us consider the y-coordinate of this dot as it moves around the circle. Initially, when t = 0, the dot is at (1,0) so that y(0) = 0. However, as the dot moves upward along the circle the value of y(t) will increase. When the dot reaches the top of the circle, at the point (0,1), its y-coordinate y(t) = 1. But then, as the dot moves further around the circle, the value y(t) will decrease, approaching zero as the dot approaches the point (-1,0) where the circle intersects the negative x-axis. If the dot continues to travel around the circle, its y-coordinate will decrease further, until y(t) = -1 when the dot is at the bottom of the circle. Finally, as the dot approaches its initial position, y(t) will approach 0 again. Then, this periodic pattern will be repeated exactly as the dot goes around the circle a second time, and again for a third completed circling, etc. This periodic process has a period P that is the circumference of the unit circle, thus $P = 2\pi$. Thus the function y(t) described by the moving dot has the properties

$$y(0) = 0 \quad \text{and} \quad y(t + 2\pi) = y(t) \quad \text{for any number t.}$$

In a similar fashion we can consider the function x(t) that is described by the x-coordinate of the dot when it is t-units around the circle. This function is similarly periodic with period 2π. At the initial point x(0) = 1. As the dot moves in the counter-clock-wise direction the value of x(t) decreases, reaching zero when the dot reaches the top of the circle, and then becoming negative as it decreases to -1 when the dot reaches the point (-1,0) where the circle intersects the negative x-axis. As t increases, the dot moves to the circle bottom point where x(t) = 0 again, and finally it increases back to 1 as the dot approaches the initial point. The function x(t) satisfies

$$x(0) = 1 \quad \text{and} \quad x(t + 2\pi) = x(t) \quad \text{for all t.}$$

We know that the circumference of the unit circle is 2π. Hence, we know the distance from the point (1,0) around the circle to points located at the circles intersection with the coordinate axis and the points where the lines y = x and y = -x intersect the circle; these distances are fractions of the circumference. For instance the distance to the point on the negative y-axis is 3/4 of the circumference, $(3/4) 2\pi = 3\pi/2$. The following table lists these points, their t-distance around the circle, and their corresponding coordinates x(t) and y(t), where we have estimated the approximate coordinates of the points on the lines y = ± x.

Table 1.5.1
Coordinates of points around the Unit Circle.

t	0	$\pi/4$	$\pi/2$	$3\pi/4$	π	$5\pi/4$	$3\pi/2$	$7\pi/4$	2π
x(t)	1	0.7	0	-0.7	-1	-0.7	0	0.7	1
y(t)	0	0.7	1	0.7	0	-0.7	-1	-0.7	0

If we sketch these points on a graph, and assume that a dot's coordinates change in a continuous fashion as the dot moves around the circle, then it seems reasonable to connect these points to form a rough sketch of the two functions x(t) and y(t) over the interval $[0,2\pi]$.

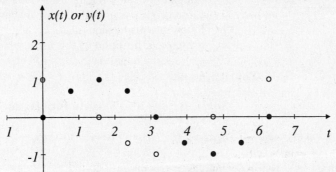

The coordinates x(t) (solid circles ●) and y(t) (open circles O) for t = 0, $\pi/4$, ... , 2π using the data from Table 1.5.1

Notice that the scale units on the t-axis are the same as the scale units on the x(t) or y(t) axis.

The coordinates do not change linearly as t increases. If the coordinates change continuously then the graphs of x(t) and y(t) will be smooth curves passing through these points.

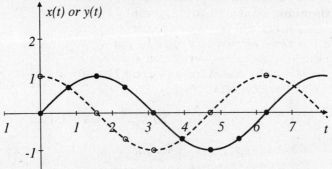

The coordinate functions for the unit circle: x(t) the dashed curve and y(t) the solid curve.

You may recognize these curves. Yes, they are the graphs of the trigonometric functions sin(t) and cos(t). The x-coordinate function gives the cos(t) and the y-coordinate function gives the sin(t). We have used t as the independent variable so that we could refer to the x and y coordinates without confusion. However, the variables x and θ (with the variable θ (the Greek letter theta) are often the independent variable, in which case you will encounter functions like y = sin(x) and y = sin(2θ). These are the same functions that you learned in relation to ratios of sides of triangles. At the end of this section these triangular relationships are reviewed. If you think of circles when you encounter trigonometric functions you will think of them as periodic functions.

TRIGONOMETRIC FUNCTIONS

Sin(t) and Cos(t)

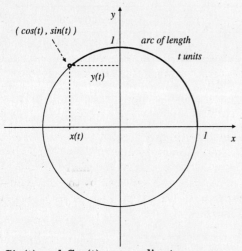

Sin(t) and Cos(t) as coordinates on the circle $x^2 + y^2 = 1$.

In view of the above discussion we see that the elementary trig functions sin(t) and cos(t) can be defined as coordinates on a unit circle, t-units around the circle. This definition does not require the introduction of any triangle. In this subsection we will consider the formation of more general sine and cosine functions by scaling and shifting the basic functions. We begin by assuming the functions sin(x) and cos(x) are periodic functions with period P = 2π and their graphs are as illustrated above. Our objective in the following discussion is to be able to easily recognize the graph of a function of the form y = b + A sin(m(x - a)). We proceed by considering the effect of each parameter in this equation.

Scaling Sine and Cosine functions.

Scaling the t-variable: y = sin(mt).

If the radian measure t is scaled by a factor m > 0, i.e. multiplied by a constant like 3t, the values of the functions cos(mt) and sin(mt) will simply be the x and y coordinates of a point m · t units around the unit circle. Consequently, the graph of a scaled sine or cosine function has a similar periodic form as the graph of the unscaled function.

However, the period will be different. As t increases the dot at position m·t along the circle moves either much faster, when m > 1, or much slower, when m < 1. This changes the period of the scaled function. The dot will return to the initial point (1,0) when m·t = 2π. Thus.

The functions sin(mt) and cos(mt) have period P = 2π/m.

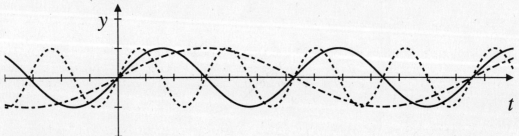

The graphs of sin(t) - solid curve, sin(2t) - dotted curve, and sin(0.5 t) - dash-doted curve.

One of the most common choices for a scaling constant is one that is a multiple of π. If the constant m = kπ then the graph of sin(kπt) will intersect the t-axis when t is a multiple of 2/k, since the period of $y = \sin(k\pi\, t)$ is $P = 2\pi/k\pi = 2/k$

For instance, $y = \sin(3\pi/2\, t)$ has the scaling constant $m = \pi/k$ for $k = 2/3$. Thus, its graph will cross the t-axis at $t = 0, 2/3, 4/3, 6/3, \ldots$ and when $t = -2/3, -4/3, \ldots$ The curve in the graph at the right has period 2. Assuming it is the graph of $y = \sin(kx)$, we conclude that $k = 1$. In this case the horizontal axis intercepts are simply the integer values.

Amplitude; Scaling the sine and cosine functions.

Multiplying a sine or cosine function by a constant in effect scales the y-axis. Traditionally, the constant is the capital letter A, i.e., as $y = A \cos(t)$. It indicates the **amplitude** or "height" of the scaled curve. As a height is positive and the coefficient A can be negative the amplitude is its absolute value.

The graph of $y = \sin(\pi x)$.

The **amplitude** of $y = A \sin(t)$ or $y = A \cos(t)$ is $|A|$.

The graph of $y = A \sin(t)$, for $A > 0$, has the same period and crosses the t-axis at the same points as the graph of $y = \sin(t)$, but it is either stretched away from the t-axis, when $A > 1$, or compressed toward the t-axis when $0 < A < 1$. If A is negative, the graph will appear as the mirror image of the graph of $\sin(t)$ inverted through the t-axis, and then stretched or compressed by the factor $|A|$. Illustrated below are the graphs of the sine function for $A = 1$, the basic curve, $A = 2$, which stretches the graph, and $A = 0.5$, which compresses the graph.

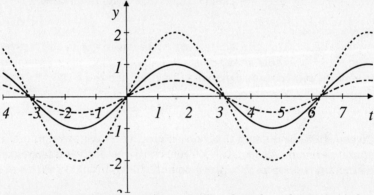

The graphs of $\sin(t)$ - solid, $2 \sin(t)$ - dotted, and $0.5 \sin(t)$ dot-dashed.

Shifted Sine and Cosine functions.

The functions $\sin(x)$ and $\cos(x)$, like any other function, can be shifted horizontally by replacing the variable x by a shifted form $x - a$. They can be shifted vertically by replacing the variable y by the shifted form $y - b$, or equivalently by adding the shift constant b to the trig function after it is evaluated.

A ***horizontally shifted*** sine or cosine function has an equation of the form

$$y = \sin(x - a) \quad \text{or} \quad y = \cos(x - a)$$

Shifting the graph of a trig function does not affect its period or its amplitude.

The equation of the curve y = sin(mx) shifted to the right a units is given by

$$y = \sin(m(x - a))$$

Thus to determine the horizontal shift constant in a function sin(mx + K) the scaling constant must be factored to give sin(m(x - K/m)). The shift will be K/m units to the right if K > 0.

In many mathematical applications, such as in physics when light, sound, or energy waves are being discussed, the general formulas are given using only sine functions. This can be accomplished because cosine functions can always be expressed as horizontally shifted sine functions. In particular, the cosine function can be expressed as the sine function shifted either to the left $\pi/2$ units or to the right $3\pi/2$ units, remember, since it is periodic these result in the same effect:

$$\cos(x) = \sin(x + \pi/2) \quad \text{or} \quad \cos(x) = \sin(x - 3\pi/2).$$

Similarly,
$$\sin(x) = \cos(x - \pi/2) \quad \text{or} \quad \sin(x) = \cos(x + 3\pi/2)$$

A *vertically shifted* scaled sine or cosine function has an equation of the form

$$y = b + \sin(x) \quad \text{or} \quad y = b + \cos(x)$$

Shifting a graph vertically does not change its period or its amplitude.

Graphing the most general shifted and scaled sine function: y = b + A sin(m(x - a)).

The most general *shifted and scaled* sine function has the form

$$y = b + A \sin(m(x - a))$$

where a and b are shift parameters that can be positive or negative, and the scale parameters m and A (the amplitude) can, without loss of generality, be assumed to be positive numbers. The graph of this function is derived from the graph of sin(x) through a series of scalings and shifts.

1. First the x-coordinate is scaled by the factor m.
   This compresses or stretches the graph of sin(x) to form the graph of sin(mx) which is a similar periodic curve with period $P = 2\pi / m$.

2. Next, the scaled graph of sin(mx) is shifted horizontally a units.
   This in effect moves the scaled graph so that the point where the graph of sin(mx) crosses the origin becomes the point $(a,0)$ on the graph sin(m(x - a)).

3. Next, the scaled horizontally shifted graph of sin(m(x - a)) is scaled in the y-direction by multiplying by the *amplitude* A.

   Relative to the graph of sin(m(x - a)), the graph of A sin(m(x - a))
   is stretched away from the x-axis if A > 1 and is compressed towards the x-axis if A < 1.

4. Finally, the graph of the x-scaled, x-shifted, y-scaled graph function sin(m(x - a)) is shifted vertically. This y-shift moves the graph up b units if b is positive, or downward $|b|$ units if b is negative.

Section 1.5 — Periodic and Trigonometric Functions

Identifying the parameters of a general sine function.

A "math-skill" that is required to intelligently "read" graphs is to discern the specific numerical values of parameters in a general function form that corresponds to a given graph. Assume that you are presented the following graph and are told that it is the graph of a general sine function. To establish the specific function you must identify the shift and scaling parameters.

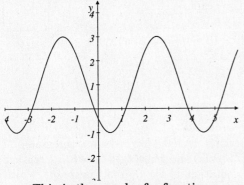

This is the graph of a function
$y = b + A \sin(m(x - a))$

Method to identify the parameters a, b, m and A from a sketch of $y = b + A \sin(m(x - a))$.

Identify the coordinates of key points on the graph.

First, establish the maximum y-coordinate, y_{max}, and the minimum y-coordinate, y_{min}. Then the vertical shift is the midpoint between these two numbers,

$$b = (y_{max} + y_{min})/2$$

and the amplitude is one half their difference,

$$A = (y_{max} - y_{min})/2$$

The period P, and hence the horizontal scale factor $m = 2\pi/P$, is found as the difference between the x-coordinates of two consecutive points where the graph passes through its maximum, (or two consecutive points where the graph passes through its minimum).

Alternatively, the period can be found by drawing any horizontal line, $y = C$, where $y_{min} < C < y_{max}$. Then choose any three consecutive points where this line intersects the graph. Let the x-coordinates of the leftmost and right-most points be x_1 and x_2. (The middle point is ignored.) Then, the function's

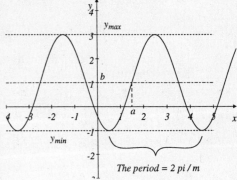

The period = $2\pi/m$

$$\text{Period} = x_2 - x_1 \qquad m = 2\pi/(x_2 - x_1)$$

Finally, the x-shift parameter a is found from the intersection of the graph with the line $y = b$, which corresponds to the shifted x-axis. a can be chosen as the x-coordinate of any point where the graph crosses $y = b$ increasing, as the graph of $\sin(x)$ appears at the origin. Normally, a is chosen as the first such positive x-coordinate.

Example 3 — Identifying a sine graph.

Problem Determine the general sine equation corresponding to the graph at the right.

Solution The maximum y-coordinate is $y_{max} = 1$ and the minimum y-coordinate is $y_{min} = -2$. Therefore,

$$b = (1 + -2)/2 = -0.5$$

and

$$A = (1 - -2)/2 = 1.5$$

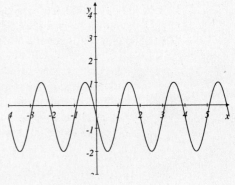

The period can be determined by observing the x-axis intersects the graph at three consecutive points with x-coordinates 1.25, 1.875, and 3.125. Using the first and third points we find

P = 3.125 - 1.125 = 2   and hence the scale constant m = $2\pi/2 = \pi$

The horizontal shift constant is identified as the first positive x-coordinate where the line y = -0.5 crosses the graph; $a = 1$. Thus the equation of the graph is

$$y = -0.5 + 1.5 \sin(\pi(x - 1))$$

The other elementary trig functions.

There are four other **elementary** trigonometric functions that are formed from the sine and cosine functions. First, there are the reciprocal functions, $1/\sin(x)$ and $1/\cos(x)$. These functions are given individual names, the *cosecant*, written in function notation as $\csc(x)$, and the *secant*, which is written as $\sec(x)$.

The **cosecant** function:

$$\csc(x) \equiv 1/\sin(x)$$

The $\csc(x)$ is not defined when $\sin(x) = 0$. These are at the values $0, \pm\pi, \pm 2\pi, \ldots$

Thus, the Natural Domain of the cosecant function is

$$D_{\csc} = \{x \mid x \neq n\pi\}$$

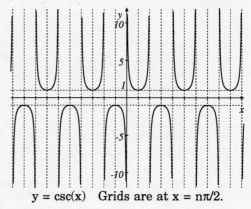

$y = \csc(x)$ Grids are at $x = n\pi/2$.

The **secant** function:

$$\sec(x) \equiv 1/\cos(x)$$

The secant function is not defined when $\cos(x) = 0$. This occurs at the points that are integer multiples of π plus $\pi/2$, i.e., at the values $x = \pm\pi/2, \pm 3\pi/2, \pm 5\pi/2, \ldots$.

Thus the Natural Domain of $\sec(x)$ is

$$D_{\sec} = \{x \mid x \neq (n + 1/2)\pi\}$$

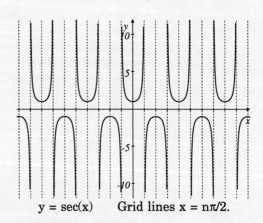

$y = \sec(x)$ Grid lines $x = n\pi/2$.

Since $|\sin(x)| \leq 1$ and $|\cos(x)| \leq 1$, their inverses are greater than or equal to one:

$$|\csc(x)| \geq 1 \quad \text{and} \quad |\sec(x)| \geq 1$$

The other two elementary trigonometric functions, the tangent and cotangent, are the quotients of the sine and cosine functions.

The **tangent** function:

$$\tan(x) = \sin(x) / \cos(x)$$

The tangent function is not defined when $\cos(x) = 0$. Thus its Natural Domain is the same as that of the secant function:

$$D_{\tan} = \{ x \mid x \neq (n + 1/2)\pi \}$$

$$\lim_{x \to \pi/2-} \tan(x) = \infty$$

$$\lim_{x \to \pi/2+} \tan(x) = -\infty$$

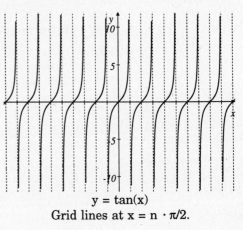

$y = \tan(x)$
Grid lines at $x = n \cdot \pi/2$.

The **cotangent** function:

$$\cot(x) = \cos(x) / \sin(x)$$

The Natural Domain of the cotangent function is the same as that of the cosecant function, since they both involve division by $\sin(x)$:

$$D_{\cot} = \{ x \mid x \neq n \cdot \pi \}$$

$$\lim_{x \to \pi-} \cot(x) = -\infty$$

$$\lim_{x \to \pi+} \cot(x) = \infty$$

$y = \cot(x)$
The grids lines are $x = n \cdot \pi/2$.

Note that $\cot(x) = 1/\tan(x)$.

Each of these four trigonometric functions is periodic. Sec(x) and csc(x) have period 2π. However, tan(x) and cot(x) have period π, as can be seen in the above graphs. Their graphs are not finite as the function values become infinitely large as their denominators approach zero. The x-values where the denominators are zero are referred to as *singular values* or sometimes as *singular points* and are the values at which the functions are not defined. As the variable x approaches one of these singular values the function values become very large, either as positive or as negative numbers.

Because the functions sec(x), csc(x), tan(x), and cot(x) are defined in terms of sin(x) and cos(x), any function involving these function can be written in an equivalent form just using the basic functions sin(x) and cos(x).

Triangular Relationships.

Most students first encountered trigonometric functions as the ratios of sides of a right triangle. These relationships are consistent with the *circular* definitions of the sine and cosine given above. While the following triangle relationships are usually introduced with an angular variable θ measured in *degrees*, in the *circular* definition the variable t is measured in *radians*, because the unit of measuring distance around the circle is the same unit (physically) as the radius of the *unit circle*.

The default unit for trigonometric functions in this text will be *radians*.

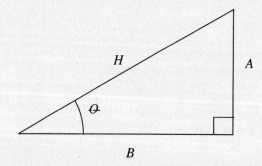

Consider the right triangle illustrated above with an acute angle θ having a hypotenuse length H, the length of the angle θ's adjacent side B and the length of its opposite side A. Then the six basic triangular formulas for the fundamental trigonometric functions are :

Triangle Identities

$$\sin(\theta) = A/H \qquad \csc(\theta) = H/A$$

$$\cos(\theta) = B/H \qquad \sec(\theta) = H/B$$

$$\tan(\theta) = A/B \qquad \cot(\theta) = B/A$$

There are many identities among trigonometric functions, however in this text we will only utilize the most basic relationships that are based on the Pythagorean identity $H^2 = A^2 + B^2$. When H = 1 the sides lengths are B = cos(θ) and A = sin(θ), and thus

Pythagorean Identities

$$1 = \sin^2(\theta) + \cos^2(\theta) \qquad \text{or} \qquad \sin^2(\theta) = 1 - \cos^2(\theta)$$

$$\sec^2(\theta) = 1 + \tan^2(\theta) \qquad \text{or} \qquad \tan^2(\theta) = \sec^2(\theta) - 1$$

$$\csc^2(\theta) = 1 + \cot^2(\theta) \qquad \text{or} \qquad \cot^2(\theta) = \csc^2(\theta) - 1$$

These identities will be used later to simplify trigonometric expressions. For example, the term $\sin^3(x)$, which means $[\sin(x)]^3$, can be expressed as

$$\sin^3(x) = \sin(x)\sin^2(x) = \sin(x)[1 - \cos^2(x)]$$

Exercise Set 1.5

1. List five naturally occurring periodic processes and indicate the basic period of each.

2. Sketch the graph of the function $y = f(x)$ assuming that $f(x)$ is periodic with period $P = 2$ and is defined over the interval [0,2) by the indicated function. Sketch the graph for $-2 \leq x \leq 6$.

   a) $y = x$ b) $y = 2 - x$ c) $y = 1 - x$ d) $y = |x - 1|$

   e) $y = \begin{cases} 0 & \text{if } 0 \leq x < 1 \\ 1 & \text{if } 1 \leq x < 2 \end{cases}$ f) $y = \begin{cases} x & \text{if } 0 \leq x < 1 \\ 2 - x & \text{if } 1 \leq x < 2 \end{cases}$

   g) $y = x^2$ h) $y = (x - 1)^2$ i) $y = x(2 - x)$ j) $y = \sqrt{x}$

3. Using a graph of the unit circle $x^2 + y^2 = 1$, estimate the values of the given numbers assuming the arguments are in radians. Compare your estimates with the values given by a scientific calculator.

   a) $\cos(0)$ b) $\sin(0)$ c) $\sin(1)$ d) $\cos(1)$ e) $\cos(2)$ f) $\sin(2)$

   g) $\cos(\pi/4)$ h) $\sin(\pi/4)$ i) $\tan(\pi/4)$ j) $\sec(\pi/3)$ k) $\csc(\pi/3)$ l) $\cot(\pi/6)$

4. Sketch the graph of the indicated functions.

   a) $y = \cos(x)$ b) $y = 2\cos(x)$ c) $y = 2\cos(x) + 1$ d) $y = -2\cos(x) + 1$

   e) $y = 2\cos(x) - 2$ f) $y = 1/2 \cos(x) - 2$ g) $y = 2\sin(x) + 1$ h) $y = 1/2 \sin(x) - 2$

5. Sketch the graph and determine the period P of the indicated function.

   a) $y = \sin(x)$ b) $y = \sin(x + \pi)$ c) $y = \sin(x + \pi/2)$ d) $y = \cos(x + \pi)$

   e) $y = \cos(2x)$ f) $y = \cos(1/2\, x)$ g) $y = \cos(x - \pi/4)$ h) $y = \cos(1/2\, x - \pi/4)$

   i) $y = \sin(\pi x)$ j) $y = 3\cos(\pi/2 - x)$ k) $y = \sin(-x)$ l) $y = \cos(-x)$

   m) $y = \sin(\pi x + \pi/2)$ n) $y = \sin(\pi x - \pi)$ o) $y = \cos(\pi x/3)$ p) $y = \cos(3\pi x)$

6. Let $f(x) = \cos(x)$, $g(x) = x^2 + 1$, $h(x) = \sin(x)$, and $k(x) = x - \pi$. Determine the following composition functions and evaluate each function at the indicated point x_0.

   a) $(f \circ g)(x)$, $x_0 = 0$ b) $(g \circ f)(x)$, $x_0 = 0$ c) $(h \circ k)(x)$, $x_0 = \pi$ d) $(g \circ h)(x)$, $x_0 = 3\pi/2$

   e) $(g \circ f \circ k)(x)$, $x_0 = \pi/2$ f) $(h \circ g)(x)$, $x_0 = 0$ g) $(g \circ f)(x)$, $x_0 = \pi$ h) $(f \circ h)(x)$, $x_0 = 0$

   i) $(h \circ f)(x)$, $x_0 = 3\pi/2$

7. Sketch the graph of the given functions and indicate the following: (i) the domain of the function; (ii) the period of the function.

   a) $y = 0.5 \tan(x)$ b) $y = \tan(2x)$ c) $y = \tan(x/2)$ d) $y = \cot(1/3 x)$ e) $y = 2\sec(x)$

   f) $y = \tan(\pi x)$ g) $y = \cot(\pi x/2)$ h) $y = \sec(\pi x)$ i) $y = \csc(x - \pi/2)$ j) $y = \tan(\pi(x - 1/2))$

8 Determine the general sine equation of the given curve.

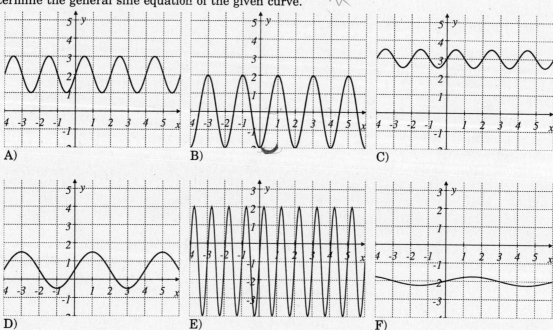

9 Express the equation for each of the curves in the previous exercise as a shifted and scaled cosine.

For problems 10-12, sketch the four curves on the same graph.

10 a) $\sin(0.5\pi x)$ b) $\sin(0.5\pi(x-1))$ c) $2\sin(0.5\pi(x-1))$ d) $1.5 + 2\sin(0.5\pi(x-1))$

11 a) $\cos(3\pi x)$ b) $\cos(3\pi(x-1))$ c) $1.5\cos(3\pi(x-1))$ d) $-2 + 1.5\cos(3\pi(x-1))$

12 a) $\cos(0.5\pi x)$ b) $\sec(0.5\pi)$ c) $\tan(0.5\pi x)$ d) $-\sin(0.5\pi x)$

13 The volume of blood pumped from the heart is a periodic function of time. During the systolic phase of a heartbeat, the heart contracts and squeezes or pumps blood into the arterial system. During the diastolic phase of a heartbeat, the muscles of the heart relax, the heart fills with blood, and no blood is pumped. Assume the length of the diastolic phase is 2/3 the length of the systolic phase.
   a) What is the period of the heartbeat?
   b) If the volume of blood pumped during the systolic phase has the form of a sine function, construct a function V(t) which would describe the volume of blood pumped over one cycle of a heartbeat.
   c) Graph the volume function, V(t) of part (b), over several heartbeats.

14 Using the Pythagorean Identities and the Triangle Identities to find the following:

   a) $\cos(\theta)$ if $\sin(\theta) = x/2$ b) $\sin(\theta)$ if $\tan(\theta) = x/3$ c) $\sec(\theta)$ if $\sin(\theta) = 2x$

   d) $\tan(\theta)$ if $\sin(\theta) = x/4$ e) $\cos(\theta)$ if $\tan^2(\theta) = x^2 + 1$ f) $\csc(\theta)$ if $\cos(\theta) = x/2$

15 Radiant energy is described by a sinusoidal function indicating the energy propagation as a function of time. This function, called a wave function, has the form
$$f(t) = A\sin(2\pi C/W \cdot t)$$
where W is the wavelength and A is the amplitude of the wave. The period of this function is P = W/C, the wavelength divided by the speed of light, C. The frequency, f, of a wave is the number of periods per unit time, usually denoted as cycles per second (cps).
$$f = C/W, \quad C = \text{speed of light}, \quad C \approx 3 \cdot 10^{10} \text{ cm/sec}$$

The nature of radiant energy is determined by its wavelength as indicated in the following table. Note that low frequency corresponds to long wavelength.

Low frequency radio	10^4
	10^2
Ultra-high frequency radio	0
	10^{-2}
Radar microwaves	10^{-4}
Visible light	10^{-6}
	10^{-8}
X-rays	10^{-10}

a) Sketch the graph of the function f(t).

b) What is the frequency of a wave with wavelength W = 0.0004 cm?

c) Sketch the graph and give the equation of a wave function with frequency of 3 cps and intensity f(t) varying between 4 and 8.

If the wavelength of visible light varies from 4000 Å to 7000 Å where Å is the symbol for the angstrom unit; one Å = 10^{-8} cm = 10^{-10}m.

d) What is the range of the frequency of visible light?

e) Sketch the "wave" associated with a yellowish-orange light having wavelength 6000 Å if the amplitude of the wave is $\sqrt{2}/L$, where L is the basic period of the wave.

16 Blood pressure varies sinusoidally between 80 (mm Hg) and 120 (mm Hg). Assume the period of the blood pressure is the same as that of a heartbeat but with a lag of one-half period. Give an equation for the blood pressure of a person with a) 70 heartbeats/minute; b) 80 heartbeats/minute; c) 80 heartbeats/minute but high blood pressure which is 20 (mm Hg) above normal.

17 Set up and graph a sine function that could be used to model the average daily temperature over several years. What is the period and amplitude of your function? (Use your local temperature.)

18 The vibration of a string may be described by sine functions. If the string is of fixed length L, such as a violin or guitar string, then the "fundamental" tone of the string is described by a sine function of period 2L. Its graph describes the string over the interval [0,L] corresponding to the violin. The first overtone is described by a sine function of period 2L/2. The second overtone is described by a sine function of period 2L/3, and so forth.

a) Give the equation of the sine function associated with the fundamental tone and the first three overtones of a string of length L = π vibrating with an amplitude A = 2.

b) On the same graph sketch the functions of part (a) for 0 ≤ x ≤ L.

19 Is the function $y = \sin(t)e^{-t/2}$ periodic? Why would this be called a "damped oscillation" curve? Sketch the graph of this curve over the interval [0,4π].

20 Sketch the graph of $y = \sin(\pi/x)$ over the interval [1/4, 4]. What does the graph look like over the interval (0, 1/4)?

21 Sketch the graph of $y = x + \sin(x)$.

22 Think about the graph of $y = \cos((\pi/2)\sin(x))$. What will be its maximum and minimum y-values? Will it be periodic? Sketch its graph over the interval [0,4π].

Section 1.6 Form, Scale, and Transformations

INTRODUCTION

Graphs provide visual images of a function that allows the viewer to understand how the function changes without requiring algebraic or calculus manipulation. However, graphs provide only a limited amount of information about a function, illustrating it in a *window* that focuses on a particular interval of x-values. The function's behavior or characteristics outside this window are not illustrated and hence unknown. If you only looked at the box in the graph at the right you would miss the function's behavior at $x = 4$. Indeed, one of the real applications of Calculus that will be developed later in this text is to help identify the *windows* in which a function has extreme values or the nature of its graph changes.

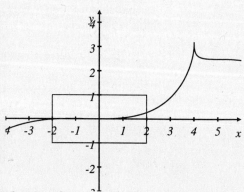

Visual information can be misleading! In this section we explore the affect of axis scaling on the appearance of graphs. In particular, we shall see how the graphs of specific types of functions can be made to appear linear by choosing scaled coordinate axes that are not uniform, and how graphs with large ranges or domains can be represented using Log transformations. This section is not intended to provide special graphing techniques. This will be done in Chapter 4 using the tools of differential calculus.

A frequent mathematical problem encountered by researchers is to determine the parameters of an equation corresponding to a graph. When only specific *data points* are provided for the graph, the problem of determining the associate model function is known as *parameter estimation*. This is a principal objective of the branch of mathematics know as *Statistics*. When the model equation is linear, $y = mx + b$, the parameters can be estimated by the *least squares* procedure that is actually available on most scientific calculators. When the data are linear on a transformed coordinate system an inverse-transform must be made to establish the corresponding model equation; for linear relationships on log-log and semi-log graphs this yields allometric and exponential models.

SCALING OF COORDINATE AXES

The standard x-y coordinate system normally has the same *uniform* scale on each axis. A uniform scale has the same physical distance on the graph between tick marks and the unit difference between successive ticks is the same. Making the unit scales on both axis the same is essential to ensure that area is preserved, i.e., the physical area of a region on the graph is proportional to the true area of the region; what you visually see is the area that would be calculated. In practice, however, often the same scale is not used on both axes. The x and y variables may have different physical or biological meanings and hence different **units**, e.g., one may be a distance, with unit cm, and the other a volume, with a liter or micro-liter, $1 \ \mu l = 10^{-6}$ liter unit. In talking about scales we will refer to the **scale** of the *units* associated with tick marks on the axis and the **physical scale** which is the actual distance between tick marks as measured with a ruler. Scaling a graph can be achieved in two ways, by unit **scaling** in which the values associated with tick marks are set and by physical scaling in which the distance between tick marks is varied.

Mathematically, scaling is achieved by a transformation, which can be indicated by a mapping of one set of variables into another:

$$(x,y) \rightarrow (u,v)$$

that is indicated by giving two transformation functions: u(x) and v(y). For simplicity, we first consider only a transformation of the x-coordinate. The three basic scaling transformations used in applications are:

Homogeneous scaling: $x \to u = k \cdot x$, k a constant.

Logarithmic scaling: $x \to u = \log(x)$.

Reciprocal scaling: $x \to u = 1/x$.

Homogeneous scaling.

The transformation $x \to u = kx$ for $k > 0$ corresponds to an expansion of the x-scale when $k > 1$ and to a compression when $0 < k < 1$. For instance, the transformation $x \to u = 2x$, illustrated at the left below, results in a stretching of the x-scale so that the visual distance between image points on the u-axis is doubled the distance on the x-axis. Whereas, the transformation $x \to u = x/3$, illustrated on the right, results in a contraction of the x-scale so that the visual distance between image points on the u-axis is one-third the distance between the corresponding points on the x-axis.

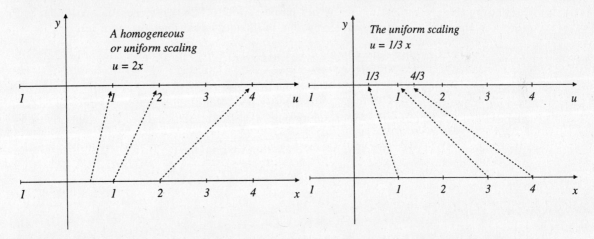

Consequently, to spread out small or close values a constant $k > 1$ is chosen and to have a large range of values appear close the scale is compressed by choosing $k < 1$.

Observe that in these illustrations the same physical scale has been used to represent one-unit on both the original x-axis and on the scaled u-axis. Think of this as the rulings on a printed sheet of graph paper with 1cm squares. When plotting data, a transformation can be made transparent by simply indicate different units on the axis corresponding to tick marks or ruled grids. e.g., as

or as

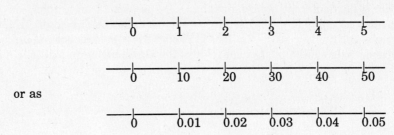

If the scaling factor k is negative, the transformation x → u = kx reverses the orientation of the axis in addition to either stretching or contracting the scale.

Positive x-values transform to negative u-values when k < 0. For example, the transformation x → u = -3x transforms or maps x = 1 to u = -3 and x = -0.5 to u = 1.5. Notice that the reverse of this transformation is u → x = -u/3.

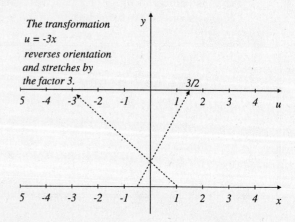

The transformation u = -3x reverses orientation and stretches by the factor 3.

Example 1   Transformation of units.

Problem   Give the scaling functions that a) convert the variable θ with angular degree units to r with radian units, and b) transform the temperature variable t °C to T °K (degree Kelvin).

Solution   To solve each problem you must know *a priori* the relationships between the indicated units.

a) The scaling from angular degrees to radians relies on the fact that the radian measurement of the distance around a unit circle is 2π while the angle measurement corresponding to the "angle" formed by the arc consisting of the full circle is 360°. The transformation equation is found by equating the ratios of each variable symbol to its equivalent measurement unit value:

$r/2\pi = \theta/360$ which implies the transformation $\theta \to r = (2\pi/360) \cdot \theta$

b) The transformation from degrees Celsius to degrees Kelvin is not a scaling transformation but is a *shift transformation*. It does not require a scaling, i.e., a compression or expansion of the physical distance between corresponding points on the t and T axes. The relationship is

t °C → T °K = t + 273.15   (approximately)

☑

Apparent slopes.

The **apparent slope**, or **physical slope,** which is the slope you "see", is the *measured change* in the y-coordinates divided by the *measured change* in the x-coordinates between any two points on the line. This physical change is found by measuring with a ruler the actual distances on the graph, as sketched, ignoring the units on the axes. The ratio of these physical distances will be the same even if the graph is enlarged or shrunk, as by a photocopier.

The appearance of graphs will vary when different coordinate scaling systems are employed. This is particularly noticeable when a line is plotted on graphs with different uniform scales; it appears as a line in each case but its apparent slope can be quite different. In the following three graphs the line is y = -1 + 2x. The graph on the right has normal uniform scales that are the same for both axis while the physical-scale of the x-axis has been modified for the left and center graphs.

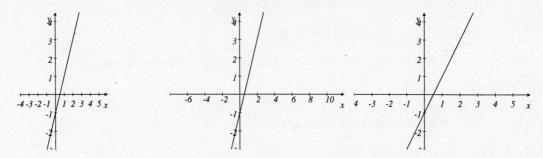

The graphs of y = -1 + 2x with different physical scales on the x-axis.

In the above graphs, the line sketched in the right graph has the *physical* or *apparent slope* of 2, whereas this *physical slope* is 4 for the left and center graphs. The center and right graphs have the same physical tick marks but different units on the x-axis. The left graph has the same units indicated as the right graph but the physical distance between x-ticks is only half of that in the right graph.

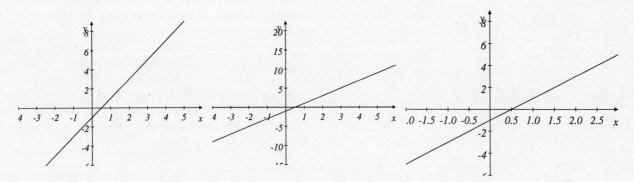

In the above sequence of graphs the same line is sketched with different scalings of the y-coordinates. The physical distance between ticks is the same in all the graphs. In the unit-value associated with a y-tick is 2 for the left and right graphs and 5 for the center graph. The x-unit value for the right graph is half that for the other two. The line in the left graph has an *apparent slope* of 1, i.e., a 45° inclination to the x-axis. In the center panel the *apparent slope* is 2/5 and in the right panel it is 1/2. (The center and right are hard to distinguish at first glance.)

Most people interpret the slope of a line by its *apparent slope*; they simply use the visual perception and do not read the scales and consequently never calculate slopes numerically. Thus scaling is the first step in *deceptive representation* of data and graphs. Yet, it has its good uses. In science it is essential to scale to be able to reflect independent and dependent variables with different units of measurement and widely different numerical values.

Logarithmic scaling.

Logarithmic scaling, for short, Log scaling, is utilized when the data or variables that must be considered range over several orders of magnitude. Recall that the *magnitude* of a number is the exponent of 10 when the number is expressed in scientific notation. For instance, $0.03 = 3 \cdot 10^{-2}$ has magnitude -2 and $56,321 = 5.6321 \cdot 10^4$ has magnitude 4 so these two numbers "differ" by $4 - (-2) = 6$ orders of magnitude. A log transformation is only valid for positive numbers since $D_{Log} = (0, +\infty)$.

The four basic features of the log transformation $x \rightarrow u = \log(x)$ are as follows.

1. The half-axis $x \in (0, +\infty)$ is transformed into the entire real line $u \in (-\infty, +\infty)$.

2. $x = 1$ transforms to $u = 0$.

3. For $x > 1$ the transformed values are positive and "compressed".
   For instance, $x = 100 \to u = 2$ and $x = 10^5 \to u = 5$.

4. For x in $(0,1)$ the transformed values are negative and "expanded".
   For example, $x = 1/10 \to u = -1$ and $x = 10^{-8} \to u = -8$.

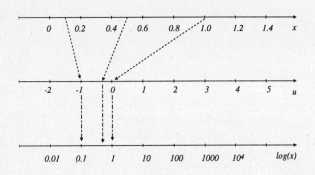

Powers of 10 are easily transformed, such as

$$x = 0.001 \to u = \log(0.001) = -3$$

and

$$x = 100{,}000 \to u = \log(100{,}000) = 5$$

Similarly, using properties of logarithms we see that x-values whose ratios are powers of ten will be transformed into u-values that differ by an integer value, e.g., $5/50 = 10$ so

$$\log(50) - \log(5) = \log(50/5) = \log(10) = 1$$

The log transformation $x \to u = \log(x)$ is illustrated at the left for three different orders of x-values. Notice that the tick marks are identically spaced on the middle u-axis and the bottom $\log(x)$ axis, however, the labels or unit-values are indicated differently.

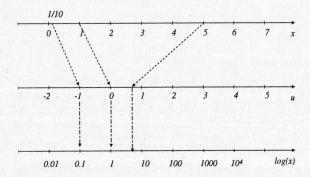

The labels on the $\log(x)$ axis are actually not the values of $\log(x)$ but the values of x. This is a typical type of labeling that you encounter if you selected "log scaling" on a spread-sheet or graphical plotting option.

On a log-scale the ticks are evenly spaced corresponding to log units and the numbers represented by the labels increase by a factor of 10 at each major tick, called a *cycle*.

It is difficult to directly plot points between the tick marks on the $\log(x)$ scaled lines. For instance, to plot the points $\log(0.5)$, $\log(5)$, or $\log(50{,}000)$ in the different panels.

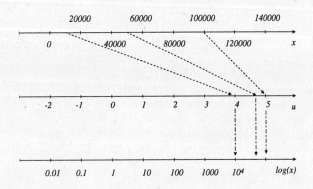

This is because the numbers $\log(2)$, $\log(3)$, $\log(4)$, $\log(5)$,... are not evenly spaced along a log-axis. $\log(5)$ for instance is about 7/10 of the way between $\log(1) = 0$ and $\log(10) = 1$.

To facilitate easier plotting of data on a logarithmic scale additional grid points corresponding $\log(n)$ for integer values of n can be included. These <u>are not uniformly spaced</u>!

The following log-scaled number line illustrates the variable distance between interior ticks, marked 2, 3, 4,5, and 8 . (The numbers 6, 7, and 9 were not indicated simply because the tick marks are too close and the numbers would have overlapped.)

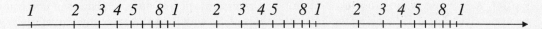

Notice that the major ticks, all labeled 1, are the same uniform distance apart; these correspond to successive powers of 10. For discussion, we refer to the distance between two successive 1-ticks as 1 P.U. (physical unit); you can measure this with a ruler. For each cycle, the physical distance between a 1-tick and a 2-tick is $\log(2)$ P.U. ≈ 0.30103 P.U. The physical distance from a 1-tick to the 3-tick is $\log(3)$ P.U. ≈ 0.4771213 P.U. The distance between a 2-tick and the following 3-tick is less that the distance between the 1-tick and 2-tick; it is the distance

$$\log(3) - \log(2) = \log(3/2) \approx 0.1760913 \quad \text{P.U.}$$

Why are there several points labeled 1? Why are the cycles 1, 2, 3, ... labels repeated? The answer is that the 1-ticks correspond to successive powers of 10, the log-line can indicate numbers differing by 3-orders of magnitude. For particular graphs the range of x-values will vary. For instance, in one graph you may wish to plot the concentrations of proteins in blood, which would likely range from nano-molar (nM), 10^{-9}, to micro-molar (µM), 10^{-6}. Referring to the above log-scaled line, in this instance the four successive 1-ticks on the log-axis, from left to right, would then correspond to 10^{-9}, 10^{-8}, 10^{-7}, and 10^{-6}. Then, the 5-tick to the left of the right-most 1-tick would correspond to 0.5 µM = 500 nM or $5 \cdot 10^{-7}$ while the middle 2-tick would represent the number 20 nM = 0.02 µM or $2 \cdot 10^{-8}$. In a different situation you may wish to plot populations of countries, ranging in values from a million to a billion, 10^6 to 10^9. Then the four 1-ticks would represent one million, 10^6, 10 million, 10^7, etc. and the population of Canada, about 27 million, would correspond to a point between the second 2- and 3-ticks.

Example 2 **Plotting data on a log scale.**

Problem How many cycles would it take to plot the following data on a log-scale axis?

Isotope	^{3}H	^{14}C	^{36}Cl	^{85}Kr	^{59}Fe	^{131}I
Half-life	12.26 y	5760 y	$3.1 \cdot 10^5$ y	10.6 y	45 d	8 d

Solution First, all data must be put into the same real unit. Converting the data given in days (d) to year units (y), we find the half-life for iron ^{59}Fe, is ≈ 0.123 y and for iodine ^{131}I is ≈ 0.0219 y. Thus these data can therefore be plotted on a line with log-scales ranging from 10^{-2} to 10^6. Using the powers of 10, we see that this would require 6 - (-2) = 8 cycles on the log-scale. This would require a log-scaled line with 9 equally spaced "1" tick marks.

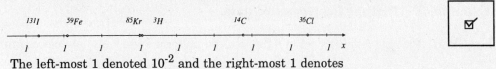

The left-most 1 denoted 10^{-2} and the right-most 1 denotes 10^6.

LOG-LOG GRAPHS

When both axes of a coordinate system are scaled logarithmically the resulting coordinate system is referred to as a **log-log** system. The representation of data or graphs of equations can be arrived at in two ways using log-log coordinates. One way is to algebraically transform x-y data into u-v data by using the two logarithmic transformations

$$x \to u = \log(x) \quad \text{and} \quad y \to v = \log(y).$$

Then, the transformed points can be plotted on a uniform u-v coordinate system. For instance, the x-y point (2, 1000) would be transformed into the u-v point $(\log(2), \log(1000)) = (0.3, 3)$(approximately).

A second way, which is the physical analogy of the first method, is to physically re-scale the coordinate axes to have distances measured in logarithmic units. Commercial **log-log graph paper** has this done for you. Because the scales are not uniform it would be difficult to plot points on such a graph if only tick marks were indicated. Normally scale-lines are provided so that you can accurately plot points off the axes. Commercial graph paper will have various "cycles" indicated. A cycle consists of a series of ticks labeled 1, 2, 3, ..., 9, 1 corresponding to numbers in an interval $[10^n, 10^{n+1}]$. For instance, on a "2 × 3 Cycles" log-log graph you could plot on the horizontal axis x-values over two orders of magnitude, say from 10 to 1000, and on the vertical axis y-values over three orders of magnitude could be plotted, such as from 0.0001 to 0.1.

Remember the ticks are physically log-distances apart, but are labeled with un-scaled x-values. When you plot a point (x,y) on a log-log scale, the physical distances are not x and y units from the "origin", which is $(\log(1), \log(1)) = (0,0)$, but are instead $\log(x)$ and $\log(y)$ physical units from the "origin" which most commonly is not even on the graph! For instance, the point corresponding to (10, 100) is physically at a position $(\log(10), \log(100)) = (1,2)$. If the point $(\log(1), \log(1))$ is not indicated on the log-log graph because you are plotting either very large or very small numbers this distance relationship will not be apparent. Furthermore, if different magnitudes are plotted on the two axis the result is a distorted graph in which the physical distances and the theoretical distances are different, this is similar to the situation where different units and physical grid distances are employed on an x-y coordinate system.

This may all seem complex, but it is really simple. As a consequence of the tick marks and lines being logarithmically placed along the axis, to plot points on log-log graph paper, you do not need to evaluate any logarithm, this has been done for you as long as you have values that fall on the grids. If you do not have log-log paper, you can still plot logarithmically scaled data by evaluating the log-values and plotting these on an ordinary u-v axis system with uniform scales.

Example 3   Plotting log-log scaled data.

Problem   Plot the following set of data on a u-v uniform system and on log-log graph paper.

x-y-data set: {(1,2),(2,10),(10,5),(50,10),(100,5),(200,2),(1000,10)}.

Solution   a) To graph these points on a uniformly scaled graph we make the change of variable

$$x \to u = \log(x) \quad \text{and} \quad y \to v = \log(y)$$

This transformed the x-y-data set into a

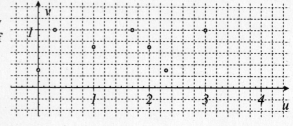

u-v-data set:
{(0, 0.3), (0.3, 1), (1, 0.7), (1.7, 1), (2, 0.7), (2.3, 0.3), (3, 1)}.

b) Alternatively, the original data can be plotted on commercial log-log graph paper. To emphasize the equivalence of the two approaches, rather than using commercial log-log paper, the data has been sketched on a graph with log-scales and grid-lines indicated.

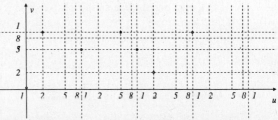

The u-axis "1"s correspond to x ranging from $1 = 10^0$ at the left to $x = 10^4$ at the right. The left-most "2" represents 2, the next "2" represents 20, and the right-most "2" represents 2000.

Notice that the graph axes, the arrows, are identical to the previous graph in physical dimensions. Hence, the points are plotted in exactly the same physical position. If you cut out one graph and overlaid it on the other the axes and points would match. The only difference is the position and labeling of the scale ticks.

MODEL PARAMETER IDENTIFICATION

Linear Log-Log Graphs.

The graph of a allometric function $y = m \cdot x^r$ is linear on a log-log coordinate system.

This can be seen by taking the log of each side of the equation and simplifying:

$$\log(y) = \log(m \cdot x^r) = \log(m) + r \log(x)$$

Setting $u = \log(x)$ and $v = \log(y)$ and defining $M = \log(m)$ this can be written as the linear equation

$$v = M + r \cdot u$$

Note that the v-intercept is $M = \log(m)$ corresponds to $u = 0$ or $x = 1$.

For instance, a graph of the equation $y = 3x^{1/2}$ on log-log graph paper is a line given by

$$\log(y) = 1/2 \log(x) + \log(3)$$

IDENTIFYING PARAMETERS OF AN ALLOMETRIC FUNCTION $m \cdot e^r$.

If experimental data are expected to have an *allometric* relationship, i.e. a power relationship, then when it is plotted on a log-log graph the data should lie on a straight line. Experimental data normally has variability and will not all lie on a line, but statistically we can select a line that "best" fits the data. From this line we can extract the numerical values of the parameters m and r.

The constant M is the v-coordinate where the data line crosses the v-axis. The v-axis is the vertical line through the point labeled 1 on the $u = \log(x)$ axis. Remember $x = 1 \rightarrow u = \log(1) = 0$. The model coefficient parameter is then

$$m = 10^M$$

The slope r of the log-log line is calculated from any two points on the line. If two points are given as numerical values, say (u_1, v_1) and (u_2, v_2), then

$$r = [v_2 - v_1] / [u_2 - u_1]$$

If the two points are specified by giving their corresponding x-y coordinates, as (x_1, y_1) and (x_2, y_2), then the slope r is calculated as the physical slope of the line, i.e., as

$$r = [\log(y_2) - \log(y_1)] / [\log(x_2) - \log(x_1)]$$

Example 4   **Determining model parameters for $y = mx^r$.**

Problem   Assume that a set of (x,y) data is plotted on a log-log graph and the line that best fits the data passes through the u-v points corresponding to the (x,y) data points A = (1,9) and B = (6,3). Sketch this line on a log-log graph and determine the parameters of the corresponding x-y model $y = mx^r$.

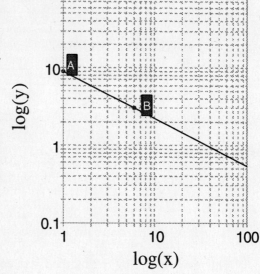

Solution   The corresponding points on the log-log graph are technically $(\log(1), \log(9)) \approx (0, 0.954)$ and $(\log(6), \log(3)) \approx (0.778, 0.477)$. As the graph has logarithmically scaled grids the x-y values can be plotted directly. The line through these two points has the u-v equation

$$v = M + r \cdot u$$

The v-shift parameter M is the $v = \log(y)$ coordinate of the point where $u = \log(x) = 0$. This is the point corresponding to the data point A in this example. Hence, $M = \log(9)$ and the model coefficient is then

$$m = 10^{\log(9)} = 9$$

The physical slope of this line is the exponent r of the allometric model. It is computed as

$$r = \frac{\log(9) - \log(3)}{\log(1) - \log(6)} = \log(9/3)/\log(1/6) = \log(3)/[-\log(6)] = 0.4771 / -0.7782$$

Thus the data fit the allometric model $y = 9x^{-0.613}$.

Example 5   **Determining the equation of a line on a log-log graph.**

Problem   Determine the x-y equation of the line in the graph at the left.

Solution
A line on a log(x)-log(y) graph has an x-y equation of the form $y = mx^r$. The first step is to identify particular points on the graph. As the line passes through $(\log(1), \log(0.7))$ we know that $m = 0.7$. To get the exponent we observe that the graph passes through the point $(\log(60), \log(30))$. Hence, using this and the "intercept" gives

$$r = [\log(30) - \log(0.7)] / [\log(60) - \log(1)] \approx 0.91782$$

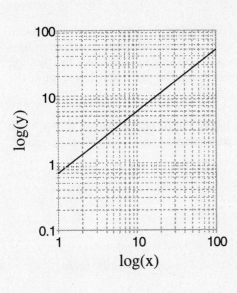

Rounding r to 0.9 gives the model equation
$$y = 0.7x^{0.9}$$

☑

Caution

If you are given data for which the scale on the log-log graph do not include the log(x) value $0 = \log(1)$, then the scale constant m or equivalently $M = \log(m)$ can not be directly read off the graph.
In this case, if you know r and a single point $(\log(x_0), \log(y_0))$ on the graph, then the value of m is found as

$$m = y_0 / x_0^r$$

Example 6   **Determining constants when log(x) = 1 is not included on the log(x) axis.**

Problem   Determine the x-y equation of the line in the graph.

solution   The equation has the form $y = mx^r$. As 1 is not on the log(x) axis we first determine r. To do this we pick two points on the line. Any two would give the same slope. Using the points

$(\log(10), \log(20))$ and $(\log(1000), \log(5))$

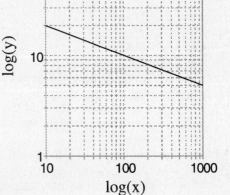

gives
$$r = [\log(5) - \log(20)] / [\log(1000) - \log(10)]$$
$$= -\log(4)/\log(100) \approx -0.60206 / 2 = -0.30103$$

Using this value and the first data point gives
$$m = 20 / 10^r = 40$$

as $r = -1/2 \log(4) = \log(4^{-1/2}) = \log(1/2)$, and hence $10^r = 10^{\log(1/2)} = 1/2$. The solution is that the line is the graph corresponding to the

$$y = 40 \, x^{\log(1/2)} \approx 40 \, x^{-0.3}$$

☑

Example 7   **An allometric model of sensory stimulation.**

Background   In studying sense receptors (eyes, ears, nose, and taste buds) in humans, it is necessary to establish a psycho-physical scale relating the perceived intensity, R, to the stimulus intensity S. The basic relation is thought to be an allometric function

$$R = cS^k$$

where c and k are parameters. The parameter k is the most interesting, as it distinguishes between different sense modalities. For example, the value of k is approximately 0.33 for a visual brightness test and 0.60 for an odor test. Estimated values of k and c are usually obtained by plotting experimental observations on a log(R) - log(S) graph and determining the slope and intercept of the line that best fits the data points.

Problem A visual test was performed and the "best fit" line on a log(R)-log(S) graph passed through the points (log(2), log(1)) and (log(8), log(16)). Estimate the parameters c and k for this experiment.

Solution The exponent k in the model equation is found as

$$k = slope = \frac{\log(R_1) - \log(R_2)}{\log(S_1) - \log(S_2)}$$

for any two points (S_1, R_1) and (S_2, R_2). Using $(S_1, R_1) = (2,1)$ and $(S_2, R_2) = (8,16)$ gives

$$r = \log(16)/\log(4) = 2\log(4)/\log(4) = 2$$

The scale coefficient c, in this model, is then given by

$$c = R_2 / S_2^r = 16 / 8^2 = 1/4$$

The data fit a simple quadratic model: $R = 0.25 S^2$.

SEMI-LOG GRAPHS

When one axis of a graph is logarithmically scaled and the other is not the graph is called a **semi-log graph**. Semi-log graphs can have either axis log-scaled, however most commonly the x-axis is uniformly scaled and the vertical axis is given a log(y) scale. Semi-log graph paper has units on one axis measured in logarithmic distance while a normal uniform unit is used on the other axis.

Linear Graphs on Semi-log Scales

The graph of any exponential function $y = c\, b^{kx}$ on a semi-log graph is linear.

Consider the general exponential function with base b:

$$y = c\, b^{kx}$$

where c and b are positive numbers and k is an arbitrary constant. Taking the logarithm of both sides of this equation results in the equation

$$\log(y) = \log(c\, b^{kx}) = \log(c) + kx \log(b)$$

Making a log transformation of only the y-variable, $y \to v = \log(y)$, results in the linear equation

$$v = C + mx$$

with slope $m = k \cdot \log(b)$ and the v-intercept is $C = \log(c)$. The three model parameters b, c, and k can not be uniquely determined from the two linear parameters C and m. The choice of the base is never unique because of the equivalence of exponential functions. Normally, the base is arbitrarily taken to be either e or 10.

Section 1.6 — Form, Scale and Transformations

Example 8 Plotting an exponential function on a semi-log graph.

Problem Sketch the graph of $y(x) = 4 \cdot 10^{0.2x}$ for $x \in [0, 14]$ on a semi-log coordinate system. Use the graph to estimate $y(3)$ and $y(11)$.

Solution As we know that the x-log(y) graph will be linear, we need only plot two points to plot it. These are chosen as points at which the function is easily evaluated. Obvious choices are for $x = 0$ and $x = 10$: $y(0) = 4$ and $y(10) = 4 \cdot 10^2 = 400$.

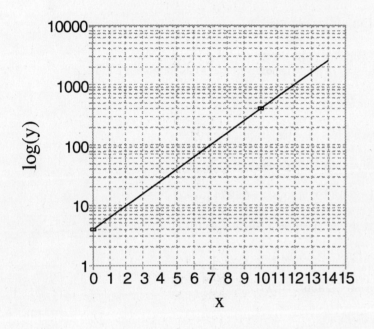

The points $(1, \log(4))$ and $(10, \log(400))$ have been plotted on the graph and a straight line passed through them. From this line we estimate the values of the function at $x = 3$ and 11. It appears that

$$\text{at } x = 3 \quad \log(y) = \log(15)$$

Thus, we estimate that

$$y(3) = 4 \cdot 10^{0.6} \approx 15$$

As its actual value is approximately 15.9 and the error is about 6%.

Similarly, from the line, we estimate

$$y(11) = 4 \cdot 10^{2.2} \approx 600$$

As $y(11) \approx 634$ this estimate has about the same % error.

Parameter Identification for exponential curves plotted on semi-log graphs.

The exponential curve $y = cb^{kx}$ will plot on a semi-log graph as a straight line for any choice of the base b. Given a line on a semi-log coordinate system how do we identify the specific model parameters? Assume that the linearized model equation is

$$\log(y) = \log(c) + mx$$

The slope constant m. The constant m is the theoretical slope of the line, defined by the ratio of the change in the log(y)-coordinates to the change in the x-coordinates for any two points on the graph.

$$m = \frac{\log(y_1) - \log(y_2)}{x_1 - x_2}$$

for any two points $(x_1, \log(y_1)), (x_2, \log(y_2))$ on the line.

The exponent k. For a specific base b, the exponent is given by $k = m/\log(b)$.

When $b = 10$, $k = m$. Since, $\log(10) = 1$.

When $b = e$, $k = m/\log(e)$ Or, equivalently, $k = m \cdot ln(10)$ since $\log(e) = ln(e)/ln(10) = 1/ln(10)$.

The coefficient c. The shift parameter log(c) is read directly from the graph as log(c) is the log(y)-axis intercept. Alternatively, if x = 0 is not included in the graph, $c = y/b^{kx}$ for any point (x,y) on the line.

Caution
The appearance of the line may be deceiving. The "physical slope" of the line, that you could measure with a ruler, is generally different than the value of m.

Example 9   **Determining exponential model parameters.**

Problem   A set of data, when plotted on a log(y)-x semi-log graph all lie on a straight line. If two of the data points are (1,2) and (5,6) what is the corresponding model exponential equation $y = ce^{kx}$?

Solution   First, the slope of the line is calculated.

$$m = [\log(2) - \log(6)] / [1 - 5] = \log(1/3)/(-4) \approx 0.1193$$

Using log(e) ≈ 0.4343, the exponent constant k is given by

$$k = m/\log(b) \approx 0.1193 / 0.4343 \approx 0.2747$$

The constant c is found by substituting the value of y and the point (1,2):

$$c \approx 2/e^{0.2747} = 1.5196$$

Thus the corresponding model equation is (approximately)

$$y = 1.5196 \, e^{0.2747x}$$

RECIPROCAL SCALES

Reciprocal transformations.

A reciprocal transformation $x \to u = 1/x$ is defined for $x \neq 0$. This transformation inverts the magnitude of numbers, making large values small and small values large. The basic properties of the reciprocal transformation are as follows.

1. $x \in (0,1) \to u \in (1,+\infty)$ and $x \in (1,+\infty) \to u \in (0,1)$
   $x \in (-1,0) \to u \in (-\infty,-1)$ and $x \in (-\infty,-1) \to u \in (-1,0)$

2. x = -1 and x = 1 are not changed by the transformation $1 \to 1$ and $-1 \to -1$.

3. The orientation of positive (or negative) points is reversed.

For example.   If $x_1 = 2$ and $x_2 = 3$, then $x_1 < x_2$ and their reciprocals $u_1 = 1/x_1 = 1/2$ and $u_2 = 1/x_2 = 1/3$ are reversed in size, $u_1 > u_2$.

Reciprocal transformations are often used to linearize hyperbolic functions. The basic *rectangular hyperbola* is the graph of $y = 1/x$. Clearly, the transformation $x \to u = 1/x$ transforms this equation into the linear equation $y = u$. The more general shifted hyperbola equation is usually written in the form

Section 1.6 — FORM, SCALE AND TRANSFORMATIONS

$$(y - b)(x - a) = c$$

A special case has the constant $c = a \cdot b$. Then the hyperbola has the form

$$y = bx / (x - a)$$

Inverting both sides of this equation gives

$$1/y = (x - a)/(bx) = x/(bx) - a/(bx) = 1/b + (-a/b)(1/x)$$

This equation is linear in the reciprocals $1/x$ and $1/y$. Therefore, the double reciprocal transformation,

$$u = 1/x \quad \text{and} \quad v = 1/y$$

changes the equation of the hyperbola into the equation of a line:

$$v = 1/b + (-a/b)\, u$$

The v-intercept is the constant $1/b$ and the slope is the negative ratio of the constants, $-b/a$. Note that you will never have an (x,y) data point giving the v-intercept since this occurs when $u = 0$, and therefore when $x = 1/u$ approaches ∞.

Example 10 — The Lineweaver-Burk plot of the Michaelis-Menten Equation.

Background The Michaelis-Menten equation is a basic formula in *Biochemistry* that indicates the rate of an enzyme activation of a protein molecule. The equation involves the rate of reaction v, the maximum reaction rate constant V_{max}, the Michaelis-Menten constant K_m, and the concentration of the substrate S. It is the hyperbolic equation

$$v = S \cdot V_{max} / [K_m + S] \qquad \textbf{Michaelis-Menten Equation}$$

Problem Determine the linear equivalent of the Michaelis-Menten equation, called the **Lineweaver-Burk** equation, by taking a double reciprocal transformation.

Solution Inverting both sides of the Michaelis-Menten equation gives

$$1/v = [K_m + S]/[S \cdot V_{max}]$$

Expressing the right side as a sum of two fractions gives

$$1/v = K_m / [S \cdot V_{max}] + S / [S \cdot V_{max}]$$

Canceling and regrouping the constants gives

$$1/v = 1/V_{max}] + [K_m / V_{max}]\, 1/S \qquad \textbf{Lineweaver-Burk Equation}$$

The model parameters, K_m and V_{max} are estimated by plotting experimental data recording reaction velocities at different substrate levels on a double reciprocal graph. A line is fit to the data, and then the parameters are found from the equation of the line.

Exercise Set 1.6

1. Draw parallel number lines to illustrate the indicated homogeneous scaling.
   a) $x \to u = 2.5x$ b) $x \to u = 0.25x$ c) $x \to u = -x$ d) $x \to u = -10x$

2. Draw parallel number lines to illustrate the indicated linear scaling.
   a) $x \to u = x - 2$ b) $x \to u = 2x + 3$ c) $x \to u = 5 - x$ d) $x \to u = x/2 - 1$

3. Draw parallel number lines to illustrate the indicated non-linear scaling.

   a) $x \to u = |x|$ b) $x \to u = x^2$ c) $x \to u = \sin(\pi x)$ for $-2 \le x \le 4$ d) $x \to u = \tan(\pi x/2)$ $|x|<1$

   e) $x \to u = \sqrt{x}$ f) $x \to u = 1/x$ $x \ne 0$ g) $x \to u = \text{SQRT}(4 - x)$ h) $x \to u = ln(x)$ $x > 0$

4. What is the actual slope of a line whose physical slope is 1, i.e. it makes a 45° angle with the horizontal axis for the indicated conditions? DRAW A SKETCH OF THE AXES SYSTEM AND THE LINE PASSING THROUGH THE ORIGIN.
   a) The tick marks on both axis are the same distance apart and have the same scale units.

   b) The tick marks on both axis are the same distance apart and the y-scale units are 2 per tick while the x-scale units are 1 per tick.

   c) The tick marks on both axis are the same distance apart and the y-scale units are 1 per tick while the x-scale units are 3 per tick.

   d) The tick marks on the y-axis are twice as far apart as those on the x-axis and have the same scale units.

   e) The tick marks on the y-axis are twice as far apart as those on the x-axis and the y-scale units are 2 per tick while the x-scale units are 1 per tick.

   f) The tick marks on the y-axis are twice as far apart as those on the x-axis and the y-scale units are 2 per tick while the x-scale units are 1 per tick.

   g) The tick marks on the y-axis are twice as far apart as those on the x-axis and the y-scale units are 1 per tick while the x-scale units are 3 per tick.

   h) The tick marks on the y-axis are one half as far apart as those on the x-axis and have the same scale units.

5. Plot all of the points on the same log-scaled number line.

   a) 1, 5, 20, 55, 90   b) 0.01, 0.005, 0.7, 0.05   c) $10^5, 10^6, 3 \cdot 10^5, 2 \cdot 10^6$

6. A set of data plotted on a log-log system was fit to a straight line passing through the data points (2,5) and (55,210). What is the equation that describe the x-y relationship of the data?

7. A set of data plotted on a semi-log system was fit to a straight line passing through the data points (2,5) and (55,210). What is the equation that describes the x-y relationship of the data?

8. A set of data plotted on a log-log system was fit to a straight line passing through the data points (0.05,0.0005) and (0.9, 0.000001). What is the equation that describes the x-y relationship of the data?

9. A set of data plotted on a double reciprocal axis system was fit to a straight line passing through the data points (2,5) and (5,10). What is the equation that describes the x-y relationship of the data?

10 (See Example 10.) Construct a Lineweaver-Burk plot of the substrate titration data for S and v: (10, 0.1), (25, 0.5), (50, 1), (100, 1.1) and (200, 1.2). Draw a line that visually approximates the data points and from this line infer the values of the Michaelis-Menten parameters K_m and V_{max}. Then write the equivalent Michaelis-Menten equation. HINT: make your 1/S axis have small grid units.

11 Determine the x-y equation of the indicated line

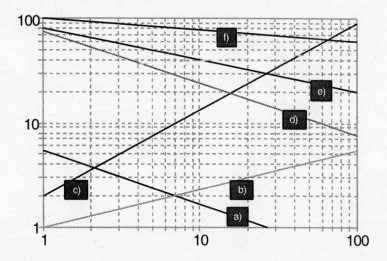

12 For each of the functions graphed in the previous exercise estimate the values of y(5) and y(10).

13 Determine the x-y equation of the indicated line

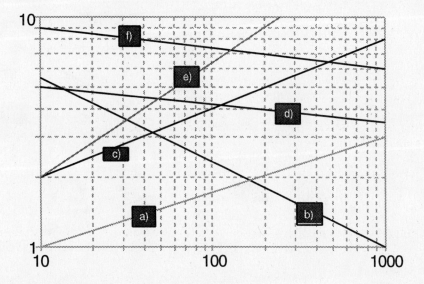

14 For each of the functions graphed in the previous exercise estimate the values of y(60) and y(200).

15 Determine the x-y equation of the indicated line

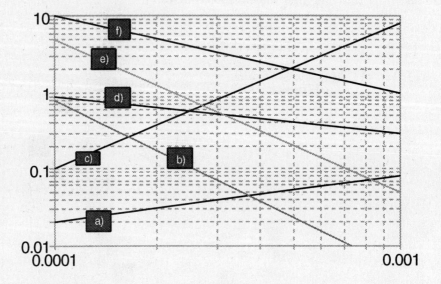

16 For each of the functions graphed in the previous exercise estimate the values of $y(5 \cdot 10^{-4})$ and $y(7.5 \cdot 10^{-4})$.

17 Determine the x-y equation of the indicated line

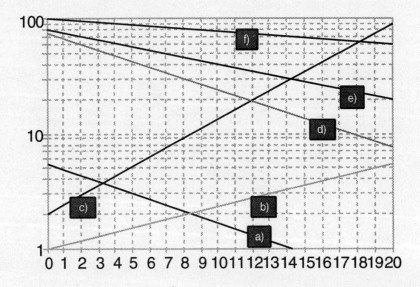

18 For each of the functions graphed in the previous exercise estimate the values of $y(5)$ and $y(10)$.

19 Determine the x-y equation of the indicated line

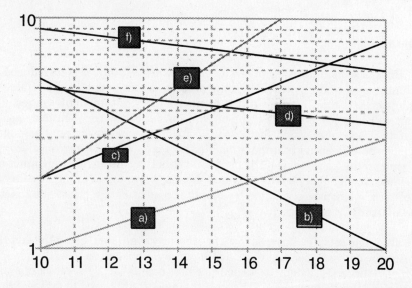

20 For each of the functions graphed in the previous exercise estimate the values of y(12) and y(15).

21 Determine the x-y equation of the indicated line

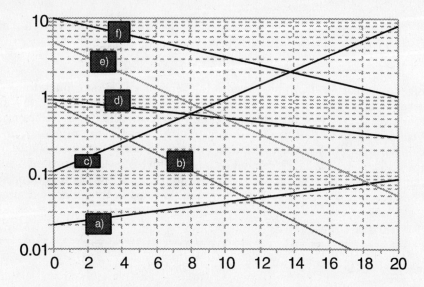

22 For each of the functions graphed in the previous exercise estimate the values of y(8) and y(15).

Section 1.7 Limits and continuity.

Introduction.

In this section the concept of a *limit of a function's values* is introduced and used to define when a function is *continuous*. The limit process is the foundation of Calculus. The mathematical definition of a limit involves a test to ensure that the function's values, f(x), are all close to some limit value, L, when the independent variable, x, is close enough to a specific number, b. Showing that this test is satisfied for a specific function f(x) can be challenging and generally requires considerable algebraic manipulation. In this text, we shall forego such exercises whenever possible, relying on the knowledge of the past 200 years that the elementary functions (as introduced in the previous sections) are *continuous* and, hence, when f(x) is an elementary function, such as e^x, its limit at x = b can be evaluated by simply substituting x = b into the function. Thus, for example, the limit of e^x at x = 3 is just e^3.

Limits of complex functions will be similarly evaluated, using the fact that the limit process and algebraic manipulations, such as adding or dividing, can be interchanged. However, in the case of division there is a major obstacle. We can not divide by zero! Unfortunately, one of the most important limits involved in Calculus, the limit that is used to define a derivative (see Chapter 3) will involve the limit of a quotient in which the denominator goes to zero. The "trick" of calculus theory is to demonstrate that this situation, dividing by zero, can be avoided by manipulating the function prior to taking the limit so that the "zero" part of the denominator cancels and then the limit can be evaluated.

LIMITS OF A FUNCTION.

The expression "the limit, as x approaches b, of f(x) is L" is written mathematically as

$$\lim_{x \to b} f(x) = L$$

This states that f(x) is always close to L whenever x is close enough to b. Mathematically, being "close" is indicated by the distance between two numbers, i.e., by the absolute value of their difference. Thus, if we introduce a tolerance value, ε (the Greek letter epsilon), to say that f(x) is as closer than ε to L is stated by the inequality $|f(x) - L| < \varepsilon$. Similarly, an inequality of the form $|x - b| < \delta$ (where δ is the Greek letter delta) indicates that the distance between x and b is less than δ. Thus, in this case x can not be larger than b plus the increment δ and it can not be smaller than b minus δ; $b - \delta < x < b + \delta$. To indicate that x ≠ δ we can require that the distance from x to δ is positive, $0 < |x - b|$. These concepts are used in the formal definition of a limit.

DEFINITION OF A LIMIT.

$$\lim_{x \to b} f(x) = L$$

if for any positive number ε there exists an number δ such that

$$|f(x) - L| < \varepsilon \quad \text{whenever} \quad b - \delta < x < b \quad \text{or} \quad b < x < b + \delta$$

Notice that the definition of a limit of f(x) at x = b does <u>not require</u> that f(b) be defined. To "prove" that a specific limit exists you must assume that ε is an arbitrary number and determine a dependent number δ such that the inequality condition of the above definition is valid. For example, to show that $\lim_{x \to b} 3x + 1 = 7$, we begin with the inequality

$$|f(x) - L| = |3x + 1 - 7|$$

Rearranging and factoring this becomes

$$|3x - 6| = 3|x - 2|$$

Thus, if we choose $\delta = \varepsilon/3$, then whenever $|x - 2| < \delta = \varepsilon/3$

$$|f(x) - L| = 3|x - 2| < 3\delta = 3\varepsilon/3 = \varepsilon$$

This type of argument is required each time a limit is "proven". To avoid this, mathematicians prove Theorems, general statements that are always true provided their hypothesis are valid, that indicate how to evaluate limits for specific types of functions. In this text, we do not generally state Theorems, preferring in most cases to indicate Rules, and we do not "prove" these. Although, we will sometimes offer Rationales of why particular rules hold.

The general limit, $\lim_{x \to b}$, is a two-sided limit. The values of x approach the number b from "above" or "the right" when $b < x$ and x gets closer to b. The values of x approach the number b from "below" or "the left" when $x < b$ and x gets closer to b. Sometimes we must consider limits when the x-values must be restricted to always being greater than or always being less than the number b. In this case we utilize one-sided limits written in the form

$$\lim_{x \to b+} f(x) \quad \text{and} \quad \lim_{x \to b-} f(x)$$

These are read as "the limit as x approaches b from the right of f(x)" and "the limit as x approaches b from the left of f(x)", respectively.

DEFINITION OF ONE-SIDED LIMITS.

$$\lim_{x \to b+} f(x) = L \qquad \textbf{Limit from the right.}$$

if for any positive number ε there exists an number δ such that

$$|f(x) - L| < \varepsilon \quad \text{whenever} \quad b < x < b + \delta$$

$$\lim_{x \to b-} f(x) = L \qquad \textbf{Limit from the left.}$$

if for any positive number ε there exists an number δ such that

$$|f(x) - L| < \varepsilon \quad \text{whenever} \quad b - \delta < x < b$$

Of course, limit variables other than x will be used. We will see limits like

$$\lim_{h \to 0} [(x + h)^3 - x^3]/h \quad \text{and} \quad \lim_{t \to 2+} [\![t]\!]$$

Limits and algebra.

The process of taking a limit can be interchanged with arithmetic operations as long as the resulting expression can be evaluated. This requires that the limits involved are all finite numbers and that division by zero does not occur. We will therefor be able to evaluate the limits of complicated functions by evaluating the limits of sub-terms and then combining these sub-limits. The following Rules are stated for a normal two-sided limit, but also apply if $x \to b$ is replaced by $x \to b+$ or $x \to b-$.

Assume that $\lim_{x \to b} f(x) = A$ and $\lim_{x \to b} g(x) = B$. Then.

The Constant Multiple limit Rule:

$$\lim_{x \to b} k \cdot f(x) = k \cdot \lim_{x \to b} f(x) = kA \quad \text{for any constant } k$$

The Sum/Difference limit Rule:

$$\lim_{x \to b} [f(x) \pm g(x)] = \lim_{x \to b} f(x) \pm \lim_{x \to b} g(x) = A \pm B$$

The Product limit Rule:

$$\lim_{x \to b} [f(x) \cdot g(x)] = \lim_{x \to b} f(x) \cdot \lim_{x \to b} g(x) = AB$$

The Quotient limit Rule:

$$\lim_{x \to b} [f(x) / g(x)] = \lim_{x \to b} f(x) / \lim_{x \to b} g(x) = A/B \quad \text{if } B \neq 0$$

These Rules allow the evaluation of a limit of a complicated function by taking the limits of subfunctions and then combining these. For instance,

$$\lim_{x \to 3} x^2 \sin(2x) / \{1 - 5e^{-0.1x}\}$$

$$= [\lim_{x \to 3} x^2] \cdot [\lim_{x \to 3} \sin(2x)] / \{[\lim_{x \to 3} 1] - 5 \lim_{x \to 3} e^{-0.1x}\}$$

The sub-limits can be evaluated simply by substituting 3 for x as each term involves a basic elementary function, and these are *continuous* at x = 3. Thus the limit is given by

$$\lim_{x \to 3} x^2 \sin(2x) / \{1 - 5e^{-0.1x}\} = 9 \sin(6) / \{1 - 5e^{-0.3}\}$$

CONTINUOUS FUNCTIONS.

Continuous functions are functions whose graphs can be sketched without lifting the pencil from the paper. To be continuous at a point x = c, f(c) must be defined and be equal to $\lim_{x \to c} f(x)$. The concept of continuity is a two-sided concept, since $\lim_{x \to c}$ is a two-sided concept. One-sided continuity can be defined by utilizing the one-sided limits $\lim_{x \to c+} f(x)$ and $\lim_{x \to c-} f(x)$.

DEFINITION OF A CONTINUOUS FUNCTION.

A function f(x) is **continuous at a point x = c** if

$$\lim_{x \to c} f(x) = f(c)$$

f(x) is **continuous on an open (a,b)**
if it is continuous at x = c for each c ∈ (a,b).

f(x) is **continuous on a closed interval [a,b]**
if it is continuous on (a,b) and if

$$\lim_{x \to a+} f(x) = f(a) \quad \text{and} \quad \lim_{x \to b-} f(x) = f(b)$$

The definition of continuity at x = c requires that three conditions be met:

1. f(c) must exist, i.e., f must be defined at x = c.

2. $\lim_{x \to c} f(x) = L$ must exist.

   In particular both $\lim_{x \to c+} f(x) = L$ and $\lim_{x \to c-} f(x) = L$

3. The limit L must be the function's value: L = f(c).

To ascertain if a function is continuous at a given number c, we generally ask the reverse question "Is f(x) not continuous at x = c ?" We look for a violation of one of the three conditions given above; if none is found, then the function is continuous.

Elementary functions are continuous at each point interior to their domain of definition.

x^n e^x sin(x)   and   cos(x) are continuous for all x.

ln(x)   and $\sqrt{x}$ are continuous for x > 0.

tan(x)   and   csc(x) are continuous for x ≠ π/2 ± nπ.

cot(x)   and   sec(x) are continuous for x ≠ ± nπ.

Example 1 Evaluating limits of continuous functions.

Problem Evaluate $\lim_{x \to 2} [x^3 - 2x + 3] / [e^x - ln(x)]$.

Solution The algebra rules are applied to reduce the problem to evaluating the limit of individual elementary functions. Applying the Constant Multiple and the Sum/Difference limit Rules, the limit of the polynomial is taken term-wise, thus the limit of the numerator is:

$$\lim_{x \to 2} [x^3 - 2x + 3] = \lim_{x \to 2} x^3 - 2 \lim_{x \to 2} x + \lim_{x \to 2} 3 = 8 - 2 \cdot 2 + 3 = 7$$

The limit of the denominator is split into a difference of two limits:

$$\lim_{x \to 2} [e^x - ln(x)] = \lim_{x \to 2} e^x - \lim_{x \to 2} ln(x) = e^2 - ln(2)$$

Applying the Quotient Limit Rule gives

$$\lim_{x \to 2} [x^3 - 2x + 3] / [e^x - ln(x)] = \lim_{x \to 2} [x^3 - 2x + 3] / \lim_{x \to 2} [e^x - ln(x)] = 7/[e^2 - ln(2)]$$

OBSERVATION:

If $f(x)$ is continuous at $x = c$, then

$$\lim_{x \to c^-} f(x) = f(c) \quad \text{and} \quad \lim_{x \to c^+} f(x) = f(c)$$

This is used to determine if a *split function* $y(x)$ that is defined by one equation $y = f(x)$ for $x > c$ and by another equation $y = g(x)$, for $x \leq c$ is actually continuous at $x = c$. If $f(x)$ and $g(x)$ are themselves continuous at c the one-sided limits of $y(x)$ at $x = c$ are just

$$\lim_{x \to c^-} y(x) = g(c) \quad \text{and} \quad \lim_{x \to c^+} y(x) = f(c)$$

If these two limits are different then $y(x)$ is not continuous at c.

Example 2 Evaluating limits and continuity of a "split" function.

Problem Where is the following function $y(x)$ continuous?

$$y(x) = \begin{cases} x^2 & \text{if } x < 0 \\ 5x - 2 & \text{if } 0 \leq x < 2 \\ x^3 & \text{if } x \geq 2 \end{cases}$$

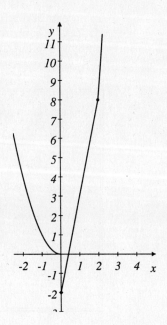

Solution This is a split function whose component functions are elementary functions so they are continuous for all x. Thus, $y(x)$ will be continuous at all x with the possible exception of the x-values where the function's definition changes, $x = 0$ and $x = 2$. These must be checked using the above observation.

At $x = 0$. $\lim_{x \to 0^-} y(x) = \lim_{x \to 0^-} x^2 = 0^2 = 0$

$\lim_{x \to 0^+} y(x) = \lim_{x \to 0^+} 5x - 2 = 5 \cdot 0 - 2 = 2$

Both one-sided limits exist, but, they are not equal. Hence, $\lim_{x \to 0} y(x)$ <u>does not exist</u>; y(x) is <u>not continuous</u> at x = 0.

At x = 2. $\lim_{x \to 2^-} y(x) = \lim_{x \to 2^-} 5x - 2 = 5 \cdot 2 - 2 = 8$

$\lim_{x \to 2^+} y(x) = \lim_{x \to 2^+} x^3 = 2^3 = 8$

As these limits are equal, $\lim_{x \to 2} y(x) = 8$. The function's value $y(2) = 2^3 = 8$ is the same as the limit. Consequently, y(x) is continuous at x = 2.

y(x) is continuous on the intervals (-∞,0) and (0,∞).

☑

Example 3 A graph with a "hole".

Problem If $y(x) = [x^2 - 2x - 3]/[x - 3]$ what is $\lim_{x \to 3} y(x)$?

Solution The function y(x) is not defined at x = 3 since its denominator is then zero. However, the $\lim_{x \to 3}$ considers what happens as x approaches 3. For x ≠ 3, y(x) can be simplified by factoring the numerator and canceling:

$[x^2 - 2x - 3]/[x - 3] = (x - 3)(x + 1)/(x - 3) = x + 1$

Hence,

$\lim_{x \to 3} y(x) = \lim_{x \to 3} [x^2 - 2x - 3]/[x - 3] = \lim_{x \to 3} x + 1 = 3 + 1 = 4$

The limit exists, even though the function is not defined at the limit point. The graph of $y = [x^2 - 2x - 3]/[x - 3]$ is the line y = x + 1 except at x = 3 where a "hole" occurs. This is often illustrated by drawing an open circle on the line at the undefined point.

☑

The previous example illustrates an important type of limit. When we study *derivatives* in subsequent Chapters we will encounter limits that have this same feature: the denominator is zero at the limit point. The method of evaluating the limit is always the same. The factor that causes the zero in the denominator is factored from the numerator and, before the limit is taken, canceled with the denominator term leaving a continuous function. The limit is then evaluated by simply evaluating the remaining function at the limit point.

Algebra of continuous functions.

Because the limit process interchanges with algebraic operations when the limits exist, and for continuous functions the limits do exist, by definition, we can conclude that algebraic combinations of continuous functions are also continuous. The usual *caveat* about division by zero must be included.

> If f(x) and g(x) are continuous at x = c then so is
>
> f(x) ± g(x),   f(x) · g(x),   and f(x)/g(x) provided g(c) ≠ 0

For the composition of two functions the second function must be continuous at the value of the first:

> f∘g(x) is continuous at x = c
> if g(x) is continuous at x = c and f(x) is continuous at x = g(c).

If f(x) and g(x) are continuous everywhere then

$$\lim_{x \to c} f(g(x)) = f(\lim_{x \to c} g(x)) = f(g(c))$$

The Extreme Value Theorem.

A function may be unbounded on an interval, e.g., y = 1/x on the open interval (0,2); as $x \to 0+$ $y \to \infty$, i.e., y becomes very large as x approaches zero. This can not happen when the function is continuous. In particular, on a closed interval [a,b] a continuous function must reach its maximum and minimum values at points in the interval. This is stated as a theorem without a proof.

The Extreme Value Theorem.

> If f(x) is continuous on [a,b] then there exist numbers x_{min} and x_{max} in [a,b] such that
>
> $$f(x_{min}) \leq f(x) \leq f(x_{max}) \quad \text{for all } x \in [a,b]$$

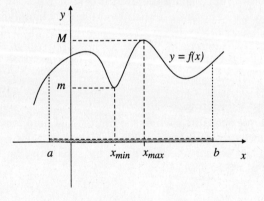

We shall denote the minimum and maximum values by m and M, respectively:

$m \equiv f(x_{min})$ is the minimum of f(x) on [a,b].

$M \equiv f(x_{max})$ is the maximum of f(x) on [a,b].

There may actually be many x-values at which the function attains its minimum or maximum values. For instance, on the interval [0, 10] f(x) = cos(πx) has the maximum M = 1, which is its value whenever x is an even integer between 0 and 10. Its minimum is m = -1, which occurs whenever x is an odd integer in the interval.

An important application of *calculus* will be to determine a function's maximum and minimum values and the x-values at which they occur.

The Intermediate Value Theorem.

Does the equation $x - ln(x) = 2$ have a solution?
This can be answered using the following *Intermediate Value Theorem*. It states that a continuous function on a closed interval assumes ever y-value between its maximum and minimum values.

The Intermediate Value Theorem.

> Assume f(x) is continuous on [a,b]. If Y is any number between the minimum m and the maximum M of f on [a,b] then there exist a value $c \in$ [a,b] such that
>
> $$f(c) = Y$$

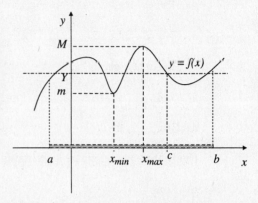

There may be more than one number c at which the function has the value Y. In the sketch at the right there are actually 5 possible choices for c for the illustrated Y-value. Although only one has been labeled on the x-axis, each x-coordinate where the graph of f(x) crosses the horizontal line y = Y satisfies the Intermediate Value Theorem (IVT). In practice, to determine c one must be able to solve the equation f(c) = Y. However, when the equation is complex, the IVT assures us that a solution exists.

Example 4 **Knowing that an equation has a solution.**

Problem Does the equation $x - ln(x) = 2$ have a solution?

If so, can you give an interval [a,b] that contains the solution?

Solution The function $f(x) = x - ln(x)$ is continuous for $x > 0$ since it is the difference of two continuous functions. Consider this function on the closed interval [0.1 , 1]. The minimum and maximum value of f(x) satisfy

$m \leq f(x) \leq M$ for any x in the interval.

In particular, when x is the right endpoint, $x = 1$, we have

$m \leq f(1) = 1 - 0 = 1$

and, similarly, when x is the left endpoint, $x = 0.1$,

$M \geq f(0.1) = 0.1 - ln(0.1) = 0.1 + ln(10) > 0.1 + 2.3 = 2.4$

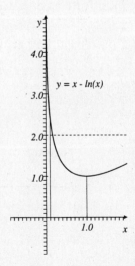

since $ln(10) \approx 2.303$. Consequently, we can conclude that the number 2 on the right of the original equation is between the minimum and maximum values of the function f on the interval [0.1,1]. Choosing Y = 2 in the Intermediate Value Theorem, its conclusion is that there is a number c, with $0.1 \leq c \leq 1$, for which f(c) = 2. Unfortunately, the theorem does not tell us how to find this number. The number x = c is then the root of the equation $x - ln(x) = 2$.

Exercise Set 1.7

1. Determine the intervals on which the given function is continuous.

   a) $f(x) = ln(x) + x$
   b) $g(x) = e^{3x} - e^{-3x}$
   c) $f(x) = \text{SQRT}(x-2) * \sec(\pi x)$
   d) $y = x + 3/(x-2)$
   e) $f(x) = \sqrt{x} \cdot \tan(x)$
   f) $f(x) = (x-2)^{0.75}$

2. Evaluate the given limit using the continuity of the elementary functions.

   a) $\lim_{x \to 3} x^2 - 5x + 1$
   b) $\lim_{x \to -1} e^x(1 - 1/x)$
   c) $\lim_{x \to 0} \sin(x^2 - 1)$
   d) $\lim_{x \to 2} x^2/(5x - 8)$
   e) $\lim_{x \to 1} e^{ln(x)}$
   f) $\lim_{x \to 3} x2^x$

3. Evaluate the one-sided limit using the continuity of the elementary functions.

   a) $\lim_{x \to 3^+} x^2 - 9$
   b) $\lim_{x \to 1^-} x/(1-x)$
   c) $\lim_{x \to 1^+} (x^2 - 1)/(x-1)$
   d) $\lim_{x \to 2^+} (x^2 - 4)/(4x - 8)$
   e) $\lim_{x \to 1^+} ln(x^2)$
   f) $\lim_{x \to 0^-} (x^3 + 3x^2)/x^2$

4. For each function i) Sketch the graph of the function. ii) Determine the limit from the left and from the right at each "split" x-value. iii) Indicate where the function is continuous.

   a) $y(x) = \begin{cases} |x| & \text{if } x < 1 \\ 1/x & \text{if } x \geq 1 \end{cases}$
   b) $y(x) = \begin{cases} |x - 1| & \text{if } x < 1 \\ 1 - x & \text{if } x \geq 1 \end{cases}$
   c) $y(x) = \begin{cases} \sqrt{x} - 1 & \text{if } x > 1 \\ \sqrt{1 - x} & \text{if } x \leq 1 \end{cases}$
   d) $y(x) = \begin{cases} 2x - 1 & \text{if } x < 0 \\ 1 - x & \text{if } x \geq 0 \end{cases}$
   e) $y(x) = \begin{cases} 2x & \text{if } x < 1 \\ \sqrt{x} & \text{if } 1 \leq x \leq 4 \\ 4 - x^2 & \text{if } x > 4 \end{cases}$
   f) $y(x) = \begin{cases} 2x - 1 & \text{if } x < 1 \\ \sqrt{x} & \text{if } 1 \leq x \leq 4 \\ x^2 - 4x & \text{if } x > 4 \end{cases}$
   g) $y(x) = \begin{cases} -3 & \text{if } x < -2 \\ [\![x]\!] & \text{if } -2 \leq x \leq 4 \\ 5 - 0.5x & \text{if } x > 4 \end{cases}$
   h) $y(x) = \begin{cases} \sin(\pi x) & \text{if } x < 1 \\ x^3 & \text{if } 1 \leq x < 2 \\ 4\sqrt{2x} & \text{if } x > 2 \end{cases}$

5. Evaluate the given limit or indicate why you can not.

   a) $\lim_{h \to 0} (3h^2 - 2h + 7)(8h - 9h^2 + 2)$
   b) $\lim_{x \to 2} [x^2 - 2x - 6] + (x - 3)/(x + 2)$
   c) $\lim_{t \to 1} (t^2 - 2t + t^5)/(t^2 - 8t - 9)$
   d) $\lim_{h \to 0} (h + 2h^4)/(h^2 + 2h)$
   e) $\lim_{x \to 3} (x^2 - 2x + 6)/(x - 3)$
   f) $\lim_{h \to 0} (h \sin(h) + h^2)/h$

6. Which of the given equations has a solution in the interval $[-1, 2]$? Why?

   a) $x^2 - 2 = 0$
   b) $e^x - 10x = 0$
   c) $xe^x = 2$
   d) $\cos(x^2) = x$
   e) $x^5 - 3x + 1 = 0$
   f) $e^x = ln(x + 2)$
   g) $x - \tan(\pi x) = 0$
   h) $x^3 = \sqrt{x}$

Chapter 2 Discrete Models.

Some things change discretely, like the increase in populations as the result of individual births or deaths, while others change continuously, like the passage of time. Many events that may seem continuous can be modeled discretely; indeed, the discrete models introduced in this chapter form the basis of most computer algorithms used to numerically model continuous process. Discrete models use difference equations to indicate the way a dynamical system changes at discrete time points. In this chapter we investigate the graphical representation of difference equations and the solution of basic equations. This provides a foundation on which the study of Calculus is built.

Section 2.1 An Introduction to Sequences and Series 106
Sequences.
Series.
Closed forms of special series.
Infinite series.

Section 2.2 Discrete Dynamical Systems 129
Introduction to Dynamical Systems.
Difference Equations.
The forward difference operator Δ.
Graphical representation of dynamical systems.
Equilibriums and limiting behavior of dynamical systems.

Section 2.3   Graphing First-Order Dynamical Systems.   139
Graphing Discrete Dynamical Systems.
Equilibriums and Stability.
The Equilibrium of Linear First-Order Difference Equations.
Steady States.
Cyclic solutions.
Using Difference Equations to Solve Non-linear Equations.

Section 2.4   Linear First-Order Difference Equations. 151
The homogeneous difference equation: $x_{n+1} = a \cdot x_n$
The non-homogeneous equation: $x_{n+1} = a \cdot x_n + b$
Non-linear first order difference equations.
Equilibriums and stability of first order systems.

Section 2.5   Higher-Order Difference Equations 164
Second-Order Difference Equations
Homogeneous 2nd-order equations.
Developing the General Solution when λ_1 and λ_2 are *complex* roots. (OPTIONAL)
Equilibriums and Steady-States.
Nonhomogeneous 2-nd Order Difference Equations.

Section 2.1 An introduction to Sequences and Series

Sequences

A **sequence** is simply a set of objects that are ordered, their order being associated with the order of natural numbers and integers. The term "sequence" is applied to many types of objects, like the sequences of days in the week, or the sequence of dives required in olympic diving competition. We will focus on mathematical sequences where the objects are numbers or variables that represent real numbers. Mathematical sequences are actually special types of functions, whose domain's are either the set N of all natural numbers or a finite set of natural numbers. When a sequence's domain is the set of all natural numbers it is said to be an *infinite sequence*. When its domain is a finite set of numbers, e.g., { 1, 2, 3, 4}, the sequence is said to be a *finite sequence*. The objects or terms of a sequence will be referred to using letters with subscripts indicating their ordinal position in the sequence, e.than standard function notation.

Finite Sequences

Finite sequences are often simply indicated as sets of numbers, with the order assigned to the numbers being their number when counting from left to right. For instance

$$\{ 5, 3, 1 \}$$

is a finite sequence with three terms; 5 is the first term, 3 the second, and 1 the third term. If this sequence is given a letter name, say b, then it could be written as $\{ b_1, b_2, b_3 \}$ where $b_1 = 5$, $b_2 = 3$, and $b_3 = 1$. When a finite sequence has many terms it is often tedious or impractical to list each term in this way. When the sequence terms have a consistent pattern, an alternative notation lists the first few terms of a sequence, enough to establish the pattern, followed by three periods ... to indicate that terms are being omitted, and finally the last term or two of the sequence. For instance, to list the sequence of the first 100 even numbers we write

$$\{ 2, 4, 6, 8, ... , 198, 200 \}$$

To fill in the omitted terms we must "recognize the pattern". Mathematically, this means that we must have a function or formula that tells us how to evaluate each sequence term. Instead of using standard function notation such as f(n), a subscript notation is used to represent sequence terms, such as f_n or X_n, which is read as "X sub n". The subscript n is called an *index* and represents a natural number; which natural number will be indicated by the **range** of the index specified by notation such as "n = 1, 100", which means that the sequence would begin with n = 1, then n = 2, then n = 3, and continue to increment n by 1 until reaching n = 100, the last term. A term with a variable subscript, e.g., X_n of X_i, is called a *general* term of the sequence, while *specific* sequence terms have numerical subscripts, such as X_3 or X_{21}. The letters used to represent the index numbers traditionally are i, j, k, l, m, and n. We will use the n and i most frequently. A finite sequence with N terms can be represented by the *sequence notation*

$$\{a_i\}_{i = 1,N} \equiv \{ a_1, a_2, a_3, a_4, ..., a_{N-1}, a_N \}$$

The first term of this sequence is a_1, the second term a_2, and the k-th term of this sequence is the term a_k. Each term a_i represents a real number which can be specified directly, like $a_4 = 20$, or by a formula, such as

$$a_i = i + i^2$$

from which a_4 could determined by substituting 4 for i and evaluating the formula

$$a_4 = 4 + 4^2 = 20$$

Sometimes we will consider only part of a sequence, called a *subsequence*, which will be like a sequence but with a restricted set of indices. These may be a consecutive set such as the subsequence

$$\{b_i\}_{i = 3,8} = \{ b_3, b_4, b_5, b_6, b_7, b_8 \}$$

or nonconsecutive, such as every third term of a sequence $\{y_n\}_{n = 1,30}$ which would be the subsequence

$$\{ y_3, y_6, y_9, \ldots, y_{30} \} \text{ or, equivalently, } \{y_{3i}\}_{i = 1,10}$$

This raises a problem of semantics, that is how we speak about sequences. The term y_3 is obviously the first term of the above subsequence, yet it is the third term of the original sequence. This same problem occurs when the index of a sequence does not start at 1. In many situations the index is associated with progressive events, such as biological generations or cycles. It is often convenient in some situations to consider the index over the range of numbers from 0 to N. In this case, the first sequence term is actually the term with index 0. For instance, a sequence of terms p_i

$$\{p_i\}_{i = 0, 20} \equiv \{ p_0, p_1, p_2, \ldots, p_{20} \}$$

has p_2 as its third term. To avoid confusion you must observe the range of the index and be careful with the adjectives used to describe sequence terms. Later, we will allow integer indices, so that a sequence could start with the term A_{-5}.

Normally, we refer to a sequence term by its index, as a number or symbol, without distinguishing its relative position in a sequence or subsequence. The first term occurring in a sequence is called the *initial term*, no matter what its index number.

Specifying sequences.

The formula specifying a sequence can be indicated either explicitly or implicitly. An explicit formula is simply a function equation, like

$$E_n = 2n$$

This indicates the nth even number. Specific sequence terms are found by replacing n in the general sequence term E_n by a numerical index value.

When a sequence is listed with the " ... " notation it is assumed that every reader can provide an explicit formula to generate the omitted terms. Unfortunately, this there are in theory infinitely many different formulas that would give exactly the same first three sequence terms. For instance, the even number sequence $\{ 2, 4, 6, 8, \ldots, 2n, \ldots, 198, 200 \}$ could, alternatively, be indicated by an implicit equation that must be solved to find the explicit function for each sequence term. Implicit equations can be simple algebraic equations, such as

$$3E_n = 2n + 2E_n$$

from which we solve for E_n to obtain $E_n = 2n$. Another type of implicit equation involves more than one sequence term, that is terms with different indices, such as

$$E_n = E_{n-1} + 2$$

This type of equation is called a *difference equation*, and it is the type of equation that arises when discrete phenomena are modeled. The remaining sections of this Chapter are devoted to studying the solution of difference equations, which often are not unique.

Example 1 **Finite sequences: specified by formulas.**

Problem (a) What is the sequence $\{d_n\}_{n=1,5}$ specified by $d_n = 2n + 1$? (b) What would be a different formula specifying this same sequence? (c) If the second formula has index variable k what is the relationship between k and the original index n?

Solution (a) The sequence terms are evaluated by substituting for n the numbers 1, 2, 3, 4, and 5. For instance, n = 3 gives $d_3 = 2 \cdot 3 + 1 = 7$. Thus the sequence is

$$\{ 3, 5, 7, 9, 11\}$$

(b) Another formula for this sequence is $a_k = 3 + 2k$ where the index k ranges from 0 to 4; the corresponding sequence is then

$$\{a_k\}_{k=0,4}$$

Observe that the two formulas are related. If we equate the formulas, we have

$$3 + 2k = 1 + 2n$$

(c) Solving for the index k, we find that $k = n - 1$. That is, the formulas are equivalent with a "shift" of the indices. Note that the shift also requires a shift in the index range. The index n goes from 1 to 5, thus the index k goes from 1 - 1 to 5 - 1, i.e., from 0 to 4.

Sequence algebra.

Sequences can be combined algebraically. They can be added, subtracted, multiplied or divided, to form new sequences. These operations are performed **term wise**, this means that the arithmetic operation is applied to combing terms of two sequences with the same index values. For a common range of the index i we can formally write:

$$\{a_i\} \bigcirc \{b_i\} \equiv \{a_i \bigcirc b_i\}$$

where the range of the indices has been omitted to simplify the notation and in the $\bigcirc$ you can substitute any of the four operations, + , - , *, or /. However, in the case of division we can not divide by zero! Thus the term a_i / b_i is not defined if $b_i = 0$.

The other algebraic operation with sequences is called *scalar multiplication*. This is simply multiplying a sequence by a constant. If k is a constant then multiplying a sequence by k means to multiply each term by k:

$$k \cdot \{a_n\} \equiv \{ k \cdot a_n \}$$

Example 2 **Sequences algebra.**

Problem Given the sequence $A = \{i^2\}$, $B = \{3i\}$ and $C = \{ i - 4 \}$ for i = 1 to 5.
What are the sequences 7B, A·B, A - C, and A/C?

Solution The sequence 7B is $7\{ 3i \} = \{ 7 \cdot 3i \} = \{ 21i \}$. Similarly,

$$A \cdot B = \{ i^2 \}\cdot\{ 3i \} = \{ i^2 \cdot 3i \} = \{ 3i^3 \}$$

The difference between the sequences A and C is

$$A - C = \{i^2\} - \{i - 4\} = \{i^2 - (i - 4)\} = \{4 - i + i^2\}$$

Each of these combined sequences is defined for i = 1,5. However, the quotient sequence

$$A/C = \{i^2\} / \{i - 4\} = \{i^2 /(i - 4)\}$$

is not defined when i = 4. This quotient is only defined for i = 1,3 and for i = 5.

Special Sequences

Certain sequences arise frequently and are referred to by names. These sequences are often characterized by implicit difference equations relating adjacent terms. These difference equations may be satisfied by many different sequences, but these will all have the same basic form that involves arbitrary constants. The *general form* of the solution is the sequence form with unspecified constants and the general sequence index. Replacing the constants by numbers gives *specific solutions*.

The Arithmetic Sequence.

An **arithmetic sequence** $\{a_n\}$ is characterized by the fact that the <u>difference</u> between any two consecutive terms is always the same constant. Considering the general term a_n, the next term of the sequence has index n + 1, i.e. is a_{n+1}. If the difference constant is the number d then the arithmetic sequence satisfies the equation

$$a_{n+1} - a_n = d \qquad \textbf{Constant Difference Equation}$$

This difference is often denoted using the **forward difference operator,** Δ (the capital Greek letter Delta), which is defined by

$$\Delta a_n \equiv a_{n+1} - a_n$$

Arithmetic sequences are thus the solutions of the difference equation

$$\Delta a_n = d$$

We can generate the terms of a solution iteratively, starting with an initial term and using the difference equation to find the next term. To do this we rearrange the constant difference equation to express a_{n+1} as a function of a_n:

$$a_{n+1} = a_n + d$$

Applying this formula iteratively, for n = 1, n = 2, ... gives

$$a_2 = a_1 + d$$

$$a_3 = a_2 + d = a_1 + d + d = a_1 + 2d$$

$$a_4 = a_3 + d = a_1 + 2d + d = a_1 + 3d$$

$$a_5 = a_4 + d = a_1 + 3d + d = a_1 + 4d$$

From these calculations we see a pattern emerging. The term a_5 is the term a_1 plus 4d. The next term following this pattern would be $a_6 = a_1 + 5d$. The index on the left is 6 and the formula has d multiplied by 5, which is the index minus one. The general form for this sequence is formed by replacing 6 with n and 5 with n-5:

$$a_n = a_1 + (n - 1)d \qquad \textbf{Arithmetic Sequence Formula}$$

If the subscript on initial term is omitted and the sequence is simply indicated by the formula

$$a_n = a + (n - 1)d$$

which is a linear function of the index n. An alternative form, is

$$a_n = b + d \cdot n \qquad \text{where} \qquad b = a - d$$

Example 3 **Determining the parameters of an arithmetic sequence.**

Problem Determine the formula for the sequence { 7, 10, 13, ... , 34 }.

Solution The pattern indicated by the first three terms, 7, 10, 13, suggest a sequence with a constant difference of d = 3 between each term. The initial term is $a_1 = 7$. The sequence formula is thus

$$a_n = 7 + (n - 1)3$$

Denoting the index of the last term by N, its value is found by solving $a_N = 34$:

$$a_N = 7 + (N - 1)3 = 34 \implies 3N = 30 \text{ or } N = 10$$

Notice that the sequence can be algebraically simplified to

$$a_n = 4 + 3n$$

☑

Geometric Sequence

A **geometric sequence** is characterized by the fact that the <u>ratio</u> of two consecutive sequence terms is always the same constant. If the constant ratio is the number r then the implicit equation describing a geometric sequence is

$$g_{n+1} / g_n = r$$

Solving this equation for g_{n+1} gives the iteration equation

$$g_{n+1} = r \cdot g_n$$

Beginning with the initial sequence term g_1 subsequent sequence terms can be generated as follows:

$$g_2 = r \cdot g_1$$
$$g_3 = r \cdot g_2 = r \cdot r g_1 = r^2 g_1$$
$$g_4 = r \cdot g_3 = r \cdot r^2 g_1 = r^3 g_1$$

$$g_5 = r \cdot g_4 = r \cdot r^3 g_1 = r^4 g_1$$

Inferring from this a pattern, we see that the general term is given by

$$g_n = r^{n-1} g_1$$

Setting the constant $g = g_1$ we have the general formula for the nth term:

$$g_n = g \cdot r^{n-1} \qquad \textbf{Geometric Sequence}$$

An alternative form for the geometric sequence is

$$g_n = (g/r) \cdot r^n \qquad \textbf{Geometric Sequence}$$

Notice that the sequence

$$\{g, gr, gr^2, gr^3, \ldots, gr^N\} = \{ g r^{n-1} \}_{n=1, N+1}$$

actually has $N + 1$ terms. A geometric sequence can be formed by simply evaluating an exponential function $f(x) = g \cdot r^{x-1}$ at the integer values $x = 1, 2, 3$, and so on. Thus, it is natural to alternatively refer to geometric sequences as *exponential sequences*.

Example 4   **Determining the parameters of a geometric sequence.**

Problem   Determine the formula for the geometric sequence $\{ 3, 12, 48, \ldots, 3{,}072 \}$.

Solution   The ratio of successive terms is $12/3 = 48/12 = 4$. Thus $r = 4$. The initial term is $g_1 = 3$. To determine the range for the index we solve for N in the equation

$$g_N = 3 \cdot 4^{N-1} = 3072$$

This gives $4^{N-1} = 1024$. Recognizing that 1024 is 2^{10} and that

$$4^{N-1} = (2^2)^{N-1} = 2^{2N-2}$$

we must have $2N - 2 = 10$ or $N = 6$. Thus the sequence is

$$\{ 3 \cdot 4^{n-1} \}_{n=1,6}$$

☑

Example 5   **Determining the equation of a geometric sequence from two terms.**

Problem   What is the value of g_2 of the geometric sequence $\{ g_n \}$ if $g_7 = 25$ and $g_5 = 10$?

Solution   Assume the sequence has the form $g_n = g r^{n-1}$. The given terms are

$$g_7 = g r^6 \qquad \text{and} \qquad g_5 = g r^4$$

The ratio constant r is found from the ratio of these two terms. Numerically, $g_7/g_5 = 25/10 = 2.5$. Using their expression in terms of r, this ratio can be simplified to give

$$g_7/g_5 = (g r^6) / (g r^4) = r^2 = 2.5$$

Solving, we find $r = 2.5^{1/2}$. The initial term g is found by substituting r into one of the terms:

$$g_5 = g\, r^4 = g\, (2.5^{1/2})^4 = g\, 2.5^2 = 6.25g = 10$$

Hence, $g = 10/6.25 = 1.6$. Finally, the required sequence term is

$$g_2 = g \cdot r = 1.6 \cdot 2.5^{1/2} \approx 2.53$$

Power Sequences

Another common type of sequence involves the index raised to a fixed power, p. The power sequences result from simply evaluating the power function $y = x^p$ at integer values of x. These sequences have the general form

$$x_n = n^p \quad n = 1, 2, 3, \ldots$$

If $p = 1$ the sequence is simply the sequence of natural numbers, $\{1, 2, 3, 4, \ldots\}$. If $p = 2$, it is the sequence of the squares of the natural numbers, $\{1, 4, 9, 16. \ldots\}$. The implicit relationship involving successive terms of a power sequence $\{x^p\}$ involves the p-th root of successive terms and has the form

$$(x_{n+1})^{1/p} = (x_n)^{1/p} + 1$$

Infinite sequences.

An **infinite sequence** is a sequence that has a term associated with every positive integer. The index for an infinite sequence is said to go to *infinity*, although ∞ is not an integer; this means that there is no upper limit for the index range. Infinite sequences are denoted as

$$\{a_i\}_{i=1,\infty} \equiv \{a_1, a_2, a_3, \ldots\}$$

There is no "last term" of an infinite sequence. Expressed in set notation, infinite sequences are denoted using the three dots convention followed by the right brace, indicating that the established pattern continues for ever. Thus, an infinite geometric, or exponential, sequence with $r = 1/2$ is

$$\{(1/2)^i\}_{i=1,\infty} = \{1/2, 1/4, 1/8, \ldots\}$$

Although infinite sequences do not have last terms, an important mathematical question is What are the values of the infinite sequence terms x_n when n becomes very large? If they approach a constant L, this is indicated as a limit

$$\lim_{n \to \infty} x_n = L$$

which is read "The limit as n goes to infinity of x sub n equals L." This indicates that all the terms of $\{x_n\}_{n=1,\infty}$ are close L as the index n becomes large. Whether such limits exist depends on the particular sequence; when a sequence has a limit, it is said to *converge* to the limit. For instance, the sequence of reciprocals $\{1/n\}$ converges to zero:

$$\text{if } x_n = 1/n \text{ then } \lim_{n \to \infty} x_n = \lim_{n \to \infty} 1/n = 0$$

If the $\lim_{n \to \infty} x_n$ does not exist, the sequence is *not convergent*. A special type of non-convergent sequences ar those for which the sequence terms become greater and greater as the index increases.

If the sequence terms become greater than any specified number then the sequence is said to *diverge to infinity*. In this case it is said that "$\lim_{n \to \infty} x_n = \infty$". For instance, the infinite power sequence with exponent p = 2 diverges to infinity:

$$\text{if } x_n = n^2 \text{ then } \lim_{n \to \infty} x_n = \lim_{n \to \infty} n^2 = \infty$$

In this text formal mathematical tests to determine if a sequence will have a limit are not considered. Intuitive notions are used to determine whether sequence terms approach a finite limit value or become infinite, without introducing rigorous proofs.

Series

A series is simply a sum of numbers or sequence terms. Associated with each finite sequence is a **finite series**, which is the sum of the sequence terms. For instance, the sum of the reciprocals of the first ten integers is the sum of $a_i = i^{-1}$ for i = 1 to 10:

$$a_1 + a_2 + a_3 + \ldots + a_9 + a_{10} = 1/1 + 1/2 + 1/3 + \ldots + 1/9 + 1/10.$$

In this expression we have used the ... convention to omit explicit listing of all the terms in the series. Such summations can be expressed more compactly by using "sigma notation", using the capital Greek letter sigma, Σ, to indicate a sum:

$$\Sigma_{i=1,10} \, a_i \qquad \text{SIGMA NOTATION}$$

This is read "the summation of a_i, from i = 1 to i = 10". The "i" in the expression is called an **index** and the number 1 is the *lower limit* and the number 10 is the *upper limit* of summation. This notation indicates that expressions a_i should be added together for i = 1, i = 2, i = 3, ... , up to i = 10. The sum of the first ten reciprocals is expressed in Sigma Notation as

$$\Sigma_{i=1,10} \, 1/i$$

Sometimes the index and its lower limit are placed beneath the Σ and the upper limit above the Σ:

$$\sum_{i=1}^{10} a_i$$

When writing series by hand this notation is easily formed and clearly distinguishes the upper and lower limits for the index. However, it is more difficult to write with ordinary word processors or typewriters and requires more space to set in type. In engineering texts the index and subscripts are often not indicated at all, being implied by "reserved index symbols". For instance, the series $\Sigma \, z_k$ has index k and its range would be an assumed common range, indicated previously in the text. We will sometimes omit the index notation when it is not critical to the discussion, or the to indicate relationships that are true for any index range.

Example 6 **Using Sigma notation.**

The following expressions illustrate the use of Sigma Notation.

$$\Sigma_{i=1,5} \, b_i = b_1 + b_2 + b_3 + b_4 + b_5$$

$$\Sigma_{j=3,5} \, A_j = A_3 + A_4 + A_5$$

$$\Sigma_{n=0,3} \, (10 + n) = (10 + 0) + (10 + 1) + (10 + 2) + (10 + 3)$$

$$\Sigma_{i=k,k+3} \, f(x_i) = f(x_k) + f(x_{k+1}) + f(x_{k+2}) + f(x_{k+3})$$

$$\Sigma_{i=-2,3}\,(2^i) = 2^{-2} + 2^{-1} + 2^0 + 2^1 + 2^2 + 2^3$$

Example 7   Evaluating finite series by expansion.

Problem   The given series is evaluated by first *expanding* and then adding the terms.

a) $\Sigma_{i=3,5}\, i^2 = 3^2 + 4^2 + 5^2 = 50$

b) $\Sigma_{j=1,4}\, 2 \cdot 3^{j-1} = 2 \cdot 3^0 + 2 \cdot 3^1 + 2 \cdot 3^2 + 2 \cdot 3^3$

$\qquad = 2[3^0 + 3^1 + 3^2 + 3^3] = 2 \cdot 40 = 80$

c) $\Sigma_{n=1,4}\, (2^n - n) = (2 - 1) + (2^2 - 2) + (2^3 - 3) + (2^4 - 4)$

$\qquad = (2 + 2^2 + 2^3 + 2^4) - (1 + 2 + 3 + 4) = 30 - 10 = 20$

d) $\Sigma_{i=1,4}\, [2 + (i-1)3] = [2 + 0] + [2 + 3] + [2 + 6] + [2 + 9] = 26$

e) The sum of the first five odd integers can be expressed as a series

$$\Sigma_{i=1,5}\,(1 + 2(i-1)) = 1 + 3 + 5 + 7 + 9 = 25$$

Observe, however, that this sum could also be expressed as the series

$$\Sigma_{j=0,4}\,(1 + 2j) = 1 + 3 + 5 + 7 + 9 = 25$$

These indices of these two series are related by the transformation $j = i - 1$. Their index limits are related: $i = 1$ corresponds to $j = 0$, and $i = 5$ corresponds to $j = 4$.

Algebra of Finite Series

ALGEBRA RULES FOR SERIES

Given two compatible series, $\Sigma\, a_i$ and $\Sigma\, b_i$, having the same index range,

$$(\Sigma\, a_i) + (\Sigma\, b_i) \equiv \Sigma(a_i + b_i) \qquad \text{Sum Rule}$$

For any constant k,

$$k(\Sigma\, a_i) \equiv \Sigma\,(ka_i) \qquad \text{Distributive Rule}$$

The Sum Rule requires that the series have the same index range.

CAUTION   The product of two series is not equal to the series formed by the term-wise product of the two sequences:

$$\Sigma\, a_i \;*\; \Sigma\, b_i \neq \Sigma\, a_i * b_i$$

For instance, if $a_i = i$ and $b_i = 2^i$, then

$$\Sigma_{i=1,2}\, a_i = 1 + 2 = 3 \quad \text{and} \quad \Sigma_{i=1,2}\, b_i = 2 + 4 = 6$$

The product of these two series is

$$\left[\Sigma_{i=1,2}\, a_i\right] \times \left[\Sigma_{i=1,2}\, b_i\right] = 3 \cdot 6 = 18$$

But, the series corresponding to the term-wise product sequence $a_i \cdot b_i$ is not the same:

$$\Sigma_{i=1,2}\, (a_i \cdot b_i) = 2 + 8 = 10$$

Example 8 **Scalar multiplication and addition of series.**

Problem Perform the indicate sum or product for the series

$$A = \Sigma_{i=1,4}\, i, \quad B = \Sigma_{i=1,4}\, i^2, \text{ and} \quad C = \Sigma_{i=1,5}\, i - 2$$

a) $3A = 3 \cdot \Sigma_{i=1,4}\, i = \Sigma_{i=1,4}\, 3i$

b) $A + B = \Sigma_{i=1,4}\, i + \Sigma_{i=1,4}\, i^2 = \Sigma_{i=1,4}(i + i^2)$

c) $A - 7B = \Sigma_{i=1,4}\, i - 7 \cdot \Sigma_{i=1,4}\, i^2 = \Sigma_{i=1,4}(i - 7i^2)$

d) The sum of series A and C cannot be expressed using the addition rule, as they are not compatible. The index of C ranges from 1 to 5 while that of A goes from 1 to 4. However, the finite series A and C have numerical values, $A = 10$ and $C = 5$, and can be added: $A + C = 15$. Their sum can be formed symbolically by writing C as a series with index range from 1 to 4 plus the term with $i = 5$:

$$A + C = \Sigma_{i=1,4}\, i + \Sigma_{i=1,4}\, (i - 2) + (5 - 2) = \Sigma_{i=1,4}\, (2i - 2) + 3$$

The technique employed in d) of the above example is referred to as "splitting" a series. Basically, a series can be split into the sum of two series with index ranges that make up the total range of the original series. The simplest way to split a series is to break it into two consecutive sum. However, sometimes more complicated methods are used, such as writing one sum over all even indices and the other over all odd indices.

SPLITTING SERIES

If j, k, and n are integers satisfying $j < k < n$, then

$$\Sigma_{i=j,n}\, a_i = \Sigma_{i=j,k}\, a_i + \Sigma_{i=k+1,n}\, a_i$$

Example 9 **Splitting series.**

Problem The following illustrates splitting a series into two.

$$\Sigma_{i=0,10}\, i^2 = \Sigma_{i=0,5}\, i^2 + \Sigma_{i=6,10}\, i^2$$

$$\Sigma_{i=-4,4}\, \sin(i\pi) = \Sigma_{i=-4,-1}\, \sin(i\pi) + \Sigma_{i=0,4}\, \sin(i\pi)$$

$$\Sigma_{i=0,10}\, a_i = a_0 + \Sigma_{i=1,9}\, a_i + a_{10}$$

If two series have different index ranges they can not be directly added. However, sometimes they can be combined by splitting them into series with common index ranges plus the other terms or series. Then, the series with common ranges can be added to form a single series; the other terms can not be combined.

Example 10 **Splitting series to combine two series with different index ranges**.

Problem Add the series $\Sigma_{i=1,100}\, a_i$ and $\Sigma_{i=3,110}\, b_i$.

Solution The a_i series index ranges from 1 to 100 while the b_i range is from 3 to 110. To combine these series we identify their largest common index range, from 3 to 100. Then both series are split to give one series with range 3 to 100 plus another term or series:

$$\Sigma_{i=1,100}\, a_i = a_1 + a_2 + \Sigma_{i=3,100}\, a_i$$

$$\Sigma_{i=3,110}\, b_i = \Sigma_{i=3,100}\, b_i + \Sigma_{i=101,110}\, b_i$$

Thus, the sum of the two series is

$$\Sigma_{i=1,100}\, a_i + \Sigma_{i=3,110}\, b_i = a_1 + a_2 + \Sigma_{i=3,100}\, (a_i + b_i) + \Sigma_{i=101,110}\, b_i$$

CLOSED FORMS OF SPECIAL SERIES

Series are evaluated by adding their terms. For some sequences, the sum of their corresponding series can be expressed as a function of the upper and lower limits. Such functions are called "closed forms" for the series. We will need some special closed forms to evaluate *definite integrals* in Chapter 5. In this section, the closed forms of eight of the simplest and most commonly encountered special series are given: for power series, with terms i^r and index i ranging from 1 to N for powers r = 0, 1, 2, and 3; for arithmetic and geometric series; and for two special types of series. The development of closed forms for series has provided a challenge to the best mathematicians for over 300 years. Consequently, you should not be dismayed if some of the following formulas do not seem obvious. The formulas are given without "proof". You should confirm their validity by evaluating the series by expansion for low values of N, say N = 2, 3 and 4.

I. The **constant series**: $\Sigma\, a_i$, in which each term a_i is a constant, k. Since the sum of a number k added to itself N times is $k \cdot N$, the constant series is easily evaluated:

SUM OF A CONSTANT

$$\Sigma_{i=1,N}\, k = \underbrace{k + k + \ldots + k}_{N\text{ - times}} = k \cdot N$$

II. The **sum of consecutive integers**: $\Sigma\, a_i$ in which $a_i = i$ is the sum of the first N integers. This form is normally introduced in High School algebra courses:

SUM OF CONSECUTIVE INTEGERS

$$\Sigma_{i=1,N}\, i = 1 + 2 + 3 + \ldots + N = N(N+1)/2$$

III. The **sum of the squares of successive integers**: $\Sigma\, i^2$. The closed form of this series is given by

SUM OF SQUARES

$$\Sigma_{i=1,N}\, i^2 = 1 + 4 + 9 + \ldots + N^2 = N(N+1)(2N+1)/6$$

IV. The **sum of the cubes of successive integers**: $\Sigma\, i^3$.

SUM OF CUBES

$$\Sigma_{i=1,N}\, i^3 = 1 + 8 + 27 + \ldots + N^3 = N^2(N+1)^2/4$$

This is not a mistake! The sum of the cubes of the first N integers is the square of the sum of the first N integers. This should make this form easy to remember.

V. The **arithmetic series** has the form $\Sigma\, a_n$, where a_n is an arithmetic sequence. The series terms can be expressed either as

$$a_n = a + d(n-1) \quad \text{or as} \quad a_n = (a-d) + dn$$

so that the first term is $a_1 = a$ and the difference in successive terms is d. This series is just the sum of the constant series with k = a - d and d times the sum of the consecutive integers. Combining the above formulas gives

SUM OF LINEAR TERMS

$$\Sigma_{n=1,N}\, [a + (n-1)d] = (a-d)N + d \cdot N(N+1)/2$$

VI. The **geometric series** is the series $\Sigma\, g_n$, corresponding to a finite geometric sequence with the nth term $g_n = g \cdot r^{n-1}$. It is useful to see how the closed form of the geometric series is derived. First, let S denote the value of the geometric series for the index n = 1 to N:

$$S = \Sigma_{n=1,N}\, g \cdot r^{n-1}$$

Then, multiplying S by the constant r and rearranging using the distributive law gives

$$r \cdot S = r \cdot \Sigma_{n=1,N}\, g \cdot r^{n-1} = \Sigma_{n=1,N}\, g \cdot r^n$$

Subtracting the series $r \cdot S$ from the series for S gives

$$S - r \cdot S = \Sigma_{n=1,N}\, [g \cdot r^{n-1} - g \cdot r^n]$$

The left side is just $(1-r)S$. Expanding the series on the right gives

$$(1-r)S = (g - g \cdot r)_{n=1} + (g \cdot r - g \cdot r^2)_{n=2} + (g \cdot r^2 - g \cdot r^3)_{n=3} + \ldots + (g \cdot r^{N-1} - g \cdot r^N)_{n=N}$$

$$= g - g \cdot r + g \cdot r - g \cdot r^2 + g \cdot r^2 - g \cdot r^3 + \ldots + g \cdot r^{N-1} - g \cdot r^N$$

After removing the parentheses, it is clear that terms $-g \cdot r$ and $+g \cdot r$ cancel; similarly, $-g \cdot r^2$ and $+g \cdot r^2$ cancel. This is also the case for each term $g \cdot r^n$ for $n = 1, 2,..., N - 1$. After canceling, all that remains is the first and last terms, g and $-g \cdot r^N$. Thus

$$(1 - r)S = g - g \cdot r^N$$

Solving for S gives the formula for the sum of a geometric series for $r \neq 1$:

$$\sum_{n=1,N} g \cdot r^{(n-1)} = g[1 - r^N] / (1 - r) \qquad \text{GEOMETRIC SUM Form 1}$$

When $r = 1$, the geometric series is really a constant series with $k = g$.
Geometric sequences, and hence, geometric series can be expressed in a variety of forms. Furthermore, it is sometimes desirable to sum geometric series over different index ranges. A simple formula can be given to handle most of the various forms. consider a series of the form

$$S = C \sum_{n=B,L} r^{n+s}$$

where C is a constant, B is the beginning index value and L is the last index value, and s is a shift of the exponent (s may be positive or negative). Then, the value of S is C times the series

$$\sum_{n=B,L} r^{n+s} = [r^B - r^{L+s+1}] / (1 - r) \qquad \text{GEOMETRIC SUM Form 2}$$

VII. The **collapsing series** is a particular type of series that is frequently encountered; it is similar to the series $(1 - r)S$ that arose when deriving the closed form for the geometric series. A collapsing series can be written so that each term is a difference between successive terms of a sequence; the general forms of collapsing series are

$$\sum_{i=n,N} (a_i - a_{i-1}) \qquad \text{or} \qquad \sum_{i=n,N} (a_{i+1} - a_i).$$

Collapsing series are easy to evaluate since, as their name suggests, when expanded they collapse through cancellation of positive and negative terms to leave only the difference of two terms:

$$\sum_{i=n,N} (a_{i+1} - a_i) = (a_{n+1} - a_n) + (a_{n+2} - a_{n+1}) + (a_{n+3} - a_{n+2})$$
$$+ ... + (a_N - a_{N-1}) + (a_{N+1} - a_N)$$

Rearranging the terms, this series can be expressed as:

$$-a_n + (a_{n+1} - a_{n+1}) + (a_{n+2} - a_{n+2}) + ... + (a_N - a_N) + a_{N+1}$$

Thus,

$$\sum_{i=n,N} (a_{i+1} - a_i) = a_{N+1} - a_n \qquad \text{COLLAPSING SUM}$$

Similarly, the closed form of the other type of collapsing series is

$$\sum_{i=n,N} (a_i - a_{i-1}) = a_N - a_{n-1} \qquad \text{COLLAPSING SUM}$$

VIII. An **alternating series** involves terms that alternate between positive and negative values. The simplest alternating series is a sum of powers of -1. Since

$$(-1)^i = 1 \text{ if i is even, and } (-1)^i = -1 \text{ if i is odd,}$$

expanding such a series, grouping the terms in pairs of +1 and -1, the closed form of the simplest alternating series is:

$$\Sigma_{i=1,N} (-1)^i = \begin{cases} -1 & \text{if n is even;} \\ 0 & \text{if n is odd.} \end{cases} \quad \text{AN ALTERNATING SERIES}$$

Example 11 **Evaluating series using closed form expressions.**

Problem Use the above formulas to express the given series in a closed form. If necessary, split the series or shift the index to obtain the appropriate indices for the closed form equation.

a) Sum of a constant:

$$\Sigma_{n=1,7} C = 7 \cdot C$$

To evaluate $\Sigma_{n=4,7} C$ we use the splitting rule; the series from 1 to 7 is the sum of the series from 1 to 3 and the series from 4 to 7. Rearranging, we write

$$\Sigma_{n=4,7} C = \Sigma_{n=1,7} C - \Sigma_{n=1,3} C = 7C - 3C = 4C$$

b) Sum of consecutive numbers:

$$\Sigma_{n=1,100} n = 100(101)/2 = 5050$$

$\Sigma_{n=6,20} n$ can be evaluated as the difference of two series:

$$\Sigma_{n=6,20} n = \Sigma_{n=1,20} n - \Sigma_{n=1,5} n = [20 \cdot 21/2] - [5 \cdot 6/2] = 210 - 15 = 195$$

A series that involves a sum of terms can be evaluated as the sum of two series. For instance

$$\Sigma_{i=1,6} (3i + 2) = 3\Sigma_{i=1,6} i + \Sigma_{i=1,6} 2 = 3(6 \cdot 7/2) + 2 \cdot 6 = 75$$

c) Sum of squares:

$$\Sigma_{i=1,10} i^2 = 10 \cdot 11 \cdot 21/6 = 385$$

Using the sum and scalar product rules, a series of quadratic function of the index is evaluated as

$$\Sigma_{k=1,5} (3k^2 - k + 2) = 3\Sigma_{k=1,5} k^2 - \Sigma_{k=1,5} k + \Sigma_{k=1,5} 2$$

$$= 3 \cdot [5 \cdot 6 \cdot 11/6] - [5 \cdot 6/2] + [2 \cdot 5]$$

$$= 165 - 15 + 10 = 160$$

d) Sum of cubes:

$$\Sigma_{i=1,10} i^3 = 10^2(11)^2/4 = 3025$$

e) Sum of an arithmetic series with a = -3, d = 5, N = 6,

$$\Sigma_{n=1,6} [-3 + (n-1)5] = = (-3-5)6 + 5 \cdot 6 \cdot 7/2 = -48 + 105 = 57$$

Or, using the addition formula,

$$\Sigma_{n=1,6} [-3 + (n-1)5] = \Sigma_{n=1,6} (-8 + 5n)$$

$$= \Sigma_{n=1,6} -8 + 5\Sigma_{n=1,6} n$$

$$= -8 \cdot 6 + 5 \cdot 6 \cdot 7/2 = 57$$

f) Sum of geometric series with initial term g = 6 and ratio r = 1/2:

$$\Sigma_{n=1,10} 6(1/2)^{n-1} = 6[1 - (1/2)^{10}/1 - (1/2)]$$

$$= 6 \cdot (1 - (1/2)^{10}/(1/2) = 12(1 - (1/2)^{10}) \approx 11.98828$$

g) Collapsing series: For any sequence $\{X_i\}$,

$$\Sigma_{i=1,10} (X_{i+1} - X_i) = X_{11} - X_1$$

$$\Sigma_{i=7,10} (X_{i+1} - X_i) = X_{11} - X_7$$

h) The sum of the geometric series $\Sigma_{n=0,6} 5 \cdot 3^{n-2}$ is evaluated by first taking the constant 5 outside the summation and then applying the Geometric Series closed Form 2. In this case, r = 3, the shift is s = -2, the initial index is B = 0 and the ending index is L = 6. The formula gives

$$5 \Sigma_{n=0,6} 3^{n-2} = 5 [3^{0-2} - 3^{6-2+1}]/[1-3]$$

$$= -(5/2)[3^{-2} - 3^5] = -607.\underline{2} - (5/2)[1/9 - 243] = -607.\underline{2}$$

☑

Example 12 Modeling Population Growth by collapsing series.

Background Let P_i denote the size of the i-the generation of a given population. The change in population size from the i-th to the (i + 1)st generation is

$$\Delta P_i = P_{i+1} - P_i$$

The total change in the population size from the first to the N-th generation is the sum of the changes $\Delta P_i = 1, 2, \ldots, N-1$. Expressed in series notation, this would read

(Total change) $TC(N) = \Sigma_{i=1,N-1} \Delta P_i = \Sigma_{i=1,N-1} [P_{i+1} - P_i]$

This is a collapsing series and thus the total change is the end population minus the starting population, as expected:

$$TC(N) = P_N - P_1$$

Normally, when modeling a population, the size of the N-th generation is not known. Frequently, the object of modeling the population is to predict P_N. This can be done by solving the above equation for P_N and expressing the total change as a series.

$$P_N = P_1 + TC(N) = P_1 + \Sigma_{i=1,N-1} \Delta P_i$$

If the change ΔP_i is known, from experimental data or from information about the population's reproductive and mortality rates, then this equation can be used to obtain P_N.

Problem For each of the given models, determine the size of the 50th generation.

(a) A population is initially 10, changes from the n-th to the (n + 1)st generation by a reproduction that is linear with the generation, 5n, and is subject to a constant "harvest" of 6 per generation.

(b) A population that is initially 10 and grows at a geometric rate of 2^n.

(c) A population that grows "logarithmicaly" is modeled by $\Delta P_n = ln(n)$ and $P_1 = 10$?

Solution (a) Denote the population size by P_n. The change between the n-th and the (n + 1)st generation is

$$\Delta P_n = 5n - 6$$

Thus the population at the 50-th generation is

$$P_{50} = P_1 + \Sigma_{n=1,49} (5n - 6) = 10 + 5[(49 \times 50)/2] - 6 \times 49 = 5841$$

(b) For this population the growth, or change ΔP_n, is given by the geometric series

$$\Delta P_n = 2^n \quad \text{and} \quad P_1 = 10$$

Using the closed Form 2 of the geometric series gives

$$P_N = 10 + \Sigma_{i=1,49} 2^i = 10 + 2(1 - 2^{49})/(1 - 2) \approx 1.1259 \cdot 10^{15}$$

(c) In this case the series is expanded as

$$P(N) = P(1) + \Sigma_{i=1,N-1} \Delta P(i)$$

$$= P(1) + \Sigma_{i=1,N-1} ln(i)$$

$$= P(1) + ln((N-1)!)$$

For instance, $ln(1) + ln(2) + ln(3) = ln(1 \cdot 2 \cdot 3) = ln(3!)$. Substituting $N = 50$, $P(1) = 10$, gives

$$P(50) = 10 + ln(50!) \approx 158.48$$

Observe that "logarithmic" growth is slower than linear growth modeled in part (a).

Finite Power Series

Series can be used to define functions. A very important class of series functions are formed as the sum of constants times powers of x, these are called **finite power series**. For instance, the series corresponding to the sequence terms $y_n = n \cdot x^n$ for n = 0, 1, ..., 4 has the form

$$\Sigma_{n=0,4}\, n \cdot x^n = 0 \cdot x^0 + 1 \cdot x + 2 \cdot x^2 + 3 \cdot x^3 + 4 \cdot x^4$$

Observe that *finite power series are really just polynomials*. The general shifted polynomial of degree N is given by the finite power series

$$P(x) = \Sigma_{n=0,N}\, a_n(x-b)^n$$

where b is a constant and $\{a_n\}_{n=0,N}$ is any sequence of N + 1 numbers. Observe also that finite power series can be viewed as the series corresponding to the product of the sequence $\{a_n\}$ and the geometric sequence $\{(x-b)^n\}$.

INFINITE SERIES

Sigma notation provides a simple way to formally extend the concept of a finite series to that of an *infinite series*. Given an infinite sequence $\{a_i\}_{i=1,\infty}$, the associated **infinite series** is a formal construct that is written as

$$\Sigma_{i=1,\infty}\, a_i$$

Recall that the infinity symbol represents a concept, not a number. The "infinite series" likewise represents the concept of a sum of the terms a_i, starting with a_1 and never terminating. The term a_∞ does not exist! An "infinite series" is not a number, it is a *symbolic form* associated with an infinite sequence, although, it is used to represent the concept of an infinite sum. This concept of an infinite sum may or may not correspond to a real number. If it does, the series is said to be *convergent*, and if it doesn't it is said to *not converge*. To define these concepts we introduce a sequence of *partial sums* of the infinite series. If this sequence converges, i.e. these partial sums approach a limit as the number of terms in the partial sum increases, then this limit is the value assigned to the infinite series and the series is said to *converge*.

EVALUATING INFINITE SERIES.

For each natural number N, the **N^{th} partial sum** of the infinite series $\Sigma_{i=1,\infty}\, a_i$ is the finite series

$$S_N \equiv \Sigma_{i=1,N}\, a_i$$

If the sequence of partial sums $\{S_N\}_{N=1,\infty}$ converges to a number A, i.e., if

$$\lim_{N \to \infty} S_N = A \quad \text{(a finite number)}$$

then the infinite series $\Sigma_{i=1,\infty}\, a_i$ is said to be **convergent**, and to **converge to the number A**.

If $\lim_{N \to \infty} S_N$ is infinite or does not exist, then the infinite series $\Sigma_{i=1,\infty}\, a_i$ **does not converge.**

If $\lim_{N \to \infty} S_N = \infty$ or $-\infty$, then the infinite series is said to **diverge to ∞ or to $-\infty$**.

Section 2.1   SEQUENCES AND SERIES

Example 13 An "infinite" geometric series.

Problem Does the infinite series $\sum_{n=1,\infty} (1/2)^{n-1}$ converge? If so, what is its value?

Solution Let S_N denote the partial sum of the series from 1 to N. Then,

$$S_1 = 1 \quad S_2 = 1 + 1/2 = 3/2 \quad S_3 = 1 + 1/2 + 1/4 = 7/4 \quad S_4 = 1 + 1/2 + 1/4 + 1/8 = 15/8$$

$$S_N = \sum_{n=1,N} (1/2)^{n-1} = 1 + 1/2 + 1/4 + 1/8 + \ldots + 1/2^{N-1}$$

S_N can be evaluated as a geometric series of Form 1 setting $g = 1$ and $r = 1/2$:

$$S_N = \sum_{n=1,N} (1/2)^{n-1} = [1 - (1/2)^N] / [1 - 1/2] = 2 - (1/2)^{N-1}$$

As N becomes larger and larger, the sum S_N becomes closer and closer to the number 2, since the terms $(1/2)^{N-1}$ becomes negligible. For instance, if $N = 101$, $(1/2)^{100} \approx 7.8 \times 10^{-31}$.

$$\lim_{N \to \infty} S_N = \lim_{N \to \infty} \{2 - (1/2)^{N-1}\} = 2$$

The sequence of partial sums $\{S_N\}$ converges to 2. Therefore, the infinite geometric series

$$\sum_{n=1,\infty} (1/2)^{n-1} \quad \text{converges to 2}$$

☑

Example 14 An "infinite series" that does not converge.

Problem Does the sequence of partial sums of the series $\sum_{n=1,\infty} 1/\sqrt{n}$ converge?

Solution The sequence of partial sums of this series is $\{S_N\}$ given

$$S_N = \sum_{n=1,N} 1/\sqrt{n} = 1 + 1/\sqrt{2} + 1/\sqrt{3} + \ldots + 1/\sqrt{N}$$

The finite series S_N does not match any of the series for which a closed forms have been given. Instead of summing this series a different approach is taken. It is compared to a second series that can be expressed in a closed form. Each term in the given series S_N is compared to the corresponding term in the second series and seen to be larger. The conclusion drawn is that the sum S_N must be greater than the sum of the corresponding second series. Finally, the closed forms of the second series are evaluated and seen to diverge, actually approaches infinity, as $N \to \infty$. It is concluded that the original sequence $\{S_N\}$ must also diverge. Since

$$1 < \sqrt{2} < \sqrt{3} < \ldots < \sqrt{N}$$

inverting each term reverses the inequalities, i.e.,

$$1/1 > 1/\sqrt{2} > 1/\sqrt{3} > \ldots > 1/\sqrt{N}$$

Consequently, each term of the series S_N is greater than $1/\sqrt{N}$. The second series which is introduced is the constant series, with $K = 1/\sqrt{N}$. It can be easily summed:

$$\sum_{i=1,N} [1/\sqrt{N}] = \sum_{i=1,N} K = K \cdot N = N/\sqrt{N} = \sqrt{N}$$

Consequently, for every positive integer N,

$$S_N > \sqrt{N}$$

As N increases S_N become larger, since it is always greater than $\sqrt{N}$, thus

$$\lim_{N \to \infty} S_N \geq \lim_{N \to \infty} \left[\Sigma_{i=1,N}\, 1/\sqrt{N} \right] = \lim_{N \to \infty} \sqrt{N} = \infty$$

The sequence of sums $\{S_N\}$ diverges to infinity as $N \to \infty$. By extension of the terminology we say that the infinite series

$$\Sigma_{n=1,\infty}\, 1/\sqrt{n} \quad diverges$$

☑

The limit in the previous example was evaluated using the basic principal that if a sequence approaches zero then the sequence of reciprocals approaches infinity:

$$\text{If } a_n > 0 \text{ and } a_n \to 0 \text{ as } n \to \infty \text{ then } 1/a_n \to +\infty$$

Example 15 Determining if an infinite series converges.

Problem Determine if the infinite series converges or diverges.

a) $\Sigma_{i=1,\infty}\, 1$ b) $\Sigma_{i=1,\infty}\, (-1)^i$ c) $\Sigma_{i=1,\infty}\, (1/i^2)$

Solutions a) For each $n = 1, 2, 3, \ldots$, the nth partial sum of this series is

$$Sn = \Sigma_{i=1,n}\, 1 = n$$

Consequently, the sequence $\{S_n\} = \{n\}$ diverges to $+\infty$ and $\Sigma_{i=1,\infty}\, 1$ is divergent.

b) The partial sums of $\Sigma_{i=1,\infty}\, (-1)^i$ are equal to either -1 or 0:

$$S_1 = -1, \quad S_2 = -1 + 1 = 0, \quad S_3 = -1 + 1 - 1 = -1 \quad \text{etc.}$$

Hence, $\{S_n\}$ does not converge and thus the series $\Sigma_{i=1,\infty}\, (-1)^i$ does not converge.

c) The N-th partial sums of the series $\Sigma_{i=1,\infty}\, (1/i^2)$ is not a simple series that can be expressed in closed form using one of our formulas. None-the-less, this infinite series does converge to the number $\pi^2/6 \approx 1.64493406$. This fact was known in 1736, but even the simplest proofs of this fact are very difficult. This convergence is illustrated by computing a few terms in the sequence of partial sums:

$$S_1 = 1/1 = 1$$

$$S_2 = (1/1) + (1/4) = 1.25$$

$$S_3 = (1/1) + (1/4) + (1/9) = 1/25 + 0.11111 = 1.36111$$

$$S_4 = (1/1) + (1/4) + (1/9) + (1/16) = 1.36111 + 0.625 = 1.42361$$

$$S_5 = S_4 + (1/25) = 1.42361 + 0.04 = 1.46361$$

$$S_6 = S_5 + (1/36) = 1.46361 + 0.02777 = 1.49138$$

$$S_7 = S_6 + (1/49) = 1.49138 +).02040 = 1.51179$$

By repeating this process, we would eventually obtain the value $S_{50} = 1.62513$ which differs from $\pi^2/6$ by about 0.0198. Using a computer to evaluate the sums gives the terms

$$S_{1000} = 1.6439345 \quad S_{2000} = 1.6444341 \quad \text{and} \quad S_{10,000} = 1.6448340718$$

It appears that this series converges very slowly to the number $\pi^2/6$. The partial sum $S_{10,000}$ is only accurate to the 3rd decimal place while its last term being added to the partial sum is $1/10,000^2 = 10^{-8}$ only adds one in the 8th decimal place. The partial sum $S_{100,000} = 1.644924068982$ is only accurate to 10^{-5}. To recognize the limit of this series by simply calculating partial sums is not practical. To determine this limit requires advanced mathematical techniques.

Infinite Power Series

One of the most important types of infinite series is the **infinite power series**, of the form

$$\Sigma_{i=0,\infty} \, a_i \, x^i$$

or, in the shifted form

$$\Sigma_{i=0,\infty} \, a_i \, (x-b)^i$$

An infinite power series may or may not converge, depending on the coefficient sequence $\{a_i\}$ and the numerical value of x. When it does converge it defines a function of x. The study of such series and their convergence is often the subject of an entire year mathematics course. We will not consider these important issues in this text. However, in later chapters we will see how ordinary functions f(x) can be represented by such power series, in a special form called a *Taylor Series expansion*. It will have the form

$$f(x) = \Sigma_{i=0,\infty} \, a_i \, (x - x_0)^i$$

where x_0 is a particular x-value. The sequence of coefficients a_i will be determined by evaluating the *derivatives* of f(x) at x_0. At this juncture let us simply list the infinite series representations of some elementary functions.

$$e^x = 1 + x + x^2/2! + x^3/3! + x^4/4! + ...$$

$$\sin(x) = x - x^3/3! + x^5/5! - x^7/7! + x^9/9! - ....$$

$$\cos(x) = 1 - x^2/2! + x^4/4! - x^6/6! + x^8/8! - x^{10}/10! + ...$$

$$1/(1+x) = 1 - x + x^2 - x^3 + x^4 - x^5 + ... \quad \text{for } |x| < 1$$

$$ln(1+x) = x - x^2/2 + x^3/3 - x^4/4 + x^5/5 - .... \quad \text{for } |x| < 1$$

EXERCISE SET 2.1

1. Evaluate the following series by expanding and then adding.

   a) $\sum_{i=1,6} (i+1)$
   b) $\sum_{i=3,10} (i^2+2)$
   c) $\sum_{k=1,10} (2k^2-k)$
   d) $\sum_{k=3,8} (k^3+(-1)^k)$

   e) $\sum_{i=-2,4} i^2$
   f) $\sum_{j=2,5} \log(j)$
   g) $\sum_{k=1,4} t^k$
   h) $\sum_{i=1,5} (t_i - t_{i-1})$

2. Write the following series as expanded sums.

   a) $\sum_{i=1,20} (1+i)^{-1}$
   b) $\sum_{k=3,7} (k+1)^2$
   c) $\sum_{j=1,N} x_j^* \Delta(x_j)$

   d) $\sum_{t=-1,5} 3^t$
   e) $\sum_{i=1,8} (i+1) - (i-1)$
   f) $\sum_{j=1,N} j \cdot \Delta x$

3. Express the following expanded sums using sigma notation.

   a) $1 + 4 + 9 + 16 + 25$
   b) $1/2 + 1/4 + 1/8 + 1/16 + \ldots + 1/1024$

   c) $5 + 7 + 9 + 11 + 13$
   d) $a_1 + b_1 + 2a_2 + b_2 + 3a_3 + b_3 + \ldots$

   e) $(a+3) + (2a+4) + (3a+5) + (4a+6)$
   f) $3 + 1 + 3 + 1 + 3 + 1 + 3 + 1$

   g) $(5-4) + (6-5) + (7-6) + (8-7)$
   h) $4 - 4 + 4 - 4 + 4 - 4 + 4 - 4$

   i) $2 + 6 + 18 + 54 + 162$
   j) $4 + 2 + 1 + 1/2 + 1/4 + 1/8$

4. Using closed forms evaluate the following series.

   a) $\sum_{n=1,30} 7$
   b) $\sum_{n=1,30} n$
   c) $\sum_{n=1,8} n^2$
   d) $\sum_{n=1,8} n^3$

   e) $\sum_{n=1,30} 2+(n-1)3$
   f) $\sum_{n=1,30} 4n$
   g) $\sum_{n=1,12} 4n^2 - n$
   h) $\sum_{n=1,12} 9+(-1)^n$

   i) $\sum_{n=1,6} (n+1)^2$
   j) $\sum_{n=1,8} n^2 - 5n + 7$
   k) $\sum_{n=1,10} 2+(n-1)^3$
   l) $\sum_{n=1,8} (4n-1)^3$

5. Evaluate the following series.

   a) $\sum_{n=1,10} 3 \cdot 2^{n-1}$
   b) $\sum_{n=1,10} 3(1/2)^{n-1}$
   c) $\sum_{n=0,10} 3 \cdot 2^{n-1}$

   d) $\sum_{n=0,10} 0.5^{n-1}$
   e) $\sum_{n=1,10} 3(-1/2)^{n-1}$
   f) $\sum_{n=1,10} 2^n$

   g) $\sum_{n=0,10} 0.1^n$
   h) $\sum_{n=0,8} -0.25^n$
   i) $\sum_{n=0,10} (-0.1)^n$

   j) $\sum_{n=0,8} 3 \cdot 0.25^n$
   k) $\sum_{n=1,25} -2(0.1)^{n-1}$
   l) $\sum_{n=1,100} 0.2^n$

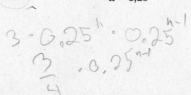

6 Evaluate the following series for the indicated parameters.

   a) $\sum_{i=1,N} a \cdot r^{i-1}$ $a = 3, r = 2, N = 7$ b) $\sum_{i=1,N} a \cdot r^{i-1}$ $a = 3, r = 0.1, N = 7$

   c) $\sum_{i=1,N} a \cdot r^{i-1}$ $a = -2, r = 6/10, N = 7$ d) $\sum_{j=3,N} a \cdot j$ $a = 2, N = 5$

   e) $\sum_{k=2,N} (k+2)^2$ $N = 7$ f) $\sum_{j=1,N} (a \cdot j + b)\Delta x$ $N = 10, a = 3, b = 2$ and $\Delta x = 10^{-5}$

7 A series can be considered as a function of its upper limit of summation. Let $S(n) = \sum_{i=1,n} a_i$

   a) Compute $S(10)$ if $a_i = 1$.   b) Compute $S(n)$ if $a_i = i + 3$.

   c) If $a_i = 2^i$, compute $S(10)$ and $S(100)$.   d) If $a_i = (i+1)h$, where h is a constant, compute $S(n)$.

   e) If $h = 1/n$ in part (d), what value does $S(n)$ have?

   f) Show that $S(n)$ is a monotone-increasing function of n if $a_i > 0$ for all i.

8 Which of the eight special series for which closed forms were given are convergent if $N \to \infty$? Indicate what restrictions must be placed on the parameters for the series to converge.

9 Evaluate the 10th partial sum $S(10)$ of the following series. Indicate whether the series appears to converge and if so what its limit might be.

   a) $\sum_{n=1,\infty} n^2$ b) $\sum_{n=1,\infty} 0.5^n$ c) $\sum_{n=1,\infty} 1/n$ d) $\sum_{n=1,\infty} \sin(n\pi/2)$

   e) $\sum_{n=1,\infty} (-2)^{-n}$ f) $\sum_{n=1,\infty} \sin(n\pi/2)/n$

10 A simple way to verify the closed form for the sum of the first N integers is to construct a rectangular region of dimensions N by N + 1 and to systematically show that the series is equal to one-half the area of the rectangle. Draw a sketch to illustrate this when N = 4. Hint: if the rectangle has height N then it can be divided in to horizontal rows of unit squares; each row can be divided so that the left part of the ith row will be the sum of i squares and the right part will be the sum of N + 1 - i squares. Add the squares on the left side of the rows and those on the right side of the rows using sigma notation.

11 A folk tale dating from early Egyptian times recounts the story of a slave who saved the Pharaoh's life. The grateful Pharaoh offered the slave a reward of his choosing. The slave seemed reasonable when he asked for one grain of wheat on the first day of the month, two grains of wheat on the second day, four grains of wheat on the third day, and so on, doubling the number of grains each day until the end of one month. The Pharaoh was overjoyed by the slave's meager request and granted it immediately. Why was he a foolish Pharaoh? If the month had 31 days, how many grains of wheat would the slave have accumulated by the end of the month?

12 A veterinarian prescribes a drug-dosage scheme for an animal in which the animal is given 10 grams of the drug on the first day and the dosage is halved on each successive day for one week. What is the total amount of drug which should be prescribed to complete this dosage scheme?

13 If a population increases in size by an amount $\Delta P(t) = t^2 - 2$ at generation t and has an initial size of 102, what will its size be after 10 generations? After N generations?

14 Give a closed form of the series $\sum_{i=n,N} (a_i - a_{i-1})$. Use n = 1 for a special case.

15 A drug is administered in 5-mg doses each day. The patient metabolizes 4.9 mg of the drug on the first day, so that 0.1 mg remains in circulation when the second dose is administered. Assume that, on successive days, the patient metabolizes one-half of the amount of the drug that was metabolized on the preceding day.

   a) How much of the drug will be metabolized on the fourth day? What will be the residual amount of drug in the patient's blood after four days?

   b) Determine a formula for the residual amount of drug in the patient after the n-th day.

   c) If the drug is given for only two days when will the patient's blood be completely free of the drug? Assume that the same metabolism process continues.

16 Write the infinite series expansions given in the text for $\sin(x)$, $\cos(x)$, e^x, $1/(1 + x)$ and $\ln(1 + x)$ using sigma notation.

17 Using the infinite series expansions given in the text and assuming that such series can be multiplied term-wise like algebra with polynomials write an infinite series expansion for the following functions.

   a) $\sin(x) + \cos(x)$ b) $\sin(x) - \cos(x)$ c) $x \cdot \sin(x)$ d) $3e^x$ e) $x \cdot e^x$

   f) $x/(1 + x)$ g) $\ln(x + 1) + 1/(1 + x)$ h) $\sin(x)/x$ i) $(1/x) \cdot \ln(1 + x)$

18 Give the infinite series expansions of the following terms by replacing x by the indicated number or variable. Simplify when possible.

   a) e^0 b) e^1 c) e^{-x} d) e^{3t} e) $\exp(x^2)$

19 Multiplication of infinite series is complicated, since like the product of polynomials, each term of one series must be multiplied times each term of the other series. However, for power series this can be done in an orderly way since the product of terms resulting in a common power of x can be grouped. Still, this presents a difficult algebraic rearrangement problem.

   a) Write the formal series expansion of $e^{t + x}$

   b) Compare the terms of degree ≤ 2 of the expansion in part a) to the corresponding terms formed from the formal product of the series for e^t times the series for e^x.

   The series for e^t is multiplied times the series for e^x, extending the multiplication of terms as we would do when multiplying polynomials:

$$1 \cdot (1 + x + x^2/2 + ...) + t \cdot (1 + x + x^2/2 + ...) + t^2/2 \cdot (1 + x + x^2/2 + ...) + t^3/3! \cdot (1 + x + x^2/2 + ...)$$

Section 2.2 Discrete Dynamical Systems.

Introduction: Discrete Dynamical Systems.

Dynamical systems are processes or measurable quantities that vary with time. A *discrete dynamical system* is a one that only changes at a discrete sequence of time-values, say t_0, t_1, t_2, t_3, ..., and remains constant over the intervals between these values. Although most biological and physical systems have many measurable aspects, in this chapter we only consider systems with one *state* variable that corresponds to a measurable aspect of the system. A mathematical description of a discrete system consists of a sequence of numerical values that indicate the state of the system. If the system's state over an interval (t_n, t_{n+1}) is denoted by X_n then the system is described by the sequence $\{X_n\}$. We will arbitrarily assume that change occurs at the end of the interval, and hence that X_n indicates the state at t_n. Normally the state of a system is governed by regular processes, so that the systems behavior can be described by a function $X(t) = F(t)$, or for a discrete model, simply a function of the index n, as $X_n = f(n)$. To establish these functions one must Model the dynamical system; this consists of forming a Dynamical Model, then analyzing this model, and solving its equations for the state X_n.

The description and analysis of discrete systems begins with a *mathematical model* of the system. A model is a mathematical analogy in which the dynamics of the system are indicated by equations involving constants, called parameters, and relationships between the state of the system at different times. A **Discrete Mathematical Model** is given by a *difference equation* that relates terms of the state-sequence indicating how to evaluate state terms using previous states or how the associated state-sequence changes. Examples of *difference equations* are:

$$x_{n+1} = 3 x_n \qquad y_{k+1} = 2y_k - y_{k-1} \qquad \Delta z_n = z_n - 1$$

The solution of a difference equation consists of a formula that explicitly expresses the general sequence term as a function of its index and model parameters (i.e., constants). The above difference equations have the following solutions:

$$x_n = x_0 \cdot 3^n \qquad y_k = [c_1 + c_2 \cdot n] \quad \text{and} \quad z_n = [z_0 - 1]2^n - 1$$

The study of difference equations and their solutions is a branch of mathematics, like the study of calculus which focuses on continuous change. A Pure mathematician might study difference equations without reference to any model, while applied mathematics normally tries to relate the mathematical aspects to models of actual systems. Our intent in this text is to emphasize the applications, although our discussions will sometimes focus on purely mathematical aspects.

Difference Equations.

Let us start this section with an example that illustrates the process of forming a difference equation from a word description.

Example 1 Forming a difference equation.

Problem Consider a sequence $\{Y_n\}$ in which each term is the average of the previous two terms. What would be the 6th term of this sequence if the first two terms are 24 and 8? Express a difference equation that this sequence satisfies.

Solution The average of two terms is one half their sum. To generate Y_6 we begin with the first two terms, $Y_1 = 24$ and $Y_2 = 8$, and sequentially apply the descriptive rule to determine Y_3, Y_4, Y_5, and Y_6:

$$Y_3 = 0.5(Y_1 + Y_2) = 16$$

$$Y_4 = 0.5(Y_2 + Y_3) = 12$$

$$Y_5 = 0.5(Y_3 + Y_4) = 14$$

$$Y_6 = 0.5(Y_4 + Y_5) = 13$$

Following this pattern we form a difference equation by introducing general subscripts. In general, the nth term of the Y-sequence is Y_n. The next term, following the nth term, has index $n + 1$, Y_{n+1}, and the term that precedes the nth term is Y_{n-1}. To write the general form of the above equations we replace the term on the left of the equal sign by Y_{n+1} and introduce the corresponding general terms into the function on the right side of the equation sign. For instance, in the last equation above identifying the subscript 6 with $n + 1$ would imply that $n = 5$ and $n - 1 = 4$. Thus the difference equation is formed by replacing Y_6 by Y_{n+1} on the left and on the right replacing Y_4 by Y_{n-1} and Y_5 by Y_n. This gives the general difference equation:

$$Y_{n+1} = 0.5(Y_{n-1} + Y_n)$$

In this difference equation the value 0.5 is a constant; if it were replaced by a symbol, say the letter p, it would be referred to as a parameter. Thus the above equation could be expressed as

$$Y_{n+1} = p(Y_{n-1} + Y_n) \quad p = 0.5$$

Difference equations are normally written in a form that expresses one sequence term explicitly as a function of one or more previous terms in the sequence. The number of sequence terms that must be known to apply the rule for generating a new term is the **order** of the difference equation.

DEFINITION OF FIRST-ORDER DIFFERENCE EQUATIONS.

The *standard form* of a **First-Order difference equation** is

$$X_{n+1} = F(X_n)$$

The difference equation is said to be **linear** if the function F is linear.

The standard **linear first-order** difference equation has the form

$$X_{n+1} = a \cdot X_n + b$$

for parameters a and b.

Higher order difference equations are similarly defined with the function $F(X_n)$ replaced by multivariable functions of more than one previous sequence term. Difference equations become more complicated as their order becomes greater. For instance, the general **Second-Order** difference equation has the standard form

$$X_{n+1} = F(X_n, X_{n-1}; P) \qquad \textbf{2}^{\textbf{nd}} \textbf{ Order Difference Eq.}$$

where **P** represents parameters, either specific numbers or unknown constants. If the function F is a linear combination of the sequence terms then the difference equation is said to be linear. The 2nd Order equation

$$X_{n+1} = 3 + 2X_n - 7X_{n-1}$$

is linear, whereas the equations

$$X_{n+1} = 3 + X_n^2 + X_{n-1} \quad \text{and} \quad X_{n+1} = 3 + X_n \cdot X_{n-1}$$

are *nonlinear*. The first has a squared term, and the second has a product $X_n \cdot X_{n-1}$. The definition of the general K^{th}-Order difference equation involves a multivariable function F of K previous sequence terms, $X_n, X_{n-1}, X_{n-2}, \ldots, X_{n-K+1}$. Notice that counting backwards, starting with the subscript n, the Kth term has the subscript n - K plus 1. The **standard form** of the Kth-**Order** difference equation is

$$X_{n+1} = F(X_n, X_{n-1}, X_{n-2}, \ldots, X_{n-K+1}; \textbf{P})$$

Difference equations of higher order do not have to actually involve each sequence term. For instance, both

$$X_{n+1} = X_n - 2X_{n-1} + X_{n-2} \quad \text{and} \quad X_{n+1} = 5X_{n-2}$$

are third-Order difference equations. They are also both linear equations. An easy direct method for calculating the order of a difference equation is to subtract the smallest subscript from the largest subscript appearing in the equation:

$$\text{Order} = \text{Largest subscript} - \text{Smallest subscript}$$

This formula can be used even when the difference equation is expressed in non-standard forms, i.e. not solved explicitly for the n + 1 term. For instance, in the difference equation $X_{n-3} = 1 - X_{n+5}/X_n$ the largest subscript is n + 5 and the smallest subscript is n - 3. Therefore, the order of the difference equation is (n + 5) - (n - 3) = 8.

Shifting indices.

Difference equations can be manipulated algebraically, like any other equation, and can also be manipulated by *shifting* their subscripts. This is a process that is useful for converting non-standard equations into the standard forms given in the above definitions. For instance

$$X_{n+1} = 3X_n(10 - X_n) \quad \text{and} \quad X_{n+1} + 3X_n^2 - 30X_n = 0$$

are different algebraic forms of the same first-Order nonlinear difference equation. The first equation has the standard form and the second is an equivalent non-standard form. Index symbols other than n may be used. If n is replaced by j in the left equation above the resulting equation is also of the standard form:

$$X_{j+1} = 3X_j(10 - X_j)$$

The indices of the standard equation can be shifted downward by one if each n is replaced by N - 1 (and hence $n + 1 \rightarrow (N - 1) + 1 = N$) to give

$$X_N = 3X_{N-1}(10 - X_{N-1})$$

The indices are shifted upward by 2 if n is replaced by N + 2, and hence $n + 1 \rightarrow N + 3$:

$$X_{N+3} = 3X_{N+2}(10 - X_{N+2})$$

A capital N was used in these shifts to help distinguish the original subscript symbol from the shifted symbol. Often this is not done and the same subscript symbol is utilized. Thus, the above shifted equations would be expressed using n as

$$X_n = 3X_{n-1}(10 - X_{n-1}) \quad \text{and} \quad X_{n+3} = 3X_{n+2}(10 - X_{n+2})$$

Example 2 Manipulating difference equations.

Problem Determine the order of the following difference equations and express them in the standard form.

a) $X_{n+3} - X_{n+2} - 5 = 0$ b) $Y_{n+2} + 4Y_n = Y_{n-5}$ c) $P_{k+1} = 0.5P_k - 20$

Solutions a) The order of the equation is the difference between the largest and smallest subscripts: $(n + 3) - (n + 2) = 1$. To put the equation into standard form, first solve for the term with the highest subscript, $n + 3$. Then, the index of this term is shifted to $N + 1$. This is done by replacing n with $N + k$ where k is found by equation the transformed highest index to $N + 1$:

$$N + k + 3 = N + 1 \implies k = -2$$

$$X_{n+3} = X_{n+2} + 5 \quad \text{then shifts to} \quad X_{N+1} = X_N + 5$$

b) The equation $Y_{n+2} + 4Y_n = Y_{n-5}$ has order $(n + 2) - (n - 5) = 7$. Its equivalent standard form is found by solving for Y_{n+2} and, as in a), shifting the indices down by $k = 1$:

$$Y_{n+2} = -4Y_n + Y_{n-5} \quad \text{which shifts to} \quad Y_{n+1} = -4Y_{n-1} + Y_{n-6}$$

c) The equation $P_{k+1} = 0.5P_k - 20$ is already in the standard form of a first-order difference equation. It does not matter that the index symbol is k instead of n and the associated sequence is $\{P_k\}$.

The difference operator Δ.

To describe the change in a sequence from one term to the next the **forward difference operator** Δ is often used. Δ is the capital Greek letter delta and it is called an "operator" because it is like a function. It indicates an algebraic process, but it operates on terms of sequences. Given a sequence $\{X_n\}$, the forward difference at the nth term is *the change from the nth to the $(n + 1)^{st}$ terms of the sequence*:

$$\Delta X_n \equiv X_{n+1} - X_n \qquad \textbf{Forward Difference } \Delta$$

To apply mathematics to problems in other disciplines one of the first steps is usually to translate a literal description of the problem into mathematical vocabulary and equations. Some key words that indicate a "forward difference" or "change equation" should be used include "difference", "change", "increase", "decrease", "grows by ", and "diminishes by " or "declines by". Sentences containing these words can usually be translated into equations involving the forward difference Δ of the quantity that is changing. Denoting the state of the system by the sequence $\{X_n\}$, the corresponding *change equation* has the form

$$\Delta X_n = \ldots\ldots\ldots\ldots\ldots$$

where the equal sign is the mathematical equivalent of the verb "is" and the right side of the equation is described in the sentence by the text following the verb.

Example 3 **Translating sentences into change equations and difference equations.**

Problem Write the **change equation** corresponding to the given sentence using the forward difference Δ and then convert this into a **standard form** difference equation.

a) The change in mass of a bacteria culture at each replication is ninety per cent of the culture's current mass.

b) During peak growing period the increase in a corn plant's height each day is three centimeters.

c) The growth in a population is two times the product of the population size and the amount that it is less than 200. $2P_n(200-P_n)$

Solution a) Denoting the mass of the culture at generation n by M_n, the change equation has the form

$$\Delta M_n = 0.9 M_n$$

Since the phrase following "is", "ninety per cent of the culture's current mass", translates into the coefficient 0.90 times the current size, i.e., the size at the nth generation. To transform this equation into the standard form of a difference equation, first substitute

$$\Delta M_n = M_{n+1} - M_n$$

to obtain $M_{n+1} - M_n = 0.9 M_n$. Rearranged algebraically, this becomes $M_{n+1} = 1.9 M_n$.

b) Assume the height of the corn is indicated by the sequence $\{H_n\}$. The change in the height is a constant, 3 cm. Thus the model equation for the change is

$$\Delta H_n = 3 \text{ cm}$$

The corresponding difference equation is $H_{n+1} - H_n = 3$, which gives the linear first-order equation $H_{n+1} = H_n + 3$.

c) If the size of the population at generation n is denoted by P_n, its change equation is

$$\Delta P_n = 2 \cdot P_n \cdot (200 - P_n)$$

Substituting $\Delta P_n = P_{n+1} - P_n$ into this equation gives

$$P_{n+1} - P_n = 2 \cdot P_n \cdot (200 - P_n)$$

which is equivalent to the nonlinear standard equation

$$P_{n+1} = P_n + 2 \cdot P_n \cdot (200 - P_n) \quad \text{or} \quad P_{n+1} = 401 P_n - P_n^2$$

Solutions to Difference Equations.

Difference equations state relationships that terms of a solution sequence must satisfy. Any sequence that *satisfies the difference equation* relations for all terms in the sequence is said to be **a solution to the difference equation**. For instance, the sequence specified by $X_n = 3^n$, this is the sequence $\{1, 3, 9, 27, ...\}$, satisfies the second-order difference equation

$$X_{n+1} = 4X_n - 3X_{n-1}$$

If you select any three consecutive terms of the sequence $\{3^n\}$ the third term will equal 4 times the second term minus 3 times the first term. For instance the three sequence terms corresponding to n = 3, 4, and 5 are 27, 81, and 243 and they satisfy the equation

$$243 = 4 \cdot 81 - 3 \cdot 27$$

More generally, substituting the general terms

$$X_{n+1} = 3^{n+1} \quad X_n = 3^n \quad \text{and} \quad X_{n-1} = 3^{n-1}$$

into the difference equation gives the equation

$$3^{n+1} = 4 \cdot 3^n - 3 \cdot 3^{n-1}$$

which is a true equation for any n, as the right side can be simplified to yield the left side of the equation:

$$4 \cdot 3^n - 3 \cdot 3^{n-1} = 4 \cdot 3^n - 3^n = (4-1)3^n = 3 \cdot 3^n = 3^{n+1}$$

You will be able to find this solution sequence using a method introduced in Section 2.4. The expression "a solution" is used because most difference equations have not one or two solutions but an infinite number of solutions, i.e., infinitely many different sequences satisfy the same difference equation. You might check that the sequence $X_n = 5 \cdot 3^n$ is also a solution of the above difference equation. The many solutions of a difference equation are normally related and can be expressed by the same formula for their general n^{th} term, however, the formula will involve a parameter or constant whose value is not mandated by the difference equation. The sequence $X_n = c \cdot 3^n$ is a solution of the above difference equation for any choice of the constant parameter c.

Example 4 Verifying solutions to difference equations.

Problem Determine if the indicated sequences is a solution of $Y_{n+1} = -Y_n + 2$.

  a) The sequence $\{0, 2, 0, 2, 0, 2, ...\}$ b) $Y_n = -1 + 2n$ c) $Y_n = 1 + 5 \cdot (-1)^n$

Solution The sequence a) is a repeating sequence that alternates between 0 and 2. If $Y_n = 0$ then $Y_{n+1} = 2$. Substituting these into $Y_{n+1} = -Y_n + 2$ results in a true equation: $2 = -0 + 2$. Consequently, the given sequence is a solution of the difference equation.

The b) sequence does not satisfy the difference equation. To show this we need only show that for at least one choice of the index n the difference equation is not a true equality when the corresponding numerical sequence terms are substituted into it. Let n = 1. Then n + 1 = 2 and by the given formula $Y_1 = -1 + 2 \cdot 1 = 1$ and $Y_2 = -1 + 2 \cdot 2 = 3$. Substituting these values into the difference equation

$$Y_{n+1} = -Y_n + 2 \quad \text{gives the false equation} \quad 3 = -1 + 2$$

To show that the c) sequence is a solution we substitute the general expressions

$$Y_n = 1 + 5 \cdot (-1)^n \quad \text{and} \quad Y_{n+1} = 1 + 5 \cdot (-1)^{n+1}$$

into the difference equation and manipulate the resulting expression to verify that both sides of the equation are equal. $Y_{n+1} = -Y_n + 2$ becomes upon substitution

$$1 + 5 \cdot (-1)^{n+1} = -(1 + 5 \cdot (-1)^n) + 2$$

which simplifies algebraically to

$$1 + 5 \cdot (-1)^n \cdot (-1) = -1 - 5 \cdot (-1)^n + 2$$

or

$$1 - 5 \cdot (-1)^n = 1 - 5 \cdot (-1)^n \quad \text{This is a true equation.}$$

☑

Difference equations are a "favorite" of computer modelers, as their direct numerical solution is easy to program. Starting with a specific numerical value for one sequence term, the following terms are found progressively by evaluating the right side of a standard form difference equation using known values to generate the next term's value. The number of terms needed to start a numerical evaluation is the order of the difference equation. This "brute force" approach is useful only when the number of calculations are feasible. For instance, if you need to know the term X_6 of a sequence and are given

$$X_3 = 10 \quad \text{and} \quad X_{n+1} = 0.5 X_n^2$$

then you can calculate $X_4 = 0.5 X_3^2 = 0.5 \cdot 10^2 = 50$. Then, $X_5 = 0.5 X_4^2 = 0.5 \cdot 50^2 = 1{,}250$ and finally, $X_6 = X_5^2 = 0.5 \cdot 1250^2 = 781{,}250$. With a calculator you might use this approach to calculate X_{10} but you would probably balk at the suggestion you could use this to calculate the term $X_{2{,}387}$. It would clearly require too many evaluations. However, with a computer this would be a simple task that could be easily programmed, even using a spread-sheet program. Sometimes numerical determinations of solutions are all that is required and this direct approach is sufficient. But, it is usually most desirable to be able to express solutions by formulas, explicit equations for the general n^{th}-term of the sequence.

Example 5 Modeling a managed resource.

Problem   A national Buffalo Park is created on the western prairies and the park management wants a predictive model of the buffalo herd to explore possible effects of management policies. Wildlife biologists estimate that normal mortality for a Buffalo herd is 5 percent each year. Assume that an initial herd is established with 85 animals and each year 38 additional buffalo are added to the herd by the park managers. Formulate a discrete model of the annual herd size. What would be the herd size after 5 years? How can this model be used to manage the herd?

Solution The mathematical model will be a difference equation. First, we introduce a herd variable; let the size of the herd at the end of the n-th year be denoted by X_n. The given information for annual changes in the herd is expressed mathematically by the forward difference model equation:

$$\Delta X_n = -\text{Deaths} + \text{New Stock}$$

$$\Delta X_n = -0.05 X_n + 38$$

A standard first-order difference equation is obtained by replacing ΔX_n with $X_{n+1} - X_n$. Solving for X_{n+1} results in the **model equation**:

$$X_{n+1} = 0.95 X_n + 38$$

Using this equation the herd size can be calculated for successive years. The initial herd size is $X_0 = 85$. Therefore,

$$X_1 = 0.95X_0 + 38 \quad = 118.75$$

$$X_2 = 0.95X_1 + 38 \quad = 150.8125$$

$$X_3 = 0.95X_2 + 38 \quad = 181.2718750$$

$$X_4 = 0.95X_3 + 38 \quad = 210.2082812$$

$$X_5 = 0.95X_4 + 38 \quad = 237.6978671$$

The above calculations were made on a calculator without any rounded off. If at each stage the values are taken as the greatest integer, i.e., rounded down to the whole animal level, the herd sizes would be 118, 150, 180, 201, and 236. While the integer values seem appropriate to describe the number of animals, the un-rounded values can be interpreted as the expected average herd size.

How to manage the herd? The above procedure indicates how to calculate the herd size for any year, provided there is sufficient time to perform the calculations. However, it would be very tedious to calculate the value of the herd size predicted by this model at the end of the Park's centennial year, X_{100}. Similarly, it would be almost impossible to answer simple questions about the predicted status of the herd. To answer quantitative questions such as "What would be the effect of changing the number of New Stock on the 5-year size of the herd?" or "How would changes in the herd's average mortality rate affect the herd size after 10 years?" we need to know the explicit equations describing the size of the herd. That is, we must be able to solve the model equation for X_n as a function of the index n and the model parameters, which in this case are the numbers 0.05, 85, and 38. A general solution of first-order equations is presented in Section 2.4.

Another type of questions is "What will be the ultimate size of the herd?" The answer mathematically is found as a limit, as n approaches infinity, of X_n. The limiting size is called the *steady-state* of the difference equation and is the quantity $\bar{X} = \lim_{n \to \infty} X_n$. Since $\bar{X}$ will also be the limit of the $(n+1)^{st}$ term, $\lim_{n \to \infty} X_{n+1} = \bar{X}$, taking the limit in the difference equation results in an algebraic equation for steady-state $\bar{X}$:

$$\lim_{n \to \infty} X_{n+1} = \lim_{n \to \infty} 0.95X_n + 38$$

$$\bar{X} = 0.95\bar{X} + 38 \quad \text{and hence} \quad \bar{X} = 760$$

The model predicts an eventual herd size of 760 buffalos. Without an explicit solution we can not answer a question such as "When will the herd reach half its expected size?"

As will be seen in the next few sections, only the simplest linear difference equations can be easily solved in functional forms. Theoretical analysis of nonlinear dynamical systems is very difficult and relies extensively on observations of computer generated numerical solutions. This is a real deficiency in our bag of mathematical tools as many relatively simple discrete dynamical systems that arise in other disciplines are nonlinear. Nonlinear problems are often approached by considering associated linear problems, study the linear problems, and then inferring characteristics of the nonlinear system from those of the associated linear system. Thus the study of simple linear systems presented in the following sections forms the basis for studying more complex dynamical systems.

Section 2.2   DISCRETE DYNAMICAL SYSTEMS   137

Exercise Set 2.2

1. Determine the value of X_6 if $X_0 = 2$ and $X_1 = 4$ and if each other term of the sequence $\{X_n\}$ is:
   a) twice the previous term.
   b) the sum of the previous term and twice the second previous term.
   c) three times the previous term minus two.
   d) one half the previous term plus three.
   e) twice the average of the two previous terms.

2. For each sequence described in Exercise 1 indicate the corresponding difference equation that describes the sequence. What is the order of the difference equation?

3. Compute X_5 if $X_0 = 3$ for the following equations:

   a) $X_{n+1} = 0.5X_n + 3$ b) $X_{n+1} = X_n - 1$ c) $X_{n+1} = X_n^2 - X_n$

   d) $\Delta X_n = 0.5$ e) $\Delta X_n = X_n$ f) $\Delta X_n = 1/X_n$

4. By shifting indices and algebraic rearrangement express each equation in the standard form and indicate its order.

   a) $X_n = 3X_{n-1}$ b) $y_{n+2} = 0.1y_{n-1}$ c) $\Delta X_n = 3X_{n-1}$ d) $\Delta y_{n+2} = 0.1y_{n-1}$

   e) $z_n = 2z_{n+1} - z_{n-1}$ f) $x_{n-3} = x_{n-5} + x_n$ g) $H_n = H_{n+1} + H_{n+2}$ h) $X_n = 3 - 2X_{n-1}$

   i) $Y_{n+2} + 3Y_{n+1} - Y_n = 2$ j) $X_n - 2X_{n+1} = 2(X_n - X_{n+1}) + 1$

5. Write the change equation corresponding to the given sentence using the forward difference Δ and then convert this into a standard form difference equation.

   a) The change in mass of a bacteria culture at each replication is ten per cent of the culture's current mass.

   b) During a drought the increase in a corn plant's height each day is 11 millimeters.

   c) The growth in a population is one half the product of the population size and the amount that it is less than 1000.

   d) The change in monthly sales of a toy is 5% of the average sales for the past year.

6. Express the given equation in the change equation form $\Delta X_n = \ldots$

   a) $X_{n+1} = 0.1X_n + 2$ b) $X_{n+1} = X_n - 1.5$ c) $X_{n+1} = X_n^2 - X_n$

   d) $X_n = 0.1X_{n-1} + 2$ e) $X_{n+2} = 2X_n$ f) $X_{n+1} = X_{n-1} - X_n$

7. Determine if the given sequence is a solution of the equation $x_{n+1} = 3x_n - 2$

   a) $x_n = 2^n + 1$ b) $x_n = 3^n + 1$ c) $x_n = 1$ d) $x_n = 1 + 3^n$

   e) $x_n = 7 \cdot 3^n$ f) $x_n = 1 - 3^n$ g) $x_n = -1$ h) $x_n = (-2)^n$

8. Determine if the given sequence is a solution of the equation $x_{n+1} = 3x_n - 2x_{n-1}$.

a) $x_n = 2^n$ b) $x_n = 3^n$ c) $x_n = 7$ d) $x_n = 7 + 2^n$

e) $x_n = 7 \cdot 3^n$ f) $x_n = 100 - 3^n$ g) $x_n = -1$ h) $x_n = (-2)^n$

9. Determine if the given sequence is a solution of the equation $x_{n+1} = 3x_n + 10x_{n-1}$.

    a) $x_n = 2^n$ b) $x_n = -5^n$ c) $x_n = 7$ d) $x_n = 7 + 2^n$

    e) $x_n = 7 \cdot (-5)^n$ f) $x_n = (-5)^n - 3^n$ g) $x_n = 10^n$ h) $x_n = (-5)^n$

10. Determine if the given sequence is a solution of the equation $x_{n+1} = 6x_n - 9x_{n-1}$.

    a) $x_n = 3^n$ b) $x_n = (-3)^n$ c) $x_n = 7n \cdot 3^n$ d) $x_n = 7 + 3^n$

    e) $x_n = (-10)^n$ f) $x_n = 3^n - n \cdot 3^n$ g) $x_n = 10^n$ h) $x_n = 6^n$

11. Assume that the normal mortality for a Buffalo herd is 22 percent each year. Assume that an initial herd is established with 100 animals and each year 25 additional buffalo are added to the herd by the park managers. Formulate a discrete model of the annual herd size. What would be the herd size after 5 years?

12. Write a change model and its equivalent standard difference equation to describe the temperature of an experimental chamber that is regulated each minute by a controller that changes the temperature by 10% of its relative difference from a "set" temperature T. Give two models, one where the term "relative" means relative to T and one where it means relative to the temperature T_n at minute n.

13. The ridership on a public transit system is optimistically projected to increase by 5% each year. Describe the change in the number of riders using a Δ-equation. Then write this as a difference equation and solve it. If the Toronto ridership was 5.2 million in 1990 what would is the projected ridership for the year 2000 using this model?

14. The ridership on a public transit system is pessimistically projected each year to decrease by 1% of the current riders but to add 2000 new riders. Describe the change in the number of riders using a Δ-equation. Then write this as a difference equation and solve it. If the Toronto ridership was 5.2 million in 1990 what would be the projected ridership for the year 2000 using this model?

15. A dog is given 136 mg of *ivermectin* to protect it against Heart Worm. If the drug is metabolized at a rate of 5% per day and to be effective there must be 30 mg of drug present, when should the dog be given another drug tablet? Answer this by modeling the amount of drug in the body and determining the first n for which $D_n \leq 30$.

Section 2.3 Graphing First-Order Dynamical Systems.

Introduction.

First-order difference equations of the form $x_{n+1} = f(x_n)$ can be very complicated and difficult to solve when the function f is nonlinear. (Solutions when $f(x) = ax + b$ will be presented in the next Section). In this section we explore graphically some of the qualitative aspects of first order difference equations. A graphical process is introduced to generate solution sequences $\{x_n\}$ using the graph of $y = f(x)$. This process demonstrates that the nature of a solutions depends on whether the graph of f is above or below the line $y = x$. If it is above, at $x = x_n$, then the change Δx_n will be positive and x_{n+1} will be greater than x_n. On the other hand, Δx_n will be negative if the graph of f is below the line $y = x$. The points where the graph of $y = f(x)$ crosses the line $y = x$ are thus critical points for the dynamical system. Their x-coordinates are called *equilibrium* or *stationary states* of the difference equation (and any associated dynamical system). At these points the system remains constant:

$$\text{if } x_n = x_E \text{ where } x_E = f(x_E) \quad \text{then} \quad x_{n+1} = f(x_n) = x_n$$

Focusing on the dynamics of solutions near an equilibrium point, we will see that some equilibriums are *attractors*, in the sense that every solution sequence $\{x_n\}$ with an initial state x_0 near x_E will converge to the equilibrium x_E. On the other hand, some equilibriums will be repulsors, and will not be the limit of solution sequence, no mater how close they may be initially. Equilibrium states classified as being *stable* or *unstable* depending on whether or not they are attractors.

Graphing Discrete Dynamical Systems.

Graphs provide an enormous amount of information via visual references. Consider a first-order discrete model of the form $x_{n+1} = f(x_n)$ where f is a function whose graph is known. A solution sequence of this difference equation could be graphed on a real number line as in the adjacent figure, where the first five sequence terms are plotted. This approach is akin to simply listing the numerical values of the solution sequence. It is only useful to illustrate a limited number of terms and provides very little information concerning the system that generated the sequence. There is a more informative alternative approach that generates solution sequences for $x_{n+1} = f(x_n)$ from the graph of f. This graphical approach provides a method of constructing the terms in a solution sequence using a series of horizontal and vertical lines. This approach allows us to visually identify equilibrium points and their stability.

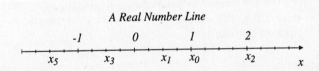

A Real Number Line

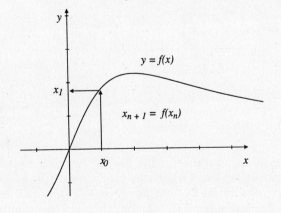

To *graph the difference equation* and its solution sequences we first plot the graph of $y = f(x)$ on a standard x-y coordinate system as in the adjacent figure. The horizontal x-axis is associated with the general sequence term x_n and the vertical y-axis is associated with the next general sequence term x_{n+1}. The value of x_{n+1} is the y-coordinate on the graph of f with x-coordinate x_n. In the figure an initial sequence term x_0 is indicated and the next term x_1 is found graphically by sketching a vertical line from x_0 on the x-axis to the graph of f. Then, a horizontal line is drawn from this intersection point to the y-axis, indicating the value $y = x_1$.

To illustrate the x_1 value on the x-axis, instead of the y-axis, it can be graphically reflected through the line $y = x$ as sketched with dotted lines in the figure at the right.

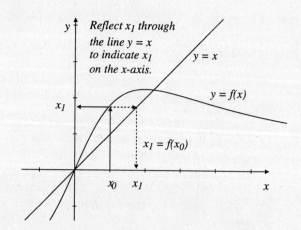

Starting at the value x_1 on the y-axis move horizontally along the line $y = x_1$ to the reflection line $y = x$, then move vertically to the x-axis at the point $x = x_1$.

The graphical processes for finding x_1 from the function's graph and then reflecting this onto the x-axis can be combined to move from the point $x = x_0$ to $x = x_1$ on the x-axis without taking the visit to the y-axis. The procedure is to move up (or down) the vertical line $x = x_0$ to the graph of $y = f(x)$, then move horizontally, as on the dotted lines in the figure, to the reflecting line $y = x$, and then move down the vertical line $x = x_1$ to the x-axis.

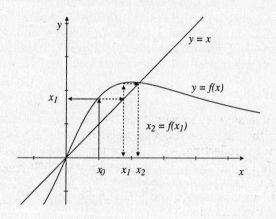

This graphical method can be used when the graph of $y = f(x)$ is known, perhaps as a graph generated by an instrument recording device, even if the actual equation for f is not known.

To find the next solution sequence term, x_2, this graphical process is repeated. The process begins at the point x_1 on the x-axis, you move to the graph of f, then to the reflecting line $y = x$, and then back to the x-axis to give the point x_2 on the x-axis. This is illustrated at the left.

The same process can be repeated three more times to generate the value x_5. Usually, when repeating this graphical process, the line to the x-axis is not sketched.

The graphical path consists of a sequence of vertical and horizontal lines, moving vertically upward or downward, as the graph dictates, to intersect the graph of $y = f(x)$, followed by a horizontal move to the reflecting line $y = x$.

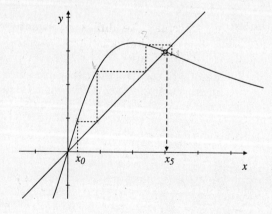

The graph at the right illustrates this process, which is euphemistically called *cob-webbing*. Assuming the scale marks are one unit, we see that the sequence starting at $x_0 = 0.25$ results after five steps in a value $x_5 \approx 3.1$ (the exact value is not important at this stage). Although not indicated explicitly in the sketch, the values of the x_is read off the graph are

$x_1 = 0.9$, $x_2 = 2.4$, $x_3 = 3.2$, and $x_4 = 2.9$

To illustrate further sequence terms would be difficult because of the limited resolution of the graphic window. The path seems to be spiraling in on the point where the function's graph crossed the reflection line. This point is called an *equilibrium* and its x-coordinate, denoted by x_E, is called **an equilibrium of the difference equation** (or dynamical system). It is the solution of the equation

$$x_E = f(x_E)$$

Difference equations may have many equilibrium values, however, linear first-order difference equations will have only one. As illustrated in the previous graph the sequence $\{x_n\}$ appears to approach x_E as n increases, that is

$$\lim_{n \to \infty} x_n = x_E$$

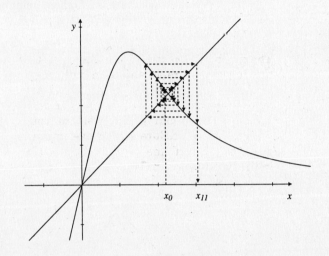

Whether or not a solution approaches an equilibrium depends on the nature of the function $f(x)$ and on the starting point of the sequence $\{x_n\}$. An equilibrium x_E is called an *attractor* if whenever x_0 is "closed" to x_E

$$\lim_{n \to \infty} x_n = x_E$$

The difference equation $x_{n+1} = f(x_n)$ corresponding to the function $y = f(x)$ sketched in the adjacent graph does not have this property. All solution sequences that start at a points x_0 near the equilibrium value x_E will move away from the equilibrium. As illustrated, when the graphical method is applied, the path spirals out from the equilibrium.

The curves $y = f(x)$ in the last two figures are similar, so why does one have an attractor equilibrium and not the other? That the slope of the graph $y = f(x)$ is negative at the intersection point, i.e., the graphs are decreasing as they cross the reflecting line $y = x$, causes the cob-webbing to spiral around the equilibrium. The major difference between the two graphs is in their slopes as they cross $y = x$. An equilibrium can not be an attractor when the slope is too steep, as in the second graph. The requirement for being an attractor is that $|\text{slope}| < 1$. In Chapter 3 we will use the *derivative* of the function f at x_E to state this condition.

Example 1   Interpreting graphs of difference equations.

Problem Consider the difference equation whose graph is indicated at the right. Use a straight edge to estimate the answers to the following questions.

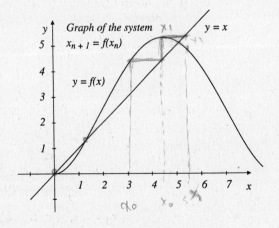

a) What are the equilibriums of the system?

b) Estimate the value x_0 for which x_1 will be the largest.

c) If $x_5 = 3$ estimate the value of x_7.

d) Sketch the graph of $y = \Delta x_n$.

e) Which x_n has the greatest Δx_n?

Note. Estimates that you make from the graph may differ slightly from the numbers given below due to variability in measurements; graphical methods have inherent accuracy limits that can not be avoided. The objective of this example is to interpret key graphical features, not to generate very precise numbers.

Solutions

a) The equilibriums are the x-coordinates where the graph crosses the reflection line $y = x$. There are three intersection points with corresponding x-coordinates of

$$x_{E1} = 0 \qquad x_{E2} \approx 1.2 \quad \text{and} \quad x_{E3} \approx 5.2$$

b) The value of x for which $f(x)$ is greatest will be the choice of x_0 giving the greatest value of $x_1 = f(x_0)$. Visually selecting the highest point on the graph $y = f(x)$ and projecting from this to the x-axis yields the approximate value $x_0 = 4.5$.

c) Using the graphical method, the terms x_6 and x_7 are constructed as indicated in the figure at the right. This construction results in the approximate value $x_7 \approx 5.3$.

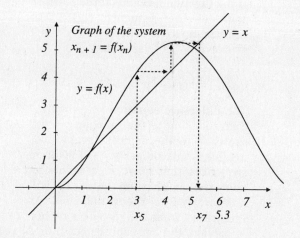

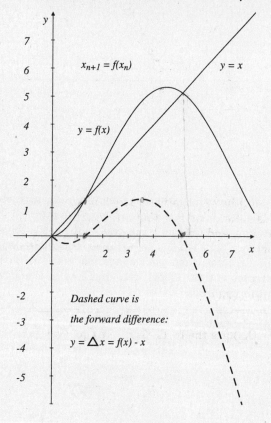

Dashed curve is the forward difference:

$y = \Delta x = f(x) - x$

d) As $\Delta x_n = x_{n+1} - x_n = f(x_n) - x_n$, the graph of the forward difference equation, $y = \Delta x$, is the dashed line $y = f(x) - x$ in the graph at the left. Notice that this graph crosses the x-axis at the equilibrium values. Δx is positive when the graph of f is above the line $y = x$ and negative when it is below.

e) The x-value for which the change Δx_n is maximum is estimated from the graph of Δx_n at the left. The highest point on the dashed curve is at $x_n \approx 3.5$, for which $\Delta x_n \approx 1.3$. This is the x-value for which the maximum <u>positive</u> change would occur. Note that a larger <u>negative</u> change occurs for x-values greater than 6. Since the original graph of the dynamical system does not illustrated negative x or x greater than 8 our discussion does not consider values in these ranges.

Equilibriums and Stability.

An equilibrium value of a difference equation is often referred to as an **equilibrium state** of the dynamical system modeled by the equation. Equilibriums play an important role in the dynamics of a systems. They are the states at which the system is actually not dynamic, not changing, but static or stationary and for this reason equilibriums are also referred to as **stationary states** of a system.

In reality, dynamical systems are subject to minor perturbations that slightly alter their states. These are generally caused by external forces, like an earthquake or changes in an instrument's temperature. When a system is at an equilibrium is perturbed, the question is whether or not it will return to the same equilibrium state. If the equilibrium x_E is an attractor, and the perturbation results in a state x_0 close to the equilibrium then the system will go through a sequence of states $\{x_n\}$ that will return to the original state:

$$\lim_{n \to \infty} x_n = x_E$$

This type of behavior is important as it *stabilizes* the system at the equilibrium state. Equilibriums are classified by their type of stability. For a simple system equilibriums will be either stable or not stable, for more complicated systems finer distinctions can be made introducing one-sided stability or cyclical stability.

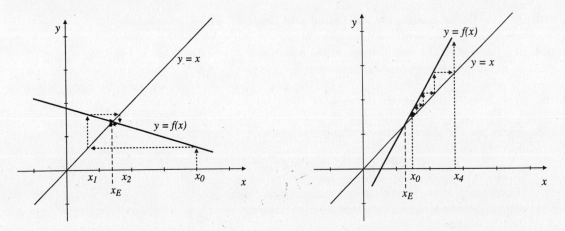

In the left panel of the above graph, the line $y = f(x)$ has a negative slope of about -0.5 and the equilibrium is stable; the cob-webbing path spirals around and into the intersection point. In the right panel, the slope of the line $y = f(x)$ is positive and greater than one, which makes the equilibrium not stable; the cob-webbing path starting near the equilibrium moves away from the equilibrium.

DEFINITION OF EQUILIBRIUMS AND THEIR CLASSIFICATION.

The **equilibriums** of the difference equation $X_{n+1} = f(X_n)$ are the roots X_E of the equation

$$x = f(x)$$

An equilibrium X_E is said to be **stable** if

$$\lim_{n \to \infty} X_n = X_E$$

for every solution sequence $\{X_n\}$ with an initial term X_0 sufficiently close X_E;

otherwise, the equilibrium is said to be **not stable** or **unstable**.

For an equilibrium to be stable **every** solution sequence that comes close to it must converge to it.

Mathematically, this means that when x_E is a stable equilibrium there is a constant δ such that whenever a solution sequence $\{x_n\}$ satisfies $|x_j - x_E| < \delta$ for some index j, then the sequence $\{x_n\}$ converges to x_E, $\lim_{n \to \infty} x_n = x_E$.

The constant δ will vary with the function f (its determination is beyond our scope).

A systems at a stable equilibrium is not critically disturbed by small random perturbations, while one at an unstable equilibrium will change when perturbed, it will either approach another equilibrium, grow infinitely large, or sometimes appear to vary chaotically. The study of **Chaos** is a quite active and exciting new area of mathematical research is concerned with studying dynamical systems whose stability characteristics change dramatically as a system parameter is varied. In *chaotic* systems unpredictable fluctuations in solutions occur as a solution that appears to converge toward an equilibrium suddenly "jumps" to distant values. Simple chaotic difference equations have been used to model phenomenon such as "infant crib death".

Example 2   **Determining equilibriums and their stability graphically.**

Problem   Determine the equilibriums of the equation $x_{n+1} = 0.5x_n(10 - x_n)$ and use a graph to decide if they are stable or not.

Solution   The equilibriums are the roots of
$$x = 0.5x(10 - x)$$
These are the roots of the rearranged equation
$$x^2 - 8x = 0$$
which are $x_{E1} = 0$ and $x_{E2} = 8$. The stability of these are not proven but are suggested by graphing the curve $y = f(x) = 0.5x(10 - x)$ and constructing solution sequence paths for initial values x_0 near each equilibrium.

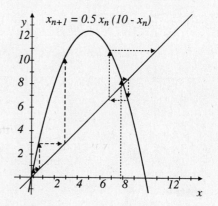

The graph of this quadratic is a parabola opening downward and intersecting the x-axis at $x = 0$ and $x = 10$. Its maximum value occurs midway between the two x-intercepts at $x = 5$, for which $y = f(5) = 12.5$. The graph at the right illustrates this curve and solution paths drawn starting near each equilibrium.

Both equilibriums appear to be unstable. Clearly for any <u>positive</u> x_n near $x_{E1} = 0$ the next value x_{n+1} will be greater since the graph of $f(x)$ is above the line $y = x$. Observe from the graph that for the dotted path starting near $x_{E2} = 8$ at $x_0 \approx 7.8$ has $x_1 \approx 8.4$, $x_2 \approx 6.7$, $x_3 \approx 10.9$ and the next value x_{10} would be found by projecting the down onto the parabola below the x-axis and off the graph window. Thus x_{10} will be a negative value. Although not illustrated, from the model equation it follows that all subsequent values would then be negative, becoming progressively larger negative numbers.

The Equilibrium of Linear First-Order Difference Equations.

With one exception, all *linear first-order* difference equations have only one equilibrium since the graph of the linear function $y = f(x) = ax + b$ and the line $y = x$ will have a single intersection point, as illustrated in the next figure. The exception occurs when the coefficient $a = 1$, in which case the graph of f is parallel to $y = x$ and, hence, $y = x$ and $y = x + b$ are either the same line (b = 0) or do not intersect (b ≠ 0).

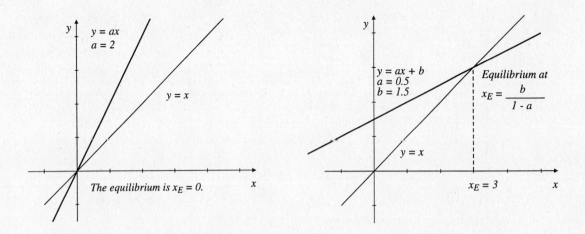

The equilibrium of a linear system is the root of $x = ax + b$, which for $a \neq 1$ is

$$x_E = b/(1-a) \qquad \textbf{Equilibrium of } \mathbf{x_{n+1} = ax_n + b}$$

If the coefficient $b = 0$ then the lines intersect at the origin and $x_E = 0$ and the difference equation is said to be **homogeneous**. If $b \neq 0$ then the linear system is said to be **nonhomogeneous** and it will have a non-zero equilibrium, as in the right panel of the above figure. A similar distinction is made for higher-order linear difference equations; homogeneous equations do not have added constants and always have $x = 0$ as an equilibrium. Homogeneous equations are used to model systems whose dynamics only depend on the system's state and not on external factors.

Steady States.

The limit of a solution sequence, if it exists, is called **the steady state of the solution $\{x_n\}$** and is denoted by x with either a double S subscript, x_{SS}, or an over-bar, $\bar{x}$.

DEFINITION OF A STEADY-STATE.

> The **steady-state** of a particular solution sequence $\{x_n\}$ of a difference equation is the limit
>
> $$x_{SS} = \lim_{n \to \infty} x_n \quad \text{if it exists.}$$

Every equilibrium is a steady state of the constant solution with $x_n = x_E$ for all n. However, a nonstable equilibrium may not be the steady state of any other solution sequence. Every steady-state is also an equilibrium. When a dynamical system has only one equilibrium then this can be the only-steady state for any solution sequence. But, if this equilibrium is not stable then most solutions will not have steady-states; they will either oscillate wildly or diverge to infinity. It is difficult to make general statements about steady-states of discrete dynamical systems. Solutions, starting from the same initial state, of very similar systems may have very different steady states or exhibit chaotic behavior. Indeed, in the study of Chaos, a standard question is "How does the steady-state of a solution change if a model parameter is varied?" For instance if the parameter a in the linear model $x_{n+1} = ax_n + 2$ is increased or decreased from $a = 1$.

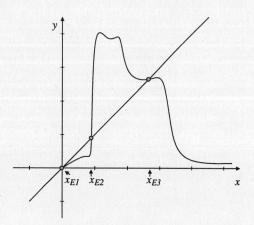
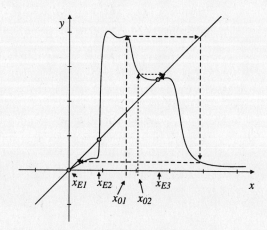

The system illustrated above has three equilibriums corresponding to the three intersections of the graphed function $y = f(x)$ and the reflection line $y = x$. As illustrated in the right panel, two solutions with nearby initial states can have different steady-states. The dotted line indicates the graphical construction of the solution with initial state x_{02}; it approaches the steady-state $\bar{x} = x_{E3}$. The dashed line, with initial state x_{01}, illustrates a solution that approaches the steady-state $\bar{x} = x_{E1} = 0$. The steady-state of a sequence with this system may be difficult to establish. You might construct several other cob-webbing paths on this graph to observe this. By reversing the graphical method, i.e., starting on the graph of $f(x)$, moving to $y = x$, and then back to the graph of $f(x)$, etc., you can establish sequence terms that preceded a given value, e.g., starting at x_n this generates x_{n-1}, then x_{n-2}, ... Using this reverse process, starting at an equilibrium, you should be able to establish solutions that start at other values and after a few steps reach the equilibrium. Such solutions then remain constant.

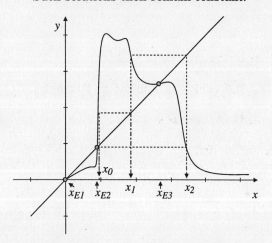

This is illustrated at the left, where the dotted line was constructed starting at the equilibrium x_{E2}, assuming $x_3 = x_{E2}$. The first point generated was $x_2 \approx 3.5$, then $x_1 \approx 1.8$, and finally $x_0 \approx 1$, which is very close to the equilibrium. In this case, x_2 is greater than the third equilibrium value but returns directly to $x_3 = x_{E2}$. Thereafter, x_4 and all further sequence terms will equal the equilibrium x_{E2}.

Observe that x_{E2} is not stable, since all solutions with $x_0 < x_{E2}$ will converge to the steady state of x_{E1}. This illustrates the fact that some solution sequences may converge to a nonstable equilibrium.

Cyclic solutions.

A natural extension of the concept of an equilibrium state is that of an *equilibrium cycle*, in which a solution repeatedly cycles through a fixed number of values. Like equilibrium states, these periodic or cyclic states can be found as the roots of an equation.

> A cyclic solution of length **n** cycles through n-values, which can be found among the roots of the equation x = g(x) where g(x) is the composition of f with itself n-times.

For instance, the numerical values of a cyclic solution of length two are found by solving x = f(f(x)); cycles of length three are found by solving x = f(f(f(x))). These equations normally become very complicated and difficult to solve. For some functions cyclic solutions do not exist, the equations have no real solutions. If a cyclic solution of length n does exist, among the roots of the corresponding equation y = g(x) you will also find all equilibrium states of the system and the values of cyclic solutions of length k where k is any divisor of n. If a ≠ -1 the linear equation $x_{n+1} = ax_n + b$ does not have any cyclic solutions. The linear equation with a = -1, $x_{n+1} = -x_n + b$, has an infinite number of cyclic solutions of length two since for any number K the sequence that alternates between K and -K + b will be a solution of the difference equation.

Example 3   **A cyclic solution of length two.**

Problem   Determine the cyclic solutions of length two for $x_{n+1} = 0.5x_n(8 - x_n)$.

Solution   The solutions are found among the roots of x = f(f(x)) where f(x) = 0.5x(8 - x). Thus
f(f(x)) = 0.5[0.5x(8 - x)](8 - [0.5x(8 - x)]) and, expanding, x = f(f(x)) becomes

$$x = 0.25x \cdot (8 - x) \cdot (0.5x^2 - 4x + 8)$$

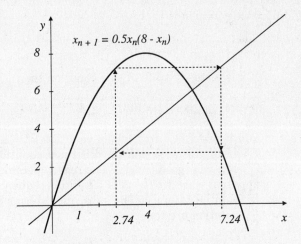

One root of this equation is x = 0, which is also an equilibrium of the difference equation. If x ≠ 0, an x can be canceled to reduce the equation to a cubic. Canceling an x and multiplying by 8, to make all the coefficients integers, an equivalent equation is

$$8 = (8 - x)(x^2 - 8x + 16)$$

or

$$0 = -x^3 + 16x^2 - 82x + 120$$

This cubic equation can be factored:

$$-x^3 + 16x^2 - 82x + 120 = (x - 6)(-x^2 + 10x - 20)$$

Its roots are thus the roots of (x - 6) = 0 and $(-x^2 + 10x - 20) = 0$. The root x = 6 gives the second equilibrium of the difference equation. The roots of the quadratic are x = 5 ± SQRT(5) which are the two values giving a non-trivial two-cycle solution sequence. You may verify by substituting into the difference equation

$$x_{n+1} = 0.5x_n(8 - x_n)$$

that when $x_0 = 5 + \sqrt{5}$ the next term is $x_1 = 5 - \sqrt{5}$ and then the next term is $x_2 = x_0 = 5 + \sqrt{5}$ and so on, alternating between $5 + \sqrt{5} \approx 7.24$ and $5 - \sqrt{5} \approx 2.74$. The geometric construction of this cyclic path is illustrated in the above figure.

Using Difference Equations to Solve Non-linear Equations.

Often the roots of an equation can not be easily found because it is too complicated. This is often the situation when the equation is non-linear. Sometimes the roots can be found using difference equations. Assume the equation has the form $F(x) = 0$. A corresponding difference equation is found by manipulating the original equation into the form $x = f(x)$. Then, consider the difference equation $x_{n+1} = f(x_n)$; its equilibrium values are the roots of the original equation. If the equilibriums are stable, then each one can be found as the steady-state of a solution sequence with an appropriate initial term x_0. Determining the equilibriums algebraically of course requires the solution of the original problem. But, numerical approximations can be found by numerically generating a solution sequence, calculating successive terms (probably on a computer) until the terms remain constant to a fixed number of decimal places. A stable equilibrium root can be approximated in this manner, however, if an equilibrium root is not stable, this approach may not lead to the root.

The difficulty in this approach is to establish the equivalent equation $x = f(x)$ so that a desired root is a stable equilibrium and then selecting an initial value that does not lead to a different root.

Example 4 An algorithm for finding Square Roots.

Problem Use a difference equation to determine the square root of a number A.

Solution The problem is to solve the equation $x^2 = A$. The approach is to determine an equivalent equation of the form $x = f(x)$. The simplest choice for this might be found by simply dividing by x to give $x = A/x$. The corresponding difference equation is then

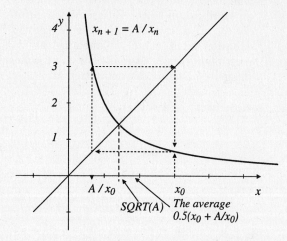

$$x_{n+1} = A / x_n$$

If $\bar{x}$ is a steady state of this equation then

$$\bar{x} = \lim_{n \to \infty} x_{n+1} = \lim_{n \to \infty} A / x_n = A / \bar{x}$$

and hence $\bar{x}$ is a root of the equation $\bar{x}^2 = A$. However, this choice for the difference equation is not a very good choice. The reason becomes apparent when we consider a solution sequence with initial term x_0. The subsequent terms of the solution are then

$$x_1 = A / x_0, \qquad x_2 = A / [A / x_0] = x_0,$$

$$x_3 = A / x_0, \qquad x_4 = A / [A / x_0] = x_0 \ldots$$

Every solution is a cyclic solution of length two. The steady-state of a cyclic solution does not exist! The only solution that will have the $\sqrt{A}$ as its steady-state is the one for which $x_0 = \sqrt{A}$.

The graphical method applied to generate the solution is seen to result in a cycle between x_0 and A/x_0. Geometrically, the equilibrium $x = \sqrt{A}$ is a point between these two values. Thus a better estimate of $\sqrt{A}$ might be given by the average of these two numbers, one half their sum. Indeed, you should verify that another form of the equation $x^2 = A$ is the equation

$$x = 0.5(x + A / x)$$

The corresponding difference equation is then

$$x_{n+1} = 0.5(x_n + A / x_n)$$

$\sqrt{A}$ is also an equilibrium of this equation, but, a stable equilibrium. As illustrated in the following figure, the graph of $y = f(x) = 0.5x(x + A/x)$ is increasing for $x > \sqrt{A}$ and hence whenever x_0 is greater than the root, the geometric solution path moves consistently towards the equilibrium point. In the figure $A = 2$ and $x_0 = 3$. This difference equation, which is known as *Newton's Method* converges to the square root very quickly. For instance, if $A = 6$ and initial term is chosen arbitrarily as A, $X_0 = 6$, then the square root of 6 is found to 9 decimal places after only 4 iterations of the difference equation:

$X_1 = (1/2)(X_0 + \{6 / X_0\}) = 3.5$

$X_2 = 1/2(X_1 + \{6 / X_1\}) = 2.607142857$

$X_3 = 1/2(X_2 + \{6 / X_2\}) = 2.454256360$

$X_4 = 1/2(X_3 + \{6 / X_3\}) = 2.449489743$

$X_5 = 1/2(X_4 + \{6 / X_4\}) = 2.449489743$

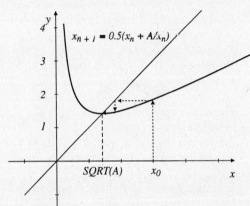

The sketch only illustrates the difference equation for positive x-values. You should sketch the function $0.5(x + A/x)$ for negative x-values to see how the negative square root is generated when x_0 is negative.

Exercise Set 2.3

1. Use a straight edge to estimate the answers to the following questions for the difference equation corresponding to the graph E7.2-1.

   a) What are the equilibriums of the system?
   b) Estimate the value x_0 for which x_1 will be the largest.
   c) If $x_5 = 3$ estimate the value of x_7.
   d) Sketch the graph of $y = \Delta x_n$.
   e) Estimate the value x_n for which the forward difference Δx_n will greatest.

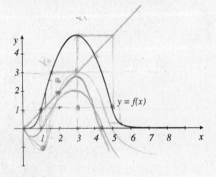

Fig. E2.3-1

2. Use a straight edge to estimate the answers to the following questions for the difference equation corresponding to the graph E7.2-2.

   a) What are the equilibriums of the system?
   b) Estimate the value x_0 for which x_1 will be the smallest.
   c) If $x_5 = 3$ estimate the value of x_4.
   d) Sketch the graph of $y = f(x_n) - 0.5$
   e) Estimate the value x_n for which the forward difference $\Delta x_n = 0.5$.

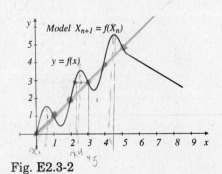

Fig. E2.3-2

3. Determine the equilibrium of the difference equation $X_{n+1} = 0.5X_n + 2$ by sketching its graph. Sketch two "paths" corresponding to initial terms $X_0 = 3$ and $X_0 = 5$. From the graph decide if the equilibrium of this equation is stable.

4 Determine the equilibrium of the difference equation $X_{n+1} = -0.5X_n + 6$ by sketching its graph. Sketch two "paths" corresponding to initial terms $X_0 = 2$ and $X_0 = 5$. From the graph decide if the equilibrium of this equation is stable.

5 Estimate the positive equilibrium of the difference equation $X_{n+1} = -X_n^2 + 9$ by sketching its graph. Sketch two "paths" corresponding to initial terms $X_0 = 2$ and $X_0 = 3$. From the graph decide if the equilibrium of this equation is stable.

6 Estimate the positive equilibrium of the difference equation $X_{n+1} = -X_n(X_n - 4)$ by sketching its graph. Sketch two "paths" corresponding to initial terms $X_0 = 1$ and $X_0 = 3$. From the graph decide if the equilibrium of this equation is stable.

7 The difference equation $X_{n+1} = -X_n(X_n - 4)$ has a cyclic solution of length two. State the equation $x = f(f(x))$ whose roots give the equilibrium and this cyclic solution. Solve for the cyclic solution. Hint: the fourth degree equation will have the factors x and $x - X_E$ where X_E is the equilibrium. Illustrate this cyclic solution on a sketch of the difference equation.

8 Repeat exercise 7 for the equation $Y_{n+1} = -Y_n(Y_n - 6)$, but this time first estimate the cycle two solutions two values from the graph and then find them algebraically.

9 a) Verify that the equation $x_{n+1} = 0.5(x_n + A/x_n)$ established in Example 4 has the steady state $\bar{x} = 3$ when $A = 9$ and $x_0 = 1$. b) What is the steady state when $x_0 = -2$?

10 In Example 4 a difference equation was used to find the square root of a number A.
   a) Using the same technique, determine a difference equation that will have the cube root of A as its equilibrium.
   b) Demonstrate that your equation has $A^{1/3}$ as its steady state by calculating the sequence term x_n for $n = 1, 2, ...$ until they do not differ in 5 decimal places, when $A = 8$ and $x_0 = 1$ and $x_0 = 4$.
   c) What is the limit of your sequence when you use $x_0 = -1$?

11 Construct a difference equation to find the root of the equation $x = 2\sin(x)$. Start with $x_0 = 1.5$ and iteratively find x_n until all your terms agree up to 1.895. What will be the next digit in the 10,000th place (4 decimal places)? How many steps did it take you to reach 4 decimal places of accuracy?

12 Consider the difference equation $x_{n+1} = Ax_n(5 - x_n)$ where A is a positive parameter.
   a) For what values of A will this difference equation only have one equilibrium, $x_{E1} = 0$? Draw a sketch to illustrate graphically why this occurs.
   b) For what value of A will the second equilibrium, $x_{E2} = 2.5$? Will this be a stable or unstable equilibrium? Draw a sketch to illustrate why.
   c) What happens to the second equilibrium as $A \to \infty$? Sketch a graph to illustrate your answer.

13 Consider the difference equation $x_{n+1} = x_n^2 + B$ where B is a parameter.
   a) For what value of B will this equation have only one equilibrium? Sketch a graph to illustrate this system.
   b) When will the system have no equilibriums? Why?
   c) For which B will zero be an equilibrium?
   d) When will the system have only positive equilibriums?

14 Consider the difference equation $x_{n+1} = B + (x_n - 1)^3$ where B is a parameter.
   a) If $B = 0$, how many equilibriums will this system have? Sketch a graph to illustrate this system.
   b) Can this system have no equilibriums? Why?
   c) For which B will zero be an equilibrium? What are the other equilibriums in this case?
   d) Could this system only have two equilibriums?

Section 2.4 Linear First-Order Difference Equations.

Introduction.

In this section the solution of the first-order linear equation $x_{n+1} = ax_n + b$ is found by generalizing the form of the first few terms $x_1, x_2, x_3, \ldots$ to establish a general term x_n. Two cases are considered, first the homogeneous case, with $b = 0$, and then the general nonhomogeneous case where $b \neq 0$. The form of the general solution for x_n is a function of the index n and the model parameters a and b. An inductive principal is used to demonstrate that these do actually provide solutions. The induction argument consists of showing the formula is valid for a starting index, say 0 or 1, and assuming it is true for all indices less than an arbitrary index n, i.e., for all indices i with $i < n$ (in particular for the index n - 1), then showing it also is valid for the index n. In these induction arguments we will use the shifted form of the linear first-order difference equation

$$x_n = ax_{n-1} + b$$

The general formula for the solutions will involve one arbitrary constant, which has been chosen as the initial value of a solution sequence with index $n = 0$. Specific solutions are formed by assigning this constant specific numerical values.

Solutions of the homogeneous equation $x_{n+1} = ax_n$.

The solution of the homogeneous equation $x_{n+1} = ax_n$ can be constructed iteratively, starting with an initial term x_0. The first four terms of this sequence are computed as

$$x_1 = ax_0$$
$$x_2 = ax_1 = a(ax_0) = a^2 x_0$$
$$x_3 = ax_2 = a(a^2 x_0) = a^3 x_0$$
$$x_4 = ax_3 = a(a^3 x_0) = a^4 x_0$$

continuing the pattern established in these evaluations, the equation for the nth term is found by replacing 4 by n in the last equation and omitting the intermediate steps:

$$x_n = a^n x_0$$

The above evaluations demonstrate this formula is valid for the indices 1, 2, 3, and 4. The formula with n replaced by n - 1 is

$$x_{n-1} = a^{n-1} x_0$$

Assuming this n - 1 formula is valid, we show that the formula gives x_n by substituting this expression for x_{n-1} into the shifted form of the difference equation and simplifying:

$$x_n = ax_{n-1} = a(a^{n-1} x_0) = a^n x_0$$

The conclusion, by the principle of mathematical induction, is that the formula is valid for all indices. If the initial term x_0 is not known it is treated as a parameter and the solution is referred to as a *general solution*. When a specific numerical value is assigned to x_0 then the solution is referred to as a *particular solution*. Often the terms are reversed in order so that the power term follows the constant.

The general solution of the
First-order Homogeneous Difference Equation $x_{n+1} = ax_n$

is the geometric sequence $\quad x_n = x_0 a^n$

Example 1   **Solving first-order homogeneous equations.**

Problem   Determine the general solution of the indicated equation and the value of the term x_{10} for the particular solution with $x_0 = 5$.

     a) $x_{n+1} = 2 x_n$ b) $x_{n+1} = 0.1 x_n$ c) $x_{n+1} = -4 x_n$

Solution   In equation a) the coefficient is $a = 2$ thus the general solution is $x_n = 2^n x_0$. The particular solution with $x_0 = 5$ is $x_n = 5 \cdot 2^n$ and thus for $n = 10$, $x_{10} = 5 \cdot 2^{10} = 5{,}120$.

The general solution to equation b) is $x_n = (0.1)^n x_0$. Setting $n = 10$ and $a = 5$, the particular solution has $x_{10} = 5 \cdot (0.1)^{10} = 0.0000000005$.

Difference equation c) has coefficient $a = -4$. Thus its general solution is $x_n = (-4)^n x_0$. Setting $n = 10$ and $x_0 = 5$, the particular solution is $x_{10} = 5 \cdot (-4)^{10} = 5{,}242{,}880$.

Example 2   **Discrete first-order population models: constant birth and death rates.**

Background   The fundamental assumption in discrete population models is that the population size varies from one generation to the next and is controlled by two processes, Births and Deaths. The simplest models assume that these processes occur at constant rates indicated by the positive parameters B and D for births and deaths, respectively. The resulting change in the population is then proportional to the size of the population at the start of a generation cycle. In mathematical vocabulary, the linear model for the change in population is thus the forward difference equation

$$\Delta X_n = B X_n - D X_n \qquad \text{Linear Population Model}$$

Notice the negative sign in front of the death coefficient indicates that the population decreases by the proportional amount.

Problem   Determine the solution of the general linear model of population dynamics.

Solution   The forward difference model is converted into a difference equation of the standard form by substituting $X_{n+1} - X_n$ for ΔX_n and solving for X_{n+1}. The equivalent equation is

$$X_{n+1} = (1 + B - D) X_n$$

This is a linear homogeneous first-order equation with $a = (1 + B - D)$. Its general solution is

$$X_n = (1 + B - D)^n X_0$$

where X_0 denotes the initial population. This is the population at the time or generation that the observer chooses to begin describing the population. Notice that if the birth rate exceeds the death rate, $B > D$, then the model predicts that the population will grow geometrically without bound.

Malthus, considering the human population of the earth, observed in 1798 that this linear model could not be valid as there seemed to be natural limitations to the size of the population, human and animal, that could exist. Yet, this basic model continues to be used to model various populations over short periods of time. It provides a "first-order approximation" of a population's dynamics. More realistic models are developed by replacing the Birth and Death rate constants by functions of the state of the population. If the species being modeled must mature for K generations before they can reproduce then the Birth rate would depend on the population X_{n-K+1} and the model equation would be a difference equation of order K.

Example 3 Compound Interest.

Background A simple and common first-order system is given by the process of compounding interest. The fundamental equation for simple interest states:

$$I = Pr$$

where I is the interest paid at the end of a period, P is the principal amount at the beginning of the period, and r is the interest rate per period. To describe the principal over several periods sequence notation is employed. Let P_0 denote the initial principal deposited and let P_n denote the principal in the account at the end of the nth interest period, after the interest has been paid. P_n is thus the principal at the start of the $(n + 1)^{st}$ interest period. Compounding of interest means that the principal increases by an amount equal to the interest paid each period and no withdrawal of principal occurs. The change in principal is described by the forward difference equation:

$$\Delta P_n = P_n r$$

Substituting $P_{n+1} - P_n$ for ΔP_n and solving for P_{n+1} leads to the equivalent difference equation describing the principal in the account:

$$P_{n+1} = (1 + r)P_n$$

The solution is thus given by

$$P_n = (1 + r)^n P_0$$

Problem What is the value of $1,000 after it is invested at 9% annual interest compounding for five years? How long will it take for this investment to double?

Solution The value of the investment after 5 years is calculated as the term $P_5 = (1 + r)^5 P_0$ with initial principal $P_0 = \$1,000.00$ and the interest rate $r = 0.09$ per year. Thus the value is

$$P_5 = (1.09)^5 \cdot 1,000 = 1,538.62.$$

The doubling time is found by solving $P_n = 2 \cdot P_0$ for n. The solution will probably not be an integer, and the required number of years will be the next higher integer value. The equation

$$(1.09)^n \cdot 1000 = 2,000$$

is solved by dividing by 1000 and taking the logarithm of both sides, using the exponent property of logarithms to move the exponent n to a coefficient of the log term, and dividing. The answer is the same if you use either the natural log, *ln*, or the log base 10, Log, on a scientific calculator:

$$n = ln(2)/ln(1.09) = Log(2)/Log(1.09) \approx 8.04323$$

Since n is greater than 8, it will take 9 years before the principal will compound to exceed $2,000. The principal after 8 and 9 years will be $P_8 = \$1,992.56$ and $P_9 = \$2,171.89$. ☑

Example 4 Effective annual interest rates.

Problem The *nominal annual interest rate* is the rate per year, but if a savings account or loan pays or charges and compounds interest more frequently, say monthly, the actual amount of interest paid over the year is can be calculated as the product of the Principal and the *effective annual interest* rate. How is the *effective interest rate* calculated?

Solution The *effective annual interest* rate is the rate that would yield the same principal (or debt) at the end of the year, assuming a single interest payment based on the principal at the start of the year, as that accumulated with the compounding of the more frequent interest payments.

If I_a denotes the simple, or nominal, annual rate of interest, the rate applicable over a period less than a year is called the *periodic interest rate*. If there are n_p interest periods per year, then the

periodic interest rate is: $I_p = I_a / n_p$

For instance, the monthly interest rate would be $I_a/12$. The principal after compounding such periodic interest payments is given by the simple interest model established in Example 3. With an initial principal P_0 at interest rate $r = I_a/n_p$, the principal after n-periods is

$$P_n = (1 + I_a/n_p)^n P_0$$

Thus, the principal at the end of one year is calculated by setting $n = n_p$:

$$P_{1\,yr} = P_{n_p} = \left(1 + \frac{I_a}{n_p}\right)^{n_p} P_0$$

The *effective annual interest rate*, denoted as I_E, is the rate that will result in the same principal $P_{1\,yr}$ after one annual payment of interest, i.e., such that

$$P_{1\,yr} = (1 + I_E) P_0$$

The effective interest rate is found by equating the right sides of the last two equations:

$$1 + I_E = \left(1 + \{I_a / n_p\}\right)^{n_p}$$

Solving for I_E results in the equation for

effective annual interest rate: $I_E = \left(1 + \{I_a / n_p\}\right)^{n_p} - 1$

If the year-end principal is known, then the effective annual interest rate is simply calculated as $I_E = (P_{1\,yr} / P_0) - 1$. When interim compounding is applied to consumer loans or credit card accounts this is the *annualized* interest rate that must be disclosed on billings. Some accounts actually use a variation of this when interest is charged from the date of purchase on credit card accounts.

To illustrate these equations, consider $1,000 invested at a nominal 10% annual interest rate for one year. Table 1 shows the corresponding effective interest rates and the accumulated principal for one year with different numbers of compounding periods. Observe that the difference between weekly and daily compounding is only ten cents over the year. You might check that compounding every hour would only add one cent to the daily interest over the year.

Table 1

Number of periods: n_p	Principal after one Year $P_{1\,yr}$	Effective annual interest I_E (per cent.)
1, annual	$1,100.00	10.0
12, monthly	$1,104.71	10.4713067
52, weekly	$1,105.06	10.5064793
365, daily	$1,105.16	10.5155781

The nonhomogeneous first-order equation $x_{n+1} = ax_n + b$.

Using the same approach employed for the homogeneous equation, we generate the sequence terms $x_1, x_2, x_3, \ldots$ until a pattern is established from which the general term is inferred. The first few terms of the solution to $x_{n+1} = ax_n + b$ beginning with the initial term x_0 are:

$x_1 = ax_0 + b$

$x_2 = ax_1 + b = a(ax_0 + b) + b = a^2 x_0 + ab + b$

$x_3 = ax_2 + b = a(a^2 x_0 + ab + b) + b = a^3 x_0 + [a^2 b + ab + b]$

$x_4 = ax_3 + b = a(a^3 x_0 + [a^2 b + ab + b]) + b = a^4 x_0 + [a^3 b + a^2 b + ab + b]$

The pattern that emerges is the n-th term x_n is $a^n x_0$ plus the sum of ba^i for $i = 0, 1, 2, \ldots$, up to $n - 1$. If we assume that this formula is valid for the index $n - 1$ the formula is

$$x_{n-1} = a^{n-1} x_0 + [a^{n-1-1} b + a^{n-1-2} b + \ldots + a^2 b + ab + b]$$

or

$$x_{n-1} = a^{n-1} x_0 + [a^{n-2} b + a^{n-3} b + \ldots + a^2 b + ab + b]$$

Substituting this expression for x_{n-1} into the difference equation $x_n = ax_{n-1} + b$ gives the equation

$$x_n = a \cdot \{a^{n-1} x_0 + [a^{n-2} b + a^{n-3} b + \ldots + a^2 b + ab + b]\} + b$$

or

$$x_n = a^n x_0 + [a^{n-1} b + a^{n-2} b + \ldots + a^2 b + ab + b]$$

Which is the form of the formula for index n. Thus, by mathematical induction, we conclude that it is valid for all n. To express this formula in a simpler form, the parameter b is factored out of the sum, which then is a geometric sum of terms a^i that can be evaluated in a "closed form" as

$$a^{n-1} + a^{n-2} + \dots + a^2 + a + 1 = \begin{cases} (1 - a^n)/(1 - a) & \text{if } a \neq 1; \\ n & \text{if } a = 1. \end{cases}$$

The general solution can then be expressed in one of two forms, depending on which closed form of the geometric series is applicable. Again we interchange the terms $a^n x_0$ to get $x_0 a^n$.

The general solution of $x_{n+1} = ax_n + b$ is

$$x_n = X_0 + b \cdot n \qquad \text{if } a = 1,$$

or

$$x_n = x_0 a^n + b \cdot [(1 - a^n)/(1 - a)] \qquad \text{if } a \neq 1.$$

An alternative form of the solution that is often more convenient to use is found by algebraically grouping the terms involving a^n and factoring to give

$$x_n = a^n \{ x_0 - [b/(1 - a)] \} + [b/(1 - a)] \qquad \text{if } a \neq 1.$$

Example 5 **Solving nonhomogeneous linear difference equations.**

Problem State the general solution and a specific solution satisfying the indicated condition.

a) $x_{n+1} = 3x_n - 1$; $x_2 = 14$ b) $x_{n+1} = 0.2x_n + 2.4$; $x_3 = 10$ c) $x_{n+1} = x_n + 2$; $x_3 = 5$

Solutions a) The parameters are $a = 3$ and $b = -1$. The general solution is

$$x_n = 3^n x_0 + (-1)(1 - 3^n)/(1 - 3) = 3^n x_0 + (1/2)(1 - 3^n)$$

The particular solution is found by setting $n = 2$ in the general solution

$$x_2 = 3^2 x_0 + (1/2)(1 - 3^2) = 9x_0 - 4$$

and solving $x_2 = 14$ for $x_0 = 2$. The particular solution is thus

$$x_n = 2 \cdot 3^n + (1/2)(1 - 3^n)$$

b) The general solution is expressed in the second form, with $a = 0.2$ and $b = 2.4$:

$$x_n = 0.2^n \{ x_0 - 2.4/(1 - 0.2) \} + 2.4/(1 - 0.2) = 0.2^n \{ x_0 - 3 \} + 3$$

Setting $n = 3$, we solve $x_3 = 5$ for x_0 to obtain the required particular solution. The solution of

$$x_3 = 0.2^3 \{ x_0 - 3 \} + 3 = 5$$

is $x_0 = 253$. Thus, the particular solution is $x_n = 253 \cdot 0.2^n + 3$.

c) The x_n coefficient is a = 1, therefor invoking the special case of the general solution. As b = 2 this gives $x_n = x_0 + 2n$. Solving $x_3 = x_0 + 2 \cdot 3 = 5$ gives $x_0 = -1$ and hence the desired particular solution is $x_n = -1 + 2n$.

☑

Example 6 Funding an annuity.

Background An *annuity* is a very important type of financial contract in which an amount of money is deposited into an interest bearing account and a fixed payment is subtracted from the account for a specified number of periods. Annuities traditionally were used for providing pre-paid retirement benefits, but are now frequently used in a variety of settings to provide guaranteed payments over a set number of periods. For example, annuities are used to spread out the payment of very high salaries of super-star athletes or top corporate executives, they are also the way most major lottery winners are paid. As we will see in this example, the amount needed to fund an annuity is often much less than the "face value" of the total amount to be paid out. The mathematical equation describing the principal in an annuity account is a forward difference equation. Denoting the principal remaining in an annuity account after the nth payment period by P_n, the model equation is

$$\Delta P_n = r \cdot P_n - A$$

where r is the periodic interest rate and A is the amount of the annuity paid out each period. (If the annuity period and the interest period are different then the interest rate is adjusted to the annuity period as the effective annual interest rate was calculated in Example 4.) Expanding the difference gives

$$P_{n+1} = (1 + r) \cdot P_n - A$$

The general form of the solution, with a = 1 + r and b = -A, gives the annuity principal after n periods:

$$P_n = (1 + r)^n P_0 - A \cdot \{1 - (1 + r)^n\} / \{1 - (1 + r)\} = (1 + r)^n P_0 - \{(1 + r)^n - 1\} \cdot A/r$$

It is assumed that the interest rate r is not zero. The alternative form for this general solution is

$$P_n = (1 + r)^n [P_0 - A/r] + A/r$$

Problem a) What is the maximum annual annuity that could be paid by a deposit of $50,000 earning interest at 8% annually for 20 years?

b) How much money must a Lottery Corporation deposit to pay a lottery winner $3,000,000, to be paid in monthly annuities of $10,000 for 25 years if the annual interest rate is 7%?

Solution a) The maximum annuity that can be paid for n periods with an initial deposit P_0 is obtained by setting the final principal to zero, $P_n = 0$, in the above equation and solving for the amount A:

The Maximum Annuity for n payments from a deposit of P_0 is

$$A = r(1 + r)^n P_0 / \{(1 + r)^n - 1\}$$

where the periodic interest rate is r. Thus, a deposit of P_0 = $50,000 at 8% annual interest, r = 0.8, will generate an annual annuity for n = 20 years of

$$A = 0.08(1 + 0.08)^{20} \cdot \$50,000 / \{(1 + 0.08)^{20} - 1\} = \$5,092.61$$

The total amount that will be paid in annuities is $20 \cdot A = \$101,852.20$.

b) To determine the initial principal needed to fund a specific annuity for a fixed number of periods the maximum annuity equation is solved for P_0 to give:

Required Initial Deposit: $P_0 = [\{(1 + r)^n - 1\} / \{r(1 + r)^n\}] \cdot A = [1 - (1 + r)^{-n}] \cdot A/r$

To fund a lottery winner of $3,000,000 in monthly annuities of $10,000 for 25 years the Lottery Corporation does not actually payout three million dollars. Instead it finances the payments as an annuity with a deposit earning interest at the simple annual interest of 7% per year. To calculate the amount of deposit needed, since the actual compounding period is one month, we first convert the annual interest of 7% into a monthly interest rate:

$$r = 0.07/12 = 0.0058333$$

The number of months required is $n = 12 \cdot 25 = 300$. Thus, the formula for Required Initial Deposit gives

$$P_0 = [\{1 - (1 + 0.0058333)^{-300}\}] \cdot 10{,}000 / 0.0058333 = \$1{,}414{,}869.03$$

This is about 47% of the putative $3 million payment.

Qualitative aspects of linear difference equations.

As introduced in the previous Section 2.3, the linear first-order equation $x_{n+1} = ax_n + b$, with $a \neq 1$, has the equilibrium value

$$x_E = b/(1 - a)$$

The steady state any solution to this equation will therefor be the equilibrium value $b/(1 - a)$, infinity, or not exist. Which occurs will depend on the coefficient a and the initial constant x_0. This is seen directly from the general solution

$$x_n = a^n \cdot [x_0 - b/(1 - a)] + b/(1 - a)$$

Recall that the limit of a^n is finite if $|a| \leq 1$ and infinite otherwise:

$$\lim_{n \to \infty} a^n = \begin{cases} \infty & \text{if } a > 1; \\ 1 & \text{if } a = 1; \\ 0 & \text{if } -1 < a < 1; \\ \text{Does Not Exist} & \text{if } a \leq -1, \end{cases}$$

However, regardless of the value of the parameter a, if the solution sequence starts at the equilibrium it will remain there. There are four possible cases for the steady-state of a solution.

The Steady State of a solution to $x_{n+1} = ax_n + b$ when $a \neq 1$ is

$$\bar{x} = \lim_{n \to \infty} x_n = \begin{cases} +\infty & \text{if } a > 1 \text{ and } [x_0 - b/(1 - a)] > 0; \\ -\infty & \text{if } a > 1 \text{ and } [x_0 - b/(1 - a)] < 0; \\ b/(1 - a) & \text{if } |a| < 1 \text{ or } x_0 = b/(1 - a); \\ \text{Does Not Exist} & \text{if } a \leq -1 \text{ and } x_0 \neq b/(1 - a). \end{cases}$$

In the third case the solution approaches the equilibrium, and in the fourth case there is no limit because powers of the negative constant a will oscillate between positive and negative values. When a = 1 the general solution is simply the arithmetic sequence $x_n = x_0 + b \cdot n$. The steady-state limit of this sequence is plus or minus infinity, depending on the sign of b.

The Steady State of a solution to $x_{n+1} = x_n + b$ is

$$\bar{x} = \lim_{n \to \infty} x_n = \begin{cases} +\infty & \text{if } b > 0; \\ -\infty & \text{if } b < 0; \\ x_0 & \text{if } b = 0. \end{cases}$$

Example 7   **Steady-states of linear difference equations.**

Problem   Utilize the appropriate case of the above limit formulas to determine the solution's steady-state for the solution of the given equation with the prescribed initial values.

Solution   a) $Y_{n+1} = 3Y_n - 2$; $Y_0 = 1$.
The initial value is the equilibrium value $\underline{Y}_E = b/(1-a) = -2/(1-3) = 1$.
Although a = 3, b = -2 the steady state is $\bar{Y} = 1$.

b) $x_{n+1} = 5x_n + 2$; $x_0 = 1$.
Since a = 5 > 1 and $x_0 - b/(1-a) = 1 - 2/(1-5) = 1.5 > 0$, $\bar{x} = \infty$.

c) $Z_{n+1} = (1/2)Z_n - 3$; $Z_0 = 2$.
As $|a| = 1/2 < 1$, the steady-state is the equilibrium value $\bar{Z} = Z_E = -3/(1 - 1/2) = -6$.

d) $Y_{n+1} = (-1/3)Y_n + 8$; $Y_0 = 12$.
As $|a| = 1/3 < 1$, the steady-state is the equilibrium value $\bar{Y} = 8/(1 - (-1/3)) = 6$.

e) $x_{n+1} = 0.25x_n + 1$; $x_0 = 2$.
As $|a| = 0.25 < 1$, the steady-state is $\bar{x} = 4/3$, which is the equilibrium $x_E = 1/(1 - 0.25)$.

f) $W_{n+1} = W_n + 2$; $W_0 = 8$.
In this case the coefficient a = 1 so that the second type of evaluation is considered. As the coefficient b = 2 is positive the steady-state is $\bar{W} = \infty$.

g) $X_{n+1} = X_n - 10$; $X_0 = 2$.
As b is negative and a = 1 the steady-state is $\bar{X} = -\infty$.

Example 8   **Repeated Drug Administration.**

Problem   A drug is administered in a dose D every Δt hours. The amount of drug in the patient's body is assumed to decay exponentially with a decay rate k. What is the amount of drug in the patient at the end of the n-th administration interval? If the dosage is 4 mg every 5 hours and the drug metabolizes at a rate of $k = ln(3)$ per hour, how much drug will be in the patient after 35 hours?

Solution   Assume that the administration time points are $t_i = i \times \Delta t$:

$$t_0 = 0, \ t_1 = \Delta t, \ t_2 = 2\Delta t, \ \ldots, \ t_n = n\Delta t, \ t_{n+1} = (n+1)\Delta t, \ \ldots.$$

The n-th administration interval is $I_n = [t_{n-1}, t_n]$.

Let B_n denote the amount of drug in the patient at the start of the n-th interval.

Let A_n denote the amount of drug in the body at the end of the n-th interval.

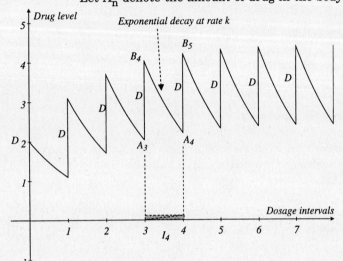

Our problem is to determine the sequence terms A_n and B_n. We will do this by establishing a difference equation that they satisfy and then solving it using the general formula.

The amount of drug at the start of an interval is the amount residual from the previous interval plus the new dosage of drug. For the n-th interval I_n:

$$B_n = A_{n-1} + D$$

On the next interval, I_{n+1}:

$$B_{n+1} = A_n + D$$

Over the interval I_n the amount of drug in the body is modeled by the exponential decay equation

$$y_n(t) = B_n e^{-k(t - t_{n-1})}$$

This is the basic exponential decay curve $y = e^{-kt}$ shifted to the right t_{n-1} units. The curve y_n is the exponential curve that decays at rate k and passes through the point (t_{n-1}, B_n). The patient's drug level at the end of the n-th interval is then the one sided limit of y_n as $t \to t_n^-$. Since the exponential function is continuous, this limit is just the value of the exponential at the limit point:

$$A_n = \lim_{t \to t_n^-} y_n(t) = B_n e^{-k(t_n - t_{n-1})}$$

But, the difference $t_n - t_{n-1} = \Delta t$, the common length of the dosage interval. Hence, the value of A_n is a constant times B_n. The constant is $C = e^{-k\Delta t}$:

$$A_n = B_n \times e^{-k \Delta t}$$

Combining this equation with the drug addition equation gives a difference equation for A_n:

$$A_n = (A_{n-1} + D)e^{-k\Delta t}$$

or

$$A_{n+1} = e^{-k\Delta t} A_n + D e^{-k\Delta t}$$

This is a linear first-order difference equation, that has the standard form

$$A_{n+1} = a A_n + b$$

with $\quad a = e^{-k\Delta t} \quad$ and $\quad b = D e^{-k\Delta t}$

Its general solution is given by

$$A_n = A_0 a^n + b(1 - a^n) / (1 - a)$$

where A_0 is the amount of drug in the patient at the start of the drug therapy. As $(e^{-k\Delta t})^n = e^{-nk\Delta t}$, this solution is

$$A_n = A_0 e^{-nk\Delta t} + D e^{-k\Delta t}(1 - e^{-nk\Delta t})/(1 - e^{-k\Delta t})$$

or

$$A_n = A_0 C^n + DC(1 - C^n)/(1 - C) \qquad C = e^{-k\Delta t}$$

The residual drug in the body at the end of each dosage interval will approach the steady state level

$$\bar{A} = b/(1 - a) = D e^{-k\Delta t}/(1 - e^{-k\Delta t}) = D/(e^{k\Delta t} - 1)$$

If the patient had no drug before the administration began we set $A_0 = 0$ and

$$A_n = D(1 - e^{-nk\Delta t})/(e^{k\Delta t} - 1)$$

For example, if the dosage is $D = 4$mg and the decay rate is $k = ln(3)$ hours^{-1} with drug administered every 5 hours, $\Delta t = 5$, the patient's lowest drug level will approach the steady state level

$$\bar{A} = D/(e^{k\Delta t} - 1) = 4/(e^{ln(3) \cdot 5} - 1) \approx 0.162895$$

This level is reached very quickly as the exponential term $e^{-nk\Delta t}$ approaches zero. For instance,

$$A_1 = 0.16460905, \quad A_2 = 0.016528646 \quad \text{and} \quad A_3 = 0.016528924$$

For $n = 3$ this exponential term $e^{-nk\Delta t}$ is only 0.000000070 or $6.969 \cdot 10^{-8}$. After 35 hours, i.e., at the end of the seventh dosage interval, the residual amount of drug in the patient would be

$$A_7 = 4C(1 - C^7)/(1 - C) \approx 0.016393165 \text{ mg}$$

as $C = e^{-ln(3) \cdot 5} = e^{-5.49} \approx 0.004115226338$. A_n is the residual amount of drug in the body at the end of the interval. The amount at the beginning of the next interval is $B_{n+1} = A_n + D$.

$$B_{n+1} = D + A_n = D + C \cdot D(1 - C^n)/(1 - C) \qquad C = e^{-k \cdot \Delta t}$$

☑

Stability Criteria for Equilibriums.

The steady state of a first-order difference equation equals the equilibrium, for any initial state x_0, only if the coefficient a is between -1 and 1:

$$\bar{x} = b/(1 - a) \text{ if } |a| < 1$$

Thus,

The equilibrium $x_E = b/(1 - a)$

of $x_{n+1} = ax_n + b$

is stable when $|a| < 1$

When $0 < a < 1$, every solution $\{x_n\}$ will converge directly to x_E. If a is negative, $-1 < a < 0$, then the solution $\{x_n\}$ will converge to x_E in an oscillatory fashion, alternating between values greater than and less than x_E. The stability of equilibriums for nonlinear systems is analyzed by replacing the system by a linear approximation (Sections 4.8 and 4.9), near the equilibrium, and applying this test for stability to the linear system. The general result is that when $x_{n+1} = f(x_n)$ has an equilibrium x_E, the equilibrium is stable when $|f'(x_E)| < 1$, where $f'(x)$ is the *derivative* of f, which is introduced in Chapter 3.

Exercise Set 2.4

1. State the general solution of the given equation and the value of the x_{10} for the indicted initial value.

   a) $X_{n+1} = -X_n$ $X_0 = 2$
   b) $X_{n+1} = X_n$ $X_0 = 2$
   c) $X_{n+1} = 2X_n$ $X_0 = 1$
   d) $X_{n+1} = 3X_n$ $X_0 = 5$
   e) $X_{n+1} = 0.1X_n$ $X_0 = 0.5$
   f) $X_{n+1} = -0.25X_n$ $X_0 = 2$

2. Determine the general solution of the given equation and x_{10} for the indicted initial value.

   a) $\Delta X_n = 0$ $X_0 = 2$
   b) $\Delta X_n = X_n$ $X_0 = 2$
   c) $\Delta X_n = 2X_n$ $X_0 = 0$
   d) $\Delta X_n = 3X_n$ $X_0 = 5$
   e) $\Delta X_n = 0.1X_n$ $X_0 = 0.5$
   f) $\Delta X_n = -0.25X_n$ $X_0 = 2$
   g) $\Delta X_n = -0.5X_n$ $X_0 = 1$
   h) $\Delta X_n = -1.3X_n$ $X_0 = 2$

3. What is the value of $500 invested at 4% simple annual interest compounded for 25 years? How long would it take this compounding deposit to reach $100,000? See Example 4.

4. What is the effective annual interest rate of a credit card account that charges 24% per year compounded monthly?

5. Some loans charge interest "in advance". The formula for this is $I = (P + I)r$ where P is the principal, r is the periodic interest rate, and I is the interest paid at the start of the period that includes interest on itself. Write a difference equation to calculate compound interest "paid in advance". Compare the interest accrued on a loan of $10,000 over a one year time at 10% annual rate when the interest is applied "in advanced" to the more common "not in advanced" interest. Assume no payments against the principal is applied. Compare the difference if the interest is compounded monthly.

6. Give the general solution of the given equation and the value of the x_5 for the indicted initial value.

   a) $X_{n+1} = X_n + 1$ $X_0 = 2$
   b) $X_{n+1} = -2X_n + 3$ $X_0 = 2$
   c) $X_{n+1} = 6X_n + 4$ $X_0 = 1$
   d) $X_{n+1} = -20X_n - 2$ $X_0 = 5$
   e) $X_{n+1} = 0.25X_n - 1$ $X_0 = 0.5$
   f) $X_{n+1} = 0.25X_n - 0.5$ $X_0 = 2$

7. Determine x_5 and the general solution of the given equation for the indicted initial value.

   a) $\Delta X_n = 1$ $X_0 = 2$
   b) $\Delta X_n = X_n - 1$ $X_0 = 2$
   c) $\Delta X_n = 2X_n - 1$ $X_0 = 0$
   d) $\Delta X_n = 3X_n + 3$ $X_0 = 5$
   e) $\Delta X_n = 0.1X_n + 0.2$ $X_0 = 0.5$
   f) $\Delta X_n = -0.25X_n - 1.5$ $X_0 = 2$

Section 2.4 LINEAR FIRST-ORDER DIFFERENCE EQUATIONS

8. Determine the steady state of the solution to the given equation and initial condition.

   a) $X_{n+1} = 0.25X_n - 1 \quad X_0 = 0.5$
   b) $X_{n+1} = 0.25X_n - 0.5 \quad X_0 = 2$
   c) $X_{n+1} = 0.5X_n - 1 \quad X_0 = 3$

   d) $X_{n+1} = 2X_n + 0.5 \quad X_0 = 2$
   e) $X_{n+1} = 5X_n - 1 \quad X_0 = 0.25$
   f) $X_{n+1} = -0.2X_n + 0.5 \quad X_0 = -1$

   g) $X_{n+1} = X_n + 0.1 \quad X_0 = 0.25$
   h) $X_{n+1} = X_n + 0.5 \quad X_0 = -1$

9. What is the maximum annual annuity that could be paid by a deposit of $100,000 earning interest at 4% annually for 25 years?

10. How much money must a Lottery Corporation deposit into an annuity account to pay a lottery winner $7,200,000, to be paid in monthly annuities of $24,000 per month for 25 years if the annual interest rate is 7%? What if the interest is compounded monthly?

11. Derive a discrete model of an epidemic in which no one recovers from the disease and each infective person has a 20% chance of infecting another person each day.

12. An ant colony increases in size each day. Assume the natural birth rate per 100 ants exceeds the natural death rate and their difference is R ants per day. A predator kills 150 ants each day. Construct a model equation describing the number of ants in the colony. What is the minimum colony size necessary for the colony not to become extinct? If the rate R = 10, how long will it take for a colony of 2000 ants to double in size?

13. Assume a population is modeled by $Y_{n+1} = r \cdot Y_n + k$.

    a) If $k = 3/2$, $r = 1.1$, and $Y_{10} = 1000$ what was its initial population Y_0?

    b) If $Y_1 = 10$ and $Y_8 = 10^4$ and $k = 0$ what is the value of r?

    c) If $\bar{Y} = 50$ and $k = 0.1$ and $\Delta Y_3 = 20$, what is Y_0?

14. A government wildlife agency wishes to manage the size of the deer population because the deer have few natural predators and efficiently reproduce so their numbers can increase beyond the size that natural food supply can sustain. Assume that the birth rate is 1.2 per doe per year and that doe and buck fawns are equally likely to be born. Assume a natural death rate of 0.2 per year.

    a) Construct a model describing the natural deer population.

    b) Construct a model with a constant removal of deer each year.

    c) If the region deer population is 3000 and 150 deer (randomly selected) are removed each year what will be the population after 10 years?

    d) What will be the steady state of this population? e) Can this population approach a steady state population of 2500, with or without hunting, if the birth and death rate parameters are altered?

Section 2.5 Higher-Order Difference Equations.

Introduction

Higher-order difference equations arise in dynamical system models with "lags" or age-structure in which maturation periods delay the influence of newborns until they reach a reproductive age. Recall that a difference equation of order K involves terms whose indices differ by K. The *standard form* of the general K^{th}-order difference equation is

$$x_{n+1} = F(x_n, x_{n-1}, x_{n-2}, \ldots, x_{n-K+1})$$

However, linear equations are normally expressed with the indices shifted up so that the lowest index is n and the highest is n + K. The *normal form* of this equation when F is linear is written as a sum of sequence terms, starting with X_{n+K} (having a lead coefficient of one) plus K - 1 lower order terms with coefficients, denoted by $A_1, A_2, \ldots, A_{K-1}$. The last term is thus $A_K X_n$:

K^{th}-order linear difference equation (Normal Form)

$$X_{n+K} + A_1 X_{n+K-1} + A_2 X_{n+K-2} + A_3 X_{n+K-3} + \ldots + A_{K-1} X_{n+1} + A_K X_n = \gamma$$

where γ is the lower case Greek letter gamma. If $\gamma = 0$ then the equation is said to be **homogeneous**. For K = 2, the normal form is usually written with coefficients a and b (rather than the subscripted A_1 and A_2):

$$X_{n+2} + a X_{n+1} + b X_n = \gamma$$

In this section we assume that the coefficients are constant, in which case a solution can be found by finding roots of an associated quadratic Auxiliary Equations. The same method can be used to solve higher-order linear constant-coefficient equations, but, entails finding the roots of higher degree polynomials. In theory, the general solution of a Kth-order difference equation will involve K arbitrary parameters, $C_1, C_2, \ldots, C_K$ that can be determined for any particular solution by specifying K sequence values. Typically, the first K sequence terms $X_0, X_1, X_2, \ldots, X_{K-1}$ are used.

SECOND-ORDER DIFFERENCE EQUATIONS

Homogeneous 2nd-order equations.

A sequential event whose change is only influenced by the current and the previous state of the event can usually be modeled by a homogeneous second-order difference equation. When there is no external influence the model will be homogeneous, and if the change is proportional to the previous states the model will be linear, resulting in a

Homogeneous Linear Second-order Difference Equation

$$X_{n+2} + a X_{n+1} + b X_n = 0 \qquad \text{H-DE}$$

If the constant b = 0 this equation reduces to a first-order homogeneous equation, $x_{n+2} + a x_{n+1} = 0$ or, shifting the indices down by one, in the standard form $x_{n+1} = -a x_n$. Thus, we would expect that the solution of the second-order equation should be similar to and an extension of the solution to this first-order equation, which is $x_n = x_0(-a)^n$. The approach to solving the second-order equation is to assume that its solution similarly involves a term raised to the power n, however, not simply the coefficient -a. An unknown parameter, λ (Lambda), is introduced. Consider the sequence defined by

$$X_n = C \lambda^n$$

where C and λ are constants to be determined.

Our objective is to find values of the constants C and λ that will make the sequence $\{X_n\}$ a solution of the difference equation. According to this formula, the $(n + 1)^{st}$ and $(n + 2)^{nd}$ terms of the sequence are:

$$X_{n+1} = C \lambda^{n+1} \quad \text{and} \quad X_{n+2} = C \lambda^{n+2}$$

Substituting these into the difference equation $X_{n+2} + aX_{n+1} + bX_n = 0$ gives

$$C \lambda^{n+2} + a \cdot C \lambda^{n+1} + b \cdot C \lambda^n = 0$$

which can be factored as

$$C \cdot \lambda^n \cdot [\lambda^2 + a \cdot \lambda + b] = 0$$

This equation is valid when one of the factors on the left is zero. If $C = 0$ or $\lambda = 0$ the solution is said to be *trivial*, since then each sequence term is zero. For any constant C, the sequence $x_n = C \cdot \lambda^n$ will be a solution sequence whenever the factor in square brackets is zero, i.e., when the parameter λ satisfies the

Auxiliary equation: $\quad \lambda^2 + a \cdot \lambda + b = 0$

The two roots of this quadratic equation are

$$\lambda_1 = (1/2)(-a + \text{SQRT}\{a^2 - 4b\})$$

and

$$\lambda_2 = (1/2)(-a - \text{SQRT}\{a^2 - 4b\})$$

For each root the sequence $\{C \cdot \lambda^n\}$ is a solution. The general solution will be a linear combination of these, which can be written as $\{C_1 \lambda_1^n + C_2 \lambda_2^n\}$ where C_1 and C_2 are arbitrary constants, one of which must be non-zero to avoid the trivial solution. This is because linear equations have the property that the sum of any two solutions is also a solution. However, there are actually three different forms of the general solution, corresponding to the three types of roots that a quadratic equation can have.

Remember that a quadratic will have either two distinct real roots, two identical or repeated roots, or no real roots. While in this text we have normally avoided reference to non-real *complex* numbers, we will have to utilize them here. The character of the roots λ_1 and λ_2 of the auxiliary equation will depend on the value of the *discriminant* $D = a^2 - 4b$. The roots are:

both real and distinct if $D = a^2 - 4b > 0$;

both real and equal if $D = a^2 - 4b = 0$;

complex conjugates if $D = a^2 - 4b < 0$.

Consequently, depending on the value of D, the general solution has one of three possible forms.

When $D = a^2 - 4b$ is positive,
the roots are real and distinct and the general solution is the sum of $C_1 \lambda_1^n$ and $C_2 \lambda_2^n$:

$$X_n = C_1 \lambda_1^n + C_2 \lambda_2^n$$

If $\lambda_1 = \lambda_2$ the sum $C_1 \lambda_1^n + C_2 \lambda_2^n$ can be expressed as a single term $C_3 \lambda_1^n$ where $C_3 = C_1 + C_2$:

$$C_1 \lambda_1^n + C_2 \lambda_2^n = C_1 \lambda_1^n + C_2 \lambda_1^n = (C_1 + C_2) \lambda_1^n = C_3 \lambda_1^n$$

We know that two parameters are required in the general solution of a second-order difference equation, but adding the two λ^n terms results in a term with only one constant. In this case the form is modified by multiplying one of the coefficients by the index n.

When $D = a^2 - 4b = 0$,
The repeated roots are $\lambda_1 = -a/2$. The general solution is then

$$X_n = (C_1 + n \cdot C_2) \lambda_1^n$$

When $D = a^2 - 4b < 0$,
the general solution, which is given below, is expressed in terms of two new parameters, r and θ. These are the *complex modulus* and *complex angle* of the *complex* roots, which can be expressed as $\lambda = re^{\pm i\theta}$. The modulus r is simply SQRT(b), but the angle θ must be found by solving the cosine equation $\cos(\theta) = -a/2r$. A discussion of how this formula is established is given following the examples illustrating the solutions of specific difference equations.

You will not need any knowledge of *complex* numbers to work the exercises.

The **general solution** of the *linear constant coefficient homogeneous second-order difference equation*

$$x_{n+2} + a x_{n+1} + b x_n = 0$$

has one of three forms, depending on the roots λ_1 and λ_2 of the

Auxiliary Equation $\quad \lambda^2 + a\lambda + b = 0$

If λ_1 and λ_2 are real and unequal:

$$X_n = C_1 \lambda_1^n + C_2 \lambda_2^n$$

If $\lambda_1 = \lambda_2$;

$$X_n = (C_1 + n \cdot C_2) \lambda_1^n$$

If λ_1 and λ_2 are complex:

$$X_n = [\, C_1 \cos(n\theta) + C_2 \sin(n\theta)\,] \, r^n$$

$$r = \sqrt{b} \quad \text{and} \quad \cos(\theta) = -a / \{\, 2\sqrt{b}\,\}$$

Example 1   **A general solution with distinct roots.**

Problem   Determine the general solution to the equation

Section 2.5 — HIGHER-ORDER DIFFERENCE EQUATIONS

$$X_{n+2} + 7 X_{n+1} + 12 X_n = 0$$

Solution The parameters in the difference equation are $a = 7$ and $b = 12$.
As the $D = a^2 - 4b = 1 > 0$ there will be two distinct real roots of the auxiliary equation

$$\lambda^2 + 7\lambda + 12 = 0$$

The two roots are $\lambda_1 = -3$ and $\lambda_2 = -4$. The general solution is therefor

$$X_n = C_1(-3)^n + C_2(-4)^n$$

Example 2 A general solution with repeated roots.

Problem Determine the general solution to the equation

$$Y_{n+2} - 6 Y_{n+1} + 9 Y_n = 0$$

Solution The equation $Y_{n+2} - 6 Y_{n+1} + 9 Y_n = 0$, has coefficients $a = -6$ and $b = 9$ and the discriminant $D = a^2 - 4b = 0$. The roots of the auxiliary equation $\lambda^2 - 6\lambda + 9 = 0$ are real and repeated:

$$\lambda_1 = \lambda_2 = 3$$

The general solution is then given by the second case:

$$Y_n = [C_1 + C_2 n]3^n$$

Example 3 A general solution with *complex* roots.

Problem Determine the general solution to the equation

$$Z_{n+2} - 2 Z_{n+1} + 4 Z_n = 0$$

Solution The equation parameters are $a = -2$ and $b = 4$. The discriminant is then negative

$$D = a^2 - 4b = -12 < 0$$

The roots of the auxiliary equation

$$\lambda^2 - 2\lambda + 4 = 0$$

are *complex* numbers: $\lambda_1 = 1 + \sqrt{3}i$ and $\lambda_2 = 1 - \sqrt{3}i$. These roots are not required to state the general solution; the required parameters are

$$r = \sqrt{b} = 2 \quad \text{and} \quad \cos(\theta) = \{-a / 2\sqrt{b}\} = 1/2$$

$$\theta = \cos^{-1}(1/2) = \pi/3 \text{ radians}$$

Caution: Do not evaluate θ on a calculator in DEGREE mode, use the RADIAN mode!

The general solution is therefore given by

$$Z_n = 2^n[\, C_1 \cos(n\pi/3) + C_2 \sin(n\pi/3)]$$

The general solution of the second-order difference equations involves two arbitrary parameters, the constants C_1 and C_2. The values of these constants are determined by specifying any two terms of a solution. The indices 0 and 1 are often used as they frequently result in the simplest algebra.

☑

Example 4 Determining Specific Solutions.

Problem   For the difference equations considered in Examples 1-3 determine the constants C_1 and C_2 so that the solution sequence has values 10 and 5 when $n = 0$ and 1, respectively.

Solution   The general solution in **Example 1** was $X_n = C_1(-3)^n + C_2(-4)^n$. Setting $n = 0$ and $n = 1$ gives two equations that are solved simultaneously for the solution constant C_1 and C_2:

$$n = 0 \quad X_0 = C_1 + C_2 = 10$$

$$n = 1 \quad X_1 = -3C_1 - 4C_2 = 5$$

Solving these equations simultaneously, we find $C_1 = 45$ and $C_2 = -35$. The particular solution is therefor

$$X_n = 45(-3)^n - 35(-4)^n$$

For **Example 2**, the general solution was $Y_n = [C_1 + C_2 n]3^n$. The specified values are

$$Y_0 = C_1 = 10 \quad Y_1 = [C_1 + C_2] \cdot 3 = 5$$

The simultaneous solutions are $C_1 = 10$ and $C_2 = -25/3$. The particular solution is thus

$$Y_n = [10 - 25n/3]3^n$$

In **Example 3** the general solution was

$$Z_n = 2^n[\, C_1 \cos(n\pi/3) + C_2 \sin(n\pi/3)]$$

The values of the trigonometric functions for $n = 0$ and 1 are evaluated first:

$$n = 0 \quad \cos(0) = 1 \quad \sin(0) = 0$$
$$n = 1 \quad \cos(\pi/3) = 1/2 \quad \sin(\pi/3) = \sqrt{3}/2$$

Therefore, the first two sequence terms are

$$Z_0 = C_1 = 10 \quad \text{and} \quad Z_1 = 2[C_1/2 + C_2\sqrt{3}/2)] = 5$$

The constants are thus $C_1 = 10$ and $C_2 = -5/\sqrt{3}$. Hence, the particular solution of the difference equation is

$$Z_n = 2^n[10 \cos(n\pi/3) - (5/\sqrt{3}) \sin(n\pi/3)]\, 2^n$$

☑

Example 5 The Fibonacci Sequence

Background The Italian scientist Leonardo Fibonacci (1170-1240) attempted to model the size of a rabbit population as a sequence. He assumed that each pair of rabbits two years old reproduced another pair. According to this model, at generation or year n, the number of rabbit pairs to be found in the next year would be the current number plus the number in the previous year. The difference equation for this model is $X_{n+1} = X_n + X_{n-1}$. Starting with a single pair that does not reproduce until the second year, he generated a solution sequence 1, 1, 2, 3, 5, 8, ... which is called a *Fibonacci Sequence*. Surprisingly, this same sequential relationship is observed in many biological situations. This simple sequence and its generalizations have stimulated a wide range of mathematical study, there is even a research journal called the *Fibonacci Quarterly* devoted to mathematical results linked to the Fibonacci Equation.

Problem Determine the general solution of the Fibonacci difference equation $X_{n+1} = X_n + X_{n-1}$ and the particular solution sequence with $X_0 = 1$ and $X_1 = 1$.

Solution First, the index of the difference equation is shifted and the equation is rearranged to the normal form of a homogeneous linear second order difference equation:

$$X_{n+2} - X_{n+1} - X_n = 0$$

The Auxiliary Equation, with $a = -1$ and $b = -1$, is $\lambda^2 - \lambda - 1 = 0$. Its roots are real and distinct, as $a^2 - 4b = 5$,

$$\lambda = 0.5(1 \pm \sqrt{5})$$

The general solution of the Fibonacci equation is thus

$$X_n = C_1 (0.5(1 + \sqrt{5}))^n + C_2 (0.5(1 - \sqrt{5}))^n$$

To obtain the solution with integer values starting with sequence terms 1, 1, we set $n = 0$ and $n = 1$ in the general solution equation, equate each expressions to 1, and solve for C_1 and C_2.

$$X_0 = C_1 (0.5(1 + \sqrt{5}))^0 + C_2 (0.5(1 - \sqrt{5}))^0 = C_1 + C_2 = 1$$

$$X_1 = C_1 (0.5(1 + \sqrt{5})) + C_2 (0.5(1 - \sqrt{5})) = 1$$

Solving, gives

$$C_1 = 0.5(1 + 1/\sqrt{5})) \quad \text{and} \quad C_2 = 0.5(1 - 1/\sqrt{5})$$

The corresponding particular solution is

$$X_n = 0.5(1 + 1/\sqrt{5})(0.5(1 + \sqrt{5}))^n + 0.5(1 - 1/\sqrt{5})(0.5(1 - \sqrt{5}))^n$$

This expression does not obviously give only integer values, yet, as you can check on a calculator with different values of n, it does generate the sequence { 1, 1, 2, 3, 5, 8, ... }.

Example 6 Model Identification: Matching a sequence to a second-order difference equation.

Background By recognizing the form of a sequence term as one of the general forms for the solution of a difference equation we can identify the parameters of the model equation for which the given sequence is a solution. The key will be to identify the roots λ_1 and λ_2 from the given sequence and then to determine the coefficients a and b for the difference equation. This is called an "inverse" or "identification" problem. Mathematically, we know that the Auxiliary Equation can be expressed as the product of factors involving its roots:

$$\lambda^2 + a\lambda + b = (\lambda - \lambda_1)(\lambda - \lambda_2)$$

Expanding the product on the right gives

$$\lambda^2 + a\lambda + b = \lambda^2 - (\lambda_1 + \lambda_2)\lambda + \lambda_1\lambda_2$$

Hence, we can **identify the model parameters**:

$$a = -(\lambda_1 + \lambda_2) \quad \text{the coefficients of } \lambda \;;$$

$$b = \lambda_1\lambda_2 \quad \text{the constant terms.}$$

Problem Determine a difference equation that will have the given sequence as a solution.

a) $X_n = 3 \cdot 2^n - 5(-1)^n$ b) $Y_n = 0.5(1 + 2n)0.3^n$

c) $W_n = \sin(\pi n/2) - \cos(\pi n/2)$ d) $Z_n = 2^n \cdot \cos(3n)$

Solution In each case we are seeking the coefficients a and b that will make the stated sequence a solution of a homogeneous difference equation.

a) The sequence X_n has the form of a general solution when the two roots are real and distinct. The fact that 3 and -5 represent the constants C_1 and C_2 is not related to the choice of the difference equation. This is made by recognizing that the terms raised to a power are the values $\lambda_1 = 2$ and $\lambda_2 = -1$. Hence, $a = -(2 - 1) = -1$ and $b = 2(-1) = -2$. The corresponding difference equation is

$$X_{n+2} - X_{n+1} - 2X_n = 0$$

b) The sequence Y_n has the form of a solution with repeated real roots: $\lambda_1 = \lambda_2 = 0.3$. Hence, the coefficient $a = -0.6$ and $b = 0.09$. The corresponding difference equation is

$$Y_{n+2} - 0.6Y_{n+1} + 0.09Y_n = 0$$

c) The sequence W_n involves trigonometric functions and thus could only correspond to a difference equation with *complex* roots of its Auxiliary Equation. There is no term raised to the power n so the radius coefficient must be $r = 1$. Consequently, as $r = b^{1/2}$, $b = 1$. The coefficient a is found by setting θ equal to the coefficient of n inside the trig functions: $\theta = \pi/2$. Then, as $\cos(\theta) = -a/(2b^{1/2})$, we solve for $a = -2b^{1/2}\cos(\theta) = -2\cos(\pi/2) = 0$. The corresponding difference equation is thus

$$W_{n+2} + W_n = 0$$

d) The sequence Z_n must similarly correspond to a model with *complex* roots. In this sequence $r = 2$, thus $b = r^2 = 4$. The inclination angle $\theta = 3$, thus

$$a = -2b^{1/2}\cos(\theta) = -4\cos(3) \approx 3.96$$

The approximate corresponding difference equation is then

$$Z_{n+2} + 3.96Z_{n+1} + 4Z_n = 0$$

Section 2.5 — Higher-Order Difference Equations

Developing the General Solution when λ_1 and λ_2 are *complex* roots. (OPTIONAL)

The following discussion is intended to show how the general solution of the equation $X_{n+2} + aX_{n+1} + bX_n = 0$ is established when the roots of the Auxiliary Equation are *complex*. The roots are *complex conjugates* and can be expressed using the symbol i, that has the property $i^2 = -1$, as

$$\lambda_1 = A + B \cdot i \qquad \text{and} \qquad \lambda_2 = A - B \cdot i$$

where the real numbers A and B are related to the model parameters a and b by

$$A = -a/2 \qquad \text{and} \qquad B = (1/2)\text{SQRT}\{4b - a^2\}$$

Notice that the sign of the terms under the square root have been reversed from those in the quadratic formula to make their sum a positive number. Also note that

$$A^2 + B^2 = b$$

The *complex* number $\lambda_1 = A + Bi$ can be graphed as a point on an x-y coordinate system with x-coordinate A and y-coordinate B. The line through the point λ_1 and the origin forms an angle of inclination with the x-axis denoted by θ. Hence, θ is called the angle of the *complex* number λ_1. Using the right triangle with angle θ having adjacent side of length A and opposite side of length B, a relationship between θ and the model parameters a and b is established as follows:

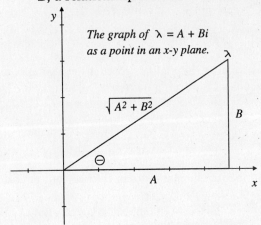

The graph of $\lambda = A + Bi$ as a point in an x-y plane.

$\cos(\theta) = $ Side Adjacent / Hypothenuse

$= A / \text{SQRT}(A^2 + B^2)$

$= -a/2 / \text{SQRT}((-a/2)^2 + [(1/2)(4b - a^2)^{1/2}]^2)$

$= -a/2 / \text{SQRT}(a^2/4 + [(1/4)(4b - a^2)])$

canceling the a^2 terms

$= -a/2 / \sqrt{b}$

thus

$$\theta = \cos^{-1}(-a / 2\sqrt{b})$$

where $\cos^{-1}$ denotes the ArcCos or Inverse Cosine function.

The modulus parameter r is the radius, the distance from the origin to the point λ_1:

$$r = \text{Hypothenuse} = \text{SQRT}(A^2 + B^2) = \sqrt{b}$$

The study of *complex* numbers and functions is an important branch of mathematics. In *complex* analysis an alternative form is used to represent *complex* numbers. A *complex* number $\lambda = A + B \cdot i$ can be represented as an exponential term with the imaginary root *i* appearing in the exponent. If r and θ are the radius and angle of the *complex* point λ in an x-y plane, then the exponential representation of λ is given by

$$\lambda = re^{\theta i}$$

Powers of *complex* numbers are easily simplified when expressed in exponential forms. Using the normal rules of exponents:

$$\lambda^n = (r \cdot e^{\theta i})^n = r^n \cdot (e^{\theta i})^n = r^n \cdot e^{n\theta i}$$

To convert a power λ^n back to the *complex* sum form, a very useful **Theorem of De Moivre** is used. Developed using trigonometric identities, it states that λ^n can be represented as r^n times a *complex* sum involving the sine and cosine of the product $n\theta$:

$$\lambda^n = r^n \cdot e^{n\theta i} = r^n \{ \cos(n\theta) + i \cdot \sin(n\theta) \}$$

Returning now to the solution of the difference equation, even though the roots λ_1 and λ_2 are *complex* numbers, the same general solution is used for real roots. The general solution, expressed with lower case coefficients c_1 and c_2 that actually represent *complex* numbers, is formally

$$X_n = c_1 \lambda_1^n + c_2 \lambda_2^n$$

Since the roots λ_1 and λ_2 have the same radius and opposite angles, θ and $-\theta$, their powers expressed in exponential form are closely related. $\lambda_1 = re^{\theta i}$ and $\lambda_2 = re^{-\theta i}$. The general solution can then be expressed as

$$X_n = c_1 \lambda_1^n + c_2 \lambda_2^n = c_1 r^n e^{n\theta i} + c_2 r^n e^{-n\theta i}$$

By De Moivre's theorem this is equivalent to the *complex* sum

$$X_n = c_1[r^n \{ \cos(n\theta) + i \cdot \sin(n\theta) \}] + c_2[r^n \{ \cos(-n\theta) + i \cdot \sin(-n\theta) \}]$$

$$= r^n \{ (c_1 + c_2) \cos(n\theta) + i \cdot (c_1 - c_2) \sin(-n\theta) \}$$

This expression can be converted to a real form, not involving the *complex* root i, by choosing c_1 and c_2 so that the sum $c_1 + c_2$ is real and the difference $c_1 - c_2$ is purely imaginary. Then, $i \cdot (c_1 - c_2)$ will be a real number. The solution of the difference equation is finally found in the stated general form

$$X_n = r^n \{ C_1 \cos(n\theta) + C_2 \sin(n\theta) \}$$

where C_1 and C_2 are arbitrary real constants, when the *complex* constants in the previous equation are chosen as the *complex* conjugates

$$c_1 = 0.5(C_1 - i C_2) \quad \text{and} \quad c_2 = 0.5(C_1 + i C_2)$$

Equilibriums and Steady-States.

An equilibrium of a second-order difference equation is a value that gives a constant solution sequence, i.e., a value such that $X_n = X_E$ for all n. Then, substituting X_E for each sequence term in the difference equation $X_{n+2} + aX_{n+1} + bX_n = 0$ gives the equation $X_E + a \cdot X_E + b \cdot X_E = 0$, which has only the solution $X_E = 0$ if $1 + a + b \neq 0$. In the special case with $1 + a + b = 0$ for number K, setting $X_1 = X_2 = K$ will result in the constant sequence $X_n = K$. Hence every constant is an equilibrium.

The equilibrium $X_E = 0$ will be stable or not stable depending on the parameters a and b. The condition for stability is that $\lim_{n \to \infty} X_n = 0$ for all solutions. In all three cases of the general solution, this limit will involve the limit of a power term, λ^n or r^n, that will either approach zero or become infinite as $n \to \infty$. As

$$\lim_{n \to \infty} \lambda^n = 0 \text{ only if } |\lambda| < 1$$

Section 2.5 — HIGHER-ORDER DIFFERENCE EQUATIONS

then a simple candidate for a particular solution is a sequence that is a constant for every term. Consider the constant sequence defined by

$$X_n = K \quad \text{then } X_{n+1} = K \quad \text{and} \quad X_{n+2} = K$$

To determine the number K these sequence terms are substituted into equation NH-DE and the resulting equation is solved for the constant K:

$$K + a \cdot K + b \cdot K = \gamma$$

Thus, if $1 + a + b \neq 0$, $\quad K = \gamma / (1 + a + b)$

If $1 + a + b = 0$, then a constant solution is not possible. In this case the next candidate for a particular solution is a constant times the index n; then

$$X_n = K \cdot n \quad X_{n+1} = K(n+1) \quad \text{and} \quad X_{n+2} = K(n+2)$$

Substituting into the difference equation and factoring out $K \cdot n$ and K gives:

$$K(n+2) + a \cdot K(n+1) + b \cdot K = \gamma$$

$$K \cdot n[1 + a + b] + K[2 + a] = \gamma$$

But, we are assuming at this point that $1 + a + b = 0$, and hence the first factored term vanishes. The remaining equation is solved for

$$K = \gamma / (2 + a)$$

Of course the same problem can occur in this case; the denominator $2 + a$ can equal zero. This can happen only when $a = -2$ and $b = 1$. In this one special case an alternative particular solution is given by

$$X_n = 0.5 \, \gamma \cdot n^2$$

These conditions and the corresponding solutions are summarized in the following.

A particular solution of the nonhomogeneous equation

$$X_{n+2} + a X_{n+1} + b X_n = \gamma \quad \gamma \text{ a constant}$$

is given by

$$X^P_n = \gamma / (1 + a + b) \qquad \text{if } a + b \neq -1;$$

$$X^P_n = \gamma \cdot n / (2 + a) \qquad \text{if } a + b = -1 \text{ and } a \neq -2;$$

$$X^P_n = \gamma \cdot n^2 / 2 \qquad \text{if } a = -2 \text{ and } b = 1.$$

Example 9 **Solving Nonhomogeneous second-order difference equations.**

Problem Determine the particular solution of $X_{n+2} + 5X_{n+1} + 6X_n = 4$ with $X_0 = 1$ and $X_1 = -1$.

Solution   The first step is to form the general solution of the equation. The associated homogeneous equation $X_{n+2} + 5X_{n+1} + 6X_n = 0$ has the Auxiliary equation $\lambda^2 + 5\lambda + 6 = 0$. The roots are $\lambda_1 = -2$ and $\lambda_2 = -3$. Thus

$$X^H_n = C_1 (-2)^n + C_2 (-3)^n$$

For this equation $\gamma = 4$ and hence a particular solution is given by

$$X^P_n = 4/(1 + 5 + 6) = 1/3$$

The general solution is therefor

$$X^{NH}_n = C_1 (-2)^n + C_2 (-3)^n + 1/3$$

To determine the particular solution with the first two values 1 and -1 we set $n = 0$ and $n = 1$ to form two equations that can be solved for C_1 and C_2.

$n = 0$ $X^{NH}_0 = C_1 (-2)^0 + C_2 (-3)^0 + 1/3 = C_1 + C_2 + 1/3 = 1$

$n = 1$ $X^{NH}_1 = C_1 (-2)^1 + C_2 (-3)^1 + 1/3 = -2C_1 - 3C_2 + 1/3 = -1$

The solution is $C_1 = 2/3$ and $C_2 = 0$. Thus the desired particular solution is

$$X_n = (2/3)(-2)^n + 1/3$$

☑

Example 10   The genetic effect of inbreeding.

Background   An issue of great interest to animal and plant breeders is the probability that offspring will have the same genetic compliment. In genetic texts the probability that two genes of an individual selected at random from the nth generation of a population are identical by descent, i.e., that they came from a common ancestor, is denoted by the α_n. The probability that they are not identical is then given by $(1 - \alpha_n)$. The sequence of probabilities $\{\alpha_n\}$ satisfies a nonhomogeneous difference equation established by indicating the probability of being different at generation n. This is the sum of the probability of being different in the two preceding generations times the chance that random mating maintains the difference. The equation is

$$(1 - \alpha_n) = [1 - (1/N)](1 - \alpha_{n-1}) + \{1/(2N)\}(1 - \alpha_{n-2})$$

where N is the mating population size that is assumed to be constant over all generations. N = 2 for simple paired matings but in botany or with the use of *in vitro* fertilization techniques N can be much larger. As N becomes larger, α_n will approach zero.

Problem   What is the general equation for the probability of identical genes in the n-th generation when N = 2 and the first two matings are not genetically identical?

Solution   First the difference equation for α_n is rearranged and the index is shifted up by 2 to obtain the equivalent equation

$$\alpha_{n+2} + [(1/N) - 1]\alpha_{n+1} - \{1/(2N)\}\alpha_n = 1/(2N)$$

This is a nonhomogeneous equation of the standard form with

$$a = (1/N) - 1 \quad b = -1/(2N) \quad \gamma = 1/(2N)$$

A particular solution of this equation, for any N, is given by the constant sequence

$$\alpha_n^p = \gamma / (1 + a + b) = 1$$

The general solution of the associated homogeneous equation

$$\alpha_{n+2} + [(1/N) - 1] \alpha_{n+1} - \{1/(2N)\}\alpha_n = 0$$

depends on the roots of the Auxiliary equation

$$\lambda^2 + [(1/N) - 1]\lambda - \{1/(2N)\} = 0$$

The roots λ_1 and λ_2 are given by

$$\lambda = (1/2)[1 - (1/N) \pm \text{SQRT}\{1 + (1/N)^2\}]$$

The general solution is therefore given by

$$\alpha_n = C_1 \lambda_1^n + C_2 \lambda_2^n + 1$$

For paired matings, with N = 2, the solution is

$$\alpha_n = C_1\{(1 + 5^{1/2})/4\}^n + C_2\{(1 - 5^{1/2})/4\}^n + 1$$

If the first two generations are not identical, then $\alpha_0 = 0$ and $\alpha_1 = 0$. Setting n = 0 and n = 1 gives the two equations:

$$\alpha_0 = C_1 + C_2 + 1 = 0$$

$$\alpha_1 = C_1\{(1 + 5^{1/2})/4\} + C_2\{(1 - 5^{1/2})/4\} + 1 = 0$$

The roots of these equations are

$$C_1 = -0.5(3/\sqrt{5} + 1) \quad \text{and} \quad C_2 = 0.5(3/\sqrt{5} - 1)$$

Thus the particular solution with $\alpha_0 = 0$ and $\alpha_1 = 0$ is

$$\alpha_n = -0.5(3/\sqrt{5} + 1)((1 + \sqrt{5})/4)^n + 0.5(3/\sqrt{5} - 1)((1 - \sqrt{5})/4)^n + 1$$

You may verify by setting n = 2, 3, and 4 that the next few sequence terms are

$$\alpha_2 = 0.25 \quad \alpha_3 = 0.375 \quad \alpha_4 = 0.5$$

These values could be found by sequentially substituting values of the α's into the original equation. The general solution is useful when considering larger n. For instance, setting n = 20 gives $\alpha_{20} = 0.983109$, which means that after 20 inbreedings of paired progeny there is greater than a 98% chance that the offspring will be genetically identical. Or, there will be less than a 2% chance they will be different.

Exercise Set 2.5

Homogeneous equations.

1. Find the general solution of the given difference equation.
   a) $X_{n+2} + 2X_{n+1} - 3X_n = 0$
   b) $X_{n+2} - 4X_n = 0$
   c) $X_{n+2} + 6X_{n+1} + 8X_n = 0$
   d) $2X_n - 3X_{n-2} = 0$
   e) $X_{n+2} + 6X_{n+1} + 9X_n = 0$
   f) $X_{n+2} - 4X_{n+1} + 4X_n = 0$
   g) $X_{n+2} - 5X_{n+1} + X_n = 0$
   h) $\Delta X_{n+1} = -\Delta X_n$
   i) $4Y_{n+2} - 3Y_{n+1} - Y_n = 0$
   j) $Z_{n+2} - 3Z_{n+1} + 2Z_n = 0$
   k) $-Y_{n+2} + Y_{n+1} + Y_n = 0$
   l) $X_{n+1} = X_{n-1} + 3X_n$
   m) $W_{n+2} + 2W_{n+1} + 3W_n = 0$
   n) $W_{n+2} + 0.04W_n = 0$
   o) $X_{n+2} + 5X_{n+1} - 4X_n = 0$
   p) $X_{n+2} - X_{n+1} + 4X_n = 0$
   q) $0.25Z_{n+2} + 3Z_{n+1} + 25Z_n = 0$
   r) $Z_{n+2} + 0.2Z_{n+1} + 0.01Z_n = 0$

2. Determine the particular solutions of $X_{n+2} - 13X_{n+1} + 40X_n = 0$ satisfying
   a) $X_0 = 2$ and $X_1 = 0$
   b) $X_0 = 0$ and $X_3 = 100$

3. Determine the particular solutions of $X_{n+2} - 6X_{n+1} + 9X_n = 0$ satisfying
   a) $X_0 = 1$ and $X_1 = 10$
   b) $X_0 = 0$ and $X_4 = 25$

4. Determine the particular solutions of $X_{n+2} - 5X_{n+1} + 9X_n = 0$ satisfying
   a) $X_0 = 1$ and $X_1 = 0$
   b) $X_0 = 0$ and $X_1 = 25$

   Hint: you will know the value of $\cos(\theta)$, remember $\sin(\theta) = \mathrm{SQRT}(1 - \cos^2(\theta))$.

5. Determine a Fibonacci sequence f_n, defined by $f_{n+1} = f_n + f_{n-1}$ for which $f_1 = 1$, $f_2 = 2$.

6. Determine a homogeneous difference equation that will have the given sequence as a solution.
   a) $X_n = 2^n + 3 \cdot 5^n$
   b) $Y_n = (1+n)3^n$
   c) $X_n = -2 \cdot 0.1^n + 3 \cdot 0.5^n$
   d) $Y_n = (1 - 2n)4^n$
   e) $W_n = \sin(\pi n/2)$
   f) $Z_n = 2^n \cdot \cos(n)$
   g) $W_n = \{\sin(\pi n/2) + \cos(\pi n/2)\}2^n$
   h) $Z_n = 2^{-n} \cdot \sin(\pi n/3)$

7. For which of the given systems is zero a stable equilibrium?
   a) $X_{n+2} - 0.25X_n = 0$
   b) $X_n + 0.09X_{n-2} = 0$
   c) $X_{n+2} - X_{n+1} + X_n = 0$
   d) $\Delta X_n = 0.5\Delta X_{n-1}$
   e) $4X_{n+2} - 3X_{n+1} - X_n = 0$
   f) $6X_{n+2} - X_{n+1} - X_n = 0$
   g) $-4X_{n+2} + 2.5X_{n+1} + X_n = 0$
   h) $X_{n+1} = X_{n-1} + 3X_n$

8. Show that $X_E = 0$ is not a stable equilibrium of $X_{n+2} + aX_{n+1} + bX_n = 0$ when $|a| > 1$ and $b > 0$.

9. For what value of the parameter b will the solution of $x_{n+2} - 3x_{n+1} + bx_n = 0$ have the solution
    a) $x_n = 5^n$
    b) $x_n = n \cdot 1.5^n$
    c) $x_n = \sin(n\pi/3) \cdot 3^n$

10. For values of k will all solutions of $y_{n+2} + ky_{n+1} + 0.5y_n = 0$ have the steady-state $\bar{y} = 0$.

11 Assume that a species normally doubles in size each generation. To control its growth, 20% of the population over one generation old is removed at the end of each year. State the model of this controlled population and determine its size after 10 years if $P_0 = 10$ and $P_1 = 20$.

Nonhomogeneous Equations.

12 Give a particular solution and the general solution for the indicated nonhomogeneous equations.

a) $X_{n+2} + X_{n+1} - 6X_n = 4$ b) $Y_{n+2} + 3Y_{n+1} - 4Y_n = 10$ c) $Z_{n+2} - 2Z_{n+1} + Z_n = 4$

d) $X_{n+2} + X_{n+1} - 2X_n = 2$ e) $Y_{n+2} - 2Y_{n+1} - 4Y_n = 3$ f) $Z_{n+2} + 3Z_{n+1} + 9Z_n = 2$

g) $Y_{n+2} - Y_{n+1} - 4Y_n = 3$ h) $Z_{n+2} + 10Z_{n+1} + 25Z_n = 2$

13 Determine the particular solution to the corresponding equation in Exercise 12 that satisfies the indicated initial conditions.

a) $X_0 = 2, X_1 = 1$ b) $Y_0 = 2, Y_1 = 1$ c) $Z_0 = 1, Z_1 = 0$ d) $X_0 = 2, X_1 = 2$

e) $Y_0 = 1, Y_1 = 2$ f) $Z_0 = 2, Z_1 = 0$ g) $Y_0 = 0, Y_1 = 1$ h) $Z_0 = 1, Z_1 = 1$

14 Determine a nonhomogeneous difference equation that will have the given sequence as a solution.

a) $X_n = 2^n + 3 \cdot 5^n + 3$ b) $Y_n = (1+n)3^n - 0.2$ c) $X_n = -2 \cdot 0.1^n + 3 \cdot 0.5^n - 2$

d) $Y_n = (1 - 2n)4^n - 1$ e) $W_n = \sin(\pi n/2) - 0.5$ f) $Z_n = 2^n \cdot \cos(n) + 7$

g) $W_n = \{\sin(\pi n/2) + \cos(\pi n/2)\}2^n - 1$ h) $Z_n = 2^{-n} \cdot \sin(\pi n/3) - 1$

15 If the daily change in mass of a bacterium growing in a culture medium equals three times the change experienced the previous day, plus five grams, determine:

a) The equation describing the daily population mass and its general solution.

b) The mass of the population on day $n = 20$ if it was initially one gram and on day 2 was 12 grams.

16 For the Genetic Inbreeding model of Example 9, determine the particular solution with $\alpha_0 = 0$ and $\alpha_1 = 1/2$ when $N = 2$. For this solution evaluate α_n for $n = 2, 3$ and 20.

17 Determine a second-order nonhomogeneous difference equation for which the given sequence $\{X_n\}$ is a solution. Determine the corresponding homogeneous equation and then the parameter γ.

a) $X_n = 2 \cdot 3^n + (-1)^n + 1$ b) $X_n = (5 - 2n)((1/2))^n - 2$

c) $X_n = 4 \cdot 2^n - 5^n - 1$ d) $X_n = 0.3(0.2)^n + (-5)^n + 0.1$

e) $X_n = [\cos(n\pi/2) + \sin(n\pi/2)]2^n + 5$ f) $X_n = [\cos(n\pi/3) + \sin(n\pi/3)](0.25)^n + 3$

18 Show that if two particular sequences $\{x^1_n\}$ and $\{x^2_n\}$ are both solutions to the nonhomogeneous equation $x_{n+2} + ax_{n+1} + bx_n = \gamma$, then the sequence $\{x^3_n\}$ defined by $x^3_n = x^1_n - x^2_n$ satisfies the associated homogeneous equation $x_{n+2} + ax_{n+1} + bx_n = 0$.

Chapter 3 Derivatives

Section 3.1 The Difference Quotient; the average rate of change over an interval 181
The Difference Quotient.
Interpretations of the Difference Quotient.
The Difference Quotient: the slope of secant lines.

Section 3.2 The Derivative. 189
Secants, Tangents, and derivatives.
Derivative notation.
Evaluating derivatives as limits.
One-sided derivatives.
When derivatives do not exist.
Infinite derivatives.
Approximating derivatives.

Section 3.3 Basic Derivative Rules 204
Specific derivative rules....- The Constant and Power Rules.
Algebraic derivative rules...
    -The Constant Multiple, Sum/Difference, Product and Quotient Rules.

Section 3.4 Derivatives of Elementary Functions 216
Derivatives of exponential functions.
Derivatives of $ln(x)$ and $\log_b(x)$.
Derivatives of trigonometric functions.
Parametric curves and derivatives.
The derivatives of $\sin(t)$ and $\cos(t)$.

Section 3.5 The Chain Rule 228
Introduction: Doing things in order.
The Chain Rule.
General Derivative Rules.

Section 3.6 Higher Derivatives and Implicit Differentiation 240
Higher derivatives.
When higher derivatives fail to exist.
Implicit differentiation.
Differentiating a function you can not solve for.
Logarithmic differentiation.

Section 3.1 The Difference Quotient:

the average rate of change over an interval.

The Difference Quotient.

The capital Greek letter *delta*, Δ, is used to indicate a change, the numerical difference between two values. For instance, Δt denotes the change in the variable t and Δw indicates the change in w. The change in a function, say $y = f(x)$, is indicated by either Δf or Δy. Differences in independent variables are sometimes referred to as **increments** of change. A difference in a function is associated with a difference in its argument variable.

DEFINITION OF ΔX AND $\Delta F(X)$.

For an independent variable x, $\quad\Delta x$ denotes a **difference** or **increment** in the variable x and is a number which may be positive, negative, or zero.

For a function $y = f(x)$ and difference Δx, $\quad$ the **difference** Δy or Δf is the **change** in the function's value from x to $x + \Delta x$:

$$\Delta y = \Delta f(x) = f(x + \Delta x) - f(x)$$

A function difference Δf depends on both the x-value and the increment Δx. For each increment Δx and specific x-value, x_0, the difference $\Delta f(x_0)$ is a number. For example, if $y = f(x) = x^2$ the difference Δy corresponding to $x_0 = 3$ and $\Delta x = 0.5$ is

$$\Delta y = f(x_0 + \Delta x) - f(x_0)$$

$$= f(3 + 0.5) - f(3) = f(3.5) - f(3)$$

$$= 3.5^2 - 3^2 = 12.25 - 9 = 3.25$$

The change induced in a function $\Delta f(x)$ relative to the change Δx in x is the ratio of these two changes. This ratio, $\Delta f / \Delta x$, is a quotient of differences which plays a central role in the development of differential calculus.

DEFINITION OF A DIFFERENCE QUOTIENT

Let $y = f(x)$. For each increment Δx, the **Difference Quotient** $\Delta f(x)/\Delta x$ is

$$\Delta y/\Delta x = \Delta f(x)/\Delta x = \{f(x + \Delta x) - f(x)\} / \Delta x$$

An alternative notation for the difference quotient that emphasizes the specific x-value at which the difference Δf should be evaluated is the symbol $\Delta f/\Delta x$ followed by a vertical bar and a subscripted indication of the x-value, for instance "x = a" or simply "a". The notation has the form:

$$\left.\frac{\Delta f(x)}{\Delta x}\right|_{x=a} = \frac{f(a + \Delta x) - f(a)}{\Delta x}$$

This notation is sometimes simply shortened to: $\left.\Delta f/\Delta x\right|_a$.

Example 1   Evaluating difference quotients.

Problem   Evaluate the difference quotient of $f(x) = x^2 - x + 2$ at $x_0 = 3$ for $\Delta x = 2$ and $\Delta x = 0.1$.

Solution   The difference quotient for $\Delta x = 2$ is

$$\Delta f(3)/\Delta x = \{f(3+2) - f(3)\}/2 = \{f(5) - f(3)\}/2 = \{22 - 8\}/2 = 7$$

The difference quotient for the same function and x-value but with $\Delta x = 0.1$ is

$$\Delta f(3)/\Delta x = \{(f(3+0.1) - f(3))/0.1\} = (8.51 - 8)/0.1 = 5.1$$

as

$$f(3.1) = (3.1)^2 - 3.1 + 2 = 8.51 \quad \text{and} \quad f(3) = 3^2 - 3 + 2 = 8$$

Example 2   A difference quotient for a negative increment.

Problem   Evaluate the difference quotient of $g(t) = t - \sqrt{t}$, at $t_0 = 3$, with increment $\Delta t = -1/2$.

Solution   To evaluate the numerator of the difference quotient we need to evaluate $g(t)$ at t_0 and $t_0 + \Delta t$, which for this problem are the numbers 3 and $3 - 1/2 = 2.5$:

$$g(t_0) = g(3) = 3 - \sqrt{3} \approx 1.267$$

and

$$g(t_0 + \Delta t) = g(3 - 1/2) = g(2.5) = 2.5 - \sqrt{2.5} \approx 0.919$$

Thus, the difference quotient is

$$\Delta g(3)/\Delta t = \{g(3 - 1/2) - g(3)\}/(-1/2) = -2\{g(2.5) - g(3)\} \approx 0.698$$

If the function and increment are specified, but a specific numerical x-value is not given, then the difference quotient can be evaluated as a function of the x-variable. Similarly, if the primary increment is not indicated as a specific number, then the difference quotient can be considered as a function of the increment. Such difference quotients are usually simplified as much as possible.

Example 3   The difference quotient of f(x) as a function of the x-value.

Problem   For $f(x) = x^2 + 1$ and increment $\Delta x = 0.2$, evaluate and algebraically simplify the difference quotient $\Delta f/\Delta x$ at $x_0 = a$.

Solution   Setting $x = a$ and $\Delta x = 0.2$ the difference quotient becomes

$$\left.\Delta f/\Delta x\right|_{x=a} = (f(a + 0.2) - f(a))/0.2$$

$$= ([(a + 0.2)^2 + 1] - [a^2 + 1])/0.2$$

$$= ([a^2 + 0.4a + 0.04 + 1] - [a^2 + 1])/0.2$$

$$= (0.4a + 0.04)/0.2 = 2a + 0.2$$

Example 4 **The difference quotient as a function of the increment.**

Problem The function that gives the area of a circular disc of radius r is $A(r) = \pi r^2$. Evaluate and simplify the difference quotient $\Delta A/\Delta r$ for an arbitrary radius $r = r_0$ and increment in the radius $\Delta r = h$.

Solution

$$\Delta A(r_0)/\Delta r = (A(r_0 + h) - A(r_0))/h$$

$$= ([\pi(r_0 + h)^2] - \pi r_0^2)/h \qquad \text{Evaluate the functions.}$$

$$= ([\pi r_0^2 + 2\pi r_0 h + \pi h^2] - \pi r_0^2)/h \qquad \text{Expand the square.}$$

$$= (2\pi r_0 h + \pi h^2)/h \qquad \text{Cancel in the numerator.}$$

$$= \{(2\pi r_0 + \pi h)h/h \qquad \text{Factor out an h.}$$

$$= 2\pi r_0 + \pi h \qquad \text{Cancel an h, if } h \neq 0.$$

The difference quotient in this example is a function of both the reference value r_0 and the increment Δr which was denoted by h.

Interpretations of the Difference Quotient

In mathematical models describing the variables are given physical or biological meanings and units and the functions take on corresponding meanings and units. For instance, if the length, l, of a rectangle is measured in meters and its width, w, in centimeters, then its area $A = l \cdot w$ has the unit $m \cdot cm = 10^{-2} m^2$. Similarly, the units of a difference quotient $\Delta f/\Delta x$ will depend on the units of the function f and of the increment Δx. The unit of $\Delta f/\Delta x$ is the ratio of the unit of f(x) to the unit of Δx. An important interpretation of the difference quotient is an "average rate of change". This interpretation is based on the principal that the total change in a function over an interval is the product of the function's "average rate of change" and the length of the interval. The length of the interval is given by the difference Δx and the total change in the function is the difference Δf.

The difference quotient as an Average Rate of Change.

Change in y = [Change in x] × [Average Rate of Change]

or, using difference notation

$$\Delta y = \quad \Delta x \times \{\Delta y/\Delta x\}$$

---

184 CHAPTER 3 DERIVATIVES

Example 5 **Average Rates of Change in Blood Pressure.**

Problem Assume that over a short period of time the blood pressure in an organ is described by the equation $P(t) = 100 - t^2$ kPa/s. The time unit is seconds, s, and the pressure unit is kilo-pascal, kPa. Determine the average rate of change in the blood pressure over the 3 second interval [2,5].

Solution The average rate of change in the blood pressure is given by the difference quotient $\Delta P/\Delta t$ evaluated at the initial time $t_0 = 2$ for the increment Δt = length of the interval [2,5], or

$$\Delta t = 5 - 2 = 3 \text{ s}$$

$$\Delta P(2)/\Delta t = [P(2 + 3) - P(2)] / 3$$

$$= [(100 - 5^2) - (100 - 2^2)] / 3$$

$$= [75 - 96]/3 = -21/3 = -7 \text{ kPa/s}$$

The average rate of change is negative, meaning the blood pressure decreased.

The vocabulary used to describe difference quotients varies in accordance with the interpretation of the independent and dependent variables. In many applications the independent variable is time. Table I lists some common interpretations of the difference quotient in various disciplines. Notice that in each case the interpretation is as an "average" value of the rate of change in the function.

Table I

Application: Interpretation of function f	Interpretation of Difference Quotient
Biology f(t) is a population size or bio-mass	$\Delta f(t)/\Delta t$ = average growth rate
Physics f(t) is the position at time t, or distance traveled	$\Delta f(t)/\Delta t$ = average velocity
Medicine C(t) is the blood concentration of a drug or toxin	$\Delta C(t)/\Delta t$ = average clearance rate
Chemistry c(x) is a molar concentration of a protein x cm down a column or gel	$\Delta c(x)/\Delta x$ = average concentration gradient
Hydrology V(t) is the storage volume of water in a lake	$\Delta V(t)/\Delta t$ = average net flow rate
Business I(t) is an inventory of goods R(t) is the revenue at time t and C(t) is a Cost Index	$\Delta I(t)/\Delta t$ = average sales rate $\Delta R(t)\Delta t$ = average rate of change in revenue $\Delta C(t)/\Delta t$ = average Inflation Rate

Example 6 **Calculating plant growth from its average rates of change.**

Problem A graduate student recorded a plant's average growth over three time intervals. The average rate of change in the plant's height over the first two days was 3 cm/d, over the next day it grew at the rate of 4 cm/d, and over the fourth to seventh days its growth rate was 9 cm/d. If the plant measured 40 cm at the start of the first day, how tall was it at the end of the seventh day?

Solution The problem involves three time intervals: [0,2], [2,3], [3,7]. Notice that in this form day 1 corresponds to the interval [0,1] and day three corresponds to the interval [2,3], so day seven is given by the interval [6,7]. The height of the plant is modeled by a function

$$H(t) = \text{height at time } t$$

The given information provides the height at time $t = 0$, $H(0) = 40$. The problem is to determine the value $H(7)$. The given average growth rates are the value of the difference quotients over the three intervals:

over [0,2], $\Delta H/\Delta t = 3$;   over [2,3], $\Delta H/\Delta t = 4$;   over [3,7], $\Delta H/\Delta t = 9$.

The height at the end of each interval is calculated by the formula:

H(end) = H(start) + [the Average Rate of Growth · Length of the time interval].

Applying this general rule to the first interval [0,2], which is of length $\Delta t = 2$, the height at $t = 2$ is given by

$$H(2) = H(0) + [3 \cdot 2] = 40 + 6 = 46 \text{ cm}$$

The second interval, [2,3], has length $\Delta t = 1$ and the formula gives

$$H(3) = H(2) + [4 \cdot 1] = 46 + 4 = 50 \text{ cm}$$

The third interval, [3,7], has length $\Delta t = 4$ and thus the formula gives the final answer

$$H(7) = H(3) + [9 \cdot 4] = 50 + 36 = 86 \text{ cm}$$

In the preceding example the units were indicated for the interim and final answers but were omitted during the calculations. *This is a common practice that will be followed throughout the text to simplify notation.*

THE DIFFERENCE QUOTIENT: THE SLOPE OF SECANT LINES

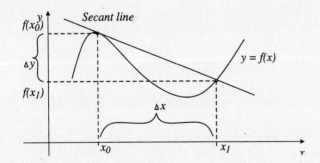

The simplest and most common geometric interpretation of the difference quotient is as the slope of a secant line. A **secant line** is a line that passes through two points on a function's graph. The **slope** of the line through the two points $(x_0, f(x_0))$ and $(x_1, f(x_1))$ is the difference quotient $\Delta f/\Delta x$ where Δx is the difference between the x-values and Δf is the change in the y-values:

$$\Delta f/\Delta x = \text{Slope of the secant line}$$

$$\Delta x = x_1 - x_0 \text{ is the change in x}$$

and

$$\Delta f = f(x_1) - f(x_0) \text{ is the change in y}$$

The increment Δx may be positive or negative, depending on how the points are labeled and thus which of the two x-values is greater. In the previous Figure, Δx is positive since $x_1 > x_0$. The graph is decreasing so the difference Δy is negative and hence the slope Δy/Δx is also negative.

DEFINITION: THE POINT-SLOPE EQUATION OF A SECANT LINE

The **point-slope equation of the secant line** through the points $(x_0, f(x_0))$ and $(x_1, f(x_1))$ is

$$y = f(x_0) + [\Delta f(x_0)/\Delta x](x - x_0)$$

The **slope** of this secant line is given by the difference quotient:

$$\Delta f(x_0)/\Delta x = \{f(x_1) - f(x_0)\} / (x_1 - x_0)$$

Example 7 Determining the equation of a secant line.

Problem Determine the equation of the secant line to the graph of $f(x) = x^3 - 2x$, passing through the points (1,-1) and (2,4).

Solution The secant line is illustrated at the right. Its equation is given in the point-slope form. We set $x_0 = 2$ and $y_0 = f(x_0) = 4$ and $x_1 = 1$ and $y_1 = f(x_1) = -1$

Then, change in x-values between x_0 and x_1 is the difference

$$\Delta x = x_1 - x_0 = 1 - 2 = -1$$

The change in the y-values, **in the same order**, is the difference $\Delta f = f(x_1) - f(x_0)$ or

$$\Delta y = y_1 - y_0 = -1 - 4 = -5$$

The slope of the secant line is the difference quotient $\Delta f/\Delta x = -5/-1 = 5$. Thus, the point slope equation of the secant line is $y = 4 + 5(x - 2)$.

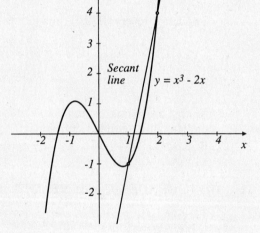

Figure 2.2

Example 8 Determining the equation of a secant line for a variable increment h = Δx.

Problem Determine the equation of the secant line passing through the point $(x_0, y_0) = (2,1)$ on the parabola $y = -0.5x^2 + 3$, and through a second point (x_1, y_1) on the parabola with x-coordinate $x_1 = 2 + h$, where h is an arbitrary increment.

Solution In the adjacent figure, the parabola is sketched with a secant line corresponding to a negative h value. To determine the secant line the slope is evaluated. First, determine the x-increment:

$$\Delta x = x_1 - x_0 = (2 + h) - 2 = h$$

The corresponding difference in the y-values is

Section 3.1 THE DIFFERENCE QUOTIENT 187

$$\Delta y = y(2 + h) - y(2)$$
$$= [-0.5(2 + h)^2 + 3] - [-0.5 \cdot 2^2 + 3]$$
$$= [-0.5 \cdot 2^2 - 0.5 \cdot 4h - 0.5h^2 + 3] - [-0.5 \cdot 2^2 + 3]$$
$$= -2 \cdot h - 0.5h^2 = -(2 + 0.5h)h$$

The secant's slope is the difference quotient

$$\Delta y/\Delta x = \{-2 \cdot h - 0.5h^2\}/h = -(2 + 0.5h)$$

The equation of the secant line through the point (2,1) is therefore

$$y = 1 - [2 + 0.5h](x - 2)$$

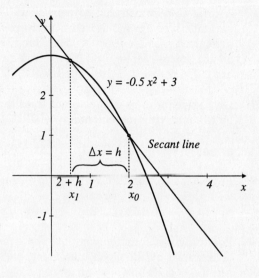

Exercise Set 3.1

1. Compute the difference quotient for the function, reference point, and increment indicated.

   a) $f(x) = 2x$; $x = 3$, $\Delta x = 0.1$
   b) $f(x) = 2x + 1$; $x = 3$, $\Delta x = 0.1$
   c) $g(x) = (x+2)/x$; $x = 1$, $\Delta x = 0.5$
   d) $g(x) = (x+2)/x$; $x = 1$, $\Delta x = -0.5$
   e) $f(x) = (x+2)/(x-2)$; $x = 1.9$, $\Delta x = 0.05$
   f) $f(x) = (x+2)/(x-2)$; $x = 1.9$, $\Delta x = 0.15$
   g) $g(x) = e^x$; $x = 0$, $h = 0.2$
   h) $f(x) = ln(x)$; $x = 1$, $h = 0.1$
   i) $g(x) = e^x$; $x = 5$, $h = 0.2$
   j) $f(x) = ln(x)$; $x = 2$, $h = 0.1$

2. Compute the indicated difference quotients using the increment $h = 0.1$. Expand and simplify as much as possible.

   a) $\Delta f(x_0)/\Delta x$; $f(x) = 2x-8$
   b) $\Delta f(t)/\Delta t \mid t=t_0$; $f(t) = t^2 + 3$
   c) $\Delta g(t)/\Delta t$; $g(t) = 1/t$
   d) $\Delta f(x_0)/\Delta x$; $\Delta x = h$, $f(x) = x^3+3$
   e) $\Delta g(z_0)/\Delta z$; $g(z) = z^{1/2}$
   f) $\Delta f(t)/\Delta t$; $f(t) = 2^t$
   g) $\Delta f(x)/\Delta x$; $f(x) = e^x$
   h) $\Delta g(x)/\Delta x$; $g(x) = e^{-x}$

3. Determine the average rate of change of the given function over the interval [1,3].

   a) $f(t) = t^2$
   b) $f(t) = t^{1/2}$
   c) $f(t) = \sin(\pi t)$
   d) $f(t) = \log_3(t)$
   e) $f(t) = 2^t$
   f) $f(t) = t(3 - t)$
   g) $g(t) = e^t$
   h) $g(t) = e^{2t}$

4. How far does a jogger run who runs 20 minutes at 1 km per hour and 10 minutes at 7 km per hour?

5. A corn plant, which was fed a growth hormone, was measured at the start of an experiment and found to be 0.8 cm in height. Over the next week of the experiment, the plant grew at an average rate of 0.15 cm per day; over the second week of the experiment, it grew at a rate of 0.05 cm per day; and, during the third week of the experiment it grew at 0.2 cm per day.
   a) Draw a graph and plot the height of the corn plant each day over the three week period.
   b) What was the height of the corn plant at the end of the three weeks?
   c) Determine the average rate of growth over the three-week period by two methods?

6 A drug is administered to a patient. The average rate of change of the drug concentration in the patient's blood is monitored over 10 minute intervals for two hours. Sketch a broken line graph to indicate a possible function describing the drug concentration if the concentration at t = 0 was C(0) = 50 µM (micro - Molar) and the average rates of change over the 12 intervals [0,10], [10,20], ... , [110,120] are 0.2, 1.5, 2.0, 1.8, 1.8, 1.6, 1.4, 0.8, 0.0, -1.0, -3.5, and -4.0 µM, respectively.

7 Give the equations of the secant lines through the point (0,3) on the graph of $y = 3 + x^4$ as determined by the following increments.

   a) $\Delta x = 1$ b) $\Delta x = 0.1$ c) $\Delta x = 0.01$

8 For each function in Exercise 3 give the equations of the secant lines determined by the point $t_0 = 1$ with the following increments.
   a) $\Delta t = 1$ b) $\Delta t = -1/2$ c) $\Delta t = 1/2$

9 For the given function compute the equation of the secant line through the point $(x_0, f(x_0))$, which is determined by the given increment, Δx.
   a) $f(x) = \sin(x)$; $x_0 = 0$, $\Delta x = \pi/2$ b) $f(x) = \tan(x)$; $x_0 = 0$, $\Delta x = \pi/4$

   c) $f(x) = \cos(\pi x)$; $x_0 = 0$, $\Delta x = -1/2$ d) $f(x) = -\sin(\pi x)$; $x_0 = 1$, $\Delta x = 1/3$

   e) $f(x) = 2\sec(x)$; $x_0 = \pi/4$, $\Delta x = -\pi/4$ f) $f(x) = \sin(x) + \cos(x)$; $x_0 = \pi/4$, $\Delta x = -\pi/4$

10 Fick's First Law describes the diffusion of a solute through tube. The one-dimensional discrete form of Fick's equation relates the average amount of a solute dissolved in a non-stirred solution that passes a position x in a time interval, this is called the average flux:
$$\Delta Q/\Delta t = -D \cdot A \cdot \Delta C/\Delta x$$
   where   Q(t) is the "flux" past a point x;
               C(x) is the solute concentration at position x along the tube;
               D is the "diffusion" constant for the solute in the fluid;
               A is the cross-sectional area of the tube.
   a) Write the form of Fick's Law for a fluid is in a tube of radius 2 and the diffusion coefficient is 0.7.
   b) If the concentration C(x) is constant, show that the diffusion is zero.
   c) If the concentration C(x) is linear show that the flux Q is also linear.

11 A fox chases a hare; initially, the fox is at the point (0,10) and the hare is at the point (10,0) of a coordinate system. Assume that the hare runs 1 unit/min in a positive direction along the x-axis. Assume that the fox runs in a straight line from its position toward the hare's position at the start of the minute, increasing its speed so that its y-coordinate decreases by one unit each minute.
   a) What will be the positions of the fox and hare after one minute? Illustrate these on a graph.
   b) What will be the positions of the fox and hare after three minutes? Illustrate these on a graph.
   c) Repeat parts (a) and (b) assuming that the fox is "smart" and runs in the direction of the point where the hare will be at the end of each minute of time.

12 Assume that the average rate of change in the concentration of a drug in a patient's blood over the nth hour interval $I_n = [n-1, n]$ is $r_n = 10^{-2}(5 - n)$ g/l. Assume that the concentration at t = 0 is C(0) = 50 mg/l. What is the concentration at t = 6 hr?

13 Compute the average rate of change of the logistics function $f(t) = 3/(1 + 2e^{-5t})$ over the interval [0,T] where T is the time at which f(t) = 2.

14 A goat bounds up a hill described by the $h(x) = 10 - (x-5)^2$. If the goat leaps one vertical unit in each bound, how many bounds will it take the goat to reach the top of the hill? What will be the longest and shortest horizontal distances covered in one jump?

Section 3.2 The Derivative

Introduction.

The principal idea underlying *differential calculus* is that if a function's graph is a smooth curve then there exists a companion function, called its **derivative**, that describes the rate of change in the curve.

Geometrically, a function f(x) is "differentiable" at x-values for which its graph has a tangent line; the slope of the tangent line is the value of the function's derivative. The tangent line provides a "linear" or "first-order" approximation of the function's graph near the "tangent point". The derivative's value at a specific point is interpreted as the rate of change in the function's values.

To define these concepts rigorously one must use limits. Differential calculus is based on the concept that a function's "instantaneous rate of change", its derivative, at a point x_0 is the limit of its "average rate of change over the interval $[x_0, x_0 + \Delta x]$" as the length Δx of the interval approaches zero. A function's average rate of change over an interval is the change in its values at the interval's endpoints divided by the length of the interval. This is the difference quotient introduced in the last section. The derivative of f is thus defined as the limit of $\Delta f/\Delta x$ as the length Δx approaches zero. Establishing derivatives by the limit process can be complicated and difficult. However, in subsequent sections simple "Derivative Rules" will be given that allow us to evaluate derivatives quickly and easily without evaluating limits.

Secants, Tangents, and Derivatives.

Tangent lines are defined at points where a graph is continuous and relatively smooth. The first tangent that you probably encountered is the tangent to a circle, which is defined as a line that touches the circle at only one point, the point of tangency. For more general curves, i.e., the graphs of functions, we can not describe a tangent line so easily. Instead, the tangent is defined as the limit of secant lines. The limit process considers the slopes of secant lines passing through a fixed point, the point of tangency, and nearby points on the curve. If, as the second point on the curve approaches the tangent point, the associated secant lines approach a single line, then this limiting line is defined to be the tangent line; its slope is defined to be the "derivative" of the function at the tangent point.

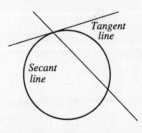

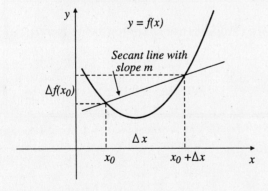

In the adjacent graph of a function f(x), the point on (x_0, y_0), where $y_0 = f(x_0)$, is a fixed reference point. Assuming that the function is defined on an interval about x_0, for each sufficiently small increment Δx we can construct a secant line through the point $(x_0, f(x_0))$ and through a second point with x-coordinate $x_0 + \Delta x$, $(x_0 + \Delta x, f(x_0 + \Delta x))$. The slope of the secant line through these two points is

$$m_S = \Delta f(x_0)/\Delta x = [f(x_0 + \Delta x) - f(x_0)] / \Delta x$$

If f(x) is continuous at x_0 then as the increment Δx approaches zero the points $(x_0 + \Delta x, f(x_0 + \Delta x))$ on the graph of f will approach the point (x_0, y_0).

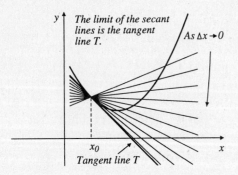

The secant lines illustrated at the left for several positive values of Δx. As the increment Δx becomes smaller the corresponding secant lines appear to approach a single line. This common limiting line is defined to be the tangent line to the graph at (x_0, y_0). The slope of a tangent line is consequently found as the limit of the slopes of the secant lines.

Slope of the Tangent Line

$$m_T = \lim_{\Delta x \to 0} \Delta f(x_0) / \Delta x$$

If this limit exists, it is a unique number, which is called *the derivative of f at x_0*.

DEFINITION OF THE DERIVATIVE.

A function f is **differentiable at x_0** if the **derivative of f at x_0** defined by

$$f'(x_0) \equiv \lim_{\Delta x \to 0} \{f(x_0 + \Delta x) - f(x_0)\}/\Delta x \quad \text{exists;}$$

If this limit fails to exist, then f is **not differentiable at x_0**.

A function is **differentiable on an open interval I** if it is differentiable at each x in I.

The **derivative of f(x)** is the function **f'** defined by

$$f'(x) = \lim_{\Delta x \to 0} \{f(x + \Delta x) - f(x)\}/\Delta x$$

at each x-value for which the limit exists.

While the ideal of a tangent line limit motivated the definition of the derivative, the limit definition of the derivative does not depend on any geometric property, it is simply algebraic. In fact, it works the other way around, the actual definition of a tangent line is given in terms of the derivative. It is the line through the point $(x_0, f(x_0))$ with slope $f'(x_0)$.

DEFINITION OF A TANGENT LINE

If f(x) is differentiable at x_0, the graph of $y = f(x)$ has a **tangent line at the point $(x_0, f(x_0))$** given by the equation

$$y = f(x_0) + f'(x_0)[x - x_0]$$

Derivative Notation

Many different notations are used to denote a derivative. The most commonly used derivative notations are:

Notation:	Which is read as:
f' or $f'(x)$	"f - prime" or "f - prime of x";
y' or $y'(x)$	"y - prime" or "y - prime of x";
dy/dx or df/dx	"the derivative of y, or f, with respect to x" or more simply as "dy, dx" or "df, dx";
$D_x y$ or $D_x f$	"the derivative with respect to x of y, or of f."
$y°$	"dy/dt" The small circle is sometimes used in the physical sciences and engineering to denote a derivative with respect to time.

When it is important to emphasize the x-value at which a derivative is being evaluated the notation will include the value explicitly, such as

$f'(3)$	"the derivative of f at x = 3" or more simply as "f - prime of 3";
$dy(x_0)/dx$ or $d/dx(y(x_0))$	"the derivative of y with respect to x at x_0" or more simply "the derivative of y at x_0";
$d[f(x)]/dx \big|_{x = -2.5}$	"the derivative of f(x) with respect to x, at x equals negative two point five."

The choice of symbols for independent variables and functions will of course vary. When a derivative is indicated by a prime it is taken with respect to the variable used with the function.

$f'(s)$	"f - prime of s" which would be understood to be the derivative of f with respect to s;
$g'(t_0)$	"the derivative of g at t_0", which would be the derivative of g with respect to t, evaluated at $t = t_0$.

Evaluating derivatives as limits.

When evaluating derivatives by the limit process we will replace the increment notation Δx with the letter "h". The use of the letter h will help avoid unnecessary algebra errors that often occur when the x-part of Δx appears next to x terms. Thus, the derivative of f(x) will be evaluated as

$$f'(x) = \lim_{h \to 0} \Delta f(x)/h = \lim_{h \to 0} \{f(x + h) - f(x)\} / h$$

The difficulty in evaluating the derivative as a limit is that the denominator of the difference quotient approaches zero. Taking the limit of the numerator and denominator separately leads to division by zero, which is not allowed. Therefore the quotient must be algebraically "simplified" to eliminate the division by h before the limit is evaluated. This is done by manipulating the numerator Δf so that an h can be factored from its terms. Prior to taking the limit $h \neq 0$ and the factored h in the numerator can be canceled with the h in the denominator.

Example 1 Evaluating a derivative as a limit.

Problem Evaluate the derivative of $f(x) = x^2 - 5$ at $x = 3$ using the limit definition. Use this to give the equation of the tangent line to the graph of f at the point $(3, f(3)) = (3, 4)$.

Solution First, evaluate and simplify the difference $\Delta f(x)$ at $x = 3$ and increment $\Delta x = h$:

$$\Delta f(3) = (f(3 + h) - f(3)) = [(3 + h)^2 - 5] - [3^2 - 5]$$

$$= [9 + 6h + h^2 - 5] - [9 - 5] = 6h + h^2 = h(6 + h)$$

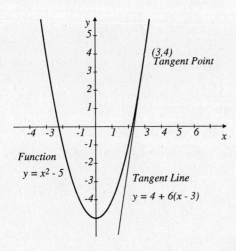

Function
$y = x^2 - 5$

Tangent Line
$y = 4 + 6(x - 3)$

Notice that an h has been factored out in the final form. When this simplified Δf is used to evaluate the difference quotient, $\Delta f/h$, this factored h will cancel with the h in the denominator. Canceling the h-terms is permissible before taking the limit to get the derivative.

$f'(3) = \lim_{h \to 0} \Delta f(3)/h$ The Definition of $f'(3)$.

$= \lim_{h \to 0} h(6 + h) / h$ From the algebra above.

$= \lim_{h \to 0} 6 + h = 6$ As $h \neq 0$ it may be canceled.

The slope of the tangent line at $(3,4)$ is thus $f'(3) = 6$. This line is sketched in the Figure at the left. The point-slope equation of the tangent line is

$$y = 4 + 6(x - 3)$$

METHOD TO EVALUATE A DERIVATIVE AS A LIMIT.

Step I *Evaluate the difference*: $\Delta f(x) = \{f(x + h) - f(x)\}$.
Expand the $(x + h)$ terms and algebraically simplify $\Delta f(x)$.

Step II *Factor an h out of $\Delta f(x)$ to obtain the form*:

$$\Delta f = \{f(x + h) - f(x)\} = h \cdot \{\text{terms involving x and h}\}$$

Step III Use the factored form of Δf, from Step II, in the numerator of the derivative limit:

$$f'(x) = \lim_{h \to 0} \Delta f(x)/\Delta x = \lim_{h \to 0} \{f(x + h) - f(x)\} / h$$

$$= \lim_{h \to 0} [h \cdot \{\text{terms involving x and h}\}] / h$$

Cancel the non-zero h in the denominator with the factored h in the numerator. Then, take the limit to determine the derivative:

$$f'(x) = \lim_{h \to 0} [\cancel{h} \cdot \{\text{terms involving x and h}\}] / \cancel{h}$$

$$= \lim_{h \to 0} \{\text{terms involving x and h}\}$$

Evaluate this limit using standard limit rules.

Evaluating derivatives as limits requires certain algebra skills and extra care to avoid clerical mistakes. Three common algebra techniques used to factor an h from the difference Δf are illustrated by the following examples.

Example 2 Derivatives of Polynomials.

Background Assume f(x) is a polynomial. To factor an h out of the difference $\Delta f(x) = f(x + h) - f(x)$:

(1) **expand the powers of (x + h)** in f(x + h),

(2) the terms in f(x + h) that involve x only, that do not have h to some power, always also appear in f(x), **subtract or cancel these x-only terms**.

(3) all remaining terms involve h, **factor out an h**.

The algebra is manageable for second or third degree polynomials using the expansions

$$(x + h)^2 = x^2 + 2xh + h^2, \quad \text{or} \quad (x + h)^3 = x^3 + 3x^2h + 3xh^2 + h^3$$

For higher powers the algebra becomes more complex but is essentially the same. The best approach to expand $(x + h)^n$, when $n \geq 4$, is to use the Binomial Theorem (See the Algebra Appendix.) If the polynomial f(x) has a term of degree n, x^n, then f(x + h) will have a term $(x + h)^n$ which when expanded will have only one term with no h-factor, it will be x^n. When you perform the subtraction f(x+h) - f(x) the constant terms and **all** of the terms x^n that do not have factors of h will cancel. This will leave only terms involving h's so that an h can always be factored out. In the next section this approach is used to establish a general rule for differentiating polynomials.

Problem Compute the derivative of $f(x) = x^2 - 5x + 7$ by the limit definition.

Solution **Step I** The difference is

$$\Delta f(x) = \{f(x + h) - f(x)\} = \{[(x + h)^2 - 5(x + h) + 7] - [x^2 - 5x + 7]\}$$

Step II Expand the (x + h) terms in f(x + h) and then subtract the f(x) terms. Then factor out an h.

$\Delta f(x) = \{ [x^2 + 2xh + h^2 - 5x - 5h + 7] - [x^2 - 5x + 7]\}$ Expand.

$= \{2xh + h^2 - 5h\}$ Subtract.

$= h(2x + h - 5)$ Factor out an h.

Step III Cancel the h term in the denominator of $\Delta f/\Delta x$ and take the limit as $h \to 0$.

$f'(x) = \lim_{h \to 0} \Delta f(x)/h$

$= \lim_{h \to 0} h(2x + h - 5) / h$ From Step II.

$= \lim_{h \to 0} 2x + h - 5$ Cancel the h.

$= 2x - 5$ Take $\lim_{h \to 0}$.

Thus, the derivative of $f(x) = x^2 - 5x + 7$ is $f'(x) = 2x - 5$.

Example 3 Derivatives of Square root functions.

Background If f(x) is the square root of a polynomial,

$$f(x) = SQRT\{P(x)\} \quad \text{where} \quad P(x) \text{ is a polynomial}$$

then Δf(x) is the difference between two square roots:

$$\Delta f(x) = SQRT\{P(x + h)\} - SQRT\{P(x)\}$$

The technique used to factor the difference is called **completing the square**. To simplify the notation, consider the difference of the square root of two terms, A and B. (In applications A and B will be polynomials.) Multiplying and dividing the difference of the square roots by the sum of the two square roots gives

$$\sqrt{A} - \sqrt{B} = [\sqrt{A} - \sqrt{B}] \cdot \left[\frac{\sqrt{A} + \sqrt{B}}{\sqrt{A} + \sqrt{B}}\right]$$

This simplifies considerably after multiplying the terms in the numerator; since the two $\sqrt{A} \cdot \sqrt{B}$ terms cancel. As $\sqrt{A} \cdot \sqrt{A} = A$, the square root disappears completely from the numerator giving:

$$\sqrt{A} - \sqrt{B} = [A - B] / \{ \sqrt{A} + \sqrt{B} \}$$

This technique converts Δf(x) to the difference of two polynomials divided by the sum of two positive square root terms. The difference can be expanded, simplified, and factored as in the previous example.

Problem Compute the derivative of f(x) = SQRT{3.5x + 7} by the limit definition.

Solution **Step I** The function difference is

$$\Delta f(x) = [f(x + h) - f(x)] = [SQRT\{3.5(x + h) +7\} - SQRT\{3.5x + 7\}]$$

Step II Complete the square for Δf(x); use the above formula with

$$A = 3.5(x + h) + 7 \quad \text{and} \quad B = 3.5x + 7$$

$$\Delta f(x) = \{[3.5(x + h) + 7] - [3.5x + 7]\} / [\, SQRT\{3.5(x + h) +7\} + SQRT\{3.5x + 7\}\,]$$

Then simplify the numerator by subtracting:

$$\Delta f(x) = 3.5h / [SQRT\{3.5(x + h) +7\} + SQRT\{3.5x + 7\}]$$

Step III Use the factored form of Δf from Step II to evaluate Δf/Δx, cancel the h's and take the limit as h → 0 to obtain the derivative.

$$f'(x) = \lim_{h \to 0} \Delta f(x)/h = \lim_{h \to 0} \big(3.5h / [SQRT\{3.5(x + h) +7\} + SQRT\{3.5x + 7\}]\big)/h$$

Cancel the h

$$= \lim_{h \to 0} 3.5 / [SQRT\{3.5(x + h) +7\} + SQRT\{3.5x + 7\}] = 3.5 / 2 \cdot SQRT\{3.5x + 7\} \; \text{Take} \lim_{h \to 0}$$

Thus,

$$D_x(SQRT(3.5x + 7) = 1.75 / SQRT\{3.5x + 7\}$$

Example 4 **Derivatives of Reciprocal functions**.

Background Assume that f(x) is the reciprocal of a polynomial, having the form

$$f(x) = 1 / P(x) \quad \text{with} \quad P(x) \text{ a polynomial.}$$

The technique required to factor the difference $\Delta f(x)$ is called **rationalizing**. This is the process of making a common denominator so that two fractions can be subtracted:

$$\{1 / P(x + h)\} - \{1 / P(x)\} = [P(x) - P(x + h)] / [P(x + h) \cdot P(x)]$$

The numerator of Δf is thereby reduced to a difference between polynomials which can be factored as in Example 2.

Problem Compute the derivative of $f(x) = 1 / (x - 3)$ using the limit definition.

Solution **Step I**: The function difference is

$$\Delta f(x) = \{f(x + h) - f(x)\} = \{[1 / ((x + h) - 3)] - [1 / (x - 3)]\}$$

Step II: The difference is "rationalized" and simplified.

$$\Delta f(x) = [1 / (x + h) - 3)] - [1 / (x - 3)]$$

$$= \{(x - 3) - (x + h - 3)\} / \{(x + h - 3)(x - 3)\}$$

$$= -h / \{(x + h - 3)(x - 3)\}$$

Step III Substituting the factored form of Δf into the difference quotient allows the h term to be canceled :

$$\Delta f(x)/h = [-h / \{(x + h - 3)(x - 3)\}] / h = -h / \{(x + h - 3) \cdot (x - 3) \cdot h\}$$

The derivative is then obtained by taking the limit as $h \to 0$:

$$f'(x) = \lim_{h \to 0} \Delta f(x)/h$$

$$= \lim_{h \to 0} -h / \{(x + h - 3) \cdot (x - 3) \cdot h\}$$

$$= \lim_{h \to 0} -1 / \{(x + h - 3)(x - 3)\} \qquad \text{Canceling}$$

$$= -1 / (x - 3)^2 \qquad \text{Since } \lim_{h \to 0} x + h - 3 = x - 3$$

Using negative exponents, this derivative can be written as

$$D_x(x - 3)^{-1} = -(x - 3)^{-2}$$

☑

One sided derivatives.

The limit involved in the definition of the derivative is a two-sided limit, we consider $h \to 0$ for both positive and negative values of h. If the h-values are restricted to being only positive or only negative, the same limit process gives what are called "one sided derivatives".

DEFINITION OF ONE-SIDED DERIVATIVES: $F'^{+}(x)$ AND $F'^{-}(x)$.

If $f(x)$ is defined on an interval $[x_0, b)$, then the **right-sided derivative of f at x_0** is

$$f'^{+}(x_0) = \lim_{h \to 0+} \Delta f(x_0)/h, \text{ if the limit exist.}$$

If $f(x)$ is defined on an interval $(a, x_0]$, then the **left-sided derivative of f at x_0** is

$$f'^{-}(x_0) = \lim_{h \to 0-} \Delta f(x_0)/h, \text{ if the limit exist.}$$

Example 5 Evaluating One-sided Derivatives

Problem Consider the multiply defined function

$$f(x) = \begin{cases} x^2 & \text{if } x < 0; \\ x & \text{if } 0 \leq x \leq 1; \\ x^2 - x + 1 & \text{if } 1 < x. \end{cases}$$

Evaluate the one-sided derivatives of f at $x = 0$ and at $x = 1$.

Solution The graph of the function f is sketched in the Figure at the right.

The function $y = f(x)$ is continuous. $f'(0)$ does not exist.

At $x = 0$.
For the left derivative, note that when $x < 0$ the first line of the definition of f gives $f(x) = x^2$. Therefore,

$$\text{if } h < 0, \quad f(0 + h) = f(h) = h^2$$

and the left-sided derivative at $x = 0$ is

$$f'^{-}(0) = \lim_{h \to 0-} [f(0 + h) - f(0)] / h$$

$$= \lim_{h \to 0-} [h^2 - 0] / h \quad = \lim_{h \to 0-} h = 0$$

To evaluate the right-derivative at $x = 0$, observe that for small positive h values, $0 < h < 1$, and hence, $f(h)$ is given by the second line of the function's definition:

$$f(0 + h) = f(h) = h$$

Therefore, at $x = 0$ the right-sided derivative is

$$f'^{+}(0) = \lim_{h \to 0+} [f(0 + h) - f(0)] / h$$

$$= \lim_{h \to 0+} [h - 0] / h \quad = \lim_{h \to 0+} 1 = 1$$

Both one-sided derivatives at $x = 0$ exist, but they are not equal:

$$f'^{+}(0) = 1 \quad \text{and} \quad f'^{-}(0) = 0$$

As we shall discuss following this example, this implies that f'(0) does not exist.

At x = 1.
To evaluate the left-sided derivative at 1, observe that for small negative values of h, $-1 < h < 0$, the sum $(1 + h)$ is less than 1 and greater than zero. Hence, $f(1 + h)$ is given by the middle line of the function's definition: $f(1 + h) = 1 + h$. Thus, the left-sided derivative at $x = 1$ is

$$f'^{-}(1) = \lim_{h \to 0^-}[f(1 + h) - f(1)] / h$$

$$= \lim_{h \to 0^-}[(1 + h) - 1] / h \quad = \lim_{h \to 0^-} 1 = 1$$

For positive h-values $(1 + h) > 1$ and $f(1 + h)$ is given by the third line of the function's definition. The right-sided derivative at 1 is then

$$f'^{+}(1) = \lim_{h \to 0^+}[f(1 + h) - f(1)] / h$$

$$= \lim_{h \to 0^+}[\{(1 + h)^2 - (1 + h) + 1\} - 1] / h \qquad \text{Evaluating } f(1 + h)$$

$$= \lim_{h \to 0^+}[\{(1 + 2h + h^2) - (1 + h) + 1\} - 1] / h \qquad \text{Expanding}$$

$$= \lim_{h \to 0^+}[1 + 2h + h^2 - 1 - h + 1 - 1] / h \qquad \text{Distributing the - sign}$$

$$= \lim_{h \to 0^+}[h + h^2] / h \qquad \text{Subtracting}$$

$$= \lim_{h \to 0^+} 1 + h = 1 \qquad \text{Canceling h's and taking the limit}$$

At x = 1, both one-sided derivatives exist and they are equal: $f'^{+}(1) = f'^{-}(1) = 1$, consequently $f'(1)$ exist and is equal to 1. ☑

When Derivatives do not exist.

Functions defined using the elementary functions, i.e., polynomials, exponentials, log, and trigonometric functions, are differentiable throughout their natural domains, with a few exceptions. However, the exceptions are important and must be recognized! Generally, for such ordinary functions, the points where a function is not differentiable are points where its definition changes form or where its tangent line becomes vertical. At the later, the derivative becomes "singular" or "infinite". This occurs when the formula for the derivative involves division and the denominator becomes zero, or when the derivative involves a function that becomes infinite, such as $\tan(x)$ at $x = \pi/2$. When a functions domain is limited, such as $SQRT(x)$ whose domain is $[0,\infty)$, the ordinary derivative can not exist at the endpoints of the domain interval. (In these cases we can sometimes establish one-sided derivatives.) To identify where a function is differentiable we instead ask "Where is it not differentiable?". To answer this you first identify the potential problem points, and then test these. One-sided derivatives can be used to ascertain whether or not a function is differentiable at an x-value where its function definition changes.

> RECALL   An ordinary two-sided limit exists if, and only if, the corresponding one-sided limits exist and are equal:
>
> $\lim_{h \to 0} g(x)$ exists   if, and only if, $\lim_{h \to 0^+} g(x) = \lim_{h \to 0^-} g(x)$

Similarly, $f'(x)$ exists if and only if both one-sided derivatives $f'^{+}(x)$ and $f'^{-}(x)$ exist and are equal.

$$f'(x_0) = L \quad \text{is equivalent to} \quad f'^{+}(x_0) = L \text{ and } f'^{-}(x_0) = L.$$

This equivalency provides two checks for the existence of a derivative.

If either one sided derivative $f'^{+}(x)$ or $f'^{-}(x)$ fails to exist then $f'(x)$ does not exist.

If $f'^{+}(x) \neq f'^{-}(x)$ then $f'(x)$ does not exist.

There are four common situations in which a derivative may not exist:

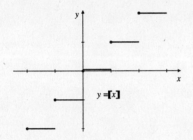

One is when the function "jumps" and is thus not continuous at x_0.

In this case one of the one-sided derivatives will be infinite. For instance, the greatest integer function $y = [\![x]\!]$ jumps at every integer x-value. An important observation is that **if $f'(x_0)$ exists then f must be continuous at x_0**. Thus, $[\![x]\!]$ is not differentiable at any integer value of x.

The second situation is when the x-value x_0 is an end-point of the domain of the function. In this case the function is not defined on an interval about x_0, and hence one of the one-sided limits can not exist. An example is given by $y = \sqrt{x}$ at $x_0 = 0$.

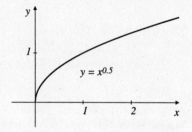

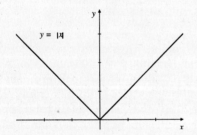

The third situation arises when the function is continuous at x_0 but the one-sided derivatives at x_0 are not equal.

This occurs when the function's graph makes a "sharp change in direction", like the graph of $y = |x|$ at $x_0 = 0$. Sometimes, this "sharp change" is not so obvious; look again at the split function $f(x)$ illustrated in Example 5 at the point $x = 0$.

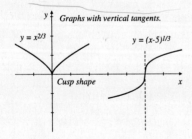

The fourth situation is when the function is continuous at x_0 and both one-sided derivatives are infinite, being either $+\infty$ or $-\infty$.

In this case the graph has a vertical tangent at the point $(x_0, f(x_0))$. The graph is said to be "cusp" shaped if the two derivatives are of different sign, i.e. $+\infty$ and $-\infty$. If they are of the same sign, then the graph is "s-shaped" when they are $+\infty$ and a "reversed s-shape" when the derivatives are $-\infty$. Infinite derivatives and vertical tangents are discussed after Example 8.

Example 6 The Absolute Value function is not differentiable at x = 0.

Problem Show that $y(x) = |x|$ is not differentiable at $x = 0$.

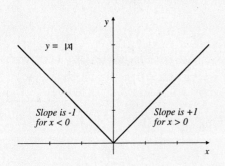

$y = |x|$

Slope is -1 for $x < 0$

Slope is +1 for $x > 0$

Solution The difference quotient $\Delta y(x)/h$ at $x = 0$ takes on two forms depending on the sign of the increment h.

$$\Delta y(0)/h = \{y(0 + h) - y(0)\} / h$$

$$= \{|h| - |0|\} / h$$

$$= \begin{cases} h/h = 1 & \text{if } h > 0; \\ -h/h = -1 & \text{if } h < 0. \end{cases}$$

Consequently, the one-sided derivatives at $x = 0$ are

$$y'^{-}(0) = \lim_{h \to 0+} \Delta y(0)/h = 1 \quad \text{and} \quad y'^{+}(0) = \lim_{h \to 0-} \Delta y(0)/h = -1$$

As these limits are not equal the derivative $y'(0) = \lim_{h \to 0} \Delta y(0)/h$ does not exist.

Example 7 The Greatest Integer function is not differentiable at integer values.

Problem Show that $y(x) = [\![x]\!]$ is not differentiable at $x_0 = 3$.

Solution The function y "jumps" at each integer value. For x just less than 3, but greater than 2, $y = [\![x]\!] = 2$, and at $x = 3$, $y = [\![3]\!] = 3$:

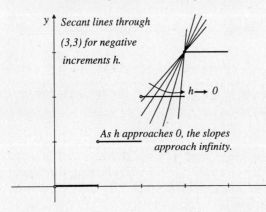

Secant lines through (3,3) for negative increments h.

$h \to 0$

As h approaches 0, the slopes approach infinity.

$$[\![x]\!] = \begin{cases} 2 & \text{if } 2 \leq x < 3; \\ 3 & \text{if } 3 \leq x < 4. \end{cases}$$

Consequently, the difference quotient $\Delta y(x)/h$ at $x = 3$ takes on two forms, depending on the sign of h. For positive h-values, $0 < h < 1$,

$$\Delta y(3)/h = \{[\![3 + h]\!] - [\![3]\!]\} / h = \{3 - 3\} / h = 0$$

For negative h-values, $-1 < h < 0$, the difference quotient is positive:

$$\Delta y(3)/h = \{2 - 3\} / h = -1 / h$$

The one sided limits required to evaluate the derivative $y'(3)$ are:

$$y'^{+}(3) = \lim_{h \to 0+} \Delta y(3)/h = \lim_{h \to 0+} 0 = 0$$

and

$$y'^{-}(3) = \lim_{h \to 0-} \Delta y(3)/h = \lim_{h \to 0-} -1/h = +\infty$$

As these limits are not equal, the derivative $y'(3)$ does not exist. The same conclusion would be reached if the value 3 were replaced by any integer n.

Example 8   **The Square Root function is not differentiable at x = 0.**

Problem   Show that $y(x) = \sqrt{x}$ is not differentiable at $x = 0$.

Solution   The square root of a negative number is not defined, the difference quotient

$$\Delta y(0)/h = \{ \sqrt{0 + h} - \sqrt{0} \} / h$$

is not defined for $h < 0$. Consequently, the limit required for the derivative does not exist, and y is not differentiable at 0. The right-sided derivative of $y = \sqrt{x}$ at $x = 0$ is $+\infty$, which is determined as follows:

$$y'^{+}(0) = \lim_{h \to 0+} \{ \sqrt{h} - \sqrt{0} \} / h = \lim_{h \to 0+} \sqrt{h} / h = \lim_{h \to 0+} 1 / \sqrt{h} = +\infty$$

Infinite derivatives.

If the derivative $f'(x_0)$ fails to exist because the $\lim_{\Delta x \to 0} \Delta f(x_0)/\Delta x$ is either $+\infty$ or $-\infty$, then derivative is said to be **infinite**. In this case, when f(x) is continuous at x_0 the function's graph has a vertical tangent line through the point $(x_0, f(x_0))$ and the shape of the graph is called an *s-shaped* or a *reversed s-shape*. A graph will also have a vertical tangent at a point where it is continuous and the one-sided derivatives are infinite and of opposite sign. In this case the slope of the graph must change sign at the point and the graph is said to be *cusp shaped*.

DEFINITION OF A VERTICAL TANGENT LINE.

If the function f(x) is continuous at $x = x_0$ and if $f'(x_0) = +\infty$ or $-\infty$, or
if the one-sided derivatives $f'^{+}(x_0)$ and $f'^{-}(x_0)$ are infinite and of opposite sign,

then   the graph of f has a **vertical tangent line at $(x_0, f(x_0))$**
and its equation is $x = x_0$.

Example 9   **Evaluating Infinite Derivatives**

Problem   Show that the derivative of $y = (x - 2)^{1/3}$ is infinite at $x = 2$. What is the tangent line to the graph at (2,0) and what shape does the graph have?

Solution   The derivative is evaluated as:

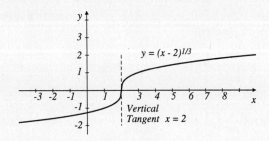

$y'(2) = \lim_{h \to 0}[y(2 + h) - y(2)] / h$
  Evaluating the function
$= \lim_{h \to 0}[(2 + h - 2)^{1/3} - 0] / h$

$= \lim_{h \to 0} h^{1/3} / h$
  Canceling the $h^{1/3}$
$= \lim_{h \to 0} 1 / h^{2/3}$
  $h^{2/3}$ is always positive!
$= +\infty$

The graph of f(x) has the vertical tangent line $x = 2$ at the point (2,0). As both the right and left derivatives are $+\infty$ the graph is "s-shaped" at this point, as illustrated above.

Approximating Derivatives.

The derivative $f'(x_0)$ can be numerically approximated by evaluating difference quotients using a small values of the x-increment h:

$$f'(x_0) \approx \Delta f(x_0)/h = \{f(x_0 + h) - f(x_0)\} / h. \qquad \textbf{Approximation of } f'(x_0)$$

This approximation consists of using the slope of a secant line as an approximation of the slope of the tangent line. The **absolute error in the approximation** is given by

$$|E(h)| = |f'(x_0) - \Delta f(x_0)/h|$$

Thus, to determine the error in an approximation it is actually necessary to know the value that is being approximated. Normally you would not know the value you are approximating. For many functions the absolute error of such approximations is proportional to the increment size, $|h|$. Thus, smaller increments will yield more accurate approximations.

Example 10 **Approximating derivatives by difference quotients**.

Problem Estimate the derivative at $x = 2$ of the functions considered in Examples 2 and 3 using the difference quotient with increment $h = 0.1$ and the absolute error of these estimates. The functions are: a) $f(x) = x^2 - 5x + 7$ and b) $y = \text{SQRT}(3.5x + 7)$ at $x_0 = 2$.

Solution a) The function's value at $x = 2$ is $f(2) = 1$; $f(2 + h) = (2 + h)2 - 5(2 + h) + 7$. Hence, the difference $\Delta f(2)$ is (after simplifying)

$$\Delta f(2) = f(2 + h) - f(2) = h^2 - h$$

The difference quotient is then

$$\Delta f(2)/h = h - 1.$$

For $h = 0.1$ this gives $\Delta f(2)/h = -0.9$. From Example 2, the value of $f'(2) = -1$. Thus, the absolute error in approximation $f'(2) \approx \Delta f(2)/h$ is $|\text{Error}| = |-1 - -0.9| = 0.1$.

b) At $x_0 = 2$ the function's value is

$$f(2) = \text{SQRT}(3.5 \cdot 2 + 7) = \text{SQRT}(14) \approx 3.741657386$$

The derivative of this function was calculated in Example 3:

$$f'(2) = 3.5 / [2 \cdot \text{SQRT}(3.5 \cdot 2 + 7)] = 1.75/\text{SQRT}(14) \approx 0.467707173$$

The difference Δf at $x = 2$ for any h is

$$\Delta f(2) = f(2 + h) - f(2) = \text{SQRT}(14 + h) - \text{SQRT}(14)$$

The calculator approximation of $\Delta f(2)/0.1$, rounded to 9 decimal places is

$$\Delta f(2)/0.1 \approx 0.464819982$$

The absolute error in the approximation $f'(2) \approx \Delta f(2)/h$ is

$$|\text{Error}| = |0.467707173 - 0.464819982| = 0.002887191$$

Exercise Set 3.2

1. Use the limit definition to evaluate the derivative of the given function at the points indicated:

   a) $f(x) = 2x + 3$ at $x_0 = 1$, $x_0 = 2$, and $x_0 = 0$ b) $g(x) = -5(x - 2)$ at $x_0 = 1$, $x_0 = 2$, and $x_0 = 3$

2. Using the limit definition, Determine the slope of the tangent line to the graph of $f(x) = x^2 + x$ at the point $(x_0, f(x_0))$ for the given x_0.

   a) $x_0 = 0$,   b) $x_0 = 1$,   c) $x_0 = 2$

3. Compute the instantaneous rate of change of $y = \text{SQRT}(x + 1)$ at the indicated x_0 using the limit process.

   a) $x_0 = 0$,   b) $x_0 = 1$,   c) $x_0 = 100$.

4. Use the limit process to determine the derivative of the given function.

   a) $y = -5x + 2$ b) $y = 3x^2$ c) $y = 1/(2x + 5)$

5. Determine the tangent line to the graph of the given function at the indicated point.

   a) $f(x) = 3x - 1$; $(1, 2)$ b) $g(x) = 3x^2 - 1$; $(0, -1)$ c) $f(t) = t^2 - 3t + 2$; $(2, 0)$

   d) $g(t) = t^2 - 4t + 4$; $(2, 0)$ e) $V(t) = 1 + 2t + 3t^2$; $(0, 1)$ f) $R(t) = \pi t^2$; $(2, 4\pi)$

   g) $y = 1/(x + 1)$; $(0, 1)$ h) $z = 1/(x^2 + 1)$; $(0, 1)$

6. For each of the following functions: (1) determine the times t at which the function is zero, and (2) compute the instantaneous rate of change of the function at these "zero values".

   a) $f(t) = 3t + 2$ b) $f(t) = t^2 - 4$ c) $M(t) = t^3 - 8$

   d) $g(t) = t^2 - t - 6$ e) $g(t) = 2t - t^2$ f) $f(t) = (t-2)^2$

7. Sketch the graph of the indicated function and use an "intuitive" approach to determine the one-sided derivatives $D_x^+ f(x_0)$ and $D_x^- f(x_0)$ at the indicated x-value.

   a) $f(x) = 2x + 1$, $x_0 = 1$ b) $f(x) = [\![x/2]\!]$, $x_0 = 3$ c) $f(x) = |3x|$, $x_0 = 0$

   d) $f(x) = x^{1/2}$, $x_0 = 0$ e) $f(x) = |3x - 1|$, $x_0 = 1/3$ f) $f(x) = x^{2/3}$, $x_0 = 0$

   g) $f(x) = \begin{cases} x^2 + 1 & \text{if } x < 1 \\ -2x + 4 & \text{if } x \geq 1 \end{cases}$ at $x_0 = 1$ h) $f(x) = \begin{cases} 2^x & \text{if } x < 0 \\ x/2 & \text{if } x \geq 0 \end{cases}$ at $x_0 = 0$

   i) $f(x) = [\![x]\!] \cdot x$, $x_0 = 2$ j) $f(x) = [\![x]\!]/x$, $x_0 = 2$

8. Show that the function $y = |x + 2|$ is not differentiable at $x = -2$.

9. Show that the function $y = [\![3x]\!]$ is not differentiable at $x = 2/3$.

10 i) Approximate the derivative $f'(x_0)$ by the difference quotient corresponding to h = 0.1, h = -0.1, h = 0.01, and h = 0.001.

ii) Compute the actual derivative $f'(x_0)$ as a limit. iii) Determine the error of each approximation in part i).

   a) $f(x) = 3x - 1$, $x_0 = 2$ b) $f(x) = x^2$, $x_0 = 2$ c) $f(x) = x^2 - 4x$, $x_0 = 2$

11 Approximate the value of $f'(x_0)$ by the difference quotient with h = 0.1 and h = -0.5 .
   USE A CALCULATOR TO EVALUATE THE FUNCTIONS.

   a) $f(x) = \sin(x)$, $x_0 = \pi/3$ b) $f(x) = \tan(x)$, $x_0 = 0$ c) $f(x) = \sin(3x)$, $x_0 = \pi$

   d) $f(x) = 10^x$, $x_0 = 0$ e) $f(x) = 10^x$, $x_0 = 1$ f) $f(x) = 2^x$, $x_0 = 0$

   g) $f(x) = e^x$, $x_0 = 0$ h) $f(x) = e^{-x}$, $x_0 = 0$ i) $f(x) = e^x$, $x_0 = 3$

   j) $f(x) = ln(x)$, $x_0 = 1$ k) $f(x) = \log(x)$, $x_0 = 1$ l) $f(x) = ln(x)$, $x_0 = 10$

12 a) Show that the different quotient of the function $f(t) = ln(t)$ has the form

$$\Delta ln(a)/\Delta t = ln[(1 + h/a)^{1/h}]$$

   b) Show that if $\lim_{\varepsilon \to 0}(1 + \varepsilon)^{1/\varepsilon} = e$ (the base of the natural log) then

$$\lim_{h \to 0} \Delta ln(a)/\Delta t = 1/a$$

   Hint: First use part a), set $\varepsilon = h/a$, and then use the first limit.

   c) Using the parts (a) and (b), what is the equation describing a population whose instantaneous growth rate is the reciprocal of the populations's size?

13 The difference quotient provides a numerical approximation of $f'(x)$ for any increment h. However, a much more accurate approximation can be made by averaging the difference quotients corresponding to positive h and negative h. This results in a formula called the **Central Difference Approximation of $f'(x)$**.

   a) Show that the average of the difference quotients $\Delta f/\Delta x$ with $\Delta x = h$ and with $\Delta x = -h$ is equal to the central difference

$$\delta f/h \equiv \{f(x + h) - f(x - h)\}/ 2h$$

   For the functions, b) $f(x) = x^2 - 5x + 7$ and c) $y = \sqrt{3.5x + 7}$
   use the Central Difference Approximation $f'(x) \approx \delta f/h \equiv \{f(x + h) - f(x - h)\}/ 2h$ to estimate the derivative $f'(x_0)$ at $x_0 = 2$ for h = 0.5, h = -0.1 and h = 0.001. Calculate the absolute errors of these approximations using the derivatives calculated in Examples 2 and 3. Compare the errors in these approximations to the corresponding errors using the simple difference quotient approximation of f', as in Example 10.

Section 3.3 Basic Derivative Rules

INTRODUCTION

In this section basic differentiation Rules are introduced. These Rules are of two types:

Specific Rules for differentiating particular functions.

Algebraic Rules for differentiating algebraic combinations of functions.

A *Rationale* is given for each Rule, which is intended to help you understand why the Rule has the form stated. For the more complex Rules the *rationale* may also provide a logical association to help you remember the Rule. These *Rationales* are not intended to be formal mathematical "Proofs", which may be found in more theoretical or advanced texts.

Specific Rules for differentiating a constant function $y = c$ and power functions $y = x^r$ are introduced first. Specific Rules for differentiating the exponential, logarithm and trigonometric functions will be introduced in the next section. Then, the basic Algebraic Rules are introduced for differentiating a constant multiple of a function, and the combination of two functions by addition/subtraction, multiplication and division. Combining these Algebraic Rules with the specific Rule for x^n, we will be able to differentiate a large class of rational functions. Another important Rule for differentiating the composition of two functions, called the Chain Rule, is introduced in Section 3.5. With the Chain Rule each specific function differentiation Rule can be generalize. The resulting General Derivative Rules will enable us to differentiate virtually any function we can express using elementary functions.

SPECIFIC DERIVATIVE RULES

The Constant Rule.

A constant function has the form $y(x) = c$ for all x. For any h, $y(x + h) = c$ and hence $\Delta y/\Delta x = 0$, for any increment Δx, and hence the rate of change in this function is always zero.

THE CONSTANT RULE

If c is a constant, then

$$D_x c = 0$$

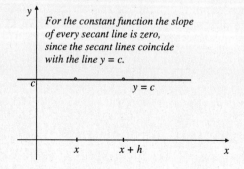

For the constant function the slope of every secant line is zero, since the secant lines coincide with the line y = c.

Rationale: The reason why this rule works is because a constant does not change; hence, its rate of change is zero! This is seen in the limit definition of the derivative. Let f be the constant function $f(x) = c$ for all x. Then,

$$f'(x) = \lim_{h \to 0} \{f(x + h) - f(x)\} / h$$

$$= \lim_{h \to 0} \{c - c\} / h$$

$$= \lim_{h \to 0} 0 / h = 0$$

Section 3.3 BASIC DERIVATIVE RULES

The Power Rule.

Power functions have the form $y = x^r$, where r is a fixed number. The derivative of $y = x^r$ is established in stages. First for positive integer powers, then for negative integers powers, then rational powers and finally for irrational powers. The important fact remains that, while the mathematical justifications for these various cases becomes more complicated, the same derivative Rule applies in all cases! We state the Power Rule for x^r and illustrate its application in examples before a *rationale* is presented for r a positive integer. *Rationales* for other types of exponents will be given later in the text.

THE POWER RULE

> For any real number r the derivative of the power $y = x^r$ is
>
> $$D_x x^r = r \cdot x^{r-1}$$
>
> provided both x^r and x^{r-1} are defined.

Example 1 Applying the Power Rule.

Problem Differentiate the given function using the Power Rule and indicate the derivative's domain.

a) x^3 b) $x^{3/4}$ c) x^{-5} d) x e) $\sqrt{x}$ f) $x^{5.7}$ g) $x^{2/3}$ h) x^π

Solutions In each case the exponent is multiplied time x raised to the exponent minus one.

a) $d/dx(x3) = 3x^{3-1} = 3x^2$, for all x.

b) If $y = x^{3/4}$ then $y' = (3/4)x^{3/4 - 1} = (3/4)x^{-1/4}$, for $x > 0$.

c) If $f(x) = x^{-5}$, then $f'(x) = -5x^{-5-1} = -5x^{-6}$, for $x \neq 0$.

d) $D_x x = D_x x^1 = 1 \cdot x^{1-1} = 1 \cdot x^0 = 1 \cdot 1 = 1$, for all x.

e) $d/dx \sqrt{x} = d/dx(x^{1/2}) = (1/2)x^{1/2 - 1} = (1/2)x^{-1/2}$, for $x > 0$.

f) $D_x x^{5.7} = 5.7 x^{5.7 - 1} = 5.7 x^{4.7}$, which is only valid for $x > 0$ since the exponent 4.7 expressed as a rational number in reduced form is $4.7 = 47/10$.

g) $D_x x^{2/3} = (2/3) x^{(2/3) - 1} = (2/3) x^{-1/3}$, for $x \neq 0$.

h) $d/dx \, x^\pi = \pi \cdot x^{\pi - 1}$ is defined for $x \geq 0$.

Rationale for the Power Rule: positive integer exponents.

Assume that n is a positive integer. The limit definition of the derivative of $y = x^n$ considers the limit of the difference quotient $\Delta y/\Delta x = \{(x + h)^n - x^n\}/h$. To evaluate this limit we must algebraically manipulate the numerator to factor an h from the difference $[(x + h)^n - x^n]$. (Review Example 2 of Section 3.2.) To accomplish this the term $(x + h)^n$ is expanded using the Binomial Theorem. To follow the argument you don't actually need to know the Binomial Theorem. The key observation is that the Binomial Theorem states that $(x + h)^n$ can be expanded as a sum of terms, each of which has the form of a constant times powers of x and h:

$$(x + h)^n = \sum c_{ij} \cdot x^i h^j$$ where i and j are positive integers such that $i + j = n$, and c_{ij} is an integer constant coefficient.

For instance, if $n = 5$, the Binomial Theorem gives the expansion

$$(x + h)^5 = x^5 + 5x^4 h + [\, 10x^3 h^2 + 10x^2 h^3 + 5x h^4 + h^5 \,]$$

where we have included in brackets all terms with h^i for $i \geq 2$. We do not need to know the details of how to calculate the coefficients in the Binomial expansion beyond the following observation. In the expansion, the term without an h, $x^n h^0 = x^n$, has coefficient 1 and the term with the h to the first power, $x^{n-1}h$, has coefficient n. Thus, the coefficient $c_{n,0} = 1$ and $c_{n-1,1} = n$, and we can write

$$(x + h)^n = x^n + nx^{n-1}h + [\text{other terms with factors } h^i \text{ for } i \geq 2]$$

At this point you might simply take this statement on faith, just as you "believe" statements of Chemist or Physicist about properties of molecules and electrons. Using this observation, the derivative of x^n is evaluated by the limit process as follows:

$d/dx(x^n) = \lim_{h \to 0} \{(x + h)^n - x^n\} / h$

$\quad = \lim_{h \to 0} \{x^n + nx^{n-1}h + [\text{other terms...}] - x^n\} / h$ Binomial Theorem

$\quad = \lim_{h \to 0} \{nx^{n-1}h + [\text{other terms...}]\} / h$ Subtracting x^n

$\quad = \lim_{h \to 0} \{nx^{n-1} + [\text{other terms with factors } h^i \text{ for } i \geq 1]\}$ Canceling h

$\quad = nx^{n-1}$ As all the terms with factors of h go to zero.

ALGEBRAIC DERIVATIVE RULES

The constant multiple Rule.

The first Algebraic Rule is for differentiating a constant multiple of a function. The Rule states that you can interchange the order of operations: the derivative of a constant times a function is just the constant times the derivative of the function.

CONSTANT MULTIPLE RULE

If $f(x)$ is a differentiable function and c is a constant, then

$$D_x\, c \cdot f(x) = c \cdot D_x\, f(x) = c \cdot f'(x)$$

Example 2   **Derivatives of functions with constant coefficients.**

Problem   Use the Power Rule and the Constant Coefficient Rule to differentiate the given function.

Solutions   a) $d/dx(5x^2) = 5 \cdot d/dx(x^2) = 5 \cdot 2x^{2-1} = 10x$

b) $d/dx(\pi x^6) = \pi \cdot d/dx(x^6) = 6\pi x^5$

c)   If $y = x^{1.2}$ then $d/dx(5y) = 5y' = 5 \cdot 1.2 x^{0.2} = 6x^{0.2}$

d) $D_x(-8x^{-5}) = -8 \cdot D_x x^{-5} = 40 x^{-6}$

Section 3.3   BASIC DERIVATIVE RULES

Rationale **for the Constant Multiple Rule.**

Using the limit definition of the derivative of $c \cdot f(x)$, and properties of limits,

$$D_x c \cdot f(x) = \lim_{h \to 0} \{c \cdot f(x + h) - c \cdot f(x)\} / h$$

$$= \lim_{h \to 0} c\{f(x + h) - f(x)\} / h \qquad \text{Factoring the constant.}$$

$$= \lim_{h \to 0} c \cdot \lim_{h \to 0} \{f(x + h) - f(x)\} / h \qquad \text{The limit of a product is the product of the limits.}$$

$$= c \cdot f'(x) \qquad \text{The } \lim_{h \to 0} c = c \text{ and by definition } \lim_{h \to 0} \{f(x + h) - f(x)\} / h = f'(x).$$

The Sum/Difference Rule.

The next Rule is like a "distributive law": $A \cdot (B \pm C) = (A \cdot B) \pm (A \cdot C)$, only multiplication by A is replaced by differentiation. It also allows the interchange of the order in which the addition and differentiation are done. It states that if two functions are differentiable then so is their sum and difference, and the derivative of a sum (or difference) is just the sum (or difference) of their derivatives.

THE SUM / DIFFERENCE RULE

If $f'(x)$ and $g'(x)$ exist then

$$D_x (f(x) + g(x)) = D_x f(x) + D_x g(x)$$

and

$$D_x (f(x) - g(x)) = D_x f(x)/dx - D_x g(x)$$

In short form:

$$(f + g)' = f' + g' \quad \text{and} \quad (f - g)' = f' - g'$$

Example 3   **Applying the Sum/Difference Rule to differentiate polynomials.**

Problem   Use the Sum/Difference Rule and the Constant Multiple Rule and the Power Rule to differentiate the given polynomial type function *term-by-term*.

   a) $y = 4x^2 + 5x - 6$ b) $g(x) = 11x^4 - 6x^3 + 77$ c) $y = 3x^{-5/2} + 2x^\pi$

Solution   a)   If $y = 4x^2 + 5x - 6$ then dy/dx is given by

$$y' = d/dx(4x^2 + 5x - 6) = d(4x^2)/dx + d(5x)/dx - d(6)/dx$$

$$= 4 \cdot 2x + 5 \cdot 1 - 0 = 8x + 5$$

   b)   If $g(x) = 11x^4 - 6x^3 + 77$, then

$$g'(x) = d/dx(11x^4 - 6x^3 + 77) = d(11x^4)/dx - d(6x^3)/dx + d(77)/dx$$

$$= 11 \cdot 4x^3 - 6 \cdot 3x^2 + 0 = 44x^3 - 18x^2$$

This same procedure can be applied to generalized *polynomial-like* functions which are composed of sums of power terms which may have non-integer exponents.

c) If $y = 3x^{-5/2} + 2x^{\pi}$ then y' is evaluated term-by-term as

$$y' = D_x(3x^{-5/2} + 2x^{\pi}) = D_x(3x^{-5/2}) + D_x(2x^{\pi})$$

$$= 3 \cdot (-5/2)x^{-7/2} + 2 \cdot \pi x^{\pi - 1} = -(15/2)x^{-7/2} + 2\pi x^{\pi - 1}$$

☑

Rationale for the Sum/Difference Rule.

The derivative of the sum $u(x) = f(x) + g(x)$ is given by

$u'(x) = \lim_{h \to 0}\{u(x + h) - u(x)\} / h$

$= \lim_{h \to 0}\{[f(x + h) + g(x + h)] - [f(x) + g(x)]\} / h$ Substitute $f + g$ for u

$= \lim_{h \to 0}\{[f(x + h) - f(x)] + [g(x + h) - g(x)]\} / h$ Rearrange

$= \lim_{h \to 0}\{[f(x + h) - f(x)] / h + [g(x + h) - g(x)] / h\}$ Use $\{A + B\}/h = A/h + B/h$

$= \lim_{h \to 0}\{[f(x + h) - f(x)] / h\} + \lim_{h \to 0}\{[g(x + h) - g(x)] / h\}$ Limit of sum = sum of limits

$= f'(x) + g'(x)$ Definition of f' and g'

Example 4 The rate of change in the Wind Chill.

Background Weather forecast in the winter frequently include references to "wind-chill temperatures". **Wind chill** is the additional cooling, the rapid heat exchange, that occurs when skin is exposed to wind. The **Wind Chill Temperature** is determined from a **Wind Chill Index**, Q, which is a function that has been empirically determined in experiments. The function Q involves constants determined from measurements of water evaporation at different temperatures and wind speeds:

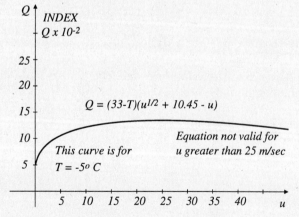

The Wind Chill Index $Q(u,T)$ for $T = 5°$.

Wind Chill Index

$$Q(u,T) = (33 - T) \cdot (10\sqrt{u} + 10.45 - u)$$

where T is the air temperature (degrees Celsius), u is wind speed (meters per second), and the constant 33 is the normal human skin temperature. The units of Q are kcal/m^2/hr. In the figure at the left, the Index $Q = Q(-5,u)$ is sketched as a function of wind speed u at constant temperature $T = -5°$.

The number that you are likely to hear on a weather forecast is not the Wind Chill Index but the **Wind Chill Temperature**, T_{WC}, which indicates "how cold it feels in the wind."

For a given air temperature, T_0, and the wind speed, u_0, one can calculate the Wind Chill index Q_0. The corresponding Wind Chill Temperature is the temperature T_{WC} that would give the same index value when the wind speed is "moderate", which is taken as the "calm" day wind speed of $u_c = 2.2$ m/sec (5 MPH). Equating the index function values, the equation $Q(u_c, T_{WC}) = Q(u_0, T_0)$ is solved for T_{WC}, this gives the formula for

$$T_{WC} = 33 - k\, Q(u,T) \qquad \text{\textbf{Wind Chill Temperature}}$$

where $k = (10\sqrt{U_c} + 10.45 - u_c)^{-1} \approx 0.043$. The Wind Chill Temperature, as a function of wind speed corresponding to an ambient temperature of $T_0 = -5°$ is sketched in the following figure.

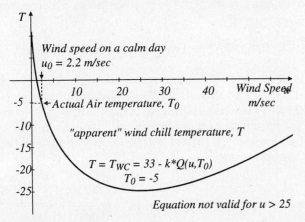

The Wind Chill Temperature as a function of wind speed u for constant temperature T.

Problems

(a) At a fixed temperature T what is the rate of change in the index Q as the wind speed u increases?

(b) What is the rate of change in the Wind Chill Temperature when the wind begins to gusts if the temperature is -5° and the wind speed is 35km/hr?

(c) How much colder would it seem under these conditions if you were exposed to a wind gust of 7 km/hr?

Solution (a) The rate of change in Q is computed as the derivative of Q with respect to u. The temperature T is treated as a constant. As $(33 - T)$ is a constant the derivative is

$$dQ/du = (33 - T)\, d/du(10\sqrt{u} + 10.45 - u) = (33 - T)\cdot[10\cdot(1/2)u^{-1/2} + 0 - 1] = (33 - T)(5u^{-1/2} - 1)$$

Notice that the derivative $Q'(u)$ is positive for wind speeds $u < 25$ m/sec (90 km/h). The empirical equation for Q is not applicable for greater wind speeds.

(b) The rate of change in the Wind Chill Temperature is the derivative, with respect to changes in wind speed:

$$D_u T_{WC} = d(33)/du - k\cdot dQ/du = 0 - k\cdot(33 - T)(5u^{-1/2} - 1) \approx -0.043(33 - T)(5u^{-1/2} - 1)$$

From this equation we see that the rate of change in the Wind Chill Temperature is negative for ambient temperatures $T < 33°$ and wind speeds below 25m/sec. This confirms that the Wind Chill Temperature decreases as the wind speed u increases.

For the given conditions we set $T = -5$ and convert the 35km/hr to $u = 9.72$ m/sec by multiplying 35 by 1000 (meters per kilometer) and dividing by 3600 (seconds per hour). The rate of change in the Wind Chill Temperature is then

$$T_{WC}' = -0.043(33 + 5)(5\cdot(9.72)^{-1/2} - 1) -0.9865 \approx -1°/(m/s)$$

(c) The apparent change in the temperature would be given by the difference ΔT_{WC}. In this example we approximate this change using the derivative. The approximate relationship is $T_{WC}' \approx \Delta T_{WC}/\Delta u$. We solve this for the change ΔT_{WC}. Thus, to a *first order approximation* a 7 km/hr gust would instantly drop the *effective temperature* by the amount equal to the product of the instantaneous rate of change times the increment of change in wind speed:

$$\Delta T_{WC} \approx \Delta u \cdot T_{WC}' = (7000/3600)\cdot -0.9865 \approx -1.92°$$

This confirms that a gust of wind on a cold day makes it feel very much colder.

The Product Rule

The derivative of a product of two functions must reflect how the product changes when each of the functions change. First, consider the change in the product A · B of two numbers when the number A is incremented by a change Δa and B is incremented by a change Δb. The product of the incremented numbers is $(A + \Delta a)(B + \Delta b)$, and hence the change in the product induced by the two increments is

$$(A + \Delta a)(B + \Delta b) - AB = \{AB + A \cdot \Delta b + B \cdot \Delta a + \Delta a \cdot \Delta b\} - AB = A \cdot \Delta b + B \cdot \Delta a + \Delta a \cdot \Delta b$$

The difference in the product is thus the sum of two *first order* terms, $A \cdot \Delta b$ and $B \cdot \Delta a$, and one *second order* term $\Delta a \cdot \Delta b$. Later, we shall indicate a *rationale* for the following Product Rule. In it, the limit definition of $D_x[f(x)\,g(x)]$ will involve a similar difference; in the limit as $h \to 0$ the first order terms will give the derivatives of f and g times the other function while the second order term will go to zero. Thus, the derivative of a product is a sum of two terms that each are the product of one function and the derivative of the other function.

THE PRODUCT RULE

If the functions f(x) and g(x) are differentiable then the product function $y = f(x)g(x)$ is differentiable and

$$d(f(x)g(x))/dx = f(x)g'(x) + g(x)f'(x)$$

Or, in a short form:

$$(f \cdot g)' = f'\cdot g + f \cdot g'$$

Notice that the Product Rule is *symmetric*. By this we mean that the same rule results if you interchange the symbols f and g. This is as you should expect. Since multiplication is commutative, $f \cdot g = g \cdot f$, there is no way to distinguish the contribution to a product as coming from the first term or the second term. Also note that the order of multiplication or of addition can be commuted, e.g. we can equivalently use

$$(fg)' = f'g + g'f \quad \text{or} \quad (fg)' = g'f + f'g$$

Example 5 **Applying the Product Rule to evaluate derivatives.**

Problem Use the Product rule to differentiate the following functions.

a) $y = (x^4 - 2)(x^5 + 3)$ and   b) $y = (x + 3x^2)(4 - x^{-2})$

Solution (a) Denoting the first term as f and the second as g we first compute the derivatives of f and g separately.

$$f(x) = x^4 - 2 \text{ and thus } f'(x) = 4x^3$$

$$g(x) = x^5 + 3 \text{ and thus } g'(x) = 5x^4$$

Applying the Product Rule gives:

$$y' = f(x)g'(x) + g(x)f'(x) = (x^4 - 2)\cdot 5x^4 + (x^5 + 3)\cdot 4x^3$$

At this point the derivative has been evaluated. However, normally you are expected to simplify such expressions. Multiplying out the terms and rearranging them in ascending powers of x leads to the derivative

$$y' = 5x^8 - 10x^4 + 4x^8 + 12x^3 = 12x^3 - 10x^4 + 9x^8$$

b) First, recognize that y is the product of two functions and then compute their individual derivatives. Set

$$f(x) = x + 3x^2 \quad \text{and} \quad g(x) = 4 - x^{-2}$$

Then

$$f'(x) = 1 + 6x \quad \text{and} \quad g'(x) = 2x^{-3}$$

Applying the Product Rule the derivative of $y = f \cdot g$ is

$$y' = (1 + 6x)(4 - x^{-2}) + (x + 3x^2)(2x^{-3})$$

Multiplying the terms and simplifying gives

$$y' = 4 + 24x + x^{-2}$$

Example 6 Using the Product Rule to determine a tangent line.

Problem Determine the tangent line at the point (1,4) to the graph of

$$y = (x^2 + 1)(x^{-3} + 1)$$

Solution The function y is the product of the two functions:

$$f(x) = x^2 + 1 \quad \text{and} \quad g(x) = x^{-3} + 1$$

The derivatives of these functions are

$$f'(x) = 2x \quad \text{and} \quad g'(x) = -3x^{-4}$$

By the Product Rule,

$$y' = f' \cdot g + f \cdot g' = 2x(x^{-3} + 1) + (x^2 + 1)(-3x^{-4}) = 2x^{-2} + 2x - 3x^{-2} - 3x^{-4}$$

The slope of the tangent line is then $y'(1) = -2$. The Point-Slope equation of the tangent line is $y = 2 - 2(x - 1)$.

CAUTION: Be careful when simplifying derivatives.

Frequently students correctly differentiate functions only to make algebraic mistakes when "simplifying" the answer. Try to be careful at this stage and if you have algebra problems seek help immediately.

Rationale for the Product Rule:

The Product Rule is derived by applying the limit definition of the derivative. The algebra used to rearrange the difference quotient is essentially that used in the consideration of the numerical product A·B discussed at the beginning of this subsection. We begin by considering the difference quotient $\Delta f \cdot g / \Delta x$ for the product of two differentiable functions f(x) and g(x). Observe that since

$$\Delta f(x) = f(x + \Delta x) - f(x) \quad \text{and} \quad \Delta g(x) = g(x + \Delta x) - g(x)$$

rearranging gives

$$f(x + \Delta x) = f(x) + \Delta f(x) \quad \text{and} \quad g(x + \Delta x) = g(x) + \Delta g(x)$$

Thus, substituting these into the difference $\Delta f \cdot g$ and simplifying gives

$$\Delta f \cdot g = f(x + \Delta x) \cdot g(x + \Delta x) - f(x) \cdot g(x)$$

$$= [f(x) + \Delta f(x)][g(x) + \Delta g(x)] - f(x) \cdot g(x)$$

$$= f(x) \cdot \Delta g(x) + g(x) \cdot \Delta f(x) + \Delta f(x) \cdot \Delta g(x)$$

The limit definition of the derivative of f·g is then evaluated as

$$\lim_{\Delta x \to 0} \Delta f \cdot g / \Delta x = \lim_{\Delta x \to 0} \{f(x) \cdot \Delta g(x) + g(x) \cdot \Delta f(x) + \Delta f(x) \cdot \Delta g(x)\} / \Delta x$$

$$= \lim_{\Delta x \to 0} f(x) \cdot \Delta g(x)/\Delta x + \lim_{\Delta x \to 0} g(x) \cdot \Delta f(x)/\Delta x + \lim_{\Delta x \to 0} \Delta f(x) \cdot \Delta g(x)/\Delta x$$

$$= f(x) \cdot \lim_{\Delta x \to 0} \Delta g(x)/\Delta x + g(x) \cdot \lim_{\Delta x \to 0} \Delta f(x)/\Delta x + \lim_{\Delta x \to 0} \Delta f(x) \cdot \Delta g(x)/\Delta x$$

By definition the first two limits give the derivatives of g and f, respectively. The third limit is zero, which follows from an important observation that a function must be continuous at each point where it is differentiable. This can be rigorously proven and is stated as a theorem.

THEOREM. **DIFFERENTIABLE FUNCTIONS ARE CONTINUOUS.**

If $f'(x_0)$ exists then f(x) is continuous at x_0.

This theorem states that if a function's graph is smooth enough that it has a tangent line at a point (x,f(x)) then it's graph can not "jump" at this point.

In particular, since f'(x) exists, we know from the above theorem that

$$\lim_{\Delta x \to 0} f(x + \Delta x) = f(x) \quad \text{or} \quad \lim_{\Delta x \to 0} \Delta f(x) = \lim_{\Delta x \to 0} \{f(x + \Delta x) - f(x)\} = 0$$

Hence,

$$\lim_{\Delta x \to 0} \Delta f(x) \cdot \Delta g(x)/\Delta x = \lim_{\Delta x \to 0} \Delta f(x) \cdot \lim_{\Delta x \to 0} \Delta g(x)/\Delta x = 0 \cdot g'(x) = 0$$

Using this limit we find the derivative of the product $f(x) \cdot g(x)$ as

$$D_x f(x) \cdot g(x) = \lim_{\Delta x \to 0} \Delta f \cdot g / \Delta x$$

$$= f(x) \cdot \lim_{\Delta x \to 0} \Delta g(x)/\Delta x + g(x) \cdot \lim_{\Delta x \to 0} \Delta f(x)/\Delta x + \lim_{\Delta x \to 0} \Delta f(x) \cdot \Delta g(x)/\Delta x$$

$$= f(x) \cdot g'(x) + g(x) \cdot f'(x) + 0$$

The Quotient Rule.

Mathematically, division is just a special form of multiplication:

$$A/B = A \cdot B^{-1}.$$

Consequently, we should be able to derive a Quotient Rule from the Product Rule. However, we should expect the Quotient Rule to be some what different from the Product Rule in one respect. Because a quotient is not symmetric, A/B is different from B/A, the Quotient Rule should not be symmetric. At this point we present the Quotient Rule and defer giving a *rationale* for it until Section 3.5.

THE QUOTIENT RULE

If f and g are differentiable functions and $g(x) \neq 0$, then

$$D_x f(x)/g(x)) = \{ g(x)f'(x) - f(x)g'(x) \} / [g(x)]^2$$

A short form of this Rule, for functions u and v, with $v \neq 0$, is:

$$(u/v)' = \{vu' - uv'\} / v^2$$

☺☺☺☺☺☺☺☺☺☺☺☺☺☺☺☺☺☺☺☺☺☺☺☺☺☺☺
Be sure to include the negative sign in front of
the term with the derivative of the denominator.
☺☺☺☺☺☺☺☺☺☺☺☺☺☺☺☺☺☺☺☺☺☺☺☺☺☺☺

Example 7 **Using the Quotient Rule.**

Problem Evaluate the derivatives of $y = f/g$ and $z = g/f$ for $f(x) = 3x$ and $g(x) = x^2 + 1$

Solution First, evaluate the derivatives of the two functions:

$$f'(x) = 3 \quad \text{and} \quad g'(x) = 2x$$

Applying the Quotient Rule the derivative of y is

$$D_x(3x/(x^2 + 1)) = \{(x^2 + 1)(3) - (3x)(2x)\} / (x^2 + 1)^2 = \{-3x^2 + 3\} / (x^2 + 1)^2$$

To differentiate z we use the Quotient rule with the roles of f and g reversed.

$$z' = D_x\{g(x)/f(x)\} = \{f(x)g'(x) - g(x)f'(x)\} / [f(x)]^2$$

$$= \{[3x \cdot 2x] - [(x^2 + 1) \cdot 3]\} / [3x]^2$$

$$= \{3x^2 - 3\} / 9x^2 = (x^{-2} - 1)/(3x^2)$$

Notice in this example that it is the position of the function that is important, not its name or symbol. The derivative of f/g is quite different from the derivative of g/f.

Example 8 **Applying the Quotient Rule to determine a tangent line.**

Problem The graph of $y = (1 + 2t)/(1 + t^3)$ is sketched below. Determine the tangent line to the graph at the point (2, 5/9).

Solution The equation of the tangent line requires the value of the derivative $y'(2)$. To use the short form of the Quotient Rule we introduce intermediate functions

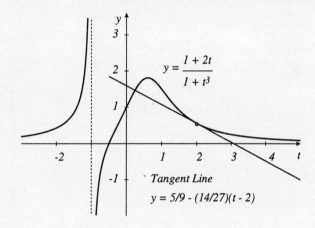

Tangent Line
$y = 5/9 - (14/27)(t - 2)$

$u = 1 + 2t$ and $v = 1 + t^3$

$u' = 2$ and $v' = 3t^2$

Then $y = u/v$ and as the derivative of u and v are:

$$y' = \{vu' - uv'\}/v^2$$

$$= \{(1 + t^3) \cdot 2 - (1 + 2t) \cdot 3t^2\} / (1 + t^3)^2$$

$$= \{(2 + 2t^3) - (3t^2 + 6t^3)\} / (1 + t^3)^2$$

$$= \{2 - 3t^2 - 4t^3\} / (1 + t^3)^2$$

The slope of the tangent line is this derivative evaluated at $t = 2$:

$$y'(2) = \{2 - 3 \cdot 2^2 - 4 \cdot 2^3\} / (1 + 2^3)^2 = -14/27$$

The equation of the tangent line is thus

$$y = 5/9 - (14/27)(t - 2)$$

You can now differentiate a large class of rational functions using the Rules given in this section. Generally, to differentiate a complex function you should apply the Algebraic Rules first and then apply Specific Rules. Algebraic Rules are applied in the same order that you would evaluate the function.

First apply the Quotient Rule, then the Product Rule, then the Sum/Difference Rule, then the Constant Multiple Rule and finally the specific Rules.

Caution! You still can not evaluate derivatives of many functions. For example, the terms $(3x + 7)^{10}$ **can not be directly differentiated with the Rules introduced in this Section.** It must first be expand to obtain a 10-th degree polynomial, which could then be differentiated term-by-term. This will be circumvented with the Chain Rule in Section 3.5.

Exercise Set 3.3

1. Determine the derivative of the given function and indicate the derivative Rules used.
   a) $f(x) = 7$
   b) $f(x) = 7x$
   c) $f(x) = x^2$
   d) $f(x) = x^{2.6}$
   e) $f(x) = x^{12}$
   f) $f(x) = x^{-5}$
   g) $f(x) = x^{-1}$
   h) $f(x) = x^{-4.6}$
   i) $f(x) = x^{3/2}$
   j) $f(x) = x^{2/3}$
   k) $f(x) = x^{1/8}$
   l) $f(x) = x^{0.6}$

2. Determine the derivative y' for the given function y.
   a) $y = 3x^2 - 2x + 1$
   b) $y = 5x^4 - 2x^5$
   c) $y = 3x^3 - 7x^{10}$
   d) $y = 1/2 x^3 - 2x^2 + 5$
   e) $y = x^{(3/2)} + 1$
   f) $y = x^{1/2} + 1$
   g) $y = x^{1/4} + x$
   h) $y = x^{1/2} + x$
   i) $y = 4x^{(3/4)} - x^2$
   j) $y = x^{1.1} - 3x^2$
   k) $y = x^{-2/3} + x^{2/3}$
   l) $y = x^{5/2} - x^{2/5}$

3. Let $f(x) = x^2 + 1$, $g(x) = 3x - 2$, and $h(x) = x^{-5}$. Evaluate the following derivatives.
   a) $D_x(f(x) \cdot g(x))$
   b) $D_x(f(x) - h(x))$
   c) $D_x(g(x) + f(x))$
   d) $D_x(h(x) \cdot g(x))$
   e) $D_x(f(x) - 3g(x))$
   f) $D_x\{1/f(x)\}$
   g) $D_x\{1/g(x)\}$
   h) $D_x\{1/h(x)\}$
   i) $D_x\{f(x)/g(x)\}$
   j) $D_x\{g(x)/f(x)\}$
   k) $D_x\{h(x)/f(x)\}$
   l) $D_x\{h(x)/g(x)\}$
   m) $D_x\{f(x) + g(x)/h(x)\}$
   n) $D_x\{[g(x) + h(x)]/f(x)\}$
   o) $D_x\{h(x) \cdot h(x)\}$

4. Evaluate the indicated derivatives assuming $f'(x) = x^2 - 2$, $g'(x) = 2x - 1$ and $h'(x) = 1/x$
   a) $D_x(f + g)(x)$
   b) $D_x(3f(x) - h(x))$
   c) $D_x(h(x) - g(x))$
   d) $D_x(h(x) + f(x) + g(x))$
   e) $D_x(f(x) - 2h(x))$
   f) $D_x(h(x) - f(x) - g(x))$

5. Evaluate the derivative y' in two ways: (1) by using the Product Rule and (2) by multiplying out the expression and then differentiating. Which is the faster method?
   a) $y = (3x + 2)(x^2 - 1)$
   b) $y = x^{1/2}(x + x^2)$
   c) $y = (t^2 - 3)(t^2 + 3)$
   d) $y = s^5(s^2 - 3s + 7)$
   e) $y = (t + 1)^2$
   f) $y = t^{3/2}(t^{1/2} + t^{5/2})$
   g) $y = (t^2 + t - 3)(t^2 - t + 3)$
   h) $y = (s^5 - 4s)(s^2 - 3s + 7)$

6. Determine the equation of the tangent line to the graph $y = f(x)$ at the indicated point (x_0, y_0).
   a) $y = x^2 - 3x$, $(-1, 4)$
   b) $y = x^3 + x^{-3}$, $(1, 2)$
   c) $y = 3x - x^3$, $(2, -2)$
   d) $y = 5x^{1/5} - x^{2/5}$, $(32, 6)$
   e) $y = x^2[3x - 5]$, $(-1, -8)$
   f) $y = x^3[1 + x^{-3}]$, $(1, 1)$
   g) $y = 3x/(1 - x^3)$, $(2, -6/7)$
   h) $y = (x^2 + 1)/(x^2 - 1)$, $(2, 5/3)$

7. In reference to the formulas of Example 4.
   a) What is the rate of change in the Wind Chill Index Q due to an increase in wind speed when $u = 2$ m/s and $T = -5°$?
   b) What is the rate of change in the Wind Chill Index Q due to an increase in temperature when $u = 2$ m/s and $T = -3°$? What is the rate of change in Q if the temperature decreases?
   c) What is the rate of change in the Wind Chill Temperature when the wind begins to gusts if the temperature is $-10°$ and the wind speed is 40km/hr?
   d) Under the conditions of c) if you were exposed to gust of 20km/hr how much colder would it seem?

Section 3.4 Derivatives of Elementary Functions

Introduction: In this section Specific Derivative Rules are introduced for the exponential and logarithmic functions and the six basic trigonometric functions. You may wish to review the properties of these elementary functions as discussed in Chapter 1.

DERIVATIVES OF EXPONENTIAL FUNCTIONS

The Rule for differentiating exponential functions will be stated with two formulas; the one that is most frequently used is the derivative of *The Exponential* function, $y = e^x$, which is a special simpler case of the more general Rule for differentiating $y = b^x$ for any base $b > 0$.

A *rationale* for the Exponential rule is provided by the following discussion, in which the general Exponential Rule is established in two stages. First, assuming that b^x is in fact differentiable, we argue that its derivative at $x = 0$ is a numerical value that depends continuously on the base number b. For each base b, this derivative is the slope of tangent line to the graph of $y = b^x$ at the point $(0,1)$. The slope of this tangent line is a unique number, which, it may surprise you to learn, has the value $ln(b)$. Hence, one way to define the natural logarithm function $ln(x)$ is as the derivative of b^x at 0, $D_x b^x |_{x=0}$. The fact that this function varies continuously with b is used to define the number "e" as the value of the base b for which the tangent line has slope 1, which is equivalent to saying that $ln(e) = 1$. The value of $D_x b^x$ for any x-value is then established using the derivative at $x = 0$ and algebra rules for exponents.

The following discussion is intended to exploit the relationship between derivatives and tangent lines to establish the Exponential Rule. It simultaneously illustrate a geometric link between exponential and logarithm functions. It provides a *rationale* for the Exponential Rule, not a proof, and could be read before or after the Examples demonstrating the use of the Exponential Rule.

The derivative of $y = b^x$ at $x = 0$.

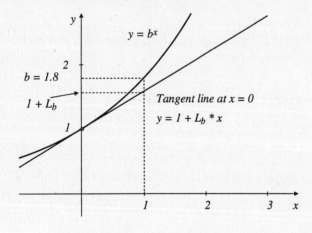

Our previous experiences with exponential functions would suggest that they are continuous and that their graphs are relative smooth for all x. Smooth enough that there is a tangent line at each point on their graph. By the definition of a tangent line,

$y'(0)$ is the slope of the tangent line to $y = b^x$ at the point $(0,1)$.

If the tangent line at $x = 0$ exists, then the limit involved in the derivative's definition

$$y'(0) = \lim_{h \to 0} \{b^{0+h} - b^0\} / h$$

must also exist. We temporarily will denote this limit's value by L_b. The subscript b emphasizes the dependence on the base b. Thus $L_b \equiv D_x b^x |_{x=0}$ and is given by the limit

$$L_b = \lim_{h \to 0} \{b^h - 1\}/h$$

since $b^{0+h} = b^h$ and $b^0 = 1$. With this notation we would write

$$d/dx(10^x)|_{x=0} = L_{10} \quad \text{and} \quad d/dx(2^x)|_{x=0} = L_2$$

In the figure above an exponential function $y = b^x$ and its tangent line are sketched, for $b = 1.8$. From the equation of the tangent line, $y = 1 + L_b x$, we know that the y-coordinate on the tangent line at x = 1 is $1 + L_b$. The number L_b can be estimated by sketching the curve $y = b^x$ and visually drawing its tangent line at (0,1). Then an estimate of L_b is $y_1 - 1$ where y_1 is the y-value on the tangent line at x = 1.

The value of L_b is zero when b = 1. This follows from the observation that the exponential with base b = 1 is simply the constant function y = 1 ($1^x = 1$). Hence, its tangent line at (0,1) is the horizontal line y = 1, which has slope zero.

In the following figure exponential curves $y = b^x$ are sketched for 6 different bases, from b = 1/4 to b = 4. The tangents to these curves at (0,1) were visually estimated and their approximate slopes estimated to give the corresponding values of L_b that are recorded in Table III.

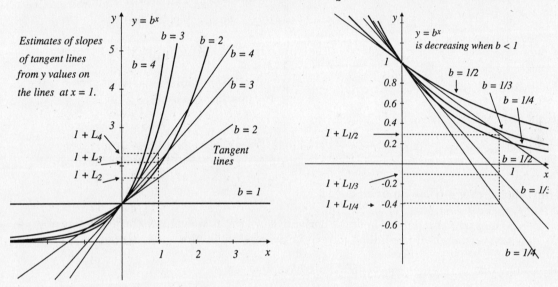

Table III Estimates of the slope L_b

Exponential base: b	1/4	1/3	1/2	1	2	3	4
Slope of tangent: L_b	-1.4	-1.1	-0.7	0	0.7	1.1	1.4

Observing these data, it would seem that the derivatives L_b have the following properties:

L_b is positive if b > 1; L_b is negative if b < 1; L_b increases as b increases.

Indeed, each of these properties can be established from the limit definition of L_b and the algebraic properties of exponential functions. The arguments in some cases are quite complex.

If the values of L_b vary continuously as b changes, then the function defined by $f(b) \equiv L_b$ for b > 0 is continuous. The Intermediate Value Theorem states that a continuous function assumes every y-value between any two function values. As b increases from 1 to 4, the values f(b) vary between $f(1) = L_1 = 0$ and $f(4) = L_4 \approx 1.4$. As the number 1 is between 0 and 1.4, it follows that for some choice of b in the interval [1,4] for which $f(b) = L_b = 1$. See the following Figure.

Furthermore, as f(b) is increasing, there is only one b value with this property. The unique number b for which $L_b = 1$ is very important. It is the number denoted by e.

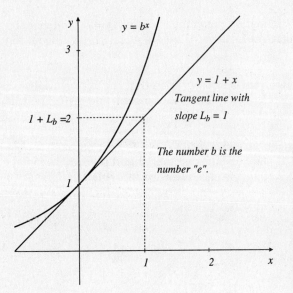

Figure 2.21

This number e is the same number you have encountered as the base of the "natural exponential" and the base of the "natural logarithm" function. This may provide your first mathematical definition of the number "e". The following definition of the number "e" replaces the value L_e with the limit that we originally introduced to define L_b.

DEFINITION. THE NUMBER e.

> The **number e** is the unique number that satisfies the identity
>
> $$\lim_{h \to 0} \{e^h - 1\} / h = 1$$
>
> The approximate value of e to 12 decimal places is
>
> $$e \approx 2.718281828459$$

The derivative of $y = b^x$ at an arbitrary x.

Using the limit definition of the derivative, we can now evaluate the derivative of $y = b^x$ at an arbitrary x-value. By definition $d/dx\, b^x$ is given by the limit

$$D_x b^x = \lim_{h \to 0} \{b^{x+h} - b^x\}/h$$

$$= \lim_{h \to 0} b^x \{b^h - 1\}/h \qquad \text{Factoring.}$$

$$= b^x \times \lim_{h \to 0} \{b^h - 1\}/h \qquad \text{As } b^x \text{ does not depend on h.}$$

$$= b^x \times L_b \qquad \text{The definition of } L_b.$$

This derivative formula indicates a very important property of the exponential function $y = b^x$. Its derivative is proportional to its own value, $y' = k \cdot y$ for some constant k and all x-values. The amazing fact is that the proportion constant is $k = L_b$, the derivative of y at $x = 0$. The derivative of the exponential function is therefor easy to determine, it is the original exponential function times the number $f(b) = L_b$.

Obviously the L_b function is important! It is required each time an exponential function b^x is differentiated. It seems natural that it should be given a special name so that we can refer to it more easily. You may be surprised to learn that the name given to this function is the *natural logarithm* function. Normally, instead of using b, x is used as the independent variable. The function $f(x) = L_x$ is of course denoted by $ln(x)$, which you know as the natural log of x. Yes, the same *ln* function that was introduce to you as the inverse of $y = e^x$ with respect to composition.

Defining functions as inverses is like a "chicken and egg" problem, "Which came first?" The following definition of $ln(x)$ is based solely on property of limits. Later in this section $ln(x)$ will be defined by its derivative, and in Section 5.4 of Chapter 5 another definition of $ln(x)$ will be given utilizing "integration", which is essentially the inverse of differentiation.

Section 3.4   DERIVATIVES OF ELEMENTARY FUNCTIONS

DEFINITION OF THE NATURAL LOGARITHM FUNCTION $LN(X)$ AS A LIMIT.

> The **natural logarithm function** $ln(x)$ is defined for $x > 0$ by
> $$ln(x) \equiv \lim_{h \to 0} \{x^h - 1\} / h$$

With the introduction of the natural log function we are now in a position to state the Exponential Rule for differentiating $y = b^x$. The important special case occurs when $b = e$, since $ln(e) = 1$, by definition!

EXPONENTIAL RULE

> For each positive number b, the derivative of $y = Exp_b(x)$ is
> $$D_x b^x = ln(b) b^x.$$
> In particular, with base e,
> $$D_x e^x = e^x$$

Example 1   **Differentiating exponential functions.**

Problem   Determine the indicated derivatives.

   a) y' if $y = 10^x$ b) $f'(3)$ if $f(x) = 2^x$ c) $D_x[2^x - 5e^x]$ d) $d/dx[x^5 \cdot 7^x]$

Solutions a) Applying the Exponential Rule with base $b = 10$:

$$y' = d/dx(10^x) = ln(10) 10^x$$

b) If $f(x) = 2^x$ then, for base $b = 2$, $f'(3) = ln(2) 2^3 \approx 0.48 \times 8 = 3.84$.

c) Using the Sum/Difference and Constant Multiple Rules :

$$D_x(2^x - 5e^x) = D_x(2^x) - 5 D_x(e^x) = ln(2) 2^x - 5e^x$$

d) Note that the function is a product of a power function, x^5, and an exponential function, 7^x. The two functions are quite different. The Power Rule is used to differentiate x^5 and the Exponential Rule is used to differentiate 7^x. Using the Product Rule and then the Power Rule and Exponential Rule the derivative is :

$$d/dx(x^5 \times 7^x) = x^5 \cdot d/dx(7^x) + 7^x \cdot d/dx(x^5)$$
$$= [x^5 \cdot ln(7) 7^x] + [7^x \cdot 5x^4] = [ln(7)x + 5] x^4 7^x.$$

Example 2 Bacteria growth and shape.

Background The relative dimensions of Rod-shaped bacterial such as *Escherichia coli (E. coli)* or *Salmonella typhimurium* depend on their intrinsic growth rate, μ, the time T between their chromosome replication and subsequent cell division, and the number C of chromosomes. The bacteria's shape may be long and thin or short and relatively thick. An indicator of shape is the ratio $S = D/L$ of the bacteria's diameter D and its length L. These shape parameters are related to the biological parameters by the equations

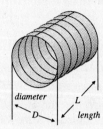

Typical rod-shape of bacteria.

$$D = K_1 2^{0.5C\mu} \quad \text{and} \quad L = K_2 2^{T\mu}$$

where K_1 and K_2 are constants.

Problem What is the rate of change in the bacteria's shape characteristic due to an increase in the growth rate?

Solution The shape characteristic is

$$S = D/L = K_1 \cdot 2^{0.5C\mu} / K_2 \cdot 2^{T\mu} = (K_1/K_2) 2^{(0.5C - T)\mu}$$

The derivative $D_\mu S$ is computed using the Exponential Rule for the independent variable μ and treating the parameters T and C as constants. The base is taken as $b = 2^{0.5C - T}$

$$S' = d/d\mu[(K_1/K_2)(2^{0.5C - T})^\mu] = (K_1/K_2)(2^{0.5C - T})^\mu \cdot ln(2^{0.5C - T})$$

which simplifies to give

$$S' = (K_1/K_2)(2^{0.5C - T})^\mu \cdot (0.5C - T) \cdot ln(2)$$

☑

From this equation it can be concluded that the bacteria's shape will vary with the changes in the intrinsic growth rate μ unless $T = 0.5C$. In this case the derivative is zero.

DERIVATIVES OF LOGARITHMIC FUNCTIONS: $ln(x)$ and $log_b(x)$.

The derivative Rules for the logarithm functions can be established either directly using the limit definition of the derivative, an approach that requires more advanced algebraic techniques than we presume of the reader, or indirectly, using the fact that the logarithm functions are the inverses of exponential functions. At this point we shall simply state the derivative Rules and later, in Section 3.6, provide a derivation employing a technique called "implicit differentiation".

$Ln(x)$ Rule

For $x > 0$, $\quad D_x ln(x) = 1/x.$

In an algebra class you were probably first introduced to the logarithm functions as inverse functions, defined by the equivalent relationships:

$$y = ln(x) \quad \Longleftrightarrow \quad x = e^y$$

$$y = \log_b(x) \quad \Longleftrightarrow \quad x = b^y$$

Previously, $ln(x)$ was defined as a limit. However, yet another definition of $ln(x)$ can be based on its derivative properties. All of these definitions are equivalent, they all define the same function. Observe that the following definition has two requirements that must be met by the ln function, one is the derivative property and the second is the value of $ln(x)$ at $x = 1$.

DEFINITION OF THE NATURAL LOGARITHM FUNCTION $LN(X)$ BY ITS DERIVATIVE.

The **natural logarithm** function $y = ln(x)$ is the unique function which satisfies

$$y' = D_x ln(x) = 1/x \quad \text{for } x > 0 \quad \text{and} \quad ln(1) = 0$$

The derivative of the general logarithmic function with base b, $\log_b(x)$, is determined from the derivative of $ln(x)$ since every logarithmic function is just a multiple of $ln(x)$. This relationship is given by the change of base formula, for any two positive numbers a and b,

$$\log_b(x) = \log_a(x) / \log_a(b)$$

When $a = e$, the $\log_a$ becomes $\log_e = ln$ and the change of base equation gives

$$\log_b(x) = ln(x) / ln(b) = [1/ln(b)]ln(x)$$

Since $1/ln(b)$ is a constant, applying the Constant Multiple Rule and the $ln(x)$ Rule to this equation gives the Logarithm Rule for differentiation.

LOGARITHM RULE FOR $LOG_B(X)$

For any base $b > 0$ and all positive x-values,

$$D_x \log_b(x) = [1 / ln(b)] \cdot 1/x = 1 / \{ln(b)x\}$$

Example 3 **Differentiating logarithmic functions.**

Problem Apply the specific derivative Rules for $ln(x)$ and $\log_b(x)$ to differentiate the given function.

a) $\log_2(x)$ b) $5 \text{Log}(t)$ c) $x^3 ln(x)$ d) $ln(x^6)$

Solutions a) For base $b = 2$: $d/dx(\log_2(x)) = 1/\{ln(2)x\}$

b) Using the independent variable t and the base $b = 10$ for $\text{Log}(x)$:

$$d/dt \, (5 \text{Log}(t)) = 5 \, d/dt \, \text{Log}(t) = 5 /[ln(10) \, t]$$

c) Using the Product Rule, $y = x^3 ln(x)$ is differentiated as

$$y' = x^3 \cdot D_x ln(x) + ln(x) \cdot D_x x^3$$

$$= x^3 \cdot 1/x + ln(x) \cdot 3x^2$$

$$= x^2 + ln(x) \cdot 3x^2 = (1 + 3ln(x))x^2$$

d) Using the exponent law for logarithms, $ln(x^6) = 6ln(x)$. It follows from the Constant Multiple Rule that

$$D_x ln(x^6) = D_x 6ln(x) = 6/x$$

☑

DERIVATIVES OF TRIGONOMETRIC FUNCTIONS

In this sub-section the Specific Rules for differentiating the basic trigonometric functions are presented. Following some examples, a *rationale* for the derivative Rules for sin(x) and cos(x) is presented that introduces an alternative approach to the traditional limit evaluation utilizing complicated trigonometric identities. It actually introduces a new concept, a "parametric curve" and the association of derivatives with the speed of a moving particle.

TRIGONOMETRIC DERIVATIVE RULES

The derivatives of the six basic trigonometric functions are:

$$D_x \sin(x) = \cos(x) \qquad\qquad D_x \cos(x) = -\sin(x)$$

$$D_x \tan(x) = \sec^2(x) \qquad\qquad D_x \cot(x) = -\csc^2(x)$$

$$D_x \sec(x) = \sec(x)\tan(x) \qquad\qquad D_x \csc(x) = -\csc(x)\cot(x)$$

CAUTION   Notice that the derivatives of the three "co.." functions, cos, cot, and csc, have a negative sign whereas the derivatives of sin, tan, and sec <u>do not</u>.

Example 4 **Differentiating trigonometric functions.**

Problem   Use the specific Trigonometric Derivative Rules to differentiate the given derivative.

a) $y = 5\cos(x)$ b) $y = t^2 + \tan(t)$ c) $y = x^3 \sec(x)$ d) $y = \sin(\theta)/\cos(\theta)$

Solutions a) $d/dx(5\cos(x)) = 5\, d/dx(\cos(x)) = -5\sin(x)$

b) If $y = t^2 + \tan(t)$ then $D_t y = y' = 2t + \sec^2(t)$

c) Using the Product Rule to differentiate $y = x^3\sec(x)$ gives

$$y' = x^3 D_x\sec(x) + D_x x^3 \cdot \sec(x) = x^3\sec(x)\tan(x) + 3x^2\sec(x)$$

d) Using the Quotient Rule, $d/d\theta\, y(\theta)$ is given by

$$y'(\theta) = \{\cos(\theta)\cdot D_\theta\sin(\theta) - \sin(\theta)\cdot D_\theta\cos(\theta)\} / \cos^2(\theta)$$

$$= \{\cos(\theta)\cdot\cos(\theta) - \sin(\theta)\cdot[-\sin(\theta)]\} / \cos^2(\theta)$$

$$= \{\cos^2(\theta) + \sin^2(\theta)\} / \cos^2(\theta)$$

$$= 1 / \cos^2(\theta) = \sec^2(\theta)$$

Since the function in Problem d) of the previous example is actually the tangent function, $y(\theta) = \sin(\theta)/\cos(\theta) = \tan(\theta)$, the example shows how the Rule for differentiating $\tan(x)$ is established using the derivatives of the sine and cosine functions. Similarly, the derivatives of $\cot(x)$, $\sec(x)$ and $\csc(x)$ can each be determined from the derivatives of $\sin(x)$ and/or $\cos(x)$ by applying the Quotient Rule.

RATIONALE FOR THE DERIVATIVE FORMULAS FOR sin(t) AND cos(t).
An alternative approach utilizing parametric curves.

The traditional approach to computing $D_t\sin(t)$ is directly by the limit definition. This requires the evaluation of several special limits that require the application of some complicated trigonometric identities. An alternative and quite different approach is given by the following. This consideration presents a different view of derivatives, introducing a parametric curve and associating the derivative of the curve's component functions with the slope of the tangent to the curve and the speed at which the curve is being generated. Several possibly new concepts are introduced in the development of these ideas, and readers with a weak mathematical background should perhaps read this through a few times without worrying about the details.

THE FOLLOWING DISCUSSION IS OPTIONAL AND NOT REQUIRED IN SUBSEQUENT SECTIONS OF THE TEXT.

The derivatives of both $\sin(t)$ and $\cos(t)$ are found simultaneously by considering the definition of $\sin(t)$ and $\cos(t)$ as *circular* functions that specify the coordinates of points on a unit circle. Points on the circle are indicated by their radian distance along the circle. Thus, we introduce the unit circle as a *parametric curve*, with points $P_t = (x(t), y(t))$ corresponding to values of a new variable t; $x(t) = \cos(t)$ and $y(t) = \sin(t)$. The derivatives of $\sin(t)$ and $\cos(t)$ are related then to the tangent line to the circle at P_t. This discussion serves as an introduction to the concept of parametric curves, which are the way paths or trajectories are described.

The Circular functions: sin(t) and cos(t).

Recall that the trigonometric functions are called "circular" functions because of their relationship to the unit circle $x^2 + y^2 = 1$. More precisely, $\cos(t)$ and $\sin(t)$ are the x and y coordinates of the point P_t located t-units around the circumference of the circle from the point $P_0 = (1,0)$. The distance is measured in a counter-clockwise direction if t is positive and in a clockwise direction if t is negative. The circle can thus be defined as a *trajectory*, consisting of the points

$$P_t = (\cos(t), \sin(t)), \text{ as } t \text{ varies.}$$

As t increases, the corresponding point P_t moves around and around the circle in a counter-clockwise direction. When a curve is specified in this way, by giving its coordinates as functions of a common variable, it is called a **parametric curve**. The common independent variable t is referred to as a **parameter**.

Parametric curves and derivatives

A **parametric curve** is specified by two functions, say f(t) and g(t) for t in some interval I. The curve, or trajectory, consists of the points (x,y) that are "generated" as t varies over I. The curve then consists of the points

$$P_t = (f(t), g(t)), \quad \text{for } t \in I$$

As t increases the associated point P_t moves along the curve, generating or tracing the curve as it goes. A parametric curve is thus a graph with something extra, it has a **direction**, indicated by the way the point moves as t increases. Not only does a parametric curve have a direction, but it also has a **speed**.

For instance, consider the unit circle given by the parametric curve $P_t = (\cos(t), \sin(t))$ for $t \in (-\infty, \infty)$. Its direction is counter-clockwise. The circle is in fact generated over and over again as t increase. It is generated once as t varies over $[0, 2\pi]$, and again as t increases through each interval of length 2π. But, the same unit circle is also generated by the curve defined by $P_t = (\cos(2t), \sin(2t))$. Notice that $x = \cos(2t)$ and $y = \sin(2t)$ satisfy $x^2 + y^2 = 1$. In this case however, the circle is traversed twice as fast, since a complete circle is generated as t varies over the interval $[0, \pi]$.

The unit circle can be generated in a clockwise direction by either the trajectory $P_t = (\cos(t), -\sin(t))$, or the trajectory $P_t = (\sin(t), \cos(t))$, in which case the point corresponding to $t = 0$ is $P_0 = (0,1)$.

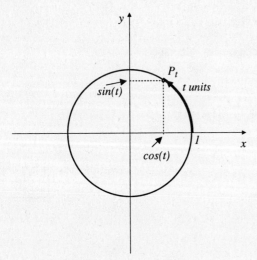

The unit circle $x^2 + y^2 = 1$.

To describe the direction and speed of a point traveling along the general parametric curve we think of the point P_t as the position of a physical particle that is moving in the x-y plane and is constrained to the curve by some force. If the force were to suddenly disappear at time t_0 then the moving particle would continue to move, carried by its momentum, in a straight line. That line is the **tangent line** to the curve at the point P_{t_0} where the restraining force vanished. At time t_0 the y-coordinate of the curve is changing at a rate $y'(t_0)$, where y' denotes $D_t y(t) = g'(t)$, and the x-coordinate is changing at a rate $x'(t_0)$, again, where $x' = D_t x(t) = f'(t)$. Dropping the subscript 0, for any number t,

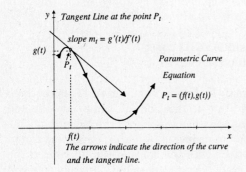

the slope of tangent line at P_t is

$$m_t = y'(t)/x'(t) = g'(t) / f'(t)$$

To establish the derivatives of the functions sin(t) and cos(t) we will work backwards from this slope equation for the parametric circle. We will want to establish the slope m_t using geometric arguments and then determine the derivatives. This however can not be accomplished without introducing the concept of speed. The following two paragraphs address the more complicated aspects of establishing for a general parametric curve the slope of the tangent line and the speed of a particle traveling along the curve.

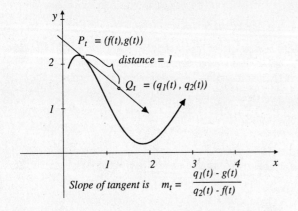

One procedure to determine the slope m_t of the tangent line at P_t involves first locating a second point on this line. Let Q_t denote the point on the tangent line one-unit away from the point P_t, in the direction along the tangent that the point generating the curve would travel were the restraining force to instantly cease. The slope m_t is the ratio of the change Δy in the y-coordinates of P_t and Q_t, to the corresponding change Δx in the x-coordinates. The coordinates of the point P_t are $(f(t), g(t))$. Denote the coordinates of the second point as $Q_t = (q_1(t), q_2(t))$. Then

$$m_t = g'(t)/f'(t) = \Delta y/\Delta x = \{q_1(t) - g(t)\}/\{q_2(t) - f(t)\}$$

We next use the algebraic property that relates the numerators and denominators of equal fractions. If A/B = C/D then there must be a constant δ such that $A = \delta \cdot C$ and $B = \delta \cdot D$. Applying this property to the two fractions that equal m_t, we conclude that the derivatives f' and g' are given by the equations

$$f'(t) = \delta[q_1(t) - f(t)]$$

and

$$g'(t) = \delta[q_2(t) - g(t)]$$

for some constant δ. Furthermore, we can show that the constant factor δ is in fact a function of the derivatives f' and g':

$$\delta = \{(f'(t))^2 + g'(t)^2)\}^{1/2}$$

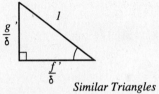

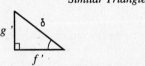

Similar Triangles

The reason why δ has this form can be seen by comparing similar right triangles in the adjacent Figure. The top triangle has hypotenuse equal to one, which is the distance from P_t to Q_t. To make the slope of the hypotenuse be the slope m_t of the tangent line we let its base length and height be

$$f'(t)/\delta = q_1(t) - f(t) \quad \text{and} \quad g'(t)/\delta = q_2(t) - g(t)$$

The lower triangle is a similar triangle obtained by multiplying all of the top triangle's side lengths by δ. Its base length is $f'(t)$ and its height is $g'(t)$ while its hypotenuse has length δ. The above formula for δ is then simply the Pythagorean equation for the length of the hypotenuse of the lower triangle. Unfortunately δ can not be determined from the point Q_t. To determine δ we must introduce a limit, this should not be surprising as we are seeking derivatives and derivatives always involve limits!

Right triangles with the same slope as the tangent line to the parametric curve $P_t = (f(t), g(t))$.

The dynamical interpretation of δ is the *speed* of the particle generating the trajectory when its at the point P_t. The speed is the rate of change in the distance P_t moves along the curve as t increases. A distance measured along the curve is called an **arc length**[†]. As the parameter t increases, say from $t = a$ to $t = b$, the point P moves from P_a to P_b and traverses a section of the curve whose arc length is denoted by

$$AL[P_a, P_b]$$

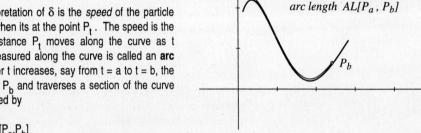

The distance along the curve between the two points is the arc length $AL[P_a, P_b]$

Unless the curve is a straight line, this arc length is greater than the Euclidian distance from P_a to P_b. The speed of the particle at P_t is then given as the derivative of the arc length with respect to the time parameter t. This is the limit of the average speed of the particle over the interval $[t, t+\Delta t]$ as the increment Δt approaches zero. The average speed is the arc length traversed in time Δt, $AL[P_t, P_{t+\Delta t}]$, divided by the time increment Δt.

$$\delta = \lim_{\Delta t \to 0} \Delta AL / \Delta t = \lim_{\Delta t \to 0} \{AL[P_t, P_{t+\Delta t}]\} / \Delta t \qquad \textbf{Speed at } P_t$$

The derivatives of sin(t) and cos(t).

The derivatives of the fundamental trigonometric functions are found by applying the above approach to the unit circle considered as the parametric curve generated by $P_t = (\cos(t), \sin(t))$. We first determine the point Q_t and then show that the speed $\delta = 1$ at each point on the trajectory. Combining these results leads to the desired derivatives.

[†] The concept of Arc Length is considered in detain in Chapter 7, Section 7.3 using integration.

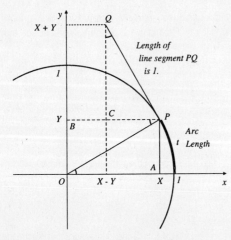

The unit circle as a parametric curve
$P_t = (\cos(t), \sin(t))$.

As P_t moves around the unit circle in a counter-clockwise direction the point Q_t, one unit along the tangent line to the circle at P_t, is as illustrated at the left. The point P_t is labeled as (X, Y), where

$$X = \cos(t) \qquad Y = \sin(t).$$

In the figure, the three right triangles OPA, POB, and QPC have the same angles (as $\angle$OPQ is a right angle) and since each has a hypothenuse of length one they are all congruent. They thus have exactly the same relative side lengths.

Using this diagram, the coordinates of the point $Q_t = (q_1(t), q_2(t))$ are

$$q_1(t) = X - Y = \cos(t) - \sin(t) \quad \text{and} \quad q_2(t) = Y + X = \sin(t) + \cos(t).$$

We now utilize the above general discussion and formulas with $g(t) = \sin(t)$ and $f(t) = \cos(t)$. First, we set the derivative of $\cos(t)$ equal to δ times the change in the x-coordinates:

$$D_t \cos(t) = \delta\{q_1 - X\} = \delta\{[\cos(t) - \sin(t)] - \cos(t)\} = -\delta \sin(t).$$

Similarly, the derivative of $\sin(t)$ is δ times the change in the y-coordinates:

$$D_t \sin(t) = \delta\{q_2 - Y\} = \delta\{[\sin(t) + \cos(t)] - \sin(t)\} = \delta \cos(t).$$

If we can establish that $\delta = 1$, then these two equations provide the basic derivative rules for the sine and cosine functions.

For this particular parametric curve δ can be obtained because it is easy to determine the arc length between points on the circle. We use the observation that the arc length $AL[P_t, P_{t+\Delta t}]$ is the distance along the circle from (1,0) to the point $P_{t+\Delta t}$ minus the distance to the point P_t. But, by definition, the point P_t is exactly t-units along the circle from (1,0) and $P_{t+\Delta t}$ is exactly $t + \Delta t$ units around the circle. Hence,

$$AL[P_t, P_{t+\Delta t}] = (t + \Delta t) - t = \Delta t$$

The speed δ is then given by the derivative of arc length

$$\delta = \lim_{\Delta t \to 0} \Delta AL/\Delta t = \lim_{\Delta t \to 0} \{AL[P_t, P_{t+\Delta t}]\}/\Delta t = \lim_{\Delta t \to 0} \Delta t/\Delta t = 1.$$

Exercise Set 3.4

1. Evaluate the indicated derivatives using the Exponential Rule:

   a) $D_x 2^x$ b) $D_x 3^x$ c) $D_x 3^{2x}$ d) $D_x x \cdot 2^x$ e) $D_x 1/3^x$

   f) $D_x 3^x \cdot 2^x$ g) $D_x[7^x - 10^x]$ h) $D_x[3^x/(1 + 3^x)]$ i) $D_x[5^x + 1/5^x]$

2. Evaluate the indicated derivatives:

   a) $D_x 4e^x$ b) $D_x[x \cdot e^x]$ c) $D_x[x/e^x]$ d) $D_x e^{3x}$ e) $D_x[1/e^x]$

   f) $D_x[5xe^{-x}]$ g) $D_x[4e^x + e^{-x}]$ h) $D_x[x^e + e^x]$ i) $D_x[e^x - e^{-x}]$

3. Use the specific logarithm Rules and the properties of logarithms to evaluate the derivatives:

   a) $D_x[3\ln(x)]$ b) $D_x[\ln(x) + 3x]$ c) $D_x \ln(x^2)$

   d) $D_x \ln(3x)$ e) $D_x \ln(2/x)$ f) $D_x \ln(4x^{-7})$

   g) $D_x[x\ln(x)]$ h) $D_x \ln(3/x^2)$ i) $D_x 1/\ln(x)$

Section 3.4 — Derivatives of Elementary Functions

4. Use the specific logarithm Rules and the properties of logarithms to evaluate the derivatives:

   a) $D_x \text{Log}(x)$
   b) $D_x \log_2(x)$
   c) $D_x \log_5(2x)$
   d) $D_x 1/\text{Log}(x)$
   e) $D_x \log_2(1/x)$
   f) $D_x[\log_5(x) - \log_3(x)]$
   g) $D_x[ln(x)\text{Log}(x)]$
   h) $D_x[ln(x) - \log_5(x)]$
   i) $D_x[ln(x)/\log_5(x)]$

5. Evaluate the indicate derivatives.

   a) $D_x[8\cos(x)]$
   b) $D_\theta[2\sin(\theta)]$
   c) $D_\theta[5\csc(\theta)]$
   d) $D_x[\cos(x) + \tan(x)]$
   e) $D_\theta[2/\sin(\theta)]$
   f) $D_\theta[2\cos(\theta) - \theta^2]$
   g) $D_\theta[\sin(\theta) \cdot \cos(\theta)]$
   h) $D_\theta[\csc(\theta) - 1/\sin(\theta)]$
   i) $d/dt[\sin(t)/t]$
   j) $D_t[t^2 \sin(t)]$
   k) $D_x[3\sin(x) + 1/\cos(x)]$
   l) $D_\theta[\theta \cot(\theta)]$
   m) $d/dx[\csc(x) + \sin(x)]$
   n) $D_t[\cos(t)/\sin(t)]$
   o) $D_x[\sin^2(x)]$

6. Differentiate the given function and evaluate the derivative at the indicated value.

   a) $y = xe^x$, at $x_0 = 0$
   b) $y = -e^x$, at $x_0 = ln(5)$
   c) $y = 3^x$, at $x = 1$
   d) $y = x/e^x$, at $x_0 = 0$
   e) $y = \sin(\theta) - \theta$, at $\theta_0 = 0$
   f) $y = t\, ln(t)$, at $t_0 = 1$

7. A particle moves along a linear path in an oscillatory fashion. Its distance at time t from a fixed point is given by the function f(t). The velocity of the point is given by the derivative f'(t). Compute its velocity at times $t_0 = 0$, $t_0 = \pi/2$, $t_0 = \pi$ when (a) $f(t) = \sin(t)$ and (b) $f(t) = t^2 \sin(t)$.

Parametric Curves.

If the curve P_t is given in parametric form $P_t = (f(t), g(t))$ then its tangent line at the point P_t has slope $m_t = g'(t)/f'(t)$. The speed that the point is traveling is

$$s = \text{SQRT}[f'^2(t) + g'^2(t)].$$

8. For the given parametric curve determine the tangent line to the curve and the speed the point is traveling at the indicated times t_0. Sketch the curve and the tangent line.

   a) The circle, $x^2 + y^2 = 4$: $P_t = (2\cos(t), 2\sin(t))$, at $t = \pi/2$.

   b) The ellipse, $x^2 + 0.04y^2 = 1$: $P_t = (\cos(t), 5\sin(t))$, at $t = \pi$.

   c) The rectangular hyperbola, $xy = 4$: $P_t = (4\sin(t), \csc(t))$, at $t = \pi/6$.

   d) The spiral $P_t = (t\cdot\cos(t), t\cdot\sin(t))$ at $t = \pi$. (To sketch this curve plot and connect the points corresponding to $t = 0, \pi/4, \pi/2, \ldots$ up to 4π to generate two rotations.)

   e) The "exponential spiral" $P_t = (e^t \cos(t), e^t \sin(t))$ at $t = 0$. (For this curve plot and connect the points corresponding to $t = 0, 0.2, 0.4, \ldots 1.0$ to sketch the initial segment of the spiral.)

9*. The derivatives of $\sin(2t)$ and $\cos(2t)$ will be easy to compute after we introduce the Chain Rule in the next section. However, they can be found by the same parametric method used to establish the derivative formulas for $\sin(t)$ and $\cos(t)$.
   a) What would be the parametric curve used?
   b) Sketch the diagram, like the one on page 225, illustrating the position P(t) on the unit circle and the corresponding tangent vector's terminus Q(t) for this curve.
   c) Carefully establish the speed term δ for this case.
   d) What are the equations for the two derivatives?

Section 3.5 The Chain Rule

Introduction.

In this section we introduce the last but probably most important algebraic derivative Rule, the **Chain Rule**, which is a rule for differentiating the composition of two function. Mathematics is a very orderly discipline. You learned early in your schooling that the order in which some arithmetic operations are performed will effect the value of a calculation. To ensure consistency, there is a hierarchy or standard order that must be followed when evaluating complicated function or algebraic expression. To differentiate complicated functions a similar sequential process must be applied, these are indicated by the Chain Rule.

With the Chain Rule we will be able to differentiate even the most complicated functions. The Chain Rule is based on the same principle of order of operations that is used to evaluate a composition of functions. The key step in applying the Chain Rule is to recognize that a complicated function can be written as the sequential composition of simpler functions. Then, the derivative of the complicated function can be evaluated utilizing the derivatives of the simpler functions. The rate of change in a composition resulting from an increase in the independent variable is actually multiplicative, reflecting the product of the consequent rates of change that occur in each function of the composition as it is evaluated.

Combining the Chain Rule with the specific derivative Rules for the elementary functions we are able to state General Derivative Rules that can be applied to differentiate elementary functions evaluated at an intermediate function $u(x)$ instead of x.

In specific problems complicated functions like $SQRT(x^2 - 6x + 1)$ or $(ln(x) + x)^3$ may be encountered. Neither of these functions can be differentiated using the derivative Rules given in the previous sections. To differentiate them we will introduce simpler functions, f and g, such that the given function has the form of a the composition function $f \circ g(x)$. For instance, $SQRT(x^2 - 6x + 1)$ is the composition $f(g(x))$ where $f(x) = SQRT(x)$ and $g(x) = x^2 - 6x + 1$. Similarly, $(ln(x) + x)^3 = f \circ g(x)$ for the functions $f(x) = x^3$ and $g(x) = ln(x) + x$. Notice that in each instance the sub-functions f and g are simple enough that their derivatives can be evaluated using the algebraic and specific Rules introduced in the previous sections of this Chapter. How the derivatives of these sub-functions are utilized to differentiate the composition function is indicated by the Chain Rule.

The Chain Rule

Differentiating a composition $f \circ g(x)$ involves a sequential process that parallels the process of evaluating the composition. Evaluating $f \circ g(x)$ can be visualized schematically as a two stage process, first, apply the function g, secondly, apply the function f:

$$x \text{ -------------> } g(x) \text{ ---------------> } f(g(x))$$

The change in a composition's value $\Delta[f \circ g]$ due to an increment Δx can be viewed similarly as a two stage process, first the increment Δx induces a change in g, Δg, and this change then induces a change in f, Δf:

$$\Delta x \text{ -------> } \Delta g = g(x + \Delta x) - g(x) \text{ --------> } \Delta f = f(g(x + \Delta x)) - f(g(x))$$

The difference quotient $\Delta[f \circ g]/\Delta x$ is the average rate of change in the composition $f(g(x))$ corresponding to the increment Δx. It is thus the average rate of change in $f(g(x))$ due to the change $\Delta g(x)$ times the average rate of change in $g(x)$ due to Δx. This statement is expressed as

$$\Delta f(g(x))/\Delta x = [\Delta f(g(x))/\Delta g(x)] \cdot [\Delta g(x)/\Delta x]$$

When the increment of change Δx approaches zero, this equation relating average rates of change leads to a similar statement relating instantaneous rates of change, i.e. derivatives; loosely,

$$D_x f(g(x)) = \lim_{\Delta x \to 0} \Delta f(g(x))/\Delta x = \lim_{\Delta x \to 0} [\Delta f(g(x))/\Delta g(x)] \cdot [\Delta g(x)/\Delta x] = D_g f \cdot D_x g$$

The corresponding derivative equation is called the Chain Rule. A more detailed *rationale* that considers the various limits involved is presented after the examples applying the Chain Rule.

THE CHAIN RULE

If f(x) and g(x) are differentiable for all x then $D_x f \circ g(x)$ exists and

$$D_x \, f \circ g(x) = f'(g(x)) \cdot g'(x)$$

In short form: If $u = g(x)$ and $y = f(u)$ then

$$dy/dx = dy/du \cdot du/dx$$

The condition that f and g are differentiable for all x can be refined to require that f be differentiable at $u = g(x)$ and $g(x)$ be differentiable at x to ensure that $f \circ g$ is differentiable at x.

The Chain Rule is <u>not</u> a symmetric Rule. When applying the Chain Rule, the order of the functions, not their names, is critical. Given functions f and g, the Chain Rule formulas for the derivative of the composition functions: $y = f \circ g$ and $z = g \circ f$ are formally different. The two derivatives are:

$$y'(x) = f'(g(x)) \cdot g'(x) \quad \text{and} \quad z'(x) = g'(f(x)) \cdot f'(x)$$

The Chain Rule is applied similarly for any composition of functions. For instance,

$$D_t H(V(t)) = H'(V(t)) \cdot V'(t)$$

where H' denotes $D_V H(V)$ and V' denotes $D_t V(t)$.

Example 1   Applying the Chain Rule.

Problem   For the given functions f and g, express the functions $y = f \circ g(x)$ and $z = g \circ f(x)$ directly as functions of x and evaluate their derivatives with respect to x using the Chain Rule.

$$f(x) = x^3 - 4x + 2 \quad \text{and} \quad g(x) = x^{1/2}$$

Solution   The functions y is given as a function of x by expanding the composition:

$$y = f \circ g(x) = f(x^{1/2}) = (x^{1/2})^3 - 4x^{1/2} + 2 = x^{3/2} - 4x^{1/2} + 2$$

In expanded form we see that y is a polynomial-like function that could be differentiated using the Power Rule and the Sum and Constant multiple Rules. However, the problem is to differentiate y using the Chain Rule. The first step is to compute the derivatives

$$f'(x) = 3x^2 - 4 \quad \text{and} \quad g'(x) = (1/2)x^{-1/2}$$

Then, by the Chain Rule

$$y'(x) = f'(g(x)) \cdot g'(x) = [3(x^{1/2})^2 - 4] \cdot (1/2)x^{-1/2}$$

$$= [3x - 4] \cdot (1/2)x^{-1/2} = (3/2)x^{1/2} - 2x^{-1/2}$$

The function z is given in terms of x as

$$z = g \circ f(x) = g(x^3 - 4x + 2) = (x^3 - 4x + 2)^{1/2}$$

To apply the Chain Rule to differentiate z we use the same derivatives f' and g'. But the Chain Rule applied to g∘f has the form

$$z'(x) = g'(f(x)) \cdot f'(x) = (1/2)[x^3 - 4x + 2]^{-1/2} \cdot [3x^2 - 4]$$

Example 2   **Using the short form of the Chain Rule.**

Problem   Compute the derivatives of 1 $y = e^{x^5}$ and 2 $w = \cos(1 - \theta^2)$.

Solution   a) The derivative of $y = e^{x^5}$ is found by introducing an intermediate variable u. Set

$$u = x^5 \quad \text{and then} \quad y = e^u.$$

The derivatives $dy/du = e^u$ and $du/dx = 5x^4$ are given by the Exponential and Power Rules. Applying the short form of the Chain Rule, the derivative of y is

$$dy/dx = dy/du \cdot du/dx = e^u \cdot 5x^4 = 5x^4 \, e^{x^5}$$

b) The function w is expressed as a composition by introducing an intermediate variable

$$v = 1 - \theta^2 \qquad \text{then} \qquad w = \cos(v).$$

The derivative of w with respect to θ is given by a short form of the Chain Rule, employing the variables v and θ in place of u and x, as

$$dw/d\theta = dw/dv \cdot dv/d\theta$$

Since $dv/d\theta = -2\theta$ and $dw/dv = -\sin(v)$, the derivative of w is

$$w' = -\sin(v) \cdot (-2\theta) = 2\theta \sin(1 - \theta^2)$$

General Derivative Rules.

The Chain Rule can be combined with each of the Specific Rules for differentiation to form General Derivative Rules. The General Rules are obtained by introducing intermediate functions u = u(x) and their derivatives. If f is an elementary function and u = u(x) is an "intermediate" function, then applying the Chain Rule to differentiate y = f(u(x)) results in the derivative

$$y' = f'(u(x)) \cdot u'(x)$$

Thus, in the list of General Derivative Rules given below, each derivative involves the product of u'(x) and another term that is a function of u(x).

GENERAL DERIVATIVE RULES

GENERAL POWER RULE

For any real number r, $\quad D_x u^r(x) = r \cdot u^{r-1}(x) \cdot u'(x)$

GENERAL EXPONENTIAL RULE

$$D_x e^{u(x)} = u'(x)\, e^{u(x)}$$

For any positive number b, $\quad D_x b^{u(x)} = ln(b)\, u'(x)\, b^{u(x)}$

GENERAL LOGARITHM RULE

$$D_x ln(u(x)) = u'(x) / u(x)$$

For any positive number b, $\quad D_x \log_b(x) = u'(x) / [ln(b)\, u(x)]$

Normally, you do not encounter a function that is expressed with the intermediate variable u = u(x) in the form indicated in the General Rules. To use these General Rules in specific differentiation problems you must first recognize the elementary function and the intermediate function u(x). The following guidelines should help you to deciding what parts of a complicated term should be selected as the intermediate function u(x).

Guide 1) Any expression (other than the independent variable) that is the argument of an elementary function, such as $2\pi x^3$ in the function $\sin(2\pi x^3)$, must be treated as an intermediate function u(x). In particular a function in an exponent, as tan(x) in the term $e^{\tan(x)}$, must be considered as an intermediate function u(x).

Guide 2) An expression within a set of parentheses is a candidate for u(x). Ask yourself "Why are the parentheses needed?" Look for repeated occurrences of the same sub-term in the complex function you are to differentiate. Normally, expressions under a square root sign or appearing to a power must be treated as an intermediate function.

General Derivative Rules

GENERAL TRIGONOMETRIC RULES

$$D_x \sin(u(x)) = \cos(u(x)) \cdot u'(x)$$

$$D_x \cos(u(x)) = -\sin(u(x)) \cdot u'(x)$$

$$D_x \tan(u(x)) = \sec^2(u(x)) \cdot u'(x)$$

$$D_x \cot(u(x)) = -\csc^2(u(x)) \cdot u'(x)$$

$$D_x \sec(u(x)) = \sec(u(x))\tan(u(x)) \cdot u'(x)$$

$$D_x \csc(u(x)) = -\csc(u(x))\cot(u(x)) \cdot u'(x)$$

A Word Of Advice

To successfully proceed with the study of Calculus it is essential that you master the use of the Chain Rule and learn to apply the General Derivative Rules.

Example 3   **Applying the Chain Rule and General Derivative Rules.**

Problem   Differentiate the expression by first identifying an intermediate function u and then applying the appropriate general derivative Rule.

     1   $(x^3 - 8x + 5)^5$ 2 $\text{Exp}(x^7 - 5x^2)$ 3 $\tan(t^3 - t^2 + 3)$ 4 $\ln(\sin(\theta))$ 5 $2^{\ln(x)}$

Solutions   For each problem we have set out the solution in a format that indicates:

     The given function;   The intermediate function u,
     the General Rule that applies;   why u was chosen;
     The derivative given by the General Rule.   and its derivative u'

     a) $y = (x^3 - 8x + 5)^5$ $u(x) = x^3 - 8x + 5$
          By General Power Rule   appears to the power 5;
          $y' = 5(x^3 - 8x + 5)^4(3x^2 - 8)$ $u'(x) = 3x^2 - 8$

     b) $y = \text{Exp}(x^7 - 5x^2)$ $u(x) = x^7 - 5x^2$
          By the General Exponential Rule   appears in the exponent;
          $y' = (7x^6 - 10x)\text{Exp}(x^7 - 5x^2)$ $u'(x) = 7x^6 - 10x$

c) $y = \tan(t^3 - t^2 + 3)$ $u(t) = t^3 - t^2 + 3$
By the General Trigonometric Rule   appears as the argument of tan;
$y' = (3t^2 - 2t)\sec^2(t^3 - t^2 + 3)$ $u'(t) = 3t^2 - 2t$

d) $y = ln(\sin(\theta))$ $u(\theta) = \sin(\theta)$
By the General Logarithm Rule   is the argument of the ln
$y' = \cos(\theta)/\sin(\theta) = \cot(\theta)$ function;
                                                                   $u'(\theta) = \cos(\theta)$

e) $y = 2^{ln(x)}$ $u(x) = ln(x)$
By the General exponential Rule   appears in the exponent;
$y' = ln(2)2^{ln(x)} \cdot 1/x$ $u'(x) = 1/x$.
$ = ln(2)2^{ln(x)}/x$

☑

Example 4   **Modeling the rate of exponential growth or decay.**

Problem   Verify that the rate of change in a population whose size is described by an exponential function $y = e^{kt}$ or $z = e^{-kt}$ is proportional to the size of the population and, hence, satisfies a Growth Model $y' = ky$ or a Decay Model $z' = -kz$, respectively.

Solution   Consider a population whose size at time t is given by $y = e^{kt}$. This equation has the form:

$$y = e^u \quad \text{with} \quad u = kt$$

Applying the General Exponential Rule, the derivative, or rate of growth, of the population size is

$$y' = e^u \cdot du/dt = e^u \cdot k = ky$$

For an exponentially decaying function the same approach is applied using $u = -kt$. The derivative of $z = e^{-kt}$ is

$$z' = e^u \cdot u' = e^{-kt} \cdot (-k) = -kz.$$

Discussion   The two Model Equations

$$y' = ky \quad \text{and} \quad z' = -kz$$

are called *differential equations* because they involve a derivative. They are also called **Dynamic Models** since they indicate the dynamics, or rates of change in the modeled populations. The functions $y(t) = e^{kt}$ and $z(t) = e^{-kt}$ are called *solutions* to the respective differential equations. In this example we started with the functions y and z and found differential equations that they satisfied. We did this by differentiating the functions using the Chain Rule. Mathematical modeling most frequently involves starting with differential equations that are established on the basis of biological, physical, or other properties of what is being "modeled". Then the differential equations are solved. Methods for solving differential equations involve the concepts of **integration** that are introduced in Chapters 5 and 6. Examples illustrating the solution of other "growth models" are introduced in Chapter 7 and Chapter 8 where general methods of solving differential equations are discussed.

☑

Differentiating complex functions frequently involves applying several different General Rules, sometimes applying the same General Rule more than once with different choices for the intermediate function. The key to dealing with complex functions is to proceed orderly. In the formulas or rules the intermediate function u(x) is a temporary variable. As each application may utilize a different intermediate u(x), the use of alternative symbols for the intermediate function is suggested. This may help you avoid errors.

Example 5   **Multiple applications of the Chain Rule; employing different intermediate variables.**

Problem   Determine the derivative of $y = (x^5 - 2x)^{1/2} + (x^3 - 1)^6$.

Solution   Applying the Sum Rule creates two terms to differentiate, each of which requires the General Power Rule. Express y as a function of two intermediate functions:

$$y = u^{1/2}(x) + v^6(x)$$

where

$$u(x) = x^5 - 2x \quad \text{and} \quad v(x) = x^3 - 1$$

Applying the Sum Rule and then the General Power Rule twice gives

$$y' = (1/2)u^{-1/2}(x)u'(x) + 6v^5(x)v'(x)$$

Substituting the expressions in terms of x for the intermediate functions and their derivatives gives:

$$y' = (1/2)(x^5 - 2x)^{-1/2}(5x^4 - 2) + 6(x^3 - 1)^5(3x^2)$$

☑

Example 6   **Nested applications of the Chain Rule; using the Chain Rule to differentiate the intermediate variable.**

Problem   Differentiate $y = (t^2 - 1)^{-2}\text{Exp}(\sin(\pi t^3))$.

Solution   Write y as a product:

$$y = f(t) \cdot g(t) \quad \text{where} \quad f(t) = (t^2 - 1)^{-2} \quad \text{and} \quad g(t) = \text{Exp}(\sin(\pi t^3)).$$

To apply the Product Rule the derivatives f' and g' must be evaluated first. The derivative of each requires applying a General Rule. f is differentiated by the General Power Rule:

$$f(t) = u^{-2}(t) \quad \text{where} \quad u(t) = t^2 - 1$$

$$f'(t) = -2u^{-3}(t)u'(t) = -2(t^2 - 1)^{-3} \cdot 2t = -4t(t^2 - 1)^{-3}$$

Applying the General Exponential Rule to differentiate g, we introduce an intermediate variable, v, and write

$$g(t) = e^{v(t)} \quad \text{where} \quad v(t) = \sin(\pi t^3)$$

$$g'(t) = e^{v(t)}v'(t)$$

However, to differentiate v(t) requires yet another application of the Chain Rule in the guise of a General Trigonometric Rule. Set $v = \sin(w)$ where $w = \pi t^3$. Then

$$v'(t) = \cos(w) \cdot w' = \cos(\pi t^3) \cdot 3\pi t^2$$

Substituting v and v' into the formula for g' gives

$$g'(t) = 3\pi t^2 \cdot \cos(\pi t^3) \cdot \text{Exp}(\sin(\pi t^3))$$

Finally, substituting the above derivatives f' and g' into the Product Rule $y' = f' \cdot g + f \cdot g'$ gives

$$y' = [-4t(t^2 - 1)^{-3}][\text{Exp}(\sin(\pi t^3))] + [(t^2 - 1)^{-2}][3\pi t^2 \cos(\pi t^3) \cdot \text{Exp}(\sin(\pi t^3))]$$

Factoring out the common exponential term and the power term $(t^2 - 1)^{-3}$ gives

$$y' = \{-4t + (t^2 - 1) \cdot 3\pi t^2 \cos(\pi t^3)\} \cdot (t^2 - 1)^{-3} \cdot \text{Exp}(\sin(\pi t^3))$$

Rationale for the Chain Rule.

The Chain Rules can be established using the limit definition of a derivative. Assume that f(x) and g(x) are differentiable. Then the derivative of the composite function $y(x) = f(g(x))$ is the limit of the difference quotient $\Delta y / \Delta x$ as $\Delta x = h \to 0$.

$$dy/dx = \lim_{h \to 0} \{y(x + h) - y(x)\} / h = \lim_{h \to 0} \{f(g(x + h)) - f(g(x))\} / h$$

The expression in the limit is symbolically manipulated by multiplying and dividing by the same non-zero term. This will not change its value. The term we multiply and divide by is the difference

$$g(x + h) - g(x)$$

The resulting expression is then grouped to form the product of two rational terms, giving

$$y' = \lim_{h \to 0} [\{f(g(x + h)) - f(g(x))\} / \{g(x + h) - g(x)\}] \cdot [\{g(x + h) - g(x)\} / h]$$

The apparent complexity of this expression is reduced by introducing an intermediate variable. Let

$$u = g(x) \quad \text{and} \quad \Delta u = g(x + h) - g(x)$$

Then, we can write $g(x + h) = u + \Delta u$. Substituting these terms into the limit equation for y' gives

$$y' = \lim_{h \to 0} \left([\{f(u + \Delta u) - f(u)\} / \Delta u] \cdot [\{u(x + h) - u(x)\} / h] \right)$$

Using the limit property that the limit of a product is the product of the limits gives

$$y' = \lim_{h \to 0} [\{f(u + \Delta u) - f(u)\} / \Delta u] \cdot \lim_{h \to 0} [\{u(x + h) - u(x)\} / h]$$

$$= df/du \cdot du/dx$$

The second limit matches precisely the form of the definition of the derivative u'(x). The first limit is seen to be the derivative dy/du after replacing the $\lim_{h \to 0}$ by $\lim_{\Delta u \to 0}$. This is justified by the observation that for a continuous function u(x),

$$\lim_{h \to 0} u(x + h) = u(x)$$

and hence $\Delta u(x) = u(x + h) - u(x)$ must approach zero as $h \to 0$.

Differentiating $g(x)^{f(x)}$.

The exponential properties $e^{ln(B)} = B$ and $ln(A^k) = k\, ln(A)$ can be used to reduce the function g^f to the exponential of a product. Ignoring the x-variable, we have

$$g^f = \exp(\,ln(g^f)\,) = \exp(\,f \cdot ln(g)\,)$$

Consequently, applying the general Exponential Rule for $e^{u(x)}$ with $u(x) = f(x) \cdot ln(g(x))$ we can differentiate $g(x)^{f(x)}$. Note that the Product Rule must be used to differentiate $u(x)$ with the Chain Rule, or General Log Rule, used to differentiate $ln(g(x))$. The resulting equation is complicated, and instead of remembering it as a Rule it is best to simply use the exponential identities in each problem.

Example 7   **Differentiating a function raised to a function.**

Problem   Differentiate $y = \sin(x)^{\tan(x)}$.

Solution   First express y as its exponential equivalent

$$y = e^{u(x)} \quad \text{where } u(x) = ln(\sin(x)^{\tan(x)})$$

Then, by the General Exponential Rule

$$y' = e^{u(x)} \cdot u'(x) = y(x) \cdot u'(x)$$

To differentiate u, express it as a product

$$u(x) = \tan(x) \cdot ln(\sin(x))$$

by the Product Rule

$$u'(x) = \tan(x) \cdot D_x ln(\sin(x)) + D_x \tan(x) \cdot ln(\sin(x))$$

The derivative of the ln term is

$$D_x ln(\sin(x)) = D_x \sin(x) / \sin(x) = \cos(x) / \sin(x) = \cot(x)$$

Thus, $u'(x) = \tan(x) \cdot \cot(x) + \sec^2(x) \cdot ln(\sin(x))$

The product $\tan(x)\cot(x) = 1$. Thus the final derivative of y is

$$y' = \sin(x)^{\tan(x)} \cdot [1 + \sec^2(x) \cdot ln(\sin(x))\,]$$

☑

Related Rates

Models or descriptions of physical or biological systems often require several variables to describe different aspects of the system. The calculus of *multivariable functions* is introduced in Chapters 9-12. However, simple systems that are described by compositions of single variable functions can be studied using the Chain Rule. Such systems are often indicated by a set of equations relating several variables. The dynamics of one variable is thus related to that of the others. The rate at which individual system-variables change, i.e., their derivatives, will thus be related, with the Chain Rule providing the relationship. For instance, if a system is governed by two equations

$$y = f(r) \quad \text{and} \quad r = g(s)$$

Then the composite system is described by the equation

$$y(s) = f \circ g(s)$$

and by the Chain Rule, its dynamic equation is

$$D_s y = f'(r) \cdot g'(s)$$

Typical "related rates" problems will specify the values of two of the three derivatives and so that the value of the third can be solved for using the dynamic equation.

Example 8 A melting snow ball.

Problem If a spherical snow ball melts uniformly so that when its radius is r_0 cm its volume is decreasing at a rate of 10 cm^3 per min and its radius is decreasing at 0.5 cm per min, at what rate is its surface area decreasing?

Solution To solve this problem we kneed to know the equations that describe the system. The two required equations give the surface area S and the volume V of the snow ball.

$$V = 4/3 \, \pi \, r^3 \quad \text{and} \quad S = 4\pi r^2$$

The variables are all considered functions of time, but the time variable is not introduced and the time is not even known. The relations between V, S, and r give the composite functions involving t. Thus when we differentiate with respect to t we have

$$dV/dt = dV/dr \cdot dr/dt \quad \text{and} \quad dS/dt = dS/dr \cdot dr/dt$$

Differentiating the equations for V and S with respect to r gives

$$dV/dr = 4\pi r^2 \quad \text{and} \quad dS/dr = 8\pi r$$

The given data imply that

$$dV/dt = -10 \quad \text{and} \quad dr/dt = -0.5 \quad \text{when} \quad r = r_0$$

with the negative sign arising from the word "decreases". The equation for dV/dt can thus be solved for r_0:

$$-10 = 4\pi r_0^2 \cdot -0.5 \quad \Longrightarrow \quad r_0 = \sqrt{5/\pi} \approx 1.26 \text{ cm}$$

Substituting this into the formula for dS/dr, we get

$$dS/dt = 8\pi \cdot \sqrt{5/\pi} \cdot -0.5 = -4\sqrt{5\pi} \approx -15.9 \text{ cm}^2/\text{min}$$

The rate of change in the surface area is -15.9 cm^2/min. It is decreasing by 15.9 cm^2 min^{-1}.

Exercise Set 3.5

1. Express $y = f \circ g(x)$ directly as a function of x and evaluate its derivative using the Chain Rule.

   a) $f(x) = x^2 + 2$; $g(x) = 2x$ b) $f(x) = 3x - 7$; $g(x) = -x + 2$ c) $f(x) = x + \sqrt{x}$; $g(x) = (x + 2)^2$

   d) $f(x) = e^x$; $g(x) = x^2 + 1$ e) $f(x) = ln(x + 1)$; $g(x) = (x + 1)(x - 1)$ f) $f(x) = x^{-1}$; $g(x) = e^{x^2}$

   g) $f(x) = (3x - 2)^3$; $g(x) = x^{1/2} - x^{-1/2}$ h) $f(x) = sin(x)$; $g(x) = 1 - x^2$

2. Express the given function as composition, $y = f \circ g$. Indicate $f(x)$, $g(x)$ and verify that your choice is valid. (The answers may not be unique.) Then use the Chain Rule to find y'.

   a) $y = (2x + 1)^2 + (2x + 1)^{1/2}$ b) $y = (2x + 1)^2 + x + 1/2$ c) $y = (x^2 + 1)exp(x^2 + 1)$

   d) $y = exp(3x - 2) + 3x - 2$ e) $y = (x + 1)(x - 1)/(1 + x^2)$ f) $y = x^2 + 6x + 9 - 2ln(x + 3)$

   g) $y = (1 - x)ln(1 - x)$ h) $y = 2 + x + e^x + exp(e^x + 1 + x)$

3. Express the function in the form $y = f(u(x))$ indicating the intermediate function $u(x)$ and the function f. Indicate which General differentiation Rules applies and use it to determine y'.

   a) $y = (3x + 1)^2$ b) $y = (2x^2 + 2)^2$ c) $y = (x^5 + 1)^3$ d) $y = (x^4 + x^3 + 1)^3$

   e) $y = (x^2 + 3x - 1)^{1/2}$ f) $y = (2x + 1)^{-1/2}$ g) $y = (x + 3)^6$ h) $y = (x + 3)^6(2x - 1)^2$

   i) $y = (2x + 1)^4 + (x^2 - 1)^3$ j) $y = (2x - 1)^2/3x^2$ k) $y = (x^3 - 2x - 8)/(5x + 1)^3$

   l) $y = (x^3 - 2x - 8)(5x - 1)^{-3}$ m) $y = (x^5 - 1)^2(x^2 - 1)^5$ n) $y = 1/(x + 1)(x - 1)$

   o) $y = 1/[(x + 1)^2(x - 1)]$ p) $y = (x^2 + 2)(x + 1)^2(2x - 1)$

4. Determine the equation of the tangent line to the graph of the function at the point $(x_0, f(x_0))$.

   a) $f(x) = (1 - x^2)^{1/2}$; $x_0 = 1/2$ b) $f(x) = sin(x^2)$; $x_0 = \pi/4$ c) $f(x) = (x - 2)^2 + (1/(x - 2))$; $x_0 = 3$

   d) $f(x) = SQRT(1 + sec^2(x))$; $x_0 = 0$ e) $f(x) = e^{2x}$; $x_0 = 0$ f) $f(x) = ln(x^3)$; $x_0 = 1$

5. Evaluate the following derivatives using the General Exponential and Logarithm Rules.

   a) $D_x e^{2x}$ b) $D_x ln(2x)$ c) $D_x exp(2x^2)$ d) $D_x ln(2x^2)$ e) $D_x exp(2x - x^3)$

   f) $D_x ln(2x - x^3)$ g) $D_x exp((x-1)^2)$ h) $D_x ln[(x - 1)^2]$ i) $D_x e^{sin(x)}$ j) $D_x 3^{2x}$

   k) $D_x Log(2x)$ l) $D_x exp_8(x^2 - 4x)$ m) $D_x 2^{ln(x)}$ n) $D_x log_8(x^5 - x^{-5})$ o) $D_x log_2[exp_8(x^3)]$

6. Evaluate the given derivative.

   a) $D_\theta ln(sin(\theta))$ b) $D_x[xe^{x^2}]$ c) $D_x[x\, ln(x^2)]$

   d) $D_\theta[sin(\theta)e^{-\theta}]$ e) $D_\theta[sin(\theta)ln(2\theta)]$ f) $D_\theta sin(e^\theta)$

   g) $D_\theta sin(ln(\theta))$ h) $D_x exp(e^x)$ i) $D_x ln(ln(x))$

   j) $D_x exp(2e^{3x})$ k) $D_\theta ln[sin(\theta)cos(\theta)]$ l) $D_\theta e^{sin(\theta)cos(\theta)}$

7   Evaluate the indicated derivative using a General Trigonometric Rule.

   a) $D_x \cos^2(x)$
   b) $D_x \cos(x^2)$
   c) $D_x \sin(1 + x^2)$

   d) $D_x (\sin(x) + 1)^3$
   e) $D_x \tan(1 - 2x)$
   f) $D_x (\cos(x) - x)^{1/2}$

   g) $D_t (\sin^2(t) + \cos^2(t))$
   h) $D_t (\sin(\sin(t)))$
   i) $D_t (\sin^5(\cos(t)))$

8   Determine the derivative $D_t y$ if

   a) $y = 3^{2t - 7}$
   b) $y = 2^{-(4 - t^2)}$
   c) $y = 5^{7^t}$

   d) $y = \log_3(3t - 5)$
   e) $y = \log_2(9 - t^2)$
   f) $y = \log_7(\log_7(t^2 - t^{-3}))$

   g) $y = \text{Log}(3^{2t - 7})$
   h) $y = \exp_3(\log_2(4t - t^{-2}))$
   i) $y = \log_8(\exp(5t^3))$

9   Use the Chain Rule to determine y'.

   a) $y = \ln(\ln(x))$
   b) $y = \sin(\sin(x^2))$
   c) $y = \ln(\ln(1/x))$
   d) $y = \sin(\sin(\sin(x)))$

10 Use the exponential properties and Chain Rule to determine y'.

   a) $y = x^x$
   b) $y = x^{\sin(x)}$
   c) $y = \sin(x)^x$

   d) $y = (x^4 - x^{-2})^{3x}$
   e) $y = x^{x^x}$
   f) $y = \ln(x)^{\ln(x)}$

11 If a spherical snow ball melts uniformly so that when its radius is r_0 cm its volume is decreasing at a rate of 8 cm^3 per min and its radius is decreasing at 2 cm per min, at what rate is its surface area decreasing?

12 If a spherical snow ball melts uniformly so that when its radius is r_0 cm its surface area is decreasing at a rate of 10 cm^2 per min and its radius is decreasing at 0.25 cm per min, at what rate is its volume decreasing?

13 If a spherical snow ball melts uniformly so that when its radius is 5 cm its volume is decreasing at a rate of 2 cm^3 per min at what rate is its radius decreasing and at what rate is its surface area decreasing?

14 If the surface area of a cube increases at 30 cm^2/s while the volume increases at 20 cm^3/s, what is the rate of increase in the side-length L of the cube?

15 If the surface area of a cube increases at 50 cm^2/s while the side-length L increase at 0.2 cm/s what is the rate of change in the volume of the cube?

16 *Poiseuille's Law* gives the volume flow through a blood vein of radius r as $V = kr^4$ for a constant k depending on pressure and viscosity. What is the percentage rate of decrease in the flow volume, $100\% \cdot V'/V$, if a person experiences trauma that contracts the vein radius by 2% per minute?

17 If a circular cylindrical shaped cell keeps a constant length:radius ratio of 20 and its volume is increasing at 100µm^3/s while its radius increases at 5µm/s what is the length of the cell?

18 Assume a rectangular ice cube melts sitting on a cold surface so that its width W and length L decrease twice as fast as its height H. If its exposed surface area is $S = 20$ cm^2 and its volume is decreasing at 4 cm^3/hr at what rate is its height decreasing? Hint: show that $V' = S \cdot H'$.

Section 3.6 Higher Derivatives and Implicit Differentiation.

Introduction

This section introduces two topics. The first is the concept of *higher order derivatives*, which result from repeatedly differentiating derivatives of a functions. The second topic is a method of differentiating called *implicit differentiation*. This is a method used to differentiate terms that involve both the variable of differentiation, x, and the dependent variable, y, without first solving for y as an explicit function of x. Implicit differentiation can be used to differentiate a product of several terms via a method called *Logarithmic Differentiation*. With this method you take the ln of the product, differentiate implicitly, and in the process avoid the complexity that arises with multiple applications of the Product Rule.

Higher Order Derivatives.

Taking a derivative is an operation or process that transforms one function into another, its derivative. Mathematicians frequently use the derivative notation D_x in this context as it is similar to "operator" notation used elsewhere in advanced mathematics, physics, and chemistry to indicate "functions" whose domains and ranges are sets of functions rather than simply numerical values. The derivative process can be visualized schematically as:

Input function: $f(x)$ $\Longrightarrow$ Differentiate: $D_x f(x)$ $\Longrightarrow$ Output Function: the Derivative $f'(x)$

Repeating this process by using the Output function f' as an Input function results in an Output of a second derivative:

Input function: $f'(x)$ $\Longrightarrow$ Differentiate: $D_x f'(x)$ $\Longrightarrow$ Output Function: the Derivative $f''(x)$

The Output function in this case is denoted by f'' and is called the **second derivative of f**. A simple example is given by $f(x) = x^4$: $f'(x) = 4x^3$ and its derivative is $f''(x) = 12x^2$.

Several different notations are used to denote the second derivative of $y = f(x)$:

Using double primes : y'' or f''

Using a superscript in parentheses: $f^{(2)}(x)$

Using the Operator Notation D_x with a superscript: $D_x^2 y$

Using the *Leibniz* differential notation d/dx with two superscripts: $d^2/dx^2 (f(x))$

NOTE: The 2 appearing as a superscript in these notations is to indicate that the derivative process has been repeated twice.
It is not an exponent that indicates a "square " or the multiplication of a term by itself.

IT <u>CAN</u> <u>NOT</u> <u>BE</u> ALGEBRAICALLY <u>CANCELED</u> <u>IN</u> <u>ANY</u> <u>WAY.</u>

Repeating the derivative process using the second derivative f'' as the Input, leads to a third function called the "third derivative"; its derivative is the "fourth derivative", and so on. The notation to denote a higher derivative is similar to that given for second derivatives. If a function is differentiated n-times, then it may be denoted with n-primes (which would be quite cumbersome for n > 3) or with an exponent-like index n, similar to the 2 in the notation for a second derivative. For higher derivatives, n > 3, the superscript notation with parentheses is the most commonly used.

Section 3.6   HIGHER DERIVATIVES AND IMPLICIT DIFFERENTIATION

DEFINITION OF HIGHER DERIVATIVES

> For a positive integer n, the **n-th derivative of f(x)** is obtained by differentiating n-times the function f(x). It is denoted by
>
> $$D^n_x f(x) \qquad d^n/dx^n f(x) \qquad \text{or} \qquad f^{(n)}(x)$$
>
> The n-th derivative of f is defined by n-nested differentiations:
>
> $$f^{(n)}(x) = d/dx(d/dx(\ldots (d/dx(f(x))) \ldots))$$
> $$\text{n-times}$$

Example 1   **Computing higher derivatives.**

Problem   Compute the first four derivatives of $f(x) = x^{10}$.

Solution   Using the Power Rule:

$$f'(x) = 10x^9$$

$$f''(x) = d/dx(f'(x)) = 90x^8$$

$$f'''(x) = f^{(3)}(x) = d/dx(90x^8) = 720x^7$$

and $\qquad f^{(4)}(x) = d/dx(720x^7) = 5040x^6$

Example 2   **Higher derivatives of polynomials are identically zero.**

Problem   Show that the fourth and all higher derivatives of $y = x^3 - 2x + 7$ are zero.

Solution   The first five derivatives of y are

$$y' = 3x^2 - 2 \qquad y'' = 6x \qquad y''' = 6 \qquad \text{and} \qquad y^{(4)} = 0 \qquad y^{(5)} = d/dx(0) = 0$$

As the derivative of the constant zero function is zero all higher derivatives exist, but are each identically zero.

As the above examples illustrate, each time a polynomial is differentiated the derivative is a polynomial of degree one less. Consequently, differentiating a polynomial of degree n n-times will result in a constant function.

> If P(x) is polynomial of degree n, then its n-th derivative $P^{(n)}(x)$ is constant and all higher derivatives are identically zero.

Example 3   **The second derivative of a product.**

Problem   Compute the second derivative of $y = x^2 e^{-0.1x}$.

Solution Applying the Product Rule, the first derivative of y is

$$y' = 2xe^{-0.1x} - 0.1x^2 e^{-0.1x}$$

The second derivative is obtained by applying the Sum/Difference Rule and then the Product Rule to each of the terms.

$$y'' = d/dx(2xe^{-0.1x}) - d/dx(0.1x^2 e^{-0.1x})$$

$$= [2e^{-0.1x} - 0.2xe^{-0.1x}] - [0.2xe^{-0.1x} - 0.01x^2 e^{-0.1x}]$$

$$= \{2 - 0.4x + 0.01x^2\}e^{-0.1x}$$

Alternatively, the first derivative could have been simplified by factoring it as

$$y' = [2x - 0.1x^2]e^{-0.1x}$$

Then, only a single application of the Product Rule is required to give y''. Try this and you will find that it is probably quicker and less prone to clerical errors.

A NOTE ON FACTORING. The exponential term can always be factor out of the derivative of a product of an exponential function and either a polynomial or trigonometric function.

When higher derivatives fail to exist.

Higher derivatives of any function can be evaluated formally, but they may not exist at specific x-values where the function and lower order derivatives are defined. One of the most common situations in which this happens is when a function has a non-integer power term $(x - x_0)^r$ where r is positive but is not an integer. The exponent of $(x - x_0)$ is decreased by one with each differentiation. Upon repeated differentiation the exponent eventually become negative and the term becomes singular at x_0, i.e. it results in division by zero when $x = x_0$. In this case higher derivative do not exist at x_0. This occurs after $n = [\![r]\!] + 1$ differentiations.

Example 4   **The Domains of higher order derivatives.**

Problem   Determine the natural domains of the higher derivatives of the function

$$y = (x - 1)^4 + (x + 2)^{3/2} - (x - 3)^{1.72} + (x + 4)^{7/3}$$

Solution   As a sum of power functions, y and its derivatives may be formally differentiated repeatedly. The first four derivatives are:

$$y' = 4(x - 1)^3 + (3/2)(x + 2)^{1/2} - 1.72(x - 3)^{0.72} + (7/3)(x - 4)^{4/3};$$

$$y'' = 12(x - 1)^2 + (3/4)(x + 2)^{-1/2} - 1.2384(x - 3)^{-0.28} + (28/9)(x - 4)^{1/3};$$

$$y^{(3)} = 24(x - 1) - (3/8)(x + 2)^{-3/2} + 0.346752(x - 3)^{-1.28} + (28/27)(x - 4)^{-2/3};$$

$$y^{(4)} = 24 + (9/16)(x + 2)^{-5/2} - 0.4438426(x - 3)^{-2.28} - (56/81)(x - 4)^{-5/3}.$$

To establish the domains of these derivatives we must check where the they are undefined. The domains depend on the domains of the sub-terms of the form $(x - x_0)^r$.

Section 3.6   HIGHER DERIVATIVES AND IMPLICIT DIFFERENTIATION

RECALL:

The domain D of the term $(x - x_0)^r$ depends on the type of number r is.

If r is an irrational number, like π or $ln(2)$, then

$$D = [x_0,\infty) \text{ if } r > 0 \quad \text{and} \quad D = (x_0,\infty) \text{ if } r < 0$$

If r is a reduced rational number: $r = m/n$ with $n > 0$, then

    if n is odd $D = (-\infty,\infty)$ if $m \geq 0$ and $D = (-\infty,x_0) \cup (x_0,\infty)$ if $m < 0$

    if n is even $D = [x_0,\infty)$ if $m \geq 0$ and $D = (x_0,\infty)$ if $m < 0$

There are four critical x-values for this function and its derivatives: 1, -2, 3, and 4.

For $x_0 = 1$: As the term $(x - 1)^r$ only occurs with positive integer exponents, this term does not restrict the domain of y or its derivatives.

For $x_0 = -2$: The term $(x + 2)^r$ occurs in each term with $r = m/2$. For y, $m = 1$, so the domain of y is restricted to $[-2,\infty)$. For the derivatives m is negative, hence their domains are restricted to $(-2,\infty)$.

For $x_0 = 3$: The term $(x - 3)^r$ appears in y with $r = 1.72$, which is the rational number 172/100 that reduces to 43/25. Thus the domain of y is not restricted by this term. The same is true for y′ as it has $r = 18/25$. However, in the second and higher derivatives r has denominator 25 but a negative numerator. Thus they are not defined at $x = 3$ and their domains are restricted to $(3,\infty)$.

For $x_0 = 4$: As y, y′ and y″ have terms $(4 - 4)^r$ with $r = m/3$ their domains are not restricted by this term. But, the third and all higher derivatives are singular at $x = 4$ because of the negative exponents of their term $(x - 4)^r$; hence, they are not defined for $x \leq 4$.

Combining the above restrictions, the Natural Domains of y and its derivatives are:

    Domain(y) = $[-2,\infty)$ Domain(y′) = $(-2,\infty)$

    Domain(y″) = $(-2,3) \cup (3,\infty)$ Domain($y^{(n)}$) = $(-2,3) \cup (3,4) \cup (4,\infty)$ for $n \geq 3$

      ☑

Implicit Differentiation.

IMPLICIT AND EXPLICIT EQUATIONS

An equation in which one variable is set equal to a function of another, like $y = f(x)$, is called an **explicit equation** for y as a function of x. Mathematical relationships that are not of this form, but consist of an equation involving two variables are called **implicit equations**. Examples of implicit equations are equations of the circle and rectangular hyperbola shown in the following figure. The implicit equations whose graphs are the illustrated curves are

$$x^2 + y^2 = 9 \quad \text{and} \quad xy = 4$$

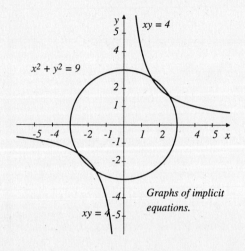
Graphs of implicit equations.

The equation for the hyperbola can be solved explicitly for y to give $y = 4/x$. The circle equation can not be solved explicitly for all y because there are two y-values for each x-value. But, if we are interested in a specific portion of the circle that is either in the top half or in the bottom half of the circle, then by suitably restricting the y-values we can give an explicit equation for that portion. You have probably done this in the past when you solved such quadratic equations in the form

$$y = \pm \text{SQRT}(9 - x^2)$$

and selected either the + sign or the - sign to refer to the top or bottom portion of the circle. As we shall see in the following examples, for more complicated implicit equations a similar process can lead to several explicit equations that describe the functional relationship of y and x for different portions of the original function's graph.

Any implicit equation involving x and y can be put in the general form

$$F(x,y) = 0 \qquad \textbf{IMPLICIT EQUATION}$$

by transferring all terms to one side of the equation. For instance,

$y = x + 4y^2$ can be written as $y - x - 4y^2 = 0$

$e^y = x^2 + y^2$ can be written as $e^y - x^2 - y^2 = 0$

The following Table IV list several examples of explicit and implicit equations.

Table IV	
Examples of Explicit Equations	Examples of Implicit Equations
$y = x^3/(1 - x^{-5})$	$y = x + 4y^2$ or $y - x - 4y^2 = 0$
$y = \sin(\pi t)$	$xy - x^2 = 0$
$\theta = e^{-3t}$	$\cos(x - y) = 0$
$z = x \sin(x)$	$e^y = x^2 + y^2$ or $e^y - x^2 - y^2 = 0$

Frequently, solving an implicit equation explicitly for one variable is very difficult or even impossible using elementary functions. There are mathematical theorems that assert the existence of solutions, i.e., of explicit equations, corresponding to a segment of a graph of an implicit equation, The basic requirement is that the graph is smooth and not a vertical line. These theorems apply only to restricted ranges of y-values. There may be several different explicit equations defined for the same interval of x-values, like $y = +\text{SQRT}(x)$ and $y = -\text{SQRT}(x)$, all corresponding to the same implicit equation, $y/x - x = 0$. However, there is no theory that tells how to find the explicit equations. Consequently, we must rely on basic algebra techniques to solve implicit equations being careful to note the domain and range of variables for which various intermediate expressions are valid.

Example 5 Solving an implicit equation explicitly.

Problem Determine the explicit equations for y as a function of x that correspond to the implicit equation

$$[(y-1)^2/x] - (y-2)^{-2} = 0$$

Solution The implicit equation must be solved for y carefully noting any multiple values when taking roots and x and y-value at which the equations are not defined. The equivalent equations are as follows:

Multiply by x and by $(y-2)^2$:
$$(y-2)^2(y-1)^2 - x = 0$$

Transfer the x term and write
the product of squares as the square of the product:
$$[(y-2)(y-1)]^2 = x$$

Take the square root,
retaining both the positive and negative options:
$$[(y-2)(y-1)] = \pm \text{SQRT}(x)$$

Multiply the y-terms and then express the
y terms as a shifted quadratic:
$$y^2 - 3y + 2 = (y - 3/2)^2 - 1/4 = \pm \text{SQRT}(x)$$

Solve for y by shifting the 1/4,
taking the ± root and shifting the 3/2:
$$y = 3/2 \pm \text{SQRT}[\,1/4 \pm \text{SQRT}(x)\,]$$

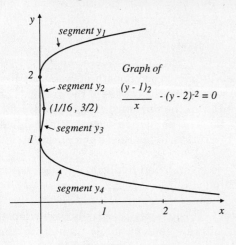

Graph of $\dfrac{(y-1)^2}{x} - (y-2)^{-2} = 0$

In this equation the two ± signs arising from taking square roots are independent of each other. Therefor four cases arise for the choice of the two signs:

$$+\;+,\qquad +\;-,\qquad -\;+,\quad \text{and}\quad -\;-$$

Each options has an implied restrictions on the domain of the function, so that the explicit function will be defined, and a different range of y-values.

For instance, when both signs are positive, the + + case, the inner term is only defined for

$$x \geq 0 \quad \text{and thus} \quad 1/4 + \sqrt{x} \geq 1/4$$

Hence, its square root is ≥ 1/2 and the corresponding y values are all ≥ 2. However, since the original implicit equation is not defined for x = 0, the equality signs can not occur. As x → ∞, the corresponding y → ∞, and the range for this case is the interval (2,∞).

The same type of analysis applied to the second case, when the signs are + -, leads to the restriction that the domain is limited to $0 < x \leq 16$ since the outer square root is not defined if $\sqrt{x} > 1/4$. Since the values of the term $1/4 - \sqrt{x}$ vary from 1/4 when x = 0 (although 0 is still not in the domain of the original equation) to 0 when x = 1/16, the range for y in this case is $3/2 \leq y < 2$. For the third and fourth cases, similar arguments give the y-ranges (1,3/2) and (-∞,1), respectively. The graph of the implicit equation is given above. In Chapter 4 graphing techniques using calculus will be introduced, that will make graphing such an equation easier. The four explicit equations for this curve, with their domains and ranges, are indicated in Table V.

Table V		
Domain	Range	Explicit Equation
$(0,\infty)$	$(2,\infty)$	$y = 1.5 + \text{SQRT}[\,¼ + \sqrt{x}\,]$
$(0, 1/16]$	$[3/2, 2)$	$y = 1.5 + \text{SQRT}[\,¼ - \sqrt{x}\,]$
$(0, 1/16)$	$(1, 3/2)$	$y = 1.5 - \text{SQRT}[\,¼ + \sqrt{x}\,]$
$(0,\infty)$	$(-\infty, 1)$	$y = 1.5 - \text{SQRT}[\,¼ - \sqrt{x}\,]$

The point (1/16, 3/2) was arbitrarily assigned to the graph of the second equation; it could have, alternatively, been included on the graph of the third equation.

Differentiating a function you can not solve for.

In practice, it is usually very difficult and often impossible to solve explicitly for a variable involved in an implicit equations. It is easy to write implicit equations that can not be solved in the ordinary sense in terms of finitely many elementary functions. For instance, we can not explicitly solve the equations

$$x + y + e^{xy} + 1 = 0 \quad \text{and} \quad \sin(x/y) + (x - y)^5 = 0$$

for y. Yet, the graphs of such implicit equations are usually relatively smooth curves that has tangent lines at most points on the curves. A tangent line will reflect the nature of the graph. The slope of the tangent line will be the value of the derivative $D_x y$ at the tangent point, where $y = f(x)$ is the (unknown) explicit function for the curve at the tangent point. In theory, the explicit equation $y = f(x)$ exists, in practice it may be impossible to express. This raises the question

"How can one find the derivative $y'(x)$ when y is given by an implicit equation without solving the equation explicitly?"

The answer is "By the process of **Implicit Differentiation**". This process is actually very simple. It consists of treating the y-variable as if it were an intermediate variable, similar to the intermediate function $u = u(x)$ employed in the Chain Rule and the General Derivative Rules. Each term of the implicit equation is differentiated by applying the standard derivative Rules, retaining the derivative y' that occurs each time y or a function of y is differentiated. The resulting equation is then solved for y'. The amazing thing is that you can always solve for y' in terms of y and x, even when the original implicit equation is impossible to solve.

When differentiating implicitly, a function $g(y)$ is differentiated by the Chain Rule:

$$d/dx(g(y)) = dg/dy \cdot dy/dx = g'(y) \cdot y' \qquad \textbf{NOTE THE } y' \textbf{ TERM !!!!!}$$

For instances,

$$d/dx(y^4) = 4y^3 \cdot dy/dx = 4y^3\, y'$$

$$d/dx(\sin(\pi y)) = \pi \cos(\pi y) \cdot dy/dx = \pi \cos(\pi y)\, y'$$

$$d/dx(e^{-3y}) = -3y' e^{-3y}$$

Section 3.6   HIGHER DERIVATIVES AND IMPLICIT DIFFERENTIATION

When x and y appear in the same expression then the algebraic differentiation Rules are applied, treating x-terms normally but y-terms as intermediate functions. For instance,

$$d/dx(x^3 y^5) = x^3 \cdot d/dx(y^5) + d/dx(x^3) \cdot y^5 \qquad \text{Product Rule}$$

$$= x^3 \cdot 5y^4 y' + 3x^2 \cdot y^5 \qquad \text{Power Rule}$$

$$= (5x^3 y^4) \cdot y' + 3x^2 y^5 \qquad \text{Algebra}$$

As another example, the quotient x/y is differentiated as

$$D_x(x/y) = \{[y \cdot D_x x] - [x \cdot D_x y]\}/y^2 \qquad \text{Quotient Rule}$$

$$= \{y - xy'\}/y^2 \qquad \text{As } D_x x = 1 \text{ and } D_x y = y'.$$

To determine the derivative y' each term of an implicit equation is differentiated in the above manner. The resulting equation will generally have several terms that may involve x's and/or y's and some terms will also have a factor of y'. It is the derivative y' that is being sought. We can always solve the differentiated equation for y', generally as a function of both x and y. If the reason for determining y' is to calculate the slope of a tangent line at a point (x_0, y_0) on the graph of the implicit equation, we can simply substitute the values x_0 and y_0 for x and y in the equation for y' to obtain the corresponding slope $y'(x_0)$.

METHOD OF IMPLICIT DIFFERENTIATION.

The Method of Implicit Differentiation
of an equation $F(x,y) = 0$ to find $D_x y$ consist of:

Step I   Differentiate with respect to x each term of the given equation $F(x,y) = 0$, treating y as an intermediate variable.

Step II   Solve the resulting equation for y':

     a) Gather all terms with a y' on one side of the equation and place all other terms on the other side.

     b) Factor out y' and divide the other side of the equation by the resulting co-factor of y' to give an explicit equation for y'.

Example 6   Evaluating Derivatives Implicitly.

Problem   Determine the derivative y' if a) $x^4 + y^5 = 10$ and b) $e^{x+y} = x \sin(y)$.

Solutions   a) Differentiate each term of the implicit equation with respect to x:

$$d/dx(x^4) + d/dx(y^5) = d/dx(10)$$

$$4x^3 + 5y^4 y' = 0$$

Solve for y':

$$y' = -4x^3/5y^4 = -(4/5)x^3 y^{-4}$$

b) The Exponential Rule is used to differentiate the left side and the Product Rule is used to differentiate the right side of the implicit equation:

$$D_x e^{x+y} = D_x[x \sin(y)]$$

$$[D_x(x+y)] \cdot e^{x+y} = D_x(x) \cdot \sin(y) + x \cdot D_x \sin(y)$$

$$[1 + y']e^{x+y} = 1 \cdot \sin(y) + x \cos(y) \cdot y'$$

To solve for y' first separate the terms that have a y' component from those that do not:

$$[e^{x+y} - x \cos(y)]y' = \sin(y) - e^{x+y}$$

Then, divide by the co-factor of y':

$$y' = \{\sin(y) - e^{x+y}\} / \{e^{x+y} - x \cos(y)\}$$

☑

Example 7   Deriving the derivative Rule $D_x ln(x) = 1/x$ by implicit differentiation.

The derivative Rule for $ln(x)$ was stated in Section 2.4 without providing a rationale for its validity. This Rule can be derived by implicit differentiation using the definition of ln as the "inverse function" of the exponential e^x, that is, that the equation

$$y = ln(x) \quad \text{is equivalent to} \quad x = e^y.$$

Problem   Determine y' if $x = e^y$ by implicit differentiation.

Solution   Differentiating implicitly, we write

$$D_x x = D_x e^y \quad \text{or} \quad 1 = y' e^y$$

Hence, solving for y' gives

$$y' = 1/e^y = 1/x$$

But, since $x = e^y$, $y = ln(x)$ and thus $y' = D_x ln(x)$ and from the above equation we conclude

$$D_x ln(x) = 1/x$$

This same approach may be used to obtain the derivative of $\log_b(x)$.

☑

Example 8   Using implicit differentiation to determine tangent lines.

Background

The implicit equation $2xy + ln(x) = x^2 y^3$ is only defied for $x > 0$. Its graph (see below) consists of two distinct curves that approach the x-axis as $x \to \infty$. The upper curve has two branches (like a squashed parabola opening to the right). Thus, for some values of x there are actually three branches of the graph of the implicit equation. This graph was generated by a computer after solving the implicit equation for y using a symbolic algebra program. This could be done since the equation is essentially a cubic in the y-variable. While you could hardly generate such a graph without a computer, you could visualize the nature of the graph for x near a particular reference value x_0 by drawing the tangent lines to the graph at the corresponding points. The three tangent lines illustrated correspond to $x_0 = 1$.

Problem Using implicit differentiation, determine the equations of the tangent lines to the curve at the three points with $x_0 = 1$.

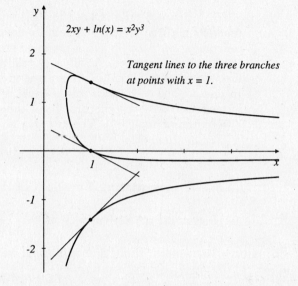

Tangent lines to the three branches at points with $x = 1$.

Solution The three points have y-coordinates satisfying the cubic equation resulting from setting $x = 1$:

$$2 \cdot 1 \cdot y + ln(1) = 1^2 y^3$$

As $ln(1) = 0$, these are the roots of

$$2y - y^3 = (2 - y^2)y = 0$$

$y_1 = \sqrt{2}$ $y_2 = -\sqrt{2}$ $y_3 = 0$

To determine the slope of the tangent lines the derivative dy/dx must be evaluated at each point. Differentiating the given equation implicitly yields:

$$D_x(2xy) + D_x ln(x) = D_x(x^2 y^3)$$

$$2x D_x y + y D_x 2x + D_x ln(x) = x^2 D_x y^3 + y^3 D_x x^2$$

$$2xy' + 2y + 1/x = 3x^2 y^2 y' + 2y^3 x$$

Solving for y' gives: $y' = \{2xy^3 - 2y - 1/x\}/\{2x - 3x^2 y^2\}$

The next step is to evaluate y' at the three points.

To compute y' at the point $(x_0, y_1) = (1, \sqrt{2})$, set $x = 1$ and $y = \sqrt{2}$:

$$y' = \{2 \cdot 1 \cdot 2^{3/2} - 2 \cdot 2^{1/2} - 1/1\} / \{2 \cdot 1 - 3 \cdot 1^2 \cdot 2^{2/2}\}$$

$$= \{4\sqrt{2} - 2\sqrt{2} - 1\}/\{2 - 6\} = \{2\sqrt{2} - 1\}/\{-4\} \approx -0.46$$

At the second point $(x_0, y_2) = (1, -\sqrt{2})$, set $x = 1$ and $y = -\sqrt{2}$:

$$y' = \{2 \cdot 1 \cdot (-2^{1/2})^3 - 2 \cdot (-2^{1/2}) - 1/1\} / \{2 \cdot 1 - 3 \cdot 1^2 \cdot (-2^{1/2})^2\}$$

$$= \{-4\sqrt{2} + 2\sqrt{2} - 1\}/\{2 - 6\} = \{2\sqrt{2} + 1\}/4 \approx 0.96$$

For the third point $(x_0, y_3) = (1, 0)$, set $x = 1$ and $y = 0$:

$$y' = \{2 \cdot 1 \cdot 0^{3/2} - 2 \cdot 0^{1/2} - 1/1\} / \{2 \cdot 1 - 3 \cdot 1^2 \cdot 0^{2/2}\} = -0.5$$

Employing the approximate numerical values for the slopes y', the three tangent lines are:

L_1: $y = \sqrt{2} - 0.46(x - 1)$ Tangent at $(1, \sqrt{2})$

L_2: $y = -\sqrt{2} + 0.96(x - 1)$ Tangent at $(1, -\sqrt{2})$

L_3: $y = -0.5(x - 1)$ Tangent at $(1, 0)$

Logarithmic differentiation.

The Product Rule is used to differentiate the product of two functions. To differentiate a product of three or more functions the Product Rule can be repeatedly applied. This time consuming approach is often prone to errors and the number of terms increases each time the Product Rule is applied. An alternative method is called **Logarithmic Differentiation**. A general formula resulting from this method is given below, but first we consider some specific examples. In this method each factor in a product is treated as an intermediate variable. A new function is formed by taking the natural logarithm of the original product. This new equation is then differentiated implicitly. The key step is to use the General Logarithm Rule:

$$D_x ln(u(x)) = u'(x) / u(x)$$

Example 9 Differentiating a product of three functions.

Problem Determine the derivative y' of $y = (1 + 5x^3)e^{7x}\cos(\pi x/2)$.

Solution To use logarithmic differentiation, we take the ln of both sides of the equation to give

$$ln(y) = ln[(1 + 5x^3)e^{7x}\cos(\pi x/2)]$$

This equation is differentiated implicitly to find y'. To simplify the process we first introduce intermediate variables to represent each factor in the product. Let

$$u = 1 + 5x^3 \quad v = e^{7x} \quad \text{and} \quad w = \cos(\pi x/2)$$

The derivatives of these intermediate functions are:

$$u' = 15x^2 \quad v' = 7e^{7x} \quad \text{and} \quad w' = -(\pi/2)\sin(\pi x/2)$$

Then, the original equation is $y = u \cdot v \cdot w$ and its logarithmic equivalent equation is

$$ln(y) = ln(u \cdot v \cdot w) = ln(u) + ln(v) + ln(w).$$

Notice that the product property of logarithms is used to change the log of the product into a sum of logs. Hence, differentiating this last equation implicitly gives

$$D_x ln(y) = D_x ln(u) + D_x ln(v) + D_x ln(w)$$

$$y'/y = u'/u + v'/v + w'/w$$

This equation is easily solved for y':

$$y' = y[u'/u + v'/v + w'/w]$$

Substituting back the x-expressions for the intermediate variables and their derivatives gives

$$y' = y[15x^2/\{1 + 5x^3\} + 7e^{7x}/e^{7x} + (-\pi/2)\sin(\pi x/2) / \cos(\pi x/2)].$$

Normally, such expression can be simplified slightly by canceling terms, e.g.,

$$y' = y[15x^2/\{1 + 5x^3\} + 7 - (\pi/2)\tan(\pi x/2)]$$

We could also substitute the explicit product for y, $(1 + 5x^3)e^{7x}\cos(\pi x/2)$, to render y' as an explicit function of x. This is generally not recommended as it makes the resulting equation more complex and normally does provide any advantage in numerical calculations of y'.

The method used in the above example can be applied to differentiate function involves products and/or quotients of three or more terms. A general formula can be stated that is applicable when y is a product of n sub-functions raised to powers. The formula is based on the observation that for any exponent r,

$$D_x ln(f^r(x)) = D_x[r \cdot ln(f(x))] = r \cdot f'(x) / f(x)$$

The general formula for Logarithmic differentiation is stated for a product of n functions each of which can be raised to a power. The functions and their corresponding exponents are indexed by subscripts.

METHOD OF LOGARITHMIC DIFFERENTIATION

If y is a product of n functions f_i raised to powers r_i,

$$y = f_1^{r_1}(x) \cdot f_2^{r_2}(x) \cdot f_3^{r_3}(x) \cdots f_n^{r_n}(x)$$

Then the derivative y' = dy/dx is given by

$$y' = y \cdot [r_1 \cdot f_1'(x)/f_1(x) + r_2 \cdot f_2'(x)/f_2(x) + \ldots + r_n \cdot f_n'(x)/f_n(x)]$$

Note: Since division is the same as multiplying with a negative exponent, Logarithmic Differentiation can be used on functions that have both multiplication and division of terms.

Example 10 **Using Logarithmic differentiation.**

Problem Differentiate the function

$$y = (x^2 - 5)^{10} (1 + e^{3x})^5 / \{(1 + 7x - x^9) \sin^3(x)\}$$

Solution The function could be differentiated by applying the Product and Quotient Rules, but this would be complicated and prone to clerical errors. The function y is the product of four intermediate functions raised to powers. It is the type of function that is best suited for Logarithmic Differentiation. The intermediate functions, their derivatives, and their exponents are set out in Table VI.

Table VI		
Function f_i	Derivative f_i'	Exponent r_i
$x^2 - 5$	$2x$	10
$1 + e^{3x}$	$3e^{3x}$	5
$1 + 7x - x^9$	$7 - 9x^8$	-1
$\sin(x)$	$\cos(x)$	-3

Using these sub-functions and their derivative, the Logarithmic Differentiation the formula gives y' directly as

$$y' = y\ [20x/(x^2 - 5)\ +\ 15e^{3x}/(1 + e^{3x})\ -\ (7 - 9x^8)/(1 + 7x - x^9)\ -\ 3\cos(x)/\sin(x)\ ]$$

The y term could be replaced by its expression as a function of x, as given in the Problem statement. Sometimes this is required. While it would appear that the resulting y' would be horrendously complicated, usually in such problems considerable simplification can be mad as each of the sum terms will have the same factors raised to either the same power or that power minus one.

☑

Exercise Set 3.6

1. Compute the third derivative of the given functions.

   a) $f(x) = 5x^2 - x + 2$ b) $f(x) = x^2 - x^{-2}$ c) $f(x) = x^{1/2}$

   d) $f(x) = x^3 + x^2 + x + 1$ e) $f(x) = x^{1/2} - x^{-1/2}$ f) $f(x) = (1 - x)^{1/2}$

   g) $f(x) = x^4 + 2x - 1$ h) $f(x) = x^{2.1} - x$ i) $f(x) = x^{-1}(1 + x)$

   j) $f(x) = (x^2 + 1)^4$ k) $f(x) = \text{SQRT}(1 - x^2)$ l) $f(x) = x(1 + x)^{-1}$

2. Evaluate the first and second derivatives at the indicated x-value.

   a) $f(x) = \exp(x^2)$, $x_0 = 0$ b) $f(x) = ln(x^2)$, $x_0 = 1$

   c) $f(x) = xe^{-3x}$, $x_0 = ln(2)$ d) $f(x) = e^{-0.1x} - e^{-0.005x}$, $x_0 = 20$

3. Compute $f''(x)$ for the given function.
   a) $f(x) = \sin(x)$ b) $f(x) = \sin(3x)$ c) $f(x) = \tan(x)$

   d) $f(x) = \cos(x^2)$ e) $f(x) = \sin(2\pi x - \pi/2)$ f) $f(x) = x \cdot \sin(x)$

4. a) Compute the first four derivatives of $y = \sin(2t)$. b) Determine the 85-th derivative $y^{(85)}(t)$.

5. Show that the function $y = \cos(2\theta) - 3\sin(2\theta)$ satisfies the differential equation $y'' = -4y$.

6. For the indicated function and reference x-value x_0, compute the coefficients of the polynomial

   $$P(x) = f(x_0) + f'(x_0)(x - x_0) + (1/2)\ f''(x_0)(x - x_0)^2 + (1/6)\ f^{(3)}(x_0)(x - x_0)^3$$

   a) $f(x) = x^3 - 1$ $x_0 = 1$ b) $f(x) = x^5 - 2x$ $x_0 = -1$ c) $f(x) = 2 - 3x^2$ $x_0 = 2$

   d) $f(x) = e^x$ $x_0 = 0$ e) $f(x) = e^{-0.1x}$ $x_0 = 0$ f) $f(x) = e^{2x}$ $x_0 = ln(3)$

   g) $f(x) = \sin(x)$ $x_0 = 0$ h) $f(x) = \sin(x)$ $x_0 = \pi/2$ i) $f(x) = \sin(3x)$ $x_0 = \pi/2$

   j) $f(x) = \cos(x)$ $x_0 = 0$ k) $f(x) = \cos(x)$ $x_0 = \pi/2$ l) $f(x) = \cos(2x)$ $x_0 = \pi/2$

Section 3.6 HIGHER DERIVATIVES AND IMPLICIT DIFFERENTIATION 253

7. Determine the domain of the function y and of all its higher derivatives:

   a) $y = (x - 3)^4$ b) $y = (x - 3)^{2.7}$ c) $y = (x - 3)^{2.72}$

   d) $y = \text{SQRT}(100 - x^2)$ e) $y = (x - 5)^{1.5} - (3 - x)^{4/3}$ f) $y = (x^{1/3} + 8)^{3/2}$

8. Determine i) the x-values at which the derivative $y'(x) = 0$, and
   ii) the x-values at which the second derivative $y''(x) = 0$:

   a) $y = x^3 - 8x + 2$ b) $y = \sin(\pi x)$ c) $y = xe^{-2x}$

   d) $y = x^3 + 27x - 1$ e) $y = \cos(2x)$ f) $y = e^{-x} - e^{-4x}$

9. Solve the given equation explicitly for y, indicating the restricted domains and corresponding ranges. Note: these are more challenging and will test your algebra skills.

   a) $(x + 2y)^{0.5} = xy$ b) $(x/y)^2 = x^2$ c) $x^2 - 2xy^4 = 1$

   d) $(x + 2y) = (xy + x - 1)/y$ e) $ln(x/y) = x^2$ f) $x^2 - 2xy^4 = y^2$

10. Determine the tangent line to the graph of the equation at the indicated point.

    a) $x^2y^3 = x - y + 9$ at $(1,2)$ b) $3x + 2y - 18x/y = 0$ at $(2,3)$ c) $(2x + x^2y)^3 = 4y^{1/2}$ at $(-1, 4)$

    d) $3xy = y/x - 4x/y$ at $(1/2,2)$ e) $x + y = e^{x-y}$ at $(0.5,0.5)$ f) $x + y^2 - 6 = ln(x/y)$ at $(2,2)$

    g) $x/(x + y) = 0.5$ at $(-2,-2)$ h) $x^y = 10x$ at $(10,2)$

11. Determine the velocity $v = x'(t)$ and the acceleration $a = x''(t)$ of a particle which oscillates about the origin on the x-axis and whose distance form the origin at time t is

    a) $x(t) = t \sin(3t)$ b) $x(t) = 5e^{\sin(t)}$ c) $x(t) = 2 - t/(t + 1)$

12. Use Logarithmic Differentiation to determine $y'(x)$:

    a) $y = (x - 1)(x^2 - 2)(x^3 - 3)$ b) $y = (3x - 1)^5(x^3 + 1)^{10}(x^2 + 6x - 1)^8$

    c) $y = (x - 1)^2(x^2 - 2)^5(x^3 - 3)^{-3}$ d) $y = (x^2 - 1)^5(x^3 - x)^2(x^{1/2} + 6x^{-1/2})$

    e) $y = (5x^2 - 1)(x^{2/3} + x^{3/2})^4/(5x + 4)^2$ f) $y = \sin(\pi x) \cos(x) e^{-3x}$

    g) $y = ln(x)e^{-2x} \sin^2(x)$ h) $y = (2 - x^3)x^2/[(x + 5)^7(2x - 1)^{1/2}]$

Chapter 4 *Applications of Derivatives.*

Overview. Calculus is a tool. It can be used to graph and analyze the features of functions and models of many real phenomena. One of its most important used is to establish optimal values of functions. In this chapter the basic applications of differential calculus are introduced with an emphasis on modeling.

Section 4.1   Qualitative Aspects of a Function: Slope and Concavity. 255
                  Using the first and second derivative to describe how a graph changes relative to the coordinate axes.

Section 4.2   Graphing with Calculus. 266
                  Introduces curve sketching using the derivative.

Section 4.3   Local Extrema: Maximum and Minimum Values 274
                  The First and Second Derivative tests are used to identify and classify extrema of a function.

Section 4.4   Optimization with constraints 282
                  Considers the extreme values of a function restricted to a specific set, and identifying the absolute maximum and minimum of a function on an interval.

Section 4.5   Optimal Resource Management 287
                  Introduces Theorems that indicate conditions which insure the existence of extreme values.

Section 4.6   Optimization: Application in various disciplines. 296
                  A kaleiptic look at optimization problems that arise in a variety of applications, with an emphasis on "word problems" and their solutions.

Section 4.7   Rolle's Theorem and the Mean Value Theorem 310
                  Introduces Theorems that indicate conditions which insure the existence of extreme values.

Section 4.8   Differentials and Linear Approximations. 317
                  Introduces the differential, df, which approximates the difference Δf. The linear approximation of functions is introduced.

Section 4.9   Taylor Polynomials and Taylor Series 324
                  Introduces higher order approximations and the representation of functions by special polynomials. The limit when the degree of the polynomials becomes infinite is an "infinite Taylor Series" that represents the function exactly.

Section 4.10   Indeterminate limits; L'Hôpital's Rule 334
                  Limits involving two functions that can not be directly evaluated because they involve concepts like the division of zero by zero, "0/0", or arithmetic with "∞", like "0 × ∞", are evaluated using a rule that considers the limits of derivatives of the functions.

Section 4.1 Qualitative Aspects of a Function: Slope, Concavity and derivatives.

Introduction.

A function's equation, y = f(x), indicates how to evaluate the function at a specific x-value in its domain, and using calculus we will be able to glean from the function information about the function's *qualitative* behavior. These are the aspects of a function, and its graph, that characterize it over a range of x-values. Qualitative features of a function are the aspects of its graph that can be visualized and used to describe the function. Graphing functions will be discussed in the next section, but first, in this section we establish the basic terms used to describe function's graphs and their relationships to derivatives.

A principal application of calculus is to determine the nature of a function's qualitative properties. Our visual perception and ability to distinguish between different graphs is incredible. We can view a graph and distinguish different curves or associate points with a straight line. How we actually perceive the appearance of a curve depends on the other visual cues that are in the field of view, the "window" as it is described in computer graphics programs. You have probably encountered the example of two line segments of equal length that appear different because of arrow-heads pointing in opposite directions at their endpoints.

In mathematics, the description of functions and their graphs are also tied to cues, mathematical cues such as the units of the variables and the orientation of numbers on a number-line, or axis. When we describe a graph we use adjectives to indicate the nature of a graph <u>relative to the axis</u> system. There are two primary sets of adjectives used: to describe the *slope* of a graph we use *increasing* and *decreasing* and to describe the *curvature* of a graph we use *concave upward* and *concave downward*. These qualitative features of a function's graph are indicated by the function's derivatives.

Qualitative features of a graph: SLOPE -- Increasing and Decreasing functions.

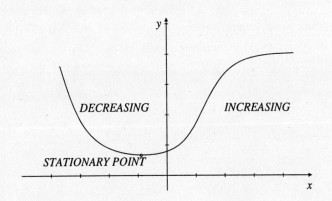

The most easily perceived feature of any graph is how it appears to change relative to the horizontal and vertical axes. When a graph rises relative to the horizontal axis, approaching the top of the graph as scanned in the positive x-direction from left to right, it is *increasing*, and it is *decreasing* if it declines towards the bottom of the graph.

The adjectives *increasing* and *decreasing* are comparative words that are defined mathematically by inequalities involving a function's values at different points in an interval.

Definition of Increasing and Decreasing on an Interval.

A function $y = f(x)$ is **increasing on an interval (a,b)** if,
for any two values x_1 and $x_2 \in (a,b)$,

$$x_1 < x_2 \text{ implies that } f(x_1) < f(x_2)$$

and $f(x)$ is **decreasing on an interval (a,b)** if,
for any two values x_1 and $x_2 \in (a,b)$,

$$x_1 < x_2 \text{ implies that } f(x_1) > f(x_2)$$

Recall that the graph of the line $y = y_0 + m(x - x_0)$ with slope m is **increasing** if $m > 0$, **decreasing** if $m < 0$, and is **horizontal** if $m = 0$.

A function is said to be **increasing (or decreasing) at specific point $(x_0, f(x_0))$** on a its graph if there is an interval containing x_0 on which the function is increasing (or decreasing).

Using the first derivative.

If $y = f(x)$ is differentiable on an interval (a,b) then its graph will have a tangent line at each point $(x_0, f(x_0))$ with $x_0 \in (a,b)$. A basic premise in Calculus is that:

The nature of a function is, to a first approximation, given by its tangent lines.

Indeed, it can be shown that a differentiable function is increasing (decreasing) at a point if and only if its tangent line at the point is increasing (decreasing). The justification of this statement involves the comparison of the sign of the difference quotient $\Delta f/h$ and the sign of its limit as $h \to 0$, which is by definition the derivative. We state this as a Theorem, a statement that is true and can be logically proven.

Theorem.

A differentiable function f, and its graph, is

increasing at a point $(x_0, f(x_0))$ if $f'(x_0) > 0$

decreasing at a point $(x_0, f(x_0))$ if $f'(x_0) < 0$

stationary at $(x_0, f(x_0))$ if $f'(x_0) = 0$

The term "stationary" is derived from the association of the derivative with the rate of change of a dynamical systems. If the derivative is zero, the dynamical system does not change or move and is "stationary" in the sense that this term is commonly used in english.

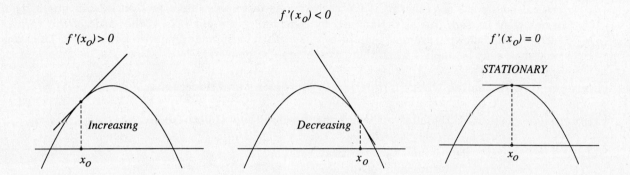

If the sign of a derivative is the same throughout an interval then the function will be either increasing, decreasing, or stationary throughout the interval. This is the basis of the following Derivative Test.

DERIVATIVE TEST FOR INCREASING AND DECREASING

> If $f'(x)$ is positive for each $x \in (a,b)$ then $f(x)$ is **increasing on (a,b).**
>
> If $f'(x)$ is negative for each $x \in (a,b)$, then $f(x)$ is **decreasing on (a,b).**
>
> If $f'(x) = 0$ for each $x \in (a,b)$ then $f(x)$ is **stationary on (a,b)**.
> It follows that $f(x) = C$ a constant throughout the interval.

To describe a graph $y = f(x)$ we will want to identify the intervals on which it is increasing or decreasing. These intervals will be found by determining the "critical points" where the graph may change its slope, from increasing to decreasing or *visa versa*. Such a change would require a change in the sign of the derivative f'. We know that for a continuous function to change from positive to negative it must pass through zero. Therefore, if the derivative f' is continuous then the sign of $f'(x)$ can only change, say from positive to negative, by passing through zero. Consequently, the only points where the function f can change from being increasing to being decreasing, or *vice-versa*, are stationary points where $f'(x) = 0$ or points where $f'(x)$ does not exist or is discontinuous.

From these observations we can determining the intervals on which a graph is increasing or decreasing. The procedure is as follows:

First, identify the **critical points of the first derivative**. These are the x-values at which the function is either stationary, $f'(x) = 0$, or at which the derivative f' is either not continuous or does not exist. If $f'(x) = 0$ on an interval include only the endpoints of this interval; this is a special case you will likely never encounter.

Next, order these critical values in increasing order, so that they partition the domain of f into subintervals. On each subinterval the sign of $f'(x)$ can not change, it is either always positive or always negative, or always zero.

Finally, apply the Derivative Test for Increasing and Decreasing on each subinterval.

Chapter 3 Application of Derivatives

Note that a simple calculator evaluation of $f'(x)$ at any x-value between two adjacent critical points will indicate the sign of the derivative throughout the subinterval. This is usually a very quick and accurate procedure. However, sometimes a more algebraic approach must be used. For instance, when the function involves unspecified parameters and the exact numerical values of the critical points are not known. If the derivative is factorable into a product (and/or quotient) of terms then its sign can be determined by considering the products of the signs of the various factors. The following example illustrates this method in a simple case with two factors. The same procedure may be applied when there are several factors.

Example 1   **Determine where a function is increasing and decreasing.**

Problem   Determine the intervals where the function $f(x) = (1/3)x^3 - (3/2)x^2 + 2x$ is increasing.

Solution   First, determine the "critical points". The stationary points are found by solving $f'(x) = 0$. Since

$$f'(x) = x^2 - 3x + 2 = (x-1)(x-2)$$

the two stationary values are

$$x_1 = 1 \quad \text{and} \quad x_2 = 2$$

$f'(x)$ is continuous and defined everywhere, thus these are the only critical values. These two numbers partition the Real line into three intervals:

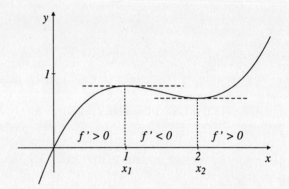

$$(-\infty, 1), \quad (1,2), \quad \text{and} \quad (2, \infty)$$

The sign of $f'(x)$ on each interval can be determined algebraically or by considering the signs of its factors, $(x - 1)$ and $(x - 2)$.

Graphically, the sign of f' on the three intervals can be illustrated by sketching three number lines with the same scales. The sign of each factor is indicated on the first two lines and the sign of their product is indicated on the bottom line.

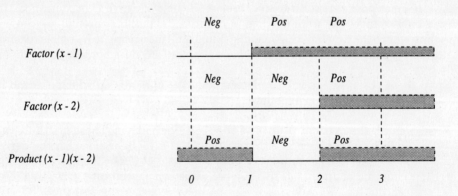

RECALL   The sign of a product is positive, if both factors are positive or both factors are negative, and is negative, if one factor is positive and the other is negative.

Consequently, we conclude that the function f is increasing on the half-lines $(-\infty,1)$ and $(2,\infty)$ and is decreasing on the interval $(1,2)$. This is consistent with our experience with cubic graphs.

Example 2 The slope of exponential functions.

Problem Determine the intervals on which the following exponential functions are increasing.

a) Exponential Growth curve $y = e^{kt}$, $k > 0$,

b) Exponential Decay curve $y = e^{-kt}$, $k > 0$,

c) Exponential absorption-decay curve $y = A(e^{-\alpha t} - e^{-\beta t})$.

Solutions a) EXPONENTIAL GROWTH function $y = e^{kt}$, k positive, is increasing for all t since its derivative,
$$y' = ke^{kt} > 0$$
is the product of the positive constant k and e^{kt}, which is always positive.

b) EXPONENTIAL DECAY function $y = e^{-kt}$, k positive, is decreasing for all t since its derivative is a product of the negative number, -k, and the positive exponential e^{-kt},
$$y' = -ke^{-kt} < 0$$

c) DIFFERENCE OF TWO EXPONENTIAL DECAYS is a function type that arises in many biological applications. Assume the coefficient A and the two decay rates, α and β, are positive, consider the graph of the function
$$y(t) = A(e^{-\alpha t} - e^{-\beta t}) \quad t \geq 0$$

The shape of this function will depend on the relative size of α and β. Assume that the term being subtracted has the greater rate of decay, i.e., $\beta > \alpha$. Then $-\beta t < -\alpha t$ for $t > 0$ and, hence, $e^{-\beta t} < e^{-\alpha t}$. Then, the function is positive, $y(t) > 0$. In this case, we will show that the graph is a curve starting at the origin that increases over an interval $(0, t_{max})$, reaching a stationary point at $t = t_{max}$, and then decreases on the interval (t_{max}, ∞). The critical value t_{max} is the root of $y'(t) = 0$. The technique for solving the equation
$$y' = A(-\alpha e^{-\alpha t} + \beta e^{-\beta t}) = 0$$
is to factor out an exponential term:
$$Ae^{-\beta t}[-\alpha e^{(\beta - \alpha)t} + \beta] = 0$$

As a product is zero only if one of the factors is zero, and the exponential $e^{-\beta t}$ is never zero, this implies that
$$-\alpha e^{(\beta - \alpha)t} + \beta = 0$$

Solving for t gives the critical value
$$t_{max} = ln(\beta/\alpha)/(\beta - \alpha)$$

Using the factored form $y'(t) = Ae^{-\beta t}[-\alpha e^{(\beta - \alpha)t} + \beta]$ it follows that the sign of y' is the same as the sign of the difference $-\alpha e^{(\beta - \alpha)t} + \beta$. If $t > t_{max}$, then this term is negative and y is decreasing. If $t < t_{max}$ then it is positive and y is increasing.

Let us repeat this with numerical values for the constants: $\alpha = 0.5$, $\beta = 2$ and $A = 5$. Then,

$$y(t) = 5[e^{-0.5t} - e^{-2t}]$$

and

$$y'(t) = 5[-0.5e^{-0.5t} + 2e^{-2t}]$$

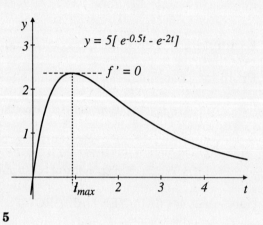

To determine the stationary value t_{max} and to determine the sign of y' for the Derivative Test, the derivative is factored

$$y'(t) = 5e^{-2t}[2 - 0.5e^{1.5t}]$$

The only stationary point for y occurs when the factor $[2 - 0.5e^{1.5t}] = 0$, or $e^{1.5t} = 4$, and hence

$$t_{max} = ln(4)/1.5 \approx 0.9242$$

Since e^t is an increasing function,

if $t < ln(4)/1.5 = t_{max}$ then $e^{1.5t} < e^{ln(4)} = 4$

and thus the factor $[2 - 0.5e^t]$ is positive. As y' is the product of this factor and the positive factor $5e^{-2t}$ it follows that $y'(t)$ is positive on the interval $(0, t_{max})$. By the Derivative Test the function y is therefore increasing on the interval $(0, t_{max})$. Employing the reverse inequality, if $t > ln(4)/1.5 = t_{max}$ a similar argument will show that $y(t)$ is decreasing on the interval (t_{max}, ∞).

Qualitative features of a graph: CONCAVITY... How a graph curves.

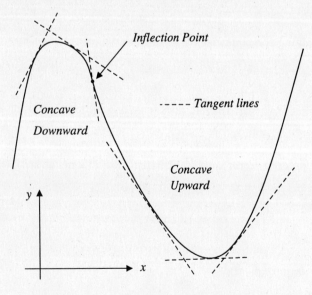

The second primary visual characteristic of a graph is how it *curves*. Mathematicians use the term *concave* to describe curving. The terms *concave upward* and *concave downward* are used to indicate the how a graph curves relative to the orientation on the vertical axis.

Curving is a local phenomena, by this we mean that the perception of a graph's curvature is influenced by its character near a specific point. The visual reference used to discern a graph's curvature at a point $(x_0, f(x_0))$ is its tangent line at that point. Whether the tangent line is explicitly illustrated or must be inferred by the viewer, our perception of curving is associated with the graphs deviation from this straight tangent line.

Visual definitions of concavity.

A graph curves upward at a point when it is above its tangent line on both sides of the point; in this case it is said to be **concave upward** at the point.

When the graph is below its tangent line on both sides of the tangent point then the graph is said to be **concave downward** at the point.

A point where the concavity changes is called an **inflection point**. The function's graph will be above its tangent line on one side of the inflection point and below it on the other side.

When sketching a curve, inflection points are the points where you usually pause briefly. This is because at these points the flowing motion of your hand must change as you arc the curve in a different direction.

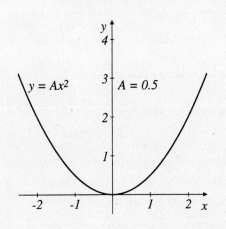

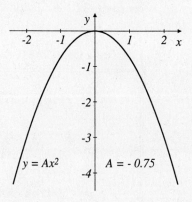

The generic example used to illustrate concavity is the parabola $y = Ax^2$.

If $A > 0$ it opens upward and is *concave upward* for all x, like the graph of $y = 0.5x^2$.

If $A < 0$, the graph opens downward and is *concave downward* for all x, like the graph of $y = -0.75x^2$.

The graph of a parabola has no inflection points.

Tangent parabolas.

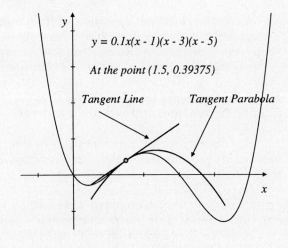

A *tangent parabola* is a natural extension of the concept of a tangent line.

The tangent parabola to the graph of f(x) at a point (c,f(c)) is the parabola P(x) characterized by:

(i) it passes through the point, $P(c) = f(c)$,

(ii) at the point of tangency it has the same slope as the curve $y = f(x)$, $P'(c) = f'(c)$, and

(iii) it has exactly the same "curvature" as the graph of f at this point, $P''(c) = f''(c)$.

Notice in the adjacent figure how the parabola approximates the curve better than the tangent line.

Definition of a Tangent Parabola

> The equation of the **tangent parabola** to the
> graph of $y = f(x)$ at the point $(x_0, f(x_0))$ is
>
> $$y(x) = f(x_0) + f'(x_0)(x - x_0) + [f''(x_0)/2](x - x_0)^2$$

Just as the slope of the tangent line indicates when a graph is increasing or decreasing at a point, the concavity of a function at a point is indicated by the concavity of its tangent parabola at that point. The tangent parabola's concavity is determined by the sign of $f''(x_0)$, which is the coefficient of the squared term. Like the sign of A in the basic reference parabola $y = Ax^2$, the tangent parabola will be concave upward if $f''(x_0)$ is positive, and it will open downward and be concave downward if $f''(x_0)$ is negative. For functions that have continuous second derivatives the concavity of the function is that of its tangent parabola and may be determined by the following test.

Second Derivative Test for Concavity

> If $f''(x_0) > 0$ then $f(x)$ is **concave upward** at $x = x_0$.
>
> If $f''(x_0) < 0$ then $f(x)$ is **concave downward** at $x = x_0$.
>
> If $f''(x_0) = 0$ then *no* **conclusion can be reached by this test!**
> In this case x_0 **may, or may not**, correspond to an inflection point.

Note that the Second Derivative Test cannot be applied if $f''(x_0)$ does not exist of if f'' is not continuous at x_0.

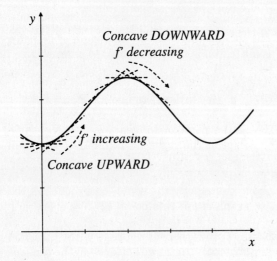

An alternative interval definition of the terms concave upward and concave downward utilizes the way the derivative changes. Remember, the second derivative is the rate of change of the first derivative!

If $f'(x)$ is **increasing on an interval** then the function f is **concave upward** on the interval.

If $f'(x)$ is **decreasing on an interval** then the function f is **concave downward** on the interval.

The graph of $y = f(x)$ is **concave upward** on an interval (a,b) when $f''(x)$ is positive for each $x \in (a,b)$, and is **concave downward** on (a,b) when $f''(x)$ is negative for each $x \in (a,b)$.

To determine the intervals on which a function is concave upward and those on which it is concave downward the first step is to identify any x-values for which the above test does not apply or is inconclusive. These are the **critical x-values** at which f' or f'' fail to exist or are discontinuous and the x-values where $f''(x) = 0$. On the open subintervals with these values as endpoints, the type of concavity of the function will not change. The concavity can be determined by evaluating $f''(x)$ at any point in the subinterval and applying the Second Derivative Test. Each of these critical x-values may correspond to an inflection point.

Test for inflections points: Assume that x_0 is a critical point at which $f''(x_0) = 0$ or does not exist. If for x in an interval containing x_0, the sign of $f''(x)$ for $x > x_0$ is the same as the sign $f''(x)$ for $x < x_0$, then $(x_0, f(x_0))$ **is not an inflection point**.

If the signs are **different**, then $(x_0, f(x_0))$ **is an inflection point**.

Example 3 Determining concavity and inflection points.

Problem   Determine the intervals on which the graph $y = (x - 5)^3$ is concave upward and downward and identify any inflection points.

Solution. The second derivative $y'' = 6(x - 5)$ is defined for all x and is zero at $x_1 = 5$.

For $x > 5$, y'' is positive, and for $x < 5$, y'' is negative. Therefore, on the interval $(-\infty, 5)$ the graph is concave downward while the graph is concave upward on $(5, \infty)$.

Observe that the graph is increasing on both intervals since $y' = 3(x - 5)^2$ is positive for $x \ne 5$.

The only candidate for an inflection point is at the critical value $x = 5$. Since the concavity changes at this value the point $(5, 0)$ is an inflection point.

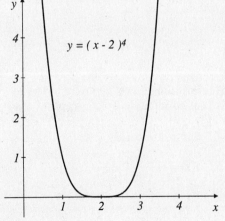

Example 4 $y'' = 0$ does not always indicate an inflection point.

Problem   Show $(2, 0)$ is not an inflection point of $y = (x - 2)^4$ although $y''(2) = 0$.

Solution   The second derivative $y'' = 12(x - 2)^2$ is defined for all x. Clearly $y''(2) = 0$, and $y''(x) > 0$ for $x \ne 2$. Therefore, the graph of this curve must be concave upward on the intervals $(-\infty, 2)$ and $(2, \infty)$.

Since the concavity does not change at $x = 2$, the point $(2, 0)$ in not an inflection point. Although $y''(0) = 0$, the graph is actually concave upward at $(2, 0)$ since the graph lies above its horizontal tangent line at this point.

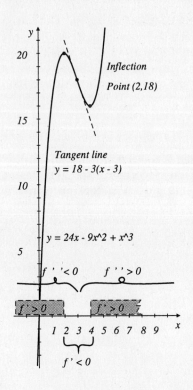

Example 5 **Describing the slope and concavity of a graph; determining the tangent at an inflection point.**

Problem   Determine the intervals on which the graph of $y = x^3 - 9x^2 + 24x$ is increasing and decreasing and the intervals on which it is concave upward and downward. What is the equation of the tangent line at the graph's inflection point?

Solution   The derivatives are

$$y' = 3x^2 - 18x + 24 \quad \text{and} \quad y'' = 6x - 18.$$

To determine where the graph is increasing and decreasing the critical points for the first derivative are found by factoring y':

$$y' = 3(x - 4)(x - 2)$$

The roots of y' = 0 are $x_1 = 2$ and $x_2 = 4$. The sign of y' is determined from the product of the signs of its factors. The factor (x - 2) is positive for x > 2, and the factor (x - 4) is positive for x > 4. Therefore,

$$y' > 0 \text{ if } x > 4 \text{ or if } x < 2$$

and

$$y' < 0 \text{ if } 2 < x < 4$$

Applying the Derivative Test, the graph of y(x) is increasing on the intervals (-∞,2) and (4,∞) and is decreasing on the interval (2,4).

The second derivative y'' has one critical point, $x_3 = 3$, y''(3) = 0. For x > 3, y''(x) = 6(x - 3) > 0 and by the Second Derivative test the graph of y is concave upward. For x < 3, y''(x) = 6(x - 3) < 0 and thus the graph is concave downward on the interval (-∞,3).

At x = 3 the Second Derivative Test is inconclusive. However, as the concavity changes at this value, the corresponding graph point (3,18) is an inflection point. The tangent line to the graph at this inflection point has slope f'(3) = -3 and its equation is

$$y = 18 - 3(x - 3) \quad \text{Tangent line at (3,18)}$$

Exercise Set 4.1

1 At the point $x_0 = 2$: i) apply the First Derivative Test to determine if the given function is increasing or decreasing, and ii) apply the Second Derivative Test to identify if the graph of the function is concave upward or concave downward.

  a) $f(x) = x^2 - 2x + 3$
  b) $f(x) = x^3 - 2x^2 + 3$
  c) $f(x) = x^2/(x - 1)$

  d) $f(x) = \{x + 1\}/\{x - 1\}$
  e) $f(x) = 1 + 1/x$
  f) $f(x) = x(1 + 1/x)$

  g) $f(x) = x^4 - 3x^3$
  h) $f(x) = \text{SQRT}(1 + x^2)$
  i) $f(x) = x^{-2}$

  j) $f(x) = e^{-x^2/2}$
  k) $f(x) = 3e^{x^2}$
  l) $f(x) - x^2 e^{-x}$

  m) $f(x) = (x^2 - 2)ln(x)$
  n) $f(x) = x(1 - ln(x))$
  o) $f(x) = ln(1/x)$

  p) $f(x) = cos(x)$
  q) $f(x) = x^2 e^{cos(x)}$
  r) $f(x) = ln(cos^2(x))$

Section 4.1 — Qualitative Aspects: Slope and Concavity

2. For the given function identify: i) The critical points where the slope may change. ii) The critical points where the concavity may change. iii) The intervals on which the function is increasing. iv) The intervals on which the function is concave downward. v) Inflections points and their tangent lines.

   a) $f(x) = x^2 - 2x + 3$
   b) $f(x) = x^3 - 2x^2 + 3$
   c) $f(x) = -x^2 - x - 1$
   d) $f(x) = x^3 - 12x + 3$
   e) $f(x) = x^3 - 12x^2 - 3x$
   f) $f(x) = (x^2 - x)^{1/2}$
   g) $f(x) = 4x^3 - 3x^2 - 2x + 6$
   h) $f(x) = (x-3)(x-1)^2$
   i) $f(x) = x^{1/2} - x$
   j) $f(x) = x^{-4} + x/32$
   k) $f(x) = x(x-2)^{1/3}$
   l) $f(x) = 2x^{5/3} - 5x^{4/3}$
   m) $f(x) = (x-2)^2(x+2)^2$
   n) $f(x) = x^4 + 0.5x + 3$
   o) $f(x) = x^4 - 8x^3 + 22x^2 - 24x + 1$

3. Determine the intervals on which the given function is decreasing.

   a) $y = 3e^{-2x}$
   b) $y = xe^{-x}$
   c) $y = x^2 e^{-x}$
   d) $y = e^{-x^2}$
   e) $y = 3(e^{-2t} - e^{-t})$
   f) $y = 3(e^{-t} - e^{-2t})$
   g) $y = [ln(x)]^2$
   h) $y = ln(x)/x$
   i) $y = x\, ln(x)$
   j) $y = sin(t - \pi)$
   k) $y = sin(\pi t)$
   l) $y = cos^2(\pi t)$

4. Determine and sketch the tangent lines and the tangent parabolas to the graph of $y = sin(x)$ at:

   a) $x = 0$
   b) $x = \pi/2$
   c) $x = 3\pi/2$

5. Sketch curves that have the indicated properties. How different can your answer appear?

   a) Its first and second derivatives are always positive.

   b) Its first derivative is always positive and its second derivative is always negative.

   c) The product of its first and second derivatives is always positive.

   d) The product of its first and second derivatives is always negative.

   e) The product of its first and second derivatives is zero when x is an integer and otherwise negative.

   f) The product of its first and second derivatives is zero when x is an integer and changes sign on successive intervals between integers.

   g) The function is continuous everywhere but its first and second derivatives do not exist at integer values, its second derivative is otherwise positive, and its first derivative is zero when x is and integer + 1/2, i.e., at 1/2, 3/2, 5/2 ...

Section 4.2 Graphing With Calculus

Introduction.

A graph provides a visual representation of a function that can quickly convey information about the function. The accuracy and details of a graph "sketch" will vary with the techniques used to plot it and how the graph is to be used.

You probably learned to graph functions by plotting points and connecting them with lines. This is not a bad method. It is the basis algorithm of most computer graphing programs. Since the resolution of a computer screen or a printer is limited, a computer can easily be programmed to "sketch" a function's graph by simply evaluating the function at sufficiently close x-values and plotting the resulting points on a coordinate system. To "sketch" a graph over a four inch interval on a standard 300 dots-per-inch laser printer this approach would only require 1200 function evaluations. However, fewer evaluations are usually sufficient. The function $f(x) = \sin(\pi x)$ is sketched below by evaluating $f(x)$ at x-values spaced $\Delta x = 0.25$, 0.1 and 0.02 apart over the interval $[0,2]$. The corresponding graph points are then connected with straight lines. Clearly, evaluating more points than the 100 x-values utilized in the $\Delta x = 0.02$ panel would not improve the graph's appearance.

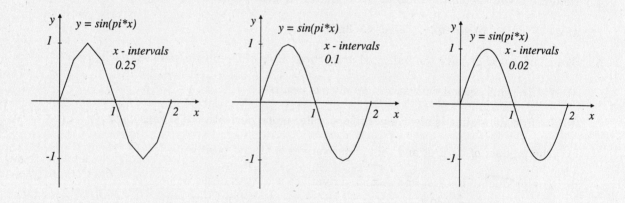

When you sketch a graph, you must decide how accurate it should be. Consider your ability to actually sketch the graph and for what purpose the graph will be used. Plotting points, like the computer generated graphs, is not a good approach for a person with limited time and calculation facilities. One objective in "curve sketching" is to convey the "essential features" of a graph with the minimum amount of computation. Considering the use aspect, in calculus courses we normally expect a "sketch" to indicate the qualitative aspects of the graph's appearance. It should be sufficiently detailed to distinguish the graph's shape and indicate "critical points" where the graph either changes its appearance or attains extreme values.

A sketch, even a "rough" sketch, should have approximately the proper slope and indicate the appropriate concavity for the function.

If a function $y = f(x)$ is twice differentiable then these features can be identified by examining the derivatives $f'(x)$ and $f''(x)$. In particular, the sign of these derivatives will provide enough information to allow the graph to be sketched with minimal function evaluation.

Using derivatives to sketch graphs.

The first step in curve sketching is to identify the critical values at which the function's characteristics may change. We will assume that the function y = f(x) is continuous and there are only a finite number of x-values (usually none) at which f'(x) or f''(x) fail to exist. For the purpose of curve sketching, the **critical values** are of three types:

CRITICAL VALUES FOR SKETCHING y = f(x)

Type I :   x-values for which f'(x) = 0;
at these values the slope of the graph may change.

Type II :   x-values for which f''(x) = 0;
at these values the concavity of the graph may change.

Type III: x-values for which f'(x) or f''(x) is not defined or is not continuous.

If f' does not exist the graph may have a "sharp corner" like y = |x| at the origin, or it may have a vertical tangent line, like the graph of $y = x^{1/3}$ at the origin.

If f'' does not exist, the graph may have a "round corner" like $y = (x - 2)^{6/5}$ at x = 2, where the graph has a horizontal tangent line but changes directions too rapidly to have a tangent parabola.

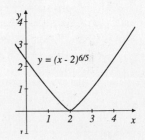

The characteristic appearance of the graph y = f(x) can only change at one of these critical values. Thus, on intervals not containing a critical point the graph's slope will not change sign and the type of concavity will not change. The signs of f' and f'' on these intervals can be found by evaluating them at <u>any</u> arbitrary "test-value" in the interval. This approach is often quicker and less prone to mistakes than methods using algebra and inequalities to analyze a function's sign over the interval.

GRAPH SKETCHING METHOD

To sketch a function y = f(x) over an interval (a,b).

STEP I   First differentiate the given function f(x) twice to determine f'(x) and f''(x).

STEP II   Identify the critical x-values where the derivatives are zero or not defined.

STEP III Determine the sign of f'(x) and f''(x) on the subintervals of (a,b) not containing critical values.

Use the First Derivative test to determine whether the graph is increasing or decreasing and the and Second Derivative test to determine if its concave upward or downward on each subinterval.

Compare the concavity on successive intervals to identify inflection points.

STEP IV   Plot key points. These include the points corresponding to critical x-values, the points (a,f(a)) and (b,f(b)) if given an interval [a,b], and other points useful to establish the scale of the graph, such as points where the graph intercepts an axis.

STEP V   "Sketch" the graph connecting these plotted points using the information from STEP III about the curve's slope and concavity. These graph segments connecting plotted points will be curves, generally not straight lines, corresponding to the function's slope and concavity; the following Table of characteristic shapes may provide a guide.

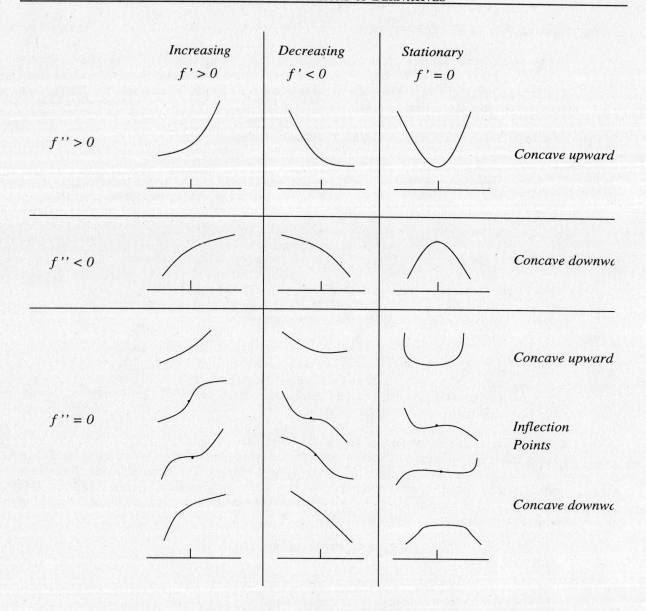

CHARACTERISTIC CURVE SHAPES

Example 1   **Applying the GRAPH SKETCHING METHOD.**

Problem   Sketch $y(x) = 3x^2 + 2x - 1$ over $[-2,1]$.

Solution   This is a simple quadratic, which you should be able to sketch without using calculus, but for illustration we sketch it applying the steps of the Graph Sketching Method.

STEP I   The derivatives are: $y'(x) = 6x + 2$ and $y''(x) = 6$

STEP II   The derivatives are defined everywhere. The only critical value is the root of $y' = 0$. We solve $6x + 2 = 0$ to obtain $x_1 = -1/3$.

STEP III   The characteristics of the graph remain constant on the subintervals $(-2, -1/3)$ and $(-1/3, 1)$.

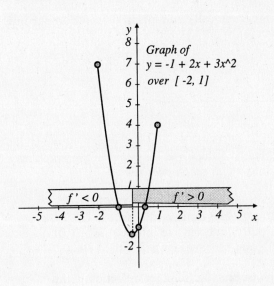

Graph of $y = -1 + 2x + 3x^2$ over $[-2, 1]$

The second derivative is the positive constant 6, so the graph is concave upward everywhere. Choosing $x = -1$ as a "test-value" in the interval $(-2,-1/3)$, $y'(-1) = -4$ is negative, thus the graph is decreasing on the interval $(-2,-1/3)$. Selecting the "test-value" $x = 0$ in $(-1/3,1)$, $y'(0) = 2$ is positive, therefor the graph is increasing on $(-1/3,1)$.

STEP IV In the figure at the left the following points are plotted:
the critical point $(-1/3, y(-1/3)) = (-1/3, -4/3)$;
the endpoints $(-2, y(-2)) = (-2, 7)$ and $(1, y(1)) = (1, 4)$;
the y-axis intercept $(0, y(0)) = (0, -1)$; and the two x-axis intercepts $(-1, 0)$ and $(1/3, 0)$ obtained by first factoring and solving $y(x) = (3x - 1)(x + 1) = 0$ for the x-values $x = 1$ and $x = 1/3$.

STEP V The graph is then sketched to pass through the plotted points. It is decreasing and concave upward on $[-2, -1/3)$ and is increasing and concave upward on $(-1/3, 1]$.

Example 2 Sketching a cubic equation.

Problem Sketch the graph of $y = f(x) = 0.2x^3 - 1.5x^2 + 3.6x$ over the interval $[-1, 5]$.

Solution Each of the five GRAPH SKETCHING steps applied as follows.

STEP I The derivatives are: $f'(x) = 0.6x^2 - 3.0x + 3.6$ and $f''(x) = 1.2x - 3.0$.

STEP II Since both derivatives are defined for all x, the only critical values are the zeros of f' and f''. These are obtained by factoring the derivatives and setting each factor equal to zero.

$$f'(x) = 0.6(x^2 - 5x + 6) = 0.6(x - 2)(x - 3) = 0$$

has two roots: $x_1 = 2$ and $x_2 = 3$. The only root of $f''(x) = 0.6(2x - 5) = 0$, is $x_3 = 2.5$.

STEP III The three critical values and the endpoints break the interval $[-1, 5]$ into four subintervals on which the character of the graph will remain constant. The sign of the derivatives on these intervals can be determined algebraically or from test-evaluations at a point in each interval.

Interval	Test Value	$f'(x)$ sign	$f''(x)$ sign
$(-1, 2)$	0	3.6 pos	-3.0 neg
$(2, 2.5)$	2.25	-0.1125 neg	-3.0 neg
$(2.5, 3)$	2.75	-0.1125 neg	3.0 pos
$(3, 5)$	4.0	1.2 pos	1.8 pos

The signs of f' and f'' on the subintervals can be graphically summarized on a number line chart like the following:

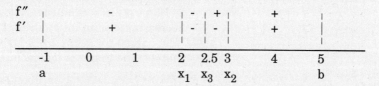

Applying the First and Second Derivative Tests gives the slope and concavity of the graph over the subintervals determined by the critical x-values. The graph over the

subinterval (-1, 2)   is increasing (f' > 0) and concave downward (f'' < 0);
subinterval (2, 2.5)   is decreasing (f' < 0) and concave downward (f'' < 0);
subinterval (2.5, 3)   is decreasing (f' < 0) and concave upward   (f'' > 0);
subinterval (3, 5) is increasing (f' > 0) and concave upward   (f'' > 0).

As the concavity changes at x = 2.5, the graph point (2.5, f(2.5)) is an inflection point.

STEP IV   The function is evaluated at the critical values and at the endpoints of the given interval [-1,5]. The evaluations on a calculator are made easier by factoring the function as follows:

$$f(x) = [(0.2x - 1.5)x + 3.6]x$$

which is evaluated by the key stroke sequence

0.2 <u>times</u> x, <u>minus</u> 1.5, <u>times</u> x, <u>plus</u> 3.6, <u>times</u> x, <u>equals</u>,

where the underlined operations indicate the appropriate calculator button to be pressed and x is the test x-value. The values are entered in a table and plotted.

x	0.0	2.0	2.5	3.0	-1.0	5.0
f(x)	0.0	2.8	2.75	2.7	-5.3	5.5

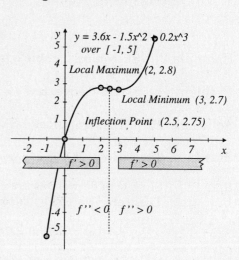

Step V   The graph is sketched using the data points and the above information. Trace the graph and notice how the motion of your hand changes at each critical point.

Example 3   **Sketching a function with a "singular" derivative; a cusp shaped curve.**

Problem   Sketch the graph of $y = (x - 7)^{2/3}$.

Solution   Apply the general Graph Sketching Method as follows:

STEP I   Calculate the derivatives.

$$y' = (2/3)(x - 7)^{-1/3} \quad \text{and} \quad y'' = (-2/9)(x - 7)^{-4/3}$$

STEP II   As neither derivative can equal zero the only critical value is $x_1 = 7$ at which y' (and hence y'') is not defined.

Section 4.2   GRAPHING WITH CALCULUS

STEP III   For $x \neq 7$, $y''(x) < 0$ since y'' is $-2/9$ times $\{(x-7)^{-2/3}\}^2$, which is positive. Thus, the graph is concave downward on both $(-\infty, 7)$ and on $(7, \infty)$. Because the cube root of a negative number is negative, the sign of y' is negative on $(-\infty, 7)$ and positive on $(7, \infty)$. This may be checked with test-values:

$$y'(-1) = (2/3)(-8)^{-1/3} = -1/3 \quad \text{and} \quad y'(15) = (2/3)(8)^{-1/3} = 1/3$$

Thus, the graph is decreasing on $(-\infty, 7)$ and increasing on $(7, \infty)$.

STEP IV   To sketch the graph, first plot the critical point $(7,0)$ and, to give an indication of the function's range, the following four easily evaluated points:

$$(-1, 4), \quad (6, 1), \quad (8, 1), \quad \text{and} \ (15, 4).$$

STEP V   The graph is sketched with a *cusp shape*. It has a point and the tangent lines to the graph approach the vertical line $x = 7$ as the tangency point $(x_0, f(x_0))$ approaches the critical point $(7, 0)$.

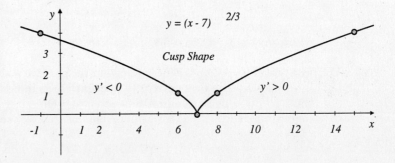

☑

Example 4   Applications to Business\Economics: Sketching Cost, Marginal Cost, and Average Costs functions.

Background   In economics the terms Cost, Marginal Cost and Average Cost refer to functions involving the number x of units or items manufactured. At a production level of x items, the **Average Cost**, $AC(x)$, of producing one item is by definition the **Cost** of x items, $C(x)$, divided by the number x of items. For the given cost function this is

$$AC(x) \equiv C(x)/x$$

The **Marginal Cost**, $MC(x)$, is defined as the rate of increase in costs when the production level x increases. The Marginal Cost is simply the derivative of the Cost with respect to the x-units:

$$MC(x) = C'(x)$$

Problem   Sketch on the same coordinate-axes the Cost, Marginal Cost and Average Cost functions for manufacturing a product given that the Cost of producing x items is:

$$C(x) = (x - 5)^2 + 10$$

Solution   This example uses the business terminology to name the functions: Cost, Marginal Cost, and Average Cost of production. The curve sketching is the same as in other examples. For the given Cost function the Average Cost function is

$$AC(x) = C(x)/x = (x^2 - 10x + 25 + 10)/x = x - 10 + 35/x$$

The corresponding Marginal Cost function is the linear function

$$MC(x) = C'(x) = 2x - 10$$

The graph of $C(x)$ is a parabola opening upward and centered at a minimum point $(5,10)$. The graph of $MC(x)$ is the line with slope 2 and x-axis intercept $x = 5$. To sketch the graph of $AC(x)$ for $x > 0$ we apply the general Graph Sketching Method over the interval $(0,\infty)$:

STEP I $AC'(x) = 1 - 35/x^2$ and $AC''(x) = 70/x^3$

STEP II   Both AC' and AC'' are undefined at $x = 0$, which is not in the interval of positive production. AC'' can not equal zero, however setting AC' to zero gives the equation

$$1 - 35/x^2 = 0$$

whose root is $x_1 = \sqrt{35} \approx 5.9$.

STEP III The graph $y = AC(x)$ is concave upward for positive x since $AC''(x) = 70/x^3 > 0$ if $x > 0$. On the interval $(0,x_1)$ the first derivative is negative, as indicated by one test-value, $AC'(1) = -34$, and thus graph is decreasing on this interval. On the remaining interval, (x_1,∞), the graph is increasing, since $AC'(x)$ is positive, as indicated by the sample evaluation:

$$AC'(10) = 1 - 35/10^2 = 0.65$$

STEP IV   Evaluating $AC(x)$ at the only critical value gives:

$$AC(\sqrt{35}) = \sqrt{35} - 10 + 35/\sqrt{35} = 2\sqrt{35} - 10 \approx 1.8$$

As the function $AC(0)$ is not defined a few additional evaluation points are selected by choosing x-values for which the arithmetic is not difficult:

$A(1) = 26$ $A(10) = 3.5$ and $A(20) = 11.75$

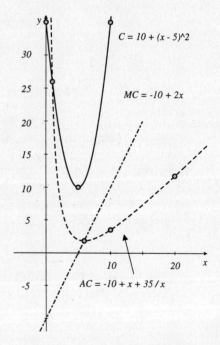

STEP V   The function $y = AC(x)$ is sketched along with the line $y = MC(x)$, and the parabola $y = C(x)$.

Notice that the graphs of $AC(x)$ and $MC(x)$ appear to intersect at the critical point $(\sqrt{35}, AC(\sqrt{35}))$. Is this a coincidence? Or, do you think it would happen for any choice of the Cost function? In the next section we will identify this critical point as the minimum point on the Average Cost curve. It would appear from this example that the minimum is where the AC curve intersects the Marginal Cost curve. Indeed, this can be shown to always be true.

Exercise Set 4.2

1. Sketch a continuous curve $y = f(x)$ over the interval $[2,6]$ that satisfies all of the given conditions
   a) i) $f(2) = 0$, $f(6) = 0$ and ii) $f(4) = 2$, $f'(4) = 0$, $f''(4) < 0$.
   b) i) $f(5) = -1$, $f'(5) = 0$, $f''(5) > 0$ and ii) $f'(3) = 0$, $f''(3) < 0$.

2. Sketch the graph of $y = f(x)$ indicating critical points where the nature of the graph changes.

   a) $y = x^2 + 4x - 12$ b) $y = (x - 1)^3$ c) $y = x^3 - 6x^2$

   d) $y = x^3 - 12x + 2$ e) $y = (1/30)(x^3 - 12x^2 + 45x + 3)$ f) $y = x^4$

   g) $y = x^4 - 9x^2$ h) $y = (x + 3)(x - 3)^2$ i) $y = x^4 - 2x^3 + 12x^2$

   j) $y = 4x^3 + 6x^2 - 24x$

3. Sketch the graph of $y = f(t)$ indicating critical points.
   a) $f(t) = t + 2/t$ b) $f(t) = 1/(1 + t^2)$ c) $f(t) = t/(1 + t^2)$

4. Sketch the graph and determine the local extrema and inflection points of the following.

   a) $y = x^{1/3}$ b) $y = x^{2/3}$ c) $y = (x - 2)^{3/5}$ d) $y = (x + 2)^{4/5}$

   e) $y = x^{1/3}(x + 2)^{1/3}$ f) $y = x^{1/3}(x - 2)^{3/5}$ g) $y = x^{2/3}(x - 2)^{3/5}$ h) $y = x^{4/3}(x - 2)^{1/5}$

5. Sketch the graph over $[0,5]$ of a) $y = xe^{-x}$ b) $y = xe^{-2x}$ c) $y = x^2 e^{-x}$

6. Sketch the graph of a) $f(t) = e^{-t^2}$ b) $y = 2te^{-t^2}$

7. Sketch the graph of the function $f(t) = 2e^{-2t} - e^{-t}$ over the interval $[0, \ln(10)]$. Hint: To find the roots of $f(t)$ use the property that $(e^A + ce^B) = e^A(1 + ce^{B-A})$.

8. Sketch the graph of the general double exponential decay curve $y = c[e^{-\alpha t} - e^{-\beta t}]$, where $c > 0$, and $\beta > \alpha > 0$ for $t \geq 0$. On what interval is the graph decreasing?

9. Sketch the graph of the logistic function $f(t) = 3/(1 + 2e^{-t})$ over the interval $[0,5]$.

10. Sketch the graph of the general logistic curve $y = B/\{1 + ce^{-\lambda Bt}\}$ when

    a) $B = 5$ $C = 2$ $\lambda = 1$ b) $B = 5$ $C = 2$ $\lambda = 0.1$ c) $B = 5$ $C = 2$ $\lambda = 10$

    d) $B = 2$ $C = 2$ $\lambda = 1$ e) $B = 5$ $C = 5$ $\lambda = 1$ f) $B = 5$ $C = 0.1$ $\lambda = 1$

    g) for unspecified constants satisfying $B > 0$, $C > 0$, and $\lambda > 0$.

11. Sketch the graph of $f(t) = 10 \exp\{-0.01 e^{-t}\}$.

12. Sketch the graph of the general Gompertz curve, $y = \exp\{-ke^{-\lambda x}\}$ for $k > 0$, $\lambda > 0$. Indicate where it is concave upward and concave downward.

13. Sketch on the same coordinate-axes the Cost, Marginal Cost and Average Cost functions (See Example 4) for manufacturing a product given that the Cost of producing x items is:
    a) $C(x) = (x - 5)^2 + 10$ b) $C(x) = x^2 - 6x + 15$ c) $C(x) = x^3 - 4x^2 + 10$

14. Sketch the graph over $[-\pi, \pi]$ and indicate the critical points and inflection points of the curve.
    a) $y = x + \sin(x)$ b) $y = \sin^2(x)$ c) $y = \sin(x)\cos(x)$ d) $y = \sin(x) - \cos(x)$

    e) $y = \sin^2(x) - \cos(x)$ f) $y = \cos(x) + \sin(x)$ g) $y = \sqrt{\cos(x)}$ h) $y = e^{\sin(x)}$

Section 4.3 Local Extrema: Maximum and Minimum Values.

Introduction: Local Maximum and Minimum of a function.

The *extrema* of a function are the maximum and minimum function values, which can be viewed "locally", in terms of function values near a reference value, or "globally", with reference to all possible functions values. The difference, by analogy, is the difference between being at the top of a hill in your home town or being atop Mt. Everest. As many functions approach infinity when $x \to \infty$, we will often want to restrict the x-values that are being considered; to do this we will refer to a "constraint set" S, which usually will be an interval.

DEFINITION OF EXTREMA

Given a function f(x) defined on a set S.

The function f has an **absolute maximum at a point $x_M \in S$** if

$$f(x) \leq f(x_M) \text{ for all } x \in S$$

The function f has an **absolute minimum at a point $x_m \in S$** if

$$f(x) \geq f(x_m) \text{ for all } x \in S$$

The function f has a **local maximum at x_0** if there is an open subinterval I of S containing the point x_0 such that

$$f(x) \leq f(x_0) \text{ for all } x \in I$$

The function f has a **local minimum at x_0** if there is an open subinterval I of S containing x_0 such that

$$f(x) \geq f(x_0) \text{ for all } x \in I$$

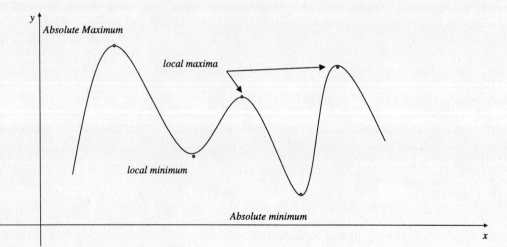

In this section the first and second derivative are used to identify local extrema. In the following sections the determination of "global" extrema over a constraint set and the issue of the existence of extrema are considered.

Determining extrema using the first derivative.

A general rule in applied mathematics, for which there are of course exceptions, is that:

> Extrema occur either at the "boundary" of the constraint set or at
> points where the characteristics of the function change.

As a graph must change direction at an extreme point it should not be surprising that a continuous function's extrema inside an interval [a,b] can occur only at critical values where the derivative is zero or does not exist. The x-values at which extrema can occur are determined from the following observations:

For a critical value not an endpoint of S.

If the graph of f(x) changes from increasing to decreasing at a critical value, $x_0 \in$ (a,b), then it has a **local maximum** at x_0, and conversely.

If the graph of f(x) changes from decreasing to increasing at a critical value, $x_0 \in$ (a,b), then it has a **local minimum** at x_0, and conversely.

For an endpoint of the constraint set S = [a,b].

For the right endpoint x = b.   If f(x) is increasing (decreasing) on an interval (c,b), then f has a local maximum (minimum) at x = b, relative to S = [a,b].

For the left endpoint x = a.   If f(x) is decreasing (increasing) on an interval (a,c), then f has a local maximum (minimum) at x = a, relative to S = [a,b].

Translating these observations into calculus terms, that is, using the derivative, results in the First Derivative Test for Extrema. It is important to note that this Test does <u>not require $f'(x_0)$ to exist!</u> It does, however, require $f'(x)$ to exist for all other x in some open interval containing x_0.

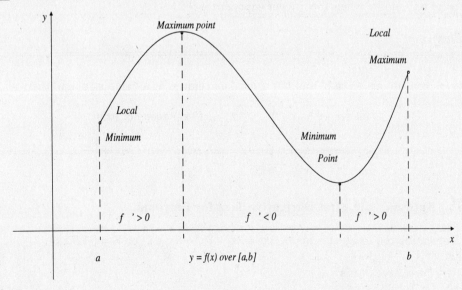

$y = f(x)$ over $[a,b]$

First Derivative Test for Extrema

Assume that $f(x)$ is continuous at $x = x_0$ and is differentiable in an open interval I containing x_0, but not necessairly at x_0. If for $x \in I$,

CASE A: $\quad f'(x) < 0$ when $x < x_0$, and $f'(x) > 0$ when $x > x_0$,
then f has a **local minimum** at x_0.

CASE B: $\quad f'(x) > 0$ when $x < x_0$, and $f'(x) < 0$ when $x > x_0$,
then f has a **local maximum** at x_0.

CASE C: $\quad f'(x)$ has the same sign for $x < x_0$ and for $x > x_0$,
then, f **does not have an extremum** at x_0.

The First Derivative Test for Extrema requires at least two function evaluations since the sign of $f'(x)$ must be known on both sides of a critical value.

To determine the local extrema of a function on a closed interval, [a,b]:

First, identify the critical x-values in (a,b) where f' is zero or fails to exist.

Then, apply the First Derivative Test to these values to determine of they correspond to local maximum or local minimum values.

Lastly, consider the derivative in the interval near each endpoint to classify it as corresponding to a local maximum or minimum.

That a continuous function f(x) can have a local maximum or minimum in the interior of the interval [a,b] only at critical points is given by Fermat's Theorem (Pierre De Fermat, 1601-1655). (He is more *famous* for the Theorem that states the equation $a^n + b^n = c^n$ can not have interger solutions a, b, and c when the integer exponent n > 2.)

Fermat's Theorem

If $f(x)$ is continuous on (a,b) and has a local extremum at a value $x_0 \; \varepsilon \; (a,b)$ then

$$\text{either} \quad f'(x_0) = 0 \quad \text{or} \quad f'(x_0) \text{ does not exist.}$$

Example 1   **Applying the First Derivative Test for Extrema.**

Problem   Determine the local extrema of $f(x) = 2x^3 - 3x^2 - 6x + 1$ on $S = [-1.5, 2.5]$.

Solution   The derivative of f is

$$f'(x) = 6x^2 - 6x - 6 = 6(x^2 - x - 1)$$

The two critical points are the roots of $f'(x) = 0$. Since the quadratic $f'(x)$ is not factorable with integer coefficients, the roots of $f'(x) = 0$ are found by applying the *quadratic formula*:

$$x_1 = 0.5(1 + \sqrt{5}) \approx 1.6$$

and

$$x_2 = 0.5(1 - \sqrt{5}) \approx -0.6$$

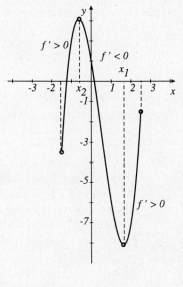

The sign of f' on the three subintervals of S that do not contain critical values, $(-1.5, x_2)$, (x_2, x_1), and $(x_1, 2.5)$, can be determined by evaluating f' at "test values" in each subinterval. We choose x-values at which the function evaluation is simple: the integer values -1, 0, and 2 are picked arbitrarily.

as $f'(-1) = 6$, f' is positive on $(-1.5, x_2)$;

as $f'(0) = -6$, f' is negative on (x_2, x_1); and

as $f'(2) = 6$, f' is positive on $(x_1, 2.5)$.

Finally, the First Derivative Test is applied.
   CASE A applies at x_2, indicating that f has a local maximum at x_2.
   At x_1, CASE B applies and f has a local minimum.
   As $f' > 0$ on $(-1.5, x_2)$, f has a local minimum at the left endpoint $x = -1.5$, relative to the set $S = [-1.5, 2.5]$.
   Since $f' > 0$ on $(x_1, 2.5)$, f has a local maximum at the right endpoint $x = 2.5$, relative to the interval S.

Example 2 Applying the First Derivative Test for Extrema at a point where the derivative does not exits.

Problem Determine the local extrema of $f(x) = x^{1/3}(x - 4)^{2/3}$ on $S = [-2, 6]$.

Solution This example is slightly more challenging since the derivative of f is more complex and more algebra skills are required to find its roots. Applying the Product Rule to differentiate f yields:

$$f'(x) = x^{1/3}(2/3)(x - 4)^{-1/3} + (1/3)x^{-2/3}(x - 4)^{2/3}.$$

To find the roots of a complicated function like this the basic technique is to express the function as a product rather than as a sum. This is done by factoring out the lowest power of each "term" that occurs in the derivative. The lowest power of x in $f'(x)$ is $-2/3$ and the lowest power of $(x - 4)$ is $-1/3$. Since

$$x^{1/3} = x \cdot x^{-2/3} \quad \text{and} \quad (x - 4)^{2/3} = (x - 4)(x - 4)^{-1/3}$$

f' can be factored as

$$f'(x) = x^{-2/3}(x - 4)^{-1/3}[(2/3)x + (1/3)(x - 4)]$$

Notice that the terms remaining in the square brackets are polynomials that can be added to give a simpler factor:

$$f'(x) = x^{-2/3}(x - 4)^{-1/3}(x - 4/3)$$

Placing the terms with negative exponents in the denominator gives

$$f'(x) = \{x - 4/3\} / \{x^{2/3}(x - 4)^{1/3}\}$$

Expressed in the last form, it is clear that f' has three critical values:

$$x_0 = 4/3, \text{ which is the only root of } f'(x) = 0;$$

and the two values at which $f'(x)$ does not exist:

$$x_1 = 0 \quad \text{and} \quad x_2 = 4$$

These three values divide the interval S into four subintervals. To apply the First Derivative Test, the sign of f' is determined on each subinterval by evaluating $f'(x^*)$ at a test value, x^*, arbitrarily chosen in each interval. The choice is usually a value for which the function evaluation is simple. We use the factored form of f' and do not need to actually determine the numerical values of these derivatives, just their signs.

Recall The sign of a fractional power of a negative number is determined by the numerator of the exponent.
With an odd numerator the term is negative, e.g., $(\text{negative})^{1/3}$ is negative.
For an even numerator the term is positive, e.g., $(\text{negative})^{2/3}$ is positive.

Interval	Test value	$f'(x^*)$
[-2,0)	$x^* = -1$	$f'(-1) = -(7/3)/\{(-5)^{1/3}\} > 0$
(0,4/3)	$x^* = 1$	$f'(1) = (-1/3)/\{(-3)^{1/3}\} > 0$
(4/3,4)	$x^* = 2$	$f'(2) = (2/3)/\{2^{2/3}(-2)^{1/3}\} < 0$
(4,6]	$x^* = 13/3$	$f'(13/3) = 3/\{(13/3)^{2/3}(1/3)^{1/3}\} > 0$

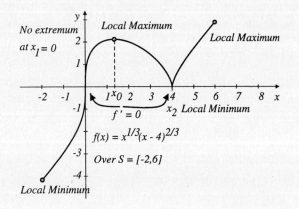

$f(x) = x^{1/3}(x - 4)^{2/3}$
Over $S = [-2, 6]$

Applying the First Derivative Test with the above information, the following conclusions are reached:

At $x = x_1 = 0$, Case C applies, f has neither a local maximum nor a local minimum since f' does not change sign at this value.

At $x = x_0 = 4/3$, Case B applies, f has a local maximum since f' changes from positive to negative at this value.

At $x = x_2 = 4$, Case A applies, f has a local minimum since f' changes from negative to positive at 4.

At the endpoints of the interval S, the definition is applied:

At $x = -2$, since $f' > 0$ on the interval (-2,0) f has a local minimum, relative to S.

At $x = 6$, since $f' > 0$ on (4,6) f has a local maximum relative to S.

Example 3 A function with an extrema at the endpoint of the constraint set.

Problem Determine the local extrema of $f(x) = 5xe^{-0.1x}$ on $S = [0,30]$.

Solution The first step is to evaluate the derivative $f'(x) = 5(e^{-0.1x} - 0.1xe^{-0.1x}) = 5(1 - 0.1x)e^{-0.1x}$

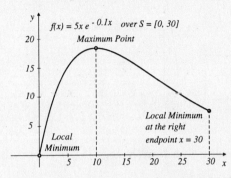

The derivative $f'(x)$ is defined for all x. Thus the only critical point is the single root $x_0 = 10$ of $f'(x) = 0$. As the exponential term $e^{-0.1x}$ is always positive, the sign of f' is the same as the sign of the factor $(1 - 0.1x)$. Hence,

$$f'(x) < 0, \text{ if } x < 10, \quad \text{and} \quad f'(x) > 0, \text{ if } x > 10$$

By Case A of the First Derivative Test, f has a local maximum at $x = 10$. At both endpoints of the constraint interval S, f has a local minimum relative to S, since f' is positive on $(0,10)$ and negative on $(10,30)$.

Using the second derivative to test for extrema.

When the function has a continuous second derivative a one-point test for extrema is available. The second derivative indicates the concavity of a graph. When a curve has a local maximum it must be concave downward and when it has a local minimum it must be concave upward. Hence, the sign of f'' will identify the type of extrema. **The following Test will not apply to endpoints of constraint sets.**

THE SECOND DERIVATIVE TEST FOR EXTREMA

Let x_0 be a critical value of $f(x)$ at which $f'(x_0) = 0$.

Case A: if $f''(x_0) < 0$, then f has a **local maximum at x_0**.

Case B: if $f''(x_0) > 0$, then f has a **local minimum at x_0**.

Case C: if $f''(x_0) = 0$, then **NO CONCLUSION CAN BE MADE**, the function f may have a maximum, a minimum, or no extremum at x_0.

Case C of the Second Derivative Test for Extrema gives NO CONCLUSION. This can be demonstrated by considering the simple power functions $f(x) = x^n$. If $n > 1$ then $f'(0) = 0$ and $f''(0) = 0$. The graph of f will have a minimum at $x = 0$ when n is even, like the parabola $y = x^2$ or $y = x^4$, $y = x^6$ ..., etc. When n is odd the graph will not have and extremum at $x = 0$, it will have an inflection point at the origin like the graphs of $y = x^3$, $y = x^5$, etc.

Example 4 Applying the Second Derivative Test for Extrema to a cubic polynomial.

Problem Determine the local extrema of

$$y = x^3 - 4x^2 + 4x + 3$$

Solution The derivatives of y are:

$$y' = 3x^2 - 8x + 4 \quad \text{and} \quad y'' = 6x - 8$$

The only critical values are the two roots of $y' = 0$; using the quadratic formula these are

$$x = \{8 \pm \text{SQRT}[64 - 48]\}/6$$

$$x_1 = 2/3 \quad \text{and} \quad x_2 = 2$$

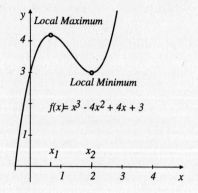

At $x_1 = 2/3$, $y'' = -4$ and Case A of the Second Derivative Test For Extrema indicates that y has a local maximum. At $x_2 = 2$, $y'' = +4$ and thus, by Case B of the Test, y has a local minimum.

Example 5 **Applying the Second Derivative Test for Extrema to a fourth degree polynomial.**

Problem Determine the local extrema of $f(x) = (x - 2)^2(x - 4)^2$.

Solution The first derivative of f is obtained by using the Product and Chain Rules:

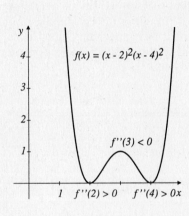

$$f'(x) = 2(x - 2)(x - 4)^2 + 2(x - 2)^2(x - 4)$$

$$= 2(x - 2)(x - 4)[(x - 4) + (x - 2)]$$

$$= 2(x - 2)(x - 4)(2x - 6)$$

The roots of $f' = 0$ are thus $x_1 = 2$, $x_2 = 4$, and $x_3 = 3$. The second derivative can be evaluated by applying the Product Rule twice:

$$f''(x) = 2[(x-4)(2x-6) + (x-2)(2x-6) + 2(x-2)(x-4)]$$

$$= 2[\{2x^2 - 14x + 24\} + \{2x^2 - 10x + 12\} + \{2x^2 - 12x + 16\}]$$

$$= 4[3x^2 - 18x + 26]$$

Applying the Second Derivative Test for Extrema at each stationary critical point:

At $x_1 = 2$: $f''(2) = +8$, f has a local minimum at $x = 2$.

At $x_2 = 4$: $f''(4) = +8$, f has a local minimum at $x = 4$.

At $x_3 = 3$: $f''(3) = -4$, f has a local maximum at $x = 3$.

Exercises Set 4.3

1. List all critical points and apply the **First Derivative Test** to determine if they correspond to extrema.

   a) $y = x^2 - 1$
   b) $y = x^2 - x$
   c) $y = x^3 - x$

   d) $y = x^3 - 12x + 6$
   e) $y = -6x^3 - 15x^2 - 36x + 3$
   f) $y = (x + 2)^2(x - 2)^2$

   g) $y = 4/(1 + 2x^2)$
   h) $y = x^{2/5}$
   i) $y = (x + 3)^{2/3}$

   j) $y = \text{SQRT}(9 + x^2)$
   k) $y = \text{SQRT}(9 - (1 - x)^2)$
   l) $y = x^5 - 5x^3$

   m) $y = xe^{-2x}$
   n) $y = e^{-x^2}$
   o) $y = x^2 e^{-x^2}$

   p) $y = e^{-2x} - e^{-4x}$
   q) $y = e^{-5x} - e^{-x}$
   r) $y = e^{-(x-2)^2}$

2. Determine the critical points of the function $y = f(x)$. Apply the **Second Derivative Test** to determine whether they correspond to local maximum or local minimum points. Determine the inflection points.

   a) $y = x - x^2$
   b) $y = (x - 2)^2$
   c) $y = (x + 2)^3$

   d) $y = (x^2 - 2)^2$
   e) $y = x(x - 2)^2$
   f) $y = (x - 2)^2(2 - x)$

   g) $y = (x^2 - 5)^{1/3}$
   h) $y = x^3(x + 2)^2$
   i) $y = (x + 7)^2(7 - x)^2$

   j) $y = (x - 2)(x - 3)(x - 4)$
   k) $y = 2x^3 - 6x^2 + 1$
   l) $y = -x/(1 + 9x^2)$

   m) $y = x^4 - x$
   n) $y = x - \sqrt{x}$
   o) $y = x - 3/x$

   p) $y = x(2 - x)^{1/3}$
   q) $y = x\,\text{SQRT}(x + 4)$
   r) $y = x^2(1 - 4/x)$

   s) $y = x^2(1 - 4x)^{2/3}$
   t) $y = xe^{-0.3x}$
   u) $y = x^2 e^{-10x}$

   v) $y = e^{-4x} - e^{-3x}$
   w) $y = 2e^{-0.01x^2}$
   x) $y = xe^{-0.09x^2}$

3. Determine the extrema of the given function over the indicate constraint interval S.

   a) $f(x) = 2 - x$ $S = [0,3]$
   b) $f(x) = 2 - x^2$ $S = [0,3]$
   c) $f(x) = (2 - x)^2$ $S = [0,3]$

   d) $f(x) = x + x^2$ $S = [-1,1]$
   e) $f(x) = 3x - x^2$ $S = [0,3]$
   f) $f(x) = -4 + 3x + x^2$ $S = [-4,4]$

   g) $f(x) = e^x$ $S = [-1,1]$
   h) $f(x) = e^{-x}$ $S = [0, ln(3)]$
   i) $f(x) = e^{-x} - e^{-2x}$ $S = [0,5]$

   j) $f(x) = xe^x$ $S = [-1,1]$
   k) $f(x) = xe^{-x}$ $S = [0, ln(3)]$
   l) $f(x) = e^{-x} + e^{-2x}$ $S = [0,5]$

   m) $f(x) = \sin(\pi x)$ $S = [-1,1]$
   n) $f(x) = \cos(\pi x/2)$ $S = [0,4]$
   o) $f(x) = \sin(\pi x)\cos(\pi x)$ $S = [-1,1]$

   p) $f(x) = \sin(x) + \cos(x)$ $S = [0,\pi]$
   q) $f(x) = \sin(x) - \cos(x)$ $S = [0,\pi]$

4. Show that if $f(x_0)$ and $f'(x)$ exists on an interval containing x_0 then the function $y = |f(x)|$ has a local minimum at x_0 even though it is not differentialble there.

5. If the derivative of f is $f'(x) = x(x - 2)(x - 3)$ at which x-values will the graph of f have local maximums or minimums? Use this to sketch the form of the graph of f.

6. If a differentiable funtion $f(x)$ has $f'(2) = 5$ and $f'(3) = -5$ will it have a local maximum? Why?

Section 4.4 Optimization With Constraints.

Introduction: Constrained Optimization.

Many applied problems that require the determination of a function's extreme values are referred to as "optimization problems". These are problems in which one wants to determine the independent variables that will lead to either a maximum or minimum of an observable *objective function*. In applications, the formulation of optimization problems normally involves constraints placed on the independent variable. The constraints are typically restrictions of the variable to "feasible" values, such as requiring a variable describing a length to be positive or limiting the volume of water in a tank to the maximum capacity of the tank. The **constraint set**, **S**, for an optimization problem can be specified in many ways, the most common is by inequalities or directly as an interval. For instance, the constraint

$$S: \text{all } x \text{ satisfy } (x-5)^2 \le 4 \quad \text{or simply} \quad S = [3,7]$$

If no constraint is indicated, the default set S is the natural domain of the function.

The *constrained optimization* problem is to determine the maximum and minimum values of a function $f(x)$ for $x \in S$.

If a solution to this problem exists, then there are x-values x_M and x_m in S that satisfy

$$f(x_m) \le f(x) \le f(x_M) \quad \text{for all } x \in S$$

The **maximum of f on S** is then $M = f(x_M)$.

The **minimum of f on S** is then $m = f(x_m)$.

Not all optimization problems have solutions!
The function f may be unbounded on S, i.e. there may be a sequence $\{x_n\}$ in S for which $\lim_{n \to \infty} f(x_n) = \infty$. Alternatively, $f(x)$ may be bounded on the set S but there is no $x \in S$ at which $f(x)$ satisfies one of the equalities. These situations can occur when either the function f is not continuous or when the set S is not a closed interval. For instance,

(i) $f(x) = 1/(x - 3)$ is unbounded on the interval (1,3); $\lim_{x \to 3^-} f(x) = -\infty$.

(ii) $f(x) = x - [\![x]\!]$ (See Example 6 Section 1.3, page 48.) has an upper bound of 1, for all x, $f(x) < 1$ but $f(x) \ne 1$ for all x ; $f(x)$ is not continuous at x = integer.

(iii) $f(x) = x^2$ does not reach a maximum value for $x \in S = [0.5,1)$; $f(x) < 1$ for all x in this interval, yet $\lim_{x \to 1^-} f(x) = 1$.

(iv)   On any infinite interval, e.g., $S = [2,\infty)$, all polynomials become unbound as $x \to \infty$.

These difficulties can be avoided by restricting the function to be continuous and the constraint set S to be a closed interval. In this case the *constrained optimization* problem always has a solution. This is an important result that is stated as a theorem.

Section 4.4 OPTIMIZATION WITH CONSTRAINTS

Extreme Value Theorem

> If f(x) is continuous on a closed interval [a,b] then f attains its absolute maximum and minimum values on the interval.
>
> There exists points x_M and x_m in [a,b] such that for $a \leq x \leq b$,
>
> $$f(x_m) \leq f(x) \leq f(x_M)$$

Determining Constrained Maximum and Minimum values.

A constrained optimization problem normally seeks not local extrema but the absolute extrema of the function on the constraint set S. This can be done using the following method.

TO DETERMINE THE MAXIMUM AND MINIMUM OF f(x) ON S = [a,b].

> Find all critical values $x \in (a,b)$ for which $f'(x) = 0$ or $f'(x)$ does not exist.
>
> Evaluate f(x) at these critical values and at the endpoints, f(a) and f(b).
>
> The **maximum of f on S** is:
>
> M = the largest of the evaluated function values.
>
> The **minimum of f on S** is:
>
> m = the smallest of the evaluated function values.

Note: if a function has no critical values in the interior of the constraint set S, then its extrema must occur at the boundary of S.

Example 1   **Determining constrained extrema.**

Problem   Determine the extrema of $f(x) = x^{1/3}(x - 2)$ on $S = [-2,4]$.

Solution   The first step is to evaluate the derivative of f:

$$f'(x) = (1/3)x^{-2/3}(x - 2) + x^{1/3}$$

To determine the critical points we factor the derivative and set each factor to zero. As $x^{1/3} = x^{-2/3} \cdot x$, we can factor $x^{-2/3}$ from each term:

$$f'(x) = x^{-2/3}[(1/3)(x - 2) + x] = \{4x - 2\} / \{3x^{2/3}\}$$

One critical value is the root $x_0 = 0.5$ of $f'(x) = 0$, and the other critical value is the singular value $x_1 = 0$ at which f' is not defined. f is evaluated at each critical value and at the endpoints of S:

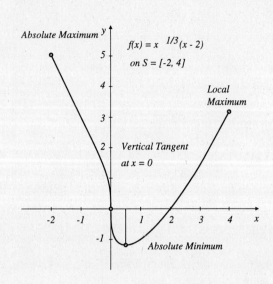

$f(-2) = (-2)^{1/3}(-2 - 2) = 4 \cdot 2^{1/3} \approx 5.04$

$f(4) = (4)^{1/3}(4 - 2) = 2 \cdot 4^{1/3} \approx 3.17$

$f(0.5) = (0.5)^{1/3}(0.5 - 2) \approx -1.19$

$f(0) = 0$

The maximum of f on S is the largest of these function values, $M = f(-2) \approx 5.04$, and the minimum of f on S is the smallest value, $m = f(0.5) \approx -1.19$.

Example 2 The extrema of a fifth degree polynomial.

Problem Determine the extrema of the function $y = 0.2x^5 - (13/3)x^3 + 36x$ on $S = [-3,4]$.

Solution The only critical points of a polynomial function will be the roots of its derivative,

$$y' = x^4 - 13x^2 + 36$$

Normally fourth degree equations are difficult to factor, but in this example, since only even powers of x occur, we can factor y'. Observe that the substitution $u = x^2$ reduces y' to a quadratic:

$$y' = u^2 - 13u + 36$$

This is a quadratic has factors $u - 9$ and $u - 4$. Hence,

$$y' = (x^2 - 9)(x^2 - 4)$$
$$= (x - 3)(x + 3)(x - 2)(x + 2)$$

The four roots of $y' = 0$ are thus $x = -3, +3, -2, +2$. The function's values at these roots are:

$y(-3) = -39.6, \quad y(-2) = -43.7\underline{3},$

$y(2) = 43.7\underline{3}, \quad y(3) = 39.6$

Evaluating the function at the right endpoint of the constraint interval $S = [-3,4]$ gives $y(4) = 71.4\underline{6}$. Based on these values the extrema of y on S are

$M = 71.4\underline{6}$ the maximum occurs at the right endpoint $x = 4$

$m = -43.7\underline{3}$ the minimum occurs at $x = -2$

Example 3 Extrema of a double exponential decay curve.

Problem Determine the extrema of $f(x) = e^{-0.02x} - e^{-0.1x}$, on $S = [0,100]$.

Solution The first step is to evaluate the derivative of f:

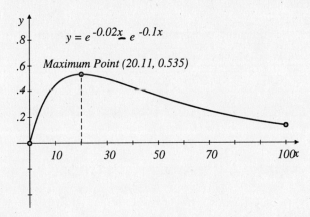

$$f'(x) = -0.02e^{-0.02x} + 0.1e^{-0.1x}$$

To find the roots of $f' = 0$, first factor f' as

$$f'(x) = -0.02e^{-0.02x}(1 - 5e^{-0.08x})$$

As the exponential $e^{-0.02x}$ is always positive, f' has only one root, obtained by solving

$$1 - 5e^{-0.08x} = 0$$

The root is

$$x_1 = \ln(1/5)/\{-0.08\} = \ln(5)/0.08 \approx 20.11$$

To determine the extrema we evaluate the function at the boundary of S, the endpoints $x = 0$ and $x = 100$, and at the critical value x_1:

$$f(0) = 0, \quad f(100) = e^{-0.02(100)} - e^{-0.1(100)} = e^{-2} - e^{-10} \approx 0.135$$

and

$$f(x_1) = e^{-0.02\ln(5)/0.08} - e^{-0.1\ln(5)/0.08} = 5^{-1/4} - 5^{-0.1/0.08} \approx 0.535$$

The maximum value of f on S is

$$M = \text{maximum } \{0, \ 0.535, \ 0.135\} = 0.535$$

and the minimum value of f on S is

$$m = \text{minimum } \{0, \ 0.535, \ 0.135\} = 0$$

Exercises Set 4.4

Each problem seeks the extrema of a function over a constraint set. In this section the emphasis is on the numerical determination of the extrema, *not graphing*. However, sketching the graphs of the functions may help you visualize the problem and give a logical check to your calculations.

1. Determine the extrema of the given function over the indicated constraint interval and identify the points where the maximum and minimum occur.

   a) $f(x) = 2x - 3$, $[0,5]$
   b) $f(x) = (2x - 1)^2$, $[0,4]$
   c) $g(x) = 5 - x^2$, $[-2,2]$
   d) $g(x) = x^3 - 3x + 1$, $[-2,2]$
   e) $f(x) = (x - 2)^3$, $[0,5]$
   f) $y = 2x^3 - 9x^2 + 12x$, $[0, 2.5]$
   g) $g(x) = 5 - x^3$, $[-1,1]$
   h) $g(x) = x^3 - 9x + 3$, $[-4,4]$
   i) $J(x) = 12x - x^3$, $[-1,1]$
   j) $g(x) = x^3 - 12x$, $[-3,3]$
   k) $y = x^2(x - 2)^2$, $[-1,1]$
   l) $y = x^2(x - 2)^2$, $[1,3]$
   m) $y = 1 - x^{2/3}$, $[-1,2]$
   n) $g(x) = \text{SQRT}\{x^2 + 2\}$, $[-1,2]$
   o) $y = x(1 - x)^{2/3}$, $[-1,2]$
   p) $y = x^{1/3}(x + 3)$, $[-8,10]$
   q) $y = x(1 - x)^{4/3}$, $[0,2]$
   r) $y = (x - 3)^{1/3}(x - 2)^{1/4}$, $[2,4]$
   s) $g(x) = e^{-2x}$, $[-1,1]$
   t) $g(x) = e^{-x^2}$, $[-1,1]$
   u) $g(x) = xe^{-2x}$, $[0,4]$
   v) $g(x) = x^2 e^{-4x}$, $[0,5]$
   w) $g(x) = \ln(5 - x^2)$, $[-2,2]$
   x) $g(x) = (x + 8)^{2/3}(x - 1)^{2/3}$, $[-8,0]$

2. Express the constraint in interval notation. Determine the extrema of the function subject to the indicated constraint.

   a) $f(x) = x(x + 2)(x - 3)$; $|x| \leq 4$
   b) $y = x^3 + x$; $(x - 2)(x + 3) \leq 0$
   c) $g(t) = t\,\text{SQRT}(1 - t^2)$; $|t| \leq 1/2$
   d) $h(s) = 2s^2 - s^4$; $|s-1| \leq 1$
   e) $R(t) = t(1/2)^t$; $|t| \leq 2$
   f) $T(x) = \ln(1 + x^2)$; $|x| \leq 1$
   g) $f(\theta) = \sin(\theta)\cos(\theta)$; $|\theta| \leq \pi/2$
   h) $f(\theta) = 1 - 2\sin(\theta)$; $|\tan(\theta)| \leq 1$
   i) $f(\theta) = \theta + \cos(\theta)$; $0 \leq \theta \leq 2\pi$
   j) $f(\theta) = \cos(\theta) - \sin(\theta)$; $|\tan(\theta/2)| \leq 1$

3. Determine the extrema of the given constrained functions.

   a) $f(t) = t + 1/t$; $t > 0$
   b) $g(x) = (1 - x^2)/x$; $x > 0$
   c) $R(t) = t^2/(1 + t)$; $t > 0$
   d) $J(x) = x/(1 + x^2)$; $x \geq 0$
   e) $f(t) = \ln(t)/t$; $t > 0$
   f) $h(x) = e^{-3x} - e^{-2x}$; $x \geq 0$
   g) $f(t) = \ln(1 + 1/t)$; $t \geq 1$
   h) $y(x) = e^{-x} - e^{-2x}$; $x \geq 0$

Section 4.5 Optimal Resources Management

Resource Management

A problem that arises in a wide variety of disciplines is how to regulate the *harvesting* of a naturally reproducing resource. The resource may be as minute as cultured bacterium or as large and disperse as the grey whale population. The resource is assumed to reproduce naturally according to an established discrete model and can be harvested to provide a benefit. The problem is to establish a harvesting policy that *best* meets a set of objectives while satisfying a companion set of limitations. We assume the benefits can be clearly quantified, the objectives are specified by a function, and that *best* means to optimize the objective function subject to certain constraints, normally imposed in the form of mathematical inequalities.

To begin, consider a periodically varying resource not subject to harvesting, whose size at the start of time period n is X_n. Assume that the resource is modeled in the absence of any harvesting by a first-order difference equation (See Chapter 2.):

Natural Model Without Harvesting

$$X_{n+1} = f(X_n)$$

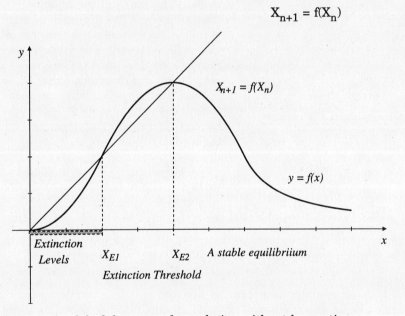

Discrete model of the *natural* population without harvesting.

Let us assume the size of the resource is positive, and negative or zero values for X_n indicate the resource has been depleted. Normally, the function f satisfies $f(0) = 0$ and $f(x) \geq 0$ for x in an interval $[0,X]$ corresponding to viable resource levels.

Recall that the resource will have *natural equilibrium values*, denoted by X_Es, determined by the intersections of the graph $y = f(x)$ and the line $y = x$. A naturally viable resource will generally have at least one non-zero equilibrium. If $f(x) = 0$, the trivial equilibrium $X_E = 0$ corresponds to no resource.

If the resource can not sustain itself when its size is too small, the trivial equilibrium will be stable. Recall, as in Section 2.2, when the graph of $y = f(x)$ is below the line $y = x$ for small x-values then the zero equilibrium is stable. As illustrated in the above figure, this occurs when $f(x) < x$ for $0 < x < X_{E1}$ where X_{E1} is the least positive equilibrium. X_{E1} then constitutes a *viability threshold*, a minimum sustainable resource level, below which the resource will naturally go to extinction.

Most biological populations depend on cross-fertilization and thus exhibit threshold levels corresponding to minimum populations needed for breeding. As the equilibrium X_{E1} is not stable in this case, the resource usually has other positive equilibrium levels that are stable. Since natural biological resources that become too large are more vulnerable to diseases and space limitations that effectively reduce their size significantly, the model function f will usually satisfy $f(x) < x$ for large x-values.

When a resource is *harvested*, i.e., its size is diminished by non-natural removal or culling, its state will depend on its natural reproductive rate between harvesting and the amount harvested. A harvesting policy will mathematically be given by a sequence $\{H_n\}$, where H_n is the amount harvested during or at the end of time period n, after the natural reproduction has occurred. A new sequence $\{Z_n\}$ is introduced to represent the size of the resource subject to harvesting. The dynamics of the harvested resource are described by the

Dynamic Model With Harvesting

$$Z_{n+1} = f(Z_n) - H_n$$

This model assumes harvesting does not affect the natural reproductive and depletion characteristics of the resource. When the harvested amount is always constant, $H_n = H$, the function $g(x) = f(x) - H$ will be use to describe the harvest model, i.e.,

$$Z_{n+1} = g(Z_n) = f(Z_n) - H \qquad \textbf{Constant Harvest Model}$$

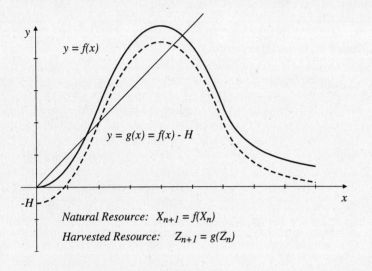

Natural Resource: $X_{n+1} = f(X_n)$
Harvested Resource: $Z_{n+1} = g(Z_n)$

In this situation the harvested resource model can be illustrated on the same x-y graph used to illustrate the dynamics of the natural resource model. The graph of the function g is the graph of $y = f(x)$ shifted downward H units. An x-value on the horizontal axis will represent the state of the resource at the start of a growing period, and can represent either X_n or Z_n, and thus the y-axis will correspond to both the natural state X_{n+1} or the harvested state Z_{n+1}, depending on the model.

A *harvesting policy* established to satisfy certain *objectives* will be subjected to constraints.

For instance, the maximum possible harvest is the size of the resource at the end of the nth time period, thus:

$$H_n \leq f(Z_n)$$

Furthermore, if $H_n = f(Z_n)$ all of the resource is harvested and the resource becomes extinct; subsequent harvest would be zero as no resource would exist. Normally, resource depletion is not an acceptable policy. Objectives are usually expressed mathematically in terms of an *objective function*, and an *optimal policy* is one that optimizes the specific objective function. A simple objective might be to maximize the amount of resource harvested, either each period or totally over a specific period. An ecologist might have the objective to maximize the size of the resource, or a dependent resource, on a continuing basis. A common business objective would seek the maximum profit derived from the harvested resource, or to maximize the return over an extended period taking into account the inflationary value of the resource. Satisfying such objectives can be difficult, however, for the simplest objectives calculus techniques can be used to establish optimal harvesting policies.

Objective: Maintaining a Constant Resource Level.

One objective is to maintain a constant resource level after each harvest. This would allow for cost effective planning of facilities for maintaining the resource, such as providing barn space for animals. A strategy to meet this objective depends on the initial resource level, Z_0, and the desired level, Z_*. If $Z_0 > Z_*$, one simply makes the first harvest to reduce the resource to Z_* and sets all subsequent harvests to equal the natural increase in the resource from Z_*.

Harvesting Policy to maintain resource at Z_* units.

$$H_0 = f(Z_0) - Z_* \quad \text{and} \quad H_n = f(Z_*) - Z_* \text{ for } n > 0$$

If $Z_0 < Z_*$, the resource can not be harvested until it naturally grows greater than Z_*. Thus $H_i = 0$ for $i = 1, 2, \ldots, N - 1$ where N is the first index for which the natural resource level exceeds the targeted level. Thereafter, the above policy is followed, $H_n = f(Z_*) - Z_*$ for $n \geq N$. Notice that this type of policy ignores the reproductive capacity of the resource except at the target level Z_*. The amount harvested is not considered in this objective. An example would be to maintain a constant dairy herd size.

Example 1 Maintaining a Constant Monkey Population.

Problem The Warden of a forest game preserve must control the size of a non-indigenous monkey population that has been introduced. The monkeys are very prolific and their initial population of 30 individuals is expected to grow according to the model $X_{n+1} = 2.2X_n$, where X_n is their number n years after being introduced. To avoid exploitation of the habitat, the number of monkeys must be kept below 500. To be on the safe side, the warden wishes to have a population of 400 monkeys. What would be the culling strategy that would achieve this most quickly?

Solution The colony should be allowed to reproduce as quickly as possible until their number exceeds 400. The model equation is a simple first-order equation with the solution

$$X_n = 30 \cdot 1.2^n$$

Setting $X_n = 400$ and solving for n gives $n = ln(400/30) / ln(1.2) \approx 14.2$. Thus the population will first exceed 400 at the end of the 15th year. The strategy is then to set

$$H_1 = 0, H_2 = 0, \ldots, H_{14} = 0 \text{ and } H_{15} = X_{15} - 400 \approx 762 - 400 \approx 62$$

and for $n \geq 16$, $H_n = 1.2 \cdot 400 - 400 = 80$

Objective: Maintain a Constant Harvest.

When the harvested resource is utilized at a constant rate, to maintain this we seek a policy that results in a constant harvest. A natural question then concerns the state of the resource for various harvest levels.

"What would be the resource size if a constant harvesting strategy was employed?"

Assume the harvest amount is H, and that $H_n = H$ for all n. The actual resource dynamics could be very erratic and in the general case can not be predicted. If the harvest amount H is too large the resource could be completely harvested after just a few harvests. However some harvest rates will result in constant resource levels. The harvested system will then be in an equilibrium, different from the resource's natural reproduction equilibrium. These *harvested resource equilibriums*, Z_Es, are the roots of the equation

$$Z_E = f(Z_E) - H$$

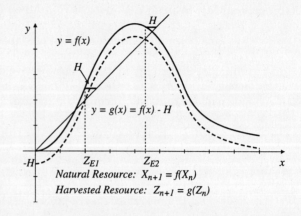

Natural Resource: $X_{n+1} = f(X_n)$
Harvested Resource: $Z_{n+1} = g(Z_n)$

Referring to the graph of the resource dynamics in a standard x-y coordinate system, the harvest equilibriums are the x-values where the graph of $g(x) = f(x) - H$ intersects the reflection line $y = x$. Thus Z_E must be a root of the equation $x = f(x) - H$. Alternatively, for a fixed harvest H, the equilibrium of the harvested resource will be a root of the equation

$$H = f(x) - x$$

The harvest amount H is graphically seen as the horizontal distance between the graph $y = f(x)$ and the line $y = x$ at an equilibrium $x = Z_E$.

There will normally only be a few, say k, positive equilibrium levels, Z_{E1}, Z_{E2}, ..., Z_{Ek} for the harvested resource. These will represent different resource levels that can each produce the same constant harvestable amount each period. Different strategies would evolve depending on secondary objectives. For instance, to maintain the smallest possible resource level one would first manage the resource to the minimum equilibrium

$$Z_{Emin} = \text{smallest of } \{Z_{E1}, Z_{E2}, ..., Z_{Ek}\}$$

However, this would probably not be a good plan. If the natural resource has a minimum sustainable threshold level, then Z_{Emin} will be a minimum sustainable threshold level for the harvested resource. It will not be a stable equilibrium. If, due to some fluctuation or perturbation in the natural biology of the resource or its environment, it under produces then the harvested resource would drop below the sustainable level. If this is not detected, harvesting could drop the resource below even its natural threshold, in which case it may never recover. Therefore, in the case of a truly natural resource, like ocean fish or whales, managing the resource to this minimum level would probably not be a good strategy.

If a resource is managed to one of the harvest equilibrium levels, and then the constant harvest is invoked, in theory the resource will remain constant. If, on the other hand, it is not initially managed, and is simply harvested, the resource will not remain constant if it is not at a Z_E equilibrium. Without being more specific about the model function f the dynamics can not be stipulated. Normally the resource levels will fluctuate and approach one of the equilibriums, a stable one. But, they may oscillate, approach limit cycles, or appear to vary chaotically.

Objective: Maximizing a Constant Harvest.

To maximize the sustainable constant harvest that can be obtained from a resource is equivalent to finding the resource level X^* for which the corresponding constant harvest level

$$H^* = \Delta X^* = f(X^*) - X^* \text{ is the largest.}$$

This is a simple optimization problem. To find the maximum of the function

$$H = f(x) - x$$

the critical x-values are found by differentiating with respect to x and setting the derivative equal to zero.

$$H' = f'(x) - 1 = 0$$

implies that

$$f'(x) = 1$$

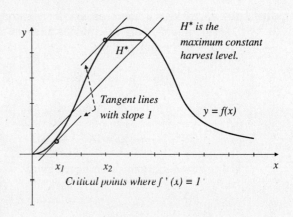

If the function f is not differentiable at a point, then it is also a critical value for the harvest function. Thus, the possible extreme resource levels are the x-coordinates of the points where the slope of the tangent line to the graph of y = f(x) has slope one, as illustrated at the right, or where the function is not differentiable. If there are several such points, some critical values, like x_1 in the figure, will correspond to "negative harvest" because the resource is actually decreasing at these states, $H = f(x_1) - x_1 < 0$. The desired state and optimal harvest level H^* will correspond to the maximum of $f(x) - x$ over all of these critical points. In bio-economic and ecology texts it is frequently assumed that the natural dynamical system function f(x) is concave downward, at least after a minimal sustainable equilibrium X_{E1}. Then, there can be at most one root of $f'(x) = 1$ (which must be greater than X_{E1}) and, by the Second Derivative Test, this root x^* corresponds to a maximum of H since $H'' = f''(x^*) < 0$. Normally, the number of critical points is small and the simplest approach is to examine the difference $f(x) - x$ for each critical point and select H^* as maximum value.

The maximum sustainable constant harvest level for a differentiable model function is

$$H^* = \text{maximum } \{ f(x) - x \mid \text{where x is a root of } f'(x) = 1\}$$

Example 2 Maximum harvest of a fish stock.

Problem Assume a fish stock under natural conditions is modeled annually by the difference equation $x_{n+1} = f(x_n)$ where the biomass (tons) of the fish in a lake is given by the split equation

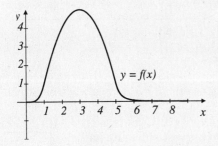

$$f(x) = \begin{cases} x^4 & \text{if } 0 \leq x < 1 \\ 1 - (x-1)(x-5) & \text{if } 1 \leq x < 5 \\ 1/(x-4)^4 & \text{if } 5 \leq x \end{cases}$$

This function and its derivative are continuous for all positive values of x. Its graph is sketched at the left. Consider the following questions.

a) If the year-5 biomass is $x_5 = 3$, what will be the x_6, x_7, and x_8 biomass levels?

b) What are the two non-zero equilibrium state levels of this resource under natural conditions?

c) If H units of this resource is harvested after it reproduces each year, what is the equation describing the resource size?

d) What will be the maximum sustainable constant harvest and what biomass will yield this harvest?

Solutions a) If $x_5 = 3$ then, as $1 \leq 3 < 5$, the next level x_6 is calculated using the middle function formula with $x = 3$, <u>not the index 5 or 6</u>,

$$x_6 = 1 - (3 - 1)(3 - 5) = 5$$

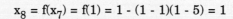

$x_7 = f(x_6) = f(5)$ is calculated using the last formula,

$$x_7 = f(x) = 1/(5 - 4)^4 = 1$$

Using the middle formula again gives

$$x_8 = f(x_7) = f(1) = 1 - (1 - 1)(1 - 5) = 1$$

The corresponding graphical method is illustrated by the arrows in Fig. 7. 25.

b) The natural equilibriums for this population are the roots of $f(x) = x$. To find these, each part of the split-function formula is equated to x and solved. When $0 \leq x < 1$ the equilibrium equation is $x = x^4$ or $x^3 = 1$, which has only the solution $x = 0$. When $1 \leq x < 5$, the equation $x = f(x)$ becomes $x = 1 - (x - 1)(x - 5)$ or $x^2 - 5x + 4 = 0$, which has two roots, $x = 1$ and $x = 4$. For $x \geq 5$ the equation $x = f(x)$ becomes $x = (x - 4)^{-4}$ which can have no solution since $1/(x - 4)^4$ is always less than 1 while x is larger than 1. The only non-zero natural equilibriums are thus $x_{E1} = 1$ and $x_{E2} = 4$, as illustrated in the graph below.

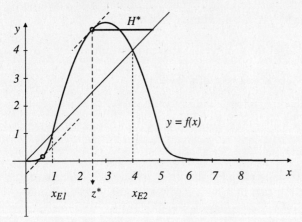

c) The corresponding harvesting model is $z_{n+1} = f(z_n) - H$. This will have three forms. For instance, if $1 \leq z_n < 5$ then $z_{n+1} = 1 - (z_n - 1)(z_n - 5) - H$.

d) The maximum constant harvest rate is found by solving $f'(x) = 1$. Since

$$f'(x) = \begin{cases} 4x^3 & \text{if } 0 \leq x < 1 \\ -2x + 6 & \text{if } 1 \leq x < 5 \\ -4/(x - 4)^5 & \text{if } 5 \leq x \end{cases}$$

When $0 \leq x < 1$, the equation $4x^3 = 1$ has the root $x = 4^{-1/3} \approx 0.63$.

When $1 \leq x < 5$, the equation $-2x + 6 = 1$ has the root $x = 2.5$.

When $5 \leq x$, the equation $-4/(x - 4)^5 = 1$ can have no solution since the left side is always negative.

The first root, $4^{-1/3}$, does not actually correspond to a positive harvest since the natural reproduction is below the replenishment level at this point (this is the left point in the graph above with tangent papallel to $y = x$). Thus, the maximum harvest level is given by the second root:

$$H^* = f(2.5) - 2.5 = 0.25 \text{ ton} \quad \text{when} \quad z^* = 2.5 \text{ ton of fish.}$$

Objective: Maximize the Harvest Profit; The Economics of Resource Farming.

In harvested resource utilization models the objective function is frequently associated with financial costs of managing the resource and the economic benefits derived from the harvested resource. The most common objective function is the "profit function" defined as the value generated by the harvested resource minus the cost of maintaining the resource and harvesting. If we assume:

> The **Cost** of maintaining resource level Z and harvesting H units is $C(Z, H)$
>
> The **Revenue** derived from H units of resource is $R(H)$;

then

> The **Profit** of harvesting H units at resource level Z is $P(Z,H) = R(H) - C(Z,H)$

At a constant harvest level, $H = f(Z) - Z$, and the profit can be expressed as a function of Z only. The critical resource levels are found as the roots of the derivative of the Profit function with respect to Z.

Example 3 Proportional Cost and Revenue.

Problem Assume that cost associated with maintaining the resource and similarly both the harvesting cost and remuneration are proportional to the resource size and amount harvested. What is the harvest level of the resource with natural dynamics modeled by $X_{n+1} = f(X_n)$ that will result in maximum profit?

Solution Assume that the maintenance expense rate is m per unit, the harvesting cost rate is h per unit, and the revenue rate is r per unit. Then the Profit function has the form

$$P = -(m \cdot Z + h \cdot H) + r \cdot H \quad \text{or} \quad P = (r - h)H - m \cdot Z$$

For a constant harvest, we know that $H = f(Z) - Z$. Thus

$$P = (r - h)(f(Z) - Z) - m \cdot Z = (r - h)f(Z) + (h - r - m)Z$$

To find the critical Z-values at which the maximum profit could occur, we differentiate P with respect to Z:

$$P' = (r - h)f'(Z) + (h - r - m)$$

The critical points are the roots of

$$f'(Z) = (r - h + m)/(r - h)$$

Normally $r > h$, as otherwise the harvest costs more than the revenue generated. Notice that for a resource with no maintenance cost, $m = 0$, the critical points are the roots of $f'(Z) = 1$, which are exactly the same values that lead to the maximum possible sustainable harvest. When $m > 0$, the critical values are where the natural reproductive function's derivative is greater than one. In the case where f is concave downward, $f'' < 0$, the first derivative f' is decreasing and consequently in this situation the optimal resource levels where $f'(Z) = (r - h + m)/(r - h)$ will be lower than the corresponding levels that maximize the harvest where $f'(Z) = 1$. ☑

Exercise Set 4.5

1. Assume a resource is modeled naturally by $x_{n+1} = f(x_n)$. Sketch the given function $f(x)$ and determine:
   i) The natural equilibrium.
   ii) The equilibrium of the resource subjected to harvesting $H = 2$ units.
   iii) If the resource level is initially 8, give a strategy to maintain a constant resource level of 6 units.
   iv) The the maximum sustainable constant harvest level.

   a) $f(x) = 0.5x + 5$ b) $f(x) = x(10 - x)$ c) $f(x) = x^2(10 - x)$

2. A renewable resource under natural conditions is modeled by a first-order difference equation $x_{n+1} = f(x_n)$ where f is given by the split equation

   $$f(x) = \begin{cases} x^2 & \text{if } 0 \leq x < 1 \\ 1 - 0.5(x-1)(x-5) & \text{if } 1 \leq x < 5 \\ 1/(x-4)^2 & \text{if } 5 \leq x \end{cases}$$

   a) Sketch a graph of this model.
   b) Calculate x_6 and x_7 if the $x_5 = 2$.
   c) What are the two non-zero steady-state levels of this resource under these natural conditions?
   d) If h units of this resource is harvested after it reproduces each year, what is the steady-state resource level that will yield the maximum constant sustainable harvest? What will be the maximum sustainable constant harvest?

3. The natural reproduction of a resource is modeled by a first-order difference equation $x_{n+1} = f(x_n)$ where f is given by the split equation

   $$f(x) = \begin{cases} 2x & \text{if } 0 \leq x < 4 \\ 8 & \text{if } 4 \leq x < 10 \\ 8/(x-9) & \text{if } 10 \leq x \end{cases}$$

   a) Sketch a graph of this model.
   b) Calculate x_6 and x_7 if the $x_3 = 2$.
   c) What is the non-zero natural steady-state level of this resource?
   d) If h units of this resource is harvested after it reproduces each year, what is the equilibrium harvested resource level when i) $h = 2$ and ii) $h = 6$?
   e) What is the maximum sustainable harvest for this resource? What will be the maximum sustainable constant harvest?

4. A renewable resource is modeled by a first-order difference equation $x_{n+1} = f(x_n)$ where f is given by the graph sketched in the figure Exer. 4.5-4.
   a) Estimate from the graph x_2 and x_3 if the $x_0 = 2$.
   b) Estimate the two non-zero steady-state levels of this resource under these natural conditions?
   c) If $h = 1$ units of this resource is harvested after it reproduces each year, estimate the equilibriums for the harvested resource.
   d) Estimate the resource level that will yield the maximum constant sustainable harvest. What will be the maximum sustainable constant harvest?

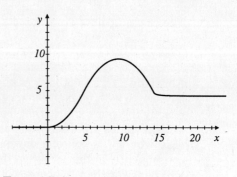

Exer. 4.5-4

5. A renewable resource size under natural conditions is modeled by a first-order difference equation $x_{n+1} = f(x_n)$ where f is given by the split equation

$$f(x) = \begin{cases} (6/25) x^2 & \text{if } 0 \leq x < 5 \\ (-6/25)(x-5)(x-15) + 6 & \text{if } 5 \leq x < 15 \\ (12/5)(x-14)^{-1} + 18/5 & \text{if } 15 \leq x \end{cases}$$

a) Sketch a graph of this model.
b) Calculate x_3 if: i) $x_0 = 2$ ii) $x_0 = 10$ iii) $x_0 = 20$
c) What is the non-zero natural steady-state level of this resource?
d) What is the equilibrium harvested resource level when H = 3 units are harvested?.
e) What is the maximum sustainable harvest for this resource? What will be the maximum sustainable constant harvest?

6. Assume that a chicken population produces according to the equation $x_{n+1} = 0.003x_n(1000 - x_n)$. If the cost of maintaining the chickens is m = 1.2¢ per chicken per day and the cost of processing a chicken is 24¢.
a) What population is required to maintain a constant harvest of 300 chickens per day?
b) What must the selling price be for the chickens harvested in part a) to break even? c) If the selling price is 98¢ what would be the maximum profit, per day, that the chicken operation could make? What would be the corresponding constant production level and the size of the chicken population?

7. Assume that a deer population has a birth rate of 1.2 per doe per year and that doe and buck fawns are equally likely. Assume a natural death rate of 0.2 per year.
a) Construct a harvesting model for the deer population.
b) If the region deer population is 5000 and 200 deer (randomly selected) are removed each year what will be the ultimate population?
c) What would be a harvesting policy that would maintain a deer population of 2500? d) Is there a maximum harvest for this population?

8. Considering the deer population described in the previous exercise, how would the model have to be altered to introduce differential culling of doe and buck deer? What would be a policy that could establish an overall population of 3000 deer, starting with a population of 4000 equally mixed doe and buck dear. Assume the natural mortality rate remains the same for both sexes.

9. Sketch the graph of a Dynamical Model $X_{n+1} = f(X_n)$ for a resource that naturally has equilibriums at $X_E = 0, 2, 4$, and 6 and $f(X) \to 0$ as $X \to \infty$. Then, sketch on the same graph the corresponding Dynamical Model with a constant harvesting of H units. Don't make the harvest H too large. Use the line y = x to identify the equilibrium values of both the natural and the harvested resource. Are these related? Explain, referring to the graph, why the stable equilibriums for the natural resource shifts downward when subjected to small amount of harvesting. What happens to unstable equilibriums? What characteristics of the graph of f will result in a decreased equilibrium value?

Section 4.6 Optimization: Applications In Different Disciplines.

Introduction: Applied Mathematics.

In this section *applied* optimization problems are considered. These have been grouped by area of application. In an introductory course not all of the applications will be considered. Your personal interest and your course emphasis will guide your selection of examples and corresponding problems. Examples are given in the following areas:

Mathematical Applications: Number Theory. Geometric optimization. Minimum Oil Boom Length

Biological Applications: Maximum growth rate; a Logistic model. The flight of a bird. Sedimentation of Red Blood Cells.

Physical Applications: Applying Fermat's Principle: Where to aim a pool shot. Hydrology: peak rainfall.

The examples each deal with *optimization*, finding minimum or maximum values of an *objective* function subject to constraints. A *routine* optimization problem, as in Sections 4.3-4, simply states a function, y = f(x), a constraint set, S, and requires a direct application of calculus to determine the extrema on the set S, if they exist. *Applied problems* are normally presented in a more indirect fashion, by describing a real or hypothetical situation using sentences and the terminology associated with the quantities or system being described. Solving an applied problem involves three steps:

(1) *Identify the problem.*   The problem must be translated into a mathematical problem, v. Variables and parameters are introduced to represent the model components. Corresponding mathematical functions and relationships are established from information that is often presented in sentences, as "word problems".

(2) *Analysis and solution.*   Apply mathematical techniques to alter the model equations and solve them for a desired variable or parameter. In optimization problems, the critical values are identified and classified to determine which correspond to maximum or minimum of an objective function on a constraint set.

(3) *Interpret the results.*   After performing the mathematical analysis, the results are presented in a **Conclusion** sentence, not just as a number, relating the answer(s) to the original problem. The results are **interpreted** in the context of the original problem. When possible, a **Discussion** of the results is given in which more general aspects of the model and its solutions are made, with possible alternative considerations.

In the real world, outside a classroom, the **Conclusion and Discussion** are the most critical part of the modeling process. You may have an excellent analysis of a difficult problem but, you must communicate your answer. It must be present in a context of the application. In business or government the **Conclusion** and **Discussion** are the kernel of a "report", what people read. In academia and research, the **Conclusion** provides the key results of a "study" or research paper. A Conclusion should be accompanied by a **Discussion** of both the actual numerical answers and of more general or theoretical aspects of the problem and its solution. It is important to identify the critical aspects of the model, its strengths and weaknesses, and to identify critical values of model parameters. Especially values at which the nature of the solution may change significantly. Such discussion could be open ended; in mathematics courses it is traditionally cut very short.

What makes applied problems difficult? The mathematics could be difficult. In real applications it often is. In introductory problems simplifying assumptions are usually made so that

Section 4.6   OPTIMIZATION: APPLICATIONS   297

the mathematics is manageable with the "tools" at hand. The real challenge is to establish the mathematical model, i.e., the variables, functions and constraints. One difficulty lies in translating English text into mathematical relationships. This can be overcome with practice, working progressively more challenging word problems. A more persistent difficulty is that applied problems frequently require *a-priori* knowledge of formulas. This required prior knowledge can make applications seem difficult to those not familiar with the discipline and its specialized terminology and "known facts".

A brief **Background** is provided for many of the examples. To work the problems it is not necessary for you to completely "understand" the background material. You should however read the background material to see how mathematical variables and functions are extracted from word descriptions.

When you consider applied problems that are outside your field, or prior experience, don't panic! Remember that you can do the mathematical analysis. Simply use the formulas indicated, and observe that the special vocabulary is usually translated to basic mathematical terms and concepts.

DISCLAIMER The examples are chosen to illustrate applied situations. Most involve simplifications of the relationships and parameters to make the problems easier. The difference between an introductory model and the cutting edge of research is frequently just the degree of complexity incorporated in a model. More accurate models may result in analysis and interpretations that are much more complicated than those presented.

MATHEMATICAL APPLICATIONS.

Number Theory Problems

These problems typically describe an algebraic property of two numbers, the variables, and ask that you determine the numbers that maximize or minimize a second algebraic combination of the numbers. The first property provides a way of eliminating one of the variables so that the expression to be optimized can be written as a function of a single variable.

Example 1   **Fixed sum with maximum product.**

Problem   Determine the two positive numbers whose sum is 200 and product is a maximum.

Solution   **Identify the problem:** Represent the numbers by two variables, say x and y. The first condition is that $x > 0$ and $y > 0$. The next condition is that their sum is 200, this is the constraint:

$$x + y = 200, \quad \text{or} \quad x = 200 - y$$

The model function to be optimized is the product, P, which depends on both variables:

$$P = x \cdot y$$

To reduce this problem to a single variable, we substitute $200 - y$ for x, giving P as a function of y only:

$$P = (200 - y) \cdot y = 200y - y^2.$$

Analysis: The mathematical problem is to find the maximum of P. Setting the derivative

$$dP/dy = 200 - 2y$$

to zero provides the critical value $y_0 = 100$. As $d^2P/dy^2 = -2$, the Second Derivative Test indicates that P has a maximum at the critical number 100. The number x corresponding to the critical y-value is $x_0 = 200 - y_0 = 100$.

Results and discussion: <u>Conclusion</u>: The two positive numbers whose sum is 200 and have the maximum product are 100 and 100. Their product is 10,000.

<u>Discussion</u>: First, notice that the two numbers are equal. Looking at the problem more closely, since the problem is symmetrical in how the numbers are utilized this is exactly what would be expected. If the variable symbols x and y were interchanged the resulting product and constraint would be the same; there is no rationale to justify making one variable larger than the other. A second observation: to generalize the problem the number 200 could be replaced by an arbitrary number N. Following the same line of analysis would yield a maximum product $P_{max} = N^2/4$ when the two numbers are both N/2.

Geometry problems.

These problems seek optimal dimensions for geometric figures. To work them you need to know specific formulas for areas, perimeters, volumes or surface areas. It is usually very helpful to visualize the problem by sketching the figure.

Example 2   **Maximum volume of a box.**

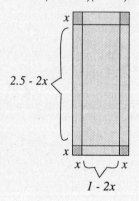

Sheet 1 by 2.5 meters.

Cut out corners x by x.

Fold to make a box with volume V.

$V = (2.5 - 2x)(1 - 2x)x$

Problem An open box is to be made by cutting the corners out of a sheet of plastic that is 1 by 2.5 meters. After removing the corners the plastic will be heated and the sides bent upward to form the box. What is the maximum volume of a box that is constructed this way?

Solution **Identify the problem:** First, sketch the sheet of plastic and indicate the corner cuts. Notice that the cut must be the same depth, x meters, on each side so that when folded the top of the box will be level. The volume of the box will be

$$V = \text{length} \cdot \text{width} \cdot \text{height}.$$

The height will be the depth of the cut, x. The width and length will be the width and length of the sheet minus the cuts at each end, $1 - 2x$ and $2.5 - 2x$. Thus

$$V = (2.5 - 2x) \cdot (1 - 2x) \cdot x$$

Analysis:

The mathematical problem is to find the maximum of V. A constraint set is implied by the physical limitations of the sheet: $0 \leq x \leq 0.5$. As $V = 0$ at both endpoints of the constraint set, by Rolle's Theorem V has an extrema in the interval (0, 2.5). The critical x-values are the roots of $V' = 0$. Using the quadratic formula to solve

$$V'(x) = 12x^2 - 14x + 2.5 = 0$$

the two roots are:

$$x_1 = \{7 - \sqrt{19}\}/12 \approx 0.22 \qquad x_2 = \{7 + \sqrt{19}\}/12 \approx 0.94$$

Only the smaller root is feasible since x must be less than half the width of the sheet. The second derivative $V''(x) = 24x - 14$ is negative for $x < 7/8$. By the Second Derivative Test, x_1 corresponds to a local maximum for V.

Results and discussion:
Conclusion: The maximum volume of a box that can be constructed out of a 1 by 2.5 m sheet of plastic is $V_{max} = V(x_1) \approx 0.254$ m^3.
Discussion: The dimensions of the box will be approximately,
              22cm high by 56cm wide by 206cm long.

☑

Example 3   Minimum length for a rectangular oil boom.

Problem   An oil spill covers 10,000 m^2 of water. To contain it a rectangular boom is utilized that has a cross link connecting the longer sides to provide stability. What is the minimum total length of boom required to enclose the oil?

Solution   **Identify the problem:** First, sketch the rectangle and the cross link component. Label the two dimensions of the boom as x and y. The cross member has the length of the shorter sides, which is chosen to be x. The two equations involved are the enclosed area formula, $A = x \cdot y = 10^4$ and the total boom length equation

$$L = 3x + 2y$$

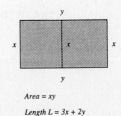

Area = xy
Length $L = 3x + 2y$

Analysis: To reduce the problem to one of a single variable, first solve the area constraint equation for $y = 10^4/x$. Then substitute this expression for y in the length formula:

$$L = 3x + 2 \cdot 10^4/x$$

To find the critical x-values set

$$L' = 3 - 2 \cdot 10^4/x^2 = 0$$

The only positive root is $x_0 = \sqrt{2/3} \, 10^2$. The other dimension would be $y_0 = \sqrt{3/2} \, 10^2$. Since

$$L'' = 4 \cdot 10^4/x^3 > 0 \quad \text{for all } x > 0$$

by the Second Derivative Test L has a minimum at x_0.

Results and discussion:
Conclusion: The minimum total length of boom require to form a rectangular closure with one cross brace to contain the oil spill is

$$L_{min} = 3\sqrt{2/3} \cdot 10^2 + 2 \cdot \sqrt{3/2} \cdot 10^2] = 2\sqrt{6} \cdot 10^2 \approx 489.9 \text{ m}$$

Discussion: The dimension of the rectangular containment boom would be

x_0 by $y_0 = \sqrt{2/3} \cdot 10^2$m by $\sqrt{3/2} \cdot 10^2$m   or approximately 81.7 by 122.5 m

☑

This optimal configuration saves only about 10m of boom over the more simply implemented 100m · 100m square configuration, which would require 500m of boom (including the cross-brace).

BIOLOGICAL APPLICATIONS

Example 4 Maximum growth rate; a Logistic Model.

Background Probably the most widely used model of "limited" growth is given by the Logistic Model. Denoting the size of a population at time t by $y = y(t)$, its rate of growth is the derivative $y'(t)$. The Logistic Model gives the growth rate as a quadratic function $f(y) = r(1 - y/M)y$.

Logistic Model $y' = f(y) = r(1 - y/M)y$

The population's intrinsic individual reproduction rate is r units/unit time. A population satisfying such a model will approach a maximum population size $y_{max} = M$, since the factor $(1 - y/M) \to 0$ as the size $y \to M$. Biologist are often interested in knowing at what stage does the population experience its maximum rate of growth?

Problem Determine the maximum rate of growth of a population described by the Logistic Model with an individual reproduction rate of 1.2 per year and a saturation level of $M = 300$ individuals.

Solution **Identify the problem:** The Logistic equation for this population is

$$y' = f(y) = 1.2(1 - y/300)y$$

We are asked to find the maximum rate of growth, i.e., the maximum of $f(y)$.

Analysis: The growth rate is a quadratic function of y. It has one critical point found by setting $f'(y) = 1.2[1 - (2/300)y] = 0$. The root $y_{max-growth} = 150$ must give a maximum value since $f''(y) = -2.4/300$ is negative for all y.

Results and Discussion:
Conclusion: The maximum growth rate for the population is $f(150) = 90$ individuals per year.

Discussion: The maximum growth rate was found but not the time at which it would occur. It is easily shown using the same method that in general the maximum growth rate will occur when the size y is one-half the maximum population size: $y_{max-growth} = 0.5 M$. The maximum growth rate will in this case be $f_{max} = 0.25 r \cdot M$

Example 5 A model of Bird flight: Optimal speed for gliding.

Background The flight of a gliding bird is achieved by the balancing of the lift generated by the bird's wings and the drag on the wings induced by gravity. The drag is a sum of two terms, associated with friction and "vortex shedding". An equation for the drag, D, as a function of air speed, U, is

$$D = SC_d U^2 + \{L^2 / (SAU^2)\}$$

where the parameters are S, a factor of wing area, A, an aspect ratio that indicates the relative length and width of the wing, C_d, the drag coefficient, and L, the lift parameter.

Problem Assuming all parameters are constant, what gliding speed U is optimal to minimize the drag?

Solution **Identify the problem:** Mathematically, the problem is to determine the speed U_{min} at which D has a minimum.

Analysis: The derivative of D with respect to U is calculated by treating all other parameters as numerical constants,

Section 4.6 — OPTIMIZATION: APPLICATIONS

$$D' = 2SC_d U - \{2L^2 / (SAU^3)\}$$

and the second derivative of the drag function is similarly computed:

$$D'' = 2SC_d + \{6L^2 / (SAU^4)\}$$

Only positive air speeds are viable. The critical speeds are the roots of $D'(U) = 0$:

$$2SC_d U - \{2L^2 / (SAU^3)\} = 0 \implies 2SC_d U^4 - \{2L^2 / (SA)\} = 0$$

Hence, the only positive critical root is

$$U_{min} = \{L^2/(S^2 C_d A)\}^{1/4}$$

As all of the parameters S, C_d, A, and L are positive, $D'' > 0$ for all positive U. Therefore, by the Second Derivative Test, D has a minimum at the critical speed U_{min}.

Results and discussion:
Conclusion: The optimal gliding air speed is $U_{min} = \{L^2/(S^2 C_d A)\}^{1/4}$.

Discussion: The results indicate that the optimal air speed actually increases only as the square root of the lift generated, i.e., for all other terms remaining constant $U_{min} \propto L^{1/2}$. There are however other facets of flight. When a bird glides, its lift must counter-balance the downward force due to gravity. The equation for lift is $L = W \cos(\Theta)$, where Θ is the angle of inclination of the glide path to the horizon and W is the bird's weight. Considering this aspect, it is seen that a heavier bird will have a greater lift component, L, and hence a faster optimal air speed when flying at the same angle of inclination. For instance a bird with identical dimensions but weighing four times as much will glide twice as fast when minimizing drag (because of the square root). Could this explain why Condors are very large? ☑

Example 6   Sedimentation Rate of Red Blood Cells.

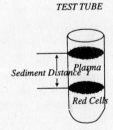
TEST TUBE

Background   A clinical procedure used to monitor various diseases is the Sedimentation Rate of red blood cells. This is determined by placing a blood sample in a vertical test tube at 22-27°C. The red cells settle to the bottom of the tube leaving a clear plasma at the top. The height Y of this top clear layer is measured over time. The rate of increase in Y is the "Sedimentation Rate". The distance Y approaches a limiting value, Y_{max}, after a long period of time. A semi-empirical model (i.e., one based in part on observations of data and in part on theory) used by hemo-rheologists and found in clinical chemistry handbooks is

$$Y(t) = Y_{max} / \{1 + \exp[-(\ln(t) - \ln(t_{50}))/b]\}$$

where b and t_{50} are positive constants that reflect the physiological state of the blood, and hence the health of the individual.

Problem   If the parameter $b = 1/2$ at what time is the red cell Sedimentation Rate a maximum?

Solution   **Identify the problem:** Mathematically, the problem is to determine the time t_{max} at which the maximum Sedimentation Rate occurs. Denoting the Sedimentation Rate by SR, it is the derivative of Y, $SR = Y'$. The distance function given above is found in hematology books. For our purposes it is more convenient to rewrite the equation in the form

$$Y = Y_{max}/\{1 + (t/t_{50})^B\}$$

with an alternative parameter $B = -1/b$. Then the Sedimentation Rate is

$$SR = dY/dt = -\{Y_{max}(B/t_{50})(t/t_{50})^{B-1}\}/\{[1 + (t/t_{50})^B]^2\}$$

Analysis: The function to be optimized is SR with $B = -2$, since $b = 1/2$. This function looks complicated but can be simplified by re-scaling the time variable. Set

$$x = t/t_{50}$$

in the SR equation and $B = -2$ to get the equation

$$SR = -Cx^{-3}(1 + x^{-2})^{-2}$$

where the constant $C = Y_{max}(B/t_{50})$. To identify the critical x-values, and thereby the critical t-values, SR is differentiated with respect to the variable x, using the Product Rule:

$$dSR/dx = SR' = -C\{-3x^{-4}(1 + x^{-2})^{-2} - 2x^{-3}(1 + x^{-2})^{-3}(-2x^{-3})\}$$

To find the critical values SR' is rearrange so that it is a product instead of a sum of terms. Using the standard procedure, we factor out the lowest powers of x and of $(1 + x^{-2})$. In this case, these are x^{-4} and $(1 + x^{-2})^{-3}$. The factored form of SR' is thus

$$SR' = -Cx^{-4}(1 + x^{-2})^{-3}\{-3(1 + x^{-2}) + 4x^{-2}\}$$

$$= -Cx^{-4}(1 + x^{-2})^{-3}\{-3 + x^{-2}\}$$

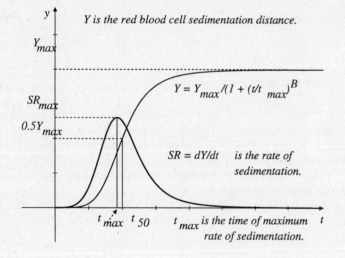

As the derivative is defined for positive x, the only critical values are the roots of $SR'(x) = 0$. Observe that SR' is the product of a constant and three factors. The first two factors are positive for all positive x. Setting the third factor, in braces, equal to zero gives

$$-3 + x^{-2} = 0$$

and solving for x gives the only positive root of $SR' = 0$, $x_0 = \sqrt{1/3}$. Therefor, because of the leading negative sign, SR' is positive for $x < x_0$ and negative for $x > x_0$. By the First Derivative Test, SR has a local maximum at x_0. As there are no other critical x-values, this is the only extremum for positive x. Converting back to t gives the maximum occurring at

$$t_{max} = t_{50} \cdot x_0 = t_{50}/\sqrt{3}$$

Results and discussion:

Conclusion. The maximum sedimentation rate occurs when at $t_{max} = t_{50}/\sqrt{3} \approx 0.577\, t_{50}$.

Discussion: At time $t = t_{50}$ the sedimentation distance $Y(t)$ is equal to one half its maximum length, for any choice of B:

$$Y(t_{50}) = Y_{max}/\{1 + (t_{50}/t_{50})^B\} = Y_{max}/2$$

Thus for the parameter b = 1/2 we found that the maximum rate of sedimentation occurs roughly at a little over half the time that it takes for the red blood cells to settle half way. Therefor, recognizing that some variability in rates would occur with different samples, an adequate time for a test to be run to measure SR would be the upper limit, t_{50}. Presumably a "normal" or standard value for this parameter can be estimated from a large number of samples and the used as a time interval for future testing.

PHYSICAL APPLICATIONS

Example 7   **Applying Fermat's Principle: Where to aim a pool shot.**

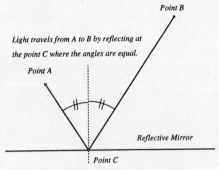

Light travels from A to B by reflecting at the point C where the angles are equal.

The lines give the shortest reflected path from A to B.

Background   The French mathematician Pierre De Fermat (1601-1665) stated that light travels by the shortest possible path between two points. If the light travels from a point A to a point B by reflection in a mirror, then it will travel to a point C on the mirror's surface and then to the point B. Fermat found that the angles of "incidence" and "reflection" at the point C must be the same.

This same principal is applied to decide where to aim a "pool" shot that must carom off a side cushion of the pool table before hitting the target ball. The balls are viewed as points, and the cue ball must travel like light in a straight line to a target point C on the cushion and reflect off the cushion at the same angle as it hits. The cushion target point C is chosen to minimize the distance traveled.

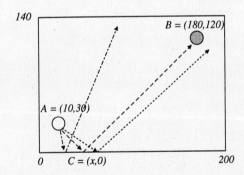

Problem   Assume a pool table with dimensions 200 by 140 cm is modeled by a rectangle with opposite corners at the origin and the point (200,140). If a cue ball is at the point A = (10,30) and you wish to hit a target ball at the point B = (180,120), at what point C on the cushion along the x-axis should the cue be shot?

Solution   Identifying the problem: The situation is illustrated in Figure 3.35. The problem is to find the x-coordinate of C so that the distance from A to C plus the distance from C to B is minimized. The total distance D is the sum of

$$D_1 = \text{Distance AC} = \text{SQRT}((x - 10)^2 + 30^2)$$

and

$$D_2 = \text{Distance CB} = \text{SQRT}((x - 180)^2 + 120^2)$$

The critical points of D are the roots of

$$D' = D_1' + D_2' = 0$$

Analysis: The algebra for this problem appears complex if you plow straight forward with the functions given. However, this and similar problems have the same structure which can be considered more easily by representing the terms under the square root as functions. Let

$$y_1 = (x - 10)^2 + 30^2 \quad \text{and} \quad y_2 = (x - 180)^2 + 120^2$$

Then, $D = \sqrt{y_1} + \sqrt{y_2}$ and

$$D' = \{ y_1' / 2\sqrt{y_1} \} + \{ y_2' / 2\sqrt{y_2} \} = \{ y_1' \cdot \sqrt{y_2} + y_2' \cdot \sqrt{y_1} \} / \{ 2 \sqrt{y_1} \cdot \sqrt{y_2} \}$$

Setting $D' = 0$ is equivalent to

$$y_1' \cdot \sqrt{y_2} + y_2' \cdot \sqrt{y_1} = 0$$

The square roots can be eliminated by first transposing one term to the other side of the equal sign, squaring both sides and then transferring back across the equal sign:

$$y_1' \cdot \sqrt{y_2} = - y_2' \cdot \sqrt{y_1}$$

$$y_1'^2 \cdot y_2 = y_2'^2 \cdot y_1$$

$$y_1'^2 \cdot y_2 - y_2'^2 \cdot y_1 = 0$$

Substituting into this equation y_1 and y_2 as given above with

$$y_1' = 2(x - 10) \quad \text{and} \quad y_2' = 2(x - 180)$$

gives the equation for the critical x-value that minimizes the distance:

$$4(x - 10)^2 \cdot [(x - 180)^2 + 120^2] - 4(x - 180)^2 \cdot [(x - 10)^2 + 30^2] = 0$$

Multiplying the squared factors times the two terms in the square brackets reveals an amazing and helpful situation, the product of the two squared terms cancel each other:

$$4(x - 10)^2 \cdot (x - 180)^2 + 4(x - 10)^2 \cdot 120^2 - 4(x - 180)^2 \cdot (x - 10)^2 - 4(x - 180)^2 \cdot 30^2$$

$$= 4(x - 10)^2 \cdot 120^2 - 4(x - 180)^2 \cdot 30^2 = 0$$

Canceling a common factor of $4 \cdot 30^2$, as $120^2 = 4^2 \cdot 30^2$, this reduces to the equation

$$(x - 10)^2 \cdot 4^2 - (x - 180)^2 = 16(x^2 - 20x + 100) - (x^2 - 360x + 32400) = 0$$

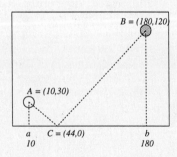

Adding common powers of x and factoring out 5 gives the factorable quadratic equation

$$5 [3x^2 + 8x - 6160] = 5(x - 44)(3x + 140) = 0$$

You could also obtain these roots using the Quadratic Equation. The root $x = -140/3$ is negative and not feasible.

The solution is thus $x = 44$.

Results and discussion.

Conclusion. To hit the target ball at the position (180,120) the que ball at position (10,30) should be aimed at the point C = (44,0) on the x-axis cushion.

Discussion: The model assumed that the cue ball has no spin which would cause it to bounce off the cushion at a reflection angle different from the incidence angle. That the incidence angles are equal can be deduced from the similar right triangles ACa and BCb indicated in the sketch. These are right triangles with the same ratio of the lengths of their side equal:

$$(44 - 10)/30 = (180 - 44)/120$$

☑

Example 8 A Hydrology model: Maximum rainfall.

Background   Models of precipitation utilize the diameter, D, of the water droplets in a cloud as an independent variable. The number of droplets per unit cloud volume is assumed to be an exponential decaying function of the diameter of the droplets:

$$N(D) = N_0 e^{-cD}$$

where the decay rate c and coefficient N_0 are dependent on the cloud conditions. The amount of water, W, in a unit volume of cloud consisting of droplets of a given diameter D is called the "water-equivalent mass" due to droplets of diameter D. This water mass is modeled by the formula

$$W(D) = \text{number of droplets} \cdot \text{volume per droplet} \cdot \text{density of water}.$$

The volume is $v = (4/3)\pi r^3 = (\pi/6)D^3$, hence

$$W(D) = (p\pi/6)D^3 \cdot N(D)$$

where p is the liquid density. The corresponding rainfall, called "mass precipitation", of droplets of diameter D is the product of the water mass W(D) and the net downward velocity of the droplets:

$$R(D) = W(D)[V_T(D) - V_u]$$

where V_u is the updraft velocity and $V_T(D)$ is the terminal velocity due to gravity. The simplest assumption is that $V_T(D)$ is proportional to D with a coefficient α, $V_T(D) = \alpha D$. The model is further simplified by assuming that there is no updraft, i.e., that $V_u = 0$. Then

$$R(D) = W(D) \cdot \alpha D = \alpha(p\pi/6)D^4 \cdot N_0 e^{-cD}$$

Problem   What size droplets correspond to peak water-equivalent mass?
What size droplets provide the peak precipitation?

Solution   **Identify the problem:** The problem is to determine the diameters D at which W(D) and R(D) are maximum. To simplify the notation we first lump all of the parameters that are multiples: set $k = pN_0\pi/6$. Then the water-equivalent mass equation is

$$W(D) = kD^3 e^{-cD}$$

And the precipitation mass equation is

$$R(D) = \alpha k D^4 e^{-cD}$$

Analysis: To determine the critical D-value, differentiate W with respect to D:

$$W'(D) = 3kD^2 e^{-cD} - ckD^3 e^{-cD} = kD^2[3 - cD]e^{-cD}$$

The only critical value for the water droplet diameter is

$$D_W = 3/c$$

Since $W'(D)$ is the product of two positive terms, kD^2 and e^{-cD}, and the linear factor $[3 - cD]$, it is clear that

$$W'(D) < 0 \quad \text{if} \quad D > D_W$$

and

$$W'(D) > 0 \quad \text{if} \quad D < D_W$$

By the First Derivative Test $W(D)$ has a maximum at D_W.

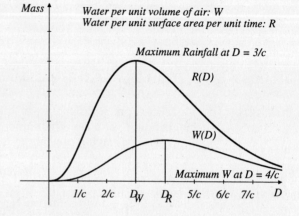

The equation for $R(D)$ is almost identical in form to that for $W(D)$; the differences being the exponent 3 is replaced by 4 and an extra multiplication coefficient, α, is included. Thus, by a similar process we find the derivative $R(D)$ with respect to D is

$$R'(D) = \alpha k D^3 [4 - cD] e^{-cD}$$

The only critical value where $R'(D) = 0$ is $D_R = 4/c$. As R' and W' have the same form, an application of the First Derivative Test, similar to that used above for W', assures that R has a maximum at D_R.

Results and discussion:
Conclusion: The peak water-equivalent mass occurs at droplets of diameter 3/c, whereas, the peak precipitation mass occurs at droplets of diameter 4/c.

Discussion: The peak water-equivalent mass does not give the maximum rainfall because the velocity of downfall is assumed proportional to the size of the droplets. Remember that as the droplet size increases their numbers per unit volume decreases. Thus slightly larger droplets, while presenting less than maximum water content in a unit volume of air, actually can deposit more water on the ground. Notice that the optimal diameters are scaled by the reciprocal of the decay rate c; as the decay rate increases the critical diameters decrease.

Exercises Set 4.6

Mathematical Applications

1. Determine the two positive numbers whose sum is a minimum and whose product equals 10.

2. What numbers A and B satisfy $3A + 2B = 100$ and have the maximum value for $A \cdot B$?

3. Determine the two numbers A and B for which $A^2 + 4B^2$ is a minimum and $A \cdot B = 1$.

4. What is the largest area of a rectangle with perimeter 10?

5. What is the minimum amount of wood required to make a rectangular open-top box with a square base that has a volume of $3.5 m^3$?

6. A condominium wishes to fence a garden area that will provide 10 rectangular plots with equal dimensions. Assume 300m of fence material is used. What is the maximum area of each individual garden plot if   a) the plots are all side by side?   b) the plots are in two rows of 5 sharing a central fence? Hint: draw a diagram! Is there a different arrangement that would give a greater area?

7. The cost of building a cylindrical silo increases linearly as the height, H, increases and as the square of the radius, r. Assume the total cost is $250 per m of height plus $500 per square meter of the base. What are the dimensions that minimize the cost of constructing a silo that has volume $V = \pi \cdot 3000 m^3$?

8. Determine the point on the line $y = 3 + 2x$ that is closest to the origin.

9. What is the point on the parabola $(x - 1)y = 1$ that is closest to the point $(1,0)$?

10. What is the largest rectangle that can be inscribed in the ellipse $(x/3)^2 + (y/2)^2 = 1$

11. Determine the minimum and maximum areas of all right triangles with perimeter 10.

12. The yield of a crop is a function of the time between planting and harvesting. If the yield is given by $y(t) = (t - 30)(50 - t)$ if $30 \leq t \leq 50$ and $y = 0$ if $t < 30$ or $t > 50$. What is the maximum yield and on what harvest day t will it be reached?

13. A farmer has 10 kilometers of fence and wishes to fence a rectangular field with one fence dividing the field in the middle. What is the largest field he or she can fence? Give the largest area and the dimensions of the field.

Biological Applications

14. A cell culture has a rate of growth $r(t) = 36t - t^2$. What is its maximum rate of growth? When is the maximum size of the culture reached?

15. Determine the maximum rate of growth of a population described by the Logistic Model with an individual reproduction rate of 3 per year and a saturation level of $M = 1000$ individuals.

16. Assume that an animal has the shape of a rectangular solid with width equal one-half of its length and height equal to one-half of its width.
    a) Give an equation representing the animal's volume minus its surface area as a function of its width.
    b) What are the dimensions of the animal for which its volume minus its surface area as a function of its width is the smallest possible value?

17. What is the minimal drag gliding speed U of a bird whose total drag, D, is $D = 30U^2 + \{L^2/(5U^2)\}$, if its lift component $L = 150$.

18. The sedimentation distance of a person's Red Blood Cells t min after sampling was observed to follow the function $Y(t) = 3.5 /\{1 + \exp{-[(\ln(t) - \ln(200))/0.2]}\}$ cm.

    By this equation, what is the maximum rate of sedimentation for this person and at what time did this occur?

19. A peach orchard is planted on a 10-acre site. The mature production of each tree is 300 lb when fewer than 65 trees are planted per acre. However, when the number of trees per acre exceeds 64, overcrowding occurs and the yield per tree is $(300 - (n - 64))$ lb, where n is the number of trees per acre.

    What is the number of trees that should be planted on the site to maximize harvest yield?

20. The speed of air escaping through your trachea when you cough is a function of the force due to chest contraction and the size of the trachea. If the trachea radius is r the speed of air can be described by the function $v(r) = F[1 + r^2(r_0 - r)]$, where the parameter F represents the force component and a normal velocity unit and r_0 is a normal trachea radius.

    What is the maximum air speed for $0 \leq r \leq r_0$?

21. The rate of blood flow through a tube of radius R is a function of the blood pressure P, the resistance pressure B and physical factors such as the density of the blood and stress parameters which we represent by parameters A and C. The flow rate, traditionally denoted by Q, is given in a simplified model as the difference between forward flow and back flow: $Q = -A \cdot P \cdot R^4 + C \cdot B \cdot R^3$.

    a) As a function of radius, what is the maximum flow rate?
    b) What is the maximum flow rate if the pressure P is a function of R, $P = 1/R^n$?

22. The shape of blood cells that are not spherical has been modeled as the solid obtained by revolving "Cassini's oval" about the x-axis. Cassini's equation is $y = B[\text{SQRT}(c^4 + 4a^2x^2) - a^2 - x^2]^{1/2}$. The biological constants a and c represent volume factors with $\sqrt{2}a > c > a > 0$. B is the strain energy for surface bending.

    What are the local maximum and minimum values for the "radius" y? What are the absolute maximum and minimum values for the radius y?

23. When an animal runs, its center of gravity moves up and down in a horizontal plane. These movements correspond to two phases of motion - that is, a "stepping" phase, where the animal's legs propel it forward, and a "floating" phase, where is it being carried by its momentum, which brings its legs forward to begin another stepping phase. The distances over which these phases occur are denoted by S and F. The "Gait" ratio $G = F/S$ is a measure of how smooth or jerky the animal's motion appears. It can be shown that G is related to the power, P, necessary for the animal to run at a speed, V. This relationship is given by

    $$P = AG\, L^4/V + B\, V^3 L^2/\{1+G\}$$

    where L is the length of the animal and A and B are constants.

    a) Show that the optimal gait G, which minimizes the required power, P, is an increasing function of the velocity, V, and a decreasing function of the animal's length, L.
    b) How would the optimal value of G be related when a horse walks, trots, canters, and gallops?
    c) How would the optimal values of G for a mouse, a dog, a cow, and an elephant be related?

Physical Applications

24 A photo-developer wishes to give the biggest prints possible for a fixed cost. Assume the cost is 10¢ per 100 cm^2 of the print area plus 2¢ per 100 cm^2 of trim area. If 10 mm must be trimmed off each end and 5 mm trimmed off each side of the print, what are the dimensions of the least cost print with a total area of 1200 cm^2?

25 Assume a pool table with dimensions 200 by 140 cm is modeled by a rectangle with opposite corners at the origin and the point (200,140). If a cue ball is at a point A = (10,30) and wishes to hit a target ball at the point B = (180,120). (Refer to Example 7.)

a) At what point C on the side cushion parallel to the x-axis should the cue be shot?

b) At what point C on the side cushion along the y-axis should the cue be shot?

c) At what point C on the side cushion parallel to the y-axis should the cue be shot?

26 A "Steel-Woman" race consists of three "legs", one the contestant runs, one is swimming, and one is covered on a bicycle. Assume the last two "legs" involve starting at a point A on one shore of a lake, swimming to a point B on an opposite parallel shore of the lake, and then running to a point C further along that lake shore. The contestant has a choice of where to swim to, i.e. the point B. Sketch a diagram to illustrate this situation. Assume the point A is the origin and the far shore is the line y = 4. If the point C is (10,4) and a contestant can run 5 times faster than she can swim, to what point B should she swim to minimize her total time for these two "legs"?

27 A model of cloud water retention and rainfall describes the water-equivalent cloud mass by
$W(D) = kD^3 e^{-cD}$ and rainfall mass by $R(D) = \alpha k D^4 e^{-cD}$, where D is a diameter of the water droplets and the parameters k and c reflect weather-physical constants. (See Example 8.) Assume that k = 2, c = 0.05, and α = 0.2.
a) For what diameter D will the water content W be a maximum?
b) For what diameter D will the rainfall R be a maximum?
c) For what diameter D will the water content minus the rainfall be a maximum?

28 What are the dimensions of a rectangular beam that is to be cut from a round log of radius r = 30cm that maximizes the "stiffness" factor $S = kwh^3$, where w = width, h = height, and k is a constant?

29 In chemistry the partial molar volume of a substance in solution is modeled in terms of its "molality", denoted by m. For common salt, sodium chloride, in water at 25°C at 1 atmosphere pressure in 1000g of water the partial molar volume is $V = 18.08 - 15.977 m^{3/2} + 0.002151 m^2$. At what molality m is the this volume a minimum?

30 The standard formula P·V = R·T that relates the pressure P, volume V and temperature T of a "perfect" gas by a constant R. An extension to "imperfect gases" is given by the

Berthelot equation $P \cdot V = R \cdot T[1 + (9PT_c/128P_c T)(1 - 6(T_c/T)^2)]$

where T_c and P_c are the critical temperature and pressure at which deviation from "perfect" behavior occurs.
For fixed temperature and pressure, what critical temperature T_c gives the maximum volume V?

31 Heat is absorbed or transferred from a body at a rate proportional to the difference in the body temperature T_b and the surrounding external temperature T. The equation for heat transfer is $Q = C(T - T_b)$. The constant C is the "heat capacity" of the body. As in most chemistry or physics laws, this formula is "ideal" and the actual term C is a function of the temperature T. One model for variation of temperature gives the heat capacity as a function of the form $C_p = \alpha + \beta T + \gamma T^{-2}$.
What is the minimum value for the capacity C_p in terms of the constants α, β, and γ?

Section 4.7 Rolle's Theorem and the Mean Value Theorem.

Introduction.

In this section four Theorems are stated. These are statements that have hypotheses and conclusions. Mathematicians **prove** Theorems by demonstrating that the conclusion is True whenever the hypothesis is True. Arguments are presented as Rationales for these Theorems but we avoid the rigor of a Proof that you will find in theoretical texts.

Rolle's Theorem.

It is often useful to know that an optimization problem has a solution on the interior of a constraint set. A theorem that is very simple in its requirements and guarantees the existence of a extreme value for a function on an interval is **Rolle's Theorem**. This theorem is deceptively powerful and is used in all areas of mathematics, from the simplest applications to the forefront of research.

Rolle's Theorem is based on the common adage "What goes up must come down." The Theorem basically states that a continuous curve that starts from the x-axis and returns to the x-axis must at least once reach an extreme point where it ceases its movement away from the axis and begins to return. The requirement of continuity is needed to avoid curves that "jump". Furthermore, if the curve is the graph of a differentiable function then at this extreme point its derivative must be zero.

Rolle's Theorem

Assume f(x) is continuous on [a,b] and f'(x) exists for each point x ∈ (a,b). If f(a) = 0 and f(b) = 0 then there exists a point c ∈ (a,b) such that

$$f'(c) = 0$$

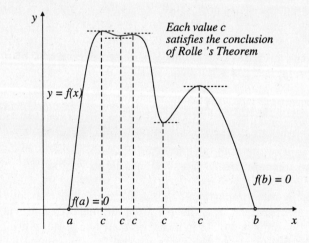

Each value c satisfies the conclusion of Rolle's Theorem

The graph illustrates Rolle's Theorem; the sketched curve is smooth and continuous. It represents the graph of a differentiable function f(x). f(a) = 0 and f(b) = 0. The x-coordinate of each point where the tangent line is horizontal could be the number c asserted to exist by Rolle's Theorem. As illustrated, there may be more than one x-value satisfying the conclusion.

Rolle's Theorem is usually stated only with the conclusion that f'(c) = 0. We have included the important corollary to this is the fact that at some value c satisfying this conclusion the function f must also assume its extreme value.

For many applied problems the existence of the extrema is more important than knowing that the derivative is zero at the point. Rolle's Theorem, like many "existence" theorems in mathematics, tells you that something exists, but it does not tell you how to find it!

Historic Comment: It is perhaps ironic that this theorem is named after Michel Rolle, a French artillery Lieutenant whose occupation was to compute trajectories for cannon fire. Although he stated the result in a mathematical treatise in 1691 while trying to prove that between every two zeros of a polynomial there must be a zero of a second polynomial, Rolle did not actually understand calculus and never realized that the second polynomial was the derivative of the first polynomial. Yet, Monsieur Rolle seems destined to have his name perpetuated in Calculus books for ever.

Applying Rolle's Theorem. Solving Equations: How many solutions are there?

Solving an equation $f(x) = 0$ by algebraic means can be very difficult or impossible. However, approximate roots can often be found numerically. Important questions to ask are
"Does the equation have a real solution?"
If so, "On what interval will there be a solution?"
And "How many roots are there?"

Sometimes these can be answered using Rolle's Theorem. The following statement is called a *corollary* as its validity follows from Rolle's Theorem.

Corollary 1. *Assume $f(x)$ is differentiable on an interval I. If the derivative $f'(x)$ is never zero on the interval, then the equation $f(x) = 0$ can have at most one root in the interval.*

Rationale The justification of this corollary utilizes a logical argument that is called a *proof by contradiction*. We assume its hypothesis is true and its conclusion is false and then show that this results in a contradiction, something that can not be. Logically, this means our assumption about it being false is actually false, it must be true.

Assume that the differentiable function $f(x)$ has two roots in I. Label these a and b. Then f is continuous and differentiable on [a,b] and $f(a) = 0$ and $f(b) = 0$. f satisfies the hypothesis of Rolle's Theorem on the interval [a,b]. The conclusion of Rolle's Theorem then assert the existence of a point c in the interval [a,b] for which $f'(c) = 0$. But, this contradicts the hypothesis of the corollary that $f'(x) \neq 0$ on the interval I. The contradiction means that the assumption that there are two roots must be false. Thus, at most one root can exist.

Since a continuous function's graph over an interval I = [a,b] would have to cross the x-axis twice if it started and ended on the same side of the axis, we can conclude that there cannot be a root of f in the interval if the conditions of Corollary 1 hold and f has the same sign at both endpoints of I. If the signs of f differ at the two endpoints then there must be a root, but only one. This observation is also stated as a corollary.

Corollary 2. *Assume that $f(x)$ is continuous on I and $f'(x) \neq 0$ on I.*

(i) If the function values $f(x)$ have the same sign at both endpoints of I, then there is no root of $f(x) = 0$ in I.

(ii) If the function values have different signs at the two endpoints of I, then there is a unique root of $f(x) = 0$ in I.

Example 1 **How many roots does a cubic equation have?**

Problem How many real roots are there of the cubic $f(x) = x^3 + x + 1$?

Solution   We know that a cubic polynomial should have either one or three real roots. Corollary 1 is applied to determine which. f(x) is continuous and differentiable for all x. Observe that $f'(x) = 3x^2 + 1$ is always positive. Hence, f satisfies the hypothesis of Corollary 1 and we conclude that there can be at most one real root of $x^3 + x + 1 = 0$.

Since f(0) = 1 and f(-1) = -1, by conclusion (ii) of Corollary 2 there is exactly one root in the interval (-1,0).

Example 2   **Examining the roots of a Quartic equation.**

Problem   How many real roots does the equation $y(x) = x^4 + 2x^3 + 3x^2 - 5 = 0$ have?

Solution   The derivative of the quartic function is $y' = 4x^3 + 6x^2 + 6x$ is positive for all positive x, since all of its coefficients are positive. Hence, by Corollary 1, there can be at most one positive root of this equation. Since y(0) = -5 and y(1) = 1, by conclusion (ii) of Corollary 2, there is a unique positive root between 0 and 1.

Similarly, for negative x, y' is always negative. This follows since $y' = x(4x^2 + 6x + 6)$ is the product of x, which is negative, and the positive quadratic factor

$$4x^2 + 6x + 6 = 4(x + 3/4)^2 + 15/4$$

Consequently, again by Corollary 1, there can be only one negative root of y.

The Mean Value Theorem.

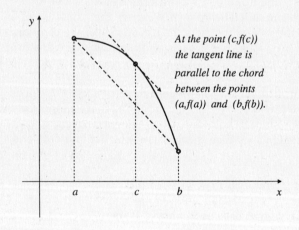

At the point (c,f(c)) the tangent line is parallel to the chord between the points (a,f(a)) and (b,f(b)).

A very useful result that is also established by a special application of Rolle's Theorem is called the **Mean Value Theorem** (this is often abreviated as **MVT**). This theorem states in mathematical terms what most people accept as an obvious physical fact:

"If an object, mathematically a point, moves from one position to another in a continuous fashion, then at some point along its path it must travel in the direction of the end point from the starting point."

The mathematical translation of this statement involves two "directions". One is the slope of the chord or straight secant line from the starting to the ending point. The other is the "direction" that the moving point takes.

If the path traveled by the point is the graph of a function, say y = f(x), then the "direction" of the moving point is the slope of the tangent line to the graph. This is the value of the derivative f'(x).

The Mean Value Theorem

> If $y = f(x)$ is continuous on $[a,b]$ and differentiable on (a,b), then there exists a number $c \in (a,b)$ such that
>
> $$f'(c) = \{f(b) - f(a)\} / \{b - a\}$$

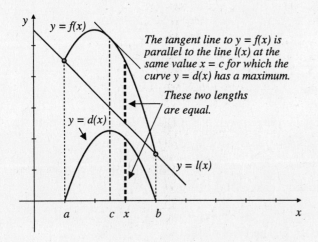

The tangent line to $y = f(x)$ is parallel to the line $l(x)$ at the same value $x = c$ for which the curve $y = d(x)$ has a maximum.

These two lengths are equal.

Rationale.

The Mean Value Theorem is a generalization of Rolle's Theorem and is established by applying Rolle's Theorem to a new function $d(x)$, defined as the vertical difference between the graph of f and the secant line $l(x)$ that passes through the terminal points $(a, f(a))$ and $(b, f(b))$. The equation of the secant line is

$$l(x) = f(a) + m(x - a)$$

where the slope is

$$m = \{f(b) - f(a)\}/\{b - a\}$$

As the curve and the secant intersect at the endpoints, $x = a$ and $x = b$.

$$d(a) = 0 \quad \text{and} \quad d(b) = 0$$

The function $d(x) = f(x) - l(x)$ is differentiable, since both $f(x)$ and $l(x)$ are differentiable on $[a,b]$. Hence, $d(x)$ satisfies the hypothesis of Rolle's Theorem and we can conclude that there exists a number c in (a,b) at which $d'(c) = 0$. By the Derivative Sum/Difference Rule

$$d'(x) = f'(x) - l'(x)$$

Thus, at the unknown number c,

$$0 = d'(c) = f'(c) - l'(c) \quad \text{or} \quad f'(c) = l'(c)$$

But the derivative of the linear function $l(x)$ is a constant, $l'(x) = m$ for all x, hence, the last equation gives the conclusion of the theorem:

$$f'(c) = \{f(b) - f(a)\}/\{b - a\}$$

Example 3 **Applying the Mean Value Theorem and solving for c.**

Problem At what point does the rate of change in the function $f(x) = 3x^3 - 6x + 1$ equal its average rate of change over the interval $[0,1]$?

Solution The Mean Value Theorem can be applied to the function f on the interval $[0,1]$. The theorem asserts that there is a number c at which $f'(c)$ equals the average rate of change.

$$f'(c) = 9c^2 - 6$$

and the average rate in f over the interval [0,1] is the slope of the secant line between the points on the graph of f at x = 0 and x = 1:

$$m = \{f(1) - f(0)\}/\{1 - 0\} = -2 - 1 / 1 = -3$$

Solving $f'(c) = m$ for c gives the equations

$$9c^2 - 6 = -3, \quad c^2 = 3/9, \quad \text{and thus} \quad c = \pm\sqrt{1/3}$$

Only the positive root is in the given interval. Hence the point is $c = \sqrt{1/3}$.

Example 4 Applying the Mean Value Theorem.

Problem   If an object moves D meters in T seconds its average speed is $\bar{S} = D/T$. Does the object actually travel at this speed?

Solution If it moves continuously the answer is "Yes." Assume that the distance the object travels in the time interval [0,t] is a differentiable function of t, y(t). The distance function satisfies the boundary conditions

$$y(0) = 0 \quad \text{and} \quad y(T) = D, \text{ the total distance.}$$

Then, by the Mean Value Theorem, at some time t = c in the interval (0,T)

$$y'(c) = \{y(T) - y(0)\} / \{T - 0\} = \{D - 0\} / \{T - 0\} = D/T$$

This illustrates the use of a Theorem to confirm our intuition. Clearly, the object must travel at even greater speeds to obtain this average speed.

If y' = 0 then y is a constant function.

The derivative Rule that was the easiest to prove is the rule:

If $y(x) = c$, a constant, then $y'(x) = 0$.

In subsequent Chapters, when we consider Integrals, we will need the converse of this rule.
Zero Derivative Theorem.

> If $y' = 0$ on an interval I, then $y(x) = c$, a constant.

Rationale.   The Zero Derivative Theorem is established by using the Mean Value Theorem. Again, we use a "proof by contradiction." We begin by assuming that the Theorem is false. Next, we demonstrate that this leads to a contradiction. The contradiction then implies that our assumption is false, i.e., a double negative, it is false that the theorem is false. Hence, the theorem must be true. The actual rationale goes as follows.

Assume the function y is not constant on the interval I and $y' = 0$ on the interval. If y is not constant, there must be two x-values in I, say $x = a$ and $x = b$ with $a < b$, at which the function has different values, $y(a) \neq y(b)$. Next, apply the Mean Value Theorem to y over the interval [a,b]. The conclusion is that for some value $c \in (a,b)$

$$y'(c) = \{y(b) - y(a)\}/\{b - a\}$$

But, we assumed that $y'(x) = 0$ for all $x \in I$. In particular that $y'(c) = 0$. Hence, the last equation gives:

$$y'(c) = 0 = \{y(b) - y(a)\}/\{b - a\}$$

which implies that the numerator of the right side is zero: $y(a) = y(b)$ This contradicts the choice of a and b as values at which y(x) is different. The conclusion is thus that there are not two x-values at which y differs, hence y must have the same constant value on the entire interval I.

We end this section by stating another important theorem which states that knowing a function's derivative identifies the function only up to a constant. This is an important result that is used in the study of Integral Calculus, where the problems involve finding functions that have a specific derivative.

Equal Derivatives Theorem.

> If the derivatives of two functions are equal then the functions differ by a constant.
>
> If $f'(x) = g'(x)$ on an interval I, then $f(x) = g(x) + C$ for some constant C.

Rationale This theorem is established using the Zero Derivative Theorem. If f and g satisfy the hypothesis of the Equal Derivatives Theorem, then the function h defined by

$$h(x) \equiv f(x) - g(x)$$

is differentiable on the interval I and for all x in I

$$h'(x) = f'(x) - g'(x) = 0$$

Thus, h(x) satisfies the hypothesis of the Zero Derivative Theorem. The conclusion is that for some constant c,

$$f(x) - g(x) = h(x) = c \quad \text{or simply} \quad f(x) = g(x) + c$$

This is the conclusion of the Equal Derivatives Theorem.

Exercises Set 4.7

1. Verify that the given functions satisfies the hypothesis of Rolle's Theorem on the given interval or indicate which condition it fails to satisfy.

   a) $f(x) = 4 - x^2$, $[-2,2]$
   b) $f(x) = 3 - x^2$, $[-2,2]$
   c) $f(x) = (x + 7)(x - 3)$, $[3,7]$
   d) $f(x) = 1 - 1/x$, $[-1,1]$
   e) $f(x) = 1 - 2(1 + x^2)^{-1}$, $[-1,1]$
   f) $f(x) = (1 - e^x)\ln(x - 1)$, $[0,2]$

2. Determine if the function satisfies the Mean Value Theorem on the given interval. If it does, state the equation involving an unknown value c that must be satisfied. Solve for the value c.

   a) $f(x) = x^2 + 3x - 2$, $[0,3]$
   b) $f(x) = x(x - 2)^2$, $[0,4]$
   c) $f(x) = (x - 1)^{1/3}$, $[-1,1]$
   d) $f(x) = \text{SQRT}\{(x - 1)^2\}$, $[-2,2]$
   e) $f(x) = (x - 1)^{1/3}(x + 3)^{5/3}$, $[-3,1]$
   f) $f(x) = (x - 2)^2(x + 2)^2$, $[-2,2]$
   g) $g(x) = e^x - 1$, $[0, \ln(3)]$
   h) $g(x) = \ln(x)$, $[1, e^2]$

3. Analyze the existence and location of the roots of $x^3 - 3x^2 + x - 1 = 0$.

4. Use Rolle's Theorem to determine the number of solutions to $\ln(x) + e^x = 0$. In what interval are the roots?

5. If f and g are differentiable and $y = f(x)$ is increasing on an interval while $y = g(x)$ is decreasing on the same interval, how many solutions to the equation $f(x) = g(x)$ can exist on the interval?

6. Use the Equal Derivatives Theorem and the fact that $\sin(0) = 0$ and $\cos(0) = 1$ to prove $\sin^2(x) = 1 - \cos^2(x)$.

7. use the Equal Derivatives Theorem, to prove that $\ln(x)$ and $\ln(kx)$ for any constant k differ by a constant K. Use the fact that $\ln(1) = 0$ to establish that this constant $K = \ln(k)$.

Section 4.8 Differentials and Linear Approximations.

Introduction.

A fundamental concept in Calculus is that a "nice" function can be locally approximated by a linear function. Indeed, this is the characteristic of being differentiable! The first order approximation of a function f(x) by a tangent line through a point $(x_0, f(x_0))$ allows us to estimate values f(x) when x is close to x_0 by the y-coordinate on the tangent line at x. This y-coordinate is simply the y-coordinate at the tangent point, $y_0 = f(x_0)$, plus an increment which is called a *differential*. The differential, denoted by df, will depend on both the derivative of $f'(x_0)$ and on the difference $\Delta x = x - x_0$. The differential df is a numerical value that provides an approximation of the difference Δf. This approximation can be used to establish "linear" or "differential" approximations of functions.

Differentials.

Just as there are two types of variables, independent variables and dependent variables, there are two types of differentials. Those representing increments of independent variables are called *primary differentials*, these are denoted by the variable preceded by the letter d, e.g., dx, dt, or dv. The second type is the *differential of a function*, which is similarly denoted using the letter d followed by the function or independent variable symbol, e.g., as df or dy if y = f(x). The differential of a function df depends on a primary differential as well as the value of f'.

DEFINITION OF DIFFERENTIALS.

A **(primary) differential, dx**, of an independent variable **x** is any numerical value.

The **differential of a function y = f(x), corresponding to a primary differential dx, at $x = x_0$,** is

$$dy = df = f'(x_0) \cdot dx$$

Differentials are denoted by the variable or function symbol preceded by a "d". This notation is very similar to the Leibniz notation for the derivative, d/dx. In defining the differential dy the derivative of the function y(x) is denoted using the "prime" notation introduced by Isaac Newton. Solving the equation differential equation $dy = y' \cdot dx$ for the derivative y', gives y' as a ratio of two differentials:

$y' = dy / dx$ **The right side of this equation is the ratio of two differentials, dy and dx, that represent numerical values.**

You have already experienced differential notation, it was used in Chapter 3 to express derivative formulas in brief forms. For instance the Quotient Rule is frequently stated in a differential form:

$$d(u/v) = \{v \cdot du - u \cdot dv\}/v^2$$

where du, dv and d(u/v) are differentials of the functions u, v and thier ratio.

Example 1 Evaluating differentials.

Problem Evaluate the differential of the indicated function at the given dx and x_0.

a) Let $f(x) = -x^2 + 1$, $dx = 0.1$, and $x_0 = 3$. Then $f'(x) = -2x$ and the differential

$$df = f'(3)dx = -6 \cdot 0.1 = -0.6$$

b) Let $f(x) = e^{3x}$, $dx = 0.01$, and $x_0 = 0$. Then, $f'(x) = 3e^{3x}$ and the differential

$$df = f'(0)dx = 3e^0 \cdot 0.01 = 0.03$$

c) Let $g(x) = (x^2 - 5)^{3/2}$, $dx = -0.25$, and $x_0 = 3$. Then, $g'(x) = 3x(x^2 - 5)^{1/2}$ and the corresponding differential of g is

$$dg = g'(2)dx = 3 \cdot 3 \cdot (9 - 5)^{1/2} \cdot (-0.25) = -4.5$$

☑

Linear approximations:
using the differential df to approximate the difference Δf.

In geometrical terms the differential df is the change in the y-values on the tangent line to the graph of f at $(x_0, f(x_0))$ as the x-value changes from x_0 to $x_0 + dx$.

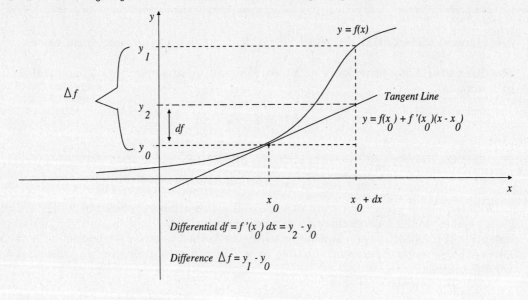

Differential $df = f'(x_0) dx = y_2 - y_0$

Difference $\Delta f = y_1 - y_0$

Recall The **difference** Δf is the change in the function's values as x changes from x_0 to $x_0 + dx$:

$$\Delta f \equiv f(x_0 + dx) - f(x_0)$$

Thus, the value of f(x) at the point $x_0 + dx$ is its value at x_0 plus the difference Δf:

$$f(x_0 + dx) = f(x_0) + \Delta f$$

Replacing Δf by the differential df in this equation gives an approximation of the function's value.

$$f(x_0 + dx) \approx f(x_0) + df$$

The right side of this approximation is precisely the y-value on the tangent line at $x = x_0 + dx$. Thus, this approximation consists of representing the function f by a linear function and is called a "first order" or "linear" approximation.

DEFINITION OF A LINEAR APPROXIMATION

> The **linear approximation of f(x) at x = a** is given by
>
> $$f(x) \approx L(x) \equiv f(a) + df = f(a) + f'(a)dx$$
> where $dx = x - a$
>
> Or, simply
>
> $$f(x) \approx f(a) + f'(a)(x - a)$$

Often complicated expressions in formulas and models can be simplified by introducing a linear approximation. This process is referred to as "linearization". To linearize a function, the first step is to identify a reference x-value, x_0, which corresponds to the number a in the above definition. It must be a value at which both the function and its derivative can be easily evaluated. Then, the linear approximation is substituted for the function. The linear approximation will generally be more accurate the closer the number x is to the reference value.

Example 2 Two linearizations of the same function.

Problem Determine the linear approximations of $f(x) = x/(2x + 1)$ at $x = 0$ and at $x = 2$. Compare the accuracy of these approximations at $x = 1$.

Solution The derivative $f'(x) = (2x + 1)^{-2}$. As $f(0) = 0$ and $f'(0) = 1$, the linear approximation of f **at x = 0** is

$$f(x) \approx L_1(x) = 0 + 1(x - 0) = x$$

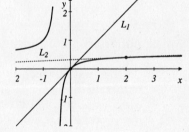

Since $f(2) = 2/5$ and $f'(2) = 1/25$, the linear approximation of f **at x = 2** is

$$f(x) \approx L_2(x) = 2/5 + (1/25)(x - 2)$$

Comparing these functions at $x = 1$, we see that

$$f(1) = 1/3 \quad \text{while} \quad L_1(1) = 1 \quad \text{and} \quad L_2(1) = 9/25$$

The approximation $f(1) \approx L_1(1)$ has an absolute error of $E = |1/3 - 1| = 2/3$.

The second approximation, $f(1) \approx L_2(1)$ has an absolute error of $E = |1/3 - 9/25| = 2/75$.

Example 3 Linear approximation of functions.

Problem Determine a linear approximation of the given function. Choose a reference value at which the function and its derivative are easily evaluated.

a) $f(x) = \sqrt{x+1}$ b) $F(x) = e^{-3x}$ c) $h(x) = ln(x^2 - 3)$ d) $g(\theta) = \cos(\theta)$

Solutions a) The function $f(x) = \sqrt{x+1}$ and its derivative $f'(x) = 0.5(x+1)^{-0.5}$ are both easily evaluated at $x = 0$. Set $x_0 = 0$. Then $f(0) = 1$, $f'(0) = 0.5$ and $dx = x - x_0 = x$. The corresponding linear approximation of f is $L(x) = f(0) + f'(0)(x - 0)$. Thus,

$$\sqrt{x+1} \approx 1 + 0.5x$$

b) $F(x) = e^{-3x}$ and $F'(x) = -3e^{-3x}$ are both easily evaluated at zero; set $x_0 = 0$. Since $F(0) = 1$ and $F'(0) = -3$, and again $dx = x - x_0 = x$, the linear approximation is

$$e^{-3x} \approx 1 - 3x$$

c) Consider $h(x) = ln(x^2 - 3)$ and its derivative $h'(x) = 2x/\{x^2 - 3\}$. Since the simplest evaluation of the natural logarithm function is $ln(1) = 0$, the reference value is chosen to make the argument $x^2 - 3 = 1$. Set $x_0 = 2$. Then, the primary differential is $dx = x - 2$ and the evaluating at x_0 gives

$$h(2) = ln(1) = 0 \quad \text{and} \quad h'(2) = 4$$

The corresponding linear approximation of h is thus

$$ln(x^2 - 3) \approx 0 + 4(x - 2) = -8 + 4x$$

Since the natural log function is **not defined** at zero this approximation clearly can not be used when $x^2 - 3 \leq 0$ or $x \leq SQRT(3) \approx 1.73$.

d) Both $g(\theta) = \cos(\theta)$ and $g'(\theta) = -\sin(\theta)$ are easy to evaluate at $\theta_0 = \pi/2$: $g(\pi/2) = 0$ and $g'(\pi/2) = -1$. The primary differential is $d\theta = \theta - \pi/2$. Thus the linear approximation of $\cos(\theta)$ at $\pi/2$ is

$$\cos(\theta) \approx 0 - 1\cdot(\theta - \pi/2) = \pi/2 - \theta$$

Numerical approximations.

Evaluating complex expressions involving elementary functions can usually be accomplished with scientific calculators. However, it is sometimes necessary to approximate the value of a complex expression directly, using simple arithmetic. A linear approximation of a complicated numerical expression is often referred to as a "differential approximation".

GENERAL METHOD FOR DETERMINING THE DIFFERENTIAL APPROXIMATION OF A NUMBER.	Application to approximate $\sqrt{51}$
1. First represent the number y to be evaluated as a function, $y = f(x)$; indicate the value x and the form of $f(x)$ and evaluate its derivative $f'(x)$.	$y = \sqrt{51} = f(x)$ $x = 51$ and $f(x) = \sqrt{x}$ $f'(x) = 1/\{2\sqrt{x}\}$
2. Next, determine a value x_0 near x such that both $f(x_0)$ and $f'(x_0)$ are easily evaluated.	If $x_0 = 49$, then $f(49) = 7$ and $f'(49) = 1/\{2\sqrt{49}\} = 1/14$

3. The primary differential is $dx = x - x_0$. The order is important!

   In this case $dx = 51 - 49 = 2$

4. Use the differential approximation

   $$f(x) \approx f(x_0) + f'(x_0)$$

   to approximate the value $y = f(x)$.

   The differential approximation is

   $$\sqrt{51} \approx 7 + (1/14) \cdot 2 \approx 7.1429$$

   A calculator value for $\sqrt{51}$ is 7.1415. This differential approximation is accurate to two decimal places.

Example 4 Approximating a cube root.

Problem Estimate the cube root of 6.

Solution Let $x = 6$ and $f(x) = x^{1/3}$. A value near 6 at which both f and $f'(x) = (1/3)x^{-2/3}$ can be easily evaluated is $x_0 = 8$. Then the primary differential is

$$dx = 6 - 8 = -2$$

The first order differential approximation is

$$6^{1/3} = f(6) \approx f(8) + f'(8) \cdot dx = 2 + (2/3)(1/4) \cdot (-2) = 2 - 1/3 = 5/3 = 1.66\underline{6}$$

The error in this approximation is about 0.15, since $6^{1/3} \approx 1.817$.

Example 5 Differential approximation of errors resulting from measurement errors.

Frequently a physical characteristic of an object is calculated from another measured quantity. If the measurement is in error, then the calculated value has an induced error. The error in the calculated value can be estimated by a differential, using the measured value as the reference.

Problem The volume of a spherical shaped cell is calculated using an average of two measurement of its diameter as viewed through a microscope. Assume that the measured diameter $D_0 = 12$ μm has a probable error of 5 percent. Estimate the probable error in the calculated cell volume. What is this potential error as a percentage of the calculated volume?

Solution The volume of a sphere as a function of its diameter is $V = (\pi/6)D^3$. Thus the estimated volume, $V(D_0)$, to four decimal places, is

$$V(12) = (\pi/6)12^3 \approx 904.7787 \text{ μm}^3$$

The true volume is $V(D)$ for some value $D = D_0 + \Delta D$, where ΔD is the actual error in measuring the diameter. The true error in our volume estimate is the difference

True Error $E = \Delta V = V(D) - V(D_0)$

However, the error can not be known without knowing the precise diameter, D! The approximate error in the calculated volume is the differential dV:

Approximate Error $\quad \hat{E} = dV = V'(D_0) \cdot dD$

The primary differential of the diameter D is given by the 5% bound for the measurement error. The error may be positive or negative, thus we consider both possibilities and set

$$dD = \pm 0.05 D_0 = \pm 0.6 \, \mu m$$

Using $V'(D) = (\pi/2)D^2$, the approximate error in the estimate is the differential

$$\hat{E} = \pm(\pi/2)(12^2) \cdot 0.6 \approx \pm 135.7168 \, \mu m^3$$

The percentage, or relative percentage, error is the ratio of the error to the true value, time 100%:

True % Error $\quad E \cdot 100\% / V = \Delta V \cdot 100\% / V = [V(D) - V(D_0)] \cdot 100\% / V$

However, to calculat this one must know the true value V. As this is not known, an approximation is given by using the calculate V, $V(D_0)$, and the differential:

Approximate %Error $\quad \hat{E} \cdot 100\% / V(D_0) = dV \cdot 100\% / V(D_0) = V'(D_0) \cdot dD \cdot 100\% / V(D_0)$

Thus, the approximate percentage error of the estimated volume is

$$\hat{E} \cdot 100\% / V(12) = \pm 135.7168 \, \mu m^3 \cdot 100\% / 904.7787 \, \mu m^3 \approx \pm 15\%$$

☑

Exercises Set 4.8

1. a) Sketch the function $f(x) = 0.5x^2$ and its tangent line at the point $x_0 = 1$. Illustrate the difference between the differential df and the difference Δf corresponding to the primary differentials $dx = 0.5$ and $dx = -1$. Calculate the absolute difference $|\Delta f - df|$ for these two values of dx.
   b) Repeat part a) for $x_0 = 4$.

2. Calculate the differential of the given function and evaluate it at the indicated reference value and primary differential.

   a) $f(x) = x^3 - 2x - 1$, $\quad x_0 = 2, \; dx = 0.1$ $\qquad$ b) $f(x) = x^3 - 2x - 1$, $\quad x_0 = 1, \quad dx = -0.1$

   c) $f(x) = x^3 - 2x - 1$, $\quad x_0 = -2, \; dx = -0.5$ $\qquad$ d) $g(x) = \text{SQRT}(25 - x^2)$, $\quad x_0 = 3, \quad dx = 0.2$

   e) $h(t) = e^{3t}$, $\quad t_0 = 0, \; dt = 0.25$ $\qquad$ f) $M(t) = -32t^2 - 1/t$, $\quad t_0 = 4, \quad dt = -0.3$

   g) $V(s) = se^{-0.02s}$, $\quad s_0 = 0, \; ds = 0.5$ $\qquad$ h) $A(s) = s/(10 + s)$, $\quad s_0 = 10, \quad ds = 2$

   i) $f(\theta) = \sin(3\theta)$, $\quad \theta_0 = \pi, \; d\theta = 0.2$ $\qquad$ j) $f(\theta) = \sin(\theta + \pi\cos(\theta))$, $\quad \theta_0 = \pi/3 \quad d\theta = 0.2$

3. Determine the linear approximation L_1 of $y = x^3 - x^{-3}$ at $x_0 = 1$ and the linear approximation L_2 of y at $x_0 = 3$. Compare these approximations at $x = 2$.

4. The accuracy of a linear approximation depends on the function, the reference point, and the primary differential. For each function compute the difference Δf, the differential df, and the absolute error $E = |\Delta f - df|$. Evaluate therse at the reference points $x_0 = 1$ and $x_0 = 10$ for three different primary differentials: $dx = 0.1$, $dx = 1$, and $dx = 10$.

   a) $f(x) = 3x^2$ $\qquad$ b) $f(x) = e^{-3x}$ $\qquad$ c) $f(x) = ln(x)$ $\qquad$ d) $f(x) = 5x/(10 + x)$

Section 4.8 — Differentials and Linear Approximations

5. Use the linear approximation $\sin(3x) \approx 3x$ to simplify the function $y = \cos(\pi/2\, \sin(3x))$.

6. Determine the linear approximations of the following functions at $x_0 = 0$. Assume k is a constant.

   a) $(x + 1)^k$ b) $\text{SQRT}(1 + kx)$ c) e^{kx} d) $\ln(kx + 1)$

   e) k^x f) $\sin(kx)$ g) $\cos(kx)$ h) $\tan(kx)$

In Exercises 7)–14) write a differential approximation of the area or volume formula used and apply it to approximate the change in the area or volume due to the indicated change in the primary variable.

7. The change in the volume of a cube with edge length x, when x = 6 mm is increased dx = 0.05 mm.

8. The change in the surface area of a cube of edge length 5 cm if the edges are decreased by 0.1 cm.

9. The change in the volume $V = \pi r^2 h/3$ of a circular cone of height h and base radius r if r is increased by 5% and h remains constant.

10. The change in the volume $V = (4/3)\pi r^3$ of a sphere of radius r when r = 5000km is increased by 1m.

11. The surface area $S = 4\pi r^2$ of a sphere of radius r, when r = 5000 km is increased by 1m.

12. The area of a circular ring with inner radius r and outer radius R, if R = 5cm is fixed and r = 3cm is decreased by 0.5cm.

13. The volume $V = s^2 h/3$ of a square based pyramid if the height h = 40m and the base edge length s increases from 45 to 50 meters.

14. The volume of a detergent box if its height:width:depth ratio remains 3:2:1 and the depth increases from 10 to 12 cm.

15. A manufacture wishes to manufacture an open square box of side length 175mm. a) How accurately must the boxes side length be measured to ensure that the resulting error in the volume is less than 5% of the theoretical volume? (Approximately.) b) How accurately should the side lengths be measured to ensure that the surface area of the exterior side of the box is within 2% of the theoretical value?

16. Nonlinear equations can be difficult to solve. An approximate solution can be obtained by linearizing an equation or part of an equation and then solving.
    a) Approximate a root of $(x - 1)e^{-2x} = 0$ by linearizing about $x_0 = 0$.
    b) Determine an approximate explicit solution $y = f(x)$ when $F(x,y) = x^2 y + \ln(x + y) = 0$ for (x,y) values near (1,0) by linearizing F(x,y) as a function of y about $y_0 = 0$ and then solving the linear equation for y.

17. Use differentials to approximate the indicated value. Note. Your answer will not be the value obtained by evaluating the number on your calculator. Use your calculator to calculate the error in your approximation and the true % relative error.

    a) $3^{1/2}$ b) $6^{-1/2}$ c) $30^{1/3}$ d) $100^{-1/3}$

    e) $4^{2.1}$ f) $2^{3.99}$ g) $25^{-0.4}$ h) $9^{-0.3}$

Section 4.9 Taylor Polynomials and Taylor Series

Introduction.

Taylor Polynomials are special polynomials that are associated with a given function at a specific reference point. These polynomials are named after the 17th century English mathematician Brook Taylor (1635-1731). The simplest Taylor Polynomial is the linear function whose graph is a tangent line to the graph of the function. A Taylor Polynomial of degree N, called an N-th Order Taylor Polynomial, will be denoted by $P_N(x)$. Taylor Polynomials are used to approximate functions. The accuracy of an approximation

$$f(x) \approx P_N(x)$$

is directly related to the degree N of P_N and the distance of the value x from a reference value x_0. As N increases the accuracy improves, and for "nice" functions, called *analytic functions*, in the limit, as $N \to \infty$, the error approaches zero. We end this section with a brief look at infinite Taylor Series.

Taylor Polynomials.

The ideal of a linear approximation of a function at a reference point can be naturally generalized to higher order approximations. For a given function $y = f(x)$ and reference value x_0, the corresponding Taylor Polynomials are defined using the values of f and its derivatives at x_0. These polynomials are "shifted" to the reference value x_0 and involve powers of the difference $x - x_0$. The coefficients of these powers are defined so that the Taylor Polynomial "matches" f and its derivatives identically at x_0.

Recall   The notation for higher order derivatives uses an exponent-like term in parentheses:

the n-th derivative of f is $\quad d^n f(x)/dx^n \equiv f^{(n)}(x)$

By definition $\quad\quad f^{(0)}(x) \equiv f(x)$

Remember how the equation of a line can be expressed in different forms. The basic slope-intercept equation

$$y = a_0 + a_1 x$$

can be expressed in the "shifted" point-slope form

$$y = b_0 + b_1(x - x_0)$$

by identifying the coefficients: $b_1 = a_1$ and $a_0 = b_0 - b_1 x_0$ or $b_0 = a_0 + a_1 x_0$. The graph in the shifted form is said to centered at the point (x_0, b_0). Normally the constant b_0 is replaced by y_0. In this text we have consistently used the shifted point-slope equation to refer to tangent lines.

Similarly, a general **N-th degree polynomial** contains only powers of x and coefficients and has the form:

$$P(x) = \Sigma_{i=1,N} \, a_i x^i = a_0 + a_1 x + a_2 x^2 + \ldots + a_N x^N$$

Remember how the basic quadratic polynomial is expressed in a shifted form, by completing the square, so that we could read off its center and easily sketch its graph, i.e., setting

$$P(x) = A + Bx + Cx^2 = b + c(x - a)^2$$

the parameters are identified as $a = -C/(2C)$, $b = A - B^2/(4C)$ and $c = C$. The function $P(x)$ is the same function no mater which form of the quadratic is used, the expanded form with coefficients A, B, and C or the shifted form with coefficients a, b, and c. The same theory applies for any degree polynomial. That is, the above N-th degree polynomial can be expressed as a **shifted polynomial** of the form

$$P(x) = \Sigma_{i=1,N} \, b_i(x - x_0)^i = b_0 + b_1(x - x_0) + b_2(x - x_0)^2 + \ldots + b_N(x - x_0)^N$$

Of course the relationship between the a and b coefficients is more complex, and depends on the shift value x_0. Taylor Polynomials have this shifted form. While any polynomial passing through a point (x_0, y_0) can be rearranged into this form, other non-polynomial functions can almost be so represented. A Taylor polynomial $P_n(x)$ can be formed from any "nice" function. The definition of a Taylor Polynomial hinges on how the coefficients, the numbers b_i, are defined. As we discuss below, a function is "nice" if the error in the approximation $f(x) \approx P_N(x)$ approaches zero as $N \to \infty$.

Consider a function f that can be differentiated repeatedly on an open interval containing x_0. The associated N-th Order Taylor Polynomial is defined as the unique *N-th degree polynomial tangent to the graph of f at x_0*. This concept of a tangent polynomial is a generalization of the concept of a tangent line and a tangent parabola, as discussed in Section 4.1. Basically, it means that the polynomial P_N and its derivatives of order 1, 2, 3, ..., N, when evaluated at x_0, have exactly the same values as the function f and its derivatives, respectively. These polynomials are expressed in the shifted form, centered at the point (x_0, y_0) on the graph of f. Let's see how this works.

For N = 1. The Taylor Polynomial P_1 is simply the polynomial whose graph is the tangent line to the graph of f at $(x_0, f(x_0))$. Thus, the equation for P_1 is

$$P_1(x) = b_0 + b_1(x - x_0)$$

where, as in the previous Section, the coefficients are

$$b_0 = f(x_0) \quad \text{and} \quad b_1 = f'(x_0)$$

Clearly, $P_1(x_0) = f(x_0)$, and since P_1's derivative is $P_1'(x) = b_1$ for all x, $P_1'(x_0) = f'(x_0)$.

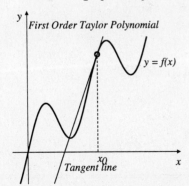
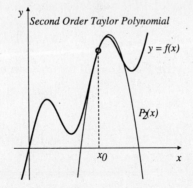
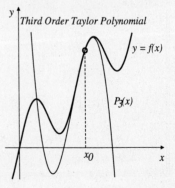

For N = 2. The second order Taylor Polynomial is the quadratic function whose graph is tangent to the graph of f at $(x_0, f(x_0))$. Its shifted form is then

$$P_2(x) = b_0 + b_1(x - x_0) + b_2(x - x_0)^2$$

Since P_2 and P_2' must also match f and f' at x_0, the constants b_0 and b_1 are the same as they were for P_1. The constant b_2 is found from the equation

$$P_2''(x_0) = f''(x_0)$$

Taking P_2's derivative twice results in the constant function

$$P_2''(x) = 2 \cdot b_2$$

Equating the last two equations and solving gives $b_2 = f''(x_0) / 2$.

For arbitrary N. Notice how P_2 is just P_1 plus an additional term. This type of relationship is true for all Taylor Polynomials. If we know P_{N-1} then P_N is found by adding one more term,

$$P_N(x) = P_{N-1}(x) + b_N(x - x_0)^N$$

In general, the coefficient b_N is found by matching the N-th derivatives:

$$D_x^N P_N(x_0) = f^{(N)}(x_0)$$

As P_{N-1} is a polynomial of degree $N - 1$, taking its derivative N-times results in zero. (Remember for instance how the fourth derivative of x^3 vanished.) Consequently, the N-th derivative of P_N is just the N-th derivative of the N-th degree term $b_N(x - x_0)^N$. But, by direct evaluation this derivative is the constant b_N times the product of the exponents each time it is differentiated,

$$D_x^N P_N(x) = D_x^N P_{N-1}(x) + D_x^N b_N(x - x_0)^N = 0 + b_N \cdot N \cdot (N-1) \cdot (N-2) \cdot \ldots \cdot 3 \cdot 2 \cdot 1$$

Thus, expressing the product as N!, the formula for b_N is found as

$$b_N = f^{(N)}(x_0) / N!$$

This formula forms the basis for the following definition.

DEFINITION OF TAYLOR POLYNOMIALS

> Assume the function $y = f(x)$ is N times differentiable at x_0. Then, the
>
> **N-th order Taylor Polynomial expansion of f(x) about x_0 is**
>
> $$P_N(x) = \sum_{n=1}^{N} \{f^{(n)}(x_0)/n!\} \cdot (x - x_0)^n$$
>
> In expanded form, the **Taylor Polynomial expansion of degree N** is
>
> $$P_N(x) = f(x_0) + f'(x_0)(x - x_0) + \{f''(x_0)/2!\}(x - x_0)^2$$
> $$+ \{f^{(3)}(x_0)/3!\}(x - x_0)^3 + \ldots + \{f^{(N)}(x_0)/N!\}(x - x_0)^N$$

Example 1 The Second order Taylor Polynomials P_2 corresponding to $\sqrt{x}$.

Problem Determine the second order Taylor Polynomial expansion of $f(x) = \sqrt{x}$ about $x_0 = 4$.

Solution The second order Taylor Polynomial is $P_2(x) = f(4) + f'(4)(x - 4) + f''(4)/2 \, (x - 4)^2$

The derivatives of f are $f'(x) = (1/2)x^{-1/2}$ and $f''(x) = (-1/4)x^{-3/2}$ These are evaluated at $x_0 = 4$ to generate the numbers needed in the coefficients:

$$f(4) = 2 \quad f'(4) = 1/4 \quad f''(4) = -1/32$$

Substituting these numerical values gives

$$P_2(x) = 2 + (1/4)(x - 4) + (-1/32)/2 \, (x - 4)^2$$

This would normally be simplified to the form $P_2(x) = 2 + (1/4)(x - 4) - (1/64)(x - 4)^2$, but no further.

Even when f(x) is already a polynomial, you can form its Taylor Polynomial at a point x_0. This is how higher order polynomials are put into the shifted form. If f(x) is a polynomial of degree K, then for N = K, $f(x) = P_N(x)$ exactly. This is also true if K < N, for then all the coefficients of terms $(x - x_0)^i$ for i = K+1, K+2, ... N are zero. In this case the Taylor Polynomial $P_N(x)$ is really a polynomial of degree K. When K > N, the Taylor polynomial $P_N(x)$ is the found by truncating, i.e., omitting, all terms of degree greater than N of the shifted form of f(x) centered at (x_0, y_0).

Example 2 A third order Taylor Polynomial corresponding to a 5th degree polynomial.

Problem Determine the 3rd order Taylor Polynomial expansion of $P(x) = 1 + x + x^5$ about $x_0 = 2$.

Solution To obtain the Taylor coefficients we require the first three derivatives of P at the point x = 2:

$$\begin{array}{ll} P(x) = 1 + x + x^5 & P(2) = 35 \\ P'(x) = 1 + 5x^4 & P'(2) = 81 \\ P''(x) = 20x^3 & P''(2) = 160 \\ P^{(3)}(x) = 60x^2 & P^{(3)}(2) = 240 \end{array}$$

Thus, the third order Taylor Polynomial expansion of P(x) about 2 is

$$P_3(x) = 35 + 081(x - 2) + (160/2)(x - 2)^2 + (240/6)(x - 2)^3$$

which simplifies to $P_3(x) = 35 + 81(x - 2) + 80(x - 2)^2 + 40(x - 2)^3$.

Example 3 A 4-th order Taylor Polynomial corresponding to $ln(x)$.

Problem Determine the 4th order Taylor Polynomial for $f(x) = ln(x)$ about $x_0 = 1$.

Solution The reference value was chosen so that the function and its derivatives are easily evaluated. $f(1) = ln(1) = 0$. The derivatives $f^{(n)}(x)$ for n = 1, 2, 3, and 4 are:

$$f'(x) = x^{-1} \quad f''(x) = -x^{-2} \quad f^{(3)}(x) = 2x^{-3} \quad f^{(4)}(x) = -6x^{-4}$$

These are evaluated at the reference value $x_0 = 1$:

$$f'(1) = 1 \quad f''(1) = -1 \quad f^{(3)}(1) = 2 \quad f^{(4)}(1) = -6$$

Using these values the 4th order Taylor expansion of $ln(x)$ about $x_0 = 1$ is

$$P_4(x) = 0 + 1(x-1)/1! + -1(x-1)^2/2! + 2(x-1)^3/3! + -6(x-1)^4/4!$$

$$P_4(x) = (x-1) - (1/2)(x-1)^2 + (1/3)(x-1)^3 - (1/4)(x-1)^4$$

Example 4 Developing Growth Models with Taylor Polynomials.

Background This example illustrates the use of a second order Taylor Polynomial to establish a limited growth model. A continuous growth model describes the rate of change in the population or substance being modeled by a differential equation. The growth rate is assumed to be a function of the size or state of the population. If $X = X(t)$ denotes the amount or size of the population at time t, a general growth model has the form

$$dX/dt = R(X)$$

One of the challenges in applied fields is to determine from observations or experiments the specific growth function R corresponding to a particular population or substance being modeled. If the form of the growth function R is totally unknown this may present an impossible task. However, one approach is to approximate R by its theoretical second order Taylor Polynomial and then to estimate the numerical values of its Taylor Polynomial coefficients from observations.

Problem Determine a second-order growth model for a general growth function R.

Solution A second order model is obtained by replacing $R(X)$ by its Taylor Polynomial $P_2(X)$. This gives a dynamic model

$$dX/dt = R(X_0) + R'(X_0)(X - X_0) + [R''(X_0)/2](X - X_0)^2$$

A primary concern in most dynamical models is to analyze the stability of the population at an **equilibrium** population size X_E. An equilibrium size X_E is defined as an X-value for which the derivative dX/dt is zero. But the derivative is equal to the Growth Rate $R(X)$, thus an equilibrium X_E is a root of

$$R(X_E) = 0 \qquad \textbf{Equilibrium Condition}$$

The dynamic growth model given by the Taylor Polynomial P_2 expanded about $X_0 = X_E$ has the form

$$dX/dt = R'(X_E)(X - X_E) + [R''(X_E)/2](X - X_E)^2$$

The equation is simplified by making a change of variables. This is accomplished by a substitution of the form $X = Y + X_E$:

$$(X - X_E) = (Y + X_E - X_E) = Y$$

The derivative $X' = Y'$ and hence the transformed growth model becomes

$$Y' = R'(X_E)Y + [R''(X_E)/2]Y^2$$

This equation is changed to a more common form by introducing two parameters:

$$\text{"The intrinsic unit growth rate"} \quad r = R'(X_E)$$

and

$$\text{"The carrying capacity"} \quad M = -2r / R''(X_E)$$

With these parameters the quadratic growth model has the form commonly referred to as the **Pearl-Verhulst** or **Logistic** growth model:

$$Y' = r(1 - Y/M)Y \qquad \textbf{Logistic Model} \quad \boxed{\checkmark}$$

Section 4.9   Taylor Polynomials and Taylor Series

Taylor Approximations.

As polynomials are easy to evaluate, Taylor Polynomials are used in numerical problems to approximate function values. In theoretical developments they are similarly used to simplify formulas. The N-th order Taylor approximation is given by

$$f(x) \approx P_N(x) \qquad \text{Taylor Approximation}$$

Its accuracy will depend on the function f, the distance $dx = x - x_0$, and the order N. An important Theorem that gives the relations between a function and its Taylor Polynomials is the following:

Taylor's Theorem

> If the (N + 1)st derivative of f, $f^{(N+1)}(x)$, is continuous on an open interval I containing the value x_0 and $P_N(x)$ is the Nth order Taylor Polynomial expansion of f about x_0, then for $x \in I$,
>
> $$f(x) = P_N(x) + R_N(x)$$
>
> where the **Remainder** function $R_N(x)$ can be expressed in the
>
> **Lagrange Remainder** form:
>
> $$R_N(x) = f^{(N+1)}(\xi) \cdot (x - x_0)^{N+1} / (N+1)!$$
>
> for some value ξ in the interval I.

Without providing a rationale for Taylor's Theorem we mention that the existence of a value ξ that gives the correct value of the remainder $R_N(x)$ is guaranteed by the Mean Value Theorem. However, there is no algorithm that tells how to determine it! To circumvent this situation, mathematicians establish error bounds for Taylor Approximations by calculating bounds for the maximum possible value of $|R_N(x)|$ when x is in an interval containing x_0.

The absolute error in the approximation $f(x) \approx P_N(x)$ is

$$|E(x)| = |f(x) - P_N(x)| = |R_N(x)| \leq M \cdot (x - x_0)^{N+1}/(N+1)! \qquad \text{Error Bound}$$

where

$$M = \text{maximum}\{|f^{(N+1)}(\xi)| : \xi \in I\}$$

I is an interval containing x and x_0

Example 5   Taylor Approximations and Errors.

Problem   Determine the Taylor Approximations of order $N \leq 4$ of $f(x) = e^{2x}$ about $x_0 = 0$. Calculate the error when using the approximations to estimate $f(1)$ and compare this to the error bound obtained from the maximum of the next higher derivative on the interval $I = [0,1]$.

Solution First, the derivatives of f are determined. They are evaluated at $x_0 = 0$ to determine the coefficients of the Taylor Polynomials; to establish the error bounds the maximum of these derivatives are required. As each derivative is a multiple of e^{2x}, which is an increasing function, the maximum over the interval [0,1] occurs at $x = 1$, giving the values indicated in the table.

n	$f^{(n)}(x)$	$f^{(n)}(0)$	maximum $f^{(n)}(x)$ on [0,1]
0	e^{2x}	1	e^2
1	$2e^{2x}$	2	$2e^2$
2	$4e^{2x}$	4	$4e^2$
3	$8e^{2x}$	8	$8e^2$
4	$16e^{2x}$	16	$16e^2$
5	$32e^{2x}$	32	$32e^2$

The first five Taylor Polynomials expansions of e^2 about $x_0 = 0$ are:

$P_0(x) = 1,$

$P_1(x) = 1 + 2x,$

$P_2(x) = 1 + 2x + 2x^2$

$P_3(x) = 1 + 2x + 2x^2 + (4/3)x^3$

$P_4(x) = 1 + 2x + 2x^2 + (4/3)x^3 + (2/3)x^4$

The Error Bounds for the approximations

$e^{2x} \approx P_n(x)$

corresponding to the maximum of the Lagrange Remainders for the approximations, are given by

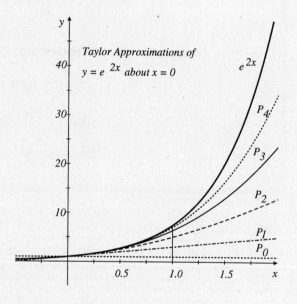

Taylor Approximations of $y = e^{2x}$ about $x = 0$

$EB_n = \max |f^{(n+1)}(\xi)| (x - x_0)^{n+1}/(n+1)! = f^{(n+1)}(1)(1 - 0)^{n+1}/(n+1)! = f^{(n+1)}(1)/(n+1)!$

The approximations $P_n(1)$ and the actual error $|E(1)|$ in the approximations, together with the error bounds derived from the Lagrange Remainder are given in the following table:

n	$P_n(1)$	$|E(1)|$	Error Bound on the interval [0,1] $f^{(n+1)}(1)/(n+1)!$
0	1	6.389056	$2e^2 \approx 14.778112$
1	3	4.389056	$2e^2 \approx 14.778112$
2	5	2.389056	$4/3\, e^2 \approx 9.8520748$
3	6.333	1.055723	$(2/3)e^2 \approx 4.9260374$
4	7	0.389056	$(4/15)e^2 \approx 1.970415$

Section 4.9 — Taylor Polynomials and Taylor Series

Observe that the actual errors are less than the Error bounds. However, the bounds do indicate the improvement of the approximation as the order n increases.

Infinite Taylor Series.

Taylor Polynomials provide an approximation of a function. If the function being approximated is relatively *nice* the accuracy of these approximations improves as the order of the approximation increases. When the error $f(x) - P_N(x) = R_N(x) \to 0$ as $N \to \infty$ on an interval I the function f is said to be an *analytic function*. Analytic functions can evaluated by an **infinite Taylor Series**, which is formally the limit of $P_N(x)$ as $N \to \infty$.

A function f(x) is **analytic on an interval I** if for all $x \in I$,

$$f(x) = \sum_{n=0}^{\infty} (f^{(n)}(x_0)/n!)(x - x_0)^n \qquad \textbf{Infinite Taylor Series}$$

*When the reference value $x_0 = 0$ the Taylor Series is called a **Maclaurin series**.*

In this subsection a brief "exposure" to infinite Taylor Series is presented. We shall not present a thorough detailed discussion of infinite series and the analysis of their convergence. Often, a whole course is devoted to the study of infinite series.

As discussed in Section 2.1 of Chapter 2, if an infinite series converges then its value is given by the limit of the sequence of "partial sums". A "partial sums" consist of the sum of the first N terms of the series. Thus, the infinite Taylor series are evaluated as

$$\sum_{n=0,\infty} f^{(n)}(x_0)(x - x_0)^n/n! \;=\; \lim_{N \to \infty} \left\{ \sum_{n=0,N} f^{(n)}(x_0)(x - x_0)^n/n! \right\}$$

The good news is that most functions you will encounter are analytic on intervals that do not contain singular points (where the function becomes infinite). Thus, the above limit does exist and in most cases is approached rapidly. In fact, your scientific calculator evaluates the elementary functions by computing partial sums of such series.

Some basic Taylor Series.

The elementary functions are analytic and can be represented by Taylor Series expansions. The exponential e^x and the trigonometric functions $\sin(x)$ and $\cos(x)$ can be expanded about $x_0 = 0$. We leave as exercises the calculations to establish the following series:

$$e^x = 1 + x + x^2/2! + x^3/3! + x^4/4! + \ldots$$

$$\sin(x) = x - x^3/3! + x^5/5! - x^7/7! + \ldots$$

$$\cos(x) = 1 - x^2/2! + x^4/4! - x^6/6! - \ldots$$

As the $ln(x)$ is not defined for $x \le 0$, the shifted function $ln(x-1)$ is expanded about $x_0 = 0$:

$$ln(x+1) = x - x^2/2 + x^3/3 - x^4/4 + \ldots \qquad \text{for } |x| < 1$$

The three dots indicate that the pattern is continued. The last three series are called "alternating" series since the sign of successive terms alternates between positive and negative.

Example 6   Determining a Taylor Series for 1/(1 + x).

Problem   Determine the infinite Taylor Series for y = 1/(1 + x) at $x_0 = 0$.

Solution   The function $y = (1 + x)^{-1}$, and each of its derivatives, is easily differentiated by the Power Rule:

$$y' = -(1 + x)^{-2}, \quad y'' = 2(1 + x)^{-3}, \quad y''' = -3 \cdot 2(1 + x)^{-4},$$

in general, the nth derivative is $y^{(n)} = (-1)^{n+1} \cdot n! \cdot (1 + x)^{n+1}$

Thus the value of this derivative at x = 0 is simply n! times an alternating $(-1)^{n+1}$ as (x + 1) = 1

$$y^{(n)}(0) = (-1)^{n+1} \cdot n! \cdot (1)^{n+1} = (-1)^{n+1} \cdot n!$$

The coefficient of the n-th power of x in the Taylor Series is this value divided by n! . Hence, the n! terms cancel and the infinite Maclaurin expansion of y is the alternating series

$$y = 1/(1 + x) = 1 - x + x^2 - x^3 + x^4 - \ldots \quad \text{for } |x| < 1$$

This series is not valid if $|x| \geq 1$ since at x = -1 the function y is not defined.

☑

A very important property of analytic functions is that they can be differentiated by differentiating their series expansions term by term.

Example 7   Differentiating Infinite Taylor Series

Problem   Show that $D_x e^x = e^x$ by differentiating its Taylor Series expansion "term by term".

Solution   Each term of the Taylor Series

$$e^x = 1 + x + x^2/2! + x^3/3! + x^4/4! + \ldots$$

is a power function whose exponent n is matched by the coefficient 1/n! . Hence

$$D_x x^n / n! = n \cdot x^{n-1} / n! = x^{n-1}/(n - 1)!$$

since

$$n/n! = \not{n} / [1 \cdot 2 \cdot 3 \cdot \ldots \cdot (n - 1) \cdot \not{n}] = 1 / [1 \cdot 2 \cdot 3 \cdot \ldots \cdot (n - 1)] = 1/(n - 1)!$$

Thus, using the D_x notation for the derivative, if we formally differentiate each term of the infinite Taylor Series for e^x we have

$$D_x e^x = D_x 1 + D_x x + D_x x^2/2! + D_x x^3/3! + D_x x^4/4! + \ldots$$
$$= 0 + 1 + x + x^2/2! + x^3/3! + x^4/4! + \ldots = e^x$$

☑

Exercises Set 4.9

1. Determine the Taylor Polynomials $P_n(x)$ expansions of $f(x) = 8 - (x-2)^4$ about $x_0 = 0$ for $n \leq 4$. Sketch the graph of f and the Taylor Polynomials over the interval $[-2,2]$.

2. Determine the second-order Taylor Polynomial of $f(x) = x^3 + x$ expanded about the point:

   a) $x_0 = 0$ b) $x_0 = 1$ c) $x_0 = -1$ d) $x_0 = 2$ e) $x_0 = 10$

3. Determine the third-order Taylor Polynomial of $f(x) = x^4 - 2x^2 + 1$ expanded about the point:

   a) $x_0 = 0$ b) $x_0 = 1$ c) $x_0 = -1$ d) $x_0 = 2$ e) $x_0 = 10$

4. Determine $P_3(x)$ for the function $f(x)$ expanded about the indicated x_0.

   a) $f(x) = 2x^2 + 3x - 2$ $x_0 = 0$ b) $f(x) = 2x^2 + 3x - 2$ $x_0 = 1$

   c) $f(x) = x^4 + 5x$ $x_0 = 0$ d) $f(x) = x^4 + 5x$ $x_0 = -2$

   e) $f(x) = e^{3x}$ $x_0 = 0$ f) $f(x) = e^{3x}$ $x_0 = -\ln(8)$

   g) $f(x) = e^{x^2}$ $x_0 = 0$ h) $f(x) = e^{-x^2}$ $x_0 = 0$

   i) $f(x) = x/(1-x)$ $x_0 = 0$ j) $f(x) = (x+1)^{1/2}$ $x_0 = 0$

   k) $f(x) = (x+1)^{1/2}$ $x_0 = 3$ l) $f(x) = 1/x$ $x_0 = 1$

   m) $f(x) = \sin(x)$ $x_0 = 0$ n) $f(x) = \sin(x)$ $x_0 = \pi$

5. Determine the Taylor Polynomial approximations of $y = e^{-x^2}$ of order $n \leq 4$ about $x_0 = 0$. Compare the error in these at $x = 1$ with the Error Bounds that are derived from the Lagrange Remainder. See Example 5.

6. Using the fact that $|\sin(x)| \leq 1$, what order Taylor Polynomial is required to approximate $\sin(x)$ to an accuracy of 0.01 on the interval a) $[-\pi/2, \pi/2]$ b) $[-\pi, \pi]$ c) $[-2\pi, 2\pi]$.

7. Derive the Taylor Series expansions about $x_0 = 0$ of the given function. Include sufficiently many terms to establish a pattern for the series.

   a) e^x b) e^{-x} c) e^{kx} d) $\ln(x+1)$

   e) $\sin(x)$ f) $\cos(x)$ g) $\sin(kx)$ h) $\cos(kx)$

8. Use the Taylor Series given in the text and differentiated "term by term" to show that:

   a) $D_x \sin(x) = \cos(x)$ b) $D_x \cos(x) = \sin(x)$ c) $D_x \ln(x+1) = 1/(x+1)$ d) $D_x e^{2x} = 2e^{2x}$

9. a) Establish the Taylor Series expansion of $\ln(x)$ about $x_0 = 1$.
   b) Establish the Taylor Series expansion of $1/x$ about $x_0 = 1$.
   c) Use term-by-term differentiation of this series to show that $D_x \ln(x) = 1/x$.

Section 4.10 Indeterminate Limits; L'Hôpital's Rule

Introduction: Limits involving infinity.

Infinity is a concept, not a number. Yet, we use ∞ as a number in limits e.g. $\lim_{x \to \infty}$. Infinity also arises when limits are evaluated. We even have an extended type of arithmetic involving ∞. We learned to evaluate limits of complex terms by considering the limits of sub-terms and then adding or multiplying these as appropriate to get the original limit. However, when the sub-term limits involve *infinity*, either as ∞ or as division by zero, *1/0*, this process often fails because our extended arithmetic can not evaluate the resulting combination of sub-limits. In this section a method is introduced that can evaluate many such limits that can not be evaluated using the basic algebra rules for limits, it is called L'Hôpital's Rule.

Evaluating Limits with extended arithmetic rules.

Infinite Sum Rule: The limit of a sum or difference of two terms is infinite if the limit of one term is finite and the limit of the other term is infinite.

Infinite Product Rule: The limit of a product of two terms is infinite if the limit of one term is finite but not zero and the limit of the other term is infinite.

Infinite Quotient Rules: The limit of a quotient is infinite if the limit of the denominator is finite but not zero and the limit of the numerator is infinite.

The limit of a quotient is zero if the limit of the denominator is infinite and the limit of the numerator is finite.

The limit of a quotient is infinite if the limit of the denominator is zero and the limit of the numerator is finite and not zero.

These *infinity* limits that can be evaluated are illustrated by the following. Each of these can be rigorously justified using the concept that to approach ∞ means that the value becomes unbounded and will become larger than <u>any</u> real number.

Assume $\lim_{x \to b} f(x) = C$ and $C \neq 0$, $\lim_{x \to b} g(x) = +\infty$, and $\lim_{x \to b} h(x) = 0$

then, the limit	Symbolically has the form:
$\lim_{x \to b} \{f(x) \pm g(x)\} = \pm \infty$	"C ± ∞" → ± ∞
$\lim_{x \to b} \{f(x) \cdot g(x)\} = \pm \infty$	"C · +∞" → ± ∞ *
$\lim_{x \to b} \{g(x) / f(x)\} = \pm \infty$	"+∞ / C" → ± ∞ *
$\lim_{x \to b} \{f(x) / g(x)\} = 0$	"C / ∞" → 0
$\lim_{x \to b} \{f(x) / h(x)\} = \pm \infty$	"C / 0" → ± ∞ #

* in these cases the sign is + if C > 0 and is - if C < 0.
\# in this case the sign depends on the sign of C and on how h(x) approaches zero.

Section 4.10   INDETERMINATE LIMITS; L'HÔPITAL'S RULE

Indeterminate limits that can not be evaluated by the extended arithmetic rules.

The objective of this Section is to evaluate *indeterminate limits* that are not covered by the above rules. These limits involve *infinity* arithmetic that can not be evaluated without further analysis.

INDETERMINATE LIMIT FORMS.

> The limit of a term is said to be **indeterminate** if taking the limits of individual sub-terms results in one of the following **indeterminate algebraic forms**:
>
> Indeterminate Quotients: $0/0$ ∞/∞
>
> Indeterminate Products: $0 \cdot \infty$ or $\infty \cdot 0$
>
> Indeterminate Difference: "$\infty - \infty$"
>
> Indeterminate Powers: 0^0 1^∞

The most obvious cases, symbolically referred to as $0/0$ or ∞/∞ limits, are ambiguous because they lead to different results in different circumstances. The ambiguity of a $0/0$ limit is best illustrated by considering the limit as $x \to 0+$ of x^n over x^k. Since $x^n/x^k = x^{n-k}$

$$\lim_{x \to 0+} x^n/x^k = \begin{cases} 0 & \text{if } n > k > 0; \\ 1 & \text{if } n = k; \\ +\infty & \text{if } 0 < n < k. \end{cases}$$

A method for evaluating Indeterminate Quotient Limits of the form $0/0$ and ∞/∞ is called *L'Hôpital's Rule*. The other indeterminate limit forms are evaluated by algebraically converting them to equivalent Quotient forms, and then applying the generalized arithmetic rules given above or using L'Hôpital's Rule.

L'Hôpital's Rule.

You have seen indeterminate limits before!
Remember the definition of the derivative?
It involves the limit of a difference quotient that has the indeterminate limit form $0/0$:

$$f'(b) = \lim_{x \to b} \{f(x) - f(b)\} / (x - b)$$

In Chapter 3, when the limit definition was used to evaluate derivatives of simple functions, division by zero was avoided by algebraic manipulations. The key step was to factor a term $(x - b)$ out of the numerator, which could then cancel the $(x - b)$ term in the denominator before the limit was taken. A similar procedure can be used to evaluate and $0/0$ limit, at least in theory. For example, the limit

$$\lim_{x \to 2} [(x - 2)/\{x^2 - 6x + 8\}]$$

has the $0/0$ form as both the numerator and denominator approach zero as $x \to 2$. The limit is easily resolved by factoring and canceling:

$$\lim_{x \to 2} [(x - 2)/\{x^2 - 6x + 8\}] = \lim_{x \to 2} [(x-2)/\{(x-2)(x-4)\}] = \lim_{x \to 2} [1/(x - 4)] = -1/2$$

This approach, factoring and canceling, is straight forward if the term going to zero can obviously be factored. For more complicated functions factoring becomes more difficult or almost impossible using ordinary algebra. Fortunately, in many cases an alternative method is available. The derivative is used to factor and cancel! How this is done is subtle, but in essence this is the result of applying l'Hôpital's Rule, which does not even mention factoring. At the end of this section a *Rationale* is given to show how l'Hôpital's Rule is related to factoring.

L'HÔPTIAL'S RULE.

If (1) the $\lim_{x \to b}[f(x)/g(x)]$ is indeterminate, of the form $0/0$ or ∞/∞,

and (2) $g'(x) \neq 0$ for all $x \neq b$ in an open interval containing b, then

$$\lim_{x \to b}[f(x)/g(x)] = \lim_{x \to b}[f'(x)/g'(x)]$$

provided the limit of $f'(x)/g'(x)$ exists.

This Rule was initially discovered by the Swiss mathematician John Bernoulli (1667-1748), but was included in the first calculus text published by the Marquis de St. Monsieur, G. l'Hôpital (1661-1704) and consequently has since been inaccurately attributed to Monsieur l'Hôpital.

NOTE: The limit in l'Hôpital's Rule can be a one sided limit and the limiting x-value b may be a finite value or infinity. Thus the regular limit $\lim_{x \to b}$ may be replaced by any of the following:

$$\lim_{x \to b^+} \qquad \lim_{x \to b^-} \qquad \lim_{x \to \infty} \qquad \lim_{x \to -\infty}$$

Another important observation is that l'Hôpital's Rule can apply even when the limit of f'/g' is not finite. If any of the *infinity* limit rules discussed earlier apply to the $\lim_{x \to b} f'(x)/g'(x)$, the resulting limit value, be it finite or infinite, applies to the original $\lim_{x \to b} f(x)/g(x)$. In particular, if the hypothesis of l'Hôpital's Rule are satisfied and

$$\lim_{x \to b} f'(x)/g'(x) = +\infty \text{ or } -\infty \qquad \text{then} \qquad \lim_{x \to b} f(x)/g(x) = +\infty \text{ or } -\infty$$

Limits of the form $0/0$ and ∞/∞.

Example 1 **Evaluating indeterminate limits of the form $0/0$.**

Problem A Evaluate $\lim_{x \to 0}(e^x - 1)/x$.

Solution The limit is indeterminate of the form $0/0$. Let $f(x) = e^x - 1$ and $g(x) = x$. By l'Hôpital's Rule the original limit is the limit of the quotient of $f'(x) = e^x$ over $g'(x) = 1$.

$$\lim_{x \to 0}(e^x - 1)/x = \lim_{x \to 0} e^x/1 = e^0/1 = 1.$$

Problem B Evaluate the one-sided limit $\lim_{x \to 4^+} \sqrt{x - 4}/(x^2 - 16)$.

Solution This limit is indeterminate of the form $0/0$. Let $f(x) = \sqrt{x - 4}$ and $g(x) = x^2 - 16$. Then

$$f'(x) = 1/\{2\sqrt{x - 4}\} \quad \text{and} \quad g'(x) = 2x$$

In this case the limit of the ratio of derivatives is infinite:

$$\lim_{x \to 4+} f'(x)/g'(x) = \lim_{x \to 4+} 1/\{2\sqrt{x-4} \cdot 2x\} = +\infty$$

because the denominator approaches zero and the numerator is the positive constant 1. Applying l'Hôpital's Rule we conclude that

$$\lim_{x \to 4+} \sqrt{x-4}/(x^2 - 16) = +\infty.$$

Example 2 Repeated applications of l'Hôpital's Rule.

Problem Evaluate $L = \lim_{x \to 0} [\{e^{2x} - 2x - 1\}/ 5x^2]$.

Solution This limit is indeterminate of the form $0/0$. Let $f(x) = e^{2x} - 2x - 1$ and $g(x) = 5x^2$. Then, as

$$f'(x) = 2e^{2x} - 2 \quad \text{and} \quad g'(x) = 10x$$

the limit of the ratio of f'/g' is

$$\lim_{x \to 0} f'(x)/g'(x) = \lim_{x \to 0} (2e^{2x} - 2)/10x$$

As f' and g' are continuous with $f'(0) = 0$ and $g'(0) = 0$ this limit is also indeterminate of the form $0/0$. In such case a second application of l'Hôpital's Rule is justified. The limit of f'/g' is determined from the limit of f''/g''. As the second derivatives are

$$f''(x) = 4e^{2x} \quad \text{and} \quad g''(x) = 10$$

$$\lim_{x \to 0} f(x)/g(x) = \lim_{x \to 0} f'(x)/g'(x) = \lim_{x \to 0} f''(x)/g''(x) = \lim_{x \to 0} 4e^{2x}/10 = 0.4$$

Caution

> Be sure that the limit satisfies the hypothesis for l'Hôpital's Rule, namely that the limit has the indeterminate $0/0$ or ∞/∞ form, before you use l'Hôpital's Rule.
>
> If you apply l'Hôpital's Rule to quotients not of the $0/0$ or ∞/∞ form your answer will be wrong!
>
> Check that the limit of the quotient is of the $0/0$ form or the ∞/∞ form before you proceed to differentiate.

Example 3 Evaluating a limit of the form ∞/∞.

Problem Evaluate the $\lim_{x \to \infty} ln(x)/\sqrt{x}$.

Solution Since both $ln(x)$ and $\sqrt{x}$ become infinite as x approaches ∞, this limit is of the indeterminate form ∞/∞. L'Hôpital's Rule is applied twice to give

$$\lim_{x \to \infty} ln(x)/\sqrt{x} = \lim_{x \to \infty} (1/x) / \{1/[2\sqrt{x}]\}$$

$$= \lim_{x \to \infty} 2\sqrt{x}/x = \lim_{x \to \infty} 2/\sqrt{x} = 0$$

Limits of the form $0 \cdot \infty$.

If $\lim_{x \to b} f(x) = 0$ and $\lim_{x \to b} g(x) = \infty$ then $\lim_{x \to b}[f(x) \cdot g(x)]$ has the indeterminate limit form $0 \cdot \infty$. Such limits are converted to equivalent limits of quotients and then l'Hôpital's Rule is applied, if warranted. There are two quotients that are equivalent to a product, $f \cdot g$, obtained by inverting either of the functions. The equivalent limits are thus

$$\lim_{x \to b}\{f(x) / [1/g(x)]\} \quad \text{or} \quad \lim_{x \to b}\{g(x) / [1/f(x)]\}$$

Example 4 Evaluating indeterminate limits of the form $0 \cdot \infty$.

Problem   Evaluate $\lim_{x \to 0+} x \cdot ln(x)$.

Solution   As $\lim_{x \to 0+} ln(x) = -\infty$, this limit is of the indeterminate form "$0 \cdot \infty$". This problem can be approached in two ways, evaluating

$$\text{as } \lim_{x \to 0+} ln(x) / (1/x) \quad \text{or} \quad \lim_{x \to 0+} x / (1/ln(x))$$

The first limit is the better choice. Applying l'Hôpital's Rule, as indicated by $\text{\textcircled{L}}$, gives

$$\lim_{x \to 0+} x \cdot ln(x) = \lim_{x \to 0+} ln(x)/(1/x) \stackrel{L}{=} \lim_{x \to 0+}\{1/x\}/\{-1/x^2\} = \lim_{x \to 0+} -x = 0$$

The reason why the second option is not a good choice becomes obvious when you try to apply l'Hôpital's Rule. If $f(x) = x$ and $g(x) = \{1/ln(x)\}$, then the derivative

$$g'(x) = D_x\{1/ln(x)\} = -\{D_x ln(x)\}/[ln(x)]^2 = -1/\{x \cdot [ln(x)]^2\}$$

The limit of the quotients f'/g' for the second choice is then

$$\lim_{x \to 0+} f'(x)/g'(x) = \lim_{x \to 0+} 1/(-1/\{x[ln(x)]^2\}) = \lim_{x \to 0+} -x[ln(x)]^2.$$

This is a more complex limit than the initial problem. Only experience can tell you which term of a product should be inverted.

Limits of the "$\infty - \infty$" form.

When a limit involves a difference between two terms that become infinite, with the same sign, then the limit is indeterminate. To evaluate such limits the difference is first converted to a product by factoring a term out of the difference. The term is chosen so that the limit of the resulting product can either be evaluated using the extended arithmetic rules, or is a limit of the form "$0 \cdot \infty$" which is then converted to a quotient so that l'Hôpital's Rule can be applied.

Example 5   Indeterminate limits of the form "$\infty - \infty$".

Problem   Evaluate $\lim_{t \to \infty} t - ln(t)$.

Solution   The difference is factored as

$$t - ln(t) = t[1 - ln(t)/t]$$

The limit of this factored form can be evaluated using an *infinity* limit rule since

$$\lim_{t \to \infty} t = \infty \text{ and } \lim_{t \to \infty} 1 - ln(t)/t = 1$$

However, to establish the second limit requires an application of l'Hôpital's Rule since $\lim_{t \to \infty} ln(t)/t$ has the ∞/∞ form. Applying l'Hôpital's Rule gives

$$\lim_{t \to \infty} ln(t) / t = \lim_{t \to \infty} (1/t) / 1 = 0$$

Consequently,

$$\lim_{t \to \infty} 1 - ln(t)/t = 1 - \lim_{t \to \infty} ln(t)/t = 1$$

The limit of the difference is thus

$$\lim_{t \to \infty} t - ln(t) = \lim_{t \to \infty} t[1 - ln(t)/t] = \lim_{t \to \infty} t \cdot \lim_{t \to \infty} 1 - ln(t)/t = \infty \cdot 1 = \infty$$

Indeterminate limits of the form 0^0 and 1^∞.

Another type of indeterminate limit can occur when evaluating limits of involving exponential functions of the form

$$\lim_{x \to b} f(x)^{g(x)}$$

The two problem limits of this type have the indeterminate forms 0^0 and 1^∞.

The form 0^0 occurs when both f and g are positive and approach zero in the limit. The problem is pinpointed by considering limits in which only the base or only the exponent varies:

$$\lim_{x \to 0+} x^k = 0 \text{ for all } k > 0$$

and, what amounts in the limit to taking the "infinite root" of a number r,

$$\lim_{x \to 0+} r^x = 1 \text{ for } 0 < r < 1$$

The general situation in which both base and exponent functions vary simultaneously is therefor not obvious. To evaluate such limits, the exponential function is algebraically converted to a product, and the equivalent limit of product has the "$0 \cdot \infty$" form. This limit is then converted to a quotient limit problem and l'Hôpital's Rule can be applied.

The conversion of f^g to a product is made by employing properties of the exp(x) and $ln(x)$ functions:

$$f(x)^{g(x)} = \exp\{ ln[f(x)^{g(x)}] \} = \exp\{ g(x) \cdot ln[f(x)] \}$$

Since the exponential function $\exp(x) = e^x$ is continuous, the limit can pass from in front of the exp function into its exponent:

$$\lim_{x \to b} f(x)^{g(x)} = \exp\{ \lim_{x \to b} g(x) \cdot ln[f(x)] \}$$

The limit on the right is indeterminate of the form $0 \cdot \infty$ in each indeterminate case:

Case 1 $\lim_{x \to b} f(x) = 0+$ and $\lim_{x \to b} g(x) = 0$

Case 2 $\lim_{x \to b} f(x) = 1$ and $\lim_{x \to b} g(x) = \infty$

CHAPTER 4 APPLICATIONS OF DERIVATIVES

In these cases, apply l'Hôpital's Rule after taking the reciprocal of one of the factors.

Example 6 A limit of the 0^0 form.

Problem Evaluate the $\lim_{x \to 0+}(5x)^x$.

Solution The exponential - logarithmic transformation of f^g is followed by applying l'Hôpital's Rule.

$$\lim_{x \to 0+}(5x)^x = \exp\{\lim_{x \to 0+}[x \ln(5x)]\}$$

$$= \exp\{\lim_{x \to 0+} \ln(5x) / (1/x)\}$$

$$= \exp\{\lim_{x \to 0+}(5/5x) / (-1/x^2)\} \qquad \text{Applying l'Hôpital's Rule}$$

$$= \exp\{\lim_{x \to 0+} -x\} = e^0 = 1$$

Caution: Don't forget to evaluate the exponential at the limit once l'Hôpital's Rule has been applied.

Example 7 A limit of the 1^∞ form. A definition of the number e.

Problem Evaluate the $\lim_{x \to \infty}(1 + 1/x)^x$.

Solution The exponential - logarithmic transformation of f^g is followed by applying l'Hôpital's Rule.

$$\lim_{x \to \infty}(1 + 1/x)^x = \exp\{\lim_{x \to \infty}[x \cdot \ln(1 + 1/x)]\}$$

$$= \exp\{\lim_{x \to \infty}[\ln(1 + 1/x) / (1/x)]\}$$

$$= \exp\{\lim_{x \to \infty}[\{(-1/x^2)/(1 + 1/x)\} / (-1/x^2)]\} \qquad \text{Applying l'Hôpital's Rule}$$

$$= \exp\{\lim_{x \to \infty}[1 / (1 + 1/x)]$$

as $1/x \to 0$

$$= e^1 = e$$

The number e in more theoretical mathematical developments is often defined by this limit:

$$e \equiv \lim_{x \to +\infty}(1 + 1/x)^x \qquad \textbf{Definition of } e$$

Rationale for l'Hôpital's Rule

The following *rationale* is intended to help you understand why l'Hôpital's Rule works. It uses some assumptions that are more restrictive than is actually required and some *slight of hand* manipulations of infinite Taylor Series. On first reading, simply follow the argument looking for the canceling, without worrying about the mathematical justification of each step.

Assume that $\lim_{x \to b} f/g$ has the indeterminate form $0/0$ and that f and g are analytic functions on an interval containing b. Then they can be represented by a Taylor Series expansion about the limit value $x = b$. For f this series has the form

$$f(x) = f(b) + f'(b) \cdot (x - b) + (1/2)f''(b) \cdot (x - b)^2 + (1/6)f^{(3)}(b) \cdot (x - b)^3 + \ldots$$

f(x) is continuous and the assumption that the limit of f/g is of the form $0/0$ means that f(b) = 0.

$$\lim_{x \to b} f(x) = f(b) = 0 \text{, since } f'(x) \text{ must be continuous at } x = b.$$

This means that the constant term in the Taylor Series expansions of f is zero. A similar argument holds for the Taylor Series expansion of g(x). Now we employ a mathematical slight of hand. The quotient f/g is expressed as the quotient of the Taylor Series expansion of f and g (the terms f(b) and g(b) are absent since they are zero):

$$\frac{f(x)}{g(x)} = \frac{f'(b) \cdot (x - b) + (1/2)f''(b) \cdot (x - b)^2 + (1/6)f^{(3)}(b) \cdot (x - b)^3 + \ldots}{g'(b) \cdot (x - b) + (1/2)g''(b) \cdot (x - b)^2 + (1/6)g^{(3)}(b) \cdot (x - b)^3 + \ldots}$$

Each term in this expression contains a factor of (x - b) to some power. Hence, this factor could be canceled. For $n \geq 2$, canceling an (x - b) reduces $(x - b)^n$ to $(x - b)^{n-1}$. These terms still contain a factor (x - b) to some power after canceling. However, the (x - b) factor will cancel completely in the first two terms, leaving only f'(b) and g'(b). The factored quotient has the form

$$\frac{f(x)}{g(x)} = \frac{f'(b) + (1/2)f''(b) \cdot (x - b) + (1/6)f^{(3)}(b) \cdot (x - b)^2 + \ldots}{g'(b) + (1/2)g''(b) \cdot (x - b) + (1/6)g^{(3)}(b) \cdot (x - b)^2 + \ldots}$$

As $x \to b$ each term with a factor $(x - b)^n \to 0$. Evaluate the limit of the factored form by evaluating the limit of each term by itself leads to l'Hôpital's Rule:

$$\lim_{x \to b} f(x)/g(x) = \lim_{x \to b} f'(x)/g'(x) = f'(b)/g'(b)$$

provided g'(b) is not zero. If both g'(b) and f'(b) are zero, then the same process could be repeated. An additional factor of (x - b) could then be canceled from the quotient of series, the remaining terms would then all still have factors of (x - b) to a power except the f'' and g'' terms. Therefore, taking the $\lim_{x \to b}$ would give the quotient f''(b)/g''(b). This would be equivalent to applying l'Hôpital's Rule twice.

Exercises Set 4.10

Evaluate the indicated limits in Exercise 1 - 9

1 a) $\lim_{x \to 1} \{x^4 + 3x^2 - 6x + 2\} / \{x^3 - 2x^2 + 5x - 4\}$

   b) $\lim_{x \to 2} \{x^4 - 4x^3 + 16x - 16\} / \{x^4 - 4x^3 + 5x^2 - 4x + 4\}$

   c) $\lim_{x \to 3} \{(x-3)^5(x^2 + 2x - 1)\} / \{(x - 3)^6(x + 2)\}$

   d) $\lim_{x \to -1} \{x^3 + 3x^2 + 3x + 1\} / \{x^4 + 5x^3 + 9x^2 + 7x + 2\}$

2 a) $\lim_{x \to 0+} \sqrt{x} / \{x^2 - x\}$ b) $\lim_{x \to 2+} \text{SQRT}(x^2 - 4) / ln(x - 1)$

   c) $\lim_{x \to \infty} e^{-x} / x^{-3}$ d) $\lim_{x \to 1} \{x^3 - 1\} / \{1 - e^{x-1}\}$

   e) $\lim_{x \to 5} \{e^{5 - x} - 1\} / (x - 5)^2$ f) $\lim_{x \to 0+} \text{SQRT}(x) / ln(x)$

3 a) $\lim_{x \to \infty} \{4x^2 - 2x + 7\} / \{x^3 - 6x + 3\}$ b) $\lim_{x \to \infty} \{5x^3 - 2x + 1\} / (x + 1)^3$

   c) $\lim_{x \to \infty} \{(x+2)(x-3)^2\} / (x+5)^4$ d) $\lim_{x \to \infty} \{5x^5 - 4x^6 + 2\} / \{6x^4 + 7x^5 + 8x^6\}$

4 a) $\lim_{x \to \infty} ln(x)/x$ b) $\lim_{x \to \infty} x^2/e^x$ c) $\lim_{x \to \infty} e^x/x^2$

   d) $\lim_{x \to \infty} \{e^x + e^{-x}\} / e^{x^2}$ e) $\lim_{x \to \infty} e^{\sqrt{x}}/\{e^x - 1\}$ f) $\lim_{x \to \infty} \exp(e^x) / x^2$

5 a) $\lim_{x \to 0} xe^{-x}$ b) $\lim_{x \to 0} xe^{1/x}$ c) $\lim_{x \to 0+} \sqrt{x}\, ln(x)$

   d) $\lim_{x \to 3} (x^2 - 9)/(e^{x/3} - e)$ e) $\lim_{x \to 0+} (1 - \cos(x))\, ln(x)$ f) $\lim_{x \to \infty} ln(x+1)/(x-1)$

6 a) $\lim_{x \to \infty} x^{-x}$ b) $\lim_{x \to 0+} x^{-x}$ c) $\lim_{x \to 0+} x^{\sqrt{x}}$

   d) $\lim_{x \to 0} (1-x)^x$ e) $\lim_{x \to 1} (ln(x))^{x-1}$ f) $\lim_{x \to \infty} x^{1/x}$

   g) $\lim_{x \to \infty} (1 + 2/x)^x$ h) $\lim_{x \to 0} (x^2 + 1)^{(x^2)}$

7 a) $\lim_{x \to 0} \{3^x - 5^x\} / x$ b) $\lim_{x \to 0} \{5^x - 3^x\} / x^2$

   c) $\lim_{x \to \infty} \{3^x - 5^x\} / x$ d) $\lim_{x \to \infty} \{5^x - 3^x\} / x^2$

8 a) $\lim_{x \to 0+} [(1/x) + ln(x)]$ b) $\lim_{x \to 0+} [1/x - e^{-x}]$

   c) $\lim_{x \to \infty} [e^x - x^2]$ d) $\lim_{x \to 0+} [1/x^2 - 1/x]$

   e) $\lim_{x \to 0+} [2/x - 2/x^2]$ f) $\lim_{x \to 0+} [5/x^2 - 5/x]$

   g) $\lim_{x \to \pi/2} [\tan(x) + 1/(x - \pi/2)]$ h) $\lim_{x \to -\infty} e^{-2x} - 3^{-x}$

9 Which of the following limits may not evaluated using l'Hôpital's Rule? For these, indicate the actual limit and the erroneous limit, which is obtained by incorrectly using l'Hôpital's Rule.

   a) $\lim_{x \to 3+} ln(x-3)/(x-3)$ b) $\lim_{x \to 2} e^{-x}/(x-2)^2$

   c) $\lim_{x \to \infty} \cos(x)/x$ d) $\lim_{x \to 0} \cos(x)/x$

   e) $\lim_{x \to 0} \sin(\sqrt{x^2}))/x^2$ f) $\lim_{x \to 0} x \cdot \sin(x^{-2})$

   g) $\lim_{x \to \infty} e^x - x$ h) $\lim_{x \to \infty} (2x)^{\sin(x)}$

10 State l'Hôpital's Rule and the conditions under which it holds for the one-sided limits
    (a) $\lim_{x \to b+}$ and (b) $\lim_{x \to b-}$.

11 State l'Hôpital's Rule when $f(b) = f'(b) = g(b) = g'(b) = 0$.

Chapter 5 *Integration*

Section 5. Areas, averages and the definite integral. ... 345

Section 5. The Definite Integral as a limit. 353

Section 5. The Fundamental Theorem of Calculus 371
and the Indefinite Integral.

Section 5. Differentiating Integral Functions; 382
The Second Fundamental Theorem of Calculus.

Section 5. General Integral Rules. 390

Section 5. The Area between curves. 398

Prologue

Calculus is divided into two major divisions, **Differential Calculus**, which has been introduced in Chapters 3 and 4, and **Integral Calculus**, which is introduced in this Chapter and explore in more detail in Chapters 6 and 7. Integral calculus focuses on *integration*, a mathematical process that in a sense is the reverse of differentiation.

The *integral*, or *antiderivative*, of a function f(x) is denoted by $\int f(x)\,dx$.

An integral is a function, it is continuous and differentiable, even if the originating function f(x) is not. The *definite integral*, which involves the *integration of a function over an interval*, is similarly denoted with the endpoints of the interval indicated as limits of integration.

The definite integral of f(x) over [a,b] is denoted by $\int_a^b f(x)\,dx$.

A definite integral evaluates as a single number, a numerical value that reflects the function's value over the entire interval of integration. To evaluate integrals a collection of Rules are introduce, that parallel the Rules of differentiation. Students are normally expected to learn these rules, however they are easily programmed into computer packages and symbolic integration is available on hand calculators. Our intentions in this text is not that you "memorize" these rules, but that you gain an understanding of the basic definition and properties of integration, to the point where you can recognize where and why integration arises in the mathematical analysis of models in other disciplines. And, that you can therefor knowledgeably utilize programs that perform integration.

Historically, the definite integral was established very early. In some sense it was known to the Greeks and Egyptians who constructed the pyramids. The definite integral is defined in terms of special sums, which we now call Riemann Sums, after the 18th century mathematician George Riemann. The numerical value of a definite integral is defined as a limit of these Riemann Sums, a limit as the number of terms being added approaches infinity. Fortunately, for many functions the definite integral can be evaluated without actually forming Riemann Sums and without evaluating limits. This is because of a result that is so important that it is called the Fundamental Theorem of Calculus, or FTC for short. The FTC *Theorem* states that if a function F(x) has f(x) as its derivative, F'(x) = f(x), then the definite integral of f(x) over an interval [a,b] is evaluated as the difference F(b) - F(a).

$$\int_a^b f(x)\ dx = F(b) - F(a)\ \text{if}\ F'(x) = f(x).$$

In this case, the function F(x) is *an antiderivative* of f(x).

The Fundamental Theorem of Calculus at first appears like an elixir, a rule that enables the evaluation of any definite integral. This illusion arises because most "text book" exercises can be solved using the FTC. However, it is easy to create functions that can not be integrated using the FTC, i.e., an antiderivative function F(x) can not be written down using the elementary functions. In many real modeling applications, the integrals that arise can not be evaluated by the FTC. In such cases the only avenue of approach is to employ numerical approximations of the integral. These are based on the limit definition of the integral, the application of which can be a tedious process. However, to understand this limit process is important, more important than learning rules that are soon forgotten. In this chapter some simple examples illustrating the evaluation of particular integrals are presented using Riemann Sums and the closed forms of specific series (introduced in Chapter 2, Section 2.1).

Our presentation introduces the basic Rules, but focuses on the application, i.e., the interpretation of integrals. Often, with their first exposure to Calculus, students learn that the area under a curve can be evaluated as a definite integral and, too often, think that every definite integral is an area. Logically, a dog is an animal with four legs but not every four legged animal is a dog. Similarly, although areas can be evaluated as integrals, not every definite integral represents an area. One of our goals is for the reader to learn other interpretations of integrals, and to learn the process, which we call the *method of calculus*, by which different the mathematical analyses of different phenomenon or models result in definite integral evaluations; these applications and models then providing alternative interpretations of the definite integral.

In this chapter the fundamentals of Integral Calculus are introduced. In Section 5.1 the *definite integral* is related to the *average value of a function* and to the *area under the graph of a function*. These associations allow us to examine relationships and properties of definite integrals without actually indicating methods to evaluate the integrals. In Section 5.2 the definite integral is defined as a limit. The limit concept is the basis of calculus, but it does not provide a practical method to evaluate definite integrals. In Section 5.3 the *Fundamental Theorem of Calculus* is introduced to give a simple way to evaluate definite integrals, utilizing *anitderivatives* or *indefinite integrals* of the function being integrated. A Second Fundamental Theorem of Calculus, which states that *integral functions*, defined with variable limits of integration are differentiable, is explored in Section 5.4. Employing the Chain Rule, in Section 5.5, the simple Integral Rules are extended to form General Integration Rules which may be applied to integrate a wide range of functions. In Section 5.6 we return to the evaluation of areas between curves.

Integration is more challenging than differentiation! If you can write down a complicated function you can probably differentiate it, however, you likely can not integrate it! Some simple looking functions can prove very difficult or even impossible to integrate. In Chapter 6, special techniques are introduced, by which some integrals that can not be directly evaluated using the General Integral Rules can be transformed into integrals that can be evaluated. For those integrals that are impossible to evaluate functionally, numerical approximation methods must be used. These are also introduced in Chapter 6. Chapter 7 contains a selection of applications, grouped by subject, that involve integration. Some of these you will be able to work through with the simplest techniques of this Chapter, while others will require methods introduced in Chapter 6.

Section 5.1 Averages, Area, and Definite Integrals.

Introduction.

The *definite integral* of a function f(x) over an interval [a,b] is a single numerical value that *reflects* the function's values f(x) for all x-values in the interval. It is indicated by an integration symbol, $\int$, with the interval's left endpoint as a subscript and its right endpoint as a superscript, followed by the function and punctuated at the end with a *differential*, dx:

$$\int_a^b f(x)\,dx$$

The definite integral represents the accumulation of the values f(x) for all x between a and b. This accumulation can be interpreted in many ways, depending on the interpretation and units assigned to x and f(x). Two interpretations, which are explored in this section, are to associate $\int_a^b f(x)\,dx$ with the average value of the function f and with the area between the graph of f and the x-axis. With these associations, some basic properties and Rules for manipulating integrals can be established. The mathematical definition of the definite integral is given in the next Section 5.2, and simple methods for evaluating definite integrals are introduced in Section 5.3.

The Definite Integral: Basic Notation.

Let us begin with a comment about notation. The symbols $\int$ is called an **integral sign**, it originated as a script German letter "ess", S for Sum, since integrals are theoretically evaluated using sums, as we will see in Section 5.2. The symbol $\int$ is <u>always</u> paired with a **differential** symbol that reflects the independent variable; if the independent variable is x, the differential is *dx*, if the independent variable is t or s the corresponding differentials are *dt* or *ds*. The function f(x) is called the **integrand**. The definite integral of f(x) over [a,b],

$$\int_a^b f(x)\,dx$$

is read literally as

"the integral, from x equal a to x equal b, of f of x, dx"

or more simply, without indicating the variable x, as

"the integral from a to b of f(x)."

The endpoints of the interval [a,b] are referred to as the **limits of integration**. The number *a* is called the **lower limit of integration** and the number *b* is called the **upper limit of integration**.

The numerical value of an integral depends on the function and the limits of integration, but in fact the integrals's value does not involve the integration variable. The integration variable x and the corresponding differential symbol dx disappear when the integral is evaluated. The integration variable is sometimes called a *dummy variable*. If the "x" in f(x) is replaced by a different variable, and the differential dx replaced by a corresponding differential, the integral value remains unchanged. For the same function f and interval [a,b], each of the following definite integrals represents the same number:

$$\int_a^b f(x)\,dx \qquad \int_a^b f(t)\,dt \qquad \int_a^b f(u)\,du$$

Thus, for instance, if $f(x) = x^3 - 7x$ and $[a,b] = [1,5]$

$$\int_1^5 x^3 - 7x\,dx \;=\; \int_1^5 t^3 - 7t\,dt \;=\; \int_1^5 u^3 - 7u\,du$$

To **evaluate** a definite integral means to determine its numerical value. We will not discuss evaluation methods at this point but using intuitive concepts about areas and averages we will discuss some basic Integration Rules that allow us to relate integrals over various intervals.

Areas and averages.

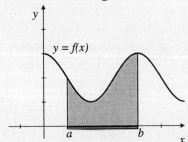

The definite integral is usually introduced for positive functions, in which case it can be interpreted as the *area between the graph of y = f(x) and the x-axis, over the interval [a,b]*. This approach utilizes intuitive notions of area and, as we shall see in Section 5.2, easily leads to the limit definition of the definite integral. However, in the world beyond calculus, applications involving definite integrals are often not concerned with areas. Indeed, when the unit associated with a function f(x) is not a measure of distance, the corresponding unit associated with the definite integral of f will not be an area!

An alternate interpretation of the definite integral is the product of "the average value of f(x) over [a,b]" and the length of the interval [a,b]. The concept of a function's *average* over an interval is a generalization of intuitive concepts of averages of numbers. The average of a set of numbers is their sum divided by how many numbers are being averaged. An average speed is simply the total distance traveled divided by the length of time spent traveling.

Exploring this analogy, consider a marathon runner who is observed at two times, 1hr and 2.5hr after the start of the race. At these times her position is 10 km and 19 km along the race course. What is her average speed over this time period? Obviously, it is (19 - 10)/ (2.5 - 1) = 6 km/hr. Did she actually run this fast throughout the time period? Of course not. She probably ran up hills slower and down faster. If she were actually monitored continuously and her speed at time t recorded as a function S(t), then the average of S(t) over [1, 2.5] would be $S^* = 6$. We know that there would be a maximum speed that she traveled, denoted S_M, and a minimum speed, S_m. Clearly we must have $S_m < 6 < S_M$, since 6 was her average speed. Did she actually ever travel at the average speed? Common sense tells us yes. Her speed varied between its minimum and maximum value and at some time she must have traveled at exactly 6 km/hr. This means there is at least one time t^* at which her speed $S(t^*) = 6$. We know that the time t^* exists but we can not determine its actual numerical value from the given information. Is this average speed associated with how far she ran over the time interval [1, 2.5]? Yes. Recall how the average was calculated. Solving for the distance covered we find

$$\text{Distance} = \text{Average Speed} \cdot \text{Elapsed Time}$$

However, it can be shown that the distance traveled can alternatively be evaluated as the definite integral of the speed function over the time interval:

$$\text{Distance} = \int_1^{2.5} S(t)\, dt$$

Notice that when the independent variable is t the integral notation is completed with the differential dt instead of dx. The Average of a function can be denoted using notation similar to that employed in computer programs, e.g., @AVG, or AVG[S,t,a,b]. We shall use

$$\text{AVG}\{S,a,b\} = \text{Average of S(t) over [a,b]}$$

Combining the last three equations, we obtain the relationship between a definite integral and the average of a function, which in this case is

$$\int_1^{2.5} S(t)\, dt = \text{AVG}\{S,1,2.5\} \cdot (2.5 - 1)$$

Definite Integral = Average * Length of Interval

Section 5.1 AVERAGES, AREAS, AND DEFINITE INTEGRALS

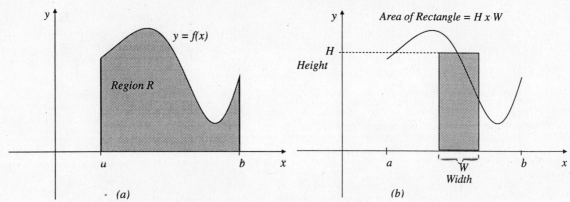

(a)   (b)

To establish the relationship between integrals, averages, and areas for more general functions we first consider a continuous function $y = f(x)$ that is positive, $f(x) > 0$ for $x \in [a,b]$. The region R under the graph of f over the interval [a,b] then has an area A. This region is bounded above by the function's graph, on the sides by the vertical lines $x = a$ and $x = b$, and below by the x-axis, as illustrated in the left figure above. In Section 5.2 we will establish that the area of the region R is given by the definite integral of f over [a,b].

$$\text{Area Under the graph of } f(x) \text{ over } [a,b] = \int_a^b f(x)\, dx$$

Our concept of area is based on an axiom, a fundamental definition, that the area of a rectangle is the product of its width, W, and its height, H,

$$A = H \times W \qquad \textbf{Area of a Rectangle}$$

Thus, if f is a constant function, $f(x) = C$ for $x \in [a,b]$, then the region under its graph is a rectangle with area $A = C \cdot (b - a)$. This tells us how to evaluate the definite integral of a constant:

Integral Rule : The integral of a constant.

> If C is a constant: $\qquad \int_a^b C\, dx = C \cdot (b - a)$

If $f(x)$ is not a constant function can the area under its graph be expressed as the area of a rectangle? The answer is yes! It is the area of a rectangle, with the x-axis interval [a,b] as its base and height equal to the average value of $f(x)$ over [a,b]. To establish this result consider the region R under the graph of f and the areas of the two rectangles indicated in Figure 4.2. Each rectangle has the interval [a,b] as its base and a height determined by the function f. The rectangle in panel (a) has height m, where

$$m = \text{ minimum value of } f(x) \text{ for } x \in [a,b]$$

This is the largest **inscribed** (meaning inside) rectangle under the graph of f. The rectangle in panel (b) has height M, where

$$M = \text{maximum value of } f(x) \text{ for } x \in [a,b]$$

This is the smallest **circumscribed** rectangle containing the region R. Observe that

$$m \le f(x) \le M \quad \text{for } x \in [a,b]$$

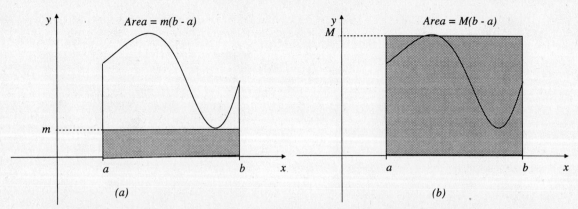

(a) (b)

From the above figure it is clear that the area A of the region R is bounded by the areas of the two rectangles:

Area of minimum Rectangle = $m(b - a) \leq A \leq M(b - a)$ = Area of maximum Rectangle

This inequality is consistent with the principle that the area of a region contained within a second region is less than or equal to the area of the second region.

For some number y^* between m and M, the area $A = y^* \cdot (b - a)$. But this is the area of a rectangle of height y^* and base [a,b]. The value y^* is none other than the average value of f over [a,b]. To show this, we define a simple function of y that gives the area of the rectangle of height y with base [a,b]:

$$A(y) \equiv (b - a) \cdot y$$

The function $A(y)$ is a linear function of the height y. It is continuous and as y increases from y = m to y = M the area function takes on each value between

$$A(m) = (b - a) \cdot m \quad \text{and} \quad A(M) = (b - a) \cdot M$$

Since the area A is between these two values, there must be a y-value y^* for which

$$A(y^*) = A$$

If we know the area A then this y-value is just

$$y^* = A/(b - a)$$

Graphically, you may visualize the rectangles with height y and base [a,b] by drawing horizontal lines, y = constant, for different values as illustrated at the right. Starting with

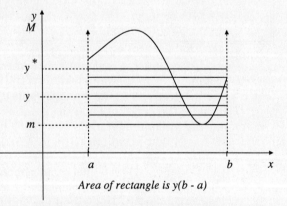

Area of rectangle is $y(b - a)$

height y = m, at which $A(m) < A$, increasing y increases the corresponding area $A(y)$. Eventually, for some y, $A(y)$ will equal A. This y-value, y^*, is the average y-value on the graph y = f(x):

$$y^* = AVG\{f,a,b\}$$

Which gives the

$$AVG\{f,a,b\} = A / (b - a)$$

Consider the areas between the curve y = f(x) and the horizontal line $y = y^*$ as illustrated at the left. Since the area under the graph of f is equal to the area of the rectangle of height y^*, the area above the

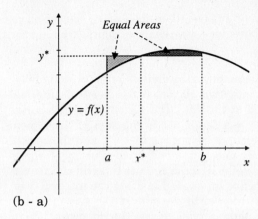

line $y = y^*$ and below the graph of f must equal the area below the line and above the graph of f. This balance of the function's values that are greater than y^* with those less than y^* agrees with the concept of the average as of a set of numbers being the number for which half the numbers are above and half below.

Replacing the area A by the integral of f over [a,b] leads to the relationship between integrals and average values of a function.

$$\int_a^b f(x)\,dx = y^* \cdot (b-a) \quad \text{or} \quad \int_a^b f(x)\,dx = \text{AVG}\{f,a,b\} \cdot (b-a)$$

There exists at least one x-value in the interval [a,b] at which $f(x) = y^*$. If x^* is one, then $f(x^*) = \text{AVG}\{f,a,b\}$, and

$$\int_a^b f(x)\,dx = f(x^*) \cdot (b-a)$$

This formula is very useful in mathematical theory. It expresses an integral as the product of the interval length and the "average" or "mean" value $y^* = f(x^*)$. This result is normally stated as a Theorem in which the unknown value x^* is usually indicated by the letter c.

THE INTEGRAL MEAN VALUE THEOREM

> If $f(x)$ is continuous on the interval [a,b],
> then there is a number $c \in [a,b]$ such that
>
> $$\int_a^b f(x)\,dx = f(c) \bullet (b-a)$$

This is the Integral theorem corresponding to the Mean Value Theorem introduced in Chapter 4, Section 4.7, that asserts the existence of a value c at which $f'(c) = [f(b) - f(a)]/(b-a)$. Observe that it is an "existence" theorem, one that tells us that something exists, but does not tell us how to find it. Such theorems are very important in developing mathematical theory but usually are seldom used in specific problems to find numerical values.

Rules about the limits of integration.

Although the properties discussed in the following are associated with areas, and hence positive functions, they are valid for any function f(x) for which the integrals exist.

Consider first the case of a "degenerate interval" [a,b], where the right and left endpoints are equal, b = a. The interval [a,a] is just a point on the x-axis. In this case, the region R under a graph of f over [a,a] is just a vertical line and hence has zero area. This is based on the axiom that a line is one-dimensional and has no area. Thus, we would conclude that the definite integral over [a,a] must be zero for any function f.

Integral Rule : Equal limits of integration.

> $$\int_a^a f(x)\,dx = 0$$

A fundamental axiom about areas is reflected in the adage "The whole is the sum of its parts."

In particular, if a region is divided into non-overlapping subregions, then the area of the region is the sum of the areas of all of the subregions. (The subregions may share a common boundary curve.)

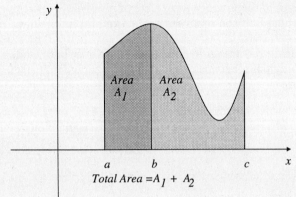

A simple example is the division of the region under a curve into two regions by a vertical line, as illustrated at the left. The region under the curve between x = a and x = c is the sum of the region under the curve from x = a to x = b and the region from x = b to x = c. The area of each of these regions is defined as the definite integral of f(x) over the respective interval. Equating the integral over the large interval to the sum of the two integrals over the subintervals gives the following Rule, which we shall refer to as the "Splitting Rule".

Integral Rule : Splitting limits of integration.

$$\int_a^c f(x)\, dx = \int_a^b f(x)\, dx + \int_b^c f(x)\, dx$$

The Splitting Rule is actually true for more general situations, for any three values a,b, and c, regardless of their order, provided the function f is integrable over each of the intervals. This Rule can be used to integrate functions that are multiply defined over different intervals.

Example 1   **Integrating a multiply defined function.**

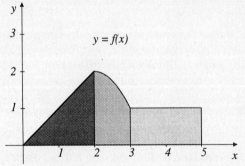

Problem   How would you determine the area under the graph of f(x) over the interval [0,5] when f is the multiply defined function

$$f(x) = \begin{cases} x & \text{if } x < 2; \\ 2 - (x-2)^2 & \text{if } 2 \le x \le 3; \\ 1 & \text{if } x > 3. \end{cases}$$

Solution   The area, illustrated at the left, is given by the definite integral

$$\int_0^5 f(x)\, dx$$

To evaluate this integral the interval [0,5] is split up at the points where the function's definition changes. Since the function is defined by different forms over the intervals [0,2], [2,3], and [3,5], these are used as the subintervals. Applying the Splitting Rule two times we write

$$\int_0^5 f(x)\, dx = \int_0^2 f(x)\, dx + \int_2^3 f(x)\, dx + \int_3^5 f(x)\, dx$$

Substituting the corresponding forms of the function on each subinterval gives

$$\int_0^5 f(x)\, dx = \int_0^2 x\, dx + \int_2^3 2 - (x-2)^2\, dx + \int_3^5 1\, dx$$

Specific Integration Rules for evaluating each of these integrals will be introduced in Section 5.3.

Section 5.1 — Averages, Areas, and Definite Integrals

When considering the area under a curve we tacitly assumed that a < b, which is because the positive direction on the x-axis corresponds to increasing x-values. This need not be the case, the "lower integration limit" can be numerically greater than the "upper integration limit". This is the case when the limits of integration are interchanged. Setting c = a in the Splitting Rule gives

$$\int_a^a f(x)\,dx = \int_a^b f(x)\,dx + \int_b^a f(x)\,dx$$

Solving for the integral from b to a and then using the fact that the integral from a to a is zero, gives

$$\int_b^a f(x)\,dx = \int_a^a f(x)\,dx - \int_a^b f(x)\,dx = -\int_a^b f(x)\,dx$$

Integral Rule : Interchange of limits.

$$\int_a^b f(x)\,dx = -\int_b^a f(x)\,dx$$

Example 2   Evaluating integrals over subintervals.

Problem   Use the Integration Rules given above to evaluate the indicated integrals when the function f(x) has the following specific integral values:

$$\int_1^3 f(x)\,dx = 5, \qquad \int_1^5 f(x)\,dx = 10, \qquad \text{and} \qquad \int_2^5 f(x)\,dx = 6.$$

(a) $\int_1^2 f(x)\,dx$ (b) $\int_2^3 f(x)\,dx$ (c) $\int_5^3 f(x)\,dx$ (d) $\int_4^4 f(x)\,dx$

Solutions   (a) Since the interval [1,5] = [1,2] ∪ [2,5],

$$\int_1^5 f(x)\,dx = \int_1^2 f(x)\,dx + \int_2^5 f(x)\,dx$$

Solving for the integral over [1,2] gives

$$\int_1^2 f(x)\,dx = \int_1^5 f(x)\,dx - \int_2^5 f(x)\,dx = 10 - 6 = 4$$

(b) Since the interval [1,3] = [1,2] ∪ [2,3],

$$\int_1^3 f(x)\,dx = \int_1^2 f(x)\,dx + \int_2^3 f(x)\,dx$$

Therefore,

$$\int_2^3 f(x)\,dx = \int_1^3 f(x)\,dx - \int_1^2 f(x)\,dx = 5 - 4 = 1$$

(c) By the Interchange of Limits Rule,

$$\int_5^3 f(x)\,dx = -\int_3^5 f(x)\,dx$$
$$= -\{\int_1^5 f(x)\,dx - \int_1^3 f(x)\,dx\} = -\{10 - 5\} = -5$$

(d) Since the limits of integration are equal,

$$\int_4^4 f(x)\,dx = 0$$

EXERCISE SET 5.1

1. Using the concept of the $\int_a^b f(x)\,dx$ as the AVG{f,a,b} · (b - a), which of the three relations ≤, =, or ≥ would make the mathematical sentence correct? Use your intuitive concept of the average of the integrand over the integration interval.

   For instance, since x^2 is positive for $x \neq 0$, the average of x^2 over any interval must be a positive number; the sentence $\int_2^5 x^2\,dx \bigcirc 0$ would be true for $\bigcirc$ replaced by ≥, as $\int_2^5 x^2\,dx \geq 0$.

   a) $\int_0^2 3x\,dx \bigcirc 0$ b) $\int_0^2 3x\,dx \bigcirc 6$

   c) $\int_0^2 3x\,dx \bigcirc \int_0^2 x\,dx$ d) $\int_0^2 3x\,dx \bigcirc 3\int_0^2 x\,dx$

   e) $\int_0^3 t^2 + 2\,dt \bigcirc 0$ f) $\int_0^3 t^2 + 2\,dt \bigcirc 2$

   g) $\int_0^3 t^2 + 2\,dt \bigcirc 12$ h) $\int_0^3 -t^2\,dt \bigcirc 0$

   i) $\int_0^3 t^2 + 4\,dt \bigcirc \int_0^3 t^2 + 2\,dt$ j) $\int_0^3 2t^2 + 4\,dt \bigcirc 2\int_0^3 t^2 + 2\,dt$

   k) $\int_1^4 x^3\,dx \bigcirc \int_1^4 x^2\,dx$ l) $\int_1^4 x^{-3}\,dx \bigcirc \int_1^4 x^{-2}\,dx$

   m) $\int_1^1 6t^2\,dt \bigcirc 0$ n) $\int_{-1}^1 x\,dx \bigcirc 0$

   o) $\int_{-4}^3 (x+1)^2\,dx \bigcirc \int_3^{-4} -(x+1)^2\,dx$

2. Express the area of the given region as a definite integral. Using geometric formulas can you state the value of the definite integral? Estimate upper and lower bounds for the integral's value by the area of the smallest rectangle containing the region and the largest rectangle inscribed within the region having the same x-axis interval as their base.

   a) Under the graph of $y = 1 + x$ over [1,3].

   b) Under the graph of $y = 10^3 x^2$ over [1,3]

   c) The top half of the region inside $x^2 + y^2 = 4$.

   d) The triangle with vertex (0,0), (3,0), and (3,6).

   e) The triangle formed by the x-axis, the line $x = 5$ and the line $y = 3(x - 2)$.

   f) Under the curve $y = 1 + \sin(x)$ over [0,2π].

3. Assume $\int_0^5 f(x)\,dx = 10$, $\int_2^5 f(x)\,dx = 7$, and $\int_3^5 f(x)\,dx = 4$. Determine

   a) $\int_0^2 f(x)\,dx$ b) $\int_0^3 f(x)\,dx$ c) $\int_3^2 f(x)\,dx$

4. Express as a single definite integral.

   a) $\int_0^2 t^2\,dt + \int_2^6 t^2\,dt$ b) $\int_2^5 (t+1)^2\,dt - \int_4^5 (t+1)^2\,dt$

   c) $\int_3^4 t + 7\,dt + \int_4^6 t + 7\,dt$ d) $\int_1^3 x^2 - 2x\,dx + \int_{-2}^1 x^2 - 2x\,dx$

   e) $\int_3^4 e^x\,dx - \int_4^3 e^x\,dx$ f) $\int_3^5 xe^x\,dx - \int_3^4 xe^x\,dx$

Section 5.2 The definite integral as a limit.

Introduction.

In this section we define the $\int_a^b f(x)\,dx$ as the limit of special types of sums, called *Riemann Sums*. Then, we apply *the method of calculus* to establish formulas to evaluate the area under a curve $y = f(x)$ over [a,b] and the average value of f(x) over [a,b]. For continuous functions these values are found to be limits, limits of Riemann Sums, having precisely the form involved in the definition of $\int_a^b f(x)\,dx$. Thus, the relationship between areas, averages and definite integrals discussed in Section 5.1 are established.

Limits of Riemann Sums are, in general, difficult to evaluate. To simplify their evaluation we introduce *uniform partition* notation. With uniform partitions integrals corresponding to areas and averages can be approximated using the *closed forms* of some simple series, introduced in Section 2.1. Taking the limit of these gives the exact value of the corresponding integrals. We shall see that this can be a difficult task, even for simple functions, and yet it is important that you understand this process. It forms is the basis for algorithms that numerically approximate integrals; a process that you may use frequently than you think in your future career, albeit, using a packaged computer program that will smooth over the nitty-gritty calculations.

To establish the integral formulas for areas and averages we use **the method of calculus**. This method consists of solving or analyzing a problem in three steps:

The method of calculus

I. The problem is solved for a **constant function**, $f(x) = c$ for $x \in$ [a,b].

II. The problem is then solved for a ***discrete* function** $f_N(x)$ that is constant on N subintervals of the interval [a,b]. This solution is the sum of terms, each like the solution for the constant function in step I. The sum is expressed in a special form involving the product of a function-related term and an increment Δx, which is the length of the subintervals.

III. For a **continuous function** f(x) the problem is solved by first constructing a sequence of discrete functions $\{f_N\}$ such that $\lim_{N \to \infty} f_N(x) = f(x)$. Then, the solution for f is found as the limit of the solutions for the discrete functions f_N, as given in step II. This limit is, by definition, a definite integral; thus, the solution is found by evaluating the integral.

When the *method of calculus* is applied, the resulting integral may be the integral of a specific function, or, as we shall see in the problem of evaluating area, it may be more complicated. In Chapter 7 many applications involving integration are introduced using *the method of calculus*. This method is very important and you should strive to understand its process so that you will recognize when integration arises in practical application problems long after you have forgotten the details and particular integration techniques that are memorized when taking a Calculus course.

Partitions and Riemann Sums.

The definite integral $\int_a^b f(x)\,dx$ is defined as a limit of special types of sums, whose general term has the form $f(x_i^*)\Delta x_i$. The first step towards defining the definite integral is to introduce the notation and vocabulary used to describe these terms. Let the interval of integration be [a,b]. To partition an interval is simply to divide the interval into subintervals. This is done in a systematic way so that the subintervals can be referred to by an index.

DEFINITION OF A PARTITION OF AN INTERVAL.

> A **partition of size N** of an interval [a,b] is an ordered set of N + 1 values, $P = \{x_0, x_1, x_2, \ldots, x_N\}$, with
>
> $$a = x_0 < x_1 < x_2 < \ldots < x_N = b.$$
>
> The partition determines N subintervals of [a,b]:
>
> $$I_1 = [x_0, x_1], \ I_2 = [x_1, x_2], \ \ldots, I_N = [x_{N-1}, x_N].$$
>
> The length of the n-th subinterval is
>
> $$\Delta x_n \equiv x_n - x_{n-1}.$$
>
> The **norm** of the partition is the maximum length of the subintervals:
>
> $$||P|| \equiv \text{maximum}\{\Delta x_n \mid i = 1, N\}$$

The seventh subinterval would be I_7, the subinterval $[x_6, x_7]$. The n-th subinterval I_n has the left-endpoint x_{n-1} and the right-endpoint is x_n, as illustrated in Figure 4.7. The index n represents any number between 1 and N, where N is the total number of subintervals. In some situations other letters will be used to represent a general subinterval. The i + 1st subinterval, I_{i+1}, would be formed by setting n = i + 1 and hence n - 1 = i, thus $I_{i+1} = [x_i, x_{i+1}]$.

A partition of [a,b] into N subintervals.

A **Riemann Sum**, named after the German mathematician George Riemann (1826 - 1866), is a sum associated with a function, a partition of an interval, and a sequence of evaluation points. In this text the evaluation points will be denoted with an asterisk and indexed with a subscript indicating the subinterval in which they lie. Corresponding to a partition P of size N we will create a set of N evaluation points

$$E = \{x_n^* \mid x_n^* \in I_n \text{ for } n = 1, 2, \ldots, N\}$$

The choice of the evaluation points is completely free, as long as

$$x_{n-1} \leq x_n^* \leq x_n$$

Consequently, there are an infinite number of different sets of evaluation points that can be considered for the same partition. Later in this section we will systematically select the evaluation points to simplify the calculations.

DEFINITION OF A RIEMANN SUM.

> The **Riemann Sum** corresponding to a function f(x), a partition $P = \{x_n\}_{n=0,N}$ of an interval [a,b], and a set of evaluation points $E = \{x_n^*\}_{n=1,N}$ has the form
>
> $$\Sigma_{n=1,N} \, f(x_n^*) \, \Delta x_n$$
>
> where
>
> $$\Delta x_n = x_n - x_{n-1}$$

The critical feature of a Riemann Sum is the form of each term, it must be the product of a function's value and the length of a subinterval. Each Riemann Sum represents a number, which depends on the function f, the partition P and the evaluation points E.

Example 1   Evaluating a Riemann Sum.

Problem   Evaluate the Riemann Sum corresponding to $f(x) = x^2 - 2x$, the interval [0,4], the partition given by $P = \{0, 1, 2, 2.5, 3, 3.25, 4\}$ and the set of evaluation points $E = \{0, 1, 2.25, 3, 3, 4\}$.

Solution   The partition P is of size $N = 6$. There is always one more partition point than the number of subintervals. The subintervals and their lengths corresponding to this partition are:

$I_1 = [0,1]$ $\Delta x_1 = 1 - 0 = 1$ $I_2 = [1,2]$ $\Delta x_2 = 2 - 1 = 1$
$I_3 = [2,2.5]$ $\Delta x_3 = 2.5 - 2 = 0.5$ $I_4 = [2.5,3]$ $\Delta x_4 = 3 - 2.5 = 0.5$
$I_5 = [3,3.25]$ $\Delta x_5 = 3.25 - 3 = 0.25$ $I_6 = [3.25,4]$ $\Delta x_6 = 4 - 3.25 = 0.75$

To evaluate the Riemann Sum the function must be evaluated at the evaluation points. These points are listed in order, so that $x_1^* = 0$, $x_2^* = 1$,..., $x_6^* = 4$. Notice that the evaluation points for the 4-th and 5-th subintervals are the same: $x_4^* = x_5^* = 3$, and that the evaluation point for the 3rd interval is not an endpoint of the interval I_3. The function values at these points are

$f(x_1^*) = f(0) = 0,$ $f(x_2^*) = f(1) = -1$ $f(x_3^*) = f(2.25) = 0.0625$

$f(x_4^*) = f(3) = 3$ $f(x_5^*) = f(3) = 3$ $f(x_6^*) = f(4) = 8$

The Riemann Sum is then

$\Sigma_{n=1,6} \, f(x_n^*)\Delta x_n = f(x_1^*)\Delta x_1 + f(x_2^*)\Delta x_2 + f(x_3^*)\Delta x_3 + f(x_4^*)\Delta x_4 + f(x_5^*)\Delta x_5 + f(x_6^*)\Delta x_6$

$= 0 \cdot 1 \; + \; -1 \cdot 1 \; + \; 0.0625 \cdot 0.5 \; + \; 3 \cdot 0.5 \; + \; 3 \cdot 0.25 \; + \; 8 \cdot 0.75 \quad\quad = 7.28125$

THE DEFINITE INTEGRAL.

To define $\int_a^b f(x) \, dx$ we consider the values of Riemann Sums corresponding to f(x) and interval [a,b] as the number N of subintervals increases. When N becomes large (think of dividing an interval into $N = 100$ or $N = 10,000$ subintervals), the corresponding Riemann Sums involve the addition of many terms (100 or 10,000). Besides the obvious computational problems, the question is "Will the values of these sums all be close to a specific number when N is very large?" By all, we mean all! For all possible choices of partitions and all possible choices of evaluation points. To indicate this we use limit notation like that used to denote the limits of sequences. However, you should be aware that this notation means much more, since as $N \to \infty$ we must include <u>all</u> possible Riemann Sums associated with partitions of size N. Don't worry, you won't actually have to this.

It can be shown that whenever the function we consider is continuous on a closed interval the limit will exist.

DEFINITION OF $\int_a^b f(x)\, dx$ AS A LIMIT.

The **definite integral of f(x) over [a,b]** is

$$\int_a^b f(x)\, dx \equiv \lim_{N \to \infty \text{ and } \max\{\Delta x^i\} \to 0} \Sigma_{i=1,N} f(x_i^*) \cdot \Delta x_i$$

if the limit exists.

The limit is taken over   all possible partitions of [a,b] of size N, $P = \{x_i\}_{i=0,N}$

and all possible choices for evaluation points $x_i^* \in [x_{i-1}, x_i]$.

The limit in the integral definition at first appears to be formidable. It is a much more complicated limit than the limits of sequences or functions, but, in practice, its generality proves very convenient. If the limit exists, then it must be the same for **all** possible partitions and evaluation points. This means that we can calculate the limit with a choice of partitions and evaluation points that makes the calculations simple. Before proceeding to evaluate integrals with this limit definition, let us first consider two problems whose solutions are given by definite integrals. For each problem we employ the *method of calculus* to establish the solution as a limit of a Riemann Sum, which by the above definition means as a definite integral.

Averages and weighted averages.

The **average** of a set of numbers is their sum divided by the number of numbers being averaged:

$$AVG\{y_1, y_2, y_3, \ldots y_N\} = \Sigma_{i=1,N}\, y_i / N \qquad \text{AVERAGE}$$

When the set of numbers includes the same number more than once then another method can be use to determine the average with fewer additions. A **weighted average** is calculated by adding the products of each distinct number multiplied by how frequently it occurs in the set being averaged. This sum is then divided by the total number of numbers in the set. However, this is just the sum of the frequencies of the distinct numbers. Formally.

If a set contains k distinct numbers, say $y_1, y_2, \ldots, y_k$,
and if the i-th number y_i occurs n_i times, then

$$AVG = \Sigma_{i=1,k}\, y_i \cdot n_i / \Sigma_{i=1,k}\, n_i \qquad \text{WEIGHTED AVERAGE}$$

The number n_i is the *weight* assigned to the i-th number. The sum in the denominator is always equal to N. The sum of the weights is just the total number of numbers being averaged.

$$\Sigma_{i=1,k}\, n_i = N$$

Example 2 **Average test score.**

Problem A class of 38 students took a test and achieved the following scores.

        18, 17, 19, 18, 14, 8, 18, 18, 16, 17, 15, 18, 19,
        17, 15, 16, 19, 18, 15, 19, 19, 16, 15, 17, 17, 19,
        16, 15, 15, 18, 17, 18, 19, 14, 19, 19, 15, 18

What is the average score on the test?

Solution A direct approach is to add up the 38 numbers and divide by 38. Alternatively, a weighted sum can be used. The seven distinct scores, their frequencies or weights, and the product of these are listed in a table:

Summary of individual test scores.

y_i	8	14	15	16	17	18	19	Total
n_i	1	2	7	4	6	9	9	38
$y_i \cdot n_i$	8	28	105	64	104	162	171	642

Using the data table, the Class Average is computed as a weighted average:

$$\text{Class Average} = [\Sigma_{i=1,7}\, y_i \times n_i]\, /\, 38 = 642 / 38 \approx 16.89$$

THE AVERAGE VALUE OF A FUNCTION.

The *average* of a function's values over an interval must be consistent with and generalize the concept of the average a set of numbers. A standard approach in mathematics is to define a concept for simple functions and then to extend this to more complicated functions. In this vein, *the average of a function f over an interval [a,b]*, denote by AVG{f,a,b}, is first defined for constant functions, then for discrete functions that have only finitely many values, and finally for continuous functions. The definition for continuous functions is given as a limit of averages of associated discrete functions, which approach the continuous function. Following the *method of calculus* this leads to the integral definition of a function's average.

The average of a constant function. STEP I.

If a function is constant on an interval, f(x) = c for all x in the interval, then intuitively its "average value" is simply the constant c. However, to bring the interval [a,b] into the discussion we observe that if f(x) = c for each x, in an intuitive sense the

"total of all the values f(x)" is the product $c \cdot (b - a)$

since the length b - a of the interval is a measure of the x-values in the interval. To get the average this product must be divided by the measure of the x-values, giving

$$\text{AVG}\{f(x) = c,\, a,b\} = c \cdot (b - a) / (b - a) = c$$

$$\text{AVG}\{f(x) = c,\, a,b\} = c \qquad \textbf{Average of a Constant function.}$$

The average of a discrete function. STEP II.

Assume the function f(x) is "discrete", and is constant on subintervals of the interval [a,b]. The average of f(x) is obtained as a weighted average of its constant values on these subintervals. To each distinct y-value, say y_i, a weight is assigned: it is the length of the interval(s) on which f(x) = y_i. The sum of the "weights" is then the sum of the lengths of the subintervals which is just the length of the interval [a,b].

More specifically, assume the function f is constant on N subintervals of an interval [a,b]. The endpoints of these subintervals are the set of x-values at which the graph of f(x) "jumps". They form a partition of the interval [a,b] which we denoted by

$$a = x_0 < x_1 < x_2 < \ldots < x_N = b$$

The function is constant on each of the N subintervals

358 CHAPTER 5 INTEGRATION

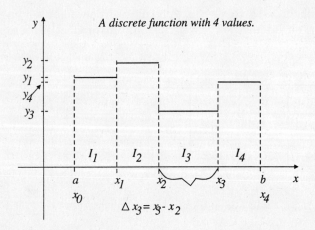

A discrete function with 4 values.

$I_1 = [x_0, x_1]$, $I_2 = [x_1, x_2]$, $I_3 = [x_2, x_3]$, ... , $I_N = [x_{N-1}, x_N]$

The n-th subinterval $I_n = [x_{n-1}, x_n]$ has length

$$\Delta x_n = x_n - x_{n-1}$$

Denote the constant value of f(x) on the subinterval I_n by y_n, then

$$f(x) = y_n \text{ if } x_{n-1} \leq x \leq x_n$$

The average of f(x) is defined to be the weighted average of these y_n values with the lengths of the corresponding subintervals as weights.

$$\text{AVG}\{f, a, b\} = \Sigma_{n=1,N} [y_n \cdot \Delta x_n] / \Sigma_{n=1,N} \Delta x_n$$

The sum in the denominator is just the total length of the interval [a,b]:

$$\Sigma_{n=1,N} \Delta x_n = b - a$$

Average of a Discrete Function f(x) over [a,b]

$$\text{AVG}\{f, a, b\} = [\Sigma_{n=1,N} y_n \cdot \Delta x_n] / (b - a)$$

Notice that the AVG{f,a,b} is given as a Riemann Sum divided by the interval length b - a.

Example 3 **Average of a discrete function.**

Problem A classical type of problem that you may have encountered as early as Jr. High School, is to determine the average speed for a trip given information about various segments of the trip. An example is the following:

"If Batman flew from Metropolis to Gotham City at 300 km/hr and immediately flew back to Metropolis at 600 km/hr what was his average speed for the round trip?

Solution The wrong approach is to average 300 and 600 to obtain the incorrect answer, 450 km/hr. The problem is that the average must be over a time interval that is not specified. To begin, assume the speed is given by a discrete function that has two values, $y_1 = 300$ and $y_2 = 600$. Assume the trip began at $t_0 = 0$, and Batman reached Gotham City at time t_1 hrs, returning to Metropolis at time t_2 hrs. Then the speed of his trip is given by the discrete function

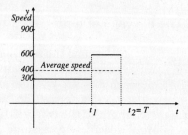

$$f(t) = \begin{cases} 300 & \text{if } 0 < t < t_1 \\ 600 & \text{if } t_1 < t < t_2 \end{cases}$$

The total time for the trip was $T = t_2$ hrs. The interval [0,T] is partitioned into two subintervals, $[t_0, t_1]$ and $[t_1, t_2]$. The average speed is the average of f over [0,T]:

$$\text{AVG}\{f,0,T\} = \{y_1 \times \Delta t_1 + y_2 \times \Delta t_2\} / \{\Delta t_1 + \Delta t_2\}$$

The next step is to establish a relationship between the times periods Δt_1 and Δt_2 and the unknown distance D between the two cities. We assume that he traveled the same distance D in each direction. Then D is the product of the average speed for each direction and the elapsed time traveled:

$$D = 300 \times \Delta t_1 \quad \text{and} \quad D = 600 \times \Delta t_2$$

Solving these for the time of each flight segment gives:

$$\Delta t_1 = D/300 \quad \text{and} \quad \Delta t_2 = D/600$$

Substituting these Δt values and the y_1 and y_2 values into the AVG equation yields an equation that can be simplified, the D terms canceling, to yields the average speed:

$$\text{AVG}\{f,0,T\} = \{300 \times D/300 + 600 \times D/600\} / \{D/300 + D/600\} = 2D / \{3D/600\} = 400$$

The average speed for the round trip was 400 km/hr.

☑

The average of a continuous function. STEP III.

Assume f(x) is a continuous function over the interval [a,b]. For the third step of the *method of Calculus*, we define a sequence of discrete functions $\{f_N\}$ such that $f_N \to f$ as $N \to \infty$. Then, we define

$$\text{AVG}\{f,a,b\} = \lim_{N \to \infty} \text{AVG}\{f_N,a,b\}$$

As the average of each discrete function $\text{AVG}\{f_N,a,b\}$ is a Riemann Sum divided by a constant, this constant times $\text{AVG}\{f,a,b\}$ is found as the limit of Riemann Sums, which, by its definition, is the definite integral of f.

More specifically, assume that f(x) is continuous on [a,b] and define for each natural number N a discrete function $f_N(x)$ as follows:

1. Choose any partition of [a,b] of size N. This creates N subintervals

   $I_n = [x_{n-1}, x_n]$ for n = 1, 2, ... ,N

2. In each subinterval I_n pick an evaluation point, x_n^*.

3. Define $f_N(x) \equiv f(x_n^*)$ if $x_{n-1} \leq x < x_n$. Don't worry about the values at the endpoints.

Each discrete function f_N approximates the function f, and thus the average of f is approximated by the average of f_N over [a,b]:

$$\text{AVG}\{f,[a,b]\} \approx \text{AVG}\{f_N,a,b\} = [\Sigma_{n=1,N} f(x_n^*) \cdot \Delta x_n] / (b-a)$$

Since the function f is assumed to be continuous, it can be shown that if the partitions are chosen so that $\max\{\Delta x_i\} \to 0$ as $N \to \infty$, then $f_N(x) \to f(x)$ for each x in the interval. Furthermore, it follows that the $\text{AVG}\{f_N,a,b\}$ approach a single constant value as $N \to \infty$. We therefore define $\text{AVG}\{f,a,b\}$ to be this limiting value.

DEFINITION OF THE AVERAGE VALUE OF F(X) OVER [A,B].

> The **average value of f over [a,b]** is
>
> $$\text{AVG}\{f,a,b\} = \lim_{N \to \infty} \left[\sum_{n=1,N} f(x_n^*) \cdot \Delta x_n\right] / (b - a)$$
>
> where $P_N = \{x_n\}$ is a partition of size N of $[a,b]$ with evaluation points
>
> $x_n^* \in [x_{n-1}, x_n]$ and $\|P_N\| \to 0$ as $N \to \infty$

Comparing this definition and the definition of the definite integral, we see that the definite integral is the product of the average function value and the length of the interval. Thus

$$\int_a^b f(x)\, dx = \text{AVG}\{f,a,b\} \cdot (b - a)$$

or

$$\text{AVG}\{f,a,b\} = \int_a^b f(x)\, dx / (b - a)$$

AREA AS A DEFINITE INTEGRAL.

As introduced in Section 5.1, we start with the definition of area based on the area of a rectangle of height H and length L, $A = L \cdot W$. Using this as our definition of area we will apply *the method of calculus* to establish a definition of the area under the graph of a continuous function. We will assume that all functions considered in the following discussion are positive, i.e. that $f(x) > 0$ on the interval $[a,b]$. The area of the region under the graph of a constant function will simply be the area of a rectangle. The area under the graph of a discrete function will be simply the sum of areas of rectangles; this sum will have the form we call a Riemann Sum. Then, the area under the graph of a continuous function will be given as a limit of areas of associated discrete functions; being the limit of Riemann Sums it will by definition be equal to the definite integral of f over $[a,b]$.

Let us introduce the notation $\text{AREA}\{f,a,b\}$ to indicate the area of the finite region bounded by the curves $y = f(x)$, $y = 0$ (the x-axis), $x = a$, and $x = b$.

We will rely on two basic principles that state axioms or assumptions about the concept of area.

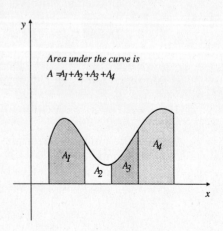

Area under the curve is
$A = A_1 + A_2 + A_3 + A_4$

Principle 1   *The whole is the sum of its parts.* This means that if a region is divided into non-overlapping subregions, then the area of the region is the sum of the area of the subregions.

Principle 2   *If a continuous function f is the limit of a sequence of functions $\{f_N\}$ then the area under the graph of f is the limit of the areas under the graphs of the functions f_N.*

Section 5.2 — The Definite Integral As A Limit

The area under a constant function. STEP I.

If $f(x) = c$, a positive constant, for $x \in [a,b]$, then the region under the graph of f is a rectangle. Its height is the function value c and its base length is the length of the interval [a,b]. Thus its area is simply the product of these two numbers.

$$\text{AREA}\{f(x) = c, a, b\} = c \cdot (b - a)$$

The area under a discrete function. STEP II.

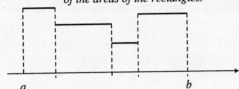

The area under a discrete function is the sum of the areas of the rectangles.

Assume that f is a discrete function that is constant on N subintervals of [a,b]. The endpoints of these subintervals forms a partition of [a,b], $\{x_i\}_{i=0,N}$. Denote the function's value on the i-th subinterval by y_i. Then

$$f(x) = y_i \text{ for } x_{i-1} < x < x_i$$

The region under the graph of f consists of N rectangles. By the first *principle* area is the sum of the areas of these rectangles. The area of the rectangle over the interval $I_i = [x_{i-1}, x_i]$ is

$$A_i = y_i \cdot \Delta x_i$$

Consequently, the area under the graph of f is

$$\text{AREA}\{f, a, b\} = \Sigma_{i=1,N}\, y_i \cdot \Delta x_i$$

The sum can be viewed as a Riemann Sum, using the index i instead of n, since for any choice of evaluation point x_i^* in the i-th subinterval, $[x_{i-1}, x_i]$, $f(x_i^*) = y_i$. This area could thus be expressed as

$$\text{AREA}\{f, a, b\} = \Sigma_{i=1,N}\, f(x_i^*) \cdot \Delta x_i$$

The area under a continuous function. STEP III.

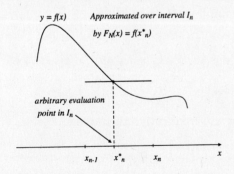

Assume that f is a continuous function on [a,b]. Then for each natural number N we approximate $f(x)$ by a discrete function $f_N(x)$ defined as follows. First, form a partition $\{x_n\}$ of [a,b] of size N, and defining $f_N(x)$ on each subinterval of the partition to equal the value of f at some point in the subinterval. If the point chosen in the n-th subinterval is denoted by x_n^* then set $y_n = f(x_n^*)$. The function f_N is then a discrete function having the form discussed above. Note that

$$f_N(x_n^*) = f(x_n^*) \text{ for each evaluation point}$$

By the second *Principle* the area under the graph of f is given as the limit of the areas under the graphs of these discrete functions.

$$\text{AREA}\{f, a, b\} = \lim_{N \to \infty} \text{AREA}\{f_N, a, b\} = \lim_{N \to \infty} \Sigma_{i=1,N}\, f_N(x_i^*) \cdot \Delta x_i$$

We formalize this result as the definition of the area under a curve.

DEFINITION OF THE AREA UNDER A CURVE.

> If f(x) is continuous and non-negative on the interval [a,b],
> then the **area under the graph of f over [a,b]** is
>
> $$A\{f,a,b\} \equiv \lim_{\Delta x \to 0 \text{ and } N \to \infty} \sum_{n=1,N} f(x_n^*) \Delta x_n$$
>
> for any partition $\{x_n\}$ and set of evaluation points $\{x_n^*\}$.

Combining this definition of the area under the graph of f with the definition of the definite integral of f over [a,b], we conclude the validity of the statement made in Section 5.1:

$$\text{Area under the graph of f over } [a,b] = \int_a^b f(x)\, dx$$

Uniform Partitions.

To simplify the notation and standardize the type of Riemann Sums we introduce the concept of a "uniform" partition. This is simply one in which each subinterval has the same length.

DEFINITION OF A UNIFORM PARTITION.

> A **uniform** partition of [a,b] into N subintervals is given by the points:
>
> $$x_n = a + n \cdot h, \quad \text{for } n = 0, 1, 2, \ldots, N$$
>
> the **mesh size of the partition, h,** is the length of each subinterval:
>
> $$\Delta x_n = h \equiv (b-a)/N.$$

For instance, a uniform partition of [3,7] into N = 5 subintervals has mesh size h = (7 - 3)/5 = 0.8. The 6 partition points are then:

$x_0 = a$	$x_1 = a + h$	$x_2 = a + 2h$	$x_3 = a + 3h$	$x_4 = a + 4h$	$x_5 = a + 5h$
$x_0 = 3$	$x_1 = 3.8$	$x_2 = 4.6$	$x_3 = 5.4$	$x_4 = 6.2$	$x_5 = 7$

To simplify the selection of function evaluation points we will frequently utilize the endpoints of the subintervals.

The **left endpoint** of the n-th subinterval I_n is $x_{n-1} = a + (n-1)h$.

The **right endpoint** of the n-th subinterval I_n is $x_n = a + nh$.

For instance, the left endpoint of I_5 is $x_4 = a + 4h$, while its right endpoint is $x_5 = a + 5h$.

The limit formulas for evaluating an integral can be simplified in form by choosing specific partitions and evaluation points. Uniform partitions of mesh size h are easily established. As the interval lengths Δx_n all equal h, the Riemann Sum associated with uniform partitions can be simplified by factoring out h. They can be standardized further by specifying the evaluation points as endpoints of the subintervals. Setting $x_n^* = x_n$, the right endpoint of I_n gives the

RIGHT ENDPOINT FORMULA

$$\int_a^b f(x)dx = \lim_{N \to \infty} h \sum_{n=1,N} f(x_n) = \lim_{N \to \infty} h \sum_{n=1,N} f(a + nh)$$

Setting $x_n^* = x_{n-1}$, the left endpoint of the subinterval I_n, gives the

LEFT ENDPOINT FORMULA

$$\int_a^b f(x)dx = \lim_{N \to \infty} h \sum_{n=1,N} f(x_{n-1}) = \lim_{N \to \infty} h \sum_{n=1,N} f(a + (n-1)h)$$

Before we apply the limit formulas to evaluate integrals let us explore how the formulas without the limits can be used to approximate integrals.

APPROXIMATING DEFINITE INTEGRALS

The $\int_a^b f(x) \, dx$ is defined as a limit of Riemann Sums, but it can be approximated by evaluating specific Riemann Sums for a finite number N. Thus

$$\int_a^b f(x) \, dx \approx \sum_{i=1,N} f(x_i^*) \cdot \Delta x_i \qquad \textbf{Approximation formula.}$$

The approximation formulas derived by using the Left and Right endpoints of a uniform partition are simple to evaluate. These approximation formulas are sometimes more easily remembered when expressed in expanded form.

Approximating the integral using right endpoints:

$$\int_a^b f(x) \, dx \approx ((b-a)/N) \, [f(a+h) + f(a+2h) + \ldots + f(b)]$$

Approximating the integral using left endpoints:

$$\int_a^b f(x)dx \approx ((b-a)/N)) \, [f(a) + f(a+h) + f(a+2h) + \ldots + f(a + (N-1)h)]$$

These approximations give the area of the rectangles illustrated in the following figures.

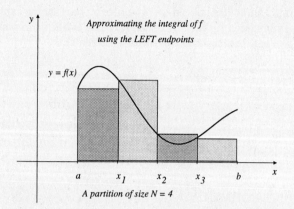
A partition of size N = 4

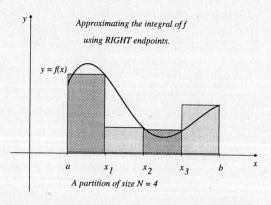
A partition of size N = 4

Example 4 Approximating an integral using right and left endpoints.

Problem Estimate $\int_0^5 2^x \, dx$ using a uniform partition of size N = 5. Compare the estimates obtained using the left and right endpoint formulas.

Solution The mesh size is h = (5 - 0)/5 = 1. The partition points are {a = 0, 1, 2, 3, 4, 5 = b}. The left endpoint approximation is thus

$$\int_0^5 2^x \, dx \approx h[f(0) + f(1) + f(2) + f(3) + f(4)]$$

$$= (1)[2^0 + 2^1 + 2^2 + 2^3 + 2^4] = 31$$

The right endpoint approximation is

$$\int_0^5 2^x \, dx \approx h[f(1) + f(2) + f(3) + f(4) + f(5)]$$

$$= (1)[2^1 + 2^2 + 2^3 + 2^4 + 2^5] = 63$$

Because the graph of f is increasing throughout the interval [0,5] the actual integral value must be between these two approximations. They provide bounds for the integral, but not very accurate bounds. In Section 5.5 a method will be introduced to evaluate this integral exactly, its actual value is 31/ln(2) or approximately 44.7.

Example 5 Approximating an integral using different size partitions.

Problem Compare the approximations of $\int_1^3 2x \, dx$ using uniform partitions of size N = 4 and N = 10. Use both the left and right endpoint formulas.

Solution First, for N = 4 the mesh size is h = (b - a)/N = (3 - 1)/4 = 0.5 and the uniform partition points are:

$x_0 = 1$ $x_1 = 1 + h = 1.5$ $x_2 = 1 + 2h = 2.0$ $x_3 = 1 + 3h = 2.5$ $x_4 = 1 + 4h = 3$

Using the left endpoint formula gives:

$$\int_1^3 2x \, dx \approx (0.5)[2(1.0) + 2(1.5) + 2(2.0) + 2(2.5)] = 7$$

Using the right endpoint formula gives the approximation:

$$\int_1^3 2x\,dx \approx (0.5)[2(1.5) + 2(2.0) + 2(2.5) + 2(3.0)] = 9$$

These are illustrated in the left panel of the following figure. Observe that because y = 2x is increasing, the actual integral value is between the upper bound given by the right endpoints and the lower bound given by the left endpoints (the shaded rectangles).

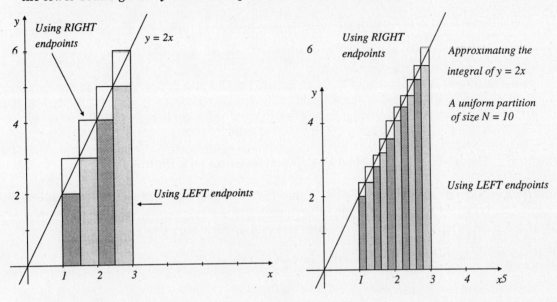

Next, consider the same integral approximated using a uniform partition of size N = 10, i.e., with mesh size

$$h = (b - a)/N = (3 - 1)/10 = 0.2$$

The uniform partition points are given by the equation

$$x_n = 1 + 0.2n \quad \text{for} \quad n = 0, 1, 2, \ldots, 10$$

Using the left endpoints, $x_n^* = x_{n-1} = 1 + 0.2(n - 1)$, the approximation is

$$\int_1^3 2x\,dx \approx (0.2)\Sigma_{n=1,10}\, 2(1 + 0.2(n - 1))$$

$$= (0.2)\Sigma_{n=1,10}\, (1.6 + 0.4n)$$

$$= (0.2)[1.6(10) + 0.4(10 \cdot 11)/2]$$

$$= (0.2)[16 + 22] = 7.6$$

Using the closed forms:
$\Sigma n = N(N+1)/2$
$\Sigma k = k \cdot N$

Repeating the same procedure with the right endpoint formula gives:

$$\int_1^3 2x\,dx \approx (0.2)\Sigma_{n=1,10}\, 2(1 + 0.2n)$$

$$= (0.2)\Sigma_{n=1,10}\, [2 + 0.4n]$$

$$= (0.2)[2 \cdot 10 + 0.4(10 \cdot 11/2)]$$

$$= (0.2)[20 + 22] = 8.4$$

These are illustrated in the right panel of the above figure. The approximations of the integral $\int_1^3 2x\,dx$ using N = 4 are 7 and 9 while the approximations for N = 10 are 7.6 and 8.4. The actual area of the trapezoid under the graph of y = 2x over the interval [1,3], which is 8. Thus the approximations with N = 10 are much more accurate than those for N = 4.

EVALUATING INTEGRALS AS LIMITS

The difficult task in evaluating an integral as a limit is to represent the Riemann Sum

$$\Sigma_{n=1,N}\, f(x_n^*)\, \Delta x_n$$

in a *closed form* as a function of N. In Section 2.1 of Chapter 2, Specific Closed Forms were given for only eight special types of sums. Using only these forms severely restricts the type of function f(x) for which you can evaluate $\int_a^b f(x)\,dx$ as a limit. The following examples illustrate this process.

Example 6   **Evaluating the integral of a linear function as a limit.**

Problem   Evaluate the integral $\int_1^3 5 + 2x\,dx$ as a limit. Use both the right and left endpoint formulas.

Solution   For an arbitrary integer N the mesh size of a uniform partition of [1,3] is

$$h = (b - a)/N = (3 - 1)/N = 2/N$$

and the partition points are

$$x_n = 1 + n \cdot h \quad \text{for} \quad n = 0,1,2,\ldots,N$$

The right endpoint evaluation points are $x_n^* = 1 + nh$ and the function is f(x) = 5 + 2x:

$$\int_1^3 5 + 2x\,dx = \lim_{N \to \infty} h \cdot \Sigma_{n=1,N}\, f(1 + nh)$$

$$= \lim_{N \to \infty} (2/N) \cdot \Sigma_{n=1,N}\, 5 + 2 \cdot (1 + n(2/N))$$

$$= \lim_{N \to \infty} 2/N \cdot \Sigma_{n=1,N}\, (7 + 4n/N)$$

$$= \lim_{N \to \infty} 2/N \cdot [7\, \Sigma_{n=1,N} 1\, + (4/N)\, \Sigma_{n=1,N} n]$$

$$= \lim_{N \to \infty} 2/N\, [7 \cdot N\, + (4/N)\, [N \cdot (N+1)/2]\,]$$

$$= \lim_{N \to \infty} 14 + 4(N+1)/N\, = \lim_{N \to \infty} 18 + 4/N\, = 18$$

The series were simplified using the close form formulas for the sum of a constant and the sum of the first N integers:

$$\Sigma_{n=1,N} 1 = N \quad \text{and} \quad \Sigma_{n=1,N} n = N(N+1)/2$$

Next we evaluate the integral using the Left endpoint formula. We use the evaluation point

$$x_n^* = x_{n-1} = 1 + h(n-1) = 1 - h + nh$$

in the general limit formula. This time we do not substitute 2/N for h until after the series have been expressed in closed form.

The integral is evaluated as

$$\int_1^3 5 + 2x\, dx = \lim_{N \to \infty} \left\{ h \cdot \Sigma_{n=1,N} f(1 - h + nh) \right\}$$

$$= \lim_{N \to \infty} \left\{ h \cdot \Sigma_{n=1,N} 5 + 2(1 - h + nh) \right\}$$

$$= \lim_{N \to \infty} \left\{ h \cdot \Sigma_{n=1,N} 7 - 2h + 2nh \right\}$$

$$= \lim_{N \to \infty} \left\{ h \cdot [(7 - 2h)\Sigma_{n=1,N} 1 + 2h\Sigma_{n=1,N} n] \right\}$$

$$= \lim_{N \to \infty} h \cdot [(7 - 2h)N + 2h \cdot N(N+1)/2] \qquad \text{Recall } h = 2/N$$

$$= \lim_{N \to \infty} 2/N \cdot [(7 - 4/N)N + 4/N \cdot N(N+1)/2]$$

$$= \lim_{N \to \infty} 14 - 8/N + 4(N+1)/N$$

$$= \lim_{N \to \infty} 18 - 4/N = 18$$

Notice that the limits are the same using either the left or right endpoints!

Example 7 **Computing the average value of a function as a limit using a uniform partition.**

Problem Determine the average value of $f(x) = 0.1x^2$ over $[0,4]$. Use a uniform partition.

Solution A uniform partition of size N has the mesh size

$$h = (4 - 0)/N = 4/N$$

The right endpoint of each subinterval is chosen as the evaluation point:

$$x_n^* = x_n = 0 + n \cdot h = 4n/N$$

The function f is evaluated at these points as

$$f(x_n^*) = 0.1(4n/N)^2 = 1.6\, n^2/N^2$$

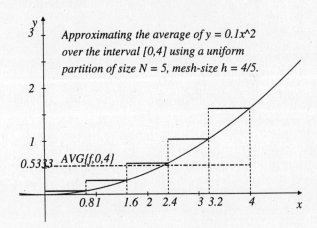

Approximating the average of $y = 0.1x^2$ over the interval $[0,4]$ using a uniform partition of size $N = 5$, mesh-size $h = 4/5$.

The discrete approximation function f_N is defined by

$$f_N(x) = 1.6\, n^2/N^2 \quad \text{if} \quad 4(n-1)/N < x < 4n/N$$

The average of f is obtained as the limit as $N \to \infty$ of the $AVG\{f_N, 0, 4\}$. We first evaluate the $AVG\{f_N, 0, 4\}$ in closed form and then take the limit of N to ∞.

$$AVG\{f_N, 0, 4\} = [\Sigma_{n=1,N} f_N(x_n^*) \cdot 4/N]/(4-0)$$

$$= 1/4\, \Sigma_{n=1,N} 1.6 n^2/N^2 \cdot 4/N]$$

$$= (1.6/N^3) \Sigma_{n=1,N} n^2 \qquad\qquad \text{$1/N^3$ can be factored out as N is a constant not involved in the summation.}$$

$$= (1.6/N^3) \cdot N(N+1)(2N+1)/6 \qquad \text{Using the closed form of the Sum of Squares series.}$$

$$= (1.6/N^3) \cdot (2N^3 + 3N^2 + N)/6 \qquad\qquad \text{Multiplying out.}$$

$$= (1.6/6)(2 + 3/N + 1/N^2)$$

Thus, since $1/N \to 0$ as $N \to \infty$, the average value is given by the limit

$$\text{AVG}\{0.1x^2, 0, 4\} = \lim_{N \to \infty} \text{AVG}\{f_N, 0, 4\}$$

$$= \lim_{N \to \infty} (1.6/6)(2 + 3/N + 1/N^2) = 3.2/6 = 0.53\underline{3}$$

☑

Example 8 **Evaluating a definite integral as a limit using uniform partitions**.

Problem Evaluate $\int_0^5 x^2\, dx$ using the limit definition of the integral; use uniform partitions and the left endpoint of each subinterval as the evaluation point.

Solution First, we form a uniform partition of [0,5] into N subintervals. The mesh size is then $h = 5/N$ and the partition points are

$$x_n = 0 + n \cdot h = 5n/N, \quad n = 1, 2, \ldots, N$$

Setting the evaluation point x_n^* equal to the left endpoint of the interval I_n, we have

$$x_n^* = 0 + (n - 1)h = (5n - 5)/N$$

The definite integral is evaluated as the limit of the associated Riemann Sums as follows:

$$\int_0^5 x^2\, dx = \lim_{N \to \infty} \Sigma_{n=1,N} ((5n - 5)/N)^2 \cdot 5/N$$

$$= \lim_{N \to \infty} \Sigma_{n=1,N} (25n^2 - 50n + 25)/N^2 \cdot 5/N$$

$$= \lim_{N \to \infty} [(125/N^3) \Sigma_{n=1,N} n^2 - 2n + 1]$$

Using the closed forms:
$\Sigma_{n=1,N} n^2 = N(N + 1)(2N + 1)/6$
$\Sigma_{n=1,N} n = N(N + 1)/2$
$\Sigma_{n=1,N} 1 = N$

$$= \lim_{N \to \infty} [(125/N^3) \cdot \{N(N + 1)(2N + 1)/6 - 2N(N + 1)/2 + N\}]$$

After simplifying

$$= \lim_{N \to \infty} (125/6)(2 - 3/N + 1/N^2) = 125/3 = 41.\underline{6}$$

As $N \to \infty$, $1/N \to 0$ and $1/N^2 \to 0$.

☑

EXERCISE SET 5.2

Averages

1 Determine the average class score on a Quiz by forming a frequency table and taking a weighted average. The individual scores are:
24, 26, 28, 23, 25, 25, 26, 27, 24, 26, 25, 25, 27, 28, 24, 25, 26, 26, 24, 26, 27, 28, 25, 27, 24, 26, 26, 25, 23, 26, 27, 24, 27, 24, 26, 24, 27, 28, 25, 23, 25, 25, 26, 28

Section 5.2   THE DEFINITE INTEGRAL AS A LIMIT   369

2   The caped Crusader drives from her Secrete Base to the Scene of the Crime, then to the villain's Hideout. After capturing the villain the crusader drives to the Slammer and then back to the Secrete Base. The Hideout is equidistant from the Scene of the Crime and the Slammer, which is one half the distance from the Base to both the Slammer and the Scene of the Crime. The Crusader manages to drive each segment at a speed 10% faster than the previous segment, and covered the first segment at 100km/h. What was the average speed of the Crusader for the entire escapade?

3   What is the average of the discrete function $f(x) = [\![x/3]\!]$ [0,10]?

4   If $f(x) = 3$ on [0,2), $f(x) = -2$ on [2,3) and $f(x) = 5$ on [3,7] what AVG{f,0,7}?

5   The discrete function f is defined by $f(x) = 1/(2n - 1)$ if $(n - 1)^2 \leq x < n^2$.
    a) Sketch a graph of f over [0,16] and find AVG(f,0,16}. b) Sketch the graph of f over $[50^2, 55^2]$; you will need different scales on the two axes. c) How long is the interval $[(n - 1)^2, n^2]$? Express the AVG{f,0,10^4} as a finite sum and evaluate it in closed form. (Note that $10^4 = 100^2$.)

6   Repeat Exercise 5 for the function $f(x) = n$ if $(n - 1)^2 \leq x < n^2$. This is the function $y = [\![1 + x^{1/2}]\!]$.

Partitions

7   Indicate a partition of the interval I into N subintervals as indicated; list the subintervals I_n and their lengths, Δt_n. The answers to (a), (e), and (g) are not unique. Also indicate, by listing or by a formula, the Left and the Right endpoints of the subintervals.

   a) I = [0,1], N = 4, max{Δt_n} = 1/2   b) I = [0,1], N = 4, uniform

   c) I = [2,5], N = 3, uniform   d) I = [2,5], N = 4, uniform

   e) I = [-1,1], N = 3, max{Δt_n} = 1   f) I = [-1,1], N = 3, uniform

   g) I = [0,3], N = 5, max{Δt_n} = 1   h) I = [0,3], N = 5, uniform

   i) I = [0,20], N = 40, uniform   j) I = [15,25], N = 50, uniform

Approximating Averages, Integrals and Areas.

8   a) Sketch the graph over [0,4] of $f(x) = x/2$. What is the area under this graph?

   b) Estimate the area under the graph of f over [0,4] using a uniform partition of size N = 4 and the Left endpoint formula. Sketch the rectangles used in your estimation.

   c) Repeat part b) using the Right endpoint formula.

   d) Repeat part c) using N = 100. You will need to give the formula for the partition points and the evaluation points. Use the Specific Closed forms for Series to evaluate the sum.

   e) Repeat part d) using the Left endpoints.

9   Estimate the definite integral of $g(t) = 3t + 7$ over [0,2]:

   a) Using right endpoints, N = 4.   b) Using left endpoints, N = 4.

   c) Using right endpoints, N = 6.   d) Using left endpoints, N = 6.

   e) Using right endpoints, N arbitrary.   f) Using left endpoints, N arbitrary.

10 Approximate $\int_0^\pi 4\sin(\theta)\,d\theta$ using a uniform partition of N = 4 subintervals. Sketch a graph to illustrate the approximation and its error.

11 Approximate $\int_0^4 f(x)\,dx$ using right endpoints for the given function and partition.

    a) P = {0, 0.5, 1, 2, 2.5, 3, 4};   f(x) = 3x - 1   b) P = uniform, N = 8;   f(x) = 1/(x + 1)
    c) P = uniform, N = 4;   f(x) = e^x d) P = uniform, N = 6;   f(x) = $ln(x + 1)$

12 Samples of an industrial effluent were analyzed and the concentration of methyl mercury in the n-th sample was recorded as m_n μg/l. Six consecutive sample values were 2, 10, 3, 5, 2, and 4? Assume that the sample periods are uniform four-hour intervals. a) What would be the average concentration of mercury in the effluent. b) Assume that the rate of effluent discharge was f(t) = 3,600t(1 - t/24) l/hr for the 24 hr period of sampling. Estimate the total amount of mercury discharged from the plant. Hint: the total mercury discharge is the product of fluid flow times mercury concentration integrated over the time interval.

13 A solute flows into a tank at a rate $Q_{in}(t)$ and flows out at a rate $Q_{out}(t)$. The amount of solute in the tank is the integral $S(T) = \int_0^T Q_{in}(t) - Q_{out}(t)\,dt$. Assume that $Q_{in}(t) = 1 + t^2$ and $Q_{out}(t) = 2t^2$

    a) When will the assimilated solute, S, reach its maximum value? Give a logical answer without using calculus.
    b) When will the outflow equal the inflow? Take this as the value T and approximate S(T) using a uniform partition of size N = 2 and t_n^* as the right endpoint of the interval, I_n.
    c) Repeat b) with N = 3 and N = 4
    d) Approximate S(T) with a uniform partition of length $\Delta t = 0.1$

14 Repeat Exercise 13 with the functions $Q_{in}(t) = t + 2$ and $Q_{out}(t) = t^2$.

15 The speed of an ultracentrifuge increases at a rate f(t) over a period [0,T].

    a) If the speed rpm(t), in revolutions per minute at time t, at time t = 0 is rpm(0) = 0, express rpm(t) as an integral and give an approximate formula for the rpm(T), T > 0.
    b) Approximate rpm(5) when $f(t) = 10e^{-t}$. Use a uniform partition with N = 5 and Right endpoints.

16 The rate of growth of most plants is a periodic function with period of one day. Assume that a corn plant grows at a rate $g'(t) = k\sin^2(\pi t)$, where t is days and k is a scaling constant that is set to 1 for simplicity.

    a) Construct a discrete approximation to the height of corn, $g(T) = \int_0^T g'(t)\,dt$.
    b) If T = 10 and g(0) = 0, what is an upper limit for the value of g(T)? Sketch a graph illustrating g(10) as an area.
    c) Approximate g(10) using a uniform partition of size N = 20. Think! Is there a quick way to make this approximation?

Integrals as limits.

17 Evaluate the given integral as a limit.

    a) $\int_0^6 3\,dx$ b) $\int_0^2 5x\,dx$ c) $\int_0^4 2x + 1\,dx$ d) $\int_2^5 7\,dx$

    e) $\int_3^{10} 4x\,dx$ f) $\int_1^4 3x - 1\,dx$ g) $\int_{-2}^5 20\,dx$ h) $\int_{-2}^0 3x\,dx$

    i) $\int_{-1}^4 5x + 7\,dx$ j) $\int_0^2 x^2\,dx$ k) $\int_{-1}^1 3x^2\,dx$ l) $\int_1^4 5x^2 - 2\,dx$

    m) $\int_{-3}^3 9 - x^2\,dx$ n) $\int_0^1 x^3\,dx$ o) $\int_1^3 x^3 - x^2\,dx$

Section 5.3 The Fundamental Theorem of Calculus and the Indefinite Integral

Introduction.

The definition of $\int_a^b f(x)\,dx$ as a limit of Riemann Sums was motivated by the fact that the solutions to many different problems are evaluated by such limits. While the limit definition fits neatly into the theoretical development of calculus, it is hardly a useful definition when one wants to evaluate specific integrals. In this section we introduce the **Fundamental Theorem of Calculus**, which provides an alternative method to evaluate definite integrals. It relates the process of integrating to the process of differentiating, thus it ties together the two main branches of Calculus.

The Fundamental Theorem allows us to integrate a function $f(x)$ if we know another function $F(x)$ such that $F'(x) = f(x)$. In this case, the integral of f over $[a,b]$ is simply the change in $F(x)$ over the same interval, $F(b) - F(a)$. The key element in this theorem is to know the function $F(x)$. A function whose derivative is $f(x)$ is called an "antiderivative" or "integral" of $f(x)$. To distinguish this function concept from the numerical value of a definite integral, the function $F(x)$ is called more precisely an "indefinite integral of $f(x)$".

The study of Integral Calculus focuses on establishing Rules and techniques to identify antiderivatives. The direct correspondence between derivatives and integrals means that for each Derivative Rule there is a corresponding Integral Rule. In this Section we state specific Integral Rules for the elementary functions. The correspondence for the sum/difference and constant multiple Algebra Rules is direct and these are also stated. However, for the Product/Quotient Rule the correspondence is more complicated and will be deferred to Chapter 6, where it forms the basis of a method called *integration by parts*.

The Fundamental Theorem of Calculus.

A concept that is fundamental to the study of dynamic or changing systems is that

> *The change in a quantity over a time period can be determined if we know the rate at which it changes.*

In function terms, this fundamental idea translates to the statement that

> *The change in the function F over an interval $[a,b]$, which is the difference $\Delta F[a,b] = F(b) - F(a)$, is determined by the value of $F'(t)$ over the interval.*

This relationship can be established using the *method of calculus*. First, it is established for constant derivatives, then for discrete derivatives, and finally using a limit for continuous derivatives. In the following discussion we will use t as the independent variable to provide an association of the mathematical concepts with our experiences of quantities varying over time.

For a Constant derivative. STEP I.

If the derivative is constant, $F'(t) = k$ for $t \in [a,b]$, then the change in the function $F(t)$ over an interval $[a,b]$ is the product of this constant rate and the length of the interval:

$$\Delta F[a,b] \equiv F(b) - F(a) = k \cdot (b - a)$$

You probably have used this relationship to compute the distance that would be traveled in a time period if the rate of movement, the speed, was constant over the time interval. A car going 90 km/h would travel $90 \cdot 3 = 270$ km in 3 hours.

For a discrete derivative. STEP II.

If the derivative F' is a discrete function, being constant over say N subintervals of the time interval [a,b], then the change in F over the whole interval [a,b] can be calculated by adding the change in F over the subintervals. If the endpoints of the subintervals, in increasing order, are the t-values t_n for n = 0, 1, ... , N, then we would calculate

$$\Delta F[a,b] = \Delta F[t_0,t_1] + \Delta F[t_1,t_2] + \ldots + \Delta F[t_{N-1},t_N]$$

Since F' is constant on each subinterval, the change over the n-th subinterval is simply the product of the subinterval length and the constant value of F' on the subinterval. The value of F' is found by evaluating F' at any point t_n^* in the subinterval. Therefore, the change in F over [a,b] is given by a Riemann Sum

$$\Delta F[a,b] = \Sigma_{n=1,N} \Delta F[t_{n-1},t_n] = \Sigma_{n=1,N} F'(t_n^*) \cdot \Delta t_n$$

In the analogy of F(t) being the distance traveled, this simply states that one calculates the total distance traveled as the sum of the distances traveled over each segment of a trip.

For a continuously varying derivative. STEP III.

If F' is a continuous function, the change ΔF can be calculated as the limit of the change ΔF_N of an associated sequence of functions $\{F_N\}$ whose derivatives F_N' are discrete approximation of F', chosen so that $\lim_{N \to \infty} F_N' = F'$. Using the same approximation process utilized to calculate the average of a function and the area under a curve, as discussed in the previous sections, this limit process will give the change in F as the limit of Riemann Sums. Hence, by definition, as the definite integral of F':

$$\Delta F[a,b] = \int_a^b F'(t) \, dt$$

This is essentially the statement of the Fundamental Theorem of Calculus.

Intuitively, this result can also be established as follows. The change ΔF is the product of the average rate of change in F and the length of the interval. Again, this is a common type of reference that is used when discussing travel distances. If you describe the speed you travelled for a trip, you probably refer to the average "speed" you traveled. You probably would calculate this as the distance traveled divided by the length of travel time. Using function notation, this relationship is

$$\text{Average Speed} = \{F(b) - F(a)\} / (b - a)$$

where,

F(t) is the distance traveled at time t,
a is the time the journey started,
b is the time the journey finished.

The distance traveled is thus

$$\Delta F[a,b] = F(b) - F(a) = \{\text{Average Speed}\} \cdot (b - a) \qquad \textbf{Net distance traveled}$$

The average speed is the average value of the derivative F' over the interval. If we denote the travel speed at time t by the function f(t), then F'(t) = f(t). Then the average speed over the journey is the average of f, which was denoted in Section 5.1 as AVG{f,a,b}, and was shown in Section 5.2 to be the definite integral of f over [a,b] divided by the difference (b - a).

$$\{\text{Average Speed}\} = \text{AVG}\{f,a,b\} = \int_a^b f(t)\, dt / (b - a)$$

Substituting this expression for the Average Speed in the previous equation, the (b - a) terms cancel to give

$$\Delta F[a,b] = \int_a^b f(t)\, dt \quad \text{or} \quad \int_a^b f(t)\, dt = F(b) - F(a)$$

This same relationship holds whenever the integrand f is the derivative of a function F and is the basis for the most important, and hence, Fundamental Theorem of Calculus. We state the theorem using x as the independent variable.

THE FUNDAMENTAL THEOREM OF CALCULUS.

If $F'(x) = f(x)$ for $x \in [a,b]$, then

$$\int_a^b f(x)\, dx = F(b) - F(a)$$

In the following we will frequently refer to the Fundamental Theorem of Calculus by the abbreviation FTC. The difference in the values of F(x) at the endpoints is sometimes denoted by a vertical bar with the left endpoint as a subscript and the right endpoint as a superscript:

$$F(x)\Big|_a^b \equiv F(b) - F(a)$$

Note: because of Word Perfect constraints the upper limit will appear shifted to the right past the lowere limit.

Example 1   **Applying the Fundamental Theorem of Calculus.**

Problem   Evaluate each of the definite integrals by "recognizing" a function whose derivative is the integrand and then applying the FTC.

(a) $\int_0^4 3x^2\, dx$ (b) $\int_1^3 2x\, 0\text{-}\, 3\, dx$ (c) $\int_{-2}^0 e^{3t}\, dt$

Solutions (a) The integrand $f(x) = 3x^2$ is the derivative of $F(x) = x^3 + 5$:

Hence, by the FTC

$$\int_0^4 3x^2\, dx = x^3 + 5 \Big|_0^4 = \{4^3 + 5\} - \{0^3 + 5\} = 64$$

(b) The integrand $f(x) = 2x - 3$ is recognized as the derivative of

$$F(x) = x^2 - 3x + 2$$

Applying the FTC the integral is evaluated as

$$\int_1^3 2x - 3\, dx = F(3) - F(1) = 2 - 0 = 2$$

(c) In this problem the independent variable is t rather than x. To apply the FTC observe that the integrand f(t) = e^{3t} is the derivative of

$$F(t) = (1/3)e^{3t}$$

The 1/3 coefficient cancels the factor of 3 that arises when the exponential term 3t is differentiated. By the FTC the value of the integral is therefor

$$\int_{-2}^{0} e^{3t}\, dt = (1/3)e^{3t}\Big|_{-2}^{0}$$

$$= (1/3)e^{3\cdot 0} - (1/3)e^{3\cdot -2}$$

$$= 1/3 - (1/3)e^{-6} \approx 0.3325\ 071$$

☑

Example 2   Finding the Average value of a function.

Problem   Evaluate the definite integral of the given function to find its "average value" over the specified interval.

  1 f(x) = 1 - 0.6x^2 over the interval [-1,1].

  2 y = $2te^{t^2}$ over the interval [0,1].

  3 T(θ) = sin(θ) over the interval [0,2π].

Solutions Using the formula A = AVG{f,a,b} = $\int_a^b$ f(x) dx / (b - a), applied to the different integrands with the corresponding integration variables, the averages are computed using the FTC.

a) In this problem the limits of integration are a = -1 and b = 1, so b - a = 2. The integrand f(x) = 1 - 0.6x^2 is the derivative of F(x) = x - 0.2x^3

Using the FTC to evaluate the integral, the average is

$$A = \int_{-1}^{1} 1 - 0.6x^2\, dx / (1 - -1) = x - 0.2x^3\Big|_{-1}^{1} / 2$$

$$= [\{1 - 0.2\cdot 1^3\} - \{-1 - 0.2\cdot(-1)^3\}] / 2 = 0.8$$

b) As $D_t e^{t^2} = 2te^{t^2}$, the average value of y over [0,1] is given by

$$A = \int_0^1 2te^{t^2} dt = e^{t^2}\Big|_0^1 = e - 1 \approx 1.718$$

c) As the derivative of -cos(θ) is sin(θ), the average of sin(θ) over [0,2π] is

$$A = \int_0^{2\pi} \sin(\theta)\, d\theta / (2\pi - 0) = -\cos(\theta)\Big|_0^{2\pi} / 2\pi$$

$$= \{-\cos(2\pi) + \cos(0)\}/ 2\pi = \{-1 + 1\}/2\pi = 0$$

☑

Antiderivatives and Indefinite Integrals.

The key step in applying the FTC to evaluate $\int_a^b f(x)\,dx$ is to find the corresponding function $F(x)$, such that $F'(x) = f(x)$. The terms **antiderivative** and **integral** are both used to describe a function F whose derivative is f.

DEFINITION OF AN ANTIDERIVATIVE OR INDEFINITE INTEGRAL

A function $F(x)$ is an antiderivative, or an integral, of $f(x)$ if $F'(x) = f(x)$.

This relationship is expressed using the *inverse* of the derivative operator D_x as

$$F(x) = D_x^{-1} f(x)$$

or using *indefinite integral* notation, without limits of integration, as

$$F(x) = \int f(x)\,dx$$

A function F is said to be <u>an</u> integral, rather than <u>the</u> integral, of a function f because it is not unique; there are infinitely many different integrals of the same function. Because the derivative of a constant is zero, adding (or subtracting) any number to a function F will give a second function with the same derivative. For instance, each of the following functions is an integral of $2x^3$:

$$F_1(x) = x^3 + 2, \quad F_2(x) = x^3 + 12, \quad F_3(x) = x^3 + -82, \quad F_4(x) = x^3$$

Notice that each function can be obtained from each other function by adding a suitable constant. For instance, $F_1(x) = F_2(x) + (-10)$. Although the integral or antiderivative F is not unique, all integrals of a given function f will differ by a constant.

If F_1 and F_2 are both integrals of f then

$$F_1 = F_2 + C \quad \text{for some constant C.}$$

DEFINITION OF THE GENERAL ANTIDERIVATIVE OR GENERAL INTEGRAL OF A FUNCTION.

The **general antiderivative** of $f(x)$ is expressed as

$$D_x^{-1} f(x) = F(x) + C$$

and the **general integral** of $f(x)$ is expressed as

$$\int f(x)\,dx = F(x) + C$$

where **C is an arbitrary constant** and F is <u>any</u> function such that $F'(x) = f(x)$.

Example 3 General Integrals.

Problem Determine the general integral of the given function.

1 $f(x) = x^3 - 2x + 7$ 2 $f(x) = e^x$ 3 $f(x) = x^{-4}$

Solutions In each case we must recognize an antiderivative of the integrand and add the arbitrary constant C to form the general integral or antiderivative.

a) An antiderivative of $f(x) = x^3 - 2x + 7$ is $F(x) = x^4/4 - x^2 + 7x$. The general antiderivative is:

$$\int x^3 - 2x + 7 \, dx = x^4/4 - x^2 + 7x + C$$

b) Since $D_x e^x = e^x$, the function $y = e^x$ is its own antiderivative. Thus

$$\int e^x \, dx = e^x + C$$

c) An antiderivative of x^{-4} is $F(x) = -x^{-3}/3$. The negative sign and the division by 3 are necessary to cancel the coefficient -3 that arises when x^{-3} is differentiated. Thus the general integral is

$$\int x^{-4} \, dx = -x^{-3}/3 + C$$

Basic Integral Rules

Which functions can you integrate?

Or, phrased differently, which functions can you recognize as a derivative?

The answer is limited by the type of functions that you can differentiate. If you have only learned to differentiate polynomials, then using the FTC you will only be able to integrate polynomials. In general, there is a duality between differentiation and integration Rules. The following table lists the basic specific Integral Rules that correspond to the specific Derivative Rules introduced in Chapter 3.

Specific Integral Rules

Type of Function	Derivative Rule	Integral Rule
Power function	$D_x x^r = r x^{r-1}$	$\int x^r \, dx = x^{r+1}/\{r+1\} + C$
Exponential Base e	$D_x e^x = e^x$	$\int e^x \, dx = e^x + C$
Natural Log	$D_x \ln(x) = 1/x$	$\int 1/x \, dx = \ln(x) + C$ **** See below.
	$D_x \ln(|x|) = 1/x$	$\int 1/x \, dx = \ln(|x|) + C$

Basic Trigonometric functions	$D_x \sin(x) = \cos(x)$	$\int \cos(x)\, dx = \sin(x) + C$
	$D_x \cos(x) = -\sin(x)$	$\int \sin(x)\, dx = -\cos(x) + C$
	$D_x \tan(x) = \sec^2(x)$	$\int \sec^2(x)\, dx = \tan(x) + C$
	$D_x \cot(x) = -\csc^2(x)$	$\int \csc^2(x)\, dx = -\cot(x) + C$
	$D_x \sec(x) = \sec(x)\tan(x)$	$\int \sec(x)\tan(x)\, dx = \sec(x) + C$
	$D_x \csc(x) = -\csc(x)\cot(x)$	$\int \csc(x)\cot(x)\, dx = -\csc(x) + C$
Exponential Base b	$D_x b^x = \ln(b) b^x$	$\int b^x\, dx = b^x / \ln(b) + C$
Logarithm Base b	$D_x \log_b(x) = 1/\{\ln(b)x\}$	$\int 1/x\, dx = \ln(b)\log_b(x) + C$

************ **The antiderivative of 1/x.** ************

The Specific Integral Rules table lists three different antiderivatives or integrals of 1/x. The functions $\ln(x)$, $\ln(|x|)$, and $\ln(b)\log_b(x)$.

The third case is equivalent to the first since the change of base formula for logarithms is $\log_b(x) = \log_a(x)/\log_a(b)$ for any positive bases a and b. Setting a = e, the natural log base, gives $\log_b(x) = \ln(x)/\ln(b)$.

Technically, the formula with the absolute value is always correct. If x > 0 then $\ln(|x|) = \ln(x)$. However, if x < 0, then $\ln(|x|)$ must be utilized since $\ln(x)$ is not define for negative x. However, there is a ruse that works. You simply write 1/x = -1/-x. Then, for instance if you need to integrate 1/x over the interval [-5,-2] you can simply write

$$\int_{-5}^{-2} 1/x\, dx = \int_{-5}^{-2} -1/(-x)\, dx = \ln(-x)\Big|_{-5}^{-2} = \ln(2) - \ln(5) = \ln(2/5)$$

If you don't know if the integration interval is negative or positive the simplest solution is to use the absolute value formula.

To simplify the notation we shall normally use the antiderivative $\ln(x)$ for the indefinite integral of $1/x$.

The $\ln(|x|)$ form will be used only in special situations to evaluate definite integrals when the argument x is negative.

Algebraic Integral Rules.

The Fundamental Theorem of Calculus gives a direct correspondence between derivatives and integrals. This carries over to the algebraic Rules for combining and manipulating the functions being differentiated or integrated. For the Constant Multiple and Sum/Difference Rules the correspondence is direct.

Sum/Difference Rule

> If f and g are integrable functions, then so are their sum and difference:
>
> $$\int f(x) + g(x)\, dx = \int f(x)\, dx + \int g(x)\, dx$$
>
> $$\int f(x) - g(x)\, dx = \int f(x)\, dx - \int g(x)\, dx$$

Constant Multiple Rule

> If k is any constant then
>
> $$\int k \cdot f(x)\, dx = k \cdot \int f(x)\, dx.$$

Examples illustrating the constant factor Rule are:

$$\int 5x^8\, dx = 5\int x^8\, dx \quad \text{and} \quad \int -3e^{2t+7}\, dt = -3\int e^{2t+7}\, dt$$

Using these two Rules the linear combinations, i.e. sums and differences of multiples, of elementary functions can be integrated using the specific rules.

Example 4   **Integrals of linear combinations of functions.**

Problem   Use the Specific Integral Rules and the Constant and Sum/Difference Rules to determine the general integrals of the given functions.

1   $f(x) = x^8 - 2x^{-1} + 7\sin(x)$ 2 $f(x) = 3\sec(x)\tan(x) - 2^x$

Solutions a) Since f is the sum of three terms its integral is found by splitting the function and evaluating three integrals:

$$I = \int x^8 - 2x^{-1} + 7\sin(x)\, dx = \int x^8\, dx - \int 2x^{-1}\, dx + \int 7\sin(x)\, dx$$

Factoring out the constants in the second and third integrals gives

$$I = \int x^8\, dx - 2\int x^{-1}\, dx + 7\int \sin(x)\, dx$$

The first integral is given by the Power Rule, the second by the Natural Log Rule and the third by the basic Rule for sin(x):

$$I = x^9/9 - 2\cdot ln(x) + 7(-\cos(x)) + C$$

b) The integral of $f(x) = 3\sec(x)\tan(x) - 2^x$ is the difference of two integrals:

$$\int 3\sec(x)\tan(x) - 2^x\, dx = 3\int \sec(x)\tan(x)\, dx - \int 2^x\, dx$$

$$= 3\sec(x) - 2^x/ln(2) + C$$

The alert reader might ask   "Why is there only one integration constant C in the above solutions? Shouldn't there be a separate integration constant with each integral?"

The answer is no. If a constant is introduced with each individual integration, then these can all be combined into one constant. For example, the integral of the quadratic polynomial $f(x) = 5 + 7x - 9x^2$ might be evaluated "term by term" introducing a different constant with each application of a Specific Integral Rule as follows:

$$\int 5 + 7x - 9x^2 \, dx = \int 5 \, dx + \int 7x \, dx - \int 9x^2 \, dx$$

$$= 5 \int 1 \, dx + 7 \int x \, dx - 9 \int x^2 \, dx$$

$$= 5(x + c_1) + 7(x^2/2 + c_2) - 9(x^3/3 + c_3)$$

$$= 5x + 7x^2/2 - 9x^3/3 + 5c_1 + 7c_2 - 9c_3$$

$$= 5x + 7x^2/2 - 9x^3/3 + C$$

where $\qquad\qquad\qquad C = 5c_1 + 7c_2 - 9c_3$

Thus, when adding or subtracting multiples of integrals, the constants can be combined as a single integral constant.

Example 5   More examples applying algebraic and specific Integral Rules.

Each of the following integrals is evaluated by applying algebraic rules to separate the elementary functions and then applying the specific integral rules. As you gain experience with this process, you will likely do some of the steps mentally and omit writing them down.

a) $\int x^9 + 4x^5 \, dx = \int x^9 \, dx + 4\int x^5 \, dx = x^{10}/10 + 4x^6/6 + C$

b) $\int x - 4x^2 + 7x^5 \, dx = \int x \, dx - 4\int x^2 \, dx + 7\int x^5 \, dx = x^2/2 - (4/3)x^3 + (7/6)x^6 + C$

c) $\int 5s^{1/2} \, ds = 5 \int s^{1/2} \, ds = 5s^{3/2}/(3/2) = (10/3)s^{3/2} + C$

d) $\int t^{0.2} + t^{-2} + t^{1/5} \, dt = t^{1.2}/1.2 + t^{-1}/-1 + t^{6/5}/(6/5) + C$

$\qquad\qquad = 0.8\underline{33}\,t^{1.2} - t^{-1} + (5/6)t^{6/5} + C$

e) $\int 4e^t \, dt = 4\int e^t \, dt = 4e^t + C$

f) $\int 4/x \, dx = 4\int 1/x \, dx = 4ln(x) + C$ (or more generally $4ln(|x|) + C$)

g) $\int 3\cos(t) - 2\sec^2(t) \, dt = 3\int \cos(t) \, dt - 2\int \sec^2(t) \, dt$

$\qquad = 3\sin(t) - 2\tan(t) + C$

EXERCISE SET 5.3

Antiderivatives and indefinite integrals.

1   Show by differentiation that both functions are antiderivatives of the same function. Find a constant C such that $F_1 = F_2 + C$.

a) $F_1(x) = x^2 - 6x + 3 \qquad\qquad F_2(x) = x^2 - 6x + 8$

b) $F_1(x) = x^{1/2} - x^{3/2} \qquad\qquad F_2(x) = x^{1/2}(1 - x) + 8$

c) $F_1(x) = (x + 3)^2$ $F_2(x) = x^2 + 6x - 12$

d) $F_1(x) = (x + 1)/(x - 1) + 3$ $F_2(x) = (x + 1)^2/(x^2 - 1),\ x \neq -1$

e) $F_1(x) = (x^2 + 3x + 2)/(x + 2)$ $F_2(x) = x - 6,\ x \neq -2$

f) $F_1(x) = ln(x)$ $F_2(x) = ln(3x)$

g) $F_1(x) = \sin^2(x)$ $F_2(x) = -\cos^2(x)$

h) $F_1(x) = e^{2x} - e^{-2x}$ $F_2(x) = e^{2x}(1 + 3e^{-2x} - e^{-4x})$

2 Give the general antiderivative of the indicated function.

a) $f(x) = x^2$ b) $f(x) = 2x - 5$ c) $f(x) = 3x^5 + 2$ d) $f(x) = x^2 - x^{-2}$

e) $f(x) = 5x^3 - 2x^5$ f) $f(x) = 6x^{-7} - 7x^6$ g) $f(x) = 3x^2 - 2/3x^4$

h) $f(x) = x^{1/2} + x^{3/4}$ i) $f(x) = 2x^{-1/2} + x^{-2}$ j) $f(x) = 6(x + 1)(x + 7)$

3 Determine the antiderivative $F(x)$ of $f(x)$ satisfying the specified condition.

a) $f(x) = x^2 - 2x - 2;\ F(3) = 2$ b) $f(x) = x^{1/2} - x;\ F(1) = 1$

c) $f(x) = 5x^5 + 2;\ F(0) = 1$ d) $f(x) = (x - 2)^2;\ F(2) = 8$

e) $f(x) = x^{-1/2} - 1;\ F(0) = 3$ f) $f(x) = \sin(x);\ F(\pi) = 0$

g) $f(x) = \sec^2(x);\ F(\pi/4) = 0$ h) $f(x) = 3e^x;\ F(0) = 1$

i) $f(x) = 3/x;\ F(1) = 2$ j) $f(x) = x - e^{-3+x};\ F(3) = 0.5$

4 Evaluate the given integral.

a) $\int 5x\ dx$ b) $\int 3x^2 - 2x + 1\ dx$ c) $\int 5x^3\ dx$

d) $\int x^{-2} - x^{-3} + 8\ dx$ e) $\int 5x^{-1/3}\ dx$ f) $\int x^{-1/2} - x^{1/2}\ dx$

g) $\int (2x + 1)^2\ dx$ h) $\int x^{-7} + 5x^2 - 2\ dx$ i) $\int 2t^{1/2}\ dt$

j) $\int (3t^{1/2} - 2)/t^{1/2}\ dt$ k) $\int x^{0.7} + x^{-0.7}\ dx$ l) $\int 4t^2(t - 3)\ dt$

5 Evaluate the given integral.

a) $\int 2e^t\ dt$ b) $\int -3e^x\ dx$ c) $\int x + e^{x+5}\ dx$

d) $\int 2/t\ dt$ e) $\int x - 1/x\ dx$ f) $\int (2x - 5)/x\ dx$

g) $\int 2^t\ dt$ h) $\int 10^x\ dx$ i) $\int 2^{x-1}\ dx$

6 Find an integral of the given function.

a) $f(\theta) = 3\cos(\theta)$ b) $f(\theta) = \sin(\theta) - 2\cos(\theta)$

c) $f(\theta) = \sec(\theta)[\sec(\theta) - \tan(\theta)]$ d) $f(\theta) = \csc(\theta)\cot(\theta)$

e) $f(\theta) = 1/\cos^2(\theta)$ f) $f(\theta) = \sin(\theta)[1 - 1/\sin(\theta) + 1/\sin^3(\theta)]$

7 For a) $f(x) = -6$ b) $f(x) = -2x + 2$ c) $f(x) = 4x^3 - 4x$

i) Determine the general "integral curve" $y = \int f(x)\ dx + C$.
ii)   Sketch the graph of the integral curve associated with $C = 0$, $C = 3$, and $C = -3$.
iii)   On the same graph, sketch the integral curve passing through $(1,3)$. Give its equation.

Section 5.3   THE FUNDAMENTAL THEOREM OF CALCULUS

Evaluating definite integrals by the Fundamental Theorem
8 Use the Fundamental Theorem of Calculus to evaluate the integral.

   a) $\int_0^2 3x \, dx$ b) $\int_1^4 2 - x^2 \, dx$ c) $\int_{-1}^3 2x + x \, dx$ d) $\int_0^{10} t^{1/2} \, dt$

   e) $\int_5^6 (t-5)^2 \, dt$ f) $\int_1^4 \sqrt{s} \, ds$ g) $\int_0^h x + 5x^4 \, dx$ h) $\int_5^6 t^3 - 2 \, dt$

9 Use the FTC to evaluate the integral.

   a) $\int_0^\pi \sin(x) \, dx$ b) $\int_0^{\pi/4} \sec^2(x) \, dx$ c) $\int_{-\pi}^\pi \cos(x) \, dx$

   d) $\int_{\pi/6}^{\pi/2} \csc(t)\cot(t) \, dt$ e) $\int_\pi^{\pi/2} 4\cos(x) \, dx$ f) $\int_0^\pi \theta + 2\cos(\theta) \, d\theta$

10 Which of the following integrals can not be evaluated using the FTC? Why?

   a) $\int_0^{10} x^2 - 1 \, dx$ b) $\int_0^{10} \text{SQRT}(x^2 - 1) \, dx$ c) $\int_0^{\pi/2} \sec^2(\theta) \, d\theta$

   d) $\int_0^{\pi/2} \sec(\theta/2) \, d\theta$ e) $\int_0^5 \ln(x) \, dx$ f) $\int_0^5 1/t \, dt$

11 Evaluate the integrals; combining them may make less work.

   a) $\int_0^6 t^2 \, dt + 3 \int_0^6 t - 1 \, dt$ b) $\int_2^5 (t+1)^2 \, dt - \int_2^5 t + 1 \, dt$

   c) $\int_3^4 t + 7 \, dt + \int_4^6 t + 7 \, dt$ d) $\int_1^3 x^2 - 2x \, dx + \int_{-2}^1 x^2 - 2x \, dx$

   e) $\int_3^4 e^x \, dx - \int_4^3 e^x \, dx$ f) $\int_3^5 7e^x \, dx + \int_5^4 7e^x \, dx$

12 Evaluate the given integral.

   a) $\int_0^1 e^x \, dx$ b) $\int_{-1}^0 e^x \, dx$ c) $\int_0^{-1} e^x \, dx$ d) $\int_0^{\ln(2)} e^x \, dx$

   e) $\int_{-1}^1 x - e^x \, dx$ f) $\int_0^{\ln(2)} -3e^x \, dx$ g) $\int_1^e 1/x \, dx$ h) $\int_1^{e^2} 1/x \, dx$

   i) $\int_{-2}^{-1} 1/x \, dx$ j) $\int_0^1 2^x \, dx$ k) $\int_{-1}^0 10^x \, dx$ l) $\int_4^5 10^x \, dx$

Average function values.
13 The marginal cost of producing x items is $MC(x) = 200 - 0.01x + 0.005x^2$. What is the average marginal cost, AMC, over the interval [50,100]? If the unit marginal cost, the marginal cost per item, is UMC = MC(x)/x, what is the average UMC, AUMC, over this same interval? Does the AMC = AUMC?

14 The rate of change in a company's capital, cash and liquid assets is its "capital flow", CF(t). A model reflecting normal fluctuations in business is $CF(t) = 0.005A \cdot \sin(t)$, where A is the total company assets at t = 0. What is the average capital flow over an interval $[0, 2n\pi]$, n an integer?

15 A crude estimate of the number of bacteria cells in a culture t hours after an experiment begins is $N(t) = 5{,}200 \cdot 2^t$. What is the average number of cells in the culture over the interval [2,5] hr?

16 A radioactive material decays exponentially. If t is measured in units of the material's "half-life" the amount present at time t is $A(t) = A_0 e^{-t}$, where A_0 is the amount of material present at time zero. What is the average amount of material present over the interval [3,7]?
    HINT: use the FTC and the fact that $D_t e^{-t} = -e^{-t}$.

17 A cyclist on the Tour d'France road race peddles up a mountain road at a speed $s(\theta) = 30(1 - \theta)$ km/h, where θ is the slope of the road. If the elevation of a road is given by $y = (x/25)^3$, x km from the starting line, what is the average cyclist's speed in reaching the check point 65 km from the starting point?

18 A projectile is delivered over a 500 km distance with its height at distance x being $h = 0.003(500 - x)x$. What is the average height of the projectile over its flight?

Section 5.4 Differentiating Integral Functions;

The 2nd Fundamental Theorem of Calculus.

Introduction.

Every continuous function is integrable.

If f is continuous on [a,b] then $\int_a^b f(x)\,dx$ exists!
Furthermore, an antiderivative of f exists!

Determining antiderivatives can be difficult, and the antiderivative of some relatively simple functions exist in theory but <u>cannot</u> be written down using finitely many elementary functions. The most frequently encountered function with this property is the exponential $f(x) = e^{-x^2}$. The antiderivative of such functions, that exist but can not express with elementary functions, can always be expressed as *integral-functions*. These are definite integrals with a variable limit of integration, such as

$$F(x) = \int_a^x f(t)\,dt$$

The Second Fundamental Theorem of Calculus, which is introduced in this section, states integral-functions like F(x) are differentiable and that $F'(x) = f(x)$.

You know that some special numbers, like π and e, are very important and that they arise naturally in many applications and mathematical problems. Because they occur often, they have been given names that we use instead of their numerical values. Similarly, functions that are very important and recur frequently in mathematics are given names. The elementary trigonometric functions are obvious examples. The *sine* function, whose value at x is denoted sin(x), is evaluated either in reference to the coordinates of points on a unit circle or by an infinite series expansion. Some important functions are integral-functions, defined by integrals with variable limits of integration. The natural logarithm function $ln(x)$ is defined later in this section as the integral-function

$$ln(x) = \int_1^x 1/t\,dt$$

In Example 4 another important integral-function is defined, one that you may not have previously encountered, the *Error Function* denoted by ERF(x):

$$\mathrm{ERF}(x) \equiv \frac{2}{\sqrt{\pi}} \int_0^x e^{-t^2}\,dt$$

This function is essential in the study of statistics as it is the basis of the Normal Probability distribution. Like $ln(x)$, the ERF(x) function can not be evaluated using polynomials or other elementary functions. You will find this function listed among the "functions" available in spreadsheet and data-base computer programs. (Many programs attach an @ to indicate a function, e.g. @ERF(x) or @LN(x).) The ERF function is evaluated by numerical algorithms that approximate the integral with a specific accuracy, it can also be found as a button on some advanced calculators.

Integral Functions: Variable limits of integration.

Assume f(x) is continuous over the interval [a,b]. Then f is continuous on the subinterval [a,x] for $a < x \leq b$. Consequently, a related function F(x) can be defined as the integral of f over the subinterval [a,x]. Traditionally, the original integrand is denoted by a "lower case" letter and its

Section 5.4 — Differentiating Integral Functions

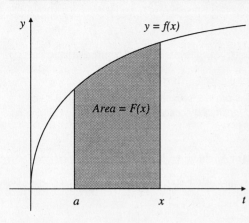

associated integral-function by the corresponding "capital" letter. Thus, we define the function F(x) by:

$$F(x) \equiv \int_a^x f(t)\, dt.$$

Note that the variable of integration is t rather than x. This change is made to avoid errors later. Technically, the integration variable is a "dummy" variable and any symbol could be used. To use x is not wrong, it simply becomes confusing sometimes to distinguish between the integration variable and the variable limit of the integral. The function F(x) can be viewed as the area under the graph of f over [a,x], when f is positive.

Example 1   Functions defined by integrals.

Problem Applying the FTC to determine the explicit form of the given integral-function.

a) $F(x) = \int_5^x t^2\, dt$ b) $G(x) = \int_0^x 7e^t\, dt$ c) $H(x) = \int_1^x t^{-3}\, dt$

Solutions Each integral is evaluated to give a function of the upper limit of integration.

a) $F(x) = \int_5^x t^2\, dt = t^3/3 \,\big|_5^x = x^3/3 - 125/3$

b) $G(x) = \int_0^x 7e^t\, dt = 7e^t\big|_0^x = 7e^x - 7$

c) $H(x) = \int_1^x t^{-3}\, dt = t^{-2}/(-2)\big|_1^x = -(1/2)x^{-2} + 1/2$

Defining the natural log function and the number e with integrals.

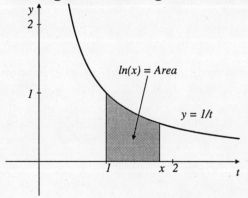

Recall that $ln(x)$ was first introduced as the inverse of the exponential function $y = e^x$, with

$$y = ln(x) \quad \text{if} \quad e^y = x$$

and

$$\exp(ln(x)) = x \quad \text{and} \quad ln(e^x) = x$$

Using these identities and the method of implicit differentiation it was shown in Section 3.6 that

$$D_x ln(x) = 1/x$$

An alternative mathematically rigorous approach, is to define $ln(x)$ as an integral-function. Since $y = 1/x$ is continuous and positive for $x > 0$, its definite integral from 1 to x exists.

ALTERNATE DEFINITION OF THE NATURAL LOGARITHM $ln(x)$

> The **natural logarithm** function is defined for $x > 0$ by
>
> $$ln(x) \equiv \int_1^x 1/t\, dt$$

A digression: Establishing the properties of $ln(x)$.

The basic properties of the *ln* function can be established from this definition using basic Integral Rules and our intuitive understanding of area. Three such properties are:

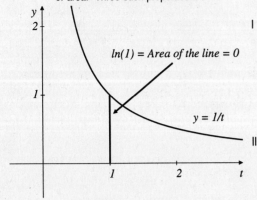

I If $x > 1$ then $ln(x)$ is positive, being the area under a positive function's graph. If $0 < x < 1$ then $ln(x)$ is negative. This is seen by applying the Interchange of Limits Rule:

$$ln(x) = \int_1^x 1/t \, dt = -\int_x^1 1/t \, dt$$

which is the negative of the area under the graph $y = 1/t$ from $t = x$ to $t = 1$.

II $ln(1) = \int_1^1 1/t \, dt = 0$

This results from the direct application of the Integral Rule for Equal limits of integration.

III The relationship $ln(1/x) = -ln(x)$.

This identity is established using the symmetry of the graph of $y = 1/t$ about the line $y = t$. Assume that $x > 1$, then $1/x < 1$ and the function values $ln(x)$ and $ln(1/x)$ each correspond to areas under the graph of $y = 1/t$ over different intervals. $ln(x)$ is the area of region over $[1, x]$ that is the sum of the two subregions with areas A_1 and A_2 as indicated in the figure.

$$ln(x) = A_1 + A_2$$

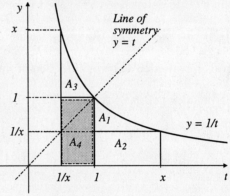

Since $1/x < 1$, $ln(1/x)$ is the negative of the area of the region beneath the $1/t$ curve over the interval $[1/x, 1]$, consisting of the subregions denoted by areas A_3 and A_4.

$$ln(1/x) = A_3 + A_4$$

The curve $y = 1/t$ is symmetric about the line $y = t$. If variables y and t are interchanged, giving the equation $t = 1/y$, and this is solved for y we get the original equation $y = 1/t$. Consequently, the two subregions with areas A_1 and A_3 are congruent, they have identical shape and size and thus $A_1 = A_3$.

The two rectangular regions have identical area, as is seen by multiplying their base and heights:

$$A_2 = (x - 1) \cdot 1/x \quad \text{and} \quad A_4 = 1 \cdot (1 - 1/x)$$

Therefore, the areas corresponding to the two integrals are equal, which gives

$$ln(1/x) = -(A_3 + A_4) = -(A_1 + A_2) = -ln(x)$$

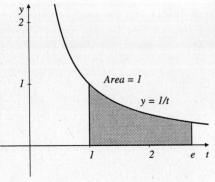

The number e.

The integral definition of $ln(x)$ provides a graphical definition of the number e. Notice that the area under the curve $y = 1/t$ over the interval $[1,x]$ increases as x increases. This area increases from zero when $x = 1$ to infinity as $x \to \infty$. As this area increases continuously, for some x-value $ln(x)$ must equal 1. This x-value is the number e. By its definition $ln(e) = 1$ and hence e is the base of $ln(x)$. By now defining "The" exponential function $EXP(x)$ as the inverse of $ln(x)$ it can be shown that $EXP(x) = e^x$.

The number *e* is the unique number satisfying

$$ln(e) = \int_1^e 1/t \, dt = 1$$

THE SECOND FUNDAMENTAL THEOREM OF CALCULUS.

Recall, that to apply the (first) Fundamental Theorem of Calculus to evaluate the integral of f(x) we must first find a function F(x) such that F'(x) = f(x). This section began with a statement that continuous functions are integrable. The Second Fundamental Theorem of Calculus states that the integral functions derived by integrating continuous functions are differentiable. It follows that **any continuous function is the derivative of another function**.

THE SECOND FUNDAMENTAL THEOREM OF INTEGRAL CALCULUS.

If the function f is continuous on [a,b] then the function defined by

$$F(x) = \int_a^x f(t) \, dt$$

is differentiable for $x \in (a,b)$ and

$$F'(x) = D_x \int_a^x f(t) \, dt = f(x)$$

To see why this is so, assume that f(x) is continuous over an interval [a,b] for $x \in (a,b)$ and define

$$F(x) = \int_a^x f(x) \, dx$$

To see that F(x) is differentiable and $D_x F(x) = f(x)$, we apply the limit definition of the derivative:

$$F'(x) = \lim_{h \to 0} \Delta F/\Delta x = \lim_{h \to 0} \{F(x + h) - F(x)\} / h$$

$$= \lim_{h \to 0} \{ \int_a^{x+h} f(x) \, dx - \int_a^x f(x) \, dx \} / h$$

To evaluate this limit the numerator must be manipulated so that an h-term can be factored out to cancel with the h in the denominator. Since the interval

$$[a, x + h] = [a, x] \cup [x, x + h],$$

$$\int_a^{x+h} f(x) \, dx = \int_a^x f(x) \, dx + \int_x^{x+h} f(x) \, dx$$

and, hence,

$$\int_a^{x+h} f(x) \, dx - \int_a^x f(x) \, dx = \int_x^{x+h} f(x) \, dx$$

Thus,

$$\Delta F(x)/\Delta x = \int_x^{x+h} f(x) \, dx / h$$

Applying the Integral Mean Value Theorem (page 349) to the integral of f over [x,x + h], we know that there exists a number c in the interval [x,x + h] at which the function f assumes its average value. To identify the number c with the interval, it is subscripted with the letter h, as c_h. Thus, the integral of f over [x, x + h] can be expressed as

$$\int_x^{x+h} f(x)\,dx = f(c_h) \cdot (x + h - x) = f(c_h) \cdot h$$

Substituting the product $f(c_h) \cdot h$ for the integral in the numerator of the difference quotient $\Delta F/\Delta x$, the h terms in the numerator and denominator can be canceled and then the limit taken to yield F'.

$$F'(x) = \lim_{h \to 0} \{f(c_h) \cdot h\}/h = \lim_{h \to 0} f(c_h)$$

As $h \to 0$, the corresponding number $c_h \to x$, since $x \leq c_h \leq x + h$ and $x + h \to x$ as $h \to 0$. Because the function f is assumed to be continuous, the limit of the function values $f(c_h)$ must be f(x). (**This is where the assumption that f is continuous is needed**.) That is,

$$F'(x) = \lim_{h \to 0} f(c_h) = f(x)$$

This is the conclusion of the **Second Fundamental Theorem of Calculus**.

Example 2 **Differentiating Integral-functions:
Applying the Second Fundamental Theorem of Calculus.**

Problem Differentiate the given integral-function. Note that the derivative is with respect to the variable appearing in the limit of the integral, **not** the integration variable.

Solutions a) $D_x \int_3^x (2t + 1)\,dt = 2x + 1$ b) $D_x \int_0^x (3t^2 + 2)^{-1}\,dt = (3x^2 + 2)^{-1}$

c) $D_x \int_1^x 1/s\,ds = 1/x$ d) $D_t \int_0^t e^{-x}\,dx = e^{-t}$ ☑

Integral-functions can also be defined with the variable in the lower limit of integration. To differentiate an integral-function where the variable appears in the lower limit of integration, such as $y = \int_x^2 t^3 - 2t\,dt$, you can interchange the limits of integration, which induces a negative sign, and then apply the Second Fundamental Theorem:

$$y = \int_x^2 t^3 - 2t\,dt = -\int_2^x t^3 - 2t\,dt$$

Therefore,

$$y' = D_x -\int_2^x t^3 - 2t\,dt = -\{D_x \int_2^x t^3 - 2t\,dt\} = -\{x^3 - 2x\}$$

More generally, integral-functions can be formed with functions of x as the upper and/or the lower limits of integration. Such functions are also differentiable.

Differentiating integral-functions of the form $\int_{g(x)}^{h(x)} f(t)\,dt$.

The Second Fundamental Theorem of Calculus can be combined with the Chain Rule to differentiate more complex integral-functions defined by integrals with functions in the limits of integration. Assume that f(x) is continuous and h(x) is differentiable. Then the function

$$y = \int_a^{h(x)} f(t)\,dt$$

is differentiable. To find its derivative let u = h(x). Then

Section 5.4   DIFFERENTIATING INTEGRAL FUNCTIONS

$$y = F(u) = \int_a^u f(t)\, dt$$

Using the Chain Rule and the Second Fundamental Theorem of Calculus,

$$D_x y = D_u F(u) \cdot D_x u = f(u) \cdot h'(x) = f(h(x)) \cdot h'(x)$$

Integral-functions defined with a function as the lower limit of integration can be similarly differentiated by applying the interchange of limits formula first and then the Chain Rule. Since an integral-function in which both integration limits are functions can be split as

$$y = \int_{g(x)}^{h(x)} f(t)\, dt = \int_{g(x)}^{a} f(t)\, dt + \int_a^{h(x)} f(t)\, dt = -\int_a^{g(x)} f(t)\, dt + \int_a^{h(x)} f(t)\, dt$$

such functions can be differentiated as

$$y' = D_x \int_a^{h(x)} f(t)\, dt - D_x \int_a^{g(x)} f(t)\, dt = f(h(x)) \cdot h'(x) - f(g(x)) \cdot g'(x)$$

DIFFERENTIATING $y = \int_{g(x)}^{h(x)} f(x)\, dx$.

If f is continuous and g and h are differentiable on $[a,b]$, then

$$y = \int_{g(x)}^{h(x)} f(t)\, dt$$

is differentiable with respect to x for $x \in (a,b)$ and

$$D_x y = f(h(x)) \cdot h'(x) - f(g(x)) \cdot g'(x)$$

Example 3   **Differentiating Integral-Functions.**

Problem   Differentiate the functions specified by integrals:

a) $\int_0^{3x+1} t^3\, dt$ b) $\int_{\sin(x)}^{0} e^t\, dt$ c) $\int_{ln(x)}^{1/x} \sin(t)\, dt$

Solutions   a) If $y = \int_0^{3x+1} t^3\, dt$ then only the upper limit is variable and

$$y' = (3x+1)^3 \cdot D_x(3x+1) = 3(3x+1)^3.$$

b) If $y = \int_{\sin(x)}^{0} e^t\, dt$, then only the lower limit is variable and hence

$$y' = -e^{\sin(x)} \cdot D_x \sin(x) = -e^{\sin(x)} \cdot \cos(x) = -\cos(x) e^{\sin(x)}.$$

Be careful to include the negative sign that multiplies the derivative term derived form the lower limit of integration!

c) The function $y = \int_{ln(x)}^{1/x} \sin(t)\, dt$ is differentiated by the general rule:

$$y' = \{\sin(1/x) \cdot D_x 1/x - \sin(ln(x)) \cdot D_x ln(x)$$

$$= -\sin(1/x)/x^2 - \sin(ln(x))/x$$

Example 4   **The ERF(x) function and the Normal Probability Distribution function.**

Background   In Statistics the function $y = e^{-x^2}$ is called a *Gaussian* density function. This function is not the derivative of an elementary function. Whenever the exponential e^{-x^2} is differentiated a factor of -2x is introduced as the derivative of the $-x^2$ in the exponent. The Gaussian function does not have this -2x factor. However, $y = e^{-x^2}$ is continuous for all x and its integral exists over any interval. A special function called the **Error Function**, denoted by ERF(x), is defined as constant $2/\sqrt{\pi}$ times the integral of the Gaussian function over the interval from 0 to x.

$$\text{ERF}(x) \equiv \frac{2}{\sqrt{\pi}} \int_0^x e^{-t^2}\, dt$$

The coefficient $2/\sqrt{\pi}$ is introduced to scale the ERF function so that it satisfies statistical laws, which we will not discuss here. The Error Function is also called a *Gaussian Distribution*. You will find this function listed, usually with the ERF name, as a standard statistical function in spreadsheet and data base programs.

The statistical distribution commonly referred to as a "bell curve" describes the **density** of many "normally distributed" numerical quantities. The standard "bell" shaped graph illustrated below is the graph of a *Normal density* with **mean** 0 and **standard deviation** 1. It is the graph of a Gaussian function with an extra coefficient of 0.5 in the exponent:

Normal Density Function

$$f(x) = \frac{1}{\sqrt{2\pi}} e^{-0.5 x^2}$$

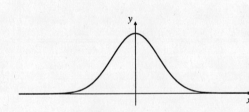

If you select at random a value X that is "normally distributed", it will have a probability of being less than a given number x. This probability is given by a "distribution" function F(x) which can be expressed as 0.5 plus an integral of the above function f. Using the integration variable t and the variable x as the upper limit of integration

$$F(x) = 0.5 + \int_0^x f(t)\, dt$$

The additive constant 0.5 term represents the area under the graph of f from $-\infty$ to 0. The integral gives the area under the graph of f from 0 to x. In the case that x < 0 it is actually the negative of the area under the curve. Combining the above equations, the Normal Distribution function F(x), with "mean" 0 and "standard deviation 1", is expressed in terms of the Error Function with the variable scaled by a factor of $1/\sqrt{2}$ as

$$F(x) = 0.5\,[1 + \text{ERF}(x/\sqrt{2})] \qquad \text{Normal Distribution}$$

The $1/\sqrt{2}$ term corresponds to the 0.5 in the exponent of f(x), since $[1/\sqrt{2}]^2 = 1/2$.

Problem   Show that the graph of the *Normal Probability Distribution* function F(x) (stated above) is increasing and is concave upward for x < 0 and concave downward for x > 0.

Solution   To determine the concavity of a function we must compute its second derivative. The first derivative F'(x) is the derivative of the constant 1/2, which is zero, plus the derivative of the integral-function $(1/2)\text{ERF}(x/\sqrt{2})$. The derivative of this ERF function is computed using the Chain Rule. Setting $u = x/\sqrt{2}$, we have

$$D_x\{(1/2)\text{ERF}(x/\sqrt{2})\} = (1/2) D_u \text{ERF}(u) \cdot D_x u$$

$D_x u = 1/\sqrt{2}$ and the derivative with respect to u is found by applying the Second Fundamental Theorem of Calculus. Using ^2 to denote the square term in the exponent, we have

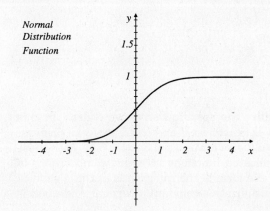

Normal Distribution Function

$$D_u ERF(u) = D_u (2/\sqrt{\pi})\int_0^u e^{t^2} dt = (2/\sqrt{\pi}) e^{u^2}$$

Thus the derivative of F is given by

$$F'(x) = D_x\, 0.5 + D_x\, (1/2)ERF(x/\sqrt{2})$$

$$= (1/2)(2/\sqrt{\pi})e^{-0.5x^2} \cdot 1/\sqrt{2} \qquad = (1/\sqrt{2\pi})e^{-0.5x^2}$$

As the exponential function is always positive the derivative F' is positive and the graph of F is increasing for all x. Differentiating F' gives the second derivative

$$F''(x) = (-x/\sqrt{2\pi})e^{-0.5x^2}$$

Again, the exponential is positive for all x so the sign of the derivative is negative for x > 0 and positive for x < 0. By the Second Derivative Test the graph of F is concave downward if x > 0 and concave upward for x < 0. The shape of the graph is thus "sigmoidal" as illustrated above. ☑

EXERCISE SET 5.4

1. Draw a sketch to illustrate the values F(3) and F(5) if $F(x) = \int_1^x 2t - 2\, dt$.

2. Draw a sketch to illustrate the values $F(\pi)$ and $F(3\pi/4)$ if $F(x) = \int_0^x \sin(t)\, dt$.

3. a) For x > 1, sketch the region whose area is $ln(x) = \int_1^x 1/t\, dt$.
   b) Using the area of rectangles, show that $ln(2) = \int_1^2 1/t\, dt$ satisfies $1/2 < ln(2) < 3/4$.
   c) Using the area of rectangles, show that $ln(3) = \int_1^3 1/t\, dt$ is less than 2.
   d) Is $ln(3) > 1$? How can you show this?

Differentiating Integral-Functions.

4. Evaluate the indicated derivative.

   a) $d/dx \int_3^x (2t^2 + t^3)\, dt$
   b) $d/dt \int_0^t 5e^x\, dx$
   c) $d/ds \int_2^s (4x^2 - 2)\, dx$
   d) $d/dx \int_0^{x^2} (t + 3)\, dt$
   e) $d/dx \int_0^{x^2 + 1} (t + 3)\, dt$
   f) $d/dx \int_1^{x^2 - 2x} (t + 3/t)\, dt$
   g) $d/dt \int_1^{(t+1)^2} (x - x^2)\, dx$
   h) $d/dx \int_{2x}^3 (1 + t)\, dt$
   i) $d/dx \int_{x^2 - 1}^5 (2t + t^4)\, dt$
   j) $d/dx \int_{x^2}^{1 + 3x} t^2\, dt$
   k) $d/ds \int_{SQRT(1-s^2)}^{SQRT(1+s^2)} (t^2 + 1)\, dt$
   l) $d/ds \int_1^{(s^2 - 2s)^6} (t - 1/t)\, dt$

5. Find $F'(\theta)$ if $F(\theta)$ is given by the integral.

   a) $\int_0^\theta \sec^2(t)\, dt$
   b) $\int_0^{\sin(\theta)} 1 + t^2\, dt$
   c) $\int_{-\theta}^\theta (1 + t^2)^{-1/2}\, dt$
   d) $\int_{\sin(\theta)}^{\tan(\theta)} 1 - x^2\, dx$
   e) $\int_\theta^{\pi - \theta} \sin(t)\, dt$
   f) $\int_{\sin(\theta)}^{\cos(\theta)} \ln(x)\, dx$

6. a) Compute the derivative of the Error Function $ERF(x) = (2/SQRT(\pi)) \int_0^x e^{-t^2}\, dt$.
   b) Write the function $y = ERF(x^{0.5})$ as an integral and find its derivative.
   c) Assume that the probability that a measurement X is less than x is given by
   $$P(X \leq x) = ERF(e^{-x}) - ERF(1 - x^3)$$
   i) Express this probability using integrals.
   What is the rate at which the probability changes when   ii) x = 0   and   iii) x = 1 ?

Section 5.5 General Integration Rules.

Introduction.

In Chapter 3 the Chain Rule was used to generalize the specific Derivative Rules. In this section the Integral counterpart of the Chain Rule is introduced and used to form General Integration Rules. These will correspond to the specific integral formulas of Section 5.3. With these General Rules we will be able to integrate more complicated functions involving compositions of elementary functions. Many functions will not match the form of the integrands in these general rules but will be close. They will only differ by a multiplicative constant. A general Method is given for determining the constant when this is the case.

The Integral Chain Rule.

The specific integral Rules for elementary functions given in Section 5.3 do not involve integrating products of functions, with the exception of two trigonometric Rules. Evaluating the integral of a product can be very difficult, but, there is a special type of product that can be integrated directly. Recall how the derivative Chain Rule works:

$$\text{If } y = F(u(x)) \text{ then } y' = D_x y = dF/du \cdot du/dx = F'(u) \cdot u'$$

The Chain Rule always results in a derivative that is the product of two terms. Consequently, when applied in reverse the Chain Rule can be used to integrate a product of two functions, but only when the functions have the specific forms of a composition and a derivative:

> One function must be a composition: $\quad f(u(x))$
>
> The other function must be the derivative of u(x): $\quad u'(x)$

Identifying the function f with the derivative of F, $F'(x) = f(x)$, we can integrate formally

$$y = F(u(x)) = \int y' \, dx = \int F'(u(x)) \cdot u'(x) \, dx$$

THE INTEGRAL FORM OF THE CHAIN RULE.

If $F'(x) = f(x)$ and $u(x)$ is an intermediate function then

$$\int f(u(x)) \cdot u'(x) \, dx = F(u(x)) + C$$

Example 1   **Integrating products using the Chain Rule.**

Problem   Evaluate each of the integrals by identifying the intermediate function u(x), its derivative u'(x), and a function f so that the integrand is of the form $f(u(x)) \cdot u'(x)$. Then, determine an anitderivative F of f and apply the Integral Chain Rule.

     1   $\int 2x(x^2 + 5)^3 \, dx$ 2 $\int \cos(x) e^{\sin(x)} \, dx$ 3 $\int 3 \, \text{SQRT}(x^3 - 3x + 2)(x^2 - 1) \, dx$

Solutions a) Whenever an integral involves a term raised to a power this term is a candidate for the intermediate function u(x). We set

$$u(x) = x^2 + 5 \quad \text{then} \quad u'(x) = 2x$$

Thus, we see that the integrand is exactly the product of u' and $f(u) = u^3$. Since x^3 is the derivative of $F(x) = \frac{1}{4} x^4$, the given integral is evaluated by the Integral Chain Rule as

$$\int 2x(x^2 + 5)^3 \, dx = \int u^3 \cdot u'(x) \, dx = \tfrac{1}{4} u^4(x) + C = \tfrac{1}{4}(x^2 + 5)^4 + C$$

b) The integrand $\cos(x) \, e^{\sin(x)}$ involves an exponential of a function term. As a rule, the term appearing in the exponent is a good candidate for the intermediate function u. Let

$$u(x) = \sin(x) \quad \text{then its derivative is} \quad u'(x) = \cos(x)$$

The function f must be the exponential function e^x, which is its own antiderivative. The integral can be evaluated directly by the Integral Chain Rule

$$\int \cos(x) e^{\sin(x)} \, dx = \int u'(x) \, e^{u(x)} \, dx = e^{u(x)} + C = e^{\sin(x)} + C$$

c) A square root is really just a power function with a fractional exponent, consequently the term under the root is a prime candidate for u(x). The function f must be the root function, $f(x) = \sqrt{x}$. We thus set

$$f(u) = \sqrt{u} \quad \text{and} \quad u(x) = x^3 - 3x + 2$$

The derivative is $u'(x) = 3x^2 - 3$. The 3 can be factored out of u' so that the integral has the required form. An antiderivative of f is

$$F(u) = u^{3/2}/(3/2) = \tfrac{2}{3} u^{3/2}$$

Therefore, applying the Integral Chain Rule gives

$$\int 3 \, \text{SQRT}(x^3 - 3x + 2)(x^2 - 1) \, dx = \int \sqrt{u} \cdot u'(x) \, dx$$

$$= \tfrac{2}{3} u^{3/2} + C = \tfrac{2}{3} (x^3 - 3x + 2)^{3/2} + C$$

General Integral Rules.

The integrals in the above example were evaluated by identifying an intermediate variable u = u(x) so that the integrand is a product of the form $f(u) \cdot u'$. Each specific Integral Rule can be generalized by substituting u(x) for the x-variable in the function being integrated and multiplying by the derivative term u'(x). TABLE I lists the resulting General Integral Rules corresponding to the specific Rules of Section 5.3. To apply these General Rules to integrate a product, say to evaluate $\int P(x) \cdot Q(x) \, dx$, it is necessary to identify the intermediate function u(x) so that one of the factors, P or Q, equals u'(x) and the other factor is an elementary function f, evaluated at u(x). Often, such integrals do not exactly match the form required for the Integral Chain Rule, but are close. They only differ by a multiplicative constant. In such cases, there is a number k such that the integrand

$$P(x) \cdot Q(x) = k \cdot f(u(x)) \cdot u'(x)$$

In which case,

$$\int P(x) \cdot Q(x) \, dx = \int k \cdot f(u(x)) \cdot u'(x) \, dx = k \cdot F(u(x)) + C$$

TABLE I GENERAL DERIVATIVE AND INTEGRAL RULES

Type of Function	General Derivative Rule	General Integral Rule
Power function	$D_x u(x)^r = r u(x)^{r-1} u'(x)$	$\int u(x)^r u'(x)\, dx = u(x)^{r+1}/\{r+1\} + C$ $r \neq -1$
Exponential Base e	$D_x e^{u(x)} = e^{u(x)} u'(x)$	$\int e^{u(x)} u'(x)\, dx = e^{u(x)} + C$
Natural Log	$D_x \ln(u(x)) = u'(x)/u(x)$	$\int u'(x)/u(x)\, dx = \ln(u(x)) + C$ or $\int u'(x)/u(x)\, dx = \ln(|u(x)|) + C$
Basic Trig functions	$D_x \sin(u(x)) = \cos(u(x)) u'(x)$	$\int \cos(u(x)) u'(x)\, dx = \sin(u(x)) + C$
	$D_x \cos(u(x)) = -\sin(u(x)) u'(x)$	$\int \sin(u(x)) u'(x)\, dx = -\cos(u(x)) + C$
	$D_x \tan(u(x)) = \sec^2(u(x)) u'(x)$	$\int \sec^2(u(x)) u'(x)\, dx = \tan(u(x)) + C$
	$D_x \cot(u(x)) = -\csc^2(u(x)) u'(x)$	$\int \csc^2(u(x)) u'(x)\, dx = -\cot(u(x)) + C$
	$D_x \sec(u(x)) = \sec(u(x))\tan(u(x)) u'(x)$	$\int \sec(u(x))\tan(u(x)) u'(x)\, dx = \sec(u(x)) + C$
	$D_x \csc(u(x)) = -\csc(u(x))\cot(u(x)) u'(x)$	$\int \csc(u(x))\cot(u(x)) u'(x)\, dx = -\csc(u(x)) + C$
Exponential Base b	$D_x b^{u(x)} = \ln(b) b^{u(x)} u'(x)$	$\int b^{u(x)} u'(x)\, dx = b^{u(x)}/\ln(b) + C$
Logarithm Base b	$D_x \log_b(u(x)) = u'(x)/\{\ln(b) u(x)\}$	$\int u'(x)/u(x)\, dx = \ln(b) \log_b(u(x)) + C$

Section 5.5   GENERAL INTEGRATION RULES

A GENERAL METHOD FOR EVALUATING INTEGRALS.

1. First, simplify the integrand and apply the Sum/Difference Integral Rule to split integrals involving sums/differences into a sum/difference of integrals.

2. If the integrand is an elementary function of x only, without a product, then it may be evaluated by one of the Specific Integral Rules presented in Section 5.3.

3. When the integrand is a product $P(x) \cdot Q(x)$, you must determine if the Integral Chain Rule can be applied. The simpler factor is usually the derivative $u'(x)$ and the other factor must then involve a composition, having the form $f(u(x))$. This can usually be recognized by parentheses, or when there is a function either in an exponent or as the argument of a *ln* or trigonometric function.

    i. Identify the functions $f(u)$ and $u(x)$ and take the derivative $u'(x)$.

    ii. Set $P(x) \cdot Q(x) = k \cdot f(u(x)) \cdot u'(x)$ and solve for the <u>constant</u> k.

    If you can not solve for k as a numerical constant, say you find $k = 2/x$ for instance, then the integral can not be evaluated by the Integral Chain Rule with the choice of $f(u)$ and $u(x)$ that was made.

Example 1   Applying the General Integral Rules.

Problem   Evaluate the integral

$$I = \int x(x^2 + 1)^4 + e^{-0.1x} + (3x - 6)/(x^2 - 4x + 5) \, dx$$

Solution   The first step is to recognize that this integral can be split into a sum of three integrals. We write

$$I = I_1 + I_2 + I_3$$

Where the three integrals are defined by

$$I_1 = \int x(x^2 + 1)^4 \, dx$$

$$I_2 = \int e^{-0.1x} \, dx$$

$$I_3 = \int (3x - 6)/(x^2 - 4x + 5) \, dx$$

Evaluating the integral I_1. The integrand is the product of x and $(x^2 + 1)^4$. The term raised to the power 4 is our choice for the intermediate function $u(x)$. Then f must be the power function $f(u) = u^4$. Proceeding to step three of the General Method, set

$$u(x) = x^2 + 1, \quad \text{and hence} \quad u'(x) = 2x$$

Next, write an equation to solve for a constant k. It has the form

$$\text{the integrand} = k \cdot u^4(x) \cdot u'(x)$$

$$x(x^2 + 1)^4 = k(x^2 + 1)^4 \cdot 2x$$

Solving for the constant k, the terms to the 4-th power cancel as do the factor of x, to give

$$2k = 1 \quad \text{or} \quad k = 1/2$$

Finally, applying the general Power Rule the integral is evaluated as

$$I_1 = k \cdot u^5/5 = (1/2)(x^2 + 1)^5/5 = 0.1(x^2 + 1)^5 + C_1$$

Evaluating the integral I_2. The integrand is the exponential of a function, -0.1x, and hence to evaluate this integral we must use the general Exponential Rule. The integrand can be considered the product of the function $P(x) = e^{-0.1x}$ and $Q(x) = 1$. The function in the exponent should be chosen as u(x):

$$u(x) = -0.1x \quad \text{and} \quad u'(x) = -0.1$$

Thus, the equation for the constant k is

$$e^{-0.1x} = k \cdot e^{-0.1x} \cdot (-0.1)$$

Canceling the exponential terms gives

$$-0.1k = 1 \quad \text{or} \quad k = -10$$

The integral is then evaluated by the General Exponential Rule as

$$I_2 = \int k\, e^{u(x)}\, u'(x)\, dx = k\, e^u = -10\, e^{-0.1x} + C_2$$

Evaluating the integral I_3. This integrand involves a rational function. The only general integral form that obviously involves a rational form is the integral of u'/u. There are three different forms of this integral listed, but we will usually use the $ln(u(x))$ formula. The function u must be the function in the denominator. Set

$$u(x) = x^2 - 4x + 5 \quad \text{then} \quad u'(x) = 2x - 4$$

The equation for k then has the form integrand = $k \cdot u'/u$:

$$(3x - 6)/(x^2 - 4x + 5) = k(2x - 4)/(x^2 - 4x + 5)$$

As the denominators are the same, we equate the numerators to form the equation

$$k(2x - 4) = (3x - 6)$$

which is solved for k = 3/2. The integral is thus evaluated by the General Natural Log formula:

$$I_3 = k ln(u(x)) = (3/2) ln(x^2 - 4x + 5) + C_3$$

Completing the solution. The final solution is the sum of the three integrals. The three integral constants are added to form a single integration constant $C = C_1 + C_2 + C_3$.

$$I = 0.1(x^2 + 1)^5 - 10e^{-0.1x} + 1.5 ln(x^2 - 4x + 5) + C.$$

Section 5.5 GENERAL INTEGRATION RULES 395

Example 2 **Integrals that can not be directly evaluated by the General Integral Rules.**

In this example we examine three integrals that **can not** be evaluate by General Integral Rules applying the General Method given above. The object of this example is to alert you to difficulties that you may encounter attempting to apply the formulas to integrals for which other techniques are required.

Problem A The integral $\int e^{x^2} dx$ can not be evaluated with elementary functions. The integrand involves an exponential with the function x^2 in the exponent. Therefor, we would expect to apply the general Exponential Rule with

$$u(x) = x^2 \quad \text{and then} \quad u'(x) = 2x$$

With this choice for $u(x)$ the equation for the constant k is

$$\text{the integrand} = k \cdot e^{u(x)} \cdot u'(x)$$

or

$$e^{x^2} = k \cdot e^{x^2} \cdot 2x$$

Canceling the exponential terms leaves the equation $k \cdot 2x = 1$, which solved for k gives

$$k = 1/2x$$

But k is not a constant! k is a function of x. Consequently, we **cannot** apply the general Exponential Rule with this choice of $u(x)$. You should ask "Will another choice of $u(x)$ work?" The answer is no. This integral cannot be evaluated by this approach, or any other approach, as an antiderivative of e to the x^2 can not be expressed by a finite number of elementary functions. (It can be expressed as an infinite series.) To demonstrate this fact is non-trivial and to prove that no such antiderivative exists requires advanced mathematical theorems.

Problem B Consider the integral $\int xe^{2x} dx$.

This integral involves an exponential with exponent 2x, and hence we attempt to evaluate it using the General Exponential Rule with

$$u(x) = 2x \quad \text{and hence} \quad u'(x) = 2$$

Proceeding as in the previous problem, an equation for k is set and found to not have a constant solution:

$$ke^{2x} \cdot 2 = xe^{2x} \quad \text{gives} \quad k = x/2$$

Again, the method fails to find the solution. This integral cannot be evaluated with this choice of $u(x)$ and applying the General Exponential Rule. However, it can be evaluated by a method called "integration by parts" that is presented in Chapter 6, Section 6.2. You may verify by computing its derivative $F'(x)$ that an antiderivative of xe^{2x} is

$$F(x) = (x/2 - 1/4)e^{2x} + C$$

Problem C Consider the integral $\int (5x^2 + 2x + 2)/(x^3 - 1) dx$

The only general formula that involves a quotient is the integral of u'/u. Setting u equal to the denominator gives

$$u(x) = x^3 - 1 \quad \text{and hence} \quad u'(x) = 3x^2$$

The equation ku'/u = integrand is

$$k(3x^2)/(x^3 - 1) = (5x^2 + 2x + 2)/(x^3 - 1)$$

As the denominators are the same, equating the numerators gives

$$3kx^2 = 5x^2 + 2x + 2 \quad \text{or} \quad k = 5/3 + 2/3x + 2/3x^2$$

Again, as k is not a constant but is a function of x this approach fails to work. The integrand does not match any of the General Integral Rules. However, using a technique called "partial fractions", which will be introduced in Chapter 6, Section 6.3, this integral can be evaluated. You can verify by differentiating it, that

$$F(x) = 3ln(x - 1) + ln(x^2 + x + 1) + C$$

is an antiderivative of the integrand.

Example 3   **Identifying the terms needed to evaluate an integral**.

In this example we focus on the terms needed to make an integrand match one of the general integral forms. Integrals are presented with blanks and the object is to determine what the blanks should be so that the integral matches a general form.

The integral Evaluates as:	General Form of the integral	The missing term is:
$\int \underline{}(x^3 - x)^5 \, dx = (x^3 - x)^6$	$\int u'(x)6u^5(x)\,dx$ $u(x) = x^3 - x$	$6u'(x) = 6(3x^2 - 1)$
$\int (2t + 3)e^{\underline{}}\,dt = e^{\underline{}}$	$\int u'(t)e^{u(t)}\,dt$ $u'(t) = 2t + 3$	$u(t) = t^2 + 3t + c$ since $u'(t) = 2t + 3$
$\int \underline{}\sin(3\pi\theta)\,d\theta = \cos(3\pi\theta)$	$\int -u'(\theta)\sin(u(\theta))\,d\theta$ $u(\theta) = 3\pi\theta$	$-u'(\theta) = -3\pi$

EXERCISE SET 5.5

For Exercises 1-12, determine an antiderivative of the given function or evaluate the indicated integral.

1 a) $y = (x + 3)^4$ b) $y = 6x(x^2 + 1)^{-3}$ c) $y = (x^2 + x + 2)^2(6x + 3)$ d) $y = (x^2 + x + 2)6x$
   e) $y = (2x^3 - 1)^5 x^2$ f) $y = (2x + 3)^4 - (2x + 3)^{-2}$ g) $y = (x^2 + 7x - 2)^{10}(2x + 7)$
   h) $y = (5x^{1/5} + 1)^4 / x^{4/5}$ i) $y = x(3x^2 + 6)^{20}$ j) $y = (3x^2 + 2)(x^3 + 2x + 7)^{-1/2}$

2 a) $2e^{2x}$ b) e^{2x} c) e^{2x+1} d) e^{-3x} e) xe^{x^2} f) $x\exp(-3x^2)$
   g) $x\exp(-3x^2+1)$ h) $\cos(x)e^{\sin(x)}$ i) $(x^2 + x + 1)\exp(2x^3 + 3x^2 + 6x - 2)$

3 a) $2/(3x + 1)$ b) $1/(2x + 1)$ c) $2x/(5x^2 + 3)$
   d) $2x^{-3}/(x^{-2} + 3)$ e) $x/(3x^2 - 4)$ f) $x^2/(x^3 + 1)$

Section 5.5   GENERAL INTEGRATION RULES   397

  g) $(x - 2)/(x^2 - 4x + 7)$ h) $(2x - 2)/(x^2 - 2x + 1)$ i) $(1 + \cos(x)) / (x + \sin(x))$

4 a) $\int 2x(x^2 - 3)^4 \, dx$ b) $\int x^3(x^4 + 5)^{-2} \, dx$ c) $\int (x^2 + 2x - 1)^2 \, dx$ d) $\int x(x^2 - 1)^3 \, dx$

5 a) $\cos(3\theta)$ b) $\sin(\theta) - \cos(2\theta)$ c) $\sec(\theta)[\sec(\theta) - \tan(\theta)]$

   d) $\theta \sin(\theta^2 + 3)$ e) $3 \sin(\pi\theta)$ f) $\tan^2(\theta - \theta^2)(1 - 2\theta)$

   g) $\sin^3(\theta) \cos(\theta)$ h) $\tan(\theta/\pi) \sec^2(\theta/\pi)$ i) $\sin^3(\theta)$

6 a) $\int 2e^t \, dt$ b) $\int e^{2x} \, dx$ c) $\int x \exp(2x^2) \, dx$ d) $\int e^{-0.01x} \, dx$

   e) $\int (x^2 + 2)\exp(x^3+6x-2) \, dx$ f) $\int 5te^{t^2} \, dt$ g) $\int \sin(\theta)e^{\cos(\theta)} \, d\theta$ h) $\int t^{-2} e^{1/t} \, dt$

7 a) $\int 2x/(x^2 + 1) \, dx$ b) $\int (t - 1)/(t^2 - 2t + 1) \, dt$

   c) $\int (x - 1)/(2x^2 - 4x + 1) \, dx$ d) $\int (2t + \cos(t))/(\sin(t) + t^2) \, dt$

   e) $\int (20x^3 + 15x^2 - 2)/(5x^4 + 5x^3 - 2x) \, dx$ f) $\int (\sin(R) + R \cos(R))/(R \sin(R)) \, dR$

8 a) $\int_5^x t^3 - 2 \, dt$ b) $\int_0^x 2s^2 - 3 \, ds$ c) $\int_1^s 2x(x^2 + 5)^3 \, dx$

   d) $\int_{-1}^s 2x(x^2 - 1)^4 \, dx$ e) $\int_{-1}^t (x + 1)^2 \, dx$ f) $\int_{-t}^1 x^2 + 1 \, dx$

   g) $\int_0^T t(3t^2 + 2)^{1/2} \, dt$ h) $\int_0^T (2t - 3)(t^2 - 3t + 6)^2 \, dt$

9 a) $\int e^t \, dt$ b) $\int e^{3t+1} \, dt$ c) $\int xe^{4x^2-2} \, dx$ d) $\int x^2 e^{x^3} \, dx$

   e) $\int (x + 3)e^{(x+3)^2} \, dx$ f) $\int e^{-x} \, dx$ g) $\int xe^{3x^2-2} \, dx$ h) $\int e^{\sqrt{x + 1}}/\sqrt{x + 1} \, dx$

10 a) $\int 3 / x \, dx$ b) $\int 1 / x^3 \, dx$ c) $\int x^2 / (x^3 - 1) \, dx$

   d) $\int (x - 1) / (x^2 - 2x + 6) \, dx$ e) $\int e^x / (1 + e^x) \, dx$ f) $\int (1 + e^x) / (x + e^x) \, dx$

   g) $\int (1 + 2x^{-3}) / (-x + x^{-2}) \, dx$ h) $\int 1 / [x(1 + \ln(x))] \, dx$

   i) $\int 1 / [x^3(1 - x^{-2})] \, dx$ j) $\int 2x / [(x^2 + 1)\ln(x^2 + 1)] \, dx$

11 a) $\int 2^x \, dx$ b) $\int 4^{x/2} \, dx$ c) $\int 3^{x+1} \, dx$ d) $\int x 3^{x^2 + 1} \, dx$

   e) $\int e^x 2^{e^x} \, dx$ f) $\int (5x^2 + 4)3^{5x^3 + 12x - 2} \, dx$ g) $\int 1/2^x \, dx$ h) $\int (1/3)^x \, dx$

12 a) $\int_0^{\ln(2)} e^{2x} \, dx$ b) $\int_{-1}^1 xe^{x^2} \, dx$ c) $\int_0^{\ln(2)} e^{-3x} \, dx$

   d) $\int_1^e 1 / x \, dx$ e) $\int_1^{e^2} 1 / x \, dx$ f) $\int_0^{10} 1 / x + 1 \, dx$

   g) $\int_0^{10} 1 / (3x + 1) \, dx$ h) $\int_2^1 3 / x \, dx$ i) $\int_e^{e^2} 1 / x \, dx$

13 Represent the given functions in the form $u^r(x)v(x)$; indicate u and v. Determine if the function has the form necessary to be integrated by the General Power Rule. If not, what would be a simple modification that would give it the proper form?

     a) $y = x(x^2 - 3)^{1/2}$ b) $y = -6(x^2 + 7x - 2)^4(2x + 7)$

     c) $y = (x^2 + 3x - 2)[x^3 + 4.5x - 6x]^5$ d) $y = (2x + 1)^{11} - (2x + 1)(2x + 1)^{-3}$

     e) $y = \sin(x^2)\cos^5(x^2)$ f) $y = e^x(1 - e^x)^{-6}$

14 Which of the two integrals is easier to evaluate? Evaluate the simpler one and indicate why the other one is more difficult.

     a) $\int_0^3 x^5(x^6 + 1)^{1/2} \, dx$ or $\int_0^3 (x^6 + 1)^{1/2} \, dx$

     b) $\int_0^\pi \sin(x^2) \, dx$ or $\int_0^\pi x \sin(x^2) \, dx$

     c) $\int_0^{\pi/2} \cos(x)\sin(x) \, dx$ or $\int_0^{\pi/2} \sin^2(x) \, dx$

Section 5.6 The Area Between Curves

Introduction.

The area under the graph of a positive function was introduced in Section 5.1 and shown in Section 5.2 to be the value of the definite integral of the function over the interval. This concept can be extended to establish the area between the graphs of any two functions. The area between the graphs of f(x) and g(x) is the integral of |f(x) - g(x)|. If the absolute value signs are omitted, the integral of the difference f(x) - g(x) can also be interpreted as an area, called a *signed* or *net* area, which allows negative "areas". This concept of a *net* amount is more often used when $\int_a^b$ f(x) - g(x) dx is interpreted not as an area but as an accumulation of mass or quantities having units associated with the functions, such as Kilo-grams or moles of a chemical substance.

The Area Between Two Curves.

Assume that f and g satisfy the inequality

$$f(x) \geq g(x) \geq 0 \text{ for } x \in [a,b]$$

Then the graph of f is above the graph of g over the interval [a,b]. The region under the graph of g is then a subregion of the region under the graph of f, as illustrated in the sketch. Consequently, the area under the graph of f must be greater that or equal to the area under the graph of g. The area of the shaded region in the figure is the **area between the graphs of f and g**. It is the area under the graph of f minus the area under the graph of g.

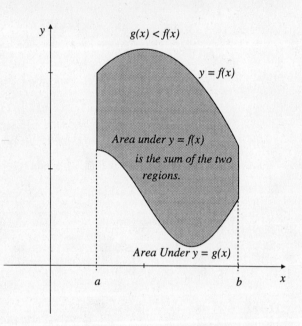

DEFINITION OF THE AREA BETWEEN TWO CURVES.

> If $f(x) \geq g(x)$ on [a,b], then the **Area between the graphs of f and g** is
>
> $$A = \int_a^b f(x)\, dx - \int_a^b g(x)\, dx = \int_a^b f(x) - g(x)\, dx$$

In the definition the condition that both f and g are non-negative was dropped. If $f(x) \geq g(x)$ for $x \in [a,b]$, then the region between the graphs of f and g over an interval [a,b] is a region whose height at any point x is given by

$$h(x) \equiv f(x) - g(x)$$

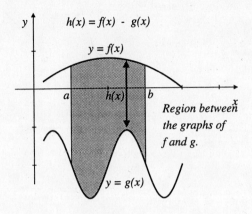

Region between the graphs of f and g.

Since $h(x) \geq 0$ the integral of $h(x)$, or equivalently of the difference $f(x) - g(x)$, must be the non-negative area under the graph of h, which is equivalent to the area between the graphs of f and g. Then, using the Sum/Difference Rule,

$$0 \leq \int_a^b h(x)\, dx = \int_a^b f(x) - g(x)\, dx$$

$$= \int_a^b f(x)\, dx - \int_a^b g(x)\, dx$$

Rearranging this inequality gives a useful integral inequality that is used repeatedly in mathematical analysis.

INTEGRAL INEQUALITY

$$\boxed{\text{If } f(x) \geq g(x) \text{ for } x \in [a,b], \text{ then } \int_a^b f(x)\, dx \geq \int_a^b g(x)\, dx.}$$

The difficulty encountered in many "area problems" is that you are given two functions and asked to determine the area between their graphs but you do not know which is the greater, or even if one is always greater than the other on the specified interval [a,b]. For simple functions that you can graph one approach that is usually helpful is to sketch the curves first. Then, it may be visually apparent which is the greater function. When the functions are more complex, you must rely on algebra to ascertain which is the upper curve. A principle that we repeatedly use is:

> If a continuous function $h(x) \neq 0$ for any $x \in [a,b]$, then $h(x)$ does not change sign in the interval [a,b]. Consequently, if $h(x) \neq 0$ for any x in the interval, then the sign of h on the interval can be determined by evaluating h at <u>any</u> particular point $c \in [a,b]$.

Thus, if the equation $f(x) = g(x)$ has no solution in the integration interval [a,b] then a quick evaluation of both functions at any fixed x-value in the interval will indicate the larger function.

Example 1 Determining areas between curves.

Problem Sketch the region between the given curves and find its area.

  a) The curves are the lines $y = 3x$ and $y = x$, over the interval [0,2].

  b) The curves are the horizontal line $y = 2$ and the curve $y = 2(1 - e^{-x})$, over [0,5].

Solutions a) A quick sketch (below right) of the two linear functions shows that the line with slope 3 is above the line with slope 1 on the interval [0,2]. The region in question is a triangular region. The upper curve is $f(x) = 3x$ and the lower curve is $g(x) = x$. The area of the triangle is

$$A = \int_a^b f(x) - g(x)\, dx$$

$$= \int_0^2 3x - x\, dx = \int_0^2 2x\, dx = x^2 \Big|_0^2 = 4$$

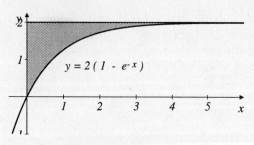
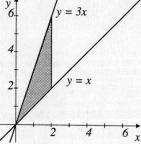

b) The graph on the left illustrates the horizontal line,

$$y = f(x) = 2$$

and the curve

$$y = g(x) = 2(1 - e^{-x})$$

which approaches the line asymptotically as $x \to \infty$. The region between them is thus below the line and above the g curve. Therefor, its area is

$$A = \int_a^b f(x) - g(x)\, dx = \int_0^5 2 - 2(1 - e^{-x})\, dx = \int_0^5 -2e^{-x}\, dx$$

$$= 2e^{-x}\big|_0^5 = -2e^{-5} + 2e^0 = 2(1 - e^{-5}) \approx 1.987$$

Example 2 Determining the area of the finite region between two parabolas.

Problem The parabolas $y = 5x - x^2$ and $y = x^2 - 5x + 4.5$ bound a finite region. What is its area?

Solution The first step is to visualize the region. A quick sketch of the curves reveals that the region R is bounded above by the curve $y = 5x - x^2$ and below by the curve $y = 4.5 - 5x + x^2$. To determine the area of R by integration the limits of the integral must be established. These are the values x_1 and x_2 indicated on the graph. To find these values the equations of the curves are solve simultaneously. Setting

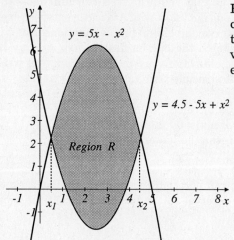

$$5x - x^2 = x^2 - 5x + 4.5$$

and gathering terms gives the quadratic equation

$$2x^2 - 10x + 4.5 = 0$$

Its two roots $x_1 = 0.5$ and $x_2 = 4.5$ are given by the Quadratic Formula:

$$x = (10 \pm \text{SQRT}(100 - 36))/4$$

The area of the region R is thus

$$A = \int_{0.5}^{4.5} (5x - x^2) - (x^2 - 5x + 4.5)\, dx$$

$$= \int_{0.5}^{4.5} -2x^2 + 10x - 4.5\, dx = -(2/3)x^3 + 5x^2 - 4.5x \big|_{0.5}^{4.5}$$

$$= [\,-(2/3)(4.5)^3 + 5(4.5)^2 - (4.5)^2\,] - [\,-(2/3)(0.5)^3 + 5(0.5)^2 - 4.5(0.5)\,]$$

Converting to fractions, this term simplifies to

$$A = [81/4] - [-13/12] = 64/3 = 31.\underline{3}$$

Areas between intersecting curves.

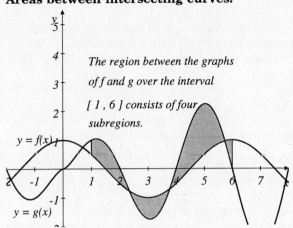

The region between the graphs of f and g over the interval [1, 6] consists of four subregions.

What is the area between two curves that do not satisfy the relationship that one is always greater than the other? If the graphs $y = f(x)$ and $y = g(x)$ have multiple intersection points for $x \in [a,b]$, with the curves crossing over, then the region between them consists of subregions over subintervals of [a,b]. One approach is to say the "area between the curves" is the sum of the areas of these subregions. This area can be expressed as the integral of the absolute value of the difference f - g.

Total Area between the graphs of f and g.

$$A = \int_a^b |f(x) - g(x)| \, dx$$

However, to evaluate this integral one still must know which function is greater to decide if the absolute value represents $f(x) - g(x)$ or $g(x) - f(x)$.

Example 3 — Computing the Total Area between intersecting curves.

Problem Determine the total area of the region between the graphs of $f(x) = 0.5(x + 1)$ and $g(x) = 2 - x^2$ over the interval [0,2].

Solution These function's graphs intersect at the point (1,1). To determine this point algebraically we set $f(x) = g(x)$ and solve for x:

$$0.5(x + 1) = 2 - x^2$$

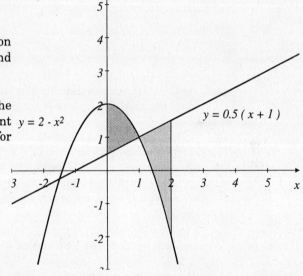

gives the quadratic

$$x^2 + 0.5x - 1.5 = (x - 1)(x + 1.5) = 0$$

whose roots are $x = 1$ and $x = -1.5$. Consequently, on the subinterval [0,1] $g(x) \geq f(x)$ and on [1,2] $f(x) \geq g(x)$.

Thus the region between the two curves consists of two subregions and the total area is found as

$$\int_0^2 |f(x) - g(x)| \, dx = \int_0^1 g(x) - f(x) \, dx + \int_1^2 f(x) - g(x) \, dx$$

$$= \int_0^1 [2 - x^2] - [0.5 + 0.5x] \, dx + \int_1^2 [0.5 + 0.5x] - [2 - x^2] \, dx$$

$$= \int_0^1 -x^2 - 0.5x + 1.5 \, dx + \int_1^2 x^2 + 0.5x - 1.5 \, dx$$

$$= -x^3/3 - 0.25x^2 + 1.5x \big|_0^1 + x^3/3 + 0.25x^2 - 1.5x \big|_1^2$$

$$= 11/12 + 19/12 = 2.5$$

Signed or Net Areas.

A second approach to interpreting the "area" between two curves that intersect and cross each other is to introduce the concept of **signed area,** which allows both positive and negative "areas". With this concept we can consider the **net area between the curves**. Arbitrarily assigning f to be the "upper" function, the integral

$$SA \equiv \int_a^b f(x) - g(x)\, dx$$

is interpreted as the **signed area of the region "below y = f(x) and above y = g(x)"**.

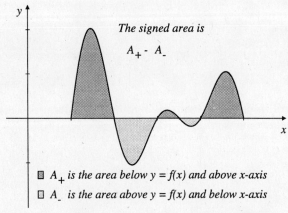

The signed area is

$A_+ - A_-$

■ A_+ is the area below $y = f(x)$ and above x-axis
□ A_- is the area above $y = f(x)$ and below x-axis

This concept can be most easily visualized if we assume that $g(x) = 0$, then its graph is the x-axis and the region between the graph of f and the x-axis consists of portions above the x-axis and portions below the x-axis. Since $g(x) = 0$, we can simply refer to this as the *signed area under the graph of f*:

$$\text{S-AREA}\{f,a,b\} = \int_a^b f(x)\, dx$$

This is the net difference of the area between the graph of f and the x-axis when the graph is above the axis minus the area when the graph is below the axis. Thus

$$\text{S-AREA}\{f,a,b\} = \int_a^b f(x)\, dx = A_+ - A_-$$

Where A_+ is the darker area indicated in the sketch and A_- is the lighter area that is below the axis. If this difference is negative it means that the area below the axis is larger than the area above the axis.

The interpretation of an integral of a difference f - g as a *net* value is more often applied when the integral is interpreted as a quantity other than an area. This occurs when the functions f and g are given units, such as volumes, masses, concentrations, etc., and the x-variable is given a commensurate unit of measure or a time unit.

Example 4   **Computing the Net Volume of blood flow**.

Problem   The volume of blood flowing through a capillary was measured. The actually flow pulsates with heart beats and the volume of flow, v(t), was given by

$v(t) = 0.5 - \sin(\pi t)$ µl/s

for $0 \leq t \leq 2$ s. (A micro liter µl is 10^{-6} l.) Negative values of the function v correspond to back flow.

What is the net flow over the two second period?
What is the average flow rate for this period?

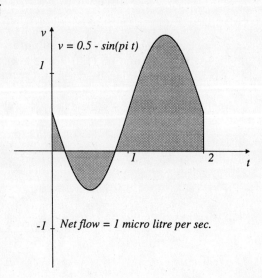

$v = 0.5 - sin(pi\, t)$

Net flow = 1 micro litre per sec.

Solution   The net flow is equivalent to the signed area under the graph of v.

$$\text{Net Flow} = \int_0^2 v(t)\, dt = \int_0^2 0.5 - \sin(\pi t)\, dt$$

$$= 0.5t + \cos(\pi t)/\pi \,\Big|_0^2 = (1 + 1/\pi) - (0 + 1/\pi) = 1 \,\mu l$$

The average flow is the integral of v divided by the length of the time period:

$$\text{AVG}\{v(t); 0, 2\} = \int_0^2 v(t)\, dt / 2 = 0.5 \,\mu l/s$$

☑

Exercise Set 5.6

1. Let $F(x) = \int_1^x f(t)\, dt$. Sketch a graph to illustrate $F(x)$ as an area. For what value of x is $F(x) = 5$?

   a) $f(t) = 3$ b) $f(t) = 3t$ c) $f(t) = 2 - t/5$ d) $f(t) = t - [\![t]\!]$

2. Sketch a graph to illustrate $F(x) = \int_1^x 1/[2t + 1]\, dt$ as an area. For what x does $F(x) = 3$?

3. Sketch the indicated region R and find its area by integration.

   a) R is the region under the curve $y = e^x - x$ over the interval [0,2].

   b) R is a triangle with the interval [3,6] as a base and an apex at (6,3).

   c) R is a triangle with the interval [3,6] as a base and an apex at (3.5,3).

   d) R is the region inside the parabola $y^2 = x$ for $0 < x < 4$.

   e) R is the diamond-shaped region with vertices at (0,0), (1,2), (3,0), and (1,-2).

   f) R is the region between the sine and cosine curves over the interval $[0, 2\pi]$.

4. For the given pair of functions, i) determine the points of intersection of their graphs and sketch their graphs; and ii) compute the area of the finite region(s) bounded by the graphs:

   a) $f(x) = x^2 + 1$; $g(x) = 5$ b) $f(x) = x^2 + 1$; $g(x) = x + 3$

   c) $f(x) = x^2 + 1$; $g(x) = 5 - x^2$ d) $f(x) = x$; $g(x) = \sin(\pi x/2)$

   e) $f(x) = \cos(x)$; $g(x) = \sin(x), |x| < \pi$ f) $f(x) = 2x^2$; $g(x) = 3x^2 - 1$

   g) $f(x) = \text{SQRT}(x)$; $g(x) = x^2$ h) $f(x) = x^3$; $g(x) = x^5$

5. Determine the area of the finite region that is:

   a) bounded by $f(x) = 2x$, $g(x) = x/2$, and $h(x) = 4 - x$;

   b) below the curve $h(x) = 1 - x$ and above the curve $g(x) = x^2$;

   c) between the curves $f(x) = \cos(x)$ and $g(x) = x^2 - \pi^2/4$ and below $h(x) = 2x/\pi + 1$;

   d) below the curve $f(x) = x(4-x)$ and above the curve $g(x) = -x$ and $h(x) = 2(x - 4)$;

   e) below the curve $f(x) = x$, above the curve $g(x) = -x$, and to the left of $y^2 = -x + 4$.

6   What is the area under $f(x) = 3^x$ over the interval $[0,2]$?

7   For what value of b is the area under $y = b^x$ over $[0,5]$ equal to 1?

8   Water flows into a tank at the rate of $5t + 3$ liters/min. It flows out at a rate of t^2 liters/min. If the tank is empty to begin with, what is its volume after 4 minutes? When will it be empty again?

Averages

9   Assume a migrating flock of geese fly for 12 hours and rest for 12 hours, and their speed $s(t)$ mph is given by
$$s(t) = A \sin(\pi t/12) \text{ for } 2n \leq t/12 < 2n + 1$$
$$\text{and } s(t) = 0 \text{ for } 2n + 1 \leq t/12 < 2n + 2 \quad \text{ for each integer n.}$$
       a) What is their average speed over a 121 hr flight period?
       b) What is the distance they would cover in 10 days?

10   If the size of a bacteria culture in a petri dish is given $A(t) = (t + 1)^{-1}\log(t+1)$ cm^2 for $t \geq 0$, what is the average size of the culture over the interval $[0,10]$?

11   The diameter of a tree trunk x meters above the ground is given by $f(x) = (50-x)$ cm.

       a) How high is the tree?   b) What is the average diameter of the tree?

12   The average percentage wage increase in settlements of unionized workers over a five year period fit the curve $y = 3 + [4e^{-0.2t} / \{10 + e^{-0.2t}\}]$. a) What was the average percentage increase over this period? b) If the average salary for this group at the start of this period was \$38,525, what was the average salary at the end of the period?

13   A tumor is estimated to be increasing in mass at a rate of $M' = 300t^5$ g/yr, $t = 30$ days at the time of diagnosis. a) What is its average rate of growth over the first trimester following its diagnosis (120 days)? b) If the tumor is 210g at time of diagnosis, what is its expected size one year later?

Chapter 6   *Methods of Integration*

Section 6.1   Integration by Substitution 406

The *method of substitution* involves the introduction of an intermediate function to evaluate integrals more complicated than those considered in Chapter 5, to which the General Integral Rules could be directly applied. When the method works, upon making the substitution the integral is transformed to an integral involving only the new variable which can then be simplified to a standard form with some algebraic manipulations. Finally, after integrating with the intermediate variable, the antiderivative of the original integral is obtained by substituting back into the original variable.

Section 6.2   Integrating Products: Integration by Parts 417

Integration by Parts is the method that is often used to integrate the product of two functions. It involves expressing one integral as a function term minus another integral. It is the integral equivalent to the Product Rule for derivatives.

Section 6.3   Integrating Rational Functions 425
The Partial Fractions Method.

Partial Fractions is the method applied to evaluate integrals with rational integrands, e.g., $\int f(x)/g(x)\,dx$. This method can be used when the denominator, $g(x)$, is the product: $g(x) = Q_1(x) \cdot Q_2(x)$. In this case the fraction $f(x)/g(x)$ can be expressed as a sum of two "partial fractions", with the Q functions as their denominators. The method works when the new partial fractions are simpler and match the integrand of a basic integration rule.

Section 6.4   Improper Integrals 434

Improper integrals are definite integrals that involve infinity, either as a limit of integration or because the integrand becomes infinite, or "singular" at a point in the integration interval. Integrals over an infinite interval, $(-\infty, b]$, $[a,\infty)$, or $(-\infty,\infty)$, are evaluated by introducing integral-functions with a finite but variable limit of integration. Then, we let this integration limit approach ∞. A similar limit method is employed when an integral is improper because of a *singularity* of the integrand, such as $\int_2^4 1/(x-2)^2\,dx$, where the integrand becomes infinite as $x \to 2$.

Section 6.5   Numerical Integration 444

Numerical integration is used to evaluate integrals of some very common functions, like a Normal Probability function or the ERF(x) function discussed in Chapter 5, Section 5.2, that can not be evaluated explicitly by antiderivative formulas. In many experimental instruments and in real-time control systems, like navigation systems, the function input is numerical data provided by sensors and the only way to evaluate integrals of such functions is numerically.

Section 6.1 Integration by Substitution.

Introduction.

The Integral Chain Rule involves the introduction of an intermediate or substitution variable. Indeed, each of the General Integration Rules is applied by making the substitution of an intermediate variable u, which is actually function u(x). As we shall see in this section, complicated integrals can sometimes be transformed to a form in which they can be algebraically simplified by the substitution method.

To evaluate an integral $\int f(x)\,dx$ by the **substitution method** involves four steps:

(i) Introduce a substitution variable, $u = u(x)$, where the expression $u(x)$ occurs in the function f(x). Solving $u = u(x)$ for x, gives x as a function of u, $x = g(u)$.

(ii) Use the corresponding differential relationship,

$$du = u'(x)\,dx \quad \text{or} \quad dx = g'(u)\,du$$

to transform the integrand and integration variable from x to u:

$$\int f(x)\,dx = \int f(g(u))\,g'(u)\,du$$

(iii) Simplify and then integrate the transformed integral with respect to the substitution variable u, yielding an antiderivative function of u.

(iv) Substitute back to the x-variable. Replace u with the function u(x) in the antiderivative of step (iii) to give an antiderivative of f(x).

The key step of course is making the "right" substitution. Experience is the greatest guide in making the choice. But, as the following discussion and examples illustrate, there are some types of integrands for which standard substitutions usually work.

Linear Substitutions: u = ax + b

The simplest substitution is used when the integrand involves linear terms of the form ax + b, where a and b are constants. The substitution is then

$$u = ax + b \quad \text{and} \quad du = a\,dx$$

or

$$x = (u - b)/a \quad \text{and} \quad dx = (1/a)\,du$$

Linear Substitution

This type of substitution is very helpful when the linear term appears to a fractional power or as the argument of an elementary function.

Example 1   **Linear Substitutions.**

Problem   Evaluate (a) $\int x\,\text{SQRT}(x - 1)\,dx$ and (b) $\int x(4x + 5)^{-1/3}\,dx$.

Solutions (a) The linear term is x - 1. Use the substitution $u = x - 1$. Then $x = u + 1$ and $dx = du$, so the transformed integral is

$$\int x\,\text{SQRT}(x - 1)\,dx = \int (u + 1)\text{SQRT}(u)\,du$$

As $\text{SQRT}(u) = u^{1/2}$, the transformed integral can be expanded algebraically and integrated using the Sum and Power Rules:

$$\int (u + 1)\text{SQRT}(u)\, du = \int u^{3/2} + u^{1/2}\, du = (2/5)u^{5/2} + (2/3)u^{3/2} + C$$

Making the substitution back into the x-variable solves the problem:

$$\int x\, \text{SQRT}(x - 1)\, dx = (2/5)(x - 1)^{5/2} + (2/3)(x - 1)^{3/2} + C$$

(b) The linear term is $4x + 5$. Make the linear substitution $u = 4x + 5$, for which

$$x = (u - 5)/4 \quad \text{and} \quad dx = (1/4)\, du$$

The integral is transformed and evaluated as follows:

$$\int x(4x + 5)^{-1/3}\, dx = \int (u - 5)/4 \cdot u^{-1/3} (1/4)\, du$$

$$= (1/16)\int u^{2/3} - 5u^{-1/3}\, du$$

$$= (1/16)[(3/5)u^{5/3} - (15/2)u^{2/3}] + C$$

$$= (1/16)[(3/5)(4x + 5)^{5/3} - (15/2)(4x + 5)^{2/3}] + C$$

☑

Nonlinear substitutions.

Integrals involving nonlinear terms can be difficult to evaluate directly. Often a substitution to replace the simplest nonlinear expression by a single variable can transform such integrals into forms that are easier to simplify. The following examples introduce standard choices for a new variable which may be employed for similar problems. Remember, normally we must be able to solve the substitution variable $u = u(x)$ for $x = g(u)$.

Example 2 **Integrals involving terms of the form $a + x^r$.**

An integrand involving shifted powers of x, having the form $a + x^r$, can often be simplified by substituting $u = a + x^r$. With this substitution we solve for x and dx as

$$x = (u - a)^{1/r} \quad \text{and} \quad dx = (1/r)(u - a)^{(1/r) - 1}\, du$$

Problem Use a substitution to evaluate $\int 1/(3 + \sqrt{x})\, dx$.

Solution The nonlinear denominator term is $u = 3 + \sqrt{x}$. Solving for x and differentiating gives

$$x = (u - 3)^2 \quad \text{and} \quad dx = 2(u - 3)\, du$$

Substituting, $1/(3 + \sqrt{x}) = 1/u$ and $2(u - 3)\, du$ for dx gives the transformed integral, which is easily evaluated after simplifying:

$$\int 1/u \cdot 2(u - 3)\, du = \int 2 - 6/u\, du = 2u - 6ln(u) + C$$

Substituting back to the x-variable gives the solution

$$\int 1/(3 + \text{SQRT}(x))\, dx = 6 + 2\, \text{SQRT}(x) - 6ln(3 + \text{SQRT}(x)) + C$$

☑

Example 3 An exponential substitution: $u = e^{h(x)}$.

Problem Evaluate $\int (2e^{3x} - 1)/(1 - e^{3x})\, dx$.

Solution As the variable x only appears in the exponential term e^{3x}, we use the substitution

$$u = e^{3x} \quad \text{and then} \quad x = (1/3)ln(u)$$

The equation for the differential dx is the derivative of $(1/3)ln(u)$ times the differential du:

$$dx = D_u(1/3)ln(u)\, du = (1/3u)\, du$$

The transformed integral is therefore

$$\int (2u - 1)/(1 - u) \cdot (1/3u)\, du = 1/3 \int (2u - 1)/(u - u^2)\, du$$

$$= -1/3 \int (2u - 1)/(u^2 - u)\, du$$

$$= -1/3\, ln(u^2 - u) + C$$

Substituting back to the x-variable, we set $u = e^{3x}$ to give

$$\int (2e^{3x} - 1)/(1 - e^{3x})\, dx = -1/3\, ln(e^{6x} - e^{3x}) + C$$

☑

Example 4 Deriving the Gompertz sigmoidal growth function.

Background The Gompertz function is used to describe populations or substances that exhibit limited growth. It involves a double exponential of the form

$$y = M e^{-ce^{-rt}} \qquad \text{GOMPERTZ EQUATION}$$

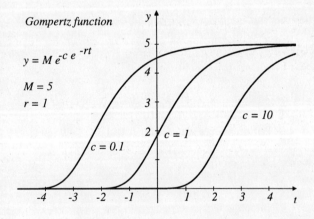

Gompertz function

$y = M e^{-c e^{-rt}}$

$M = 5$
$r = 1$

The graph of this equation is "sigmoidal" or "s-shaped" as illustrated in the sketch where the curve for three different c-values are drawn. Notice that varying c in effect creates a horizontal shift in the curve but does not change its shape or asymptotes. The parameter r is called the "natural" growth rate. As time increases, i.e. as $t \to \infty$, this function approaches a limiting size, M.

The Gompertz function is derived from a continuous growth model that is a generalization of the

Exponential Growth Model $y' = ry$

The generalization is made by replacing the constant growth rate r by the "saturating" growth rate $rln(M/y)$. Notice that as $y \to M$, $ln(M/y) \to ln(1) = 0$. The Gompertz Growth Model has the form

$$y' = rln(M/y) \cdot y \qquad \text{Gompertz Growth Model}$$

The Gompertz function is the solution of this differential equation which is found by a method called "separation of variables", which is considered in Chapter 8. Applying this method results in equating two integrals, one with respect to t and one having the function variable y as the integration variable. Integrating each separately, equating the antiderivatives, and then solving for y as a function of t gives the Gompertz function. In the following problem we evaluate these integrals.

Problem Determine the Gompertz function by integrating and solving the integral equation

$$\int -1/\{y\,ln(M/y)\}\,dy = \int -r\,dt \qquad \text{for } y < M$$

Solution As r is a constant, the integral on the right is just

$$\int -r\,dt = -rt + c$$

The y-integral looks difficult, but it can be integrated after a single substitution. Replacing the nonlinear term by a new variable u, we set

$$u = ln(M/y) \quad \text{and then} \quad du = -1/y\,dy$$

Remember that M is a constant. Hence,

$$D_y ln(M/y) = D_y[ln(M) - ln(y)] = D_y ln(M) - D_y ln(y) = -1/y$$

Separating the product in the denominator, we can directly transform the y-integral to one in the variable u, which is easily integrated:

$$\int -1/\{y\,ln(M/y)\}\,dy = \int 1/ln(M/y) \cdot -1/y\,dy = \int 1/u\,du = ln(u) + c$$

Substituting back into the y variable and equating this antiderivative to the first integral gives the following equation, in which the integration constants have been combined,

$$ln(ln(M/y))) = -rt + c$$

To obtain the Gompertz function this is solved for y. Taking the exponential of both sides:

$$ln(M/y) = e^{-rt + c} = Ce^{-rt}$$

where the new constant $C = e^c$. Taking the exponential again removes the last ln :

$$M/y = \exp(Ce^{-rt})$$

Finally, solving for y gives the Gompertz equation: $y = M \exp(-Ce^{-rt})$.

Evaluating definite integrals with substitutions.

There are two ways to evaluate a definite integral $I = \int_a^b f(x)\,dx$ when using a substitution $u = g(x)$ to find an antiderivative.

The first way is to simply use the substitution to find an antiderivative F(x) of f(x), as in the previous examples. Then apply the Fundamental Theorem of Calculus: $I = F(b) - F(a)$.

The second approach, that can sometimes save work and make the evaluation easier, is to transform the limits of integration so that the transformed integral in the substitution variable is a definite integral, which can be evaluated without substituting back to the x-variable.

The limits are normally transformed directly. If the substitution is u = u(x) then the transformed integral will have limits A = u(a) and B = u(b). i.e. if solving for x gives x = g(u) the transformation is

$$\int_a^b f(x)\, dx \quad \Longleftrightarrow \quad \int_A^B f(g(u))\, g'(u)\, du$$

Caution: this method depends on the substitution giving a one-to-one correspondence between the x-interval [a,b] and a u-interval, which we denote as [A,B].

Example 5 **Evaluating a definite integral**.

Problem Evaluate $I = \int_7^{15} x^2/\text{SQRT}(3x+4)\, dx$.

Solution Taking the first approach, we use a substitution to find an antiderivative of $x^2/\text{SQRT}(3x+4)$ and then apply the FTC. We make the non-linear substitution $u = \text{SQRT}(3x+4)$. Then

$$x = (u^2 - 4)/3 \quad \text{and} \quad dx = 2u/3\, du$$

The transformed indefinite integral is

$$\int [(u^2 - 4)/3]^2 / u \cdot (2u/3)\, du = 2/27 \int (u^2 - 4)^2\, du = 2/27 \int u^4 - 8u^2 + 16\, du$$

$$= (2/27)[\, u^5/5 - (8/3)u^3 + 16u\,]$$

Substituting back into the x-variable gives the antiderivative

$$F(x) = (2/27)[\, \text{SQRT}(3x+4)^5/5 - (8/3)\,\text{SQRT}(3x+4)^3 + 16\,\text{SQRT}(3x+4)\,]$$

Then the value of $I = F(15) - F(7)$. We leave this calculation to the reader at this point. To evaluate this you will repeatedly evaluate the square root term at one of two values, these values are precisely the limits of integration that arise if one transforms the limits of integration.

Taking the second approach, which we can do since the square root function $u(x) = \text{SQRT}(3x+4)$ is one-to-one, the limits of integration for the transformed integral are derived from the corresponding limits for the original integral:

the lower limit $x = 7 \longrightarrow u = \text{SQRT}(3\cdot 7 + 4) = \text{SQRT}(25) = 5$;

the upper limit $x = 15 \longrightarrow u = \text{SQRT}(3\cdot 15 + 4) = \text{SQRT}(49) = 7$.

Thus, the substitution results in the definite u-integral over [5,7]:

$$\int_7^{15} x^2/\text{SQRT}(3x+4)\, dx = 2/27 \int_5^7 u^4 - 8u^2 + 16\, du$$

$$= 2/27\, [u^5/5 - 8u^3/3 + 16u]\,\big|_5^7$$

$$= 2/27\, [(7^5 - 5^5)/5 - (8/3)(7^3 - 5^3) + 16(7-5)] \approx 162.005.$$

This is precisely the answer you should get when you evaluate $F(15) - F(7)$. The second approach technically requires the same numerical evaluation but omits the added complexity that arises when we substituted back to x to get F(x).

Section 6.1   INTEGRATION BY SUBSTITUTION   411

Trigonometric Substitutions for $a^2 + x^2$, $a^2 - x^2$, or $x^2 - a^2$.

Integrals involving quadratic functions can often be evaluated by using simple "trigonometric" substitutions. The reason why these substitutions "work" is because of the Pythagorean Identities:

$$\sin^2(\theta) + \cos^2(\theta) = 1, \quad \sec^2(\theta) = 1 + \tan^2(\theta) \quad \text{or} \quad \sec^2(\theta) - 1 = \tan^2(\theta)$$

These identities facilitate the conversion of a sum or difference involving a squared term into a single squared term, which usually simplifies an expression. There are three basic trig-substitutions, depending on the form of the integrand.

SIN SUBSTITUTION   For $a^2 - x^2$ use the substitution: $x = a \sin(\theta)$ and $dx = a \cos(\theta) \, d\theta$.

Then, $a^2 - x^2 = a^2 - a^2 \sin^2(\theta) = a^2[1 - \sin^2(\theta)] = a^2 \cos^2(\theta)$.

To substitute for θ after integrating use $\theta = \sin^{-1}(x/a)$.

TAN SUBSTITUTION   For $a^2 + x^2$ use the substitution: $x = a \tan(\theta)$ and $dx = a \sec^2(\theta) \, d\theta$,

Then, $a^2 + x^2 = a^2 + a^2 \tan^2(\theta) = a^2[1 + \tan^2(\theta)] = a^2 \sec^2(\theta)$.

To substitute for θ after integrating in this case use $\theta = \tan^{-1}(x/a)$.

SEC SUBSTITUTION   For $x^2 - a^2$ use the substitution: $x = a \sec(\theta)$ and $dx = a \sec(\theta)\tan(\theta) \, d\theta$.

Then, $x^2 - a^2 = a^2\sec^2(\theta) - a^2 = a^2[\sec^2(\theta) - 1] = a^2 \tan^2(\theta)$.

To substitute for θ after integrating use $\theta = \sec^{-1}(x/a)$.

The three substitutions are illustrated by right triangles having angle θ and side lengths determined by x and the constant a:

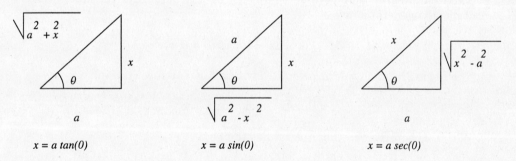

$x = a\,tan(\theta)$ $x = a\,sin(\theta)$ $x = a\,sec(\theta)$

With these substitutions the transformed integral will have θ as the variable of integration. After integrating the transformed θ-integral the antiderivtive must be converted back from the θ-variable to the x-variable. The above triangles can be used to make this conversion. Using the representation of trig functions as the ratios of sides of the triangle any of the trigonometric functions can be expressed as functions of x. For instance, if the substitution $x = 5 \sin(\theta)$ is made, $a = 5$, and if after integrating with respect to θ the antiderivative involves the term $\cos(\theta)$, we would utilize the middle triangle and the relationship that

$$\cos(\theta) = [\text{side adjacent}]/[\text{hypotenuses}]$$

to give $\cos(\theta) = \text{SQRT}(25 - x^2)/5$

Example 6 Using a sin(θ) substitution.

Problem Evaluate $\int 1/\text{SQRT}(1 - x^2)\, dx$.

Solution The constant $a = 1$ and the substitution is $x = \sin(\theta)$. Then $dx = \cos(\theta)\, d\theta$. The transformed integral is

$$\int [1/\text{SQRT}(1 - \sin^2(\theta))] \cdot \cos(\theta)\, d\theta = \int \cos(\theta)/\text{SQRT}(\cos^2(\theta))\, d\theta = \int 1\, d\theta = \theta + C$$

To substitute back to the x-variable we use the inverse sine function. This function is defined for $-\pi/2 \le \theta \le \pi/2$ by

$$\sin(\theta) = x \iff \theta = \text{Arcsin}(x) \quad \text{or} \quad \theta = \sin^{-1}(x) \textbf{ which is not } 1/\sin(\theta)!$$

Therefore, the integral is

$$\int 1/\text{SQRT}(1 - x^2)\, dx = \text{Arcsin}(x) + C$$

Example 7 Using a tan(θ) substitution.

Problem Evaluate the integral $I = \int (4x + 1)/(x^2 + 1)\, dx$.

Solution The integrand can be split int the sum of two terms

$$(4x + 1)/(x^2 + 1) = 4x/(x^2 + 1) + 1/(x^2 + 1)$$

Hence, the integral I is the sum of two integrals

$$I = \int 4x/(x^2 + 1)\, dx + \int 1/(x^2 + 1)\, dx$$

The first integral can be evaluated directly by a General Integral Rule. If $u(x) = x^2 + 1$, then it has the general form

$$\int 2u'(x)/u(x)\, dx = 2ln(u(x)) = 2ln(x^2 + 1)$$

The second integral can be evaluated with a tangent substitution. Set

$$x = \tan(\theta) \quad \text{and} \quad dx = \sec^2(\theta)\, d\theta$$

With this substitution the integral has the form

$$\int 1/(x^2 + 1)\, dx = \int 1/(\tan^2(\theta) + 1) \cdot \sec^2(\theta)\, d\theta$$

Using the identity $1 + \tan^2(\theta) = \sec^2(\theta)$, this integral becomes

$$\int \sec^2(\theta)/\sec^2(\theta)\, d\theta = \int 1\, d\theta = \theta = \tan^{-1}(x)$$

In substituting back into the x-variable we introduced the inverse tangent function denoted by $\text{Arctan}(x)$ or $\tan^{-1}(x)$. For $-\pi/2 < \theta < \pi/2$,

$$x = \tan(\theta) \iff \theta = \tan^{-1}(x) \text{ or } \theta = \text{Arctan}(x)$$

The original integral is the sum of the two integrals evaluated above,

$$I = \int (4x + 1)/(x^2 + 1)\, dx = 2ln(x^2 + 1) + \tan^{-1}(x) + C$$

Section 6.1 INTEGRATION BY SUBSTITUTION

The above examples establish the integrals

$$\int 1/\sqrt{1-x^2}\,dx = \sin^{-1}(x) \quad \text{and} \quad \int 1/(1+x^2)\,dx = \tan^{-1}(x).$$

The inverse sine and tangent functions can actually be defined by these equations. The corresponding derivative equations provide two new derivative rules to add to those established in Chapter 3.

DERIVATIVES OF $\tan^{-1}(x)$ AND $\sin^{-1}(x)$

$$D_x \operatorname{Arcsin}(x) = D_x \sin^{-1}(x) = 1/\sqrt{1-x^2}$$

$$D_x \operatorname{Arctan}(x) = D_x \tan^{-1}(x) = 1/(x^2+1)$$

Integrals involving terms of the form $1/(x^2+a^2)$ and $1/\sqrt{a^2-x^2}$, a a constant, occur frequently in applications. To evaluate them we state the corresponding General Integration Rules, which can be verified by differentiating, using the Chain Rule and the above derivative rules.

INTEGRALS OF $(A^2 + U^2(X))^{-1}$ AND $(A^2 - U^2(X))^{-1/2}$

$$\int u'(x)/(a^2 + u^2(x))\,dx = (1/a)\tan^{-1}(u(x)/a) + C \qquad \textbf{Arctan Rule}$$

$$\int u'(x)/\sqrt{a^2 - u^2(x)}\,dx = \sin^{-1}(u(x)/a) + C \qquad \textbf{Arcsin Rule}$$

Example 8 **Applying the $\tan^{-1}$ and $\sin^{-1}$ integral rules.**

Problem Evaluate (a) $\int 3/(49 + 9x^2)\,dx$ and (b) $\int 8x/\sqrt{25 - 16x^4}\,dx$.

Solution (a) To apply the general Arctan integral rule we set

$$a = \sqrt{49} = 7 \quad \text{and} \quad u = 3x$$

Then $u' = 3$ and the Arctan Rule gives

$$\int 3/(49 + 9x^2)\,dx = \int u'/(7^2 + u^2)\,dx = (1/7)\tan^{-1}(3x/7) + C$$

(b) To apply the general Arcsin rule, set $a = \sqrt{25}$ and $u = 4x^2$. Then $u' = 8x$ and the Arcsin Rule gives

$$\int 8x/\sqrt{25 - 16x^4}\,dx = \sin^{-1}(4x^2/5) + C$$

Example 9 Using a sec(θ) substitution.

Problem Evaluate $I = \int 1/[x \cdot \text{SQRT}(x^2 - 36)]\, dx$.

Solution The constant $a = 6$ and the substitution is $x = 6\sec(\theta)$. Then $dx = 6\sec(\theta)\tan(\theta)\, d\theta$. The transformed integral is

$$\int 6\sec(\theta)\tan(\theta) / [6\sec(\theta)\, \text{SQRT}(36\sec^2(\theta) - 36)]\, d\theta$$

As $\text{SQRT}(36\sec^2(\theta) - 36) = \text{SQRT}(36\tan^2(\theta)) = 6\tan(\theta)$, assuming that $0 \le \theta < \pi/2$, the integral reduces to

$$I = \int 1/6\, d\theta = \theta/6 + C = (1/6)\sec^{-1}(x) + C$$

☑

Example 10 Integrating SQRT(1 + x²)/x.

Problem Evaluate the integral $I = \int (1 + x^2)^{1/2}/x\, dx$.

Solution Making the substitution $x = \tan(\theta)$ and $dx = \sec^2(\theta)\, d\theta$

$$I = \int (1 + \tan^2(\theta))^{1/2}/\tan(\theta)\ \sec^2(\theta)\, d\theta$$

$$= \int \sec(\theta)\cdot\sec^2(\theta) / \tan(\theta)\, d\theta$$

$$= \int \sec(\theta)\cdot(1 + \tan^2(\theta)) / \tan(\theta)\, d\theta$$

$$= \int \sec(\theta)/\tan(\theta) + \sec(\theta)\tan(\theta)\, d\theta$$

$$= \int \sec(\theta)/\tan(\theta)\, d\theta + \int \sec(\theta)\tan(\theta)\, d\theta$$

As $D_\theta \sec(\theta) = \sec(\theta)\tan(\theta)$, the second integral is just $\sec(\theta)$. In the first integral the integrand is simplified by converting to sines and cosines:

$$\sec(\theta)/\tan(\theta) = \{1/\cos(\theta)\}/\{\sin(\theta)/\cos(\theta)\} = 1/\sin(\theta) = \csc(\theta)$$

Thus the integral I is given by

$$I = \int \csc(\theta)\, d\theta + \sec(\theta) + C$$

To evaluate the integral of $\csc(\theta)$ we use a mathematical ploy that is not obvious but yields an exact integral. We multiply and divide by the same term. The term[†] is the sum

$$u(\theta) = \csc(\theta) + \cot(\theta)$$

When this is done the resulting integral has the form $-\int u'/u\, d\theta = -\ln(u)$. The minus sign arises because of the minus sign in the derivatives of $\cot(\theta)$ and $\csc(\theta)$. Thus

$$\int \csc(\theta)\, d\theta = \int \csc(\theta)[\csc(\theta) + \cot(\theta)] / [\csc(\theta) + \cot(\theta)]\, d\theta$$

$$= \int [\csc^2(\theta) + \csc(\theta)\cot(\theta)] / [\csc(\theta) + \cot(\theta)]\, d\theta$$

$$\int \csc(\theta)\, d\theta = -\ln[\,\csc(\theta) + \cot(\theta)\,] \qquad \textbf{Integral of csc(θ).}$$

[†]To integrate $\sec(\theta)$ you use the term $u = \sec(\theta) + \tan(\theta)$.

Although its not needed in this example, a similar approach will render the formula

$$\int \sec(\theta)\, d\theta = -ln[\, \sec(\theta) + \tan(\theta)\,]$$ **Integral of sec(θ).**

Using this last integration, the θ antiderivative of the original integral I is given by

$$I = -ln(\csc(\theta) + \cot(\theta)) + \sec(\theta) + C$$

To substitute back into the x-variable we refer to the triangle for the tangent substitution with $a = 1$. The required identities are

$$\csc(\theta) = SQRT(1 + x^2)/x \quad \sec(\theta) = SQRT(1 + x^2) \quad \text{and} \quad \cot(\theta) = 1/x.$$

Making these substitutions finally gives the original integral as a function of x:

$$I = -ln[\, (SQRT(1 + x^2) + 1)/x\,] + SQRT(1 + x^2) + C$$

☑

EXERCISE SET 6.1

1. Evaluate the following integrals using a linear substitution.

   a) $\int 1/SQRT(x+3)\, dx$ b) $\int x/SQRT(x+3)\, dx$ c) $\int x^2 SQRT(x+3)\, dx$

   d) $\int x\, SQRT(2x+5)\, dx$ e) $\int x(-8x+2)^{1/5}\, dx$ f) $\int x(3x-5)^{0.8}\, dx$

   g) $\int x^2/SQRT(1-2x)\, dx$ h) $\int (x+2)/SQRT(x-2)\, dx$ i) $\int x/(x+3)^{1/3}\, dx$

   j) $\int x(3x-2)^{1/3}\, dx$ k) $\int x/(2x-5)^{2/3}\, dx$ l) $\int x(5x+4)^{2/3}\, dx$

2. Use the indicated substitution to evaluate the integral.

   a) $\int 5/(x - x^{4/5})\, dx$; $u = x^{1/5}$ b) $\int 1/(x + x^{3/5})\, dx$; $u = x^{1/5}$

   c) $\int 2x^5 SQRT(1 - x^3)\, dx$; $u = 1 - x^3$ d) $\int 2x^5 SQRT(1 - x^3)\, dx$; $u = SQRT(1 - x^3)$

   e) $\int 2x^8 SQRT(1 - x^3)\, dx$; $u = 1 - x^3$ f) $\int 2x^8 SQRT(1 - x^3)\, dx$; $u = SQRT(1 - x^3)$

   g) $\int x^{-2} SQRT(1/x - 1)\, dx$; $u = 1/x$ h) $\int x^{-3} SQRT(1/x - 1)\, dx$; $u = 1/x$

   i) $\int 1/[1 - e^{-x}]\, dx$; $u = e^x$ j) $\int 1/[x^{1/2} - x^{1/3}]\, dx$ $u = x^{1/6}$, then $z = u - 1$.

   k) $\int 3/(x^{1/2} - x^{1/4})\, dx$; $u = x^{1/4}$ l) $\int 3/(x^{1/2} + x^{1/3})\, dx$; $u = x^{1/6}$

3. Determine an antiderivative of the indicated function using a trig substitution.

   a) $f(x) = SQRT(1 - x^2)$ b) $f(x) = (1 + x^2)^{-1/2}$ c) $f(x) = SQRT(4 - x^2)$

   d) $f(x) = (25 + x^2)^{-1/2}$ e) $f(x) = SQRT(4 - 9x^2)$ f) $f(x) = (25 + 9x^2)^{-1/2}$

   g) $f(x) = (4 + 9x^2)^{-1}$ h) $f(x) = (4 + 9x^2)^{-1/2}$ i) $f(x) = 1/SQRT(4 - x^2)$

   j) $f(x) = 1/(3 - 2x^2)$ k) $f(x) = 1/[x\, SQRT(1 - x^2)]$ l) $f(x) = x^2/SQRT(4 + x^2)$

4 a) Verify, by differentiating both sides of the equation, that

$$\int 1/\mathrm{SQRT}(x^2 - a^2)\, dx = ln(x + \mathrm{SQRT}(x^2 - a^2))$$

b) Derive the identity of part a) by using a substitution to evaluate the integral.

5 When an integrand involves a quadratic with an x-term the quadratic must first be algebraically manipulated to obtain a form suitable for a trigonometric substitution. This process is called *completing the square*, as discussed in Chapter 1. A quadratic term with an x-term present can be rearrange to the *shifted* form

$$A x^2 + Bx + C = b + m(x - a)^2$$

where $\qquad m = A \qquad a = -B/2A \qquad$ and $\qquad b = C - (B/2A)^2$

To apply a trig substitution first make a linear substitution $u = x - a$. Note that $du = dx$. Then, make a substitution $u = m^{1/2} \sin(\theta)$ when $m > 0$ or $u = (-m)^{1/2} \tan(\theta)$ when $m < 0$. Use this technique to evaluate:

a) $\int 1/\mathrm{SQRT}(2 - x - x^2)\, dx$
b) $\int 1/\mathrm{SQRT}(3x^2 - 6x + 9)\, dx$

c) $\int x/\mathrm{SQRT}(x^2 + 2x)\, dx$
d) $\int 1/\mathrm{SQRT}(x^2 - 6x)\, dx$

6 Evaluate the given definite integral using a substitution. Indicate the limits of integration for the transformed integral.

a) $\int_0^6 1/\mathrm{SQRT}(x + 3)\, dx$
b) $\int_0^4 x/\mathrm{SQRT}(x + 12)\, dx$

c) $\int_{-1}^1 x/(x + 2)^{1/3}\, dx$
d) $\int_{-1}^1 x(4x - 5)^{1/3}\, dx$
e) $\int_{-1}^1 x/(x + 2)^{2/3}\, dx$

f) $\int_{-1}^1 x(4x - 5)^{2/3}\, dx$
g) $\int_0^6 1/\mathrm{SQRT}(81 - x^2)\, dx$
h) $\int_0^4 3/\mathrm{SQRT}(9 + x^2)\, dx$

7 Determine the area of the pie-shaped sector of the circle $x^2 + y^2 = 25$ between the points (5,0) and (3,4). Hint, draw a sketch of the region first to establish the integrands and limits of integration.

8 The marginal production rate for a milling machine t hours after it is placed in service is $MP = t/\mathrm{SQRT}(9 + t^2)$ tons/h. What is the amount of material milled by the machine over an 8h shift following start up?

9 Determine the area of the finite region bounded by $y = \mathrm{SQRT}(x^2 - x^3)$ and the x-axis.

Section 6.2 Integrating Products: Integration by Parts.

Introduction.

In this section a method called "Integration by Parts" is introduced to evaluate integrals that involve products that are not of the form $f(u(x)) \cdot u'(x)$ used in the Integral Chain Rule. To establish a rule for integrating a product we naturally look to the Product Rule for derivatives. The corresponding integration rule would require an integrand that has a sum of two products[†]:

$$D_x[f(x)g(x)] = f'(x)g(x) + f(x)g'(x) \quad \text{Product Derivative Rule}$$

$$\Updownarrow$$

$$\int \{ f'(x)g(x) + f(x)g'(x) \} \, dx = f(x)g(x) \quad \text{Product Integration Rule}$$

or, applying the Sum Rule,

$$\int f'(x)g(x) \, dx + \int f(x)g'(x) \, dx = f(x)g(x)$$

Solving this equation for the first integral gives an equation that expresses the integral of a product, albeit a special type of product, as a related product minus another integral.

INTEGRATION BY PARTS RULE

$$\int f'(x)g(x) \, dx = f(x)g(x) - \int f(x)g'(x) \, dx$$

A simpler way to write this Rule is to utilize differentials.

Recall. The differentials of the functions u and v are $du = u'(x) \cdot dx$ and $dv = v'(x) \cdot dx$.

Utilizing u and v, without indicating the variable as u(x), instead of the functions f(x) and g(x) gives an alternative "differential" or "short" form of this rule:

DIFFERENTIAL FORM: INTEGRATION BY PARTS RULE

$$\int u \, dv = uv - \int v \, du$$

[†] The constant of integration will be omitted in the integration by parts formula, for brevity. In evaluating specific integrals, the integration constant will be added at the last step in the integration process.

Integration by Parts shall be referred to as the IP method. Application of the IP method to evaluate an integral, $I = \int h(x)\, dx$ where $h(x)$ is a product of terms, involves four steps:

Step 1   Express the integrand as a product of two terms: $h(x) = u(x) \cdot v'(x)$
Then, using the differential $dv = v'(x) \cdot dx$,

$$I = \int u\, dv$$

Step 2   Determine the integral $v(x) = \int v'(x)\, dx$ and the differential $du = u'(x)\, dx$.

Step 3   Apply the IP formula to write $I = uv - \int v\, du$.

Step 4   Evaluate the integral $\int v\, du$ and substitute this into the IP formula of Step 3 to give the integral I.

The IP Rule is different than the integration Rules introduced previously. It does not evaluate an integral explicitly. Rather, it equates one integral to a product of functions **minus** another integral. It just trades one integration problem for another. How can that help? If we can't integrate one product why should we be able to integrate a different product? The answer hinges on the functions u and v'. As we know, some functions are more easily integrated than others. The Integration by Parts formula is used when the $\int u\, dv$ integral can not be evaluated directly and "works" when the second integral $\int v\, du$ can be evaluated. This occurs when u and v' can be chosen so that $u' \cdot v$ is a simpler expression than the original $u \cdot v'$. Thus the key step in applying the IP formula is to choose which factor of the original integrand should be u and which factor should be v'.

How to choose the functions u and v'.

Given an integral $\int h(x)\, dx$, the first step is to express h as a product:

$$h(x) = u(x) v'(x)$$

The objective in choosing the functions u and v' is that $v'(x)$ can be integrated and that $u(x)$ becomes simpler when differentiated, so that $\int v\, du$ can be evaluated.

NOTE   The wrong choice for u and v' can lead to a more difficult integral than you started with!

To make the choice, it is helpful to recall how functions change when they are differentiated or integrated. The following general observations may provide insight.

Polynomials   The derivative $P'(x)$ is generally simpler than a polynomial $P(x)$. Hence, polynomials are a good choice for u. Although integrating a polynomial increases its complexity, polynomials are easy to integrate and hence are also sometime chosen as v'.

Powers or Roots   Differentiating and integrating a simple power, like $x^{3/5}$, does not change its complexity. Differentiating a power term like $[g(x)]^r$ adds the factor $g'(x)$ and increases complexity. To integrate a general power term the derivative must be present; however, the integral of $v' = g^r(x) g'(x)$ is actually less complex.

Exponentials   An exponential e^{kx}, k a constant, does not change in complexity when differentiated or when integrated. It is easy to integrate and hence a good choice for v'.

An exponential with a function in the exponent, $e^{f(x)}$, increases in complexity when differentiated: $D_x e^{f(x)} = f'(x) e^{f(x)}$. However, the integral of $dv = e^{f(x)} f'(x)$ is actually simpler, as it no longer involves a product.

Section 6.2 INTEGRATING PRODUCTS: INTEGRATION BY PARTS 419

Logarithms The derivative $D_x ln(x) = 1/x$ is much simpler than $ln(x)$. The integral of $ln(x)$ is more complex, as will be seen in an Example. Log functions are usually a good choice for the u term.

Trig functions The basic trig functions $sin(x)$ and $cos(x)$ do not change their complexity when differentiated or when integrated; they are usually chosen as v'.

Based on the above observations, a general guide for choosing u an v' is:

Choose the function u(x) to be a polynomial or *ln* function, if they are factors, followed in preference by simple exponential and elementary trig functions.

Choose the function v'(x) as a product that corresponds to a general Integral Rule, e.g., $g'(x)g^r(x)$ or $g'(x)e^{g(x)}$ if such terms are present, followed, in order of preference, by simple exponentials, elementary trig functions, and then the lowest order polynomial factors.

To facilitate evaluating integrals using the IP method, it is helpful to create a template for writing down the functions u and v and their derivatives:

$$u = \underline{\hspace{2cm}} \qquad v' = \underline{\hspace{2cm}}$$

$$u' = \underline{\hspace{2cm}} \qquad v = \underline{\hspace{2cm}}$$

You first choose the functions u and v' and then compute the other two functions.

Example 1 Integrating the product of an exponential and a polynomial.

Problem Evaluate $\int xe^{2x} \, dx$.

Solution This integral does not match any of the General Integral Formulas. It involves a product so we try the Integration by Parts method. Choose the polynomial for u, set u = x. This leaves the exponential for v', $v' = e^{2x}$. The derivative of u is u' = 1, a simpler function by far. The integral of v' is $v = (1/2)e^{2x}$. The corresponding function matrix is

$$u = \underline{\ \ x\ \ } \qquad v' = \underline{\ \ e^{2x}\ \ }$$

$$u' = \underline{\ \ 1\ \ } \qquad v = \underline{\ \ e^{2x}/2\ \ }$$

The IP formula then gives

$$\int xe^{2x} \, dx = xe^{2x}/2 - \int 1 \cdot e^{2x}/2 \, dx = xe^{2x}/2 - e^{2x}/4 = (x/2 - 1/4)e^{2x} + C$$

CAUTION: Students make more mistakes using the IP formula due to carelessness than with any other rule. To avoid mistakes, write down the function matrix for each problem and **be careful to include the negative sign before the second integral.**

Example 2 Integrating a polynomial times a trig function.

Problem Evaluate $\int \theta \cos(2\theta) \, d\theta$.

Solution As the integrand is a product of two terms we try the IP method. We set $u = \theta$ and therefore v' must be $\cos(2\theta)$. We enter these into the template and then enter the derivative u' and the integral v.

$$u = \underline{\quad\theta\quad} \qquad v' = \underline{\quad\cos(2\theta)\quad}$$

$$u' = \underline{\quad 1 \quad} \qquad v = \underline{\quad \sin(2\theta)/2 \quad}$$

The IP formula then gives

$$\int \theta \cos(2\theta)\, d\theta = \theta \sin(2\theta)/2 - \int \sin(2\theta)/2\, d\theta$$

$$= \theta \sin(2\theta)/2 - [-\cos(2\theta)/4]$$

$$= \theta/2\, \sin(2\theta) + (1/4)\cos(2\theta) + C$$

Be careful of the double negative signs!

Example 3 **The integral of $ln(x)$.**

Problem Use Integration by Parts to evaluate $\int ln(x)\, dx$.

Solution Although the integrand is a single function the IP method is applied by introducing a second function that is just the constant 1. Since we are trying to integrate $ln(x)$ we can not choose this to be the derivative v', hence we set $v' = 1$ and $u = ln(x)$. Filling in the function template we have

$$u = \underline{\quad ln(x) \quad} \qquad v' = \underline{\quad 1 \quad}$$

$$u' = \underline{\quad 1/x \quad} \qquad v = \underline{\quad x \quad}$$

The IP formula gives

$$\int ln(x)\, dx = x\,ln(x) - \int (1/x)\, x\, dx$$

$$= x\,ln(x) - \int 1\, dx = x\,ln(x) - x + C$$

We can check the validity of this integral by differentiating. Using the product rule we find

$$D_x[x\,ln(x) - x + C] = x \cdot D_x ln(x) + ln(x) \cdot D_x x - D_x x + D_x C$$

$$= x \cdot 1/x + ln(x) - 1 = ln(x)$$

We can now add the integral of $ln(x)$ to our list of fundamental Integral Formulas.

INTEGRAL OF $ln(x)$

$$\int ln(x)\, dx = x\,ln(x) - x + C$$

Evaluating Definite Integrals using Integration by Parts.

When evaluating a definite integral using the IP formula the mechanics are exactly the same but both the product uv and the second integral must be evaluated at the limits of integration:

$$\int_a^b u(x)v'(x)\, dx = u(x)v(x)\Big|_a^b - \int_a^b v(x)u'(x)\, dx$$

where the notation for evaluating the product uv is defined as

$$u(x)v(x)\big|_a^b \equiv u(b)v(b) - u(a)v(a)$$

Example 4 Average dose of a drug. Evaluating a definite integral by the IP method.

Problem A drug is administered orally and the concentration in the blood is found to fit the curve

$$y = 5t \cdot e^{-0.05t} \;\; \mu\text{ Moles }/\text{l, t in hours}$$

What is the average blood concentration of the drug over the first twenty four hours following administration?

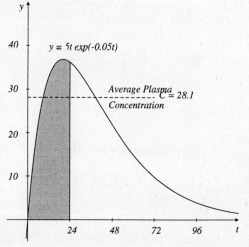

Solution The average concentration is the definite integral of the concentration curve over the time period divided by the length of the time period. The integral is

$$I = \int_0^{24} y(t)\, dt = \int_0^{24} 5t e^{-.05t}\, dt$$

Setting u equal to the polynomial, $5t$, the derivative v' must be $e^{-0.05t}$ and the function template is

$$u = \underline{\;\;\;\;\;5t\;\;\;\;\;} \quad v' = \underline{\;\;\;\;\;e^{-.05t}\;\;\;\;\;}$$

$$u' = \underline{\;\;\;\;\;5\;\;\;\;\;} \quad v = \underline{\;\;-20e^{-.05t}\;\;}$$

The definite integral is evaluated by the IP formula as

$$\int_0^{24} 5t e^{-0.05t}\, dt = -100 t e^{-0.05t}\big|_0^{24} - \int_0^{24} -100 e^{-0.05t}\, dt.$$

$$= -100 t e^{-0.05t}\big|_0^{24} - 2000 e^{-0.05t}\big|_0^{24}$$

$$= -100(t + 20) e^{-0.05t}\big|_0^{24}$$

$$= -100\, [44 e^{-1.2} - 20 e^0] \approx 674.7$$

The average blood plasma drug concentration is therefore approximately

$$C = \int_0^{24} 5t e^{-.05t}\, dt\, /\, 24 \approx 28.1 \;\mu\text{Moles/l}$$

☑

Frequently it is necessary to apply the IP method more than once to evaluate an integral. When doing so, the "dummy" functions u and v' will change with each application. If you experience confusion using the same symbols you can use other symbols for the second (or third) application.

Example 5 Repeated application of Integration by Parts.

Problem Evaluate the integral $I = \int x^2 e^{-x}\, dx$.

Solution The IP method seems appropriate because of the product. Choose u as the polynomial x^2. Then $v' = e^{-x}$ and the function template is

$$u = \underline{\;\;\;x^2\;\;\;} \quad v' = \underline{\;\;\;e^{-x}\;\;\;}$$

$$u' = \underline{\;\;2x\;\;} \quad v = \underline{\;\;-e^{-x}\;\;}$$

The IP formula then gives

$$I = \int x^2 e^{-x}\, dx = -x^2 e^{-x} - \int -2x e^{-x}\, dx$$

The resulting integral still involves a product. Hence, we apply the IP method again, this time to evaluate $I_1 = \int -2x e^{-x}\, dx$. Choosing the u term as the polynomial $-2x$ we have

$$u = \underline{\quad -2x \quad} \qquad v' = \underline{\quad e^{-x} \quad}$$

$$u' = \underline{\;-\;-\;-2\;-\;-\;-\;-\;} \qquad v = \underline{\;-\;-e^{-x}\;-\;-\;-\;-\;-\;}$$

The second integral is then evaluated by the IP rule as

$$I_1 = \int -2x e^{-x}\, dx = 2x e^{-x} - \int 2 e^{-x}\, dx$$

$$= 2x e^{-x} - (-2 e^{-x}) = 2x e^{-x} + 2 e^{-x}$$

Substituting this into the first IP equation gives

$$I = \int x^2 e^{-x}\, dx = -x^2 e^{-x} - \{2x e^{-x} + 2 e^{-x}\} = -[x^2 + 2x + 2] e^{-x} + C$$

Integral Formula for $\int \sin^m(\theta)\cos^n(\theta)\, d\theta$

Products of the form $\sin^m(\theta)\cos^n(\theta)$, where m and n positive integers, can be reduced algebraically to two types of functions. For instance, applying the Pythagorean Identity,

$$\sin^4(\theta)\cos^3(\theta) = \sin^4(\theta)\cos^2(\theta)\cos(\theta) = \sin^4(\theta)(1 - \sin^2(\theta))\cos(\theta)$$

$$= [\sin^4(\theta) - \sin^6(\theta)]\cos(\theta) = \sin^4(\theta)\cos(\theta) - \sin^6(\theta)\cos(\theta)$$

and

$$\sin^2(\theta)\cos^2(\theta) = \sin^2(\theta)(1 - \sin^2(\theta)) = \sin^2(\theta) - \sin^4(\theta)$$

The general integral $\int \sin^m(\theta)\cos^n(\theta)\, d\theta$ can thus be algebraically changed into an integral of one of two types:

Type 1 terms have the form $\sin(\theta)\cos^j(\theta)$ or $\cos(\theta)\sin^j(\theta)$ for an integer exponent j.

Integrals of Type 1 products can be integrated directly as they match the General Power Rule for Integration: $g'(\theta)g^j(\theta)$, with $g(\theta) = \sin(\theta)$ or $g(\theta) = \cos(\theta)$.

Type 2 terms have the form $\cos^i(\theta)$ or $\sin^i(\theta)$ where i is an even integer.

Integrals of Type 2 terms must be integrated using the IP method.
Each time the IP formula is applied the new integrand can be rearranged using the pythagorean identity to give a multiple of the original integrand plus a term of the same form with the exponent reduced by 2. (See the second example below.)
Repeating this process will eventually lead to evaluating an integral of $\sin^2(\theta)$ or $\cos^2(\theta)$.

Example 6 Integrating $\sin^2(\theta)$ and $\cos^2(\theta)$.

Problem Determine Integral Formulas for $\int \sin^2(\theta)\, d\theta$ and $\int \cos^2(\theta)\, d\theta$.

Solution Consider first the integral $I = \int \sin^2(\theta)\, d\theta$.

To use the IP method first write the integrand as a product: $\sin^2(\theta) = \sin(\theta)\sin(\theta)$. Then set both u and v' equal to $\sin(\theta)$. The corresponding function template is:

$$u = \underline{\quad\sin(\theta)\quad} \qquad v' = \underline{\quad\sin(\theta)\quad}$$

$$u' = \underline{\quad\cos(\theta)\quad} \qquad v = \underline{\quad -\cos(\theta)\quad}$$

The IP formula then gives

$$I = -\sin(\theta)\cos(\theta) - \int \cos(\theta) \times (-\cos(\theta))\, d\theta$$

$$= -\sin(\theta)\cos(\theta) + \int \cos^2(\theta)\, d\theta$$

$$= -\sin(\theta)\cos(\theta) + \int 1 - \sin^2(\theta)\, d\theta$$

$$= -\sin(\theta)\cos(\theta) + \theta - \int \sin^2(\theta)\, d\theta$$

The integral on the right is the original integral I that we stated with. If we replace this integral by the symbol I the above equation has the form

$$I = -\sin(\theta)\cos(\theta) + \theta - I$$

which is solved for I to give

$$I = \int \sin^2(\theta)\, d\theta = \theta/2 - (1/2)\sin(\theta)\cos(\theta) + C$$

The integral of $\cos^2(\theta)$ could be found by exactly the same method, or more simply by using the integral of $\sin^2(\theta)$ and the $\cos^2(\theta) = 1 - \sin^2(\theta)$ identity.

$$\int \cos^2(\theta)\, d\theta = \int 1 - \sin^2(\theta)\, d\theta = \theta - [\theta/2 - (1/2)\sin(\theta)\cos(\theta)] = \theta/2 + (1/2)\sin(\theta)\cos(\theta) + C$$

An alternative formula:

$$\int \sin^2(\theta)\, d\theta = 0.5(\theta - 0.5\sin(2\theta))$$

is sometimes listed in tables of integrals. This is based on the identity $\sin(\theta)\cos(\theta) = 0.5\sin(2\theta)$. However, this form of the antiderivative is difficult to convert back to x-terms when the θ-integral arose as the result of a trig substitution. Consequently, we will not use this form.

Example 7 Integrating a product of the form $\int \sin^m(\theta)\, d\theta$ for even m.

Problem Evaluate $I = \int \sin^4(\theta)\, d\theta$.

Solution Apply the IP formula would with $u = \sin^3(\theta)$ and $dv = \cos(\theta)$ and then manipulate the resulting integral and use the formula for the integral of $\sin^2(\theta)$:

$$I = \int \sin^4(\theta)\, d\theta = \int \sin(\theta)\sin^3(\theta)\, d\theta$$

$$= -\cos(\theta)\sin^3(\theta) - \int -3\cos^2(\theta)\sin^2(\theta)\, d\theta$$

$$= -\cos(\theta)\sin^3(\theta) - \int -3[1 - \sin^2(\theta)]\sin^2(\theta)\, d\theta$$

$$= -\cos(\theta)\sin^3(\theta) - \int -3[1 - \sin^2(\theta)]\sin^2(\theta)\, d\theta$$

$$= -\cos(\theta)\sin^3(\theta) - \int -3\sin^2(\theta)\, d\theta - 3\int \sin^4(\theta)\, d\theta$$

$$= -\cos(\theta)\sin^3(\theta) - \int -3\sin^2(\theta)\, d\theta - 3\,I$$

Solving for the integral I gives

$$I = -(1/4)\cos(\theta)\sin^3(\theta) + (3/4)\int \sin^2(\theta)\, d\theta$$

Now, we need only substitute the above integral for $\sin^2(\theta)$ to get:

$$I = -(1/4)\cos(\theta)\sin^3(\theta) + (3/4)[\theta/2 - (1/2)\sin(\theta)\cos(\theta)] + C$$

EXERCISE SET 6.2

In exercises 1 - 5 evaluate the given indefinite integrals using Integration by Parts.

1.
   a) $\int xe^{3x}\, dx$
   b) $\int xe^{2x+5}\, dx$
   c) $\int 4xe^{-6x}\, dx$
   d) $\int (x+3)e^{4x}\, dx$
   e) $\int xe^{kx}\, dx$
   f) $\int xe^{-kx}\, dx$
   g) $\int x^3 e^{x^2}\, dx$
   h) $\int x^5 e^{x^3}\, dx$
   i) $\int 2x^3 e^{-x^2}\, dx$

2.
   a) $\int x^2 \ln(x)\, dx$
   b) $\int (x+4)\ln(x)\, dx$
   c) $\int x^2 \ln(2x)\, dx$
   d) $\int \ln(5+x)\, dx$
   e) $\int x^{1/3}\ln(x)\, dx$
   f) $\int \ln(x^{3x})\, dx$
   g) $\int \ln(x)\ln(x)\, dx$
   h) $\int x(\ln(x))^2\, dx$
   i) $\int \ln(\ln(x))/x\, dx$

3.
   a) $\int_0^2 xe^{2x}\, dx$
   b) $\int_0^3 xe^{x-3}\, dx$
   c) $\int_{-1}^1 xe^{-0.5x}\, dx$
   d) $\int_1^e 3x\ln(x)\, dx$
   e) $\int_{-2}^{-1}(x+1)\ln(-x)\, dx$
   f) $\int_1^2 (x^2-4)\ln(x)\, dx$

4. Repeated application of the IP method may be required to evaluate these integrals.
   a) $\int x^2 e^{-3x}\, dx$
   b) $\int x^2 e^{4x-5}\, dx$
   c) $\int x^5 e^{x^2}\, dx$
   d) $\int (x^2 - 2x + 5)e^{-8x}\, dx$
   e) $\int x^3 e^{5x}\, dx$
   f) $\int (x^3 - x)e^{3x}\, dx$

5.
   a) $\int \theta \cos(3\theta)\, d\theta$
   b) $\int \theta \sin(\pi\theta)\, d\theta$
   c) $\int \theta \sec^2(3\theta)\, d\theta$
   d) $\int \theta \sin(\theta + 1)\, d\theta$
   e) $\int \theta^3 \cos(\theta^2)\, d\theta$
   f) $\int \cos(\theta)\cos(\theta)\, d\theta$
   g) $\int \sin(\theta)e^\theta\, d\theta$
   h) $\int \cos^3(\theta)\, d\theta$
   i) $\int \sin(2\theta)e^\theta\, d\theta$
   j) $\int \cos(\theta)e^{-2\theta}\, d\theta$
   k) $\int \sin(\pi\theta)e^{-3\theta}\, d\theta$
   l) $\int \theta^2 \cos(\theta)\, d\theta$

6.
   a) $\int \sin^2(3\theta)\, d\theta$
   b) $\int \sin^3(\theta)\cos(\theta)\, d\theta$
   c) $\int \sin^3(\theta)\, d\theta$
   d) $\int \sin^4(\theta)\, d\theta$
   e) $\int \tan^2(\theta)\, d\theta$
   f) $\int 1/\cos(\theta)\, d\theta$

7. Transform the integral by a substitution and then use Integration by Parts to evaluate it.
   a) $\int \sin(\sqrt{2+x}))\, dx$
   b) $\int e^{\sqrt{3-2x}}\, dx$
   c) $\int xe^{\sqrt{2-x}}\, dx$
   d) $\int e^{(x+1)^{1/3}}\, dx$

8. Calculate the average blood concentration of the drug over the first twenty four hours following oral administration if its the blood concentration is found to fit the curve (a) $y = 3te^{-0.2t}$ Moles/l, and (b) $y = 4t^2 e^{-0.3t}$ Moles/l, t in hours.

9. The value of a currency fluctuated by a factor of $y = 1 + e^{-0.002t}\cos(\pi t)$, t in days after instituting a new tax policy. What was the average fluctuation over the first two years of the new policy.

10. Determine an antiderivative of the indicated function by first making a trigonometric substitution and then using integration by parts.
    a) $f(x) = \sqrt{1 + x^2}$
    b) $f(x) = \sqrt{1 + 4x^2}$
    c) $f(x) = \sqrt{4 + x^2}$
    d) $f(x) = \sqrt{25 + 9x^2}$
    e) $f(x) = (4 - 9x^2)^{3/2}$
    f) $f(x) = (25 - 9x^2)^{-3/2}$

Section 6.3 Integrating Rational Functions: The Partial Fractions Method.

Introduction.

The "Partial Fractions Method" is a technique used to evaluate integrals of rational functions, integrals of the form

$$\int f(x)/g(x)\, dx$$

when the denominator $g(x)$ can be factored into a product of two terms:

$$g(x) = Q_1(x) \cdot Q_2(x).$$

This method is based on the fact that a fraction whose denominator can be factored can be expressed as a sum of simpler fractions.

You know that the simplest way to add two fractions is to form a common denominator by multiplying the two denominators. The numerator is then the sum of the "cross-products", each numerator multiplied by the other denominator. You learned this procedure for adding numerical fractions like the sum

$$2/5 + 1/3 = (3 \cdot 2 + 5 \cdot 1)/(5 \cdot 3) = 11/15$$

and should know that the same method works when one has rational functions. For instance

$$P_1(x)/Q_1(x) + P_2(x)/Q_2(x) = [P_1(x) \cdot Q_2(x) + P_2(x) \cdot Q_1(x)] / [Q_1(x) \cdot Q_2(x)]$$

To apply the Partial Fractions method you must work backwards from one rational function to form a sum of two rational functions. The problem can be simply illustrated using a numerical example.

Given the fraction 11/15, find the two fractions whose sum is 11/15.

This is not a skill normally stressed when teaching arithmetic. The process of expanding a fraction as a sum of partial fractions consist of

1. Factoring the denominator: $15 = 5 \cdot 3$

2. Writing two fractions whose denominators are the factors and whose numerators are unknown constants: $A/3$ and $B/5$.

3. Add these fractions and equate their sum to the original given fraction.

$$A/3 + B/5 = [5A + 3B] / 15 = 11/15$$

4. As the denominators are the same, this mean the numerators must be the same. Equate them and solve for the constants:
$$11 = 5A + 3B$$

The solutions of this equation is not unique, but, if we assume that A and B are integers and that $1 \le A < 3$ and $1 \le B < 5$, there is only one solution, $A = 1$ and $B = 2$.

The partial fraction method involves the same process applied to rational functions. We will consider quotients of polynomials.

The Method of Partial Fractions.

Assume f(x) and g(x) are polynomials and f(x)/g(x) is of **reduced form**. This means that

$$\text{degree}(f) < \text{degree}(g)$$

and that

$$f(x) \text{ and } g(x) \text{ have no common factors.}$$

Assume that g is factorable, with factors Q_1 and Q_2. Then a basic algebra theorem states that there exist two polynomials P_1 and P_2 such that

$$\text{degree}(P_1) < \text{degree}(Q_1) \quad \text{and} \quad \text{degree}(P_2) < \text{degree}(Q_2)$$

and the quotient f/g can be expressed as the sum of *partial fractions*:

$$f(x)/g(x) = P_1(x)/Q_1(x) + P_2(x)/Q_2(x)$$

Integrating this equation gives the basic formula for the **partial fractions method**.

$$\int f(x)/g(x)\, dx = \int P_1(x)/Q_1(x)\, dx + \int P_2(x)/Q_2(x)\, dx$$

Applying the Partial Fractions Method consists of algebraically determining the functions $P_1(x)$ and $P_2(x)$ and then evaluating the P/Q integrals. These functions are found by solving the equation obtained by equating the numerators:

$$f(x) = Q_2(x)P_1(x) + Q_1(x)P_2(x)$$

Just as in the numerical case, to solve this for P_1 and P_2 some additional conditions must be imposed. One important condition is that each numerator polynomial P must be of lower degree than its denominator Q. It is known that polynomials can be factored into the product of linear and/or quadratic terms. If Q is linear, $Q(x) = ax + b$, the corresponding numerator must simply be a constant, $P(x) = A$. The partial fraction P/Q will then have the form

$$A/(ax + b) \quad \text{for some (unknown) constant A.}$$

Similarly, if the factor Q is a quadratic, $Q(x) = ax^2 + bx + c$, then the corresponding numerator P will be linear, $P(x) = Ax + B$, for some unknown constants A and B. The partial fraction in this case will have the form

$$[Ax + B] / [ax^2 + bx + c]$$

In all cases the right side of the equation

$$f(x) = Q_2(x)P_1(x) + Q_1(x)P_2(x)$$

will be a polynomial involving the unknown constants, A, B, C, ... When the two Q · P terms have been multiplied and added the equation will equate the given numerator f(x) to a polynomial involving constants. This is solved for the constants by creating an associated set of equations. These are formed from the following algebraic property.

Definition of equal polynomials.   Two polynomials are equal only if they have the same degree and the coefficients of like powers are the same.

For instance, if the polynomial $3 - 2x^2 + 4x^5$ is set equal to a general polynomial,

$$3 - 2x^2 + 4x^5 = a_0 + a_1 x + a_2 x^2 + a_3 x^3 + \ldots + a_n x^n$$

Section 6.3 — The Partial Fractions Method

$$(1/5)ln(N/(5 - N)) = 3t + c$$

To solve for N, first multiply by 5 and then take the exponential of both sides to invert the *ln* function:

$$ln(N/(5 - N)) = 15t + 5c$$

$$N/(5 - N) = e^{15t + 5c}$$

Next, invert each side of this equation. This can be done to the right side by simply changing the sign of the exponent. The right side is further simplified by introducing a different constant, $C = e^{-5c}$, so that the equation becomes:

$$(5 - N)/N = e^{-(15t + 5c)} = Ce^{-15t}$$

Multiplying by N and gathering the N terms on one side:

$$5 = N + NCe^{-15t} = N(1 + Ce^{-15t})$$

Finally, dividing by the term $(1 + Ce^{-15t})$ results in the *Logistic equation*. This equation is also referred to as the *Pearl-Verhulst equation*.

$$N(t) = 5/(1 + Ce^{-15t})$$

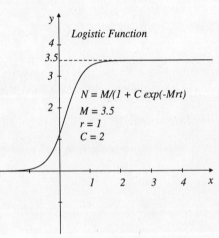

The values of the constants, $M = 5$ and $r = 3$, did not affect the determination of the unknown coefficients. If you repeated the above solution with the symbols M and r you would arrive at the same form of the solution:

$$N(t) = M/(1 + Ce^{-Mrt})$$

LOGISTIC EQUATION

☑

The above Figure illustrates the graph of the logistic function.

Reduced Integrands.

A key assumption of the Partial Fractions method is that the quotient f/g is "reduced". If this is not the case, then it must be reduced before forming the partial fractions. If f and g have a common factor then this should be canceled. If the degree of f is greater than or equal to the degree of g then you must divide g into f to represent the quotient f/g as a polynomial plus a quotient that is reduced. Recall how you reduced a numerical fraction like 27/4. Since

$$27 \div 4 = 6 \text{ remainder } 3 \quad \text{we write} \quad 27/4 = 6 + 3/4$$

For polynomials the same process works. If the degree(f) ≥ degree(g) we divide g into f. If

$$f(x) \div g(x) = h(x) \text{ with remainder } r(x), \quad \text{we write} \quad f/g = h + r/g$$

The remainder r(x) will always satisfy degree(r) < degree(g). Then, we can evaluate the integral of f/g as a sum:

$$\int f(x)/g(x) \, dx = \int h(x) \, dx + \int r(x)/g(x) \, dx$$

Example 3 Partial Fractions Requiring Division.

Problem Evaluate $I = \int (2 + 2x - 3x^3)/(1 - x^2)\, dx$.

Solution The integrand is not reduced since the degree of the numerator is 3 and the degree of the denominator is 2. Division of the denominator into the numerator gives

$$
\begin{array}{r}
3x \\
-x^2 + 1 \,\overline{\smash{)}\, -3x^3 + 0x^2 + 2x + 2} \\
-3x^3 + 0x^2 + 3x \\
\hline
-x + 2 \quad \text{The remainder.}
\end{array}
$$

So we can write
$$(2 + 2x - 3x^3)/(1 - x^2) = 3x + (2 - x)/(1 - x^2)$$

The integral is therefore
$$I = \int 3x + (2 - x)/(1 - x^2)\, dx = (3/2)x^2 + \int (2 - x)/(1 - x^2)\, dx$$

Now, the remaining integral can be evaluated using partial fractions. As
$$1 - x^2 = (1 - x)(1 + x)$$

for some undetermined constants A and B,
$$\int (2 - x)/(1 - x^2)\, dx = \int A/(1 + x) + B/(1 - x)\, dx$$
$$= A \ln(1 + x) - B \ln(1 - x) = \ln[(1 + x)^A/(1 - x)^B]$$

The numerical values of A and B are found by setting
$$2 - x = A(1 - x) + B(1 + x) = (A + B) + (B - A)x$$

Equating the coefficients of like powers gives the system of equations

$$
\begin{array}{ll}
2 = A + B & \text{coefficients of x;} \\
-1 = B - A & \text{constant terms.}
\end{array}
$$

Solving gives $B = 1/2$ and $A = 3/2$. Thus
$$I = (3/2)x^2 + (3/2)\ln(1 + x) - (1/2)\ln(1 - x) + C$$

☑

The Partial Fractions method can be applied if there are more than two factors in the denominator. There will be one "partial fraction" for each factor of the denominator.

Example 4 Integrating by Partial Fractions with three factors.

Problem Evaluate $I = \int \{2x^2 - 3x + 2\}/\{x^3 - 6x^2 + 11x - 6\}\, dx$.

Solution The denominator is factorable, with three linear factors:
$$x^3 - 6x^2 + 11x - 6 = (x - 1)(x - 2)(x - 3)$$

The integrand is then equated to the sum of the three partial fractions of the form

$$A/(x - 1) + B/(x - 2) + C/(x - 3) =$$

$$[A(x - 2)(x - 3) + B(x - 1)(x - 3) + C(x - 1)(x - 2)] / [(x - 1)(x - 2)(x - 3)]$$

The numerator can be expanded and then simplified to give the quadratic

$$(A + B + C)x^2 + (-5A - 4B - 3C)x + (6A + 3B + 2C)$$

Equating this polynomial to the numerator $2x^2 - 3x + 2$ of the integral gives a system of three equation in the undetermined coefficient A, B, and C:

$$\begin{array}{ll} A + B + C = 2 & x^2 \text{ coefficients} \\ -5A - 4B - 3C = -3 & x \text{ coefficients} \\ 6A + 3B + 2C = 2 & \text{constants} \end{array}$$

The simultaneous solution of these equations is (after some work)

$$A = 1/2 \quad B = -4 \quad C = 11/2$$

The integral is therefore

$$I = \int A/(x - 1) + B/(x - 2) + C/(x - 3) \, dx$$

$$= A\ln(x - 1) + B\ln(x - 2) + C\ln(x - 3)$$

$$= (1/2)\ln(x - 1) - 4\ln(x - 2) + (11/2)\ln(x - 3)$$

$$= \ln((x - 1)^{1/2}(x - 3)^{11/2}(x - 2)^{-4}) + \text{integration constant}$$

Repeated factors.

A special situation occurs if g(x) has a multiple factor, that is one which occurs to an integer power, like $[Q(x)]^3$. In this case the algebra is slightly more complicated. When the quotient f/g is expanded as a sum of partial fractions, there must be a partial fraction for each power of the multiple factor. For instance, if $g(x) = x^3$ the partial fraction expansion will have three terms, A/x, B/x^2, and C/x^3. You can verify by adding the fractions that

$$\{3x^2 + 2x + 1\}/x^3 = 3/x + 2/x^2 + 1/x^3$$

In general, if g(x) has a "multiple" linear factor, $(ax + b)^n$, the expansion of f/g as partial fractions will include the n fractions of the form

$$A_1/(ax + b) + A_2/(ax + b)^2 + A_3/(ax + b)^3 + \ldots + A_n/(ax + b)^n$$

There will be n undetermined coefficients $A_1, A_2, \ldots, A_n$, some of which may turn out to be zero, in which case the corresponding fraction will not appear when the partial fractions are integrated. You will not likely encounter problems with a large number of factors since the algebra required to solve for the unknown coefficients increases as the square of the number of coefficients.

Example 5   **Partial Fractions with Multiple linear factors.**

Problem   Evaluate $I = \int \{2x^2 - 4x - 5\}/\{(x - 3)(x + 2)^2\} \, dx$.

Solution The factor $(x + 2)$ occurs twice in the denominator. The integrand is expressed as a sum of partial fractions having the form

$$\{2x^2 - 4x - 5\}/\{(x - 3)(x + 2)^2\} = A/(x - 3) + B/(x + 2) + C/(x + 2)^2$$

The partial fractions are added using the common denominator $(x - 3)(x + 2)^2$. The numerator of the added fractions is

$$A(x + 2)^2 + B(x - 3)(x + 2) + C(x - 3)$$

$$= (A + B)x^2 + (4A - B + C)x + (4A - 6B - 3C)$$

Equating this polynomial to the numerator of the integrand gives a system of three equation for A, B, and C:

$$\begin{aligned} A + B &= 2 & x^2 \text{ coefficients} \\ 4A - B + C &= -4 & x \text{ coefficients} \\ 4A - 6B - 3C &= -5 & \text{constants} \end{aligned}$$

The solutions are $A = 1/25$, $B = 49/25$, and $C = -11/5$. Hence the integral is

$$I = \int (1/25)/(x - 3) + (49/25)/(x + 2) - (11/5)/(x + 2)^2 \, dx$$

$$= (1/25)ln(x - 3) + (49/25)ln(x + 2) + (11/5)/(x + 2) + \text{constant}$$

☑

Quadratic Factors.

If the denominator $g(x)$ has an irreducible quadratic factor, one that can not be factored into linear factors, then the expansion of $f(x)/g(x)$ as a sum of partial fractions must have a fraction with the quadratic as the denominator. The numerator in this case must be a linear term of the form $Ax + B$. For example if $g(x) = (x^2 + x + 1)(x + 5)$ then the partial fraction expansion of f/g will have the form

$$f(x)/g(x) = (Ax + B)/(x^2 + x + 1) + C/(x + 5).$$

Recall: You can check if a quadratic term can be factored with the Quadratic Formula. If the roots of $ax^2 + bx + c = 0$ are complex numbers then it can not be factored. Since the roots $x = \{-b \pm SQRT(b^2 - 4ac)\}/2a$ are complex when the term in the square root is negative,

$ax^2 + bx + c$ is "irreducible" and can not be factored if $b^2 < 4ac$.

Example 6 **Partial Fractions with a quadratic factor.**

Problem Evaluate the integral $I = \int \{6x^2 - 4x + 4\}/\{(x - 2)(x^2 + 1)\} \, dx$.

Solution The factor $x^2 + 1$ is irreducible. The partial fraction expansion has the form

$$\{6x^2 - 4x + 4\}/\{(x - 2)(x^2 + 1)\} = A/(x + 2) + (Bx + C)/(x^2 + 1)$$

Adding the fractions gives a fraction with the common denominator $(x - 2)(x^2 + 1)$ and numerator

$$A(x^2 + 1) + (Bx + C)(x - 2) = (A + B)x^2 + (-2B + C)x + (A - 2C)$$

Equating this to the numerator $6x^2 - 4x + 4$ gives the system of equations

$$A + B = 6 \quad x^2 \text{ coefficients}$$
$$-2B + C = -4 \quad x \text{ coefficients}$$
$$A - 2C = 4 \quad \text{constants}$$

The solution of this system is $A = 4$, $B = 2$, and $C = 0$. Thus

$$I = \int 4/(x+2) + 2x/(x^2+1) \, dx = 4ln(x+2) + ln(x^2+1) + \text{Constant}$$

EXERCISE SET 6.3

1. Evaluate the given integral using the Partial Fractions Method.
    a) $\int 1/(x^2 - 1) \, dx$
    b) $\int (2x + 1)/(x^2 - 1) \, dx$
    c) $\int (x + 4)/(x^2 + x - 2) \, dx$
    d) $\int (x + 3)/(x^2 - 2x) \, dx$
    e) $\int (3x^2 + 2)/[(x - 1)(x + 2)(x + 3)] \, dx$
    f) $\int (3x^2 + x + 2)/[(x^2 - 1)(x + 4)] \, dx$
    g) $\int (3x^2 - 2x)/[(x - 1)(x^2 - 3)] \, dx$
    h) $\int x^2/(x^3 - 3x^2 + 2x) \, dx$

2. Evaluate the given integral by first reducing the quotient.
    a) $\int (x^2 + 2)/x \, dx$
    b) $\int (x^2 + 2)/(x + 2) \, dx$
    c) $\int (x^3 - x + 1)/(x - 2) \, dx$
    d) $\int (x^3 + 1)/(x + 2) \, dx$
    e) $\int (x^2 + 1)/(x^2 - 1) \, dx$
    f) $\int (x^3 + 1)/(x^2 - 1) \, dx$
    g) $\int (x^3 + 1)/(x^2 + 3x + 2) \, dx$
    h) $\int (x^{10} - 1)/(x - 1) \, dx$

3. Evaluate the indicated integrals.
    a) $\int x/(x - 2)^2 \, dx$
    b) $\int (x^2 + 2)/x^3 \, dx$
    c) $\int (x + 1)/(x - 1)^2 \, dx$
    d) $\int (x - 1)/[x^2(x + 1)] \, dx$
    e) $\int x^3/(x - 1)^2 \, dx$
    f) $\int (x^2 + x + 1)/[(x - 1)^2(x + 2)^2] \, dx$
    g) $\int (x^2 + 1)/[(x - 1)^2(x + 3)^2] \, dx$
    h) $\int (x + 2)/[x(x - 1)^3] \, dx$

4. Determine an antiderivative of the given rational function.
    a) $(3x^2 + 1)/[(x^2 + 1)x]$
    b) $(10x^2 - 16x + 10)/[(x^2 - x + 1)(x - 2)]$
    c) $3x^2/[(x^2 - x + 1)(x + 1)]$
    d) $3/(x^3 + x)$
    e) $(2x^2 - 2)/(x^3 + x^2 + x)$
    f) $7x^3 - 1/(2x^4 - 2x)$

Section 6.4 Improper Integrals

Introduction.

Two assumptions were made when the definite integral $\int_a^b f(x)\,dx$ was defined. One was that [a,b] is a finite interval and the second was that f(x) is continuous on the interval [a,b]. Sometimes we encounter definite integrals that violate one of these assumptions, either an integral over an infinite interval or the integral of a function that is *singular* because it becomes infinite at an endpoint of integration interval. Such integrals are called **improper** or **singular** integrals.

The first type of **improper** integral occurs when the integration interval is infinite. Integrals over infinite intervals have one of the forms:

$$\int_a^\infty f(x)\,dx \qquad \int_{-\infty}^b f(x)\,dx \qquad \int_{-\infty}^\infty f(x)\,dx$$

The second type of **improper** integral occurs when the integrand is *singular* at an endpoint of the integration interval [a,b]. These integrals have the form $\int_a^b f(x)\,dx$ where either

$$\lim_{x \to a+} f(x) = \pm\infty \text{ and f is continuous on (a,b]}$$

or

$$\lim_{x \to b-} f(x) = \pm\infty \text{ and f is continuous on [a,b).}$$

For instance,

$$\int_0^1 1/\sqrt{1-x}\,dx \text{ is singular at } x = 1, \text{ since } \lim_{x \to 1-} 1/\sqrt{1-x} = \infty.$$

and

$$\int_0^1 ln(x)\,dx \text{ is singular at } x = 0, \text{ since } \lim_{x \to 0+} ln(x) = -\infty.$$

To recognize improper integrals you should be conscious of the types of singularities that can arise. The three types of singularities that you will likely encounter are:

Division by zero: This occurs when f(x) contains a rational expression and the denominator becomes zero. Look for factors like $(x - a)^r$ or $ln(x/b)$ in the denominator.

Logarithm functions: They become infinite if their argument approaches zero.

$$ln(x - a) \text{ is singular at } x = a$$

Trigonometric functions: Tan(x), Cot(x), Sec(x) and Csc(x) all have periodic singularities:
tan(x) and sec(x) are singular at values $x = \pi/2 + n\pi$,
cot(x) and csc(x) are singular at $x = n\pi$, n an integer.

Improper integrals are evaluated by a standard calculus technique, they are approximated by well defined integrals and the a limit of these approximations is taken. The approach is to introduce a new variable integration limit, say t, to replace the ∞ or the endpoint at which the integrand is singular. The integral of f over the interval with endpoint t defines an integral-function F(t). If the limit of F(t) as t approaches the singular endpoint exists, as a finite number, the integral is said to *converge* and this limit is the value of the improper integral. If the limit fails to exist the improper integral does *not converge* and does not exist. If the limit is infinite, the improper integral *diverges to infinity* and is loosely said to be "infinite".

Improper Integrals over infinite intervals.

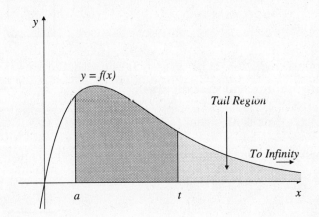

If the function f is positive, $\int_a^\infty f(x)\, dx$ can be visualized as the area under the graph of f over the infinite interval [a,∞). If we split the infinite interval [a,∞) into two intervals, [a,t] and [t,∞), then we should have

$$\int_a^\infty f(x)\, dx = \int_a^t f(x)\, dx + \int_t^\infty f(x)\, dx$$

The integral on the right represents the area of the "tail" region under the graph of f. This concept is utilized in statistics. As $t \to \infty$ the area under the graph over [a,t] should approach the area under the graph over the infinite interval [a,∞) and the area of the "tail" region should approach zero.

DEFINITION OF IMPROPER INTEGRALS OVER INFINITE INTERVALS.

> The **Improper Integrals** of f(x) over the infinite intervals [a,∞) and (-∞,b] are defined by the following limits, if they exist:
>
> $$\int_a^\infty f(x)\, dx = \lim_{t \to \infty} \int_a^t f(x)\, dx$$
>
> $$\int_{-\infty}^b f(x)\, dx = \lim_{t \to -\infty} \int_t^b f(x)\, dx$$
>
> If the limit exists, the improper integral is said to **converge**.
>
> If the limit does not exist, the improper integral is said to **diverge**.
>
> If the limit is ∞ or -∞, the improper integral is said to be **infinite**.
>
> The **Improper Integral** of f(x) over the interval (-∞,∞) is the sum
>
> $$\int_{-\infty}^\infty f(x)\, dx = \int_{-\infty}^c f(x)\, dx + \int_c^\infty f(x)\, dx$$
>
> for any number c, provided both the integrals on the right converge; otherwise the integral over (-∞,∞) is divergent.

Example 1 **A convergent improper integral over an infinite interval**.

Problem   Evaluate the improper integral $\int_0^\infty e^{-3x}\, dx$.

Solution   The integral is evaluated as a limit

$$\int_0^\infty e^{-3x}\, dx = \lim_{t \to \infty} \int_0^t e^{-3x}\, dx$$

$$= \lim_{t \to \infty} (-1/3)e^{-3x} \Big|_0^t$$

$$= \lim_{t \to \infty} -e^{-3t} + 1/3 = 1/3$$

The integral converges to 1/3.

Example 2   An improper integral over an infinite interval that diverges.

Problem Show that the improper integral $\int_{10}^{\infty} 1/\sqrt{x - 5} \, dx$ is divergent.

Solution To test if the integral converges we consider the

$$\lim_{t \to \infty} \int_{10}^{t} 1/\sqrt{x - 5} \, dx = \lim_{t \to \infty} 2\sqrt{x - 5}$$

$$= \lim_{t \to \infty} 2\sqrt{t - 5} - 0 = \infty$$

Consequently, the integral "diverges to infinity", and is said to be "infinite".

Example 3   Evaluating the Bio-availability of a drug. A two compartment model.

Background

A model used to describe a wide range of sequential processes is the *two compartment model*. The simplest two compartment model is illustrated as a flow diagram with rectangles representing the compartments and arrows indicating the direction of transfers. Compartment models are based on the assumption that a substance exists in "compartments" and leaves each compartment at a rate proportional to the amount present. In the simplest model the substance leaving the first compartment enters the second compartment and then is "eliminated" from this compartment into an external void. The amount of substance in each compartment can be determined by solving a dynamical model consisting of two differential equations, which indicate the rates of change in the amount of substance in each compartment. The simplest model is described by the equations

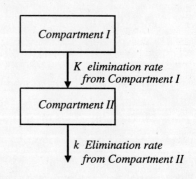

Simple Two Compartment Model

$$y_1' = -Ky_1 \quad \text{and} \quad y_2' = Ky_1 - ky_2$$

If the initial amount of substance in Compartment I is $y_1(0) = D$ and initially the Compartment II contains no substance, $y_2(0) = 0$, then the solution of this system is

Amount in Compartment I $y_1(t) = (D/K)e^{-Kt}$

Amount in Compartment II $y_2(t) = D(k/(k - K))(e^{-Kt} - e^{-kt})$

Drug distribution in humans and animals is frequently modeled using a two compartment model. The first compartment is the muscle into which the drug is injected and the second compartment is the blood plasma. The effectiveness of a drug often depends on how much drug remains in the person's blood system for a period of time. A measure of this is given by the drug's *Bioavailability*, which is also called the individual's *Total Dose*, denoted TD. The Total Dose is defined as the integral of the blood drug level over $[0,\infty)$, with 0 time being the time the drug is administered.

Problem Assume that D = 5mg of drug is injected into muscle and is absorbed into the blood at a rate of K = 0.2 hr^{-1}. The drug is eliminated by filtration through the kidney at a rate of k = 0.001 hr^{-1}. What is the individuals total dose of this drug?

Solution The blood plasma level of drug is, from the solution y_2 stated above,

$$y = c(e^{-0.2t} - e^{-0.001t})$$

where the coefficient is

$$c = 5 \cdot 0.001/(0.001 - 0.2) = 0.005/-0.199 \approx -0.0251$$

The person's Total Dose is by definition the integral of the plasma levels over all future time and is given by the improper integral

$$TD = \int_0^\infty c(e^{-0.2t} - e^{-0.001t})\, dt$$

Taking the constant c outside the integral, the improper integral is evaluated as a limit. Using s as the variable for the limit, we set

$$TD = c \cdot \lim_{s \to \infty} \int_0^s (e^{-0.2t} - e^{-0.001t})\, dt$$

$$= c \cdot \lim_{s \to \infty} (-5e^{-0.2t} + 1000e^{-0.001t}) \Big|_0^s$$

$$= c \cdot \lim_{s \to \infty} \{(-5e^{-0.2s} + 1000e^{-0.001s}) - (-5 + 1000)\} = c \cdot -995$$

since each exponential term approaches 0 as $s \to \infty$. Substituting for c its numerical value gives the total dose commitment:

$$TD = -0.005/0.199 \cdot -995 = 25$$

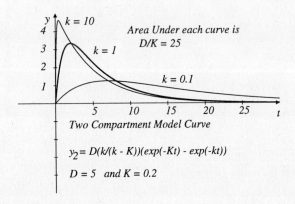

Two Compartment Model Curve

$y_2 = D(k/(k - K))(\exp(-Kt) - \exp(-kt))$

$D = 5$ and $K = 0.2$

You may have noticed that this Total Dose commitment is TD = D/K. You could verify this by integrating the general function y_2 over $[0,\infty)$. Interpreting this observation, we see that the total exposure to the drug depends on the dose and how fast it moves into the blood from the muscle, but not on how fast it is cleared from the blood. The graphs at the left illustrate the y_2 curve for three different eliminations rates and the same dose and absorption rate. Although the curves have different shapes, the area under each curve over the half line $[0,\infty)$ is 25.

Improper Integrals over finite intervals.

Improper integrals for which the integrand is singular at an endpoint are also evaluated by limits. For these integrals the corresponding region under the graph of the integrand extends to

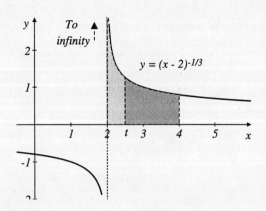

infinity in the vertical direction. For instance, the function $f(x) = 1/(x-2)^{1/3}$ is singular at $x = 2$. The integral of f over [2,4] is improper because the function becomes unbounded as x approaches 2. This improper integral does converge and represents the finite area under the graph over [2,4], even though this region extends infinitely in the positive y-direction. Splitting the interval at some $t \in (2,4)$ would give formally:

$$\int_2^4 (x-2)^{-1/3}\,dx = \int_t^4 (x-2)^{-1/3}\,dx + \int_2^t (x-2)^{-3}\,dx$$

The integral over [t,4] is simply the finite area indicated in the figure. The integral over [2,t] corresponds to the area of the region with infinite height at $x = 2$. If the area of this region is finite, as $t \to 2$ from the right this area will approach zero. Consequently, if the integral converges, the improper integral is given by the limit

$$\int_2^4 (x-2)^{-1/3}\,dx = \lim_{t \to 2+} \int_t^4 (x-2)^{-1/3}\,dx$$

As $\lim_{t \to 2+} \int_2^t (x-2)^{-3}\,dx = 0$.

DEFINITION OF AN IMPROPER INTEGRAL OVER A FINITE INTERVAL.

If $f(x)$ is continuous on [a,b) and $\lim_{x \to b^-} f(x) = \pm\infty$, then

$$\int_a^b f(x)\,dx = \lim_{t \to b^-} \int_a^t f(x)\,dx, \text{ if it exists.}$$

If $f(x)$ is continuous on (a,b] and $\lim_{x \to a^+} f(x) = \pm\infty$, then

$$\int_a^b f(x)\,dx = \lim_{t \to a^+} \int_t^b f(x)\,dx, \text{ if it exists.}$$

If the limit exists the integral **converges**.

If the limit does not exist, the integral **diverges**.

If the limit is ∞ (or $-\infty$) then the improper integral is said to **diverge to (negative) infinity**.

Example 4 **A convergent improper integral over a finite interval**.

Problem Evaluate $I = \int_3^5 (x-3)^{-2/3}\,dx$.

Solution The integral I converges if

$$\lim_{t \to 3+} \int_t^5 (x-3)^{-2/3}\,dx$$

exists. An antiderivative of $(x - 3)^{-2/3}$ is $3(x - 3)^{1/3}$. Applying the FTC to the integral over the interval $[t,5]$, the value of the improper integral is given by

$$I = \lim_{t \to 3+} 3(x - 3)^{1/3} \Big|_t^5$$

$$= \lim_{t \to 3+} 3 \cdot 2^{1/3} - 3(t - 3)^{1/3} = 3 \cdot 2^{1/3}$$

The integral converges to $3 \cdot 2^{1/3} \approx 3.78$. Thus the area of this infinite region is less than the area of a square of side-length 2.

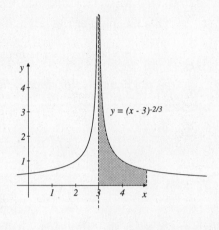

Example 5   A divergent integral over a finite interval.

Problem   Does $\int_0^1 1/x \, dx$ converge? If so to what value?

Solution   The integral is improper because the integrand approaches ∞ as $x \to 0+$. To test if the integral converges, we consider the

$$\lim_{t \to 0+} \int_t^1 1/x \, dx = \lim_{t \to 0+} ln(x) \Big|_t^1$$

$$= \lim_{t \to 0+} ln(1) - ln(t) = \infty$$

As this limit is not finite the improper integral does not converge; it diverges to ∞.

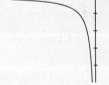

Example 6   The improper integral $\int_c^b (x - c)^{-r} \, dx$.

Problem   Because of the negative sign in the exponent, the function $y = (x - c)^{-r}$ is singular at $x = c$ for all positive numbers r. For what values of r is the integral $I = \int_c^b (x - c)^{-r} \, dx$ convergent?

Solution   Two cases must be considered. If $r = 1$ the integral diverges, as

$$\int_c^b (x - c)^{-1} \, dx = \lim_{t \to c+} ln(x - c) \Big|_t^b = \lim_{t \to c+} \{ln(b - c) - ln(t - c)\} = \infty$$

If $r \neq 1$, an antiderivative of $(x - c)^{-r}$ is $(x - c)^{-r + 1}/(1 - r)$. Thus,

$$\int_c^b (x - c)^{-r} \, dx = \lim_{t \to c+} (x - c)^{1 - r}/(1 - r) \Big|_t^b$$

$$= \lim_{t \to c+} \{(b - c)^{1 - r} - (t - c)^{1 - r}\}/(1 - r)$$

The first term $(b - c)^{1 - r}$ is constant and can thus be taken outside the limit. Thus

$$I = (b - c)^{1 - r}/(1 - r) - \{1/(1 - r)\} \cdot \lim_{t \to c+} (t - c)^{1 - r}$$

The value of the remaining limit depends on the parameter r.

If $r < 1$, then $1 - r$ is positive and $\lim_{t \to c^+}(t - c)^{1-r} = 0$

This means that the original integral converges to $I = (b - c)^{1-r}/(1 - r)$ for $0 < r < 1$.

If $r > 1$ then $1 - r$ is negative and $\lim_{t \to c^+}(t - c)^{1-r} = \lim_{t \to c^+} 1/(t - c)^{r-1} = +\infty$

as the denominator approaches zero and $r - 1 > 0$. In summary,

$$\int_c^b (x - c)^{-r} dx = \begin{cases} (b - c)^{1-r}/(1 - r) & \text{if } 0 < r < 1 \\ \text{diverges to } \infty & \text{if } r \geq 1 \end{cases}$$

☑

Integrals with singularities within an integration interval.

Improper integrals also occur when an integrand is singular at an interior point of the interval of integration. This is dealt with by splitting the integration interval at the singular point, giving two improper integrals with endpoint singularities. This type of singularity is sometimes not recognized, which can lead to erroneous evaluations.

DEFINITION OF AN IMPROPER INTEGRAL WITH AN "INTERIOR SINGULARITY".

If $f(x)$ has a singularity at $x = c$, and $a < c < b$, then

$$\int_a^b f(x) \, dx \equiv \int_a^c f(x) \, dx + \int_c^b f(x) \, dx$$

provided both improper integrals over $[a,c]$ and $[c,b]$ converge.
If one of the integrals diverges then the integral over $[a,b]$ diverges.

Example 7 **Integrating over a singularity.**

Problem   Evaluate (a) $\int_{-2}^1 1/x^2 \, dx$ and (b) $\int_0^5 (x - 2)^{-1/3} \, dx$.

Solutions   (a) A naive evaluation of the integral of $y = 1/x^2$ over the interval $[-2,1]$ would give

$$\int_{-2}^1 1/x^2 \, dx = -1/x \Big|_{-2}^1 = -1 - 1/2 = -3/2, \quad \textbf{which can not be valid!}$$

Since the integrand $1/x^2$ is always positive, its integral can not be negative. The difficulty is that this is an improper integral because the integrand $1/x^2$ is singular and becomes infinite at $x = 0$. When an interior singularity occurs, the improper integral must be split at the singular x=value to form two improper integrals having singularities only at endpoints of their integration intervals. Formally, we write

$$\int_{-2}^1 1/x^2 \, dx = \int_{-2}^0 1/x^2 \, dx + \int_0^1 1/x^2 \, dx$$

The original integral exists only if both improper integrals on the right exist. For this particular problem we can refer to the general case considered in Example 6 with $c = 0$ and $r = 2$ to conclude that $\int_0^1 1/x^2 \, dx$ does not converge, and hence the original integral can not be evaluated.

(b) $\int_0^5 (x-2)^{-1/3} dx$ is improper since the integrand $1/(x-2)^{1/3}$ is singular at $x = 2$. To evaluate this integral it is split at the singular point. We consider the limit of both split improper integrals simultaneously using the limit variables t and s.

$$\int_0^5 (x-2)^{-1/3} dx = \int_0^2 (x-2)^{-1/3} dx + \int_2^5 (x-2)^{-1/3} dx$$

$$= \lim_{t \to 2^-} \int_0^t (x-2)^{-1/3} dx + \lim_{s \to 2^+} \int_s^5 (x-2)^{-1/3} dx$$

$$= \lim_{t \to 2^-} (3/2)(x-2)^{2/3} \big|_0^t + \lim_{s \to 2^+} (3/2)(x-2)^{2/3} \big|_s^5$$

$$= \lim_{t \to 2^-} (3/2)[(t-2)^{2/3} - (-2)^{2/3}] + \lim_{s \to 2^+} (3/2)[3^{2/3} - (s-2)^{2/3}]$$

$$= (3/2)[(2-2)^{2/3} - (-2)^{2/3}] + (3/2)[3^{2/3} - (2-2)^{2/3}]$$

$$= (3/2)\{-(-2)^{2/3} + 3^{2/3}\} \approx 0.739$$

The Integral converges.

☑

Improper integrals with indeterminate limits.

Frequently the limits involved in evaluating improper integrals are "Indeterminate", having the symbolic limit forms "0/0", "∞/∞", or "0 · ∞". Evaluation of indeterminate limits was introduced in Chapter 4, Section 4.10. The following examples use the principal result of that section called L'Hôpital's Rule.

Recall.

L'Hôpital's Rule

If $\lim f(x) = 0$ and $\lim g(x) = 0$, or $\lim f(x) = \infty$ and $\lim g(x) = \infty$, then,

$$\lim f(x)/g(x) = \lim f'(x)/g'(x) \text{ provided this limit exists.}$$

Example 8 An improper integral with an "indeterminate" limit.

Problem Evaluate the improper integral $\int_0^\infty xe^{-2x} dx$.

Solution The antiderivative of xe^{-2x} is obtained using Integration by Parts, which was introduced in Section 6.2. To apply the integration by parts method, we choose functions u and v', so that $u(x)v'(x) = xe^{-2x}$. Then u' and v are determined and the IP formula is applied.
Choosing $u(x) = x$ and $v'(x)$ as the exponential term:

$$u = \underline{\quad x \quad} \qquad v' = \underline{\quad e^{-2x} \quad}$$

$$u' = \underline{\_\_ 1 \_\_\_\_\_} \qquad v = \underline{\_\_ -0.5e^{-2x} \_\_\_\_}$$

The IP formula gives the antiderivative:

$$\int xe^{-2x}\,dx = -0.5xe^{-2x} - \int -0.5e^{-2x}\,dx = -0.5xe^{-2x} - 0.25e^{-2x} = -0.25(2x+1)e^{-2x}$$

The improper integral is then evaluated by the limit of the definite integrals evaluated using the above antiderivative of xe^{-2x}.

$$\int_0^\infty xe^{-2x}\,dx = \lim_{t\to\infty} \int_0^t xe^{-2x}\,dx$$

$$= \lim_{t\to\infty} -0.25(2x+1)e^{-2x}\Big|_0^t$$

$$= \lim_{t\to\infty} -0.25\{(2t+1)e^{-2t} - (0+1)e^0\}$$

$$= -0.25 \lim_{t\to\infty} \{(2t-1)e^{-2t}\} + 0.25 = 0.25$$

To show that the limit of $(2t-1)e^{-2t}$ is zero requires the application of L'Hôpital's Rule since the limit has the indeterminate form "$\infty \cdot 0$". The term $2t-1$ goes to ∞ and e^{-2t} goes to zero as $t\to\infty$. It is evaluated by first manipulating it into an equivalent "∞/∞" form:

$$\lim_{t\to\infty} \{(2t-1)e^{-2t}\} = \lim_{t\to\infty} \{(2t-1)/e^{2t}\}$$

In this form L'Hôpital's Rule can be applied. Set $f(t) = 2t-1$ and $g(t) = e^{2t}$. Then

$$f'(t) = 2 \quad \text{and} \quad g'(t) = 2e^{2t}$$

and L'Hôpital's Rule gives

$$\lim_{t\to\infty} (2t-1)/e^{2t} = \lim_{t\to\infty} 2/2e^{2t} = 0$$

EXERCISE SET 6.4

1. Determine if the integral converges or diverges. Evaluate the integral if it converges.

   a) $\int_4^\infty 5x^{-3/4}\,dx$
   b) $\int_3^\infty x\,e^{-x^2}\,dx$
   c) $\int_2^\infty x\,e^{x^2}\,dx$
   d) $\int_0^\infty x^{4/3}\,dx$
   e) $\int_1^\infty 2/x\,dx$
   f) $\int_2^\infty x^{-3} - x^{-2}\,dx$
   g) $\int_1^\infty x^{-4/3}\,dx$
   h) $\int_0^\infty x\,e^{-10x^2}\,dx$
   i) $\int_2^\infty x^2\,e^{-x^3}\,dx$
   j) $\int_0^\infty e^{-0.3x}\,dx$
   k) $\int_2^\infty 1/(x\,ln(x))\,dx$
   l) $\int_1^\infty 6x^2 e^{-2x^3 - 1}\,dx$
   m) $\int_{-\infty}^\infty x\,e^{-x^2}\,dx$
   n) $\int_{-\infty}^\infty 5x^3 e^{-x^4}\,dx$
   o) $\int_{-\infty}^\infty x/(1+x^2)\,dx$
   p) $\int_{-\infty}^\infty x^2 e^{-x^3}\,dx$
   q) $\int_{-\infty}^\infty x^3 e^{-x^4}\,dx$
   r) $\int_0^\infty x^2/(1+x^3)\,dx$
   s) $\int_0^\infty 2/(1+e^x)\,dx$
   t) $\int_{-\infty}^\infty x/(1+x^2)^2\,dx$
   u) $\int_{-\infty}^\infty x/\text{SQRT}(4+x^2)\,dx$

2. Determine if the integral is "improper" and if so, evaluate the integral if it converges or show why it does not exist.

   a) $\int_0^1 x^{-2/3}\,dx$
   b) $\int_0^1 x^{-2}\,dx$
   c) $\int_0^1 ln(2x)/x\,dx$
   d) $\int_0^5 (x-5)^{-1/3}\,dx$
   e) $\int_0^5 (x-5)^{-4/3}\,dx$
   f) $\int_0^5 x/\text{SQRT}(25 - x^2)\,dx$
   g) $\int_0^3 5x\,log_3(x)\,dx$
   h) $\int_0^5 e^{-1/x}/x^2\,dx$
   i) $\int_1^{10} log_3(x-1)\,dx$

3. Determine if the integral is "improper". Evaluate the integral or indicate why it does not exist. You will need to use Integration by Parts to evaluate some antiderivatives.

a) $\int_0^{\pi/2} \tan(\theta)\, d\theta$
b) $\int_0^{\pi/2} \cos(\sqrt{\theta})/\sqrt{\theta}\, d\theta$
c) $\int_0^{\infty} \sin(\theta)e^{-\theta}\, d\theta$
d) $\int_0^{\infty} \cos(t)e^{-t}\, dt$
e) $\int_0^{\pi} -\sin(1/t)/t^2\, dt$
f) $\int_0^{\pi/2} \sec^2(\theta)\, d\theta$
g) $\int_0^{\infty} \cos(t)e^{\sin(t)}\, dt$
h) $\int_0^{\infty} (\sin(t) + t\cos(t))e^{-t\sin(t)}\, dt$
i) $\int_0^{\infty} (\sin(t) + t\cos(t))e^{t\sin(t)}\, dt$
j) $\int_0^{\pi/2} \ln(\sin(\theta))\cot(\theta)\, d\theta$
k) $\int_1^{\infty} (-\sin(t)/t^2 + \cos(t)/t)e^{\sin(t)/t}\, dt$
l) $\int_0^{\pi} \sin(\theta)/(1 - \cos^2(\theta))\, d\theta$
m) $\int_0^{\infty} (\cos(t) - 1)e^{\sin(t) - t}\, dt$
n) $\int_0^{\infty} \sin(\theta)e^{\theta}\, d\theta$
o) $\int_{-2}^{1} 1/(2 - t - t^2)\, dt$
p) $\int_1^{2} (t^2 + 2t)/\sqrt{-4 + 3t^2 + t^3}\, dt$

4. Indicate if the given integral is "improper", and why. Evaluate the integral, if it exists.

a) $\int_{-1}^{1} x^{-2/3}\, dx$
b) $\int_{-1}^{1} x^{-3/2}\, dx$
c) $\int_{-1}^{1} (2x - 1)^{-2/3}\, dx$
d) $\int_{-5}^{5} (x^2 - 3x + 2)/x\, dx$
e) $\int_0^{5} (x - 2)^{-.75}\, dx$
f) $\int_0^{5} x/(x^2 - 1)\, dx$
g) $\int_0^{5} x\sqrt{x^2 - 4}\, dx$
h) $\int_1^{10} (2x - 3)^{-1.5}\, dx$
i) $\int_0^{10} (x + 5)/(x - 5)\, dx$
j) $\int_{-5}^{5} (2x - 1)(x^2 - x - 2)^{-1/3}\, dx$
k) $\int_{2.5}^{3.5} (2x - 5)/(x^2 - 5x + 6)\, dx$
l) $\int_{2.5}^{5} (3x - 8)/(x^2 - 5x + 6)\, dx$
m) $\int_0^{1} \cos(\pi\theta)/(1 - \sin(\pi\theta))^{1/4}\, d\theta$
n) $\int_0^{\pi} \tan(\theta)\, d\theta$

5. What is the Bioavailability of a drug if its Plasma level over the interval $[0,\infty)$ is given by the indicated function. (See Example 3.) Assume that k is a positive constant.

a) $C(t) = 0.05e^{-kt}$
b) $C(t) = 0.25te^{-kt}$
c) $C(t) = 3.1[e^{-2t} - e^{-6t}]$
d) $C(t) = 7t^2 e^{-kt}$
e) $C(t) = -35[e^{-t} + e^{-0.5t} - 2e^{-0.1t}]$
f) $C(t) = 4.5\sqrt{3 + 1/t}/t^2$

Section 6.5 Numerical integration

Introduction: Approximating Integrals.

In scientific research and in industrial and business applications there is often a need to evaluate definite integrals for which the FTC cannot be applied. This occurs when the integrand is a complex function whose antiderivative we cannot establish or is a function that is not given by a formula, like y = f(x), but rather is generated as numerical values at specific x-values. Typically, such values are derived from experimental instruments or "real-time" data sensors. In such cases, as the Fundamental Theorem of Calculus can not be used to evaluate the integral, numerical methods are employed to approximate specific definite integrals.

In this section three formulas are introduced to numerically approximate $\int_a^b f(x)\,dx$. These formulas are derived by approximating the function f(x) by a polynomial over subintervals of the interval [a,b]. The reason for this is that the definite integral of a polynomial of degree N can be given explicitly by a numerical "quadrature" formula using N + 1 data values.

If a polynomial P(x) approximates f(x) over an interval [a,b] with an **error** e(x), then f(x) = P(x) + e(x), and

$$\int_a^b f(x)\,dx = \int_a^b P(x)\,dx + \int_a^b e(x)\,dx$$

The integral of f(x) is approximated by the integral of the polynomial P(x),

$$\int_a^b f(x)\,dx \approx \int_a^b P(x)\,dx$$

with an absolute error given by

$$|E| = \left|\int_a^b f(x)\,dx - \int_a^b P(x)\,dx\right| = \left|\int_a^b e(x)\,dx\right| \leq \max\{|e(x)|\} \cdot (b - a)$$

If the error function is small, $|e(x)| < \delta$, for some small number δ, then the absolute error will be less than δ times the length of the interval [a,b], $|E| \leq \delta(b - a)$. Thus to make an approximation more accurate two factors can be considered, how to minimize the error e(x) and how to minimize the length of the integration interval.

One way to diminish the error $|E|$ is to make the integration interval length smaller; this is achieved by employing a basic calculus technique. The original integration interval [a,b] is partitioned by a set $\{x_n\}$ into N subintervals and the original integral is the sum of the integrals over each subinterval. If each subinterval integral is approximated, presumably using approximating polynomials P_n on the subinterval $[x_{n-1}, x_n]$, and with an error proportional to the subinterval's length Δx_n, the sum of these approximations gives an approximation to the original interval:

$$\int_a^b f(x)\,dx = \sum_{n=1}^{N} \int_{x_{n-1}}^{x_n} f(x)\,dx \approx \sum_{n=1}^{N} \int_{x_{n-1}}^{x_n} P_n(x)\,dx$$

The second factor that could reduce the absolute error $|E|$ is to reduce the magnitude of the error at each point, i.e, e(x). This can be done by employing higher order polynomials to approximate the function f(x). As we shall see, the increased computational cost is offset by the increased accuracy.

Three different approximation formulas will be introduced by choosing the polynomial P(x) to be of degree 0, 1 and 2. For degree 0: P(x) = C, the formula is called the "Rectangle Rule", since the region under a constant function is a rectangle. For degree = 1: P(x) = C + Bx, and the resulting formula is called the "Trapezoid Rule", because the shape of the region under the graph of a linear function is trapezoidal. For degree = 2: $P(x) = C + Bx + Ax^2$, and the formula is called "Simpson's Rule", which is named after Thomas Simpson (1720 - 1761) (who did not originate the rule but was the first to state it in his book). It utilizes quadratics to approximate the integrand f(x) and thus is based on a formula for the area under a parabola.

The Rectangle Rule.

The simplest approximation of a function is given by a constant function. If the function $P(x) = y^*$ where y^* is the average value of f over [a,b], then as discussed in Chapter 5, Sections 5.1 and 5.2, the integral of f over [a,b] is given exactly by the integral of P(x), i.e., if y^* =AVG{f,a,b} then

$$\int_a^b f(x)\, dx = \int_a^b P(x)\, dx = \int_a^b y^*\, dx = y^* \cdot (b - a)$$

since the average value AVG{f,a,b} is defined to be the integral of f over [a,b] divided by b - a. Remember that it is this integral that we want to evaluate, we of course do not know its value and thus do not know AVG{f,a,b} *a priori*. If we randomly select a value y^* our chance of picking the average of f is very slim.

The Rectangle Rule is established by partitioning the interval [a,b] into subintervals and choosing the approximating polynomial P(x) to be a constant on each subinterval. The constant value can be chosen as the value of f at any evaluation point in the subinterval. If the partition has N subintervals, and the evaluation point in the n-th subinterval is x_n^* then

$$P(x) = f(x_n^*) \text{ for } x_{n-1} \leq x \leq x_n$$

Consequently, the integral of P(x) over $[x_{n-1}, x_n]$ is just the constant $f(x_n^*)$ times $\Delta x_n = x_n - x_{n-1}$. The resulting approximation formula becomes

$$\int_a^b f(x)\, dx = \Sigma_{n=1,N}[f(x_n^*) \cdot \Delta x_n]$$

This formula is simplified if we use a uniform partition. Recall that a uniform partition of [a,b] into N subintervals of equal length h = (b - a)/N is given by the x-values

$$x_0 = a, \quad x_1 = a + h, \quad x_2 = a + 2h, \quad \ldots, \quad x_N = a + Nh = b$$

and the n-th subinterval is

$$I_n = [x_{n-1}, x_n] = [a + (n-1)h, a + nh]$$

For a uniform partition $\Delta x_n = h$ for each n, and hence the Rectangle Approximation formula is

$$\int_a^b f(x)\, dx \approx h \cdot \Sigma_{n=1,N} f(x_n^*)$$

Two similar formulas are formed by choosing the evaluation points as the right and left endpoints of the subintervals (See the following graphs). These approximations consist of approximating the area under the graph of f by the sum of the areas of rectangles, as illustrated above. Each rule adds N function values and multiplies by the subinterval width h. We formally state these formulas as the Rectangle Rule. We will consider numerical examples after stating the other approximation formulas.

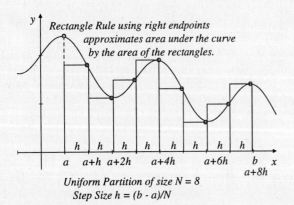
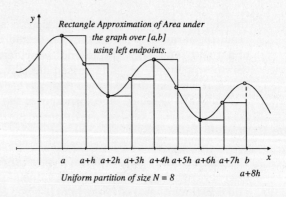

RECTANGLE RULE FOR INTEGRATION

For a uniform partition of [a,b] into N subintervals of length $h = (b-a)/N$:

$$\int_a^b f(x)\, dx \approx h[f(a+h) + f(a+2h) + \ldots + f(b)] \qquad \text{Using right endpoints}$$

$$\int_a^b f(x)\, dx \approx h[f(a) + f(a+h) + \ldots + f(a+(N-1)h)] \qquad \text{Using left endpoints}$$

The Trapezoid Rule.

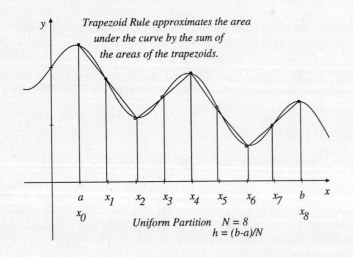

The "Trapezoid Rule" for approximating $\int_a^b f(x)\, dx$ is derived by approximating the function f by a linear function P over each subinterval of the interval [a,b]. The linear segments are chosen so that P(x) has the same values as f(x) at the endpoints of the subintervals.

The equation for P(x) over the n-th subinterval $[x_{n-1}, x_n]$ is given by

$$P(x) = f(x_{n-1}) + m(x - x_{n-1})$$

This is the point slope equation of a line passing through the point $(x_{n-1}, f(x_{n-1}))$, with slope

$$m = \{f(x_n) - f(x_{n-1})\}/h$$

The approximating polynomial P matches the function f at the endpoints of the subinterval:

$$P(x_{n-1}) = f(x_{n-1})$$

and

$$P(x_n) = f(x_n)$$

The integral of P(x) over the general interval $[x_{n-1}, x_n]$, is the area under its graph, which is the area of a trapezoid as illustrated at the right. This area is simply one half the base length times the sum of the heights at the endpoints. You can actually establish this formula by integrating the function P over the interval.

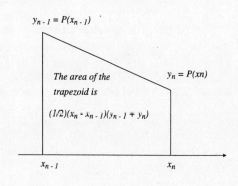

$$\int_{x_{n-1}}^{x_n} P(x)\, dx = \frac{1}{2}[P(x_{n-1}) + P(x_n)] \cdot (x_n - x_{n-1})$$

Area of a trapezoid.

The Trapezoid Rule results from summing these areas over each subinterval of a uniform partition of the integration interval [a,b] into N subintervals. Since the intervals are all of the same length the term $\Delta x_n = h$ for each n:

$$\int_a^b f(x)\, dx \approx \Sigma_{n=1,N} (h/2)[f(x_{n-1}) + f(x_n)]$$

Factoring out the h/2 term and expanding, this sum has the form

$$(h/2)\{[f(x_0) + f(x_1)] + [f(x_1) + f(x_2)] + [f(x_2) + f(x_3)] + \ldots + [f(x_{N-1}) + f(x_N)]\}$$

As each term $f(x_n)$ appears twice, except for the first and last terms, the approximation becomes

$$\int_a^b f(x)\, dx \approx (h/2)\{f(x_0) + 2f(x_1) + 2f(x_2) + \ldots + 2f(x_{N-1}) + f(x_N)\}$$

If the 2 is factored out, this sum is sometimes more easily remembered when the partition points are expressed as $x_n = a + nh$. The resulting sum is known as the Trapezoid Rule.

THE TRAPEZOID RULE FOR INTEGRATION.

$$\int_a^b f(x)\, dx \approx h[f(a)/2 + f(a + h) + f(a + 2h) + \ldots + f(a + (N-1)h) + f(b)/2]$$

Observe that the Trapezoid Rule requires exactly the same numerical data, the function values at the partition points, as the Rectangle Rule requires. But, it is normally much more accurate! In practice the Trapezoid Rule is used in preference to the Rectangle Rule.

Example 1   **Applying the Trapezoid Rule.**

Problem   Approximate $\int_0^4 x(4 - x)\, dx$ using the Trapezoid Rule for partitions of size N = 4 and N = 10.

Solutions   For N = 4, each subinterval is of length h = (4-0)/4 = 1. The function f(x) = x(4 - x). The basic formula gives

$$\int_0^4 x(4-x)\,dx \approx (1)[f(0)/2 + f(1) + f(2) + f(3) + f(4)/2]$$

$$= [0/2 + 3 + 4 + 3 + 0/2] = 10$$

For N = 10, the step size is h = (4 - 0)/10 = 0.4. The Trapezoid Rule gives

$$\int_0^4 x(4-x)\,dx \approx 0.4[f(0)/2 + f(0.4) + f(0.8) + f(1.2) + \ldots + f(3.6) + f(4)/2]$$

$$= 0.4[0/2 + 1.44 + 2.56 + 3.36 + 3.84 + 4 + 3.84 + 3.36 + 2.56 + 1.44 + 0/2]$$

$$= 0.4 \cdot 26.4 = 10.56$$

These approximations can be compared to the exact value obtained by the Fundamental Theorem of Calculus

$$\int_0^4 x(4-x)\,dx = 2x^2 - x^3/3 \;\Big|_0^4 = [2 \cdot 4^2 - 4^3/3] = 10.\underline{6}$$

☑

Simpson's Rule

The numerical approximation of $\int_a^b f(x)\,dx$ can be made more accurate by choosing an approximation function $P(x)$ that more closely resembles the function f. The next logical choice to consider, following the pattern used to form the Rectangle and Trapezoid Rules, is to choose $P(x)$ as a quadratic function over the subintervals, matching the function f at points of a partition.

The graph of $P(x) = Ax^2 + Bx + C$ is a parabola, as illustrated at the right. It is completely determined by any three points that it passes through. In particular, the area under a parabola over an interval can be expressed as a function of its y-values at the endpoints and the midpoint of the interval. The algebra of this is simplified if the points are denoted by $x = z$, $x = z + h$ and $x = z + 2h$.

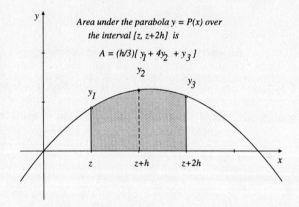

Area under the parabola $y = P(x)$ over the interval $[z, z+2h]$ is

$$A = (h/3)[y_1 + 4y_2 + y_3]$$

Consider a parabola $P(x)$ over the interval $[z, z + 2h]$. Denote the y-values at the endpoints and midpoint of the interval by

$$P(z) = y_1, \quad P(z + h) = y_2, \quad P(z + 2h) = y_3$$

as illustrated. The area under the graph of the parabola over this interval can be shown to be a simple combination of these three y-values:

$$\int_z^{z+2h} P(x)\,dx = (h/3)[y_1 + 4y_2 + y_3] \qquad \textbf{3-Point Formula}$$

Simpson's Rule is derived by approximating the function f by different parabolas over successive pairs of subintervals of a uniform partition of size 2N, so that there will be an even number of subintervals. This is illustrated on the next page. The parabolas are chosen to agree with the values of f(x) at the partition points as illustrated. In the figure, the interval [a,b] is partitioned into 2N = 8 subintervals and f(x) is approximated by N = 4 parabolas:

$f(x) \approx P_1(x)$ over $[a, a + 2h]$; $\qquad\qquad f(x) \approx P_2(x)$ over $[a + 2h, a + 4h]$;

$f(x) \approx P_3(x)$ over $[a + 4h, a + 6h]$; $\qquad\qquad f(x) \approx P_4(x)$ over $[a + 6h, b]$.

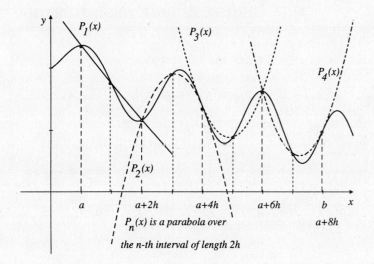

$P_n(x)$ is a parabola over the n-th interval of length 2h

Simpson's Rule approximates the integral of f by the sum of the integrals of the four parabolas over their respective subintervals:

$$\int_a^b f(x)\,dx \approx \int_a^{a+2h} P_1(x)\,dx$$

$$+ \int_{a+2h}^{a+4h} P_2(x)\,dx$$

$$+ \int_{a+4h}^{a+6h} P_3(x)\,dx$$

$$+ \int_{a+6h}^{b} P_4(x)\,dx$$

The integral of each parabola can be evaluated by the above **3-point formula**. The actual equation of each parabola is not needed, only their y-values at the endpoints, which by our construction are exactly the same as the values of f at the endpoints:

$$P(a + nh) = f(a + nh)$$

The resulting approximation of the integral of f is

$$\int_a^b f(x)\,dx \approx h/3[f(a) + 4f(a+h) + f(a+2h)] \qquad \int_a^{a+2h} P_1(x)\,dx$$
$$+ h/3[f(a+2h) + 4f(a+3h) + f(a+4h)] \qquad \int_{a+2h}^{a+4h} P_2(x)\,dx$$
$$+ h/3[f(a+4h) + 4f(a+5h) + f(a+6h)] \qquad \int_{a+4h}^{a+6h} P_3(x)\,dx$$
$$+ h/3[f(a+6h) + 4f(a+7h) + f(b)] \qquad \int_{a+6h}^{b} P_4(x)\,dx$$

Factoring out h/3 and adding like terms, the above equation reduces to one with an apparent pattern:

$$\int_a^b f(x)\,dx \approx h/3\{f(a) + 4f(a+h) + 2f(a+2h) + 4f(a+3h)$$
$$+ 2f(a+4h) + 4f(a+5h) + 2f(a+6h) + 4f(a+7h) + f(b)\}$$

Observe that this sum adds:

i) the value of f at the first and last points, f(a) and f(b);

ii) four times f evaluated at each odd partition point, $4f(a+nh)$, n = 1, 3, 5, 7;

iii) two times f evaluated at each even partition point, $2f(a+nh)$, n = 2, 4, 6.

Expressed in a compact form, this formula reads as follows:

$$\int_a^b f(x)\,dx \approx h/3\{f(a) + f(b) + 4 \cdot \Sigma_{odd\ n\ =\ 1, N-1} f(a+nh) + 2 \cdot \Sigma_{even\ n\ =\ 2, N-2} f(a+nh)\}$$

This equation is sometimes more easily remembered without using the Σ notation and just recognizing the pattern.

SIMPSON'S RULE FOR INTEGRATION.

For $x_n = a + nh$, $n = 0, 1, 2, \ldots, N$ where N is an even integer and $h = (b - a)/N$,

$$\int_a^b f(x)\, dx \approx h/3 \{f(x_0) + 4f(x_1) + 2f(x_2) + 4f(x_3) + 2f(x_4) +$$

$$\ldots + 2f(x_{N-2}) + 4f(x_{N-1}) + f(x_N)\}$$

Remember: TO APPLY SIMPSON'S RULE, THE UNIFORM PARTITION MUST HAVE AN EVEN NUMBER OF SUBINTERVALS.

Example 2 **Applying Simpson's Rule**.

Problem   Use Simpson's Rule with N = 6 to approximate $\int_0^4 x^3\, dx$.

Solution   The function is $f(x) = x^3$. The step size is $h = (4 - 0)/6 = 2/3$. Applying Simpson's Rule the integral is approximated by the sum:

$$(2/3)/3 \cdot \{f(0) + 4f(2/3) + 2f(4/3) + 4f(6/3) + 2f(8/3) + 4f(10/3) + f(4)\}$$

$$= (2/9) \cdot \{0^3 + 4(2/3)^3 + 2(4/3)^3 + 4(6/3)^3 + 2(8/3)^3 + 4(10/3)^3 + 4^3\}$$

$$= (2/9) \cdot \{0 + 32/27 + 128/27 + 864/27 + 1024/27 + 4000/27 + 64\}$$

$$= (2/9) \cdot \{6048/27 + 64\} = (2/9) \cdot 288 = 64$$

The exact value of this integral determined by the Fundamental Theorem of Calculus is:

$$\int_0^4 x^3\, dx = x^4/4 \,|_0^4 = 4^4/4 = 64$$

The last Example illustrates the fact that Simpson's Rule is "exact" for polynomials of degree three. Even though the Rule is derived using second-order polynomial approximations it always gives the exact value of the integral of a cubic polynomial.

Example 3 **Integrating with numerical data**.

Background

An experiment investigating the effects of a blood protein called "LACI"[†] on the "activity" of the blood coagulation factor VII generated numerical data as indicated in the following figures. The different panels indicate "activation" levels at 3 second intervals with and without the LACI protein added to the reaction solution. The values on the y-axis reflect the different rates at which a blood clot would form following an injury. The investigators wished to quantify the overall effects of the LACI protein

---

[†] Since this example was originally written, the name for LACI has been officially changed to Tissue Factor Pathway Inhibitor.

by computing and comparing the areas under the curves passing though the data. (The curves are not shown in the figure.)

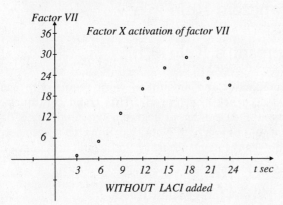
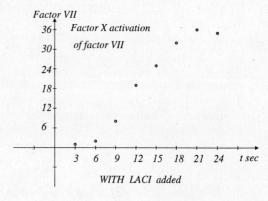

Problem Use the following tabulated data corresponding to the above graphs to estimate the areas under the two curves (formed when the dots are connected) using the trapezoid rule.

Time (seconds)	3	6	9	12	15	18	21	24
Factor VII without LACI	1	5	13	20	26	29	23	21
Factor VII with LACI	1	2	8	19	25	32	36	35

Solution The 8 data points for each experimental condition give only 7 intervals. To apply Simpson's Rule an even number is required. To achieve this we improvise and note that the graphs begin at the origin. At time $t = 0$ there is no reaction and the corresponding rate is zero. Using the "zero,zero" points the number of intervals is $N = 8$ and the step size is $h = 3$ s.

The estimated area under the curve without LACI is

$$A_{\text{No LACI}} = (3/3)\{0 + 4\cdot 1 + 2\cdot 5 + 4\cdot 13 + 2\cdot 20 + 4\cdot 26 + 2\cdot 29 + 4\cdot 23 + 20\} = 380.$$

The corresponding area with LACI present is

$$A_{\text{LACI}} = (3/3)\{0 + 4\cdot 1 + 2\cdot 2 + 4\cdot 8 + 2\cdot 19 + 4\cdot 25 + 2\cdot 32 + 4\cdot 36 + 35\} = 421.$$

Looking at the data, graphed or in the table, it may be difficult to distinguish the effect of LACI on factor VII levels. The approximate integrals indicate that over the observation period about 10% more factor VII is present when LACI is included in the assay solution.

Errors in numerical integration.

The error incurred using the different numerical integration Rules depends on the step size, $h = (b - a)/N$, and the nature of the function f. When the function f is continuous the errors of the three formulas will be proportional (expressed by the symbol $\propto$) to different powers of h:

$$\text{The error of the Rectangle Rule} \propto h.$$

$$\text{The error of the Trapezoid Rule} \propto h^2.$$

$$\text{The error of the Simpson's Rule} \propto h^4.$$

Consequently, in return for the increased complexity and amount of work required to use the Trapezoid and Simpson's Rules an increased accuracy is achieved. This trend continues with "higher order" Rules. Generally, if the polynomial $P(x)$ is a higher degree polynomial, say of degree n, then the error associated with the method would be $\propto h^{n+1}$. Note that Simpson's Rule actually does better than this general rule indicates; it gives the same accuracy that would be achieve using a "cubic" $P(x)$.

Example 4 **Comparing the accuracy of the different Integration Rules.**

Problem Compare the accuracy of the numerical integration Rules by approximating

$$\int_0^2 x\, e^{-x^2}\, dx = (1/2)e^{-x^2}\big|_0^2 = (e^0 - e^{-4})/2 = 0.49084218 \text{ (by calculator)}.$$

Using a uniform partition of size $N = 10$ of the interval $[0,2]$.

Solution The data required for these calculations is the values of the function at the partition points $x_n = 0.2n$, for $n = 0, 1, \ldots, 10$. These are indicated by $y_n = x_n \exp(-x_n^2)$. They were found using a spread-sheet program (a calculator would be sufficient) and are listed in Table I along with the "weights" required for each rule and the product of the weights times the y-values. The sums of the columns are also calculated. Using these data the estimates of the integral are obtained as follows:

Applying the Rectangle Rule:

Using left end points: $h \cdot \Sigma_{n=0,9} y_n = 0.2 \cdot 2.417032 = 0.483406$

Using right end points: $h \cdot \Sigma_{n=1,10} y_n = 0.2 \cdot 2.453663 = 0.490732$

Applying the Trapezoid Rule: $h \cdot \Sigma_{n=0,10} y_n \cdot Wt_T = 0.2 \cdot 2.435347 = 0.487069$

Applying Simpson's Rule: $(h/3) \cdot \Sigma_{n=0,10} y_n \cdot Wt_S = (0.2 / 3) \cdot 7.36337567 = 0.49089171$

The errors incurred using the different Rules are:

	Error
Rectangle Rule Using Left Endpoints:	0.007435
Rectangle Rule Using Right Endpoints:	0.000109
Trapezoid Rule	0.003772
Simpson's Rule	-0.0000495

Table I Data for Numerical Integration.

n	x_n	$y_n = x_n e^{-x_n^2}$	Trapezoid Weight Wt_T	Trapezoid Rule $Wt_T \cdot y_n$	Simpson's Weight Wt_S	Simpson's Rule $Wt_S \cdot y_n$
0	0.0	0.0	0.5	0.0000	1	0.0000
1	0.2	0.192157	1	0.1921	4	0.7684
2	0.4	0.340857	1	0.3408	2	0.6816
3	0.6	0.418605	1	0.4186	4	1.6744
4	0.8	0.421833	1	0.4218	2	0.8436
5	1.0	0.367879	1	0.3678	4	1.4712
6	1.2	0.284313	1	0.2843	2	0.5686
7	1.4	0.197201	1	0.1972	4	0.7888
8	1.6	0.123687	1	0.1236	2	0.2472
9	1.8	0.070495	1	0.0704	4	0.2816
10	2.0	0.036631	0.5	0.0183	1	0.0366
SUM		2.4532		2.4349		7.3620

Notice that for h = 0.2 in this particular example the Rectangle Rule approximation with the right endpoints is actually more accurate than the Trapezoid Rule. This is not normally the case, but it can happen. This process was repeated with N = 20 subintervals, h = 0.1. Without presenting the data the results are summarized as follows:

Rule	Estimate of Integral	Error
Rectangle, Left Endpoints	0.488069	0.0027724
Rectangle, Right Endpoints	0.4917328	-0.00890
Trapezoid	0.489901	0.000940
Simpson's	0.49084516	-0.0000029

Employing the finer partition gave better estimates. Also, the relative errors of the methods is more clearly demonstrated.

EXERCISE SET 6.5

1. For the given function f(x) estimate $\int_0^2 f(x)\, dx$ using the Rectangle Rule with:

   i)   N = 4 and right endpoints   ii)   N = 4 and left endpoints
   iii)   N = 6 and right endpoints   iv)   N = 6 and left endpoints.

   a) $3x + 5$ b) x^2 c) $(x-2)^2$ d) $x(2-x)^2$

2. If a migrating flock of geese flies with speed $s(t) = 90\sin^2(\pi t)$ km/day, where the unit of time is t days, express the distance they fly in 10 days as a definite integral. Approximate this distance using the Rectangle Rule with N = 20.

3 Estimate the average of $R(t) = 1/(t + 1)$ over [0,10] using the Rectangle Rule and $N = 5$.

4 Estimate the average value of $f(x) = 3^x$ over the interval [0,2]
   a) Using the both the left and right Rectangle Rules for $N = 4$.
   b) Using the Trapezoid Rule with $N = 4$.
   c) Using the both the left and right Rectangle Rules for $N = 8$.
   d) Using the Trapezoid Rule with $N = 8$.
   e) Using Simpson's Rule with $N = 8$.

5 Approximate $\int_0^1 f(x)\, dx$ for the given function using a uniform partition of size $N = 6$ and
   i)   the Rectangle Rules   ii) the Trapezoid Rule
   iii) Simpson's Rule.   iv) Compare the error in each approximation.

   a) $f(x) = 5e^x$ b) $f(x) = \sin(\pi x)$ c) $f(x) = ln(x + 1)$ $f(x) = e^{-x^2}$

6 Repeat Exercises 6 for $N = 10$.

7 Approximate the $\int_0^3 [\![x]\!]\, dx$ using the Rectangle Rule with a) $N = 6$ and b) $N = 9$. Compare the estimates using the left and right endpoints.

8 Estimate $\int_0^4 f(x)\, dx$ if the function f is given by the tabulated values:

x	0.0	0.5	1.0	1.5	2.0	2.5	3.0	3.5	4.0
f(x)	1.0	1.0	2	1.5	3.0	1.2	1.1	0.8	0.1

Use a) the Rectangle Rule; b) the Trapezoid Rule; c) Simpson's Rule.

9 Approximate $\int_0^2 f(x)\, dx$ if f is given by the non-uniformly spaced data:

x	0.0	0.2	0.4	0.5	0.9	1.3	1.8	1.9	2.0
f(x)	0.0	1.5	1.2	2.0	1.6	3.4	0.3	0.5	0.8

a) Using rectangular approximations. b) Using trapezoids.
Caution: The Rules employing uniform partitions do not apply. Sketch the data and the approximating polynomial over each subinterval.

An alternative way to approximate integrals.

Taylor polynomials can be used to approximate a function. (See Chapter 4 Section 4.9.) Another method of approximating an integral is by the integral of a Taylor Polynomial which approximates the integrand. The virtue of this method is that integrals of polynomials are very easily evaluated. If P_n is an nth-order Taylor polynomial for $f(x)$ about a point $x = c$, then the **Taylor Approximation of the integral** is

$$\int_a^b f(x)\, dx \approx \int_a^b P_n(x)\, dx$$

Recall that $P_n(x) = \sum_{i=1}^n \{f^{(i)}(c)/i!\}(x - c)^i$, where $f^{(i)}(c)$ is the i-th derivative of f evaluated at $x = c$.

10 Approximate $\int_0^1 e^x\, dx$ using the third-order Taylor polynomial, for e^x about $c = 0$:
    $P_3(x) = 1 + x + x^2/2 + x^3/3!$

11 Approximate $\int_1^{1.5} 1/x\, dx$
    Use a third-order Taylor approximation $P_3(x)$ of $f(x) = x^{-1}$ about $c = 1$.

12 Approximate $\int_0^1 1/(1 + x^3)\, dx$
    Use a ninth-order Taylor approximation of $f(x) = 1/(1 + x^3)$ about $c = 0$: $P_9(x) = 1 - x^3 + 2x^6 - 6x^9$

   This expansion is most easily obtained by setting $x^3 = t$, determining a third-order expansion of $y = 1/(1 + t)$ about $t = 0$, and then substituting back into the x-variable.

13 Use the fourth-order Taylor approximations of the integrand $f(t)$ about the lower limit of integration approximate to obtain a polynomial that approximates the given integral-functions.

   a) $\int_0^x \text{SQRT}(t + 1)\, dt$ b) $\int_0^x (1 + t)^6\, dt$ c) $\int_0^x e^t\, dt$
   d) $\int_0^x \sin(t)\, dt$ e) $\int_0^x \sin(t^2)\, dt$ f) $\int_0^x \cos(\pi t)\, dt$
   g) $\int_0^x \cos(t^3)\, dt$ h) $\int_1^x 1/t\, dt$ i) $\int_1^x ln(t)\, dt$
   j) $\int_0^x e^{2t}\, dt$ k) $\int_0^x e^{t^2}\, dt$ l) $\int_1^x t^{2/3}\, dt$

Chapter 7 *Applications Involving Integration*

Section 7.1 Business and Economics 457
   Maximum profit from extended operations.
   Consumer's and Producers's Surplus.
   Equipment Utility; Uses the net area between two curves Section 5.6.

Section 7.2 Volumes of Solids of Revolution 461
   The *method of calculus* is used to establish the formulas in this section.
   Volume of a Cone.
   Volume of a dimple.
   Volume of an annulus or shell of revolution.
   Volume of a solid of revolution about the y-axis.

Section 7.3 The Length of a Curve; 467
   The Area of a Surface of Revolution.
   Arc Length.
   The circumference of a circle.
   The surface area of a sphere. Uses the parametric description of a circle.

Section 7.4 Probability and Statistics 474
   Computing Probabilities for a Uniform Density.
   A finite non-uniform distribution.
   Probability of an accident with a constant "hazard rate".
   The Expected Value or Mean of a uniform distribution.
   The Expected Value or Mean of a non-uniform distribution.
   The Expected Value or Mean of a Hazard distribution. This example uses
       Integration by Parts, Section 6.2, and L'Hôpital's Rule, Section 4.10.
   Predicting the number of AIDS cases by numerical integration. This example
       use the Trapezoid Rule, Section 6.5, to integrate a Delay-integral.

Section 7.5 Chemistry, Physics, and Hydrology 483
   Enzyme Kinetics; The Integrated Michaelis-Menten equation.
   Work: the integral of force over distance.
   Hooke's Law for stretching and compressing springs.
   Distance traveled: the integral of velocity over time traveled.
   Reservoir Storage Capacity
   Fluid infiltration of porous soils: The Horton Equation.
   Fluid Pressure and Fluid Force.

Section 7.6 Physiology and Biology 495
   Metabolic Rates and Basal Metabolism.
   Cardiac Output.
   Oxygen debt due to exercise.
   Growth and Decay Models; use Partial Fractions Section 6.3.
   Ruminant Digestion; the Passage of food through a sheep's stomach; uses Improper
       Integrals, Section 6.4.

Section 7.7 Toxicology and Environmental Modeling 503
   Uptake of pollutants by freshwater fish.
   Lake contamination and purification.
   The Dose Commitment resulting from toxic exposure; uses Improper Integrals Section 5.4.

Foreword

In this chapter various applications involving integrals are introduced. Many of the Examples begin with *Background* material introducing the vocabulary of the application and indicating how a particular quantity is described by an integral. The Background presentations are intentionally brief and often the nomenclature and complexity of the models have been simplified. Due to time constraints, most introductory calculus courses will only be able to cover a selection of these topics, but you are encouraged to browse topics not directly assigned to see the common role of the integral in the different applications.

You should view each Example as a "word problem". You should read it through first to identify the variables and notation utilized, then reread it looking for Key Words that indicate mathematical functions, equations, and when an integral is required. Common Key Words that indicate an integral is to be taken include: total, average, cumulative, and sum.

While it would be nice to rigorously "model" these topics, starting from basic assumptions and "first principles" to establish functional relationships and from these deduce the model equation(s), this would require a more extensive study and take us beyond the scope of an introductory calculus text. Most Colleges and Universities have mathematical modeling courses, but these typically require at least a year of mathematics. You will find the topics introduced in this chapter "modeled" in specialized discipline courses. To find introductory modeling books in your library you might try a computer search with the Kew Words *mathematical* and *model* plus discipline words such as *costs*, for economics, or *enzyme kinetics*, for biochemistry or physiology. You may also want to inquire about mathematical modeling from your course Professor, who can probably guide you to other sources or faculty directly involved in mathematical modeling.

To follow the solutions of the Example Problems it is not necessary that you fully "understand" the Background material, or memorize the vocabulary for specific applications, it is there to be referred to if needed, the Solutions depend on applying the integration techniques of the previous Chapters.

The sections of this Chapter are organized by subject. All of the examples have in common that their solutions utilize various integration methods previously introduced in this text. Many of the integrals considered can be evaluated using the General Integration Rules of Chapter 5, a few, as indicated in the following List of Examples, require integration techniques and methods introduced in Chapter 6. To establishing the integral formulas appropriate for different problems we invariably follow the *method of calculus*, giving a formula for a constant function, then for discrete functions as Riemann Sums, and finally for continuous functions as a limit of the solutions for associated discrete functions. This limit we recognize as a definite integral, employing the definition given in Chapter 5. In some problems we solve derivative equations by separating the variables and integrating. This method is considered in more detain in Chapter 8.

Section 7.1 Business and Economics

In business and economics applications, rates of change are called *marginals* instead of derivatives. Instead of indicating the rate of change in a cost variable C as C′ economics text often used the notation MC. Similarly, for instance, the marginal profit MP denotes the derivative of the profit, P′, etc. If a marginal function is known, the corresponding function, e.g., the Cost or Profit, is determined as the integral of the marginal, up to a constant. This of course is just a statement of the Fundamental Theorem of Calculus that states that $y = \int y' \, dx$ + a constant. In many actual applications the Marginal of an economic quantity is approximated by measured or estimated by observing the response change in the quantity to a unit change in a primary variable; this gives the differential approximation of the derivative. For instance the change ΔC in the cost of a manufactured item when a primary variable, x, such as gas prices or wage levels, are increased one unit gives an approximation of the derivative C′, and thus an estimate of the marginal cost MC. The change in cost when the primary variable x changes from x_1 to x_2 is then given by the integral of this marginal cost over the interval $[x_1, x_2]$:

$$C(x_2) - C(x_1) = \int_{x_1}^{x_2} MC(x) \, dx$$

This type of reasoning is used to establish various quantities as integrals of the marginals.

Example 1   **Maximum profit from extended operations.**

Background   When a factory operates past a normal shift time there is an increase in production that translates into increased revenue. But, there is also an increase in expenses as labor costs increase and machinery requires greater maintenance. If MC(t) is the marginal cost and MR(t) is the marginal revenue associated with operating a business t hours past a normal shift, then the profit derived from t hours of overtime is the difference between the increased revenue and the increased costs. Each of these is computed as an integral of a marginal:

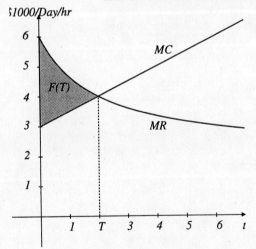

$$P(t) = R(t) - C(t) = \int_0^t MR(t) - \int_0^t MC(t) \, dt$$

or simply

$$P(t) = \int_0^t MR(t) - MC(t) \, dt$$

Graphically, P(t) is the area between the marginal revenue and marginal cost curves over the interval [0,t]. As long as the marginal revenue exceeds the marginal costs the profit increases with time. Normally, marginal costs increase with time while marginal revenues eventually decrease (overtime workers become less efficient). This means that there is an equilibrium time T where MR = MC, as sketched at the left. To operate beyond this period of time results in a loss.

Intuitively, to maximize its profit the company should operate T hrs of overtime, as it begins to loose money operating beyond this time. You can verify that T is a critical point for the function P(t); using the Second Fundamental Theorem of Calculus (Section 4.4) the derivative of the integral function P(T) is just MR(T) - MC(T). Hence,

$$P'(T) \equiv MP(T) = MR(T) - MC(T) = 0$$

That T actually corresponds to a maximum value follows from either the First Derivative Test or the 2nd Derivative Test (Section 3.3):

$$P''(T) = MR'(T) - MC'(T) < 0$$

because of the assumption that the marginal revenue is decreasing, $MR'(t) < 0$, while the marginal cost are increasing, $MC'(t) > 0$. The maximum profit is $P(T)$; as illustrated in the figure, this is the area between the curves $MC(t)$ and $MR(t)$ over the interval $[0,T]$.

Problem Consider a Lumber mill that operates two shifts and can extend the length of one shift to increase production. Assume the marginal revenue and cost for the mill operations are

$$MR(t) = 2 + 8/(t + 2) \quad \text{and} \quad MC(t) = 3 + t/2$$

with units of $1000/hour of overtime per day. What is the maximum extra profit per day that can be derived from overtime operations?

Solution First the equilibrium or break-even time T must be determined. Setting $MR = MC$ gives

$$2 + 8/(t + 2) = 3 + t/2$$

This nonlinear equation is solved by subtracting 2 from each side and then multiplying by the product of the denominators, $2(t + 2)$, and simplifying to give

$$t^2 + 4t - 12 = (t + 6)(t - 2) = 0 \implies t = 2 \text{ or } t = -6$$

Only the positive root $T = 2$ is a feasible solution. Next, the maximum extra profit is found by evaluating the integral of $MR(t) - MC(t)$ over the interval $[0,T]$:

$$P(2) = \int_0^2 [2 + 8/(t + 2)] - [3 + t/2]\, dt = -t + 8\ln(t + 2) - t^2/4 \Big|_0^2$$

$$= \{-2 + 8\ln(4) - 1\} - 8\ln(2) = 8\ln(4/2) - 3 \approx 2.545$$

The conclusion is that the mill would then earn approximately $2,545 per day by running two hours of overtime.

Example 2 Consumer's Surplus and Producer's Surplus.

Background The revenue associated with a product depends on the cost of manufacturing and the price per unit. The selling price of an item will depend on the number of units manufactured. The **supply curve**, $S(x)$, indicates the price per item that a manufacture will charge when the total production is x-units. Solving the equation $p = S(x)$ for x indicates the number of items a producer is willing to make if the selling price is p per unit. The supply function normally is an increasing function.

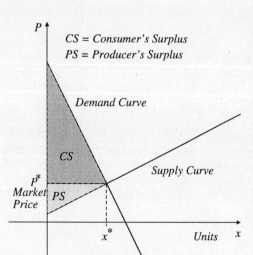

The unit price consumers are willing to pay for an item is given by the **demand curve**, $D(x)$, where x is the total number of units manufactured. Psychologically, consumers will pay a higher amount for a scarce product and will pay less for a very abundant product. Consequently, the demand function is normally a decreasing function of x.

The point (x^*, p^*) where the graphs of $S(x)$ and $D(x)$ intersect indicates the **market price** of the product, as sketched at the left. It indicates the maximum number x^* of units that can be manufactured and all will be sold at the market price p^*. The total Revenue generated at the market price is the product of the price times the amount sold:

$$\text{Market Price Revenue} = p^* \cdot x^*$$

In economic text books two quantities are used to discuss marketing conditions. The **consumer's surplus**, CS, which is the difference between what consumers would have been willing to pay for lower production levels and the market price p^*; CS is the area between the vertical line $p = p^*$ and the demand curve and is evaluate as an integral

$$CS = \int_0^{x^*} D(x) - p^* \, dx \qquad \text{Consumer's Surplus}$$

The difference between the revenue received when the production is x^* and the amount the producers would have been willing to receive when the production was less than x^* is the **producer's surplus**, PS. It is likewise defined by an integral,

$$PS = \int_0^{x^*} p^* - S(x) \, dx \qquad \text{Producer's Surplus}$$

and is the area between the supply curve and the line $p = p^*$.

Problem Determine the Consumer's Surplus and the Producer's Surplus for the linear demand and supply curves

$$D(x) = 20 - 2x \quad \text{and} \quad S(x) = 1 + x/2$$

Solution First, determine the market volume x^* by solving $D(x) = S(x)$ for $x = x^*$:

$$20 - 2x = 1 + x/2 \implies x^* = 7.6.$$

The market price is then $p^* = S(7.6) = 4.8$. Using these values gives

$$CS = \int_0^{7.6} 20 - 2x - 4.8 \, dx = 15.2x - x^2 \big|_0^{7.6} = 57.75$$

$$PS = \int_0^{7.6} 4.8 - [1 + x/2] \, dx = 3.8x - x^2/4 \big|_0^{7.6} = 14.44$$

Example 3 Equipment utility.

Problem Assume that a piece of equipment generates a marginal revenue of

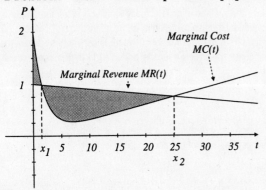

$$MR(x) = 1 - 0.01x \quad \$1000 \text{ per year}$$

when it has been in service for x years. Assume the marginal cost of running the machine is

$$MC(x) = 2e^{-0.5x} + 0.03x \quad \$1000 \text{ per year}$$

The exponential term represents the depreciation of purchase costs and the linear term represents increased cost of maintenance. These curves are illustrated at the left. They intersect at two points: $x_1 \approx 1.51$, when start-up costs are recovered and the machine starts to be profitable, and $x_2 \approx 25$ (actually 24.998), when increasing maintenance cost exceed revenues. At x_2 the machine should be replaced, as it will no longer be profitable.

What is the "equipment utility", i.e., the "total net profit" derived from the machine?

Solution The Key Word in this question is *total*; this indicates an integral of the *marginal net profit*, the marginal revenue minus the marginal cost, over the time period corresponding to the life span of the equipment:

$$\text{TNP} = \int_0^{25} MR(x) - MC(x)\, dx$$

For the specified marginal functions the TNP is

$$\int_0^{25} 1 - 0.04x - 2e^{-0.5x}\, dx = x - .02x^2 + 4e^{-0.5x} \Big|_0^{25}$$

$$= 25 - 12.5 + 4e^{-12.5} - 2 \approx 10.5$$

Thus the equipment utility value is about $10,500.

EXERCISE SET 7.1

1. A business generates sales of 200 items at a profit of $2,500. If they increase the sales to 300 and the marginal profit is $MP(x) = 0.02x$, for x units sold, what is the change in the profit?

2. What is the maximum extra profit that can be derived from extended operating hours if an extension of t hr past the regular shift results in a marginal revenue is $MR(t) = 5t(t - 4)$ $100/hr and the marginal cost $MC = 0.8t$ $100/hr?

3. A product has a Supply curve $S(x) = 0.5x$ and a Demand curve $D(x) = 2 + 3/(x + 1)$, for x units sold.
   a) What is the market price of this product, the optimal production level and the corresponding market revenue?
   b) Compute the Consumer's and Producer's surplus for this product.

4. A company's marginal cost is $MC(x) = 2x^2 - 60x + 500$, and its marginal revenue is $MR = 300 - 10x$, for x units of sales. The company currently sells 18 units per day.
   a) For what sales level x_p is the marginal profit MP a maximum?
   b) If the company changed its marketing to sell x_p units what would be the net change in the company's daily profit? (Hint: integrate MP between x_p and 18.)

5. **Future value of invested income**. While most personal savings accounts pay interest on deposits at discrete intervals, some banks and business utilize continuously compounding interest. This can be visualized as the limit of paying a simple interest rate over very short periods; the *annualized interest rate* is the equivalent interest rate that would result in the same total amount of interest if interest is paid only once at the end of the year. The reason business use these is because the mathematics is actually simpler. The future value of monies being compounded continuously is calculated as the integral of the revenue times an exponential interest function $e^{I \cdot \Delta T}$ where ΔT is the period of time from when the monies are deposited, t, and a terminal time T. If the revenue function R(t) is continuous and I is the continuous interest rate, then the future value, at the end of T years, of income generated over the period [0,T] that is continuously compounded is

$$FV = \int_0^T R(t) e^{I(T-t)}\, dt$$

Assume that a Retirement Fund arranges that a continuous income of $1.2 million per year (t in units of years) is deposited in an account paying 4% interest, compounded continuously, for a 5 year period. What will be the value of the accumulated income and interest at the end of the five years?

6. **Present value of invested future income**. If revenue is generated at a continuous rate of R(t) and the continuously compounding interest rate is I, then the present value of the future revenue over an interval [0,T] is

$$PV = \int_0^T R(t) e^{-I \cdot t}\, dt$$

The person paying the revenue R(t) could do so by depositing the amount PV in the interest account and paying out the revenue from the account. (a) What is the present value of a business that generates a continuous revenue stream of $R(t) = 40,000 + 1000t$ $/year when the continuous interest rate is I = 5%?

Section 7.2 Volumes of Solids of Revolution

A **solid of revolution** is formed or generated by rotating about the x-axis a region in the x-y plane. To do this a third dimension axis perpendicular to the x-y plane is introduced, which is sketched as a z-axis that extends toward the viewer. Rotating a curve y = f(x) around the x-axis generates a three dimensional surface or *shell* that is the surface of the *solid of revolution* corresponding to the plane region under the graph of y = f(x). To avoid complexities we assume the function f is positive on the interval of interest. This is illustrated below.

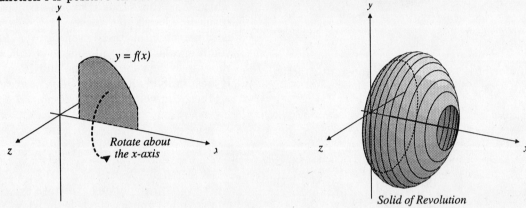

Solid of Revolution

The volume of a solid of revolution is calculated by evaluating an integral over the interval [a,b] and does not actually require the z-coordinate. The formula for the volume is established by applying the *fundamental method of calculus*:

First, the volume is established for the solid of revolution corresponding to a **constant function**.

Next, the volume is established for the solid of revolution corresponding to a **discrete function** that is constant on subintervals of an interval [a,b]. This volume is expressed as a finite sum that has the form of a Reimann Sum: each term is the product of a function value and a difference Δx.

Then, it is argued that a **continuous function** f can be approximated by a sequence of discrete functions, $\{f_n\}$, so that $\lim_{n \to \infty} f_n(x) = f(x)$.

Finally, if V_n is the volume of solid of revolution associated with the approximating function f_n, then the volume V associated with the function f is given as the limit of the sequence $\{V_n\}$. However, the limit of a sequence of Riemann Sums is by definition a definite integral. As we shall see, the integrand is actually $\pi[f(x)]^2$.

More specifically, consider first the solid of revolution associated with the constant function f(x) = c over an interval [a,b]. This solid is a horizontal circular cylinder, as illustrated in the top figure on the following page. Hence, its volume is the product of its base area times its width, which is the length of the interval [a,b]. Its width is thus $\Delta x = b - a$ and its base area is the area of the circular disk of radius c, $A = \pi \cdot c^2$. Thus, the

462   CHAPTER 7   APPLICATIONS INVOLVING INTEGRATION

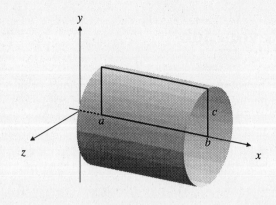

Cylinder of revolution of the constant function $f(x) = c$ over $[a,b]$.

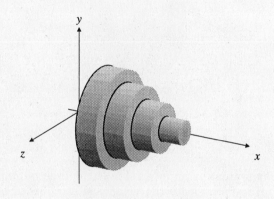

Solid of revolution of a function that is constant on four subintervals.

Volume of the Solid of Revolution corresponding to a constant function $f(x) = c$:

$$V = \pi \cdot c^2 \Delta x$$

Next, consider a discrete function $f_N(x)$ that is positive and constant over N subintervals of $[a,b]$. The solid of revolution associated with this type of function consists of N horizontal cylinders each of which has the same shape as the constant of revolution illustrated at the left. This type of solid is illustrated in the lower figurer for N = 4 subintervals. The total volume of this solid, V, is the sum of the volumes of the individual cylinders. If these individual cylinder volumes are denoted by v_n then the volume of the combined solid V_N is given by:

$$V_N = \Sigma_{n=1,N} \, v_n$$

where v_n is the volume of the n-th cylinder. Each sub-cylinder volume is calculated as its width times π times its radius squared:

If f_N is constant on the N subintervals intervals of $[a,b]$ determined by the partition

$$a = x_0, x_1, x_2, \ldots, x_N = b$$

and is defined on the n-th subinterval (x_{n-1}, x_n) by

$$f(x) = y_n \text{ for } x_{n-1} \le x < x_n$$

then the radius of the nth cylinder is y_n and its volume is

$$v_n = \pi \cdot y_n^2 \, \Delta x_n$$

where the width of this cylinder is $\Delta x_n = x_n - x_{n-1}$. Adding up these volumes gives the

Volume of the Solid of Revolution corresponding to a discrete function F_N:

$$V_N = \Sigma_{n=1,N} \, \pi \cdot y_n^2 \, \Delta x_n$$

Finally, to determine the volume V of the solid of revolution about the x-axis corresponding to a continuous positive function $f(x)$, the basic method of calculus is applied. If the function $f(x)$ is positive for x in an interval $[a,b]$ as illustrated in the left panel of the following figure, we construct a sequence of discrete functions $\{f_N\}$ such that $\lim_{N \to \infty} f_N(x) = f(x)$. To do this for each natural number N let $\{x_n\}_{n=0,N}$ be a partition of the interval $[a,b]$ into N subintervals. Define the discrete function $f_N(x)$ by

$$f_N(x) = y_n \equiv f(x_{n-1}) \quad \text{when} \quad x_{n-1} < x < x_n$$

f_N has the constant value of the function f at the left endpoint of the subinterval[†]. The volume V is then approximated by the volume V_N. If the partitions are selected so that as N increases the maximum Δx_n approaches zero, then we argue that $V = \lim_{N \to \infty} V_N$.

The solid of revolution corresponding to each discrete function consists of a series of disks having the width of the subintervals as sketched in the right panel of the figure, for N = 5. The volume of such solids is the sum of the volume of the individual disks, which are each cylinders, and is given by

$$V_N = \Sigma_{n=1,N}\, \pi \cdot y_n^2 \cdot \Delta x_n = \Sigma_{n=1,N}\, \pi \cdot [f(x_n)]^2 \cdot \Delta x_n$$

Notice that these sums have the form of a Riemann Sum, i.e. it is a sum of a function term $g(x_n^*)$ times Δx_n where the function is not f(x) but $g(x) = \pi \cdot [f(x)]^2$.

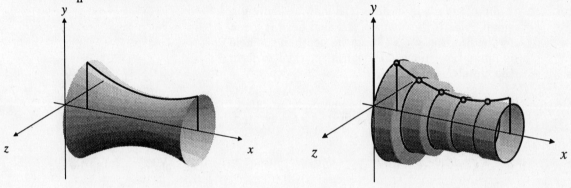

Taking the limit as N goes to infinity and the lengths Δx_n approach zero, gives the true volume V. But, with $y_n = f(x_n)$ the limit of these Riemann Sums is by definition the definite integral of $\pi \cdot f(x)^2$.

$$V = \lim_{N \to \infty} \Sigma_{n=1,N}\, \pi \cdot [f(x_n)]^2 \Delta x_n = \int_a^b \pi \cdot [f(x)]^2\, dx$$

Volume of the Solid of Revolution about the x-axis corresponding to y = f(x) over the interval [a,b] is

$$V = \int_a^b \pi \cdot [f(x)]^2\, dx$$

Example 1   The volume of a cone.

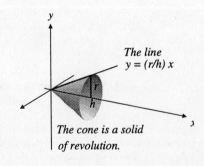

The line $y = (r/h)\, x$

The cone is a solid of revolution.

Problem   Establish the formula for the volume of a circular cone with height h and base radius r.

Solution   The cone can be visualized as the solid of revolution about the x-axis over the interval [0,h] of the linear function

$$f(x) = (r/h)x$$

[†]   The evaluation point could be any point within the subinterval. Using an endpoint normally simplifies the form of the sum.

The volume of this solid is then given by the integral

$$V = \int_0^h \pi \cdot [(r/h)x]^2 \, dx = \pi \cdot (r/h)^2 \, x^3/3 \Big|_0^h$$

$$= \pi \cdot (r/h)^2 \cdot [h^3/3 - 0^3/3] = \pi r^2 h/3$$

Example 2 Volume of a dimple.

Problem A dimple is made in a vertical slab by drilling a small hole. If the shape of the drill head is given by the function $y = x^{1/3}$ and the depth of the hole is 8mm, how much material is removed to form the dimple?

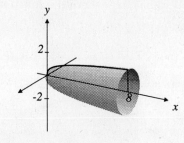

Solution The material removed is the volume of a solid of revolution over the interval $[0,8]$ of $y = x^{1/3}$. It is given by

$$V = \int_0^8 \pi \cdot [x^{1/3}]^2 \, dx$$

$$= \int_0^8 \pi x^{2/3} \, dx = \pi \cdot (3/5) x^{5/3} \Big|_0^8$$

$$= 96\pi/5 \approx 60.32 \text{ mm}^3$$

The volume created by rotation of the region under the graph of $y = x^{1/3}$ about the x-axis for $0 \leq x \leq 8$.

Two other types of solids of revolution arise in various applications. One is where the plane region between the graph of $y = f(x)$ and a horizontal line $y = y_0$ is revolved about the line. The general formula is obtained by replacing $[f(x)]^2$ by $(radius)^2$, where the *radius* is the distance between the graph $y = f(x)$ and the rotation "axis": the distance $|f(x) - y_0|$.

If the region between a curve $y = f(x)$ and the line $y = y_0$ is rotated about the line, the Volume of the corresponding Solid of Revolution is

$$V = \int_a^b \pi [f(x) - y_0]^2 \, dx$$

Another type of rotational solid is the shell formed by rotating the plane region between the graphs of two positive functions $f(x)$ and $g(x)$ about the x-axis. The volume of this solid is the difference of the volumes of revolution corresponding to $f(x)$ and to $g(x)$.

If the region between the graphs of $y = f(x)$ and $y = g(x)$ is rotated about the x-axis the Volume of the corresponding Solid of Revolution is

$$V = \pi \int_a^b [f(x)]^2 - [g(x)]^2 \, dx$$

Example 3 Volume of a shell of revolution.

Problem What is the volume of the shell formed by rotating around the x-axis the region bounded above by $y = 2x$ and below by $y = x^2$ over the interval $[0,2]$?

Solution This solid is the region between the surfaces generated by rotating the outer curve y = 2x and the inner curve y = x^2 about the x-axis. It can be viewed as the solid generated by revolving the upper curve and minus the volume generated by rotating the inner curve. The volume V is thus the difference in the volumes of the two solids of revolutions.

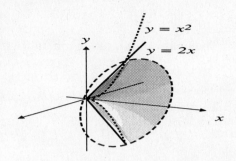

$$V = \int_0^2 \pi \cdot [2x]^2 \, dx - \int_0^2 \pi \cdot [x^2]^2 \, dx$$

$$= \pi \cdot \int_0^2 4x^2 - x^4 \, dx$$

$$= \pi \cdot \{4x^3/3 - x^5/5\} \Big|_0^2$$

$$= \pi \cdot \{32/3 - 32/5\} = 64\pi/15 \approx 13.404$$

☑

Rotation about the y-axis.

If the graph of y = f(x) is monotone increasing or monotone decreasing on an interval [a,b], then the function f is one-to-one on the interval and hence its inverse f^{-1} exists. We can thus solve the equation y = f(x) for x, x = f^{-1}(y). Rotating the curve y = f(x) about the y-axis also generates a solid of revolution, which is actually the three-dimensional region swept by the two dimensional area between the y-axis and the curve for y in an interval with endpoints A = f(a) and B = f(b); as we don't know *a priori* which of the endpoint values A or B is larger, we shall denote this y-interval by [c,d]. The volume of this Solid of Revolution about the y-axis is given by an integral of $\pi(\text{radius})^2$. In this case the radius is $|x|$, the distance between the y-axis and the graph of f. This is the distance from x = 0 (the y-axis) to x = f^{-1}(y) (the x-coordinated on the graph for a given y). As $|x|^2 = (f^{-1}x))^2$, the

**Volume of the Solid of Revolution about the y-axis
associated with the graph of y = f(x) is**

$$V = \int_c^d \pi \cdot [f^{-1}(y)]^2 \, dy$$

If a vertical rotation axis other than the y-axis is considered, say the line x = x_0, then a similar formula will apply, with the *radius* = $|f^{-1}(y) - x_0|$.

Example 4   **Volume of a solid of revolution about the y-axis.**

Problem What is the volume of the solid obtained by rotating about the y-axis the portion of the curve y = *ln*(x) with x between 1 and e^2? This is the solid region that is above the surface illustrated at the right and below the horizontal plane y = 2.

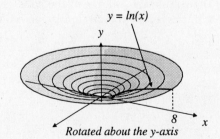

Rotated about the y-axis

Solution The radius of this solid of revolution is the distance between the y-axis and the curve. This is the value of x expressed as a function of y, radius = $x = e^y$.

The limits of integration are the y-values corresponding to $x = 1$ and $x = e^2$:

$$c = ln(1) = 0 \quad \text{and} \quad d = ln(e^2) = 2$$

Thus, the volume is

$$V = \int_0^2 \pi [e^y]^2 \, dy = \int_0^2 \pi e^{2y} \, dy$$

$$= (\pi/2) e^{2y} \Big|_0^2 = (\pi/2)(e^4 - 1) \approx 84.197$$

☑

EXERCISE SET 7.2

1 Find the volume of the solid of revolution generated by revolving the given curve y = f(x) about the x-axis over the given interval.   Sketch the region.

   a)   y = 3x + 2, over [0,2]   b) $y = e^{-2x}$, over [0,1]   c) $y = x^{1/2} + 1$, over [1,4]

   d)   y = SQRT(1 - x^2), over [0,1]   e) $y = x^2 + 2$, over [0,2]   f) $y = e^x - 1$, over [0,1]

   g)   y = sin(x), over [0,2π]   h) $y = x - x^{1/2}$, over [1,9]

2 Find the volume of the solid of revolution generated by revolving about the x-axis the region bounded by the given curves. Sketch the solid region.

   a)   y = x, $y = x^2$, x = 1, x = 4   b)   y = 1/x, y = 1, x = 1, x = 3

   c) $y = x^2 + 1$, y = 1, x = 1, x = 2   d)   y = 1, $y = \cos^{1/2}(\pi x)$, x = -1, x = 1

3 Find the volume of the solid of revolution generated by revolving the given curve y = f(x) about the y-axis for x in the given interval. These are the solid regions bounded by the surface of revolution and horizontal plane(s). These regions will thus include a portion of the y-axis. Sketch the region.

   a)   y = 3x, over [0,2]   b) $y = x^2$, over [0,1]   c)   y = 3x, over [2,4]

   d) $y = x^{1/2}$, over [0,1]   e)   y = 3x + 5, over [0,2]   f) $y = x^2 - 1$, over [1,4]

4 Determine the volume of the solid generated by revolving the curve y = 2x a) about the line y = 1 for x in the interval [1,5] and b) about the line x = 3 for x in [0,1].

Section 7.3 The Length of a Curve; the Area of a Surface of Revolution.

The length of a curve.

The length of a straight line segment connecting two points is given by the **Euclidean distance**, the length of the straight line segment between points (x_0,y_0) and (x_1,y_1) is

$$D = \text{SQRT}((x_1 - x_0)^2 + (y_1 - y_0)^2)$$

However, the length of a curved line connecting two points, called an **Arc Length**, depends on the nature of the curve and can be difficult to calculate. The formula for computing an Arc Length involves an integral and is established by applying the basic *method of calculus*. The Arc Length of a straight line is simply its Euclidean Length. Next, the Arc Length of a polygonal line, called a *piecewise linear curve*, is simply the sum of the lengths of its linear segments; this sum will have the form of a Riemann Sum. To establish the Arc Length of a continuous curved line we introduce approximations of the curved line by polygonal lines. If the approximations can be made successively more accurate, that is made to approach the original given curved line, then the length of the curved line will be the limit of the lengths of the piecewise linear approximations. The limit of the Riemann Sums will be a definite integral, its integrand will not simply be the given function but will involve the square root of one plus the derivative squared.

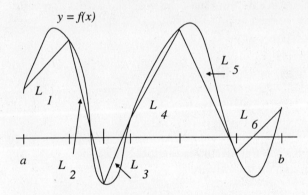

The L lines form a piecewise linear approximation of $y = f(x)$.

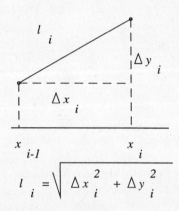

Consider a curve C that is the graph of a differentiable function $y = f(x)$ over the interval [a,b]. To approximate the length of the curve between the points $(a,f(a))$ and $(b,f(b))$ we create polygonal lines, PL_N, that approximate the curve C. To do this, let $\{x_n\}_{n=0,N}$ be a partition of [a,b] into N subintervals. The polygonal curve PL_N consists of the line segments connecting the points $(x_n,f(x_n))$ on the graph of f. This is the set of line segments $L_1, L_2, ..., L_N$ as illustrated in the graph above. The length of the i-th segment L_i, connecting the points $(x_{n-1}, f(x_{n-1}))$ and $(x_n, f(x_n))$ is

$$l_i = \text{SQRT}\{\Delta x_i^2 + \Delta y_i^2\}$$

where

$$\Delta x_i = x_i - x_{i-1} \quad \text{and} \quad \Delta y_i = f(x_i) - f(x_{i-1})$$

The length l_i can be expressed in terms of the derivative of f. Recall that the Mean Value Theorem (Section 4.7 of Chapter 4) asserts that there is an x-value in the interval $[x_{i-1}, x_i]$ at which $f'(x)$ equals the slope $\Delta y_i/\Delta x_i$ of the line segment L_i. Assuming the derivative f' is continuous, we know that there exists a value x_i^* (we will not actually need to know its value, we just need to know that it exists) that satisfies the equation

$$f'(x_i^*) = \Delta y_i / \Delta x_i$$

Solving this equation for Δy_i gives:

$$\Delta y_i = f(x_i) - f(x_{i-1}) = f'(x_i^*) \Delta x_i$$

Substituting $f'(x_i^*)\Delta x_i$ for Δy_i in the equation for the length l_i gives

$$l_i = \text{SQRT}\{\Delta x_i^2 + [f'(x_i^*)\Delta x_i]^2\} = \text{SQRT}\{1 + [f'(x_i^*)]^2\}\, \Delta x_i$$

The total length of the polygonal line PL is the sum of the lengths of the segments:

$$\text{Arc Length of } PL_N = \Sigma_{i=1,N} \text{ SQRT}\{1 + [f'(x_i^*)]^2\}\, \Delta x_i$$

This summation has the form of a Riemann Sum, each term is a function value times an increment Δx. The Arc Length of the curve C is the limit of the lengths of lines PL_N as the number of segments N increases, assuming that the Δx_is approach zero:

$$\text{Arc Length of C} = \lim_{N \to \infty} \Sigma_{i=1,N} \text{ SQRT}\{1 + [f'(x_i^*)]^2\}\Delta x_i$$

By definition, a limit of this type of Riemann Sum is a definite integral, with the expression before the Δx terms as the integrand. Thus,

The Arc Length of the curve C :

$$y = f(x) \text{ from } (a, f(a)) \text{ to } (b, f(b))$$

$$AL = \int_a^b \text{SQRT}\{1 + [f'(x)]^2\}\, dx$$

Caution: because of the SQRT function, even for simple curves the integral in the Arc Length formula can be difficult or impossible to evaluate!

Example 1 **Computing the Arc Length of a straight line.**

Problem Use the Arc length formula to compute the length of the straight line connecting (2,3) and (4,7).

Solution First, we determine that the straight line connecting these points is the graph of $y = f(x) = -1 + 2x$. Then, as $y' = f'(x) = 2$, the Arc Length formula gives

$$AL = \int_2^4 \text{SQRT}(1 + 2^2)\, dx = \int_2^4 \sqrt{5}\, dx = \sqrt{5}\, x \Big|_2^4 = \sqrt{5}[4 - 2] = 2\sqrt{5}$$

Observe that this is exactly the Euclidean distance

$$D = \text{SQRT}((4-2)^2 + (7-3)^2) = \sqrt{20} = 2\sqrt{5}$$

Example 2 **Computing the Arc Length of an allometric function.**

Problem Determine the Arc Length of the segment of the curve $y = (1/3)x^{3/2}$ over the interval [4,9].

Solution To apply the Arc Length formula, first differentiate to obtain y':

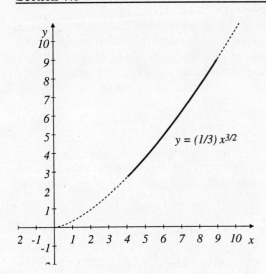

$$y' = (1/2)x^{1/2}$$

Then,

$$\text{Arc Length} = \int_4^9 \text{SQRT}(1 + [(1/2)x^{1/2}]^2)\, dx$$

$$= \int_4^9 \text{SQRT}(1 + x/4)\, dx$$

$$= (8/3)(1 + x/4)^{3/2} \Big|_4^9$$

$$= (8/3)\{(1 + 9/4)^{3/2} - (1 + 4/4)^{3/2}\}$$

$$= (8/3)\{(13/4)^{3/2} - 2^{3/2}\} \approx 8.08158$$

☑

In many physics and engineering problems it is convenient to use **differential forms** of the Arc Length formulas. In these, the derivative $y' = f'(x)$ is replaced by a ratio of differentials dy/dx or the derivative $g'(y)$ is replaced by the ratio dx/dy. The corresponding differential formulas for Arc Length are shorter to write:

$$\text{AL} = \int_a^b \text{SQRT}(1 + (dy/dx)^2)\, dx \quad \text{and} \quad \text{AL} = \int_c^d \text{SQRT}(1 + (dx/dy)^2)\, dy$$

As the differentials dx and dy can be algebraically manipulated like regular variables, these two formulas can be combined into one form if we take some liberty with the integral notation. Since, for a positive differential dx, $\text{SQRT}(1/dx^2) = 1/dx$,

$$\text{SQRT}(1 + (dy/dx)^2)\cdot dx = \text{SQRT}((dx^2 + dy^2)/dx^2)\cdot dx = \text{SQRT}(dx^2 + dy^2)\cdot dx/dx = \text{SQRT}(dx^2 + dy^2)$$

Similarly, for a positive differential dy,

$$\text{SQRT}(1 + (dx/dy)^2)\cdot dy = \text{SQRT}((dy^2 + dx^2)/dy^2)\cdot dy = \text{SQRT}(dy^2 + dx^2)\cdot dy/dy = \text{SQRT}(dy^2 + dx^2),$$

Both differential Arc Length formulas can be expressed by the same form, which is commonly encountered in applications in Physics, Engineering, and Physical Chemistry:

$$\text{AL} = \int \text{SQRT}(dx^2 + dy^2)$$

Sometimes the Arc Length is calculated using a new variable, s, which is the actual measure of length along the curve, by defining the differential of s, ds, to be the differential of arc length:

$$ds \equiv \text{SQRT}(dx^2 + dy^2) \quad \textbf{differential of arc length}$$

The Arc Length formula can be further simplified to a very brief form that involves a different type of integral called a **line integral**. Line integrals are denoted as $\oint F(s)\, ds$ where s is the arc length variable and ds is its differential. Line integrals are encountered in problems that involve calculating the flux or change in a quantity over a two-dimensional region. In advanced calculus it is shown that the flux can be evaluated as a line integral around the boundary of the region. A special situation occurs when the function being integrated is simply the constant function $F(s) = 1$, then the line integral gives the Arc Length of the curve over which it is integrated

$$\text{AL} = \oint 1\, ds$$

Establishing the limits of integration depend on the curve C and we will not discuss the general theory. However, when the curve C can be expressed as a *parametric curve* then such integrals can be expressed in terms of the parameter specifying the curve. As the integrals in the Arc Length formulas are notoriously difficult to evaluate, even for simple curves, introducing a parametric form of the curve can result in a simpler integral to evaluate. This is illustrated in the following example.

Example 3 **The circumference of a circle.**

Problem You of course know the formula for the circumference of a circle, $2\pi r$, but have you ever verified this by a different means? Determine the circumference of the unit circle $x^2 + y^2 = 1$ using an Arc Length formula.

Solution Since this is a circle of radius 1 its circumference is 2π. By symmetry, the length is twice the length of the upper half circle that is the graph of $y = f(x) = \text{SQRT}(1 - x^2)$ for $-1 \leq x \leq 1$. To apply the basic Arc Length formula we first compute $y' = -x(1-x^2)^{-1/2}$ and integrate $\text{SQRT}(1 + y'^2)$ over the $[-1, 1]$:

$$AL = \int_{-1}^{1} \text{SQRT}(1 + [x^2(1-x^2)^{-1}]) \, dx$$

This is an improper integral as the term $(1-x^2)^{-1}$ is singular, i.e., infinite at both $x = -1$ and $x = 1$. It can be algebraically simplified to

$$AL = \int_{-1}^{1} (1-x^2)^{-1/2} \, dx$$

An antiderivative of $\int 1/\text{SQRT}(1-x^2) \, dx$ can be found using the trig substitution $x = \sin(\theta)$, which leads to the Arcsin Rule of Section 6.1

$$AL = \int_{-1}^{1} (1-x^2)^{-1/2} \, dx = \text{Arcsin}(1) - \text{Arcsin}(-1) = \pi/2 - (-\pi/2) = \pi$$

Thus the circumference is 2π as expected. This result could also be obtained by representing the circle as a parametric curve. The unit circle is the trajectory of the path

$$x = \cos(\theta) \quad \text{and} \quad y = \sin(\theta) \quad \text{for } \theta \in [0, 2\pi].$$

Using these equations the differentials of x and y are

$$dx = -\sin(\theta) \, d\theta \quad \text{and} \quad dy = \cos(\theta) \, d\theta.$$

For a positive differential $d\theta$, the corresponding differential of arc length is

$$ds = \text{SQRT}(dx^2 + dy^2) = \text{SQRT}((-\sin(\theta)\,d\theta)^2 + (\cos(\theta)\,d\theta)^2) = \text{SQRT}(\sin(\theta)^2 + \cos(\theta)^2) \, d\theta = d\theta$$

The Arc Length of the circle is then given by the line integral of ds over $[0, 2\pi]$:

$$AL = \int_{\theta=0}^{\theta=2\pi} ds = \int_{\theta=0}^{\theta=2\pi} 1 \, d\theta = \theta\big|_0^{2\pi} = 2\pi$$

☑

The Area of a Surface of Revolution.

In Section 7.2 we considered the volume of a solid of revolution. We now consider the problem of determining the surface area of such solids. The surface generated by revolving the curve $y = f(x)$ about the x-axis, for x in an interval $[a,b]$, has an area denoted by SA. This area is computed via a formula that is similar to the Arc Length formula. It is established by approximating the graph of f by a polygonal line PL and considering the area of the surface of revolution of the line segments of PL.

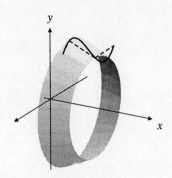
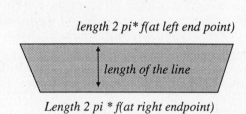

length 2 pi f(at left end point)*

length of the line

*Length 2 pi * f(at right endpoint)*

The key observation is that the surface of revolution of a line segment is a continuous ribbon that encircles the x-axis and has different radii at its two edges, if it is cut parallel to the x-axis and laid flat, it will form a trapezoid shaped strip. Its height is the length of the line segment which we denote by Δs_i. The strip's base length at each side edge is 2π times the corresponding y-coordinate. If, as in the discussion of arc lengths, the line L_i over an interval $[x_{i-1}, x_i]$ passes through the graph of $y = f(x)$ at the endpoints, then the y-coordinates at the endpoints are

$$y_{i-1} = f(x_{i-1}) \quad \text{and} \quad y_i = f(x_i)$$

As the area of a trapezoid of height h and base lengths B_1 and B_2 is $A = [B_1 + B_2]/2 \cdot h$, the area of the strip derived from rotating the line segment L_i is

$$SA_i = [2\pi \cdot f(x_i) + 2\pi \cdot f(x_{i-1})]/2 \cdot \Delta s_i = 2\pi[f(x_i) + f(x_{i-1})]/2 \cdot \Delta s_i$$

The number $y_i^* = [f(x_i) + f(x_{i-1})]/2$ is the average of the function values $f(x_i)$ and $f(x_{i-1})$. It is thus a number that is between these two numbers. If the function f is continuous, the Intermediate Value Theorem states that there is an x-value, x_i^*, in the interval $[x_{i-1}, x_i]$ at which $f(x_i^*) = y_i^*$. Therefore the surface area of the strip can be expressed as

$$SA_i = 2\pi \cdot f(x_i^*) \cdot \Delta s_i$$

By factoring out the difference Δx_i, the length of the line segment L_i can be expressed as

$$\Delta s_i = \text{SQRT}(1 + (\Delta y_i/\Delta x_i)^2) \cdot \Delta x_i$$

Combining these identities, we see that the surface area of the strip generated by rotating the line segment L_i about the x-axis is the average circumference of the strip, $2\pi f(x_i^*)$ times Δs_i

$$SA_i = 2\pi \cdot f(x_i^*) \cdot \text{SQRT}(1 + (\Delta y_i/\Delta x_i)^2) \cdot \Delta x_i$$

Adding the areas of the strips over each subinterval gives a Riemann Sum that is the area of the surface of revolution of the polygonal line PL:

$$\Sigma SA_i = \Sigma 2\pi \cdot f(x_i^*) \cdot \text{SQRT}(1 + (\Delta y_i/\Delta x_i)^2) \cdot \Delta x_i$$

In the limit, as the number of subintervals increases to ∞, these sums should approach the area of the surface of revolution of the curve C. Again, in the limit the Riemann Sum results in a definite integral. The slope term $\Delta y_i/\Delta x_i$, in the limit, becomes the derivative $f'(x)$. The radius term simply becomes $2\pi \cdot f(x)$ in the limit. The constant 2π is placed as a multiple in front of the integral.

The Area of the Surface of Revolution of y = f(x) about the x-axis is

$$SA = 2\pi \int_a^b f(x) \, \text{SQRT}(1 + [f'(x)]^2) \, dx$$

An alternative formula uses the variable y in place of f(x) and the differential form of the arc length, as introduced earlier in this section:

$$SA = 2\pi \int_a^b y \cdot \text{SQRT}(dx^2 + dy^2)$$

Another form using the differential ds of the arc length variable s, is simply

$$SA = 2\pi \int_a^b y \cdot ds$$

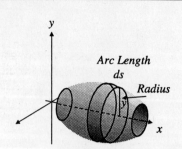

Example 4 **Computing the surface area of sphere.**

Problem Determine the surface area of the sphere of radius 2 centered at the origin.

Solution The unit sphere is the surface $x^2 + y^2 + z^2 = 4$ which is the surface of revolution about the x-axis of the function

$$y = \text{SQRT}(4 - x^2) \text{ for } -2 \leq x \leq 2$$

Since

$$y' = -x / \text{SQRT}(4 - x^2)$$

the differential of arc length is

$$ds = \text{SQRT}(1 + y'^2) = \text{SQRT}(1 + x^2 / (4 - x^2)) = \text{SQRT}(4 / (4 - x^2))$$

The surface area of the sphere is then given by

$$SA = 2\pi \int_{-2}^2 y \cdot ds = 2\pi \int_{-2}^2 \text{SQRT}(4 - x^2) \cdot \text{SQRT}(4/(4 - x^2)) \, dx$$

$$= 2\pi \int_{-2}^2 2 \, dx = 2\pi \cdot 2x \big|_{-2}^2 = 16\pi$$

EXERCISE SET 7.3

1 Determine the arc length of the given curve over the specified range.
  a) $y = (1/3)x^{3/2}$, from $x = 0$ to $x = 5$

  b) $y = (1/3)(2 + x^2)^{3/2}$, from $x = 0$ to $x = 5$

  c) $y = (e^x + e^{-x})/2$, from $x = 0$ to $x = ln(5)$ Hint: $1 + (y')^2$ is a perfect square.

  d) $y = x^4/4 + x^{-2}/8$, from $x = 1$ to $x = 4$

  e) $y = x^2/2 - ln(x)/4$, from $x = 1/4$ to $x = 5/4$.

  f) $y = \int_4^x \text{SQRT}(t^3 - 2t - 1) \, dt$, from $x = 4$ to $x = 6$

Section 7.3   THE LENGTH OF A CURVE   473

2   Determine the Arc Length of the curve that is the graph of the relationship over the indicated ranges.
     a) $y^3 = 8x^2$ from $x = 1$ to $x = 8$ b) $24xy = x^4 + 48$ from $x = 1$ to $x = 2$

     c) $e^y = \cos(x)$ from $x = 0$ to $x = \pi/3$ d) $y^2 = (x^2 + 2)^3$ from $x = 0$ to $x = 2$

3   Determine the circumference of the circle $(x - 2)^2 + (y + 3)^2 = 9$, using an Arc Length formula.

4   Determine the length of the parametric curves:
     a) $x = t^3$, $y = t^2$ for $0 \le t \le 5$ b) $x = 1 + 2t$, $y = 1 - t^2$ for $0 \le t \le 2$

     c) $x = \cos^2(\theta)$, $y = \sin^2(\theta)$ for $0 < \theta < \pi$ d) $x = \cos^2(\theta)$, $y = \cos^2(\theta)$ for $0 \le \theta \le \pi$

     e) $x = \cos^2(\theta)$, $y = \cos(\theta)$ for $0 \le \theta \le \pi$ f) $x = e^{2\theta}\cos(\theta)$, $y = e^{2\theta}\sin(\theta)$ for $0 \le \theta \le \pi$

5   Determine the area of the surface of revolution about the x-axis of the indicated curve segment. Sketch the curve and roughly sketch the surface of revolution.
     a) $y = 3x$, [0,3]   b) $y = 3x + 1$, [0,2]   c) $y = 5x^2$, [0,3]

     d) $y = x^3$, [-2,2]   e) $y = x^{1/2}$, [1,4]   f) $y = (2x - x^2)^{0.5}$, [1,2]

6   If the curve $y = f(x)$ is strictly increasing or strictly decreasing for $x \in [a,b]$, $a > 0$, and is rotated about the y-axis, the resulting surface of revolution has an area given by an integral where the radius term is simply x, the distance from the y-axis to the curve.
     Area of Surface revolved about the y-axis.

     $$SA = 2\pi \int_a^b x \sqrt{1 + [f'(x)]^2}\, dx$$

     If the curve can be expressed as a function of y, in the form $x = g(y)$, for $c \le y \le d$, then the formula for the surface area is

     $$SA = 2\pi \int_c^d g(y) \sqrt{1 + [g'(y)]^2}\, dy$$

     Use these formulas to evaluate the area of the surface obtained by revolving the given curve about the y-axis. Sketch the surface.

     a) $y = 2x$, $x \in [0,1]$ b) $y = 3x^2$, $x \in [0,3]$

     c) $y = 0.5x - 2$, $x \in [0,1]$ d) $y = 2x^{3/2}$, $x \in [0,1]$

     e) The ellipse $x^2 + 4y^2 = 4$. Hint: use the symmetry of the ellipse.

     f) $y = x^4/4 + x^{-2}/8$, $x \in [1,2]$ g) $x = 0.5(e^y + e^{-y})$, $y \in [0, \ln(3)]$

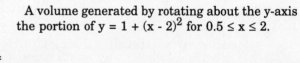

A volume generated by rotating about the y-axis the portion of $y = 1 + (x - 2)^2$ for $0.5 \le x \le 2$.

7   Consider the line with slope k, $y = kx$ over an interval [0,t] for a constant k.
     a) State the integrals that give: (i) the Arc Length of this curve, (ii) the volume of the solid of revolution of this curve about the x-axis, and (iii) the area of the surface of revolution of this curve.
     b) Evaluate the three integrals in a) and relate these to basic geometric formulas.

8   Use the approach employed in Example 5, setting the radius = R to determine the formula for the surface area of a sphere of radius R.

9   Consider the curve $y = 1/x$ over an interval [0,t].
     a) State the integrals that give: i) the Arc Length of this curve, ii) the volume of the solid of revolution formed by rotating the region under this curve about the x-axis, and iii) the surface area of this solid.
     b) Which of the integrals in part a) is easier to evaluate?
     c) If the right endpoint t is set to ∞ do the corresponding improper integrals converge? How do you relate these observations to the volume and surface area of the corresponding infinite solid of revolution?

Section 7.4 Probability and Statistics

A **Probability** is a numerical value between 0 and 1 that indicates the chance of an event occurring. In this section we consider events that are numerical and are recorded as a number x, which is called a **random variable**. We assume that the range of possible x-values is in an interval, I, which is called the **sample space**. This interval may be finite or infinite. The chances that specific values of x will occur are described by a **probability density** function. The probability that the random variable x is less than a particular x-value is given by the **probability distribution** function, which is an integral-function of the density function.

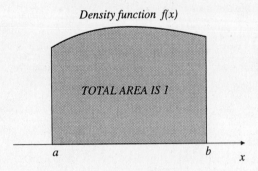

Density function f(x)

TOTAL AREA IS 1

A probability density function indicates how "probable" it is that an event will result in a specific x value. A density function f(x) must be non-negative and its integral over the sample space I must equal one:

$$f(x) \geq 0 \text{ for } x \in I \quad \text{and} \quad \int_I f(x) \, dx = 1$$

The chance that the event with density f(x) results in a value x between two numbers c and d in the sample interval, (assuming $c < d$) is defined to be the probability

$$P[\, c < x < d \,] \equiv \int_c^d f(x) \, dx$$

The probability that an outcome is less than a given value x, is given by the **probability distribution** function, F(x), defined as the integral of the density function over all values in the probability space less than x. The distribution function F(x) is thus an integral-function of the density function, with the upper limit of integration equal to x and lower limit a being the left endpoint of the interval I. Using the "dummy" integration variable t, the distribution function is defined by

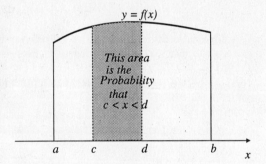

Probability Distribution Function

$$F(x) = \int_a^x f(t) \, dt$$

The distribution function F(x) is thus an antiderivative of the density f(x). Consequently, the density function corresponding to a given distribution F(x) is simply $f(x) = F'(x)$. For any number x_0 in the sample interval I,

$$P[\, a < x < x_0 \,] = F(x_0)$$

Applying the Fundamental Theorem of Calculus, the probability that the random variable x is in a subinterval [c, d] of the sample space I is

$$P[\, c < x < d \,] = \int_c^d f(x) \, dx = F(d) - F(c).$$

Since the derivative $F'(x) = f(x) \geq 0$ the distribution function F(x) is non-decreasing. At the left endpoint of the sample space I the distribution function is zero, $F(a) = 0$, and, since the total probability is 1, at the right endpoint, b, the distribution function must equal one, $F(b) = 1$. It is a simple application of the Splitting Rule for limits of integration that the probability of an event being greater than x_0 is 1 minus the probability that it is less than x_0:

$$P[\,x > x_0\,] = 1 - F(x_0)$$

Example 1   **Computing Probabilities for a Uniform Density.**

Background   An event for which every x-value in a finite interval I = [a,b] is equally likely to occur is said to be **uniformly** distributed. The **uniform density** f(x) = 1/(b - a), i.e., the density function is a constant function equal to the reciprocal of the length of the interval I.

Problem   If the random variable x is uniformly distributed on I = [3,5] what is the probability that
(a) x < 3.5,   (b) x > 3.5,   and (c) 3.5 < x < 4.75 ?

Solutions   The uniform density function is f(x) = 1/2 since the length of [3,5] is 2. In each case the answer is the integral of f over the appropriate subinterval. For (a) and (b) one of the endpoints of I must be used as a limit of integration.

(a) This probability is given by the distribution function

$$F(3.5) = P[x < 3.5] = \int_3^{3.5} 1/2 \, dx = x/2 \,\big|_3^{3.5} = [3.5 - 3]/2 = 0.25$$

(b) Using part (a),   P[x > 3.5] = 1 - P[x < 3.5] = 1 - 0.25 = 0.75

(c) $P[3.5 < x < 4.75] = \int_{3.5}^{4.75} 1/2 \, dx = x/2 \,\big|_{3.5}^{4.75} = [4.75 - 3.5]/2 = 0.625$

Example 2   **A finite non-uniform distribution.**

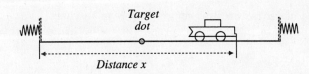

Problem   A carnival game requires the player to release a spring loaded toy car with the objective that the car will come to rest on a red dot in the middle of a 3 meter track. Because the track is curved and has spring bumper at each end the probability that the front of the car ends up at a point x-units along the track is governed by a density function of the form

$$f(x) = A(x - 1.5)^2$$

(a) What is the value of the constant A?

(b) What is the probability that any portion of a 10cm car comes to rest over a red dot of radius 2 cm at the center of the track?

Solutions   (a) The constant A is determined by the requirement that the total probability must be equal to one:

$$\int_0^3 A(x - 1.5)^2 \, dx = 1$$

Evaluating the integral and solving gives A = 4/9.

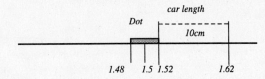

$$1 = (A/3)(x - 1.5)^3 \,\big|_0^3 = (A/3)[1.5^3 - (-1.5)^3] = 6.75A/3$$

(b) As illustrated at the left, the x-values for which a portion of the car will be over the dot are in the interval (1.48 , 1.62). The probability of a "win" is

$$P[1.48 < x < 1.62] = \int_{1.48}^{1.62} (4/9)(x - 1.5)^2 \, dx$$

$= (4/27)(x - 1.5)^3 \big|_{1.48}^{1.62} = (4/27)[0.12^3 - 0.02^3] = 0.0002548,$

The chance of winning is about 1 in 4000.

Example 3 **Probability of an accident with a constant "hazard rate".**

Background A "hazard distribution" is used to predict an event that is the first occurrence of an "accident" or "hazard". The recorded value is the time t at which the event occurs, measured from some initial zero time. Often the "zero" time is taken as the time at which the last accident occurred. Since a truly random accident is equally likely to happen at any time, and is not dependent on when the last accident happened it is common to assume that the probability that the event will occur at a time t, given it has not occurred up to time t, is the same for all time t > 0. This is referred to as a "constant hazard rate". The sample space is the interval I = [0,∞] and the probability density function is an **exponential density** function of the form

$$f(t) = \lambda e^{-\lambda t} \qquad \text{Exponential or Hazard Density Function}$$

where λ is the constant **hazard rate**.

Problem (a) Assume that a personal computer has a failure hazard rate of $\lambda = 0.3$ years. What is the probability that it will fail in the first 90 days of service?

(b) If a light bulb has a constant hazard rate of $\lambda = 2$ years for failure, and it has burnt for 3 years, what is the probability that it will burn for another 3 years?

(c) If a person with a particular virus has constant hazard rate of manifesting an associated disease of $\lambda = 1/9$, what is the time T for which the person has a 50% chance of showing symptoms of the disease in the interval [0,T]? Assume that 0 is the time the person became infected with the virus.

Solutions (a) Since the hazard rate is given in units of years, we first convert 90 days to years. For convenience we assume that 90 days corresponds to 1/4 year. The probability of failure is then

$$P[0 < t < 1/4] = \int_0^{1/4} 0.03 e^{-0.03t} \, dt = 1 - e^{-0.0075} \approx 0.0074719$$

There is about a 3 in 400 chance of failure.

(b) The probability that the bulb fails in years 3 to 6 is

$$P[3 < t < 6] = \int_3^6 2e^{-2t} \, dt = e^{-6} - e^{-12} \approx 0.0024726$$

The chance of failure is about 1 in 500.

(c) This problem requires us to find T such that F(T) = 0.5, where F is the Hazard distribution function. Setting

$$0.5 = F(T) = \int_0^T (1/9)e^{-t/9} \, dt = 1 - e^{-T/9}$$

we solve for the time

$$T = -9 \ln(0.5) \approx 6.238 \text{ years}$$

Mean values.

The **expected value** of a random variable x is the average of all possible x-values, weighted by the chance that each x-value has of occurring. If the random variable x has the density function f(x) the expected value of x is its *mean*, which is often denoted by μ, the Greek letter "mu", or by $\bar{x}$ and is evaluated as

$$\mu = \int x \cdot f(x) \, dx \qquad \textbf{Mean Value}$$

The mean is actually a weighted average of x over the sample interval I, using the density function f(x) as the weight which is given as the ratio of two integrals, $\int x \cdot f(x) \, dx / \int f(x) \, dx$, however, because f(x) os a density function, $\int f(x) \, dx = 1$.

Example 4 **The Expected Value or *Mean* of a uniform distribution.**

Problem   What is the mean value of the random variable considered in Example 1?

Solution The uniformly distributed variable of **Example 1** has a density f(x) = 1/2 for 3 ≤ x ≤ 5. Its mean is

$$\mu = \int_3^5 x \cdot 1/2 \, dx = x^2/4 \Big|_3^5 = [5^2 - 3^2]/4 = 4$$

For any uniform distribution the mean μ is the mid-point of the sample interval I. If x has the density function f(x) = 1/(b - a) for a ≤ x ≤ b, then it mean is

$$\mu = \int_a^b x \cdot 1/(b-a) \, dx = x^2/[2(b-a)] \Big|_a^b = [b^2 - a^2] / [2(b-a)] = [a+b]/2$$

as $b^2 - a^2 = (b-a)(b+a)$.

Example 5 **The Expected Value or *Mean* of a non-uniform distribution.**

Problem   What is the mean value of the random variable considered in Example 2?

Solution In **Example 2** the density function is f(x) = (4/9)(x - 1.5)² and the sample space is the interval I = [0,3] with units of meters. The mean value of this random variable is

$$\mu = \int_0^3 x \cdot (4/9)(x - 3/2)^2 \, dx$$

$$= \int_0^3 (4/9)(x^3 - 3x^2 + (9/4)x) \, dx$$

$$= (4/9)(x^4/4 - x^3 + (9/8)x^2 \Big|_0^3$$

$$= 9 - 12 + 9/2 = 3.5\text{m} \qquad \text{μ has the same units as x.}$$

Example 6 **The Expected Value or *Mean* of a Hazard distribution.**

Problem   What is the mean of a random variable with a Hazard distribution as considered in Example 3?

Solution In **Example 3** the hazard variable t was introduced with a constant hazard distribution f(t) = $\lambda e^{-\lambda t}$ over the infinite sample space I = [0,∞). As we shall now show, its mean is simply 1/λ, the reciprocal of the *hazard rate* λ. Since the sample space is infinite, the mean value is given by the improper integral

$$\mu = \int_0^\infty t \cdot \lambda e^{-\lambda t} \, dt$$

The antiderivative of $\lambda t e^{-\lambda t}$ must be determined using the method of "integration by parts" as developed in Chapter 6, Section 6.2. Introducing the intermediate variables $u = t$ and $v' = \lambda e^{-\lambda t}$, we determine the derivative $u' = 1$ and the antiderivative $v = -e^{-\lambda t}$:

$$u = t \qquad v' = \lambda e^{-\lambda t}$$

$$u' = 1 \qquad v = -e^{-\lambda t}$$

Applying the Integration by Parts formula gives the antiderivative:

$$\int t\lambda e^{-\lambda t}\, dt = -te^{-\lambda t} - \int -e^{-\lambda t}\, dt = te^{-\lambda t} - (1/\lambda)e^{-\lambda t} = (t - 1/\lambda)e^{-\lambda t}.$$

The improper integral is evaluated as a limit:

$$\mu = \lim_{T \to \infty} \int_0^T t\lambda e^{-\lambda t}\, dt$$

$$= \lim_{T \to \infty} (t - 1/\lambda)e^{-\lambda t} \big|_0^T$$

$$= \lim_{T \to \infty} [(T - 1/\lambda)e^{-\lambda T} + 1/\lambda]$$

$$= 1/\lambda + \lim_{T \to \infty} (T - 1/\lambda)e^{-\lambda T} = 1/\lambda$$

The final limit has the indeterminate form "$\infty \cdot 0$" as $T \to \infty$ since

$$T - 1/\lambda \to \infty \qquad \text{and} \qquad e^{-\lambda T} \to 0$$

To show that its limit is zero we rearrange it, putting the exponential term in the denominator, and then apply L'Hôpital's Rule, as developed in Chapter 4, Section 4.10:

$$\lim_{T \to \infty} (T - 1/\lambda)e^{-\lambda T} = \lim_{T \to \infty} [T - 1/\lambda] / [e^{\lambda T}]$$

This limit has the indeterminate form "∞/∞", so we differentiate the numerator and denominator separately, not as a quotient, and consider the limit of the ratio of the derivatives:

$$= \lim_{T \to \infty} 1 / \lambda e^{\lambda T} = 0 \qquad \qquad \text{since } e^{\lambda T} \to \infty.$$

Consequently, $\mu = 1/\lambda$. This indicates how hazard rates are estimated. One observes many accidents or failures, and takes λ to be the reciprocal of the average time between the accidents.

Example 7 **Predicting the expected number of AIDS cases by numerical integration.**

This example is based on methods used by the Center for Disease Control to predict the number of AIDS cases that are likely to occur, using data on those that had already been recorded. It is complicated by the fact that the occurrence of AIDs follows the exposure to HIV with a lag. The expected number of AIDS cases is given by an integral, however the density function is too complicated to integrate with the Fundamental Theorem of Calculus, it can only be approximated numerically. In the Solution we construct a table of values for the complicated density function and then a table of values required to approximate the integral using the Trapezoid Rule.

Even if you have not studied numerical integration, you may find it interesting to read how real data are used to approach this important problem.

Background This background is intended to motivate and explain the reasoning behind a process used to predict the number of AIDS cases that will be diagnosed some time after people are infected with HIV. The modeling is of necessity complex and consequently may prove difficult to comprehend on first reading.

It is not necessary that you understand this discussion to work the problem.

The AIDS epidemic presented a difficult problem of predicting the number of future cases expected to occur so that adequate health facilities could be established. The disease known as AIDS involves two stages. One is a clinically benign pre-AIDS state in which a person is infected and develops an antibody to the HIV virus. These individuals enter the AIDS state at some future time, diagnosed by the presence of multiple clinical symptoms. In modeling this phenomena it is assumed that the "incubation" period, the time an individual has HIV antibody before being classified as having AIDS, is a random variable with density function f(t) and a probability distribution function F(t). Then the probability that an HIV individual infected at time t = 0 becomes an active AIDS patient in an interval [a,b] is

$$P_A[a,b] = F(b) - F(a).$$

A critical component of the model is the number of individuals contacting the disease at time t. This is modeled by an Infection function denoted by

$$I(t) = \text{Number of New HIV infectives at time t}.$$

This is the function that is actually not known due to the complexity of the disease and its mode of transmission. The data that is recorded in national data registers and world wide, in Paris by the World Health Organization, is the number of the new AIDS cases reported over specific periods. The expected number of AIDS cases to be reported in a period [a,b] that resulted from infections prior to time T is the integral over the infection period, the interval [1980,T], of the infection function I(t) times the probability that these infections result in a diagnosed AIDS case. We denote this expected number as

$$NA(a,b;T) = \int_{1980}^{T} I(s) \cdot P_A\{s,[a,b]\}\, ds$$

where

$P_A\{s,[a,b]\} = $ the probability an individual is infected at time s and is diagnosed as having AIDS in the interval [a,b]}.

Relating this probability to the distribution function F is achieved by using a time-shift.

The probability that someone who becomes infected at time s develops AIDS before time t is $F(t - s)$.

Hence, the probability that an infective at time s is diagnosed in the interval [a,b] is the probability that they are diagnosed before time b minus the probability they were diagnosed prior to time a.

$$P_A\{s,[a,b]\} = F(b - s) - F(a - s)$$

Consequently, the number of AIDS cases occurring in the interval [a,b] due to infections prior to time T is given by the "delay-integral" equation

$$NA(a,b;T) = \int_{1980}^{T} I(s)[F(b - s) - F(a - s)]\, ds.$$

The lower limit of integration is chosen as a year prior to the "apparent" start of the epidemic; 1980 has been used to model North America and Europe data. This integral represents the sum of all of the new HIV infectives up to time T multiplied by the probability that they are reported as having converted in the reporting interval [a,b].

Problem On the basis of the following data estimate the number of new AIDS cases that will be reported in the year 1994-1995 due to HIV infections prior to 1987. This is the number NA(1994,1995;1987) given by the above delay-integral equation.

Use a 1988 estimate of the incubation distribution given by the "Weibull" distribution

$$F(t) = 1 - e^{-\lambda t^\alpha}$$

where $\lambda = 1/9$ is the reciprocal of the observed mean incubation time, 9 years, and $\alpha = 2.286$ was statistically estimated from observations. The estimated incidence of HIV infections in the US from 1980 to 1987 is given by a discrete step function, scaled by the population constant $N = 159{,}000$:

$I(s) = 0.0012 \cdot N$ for $1980 < s < 1981$,

$I(s) = 0.0879 \cdot N$ for $1981 < s < 1982$,

$I(s) = 0.1285 \cdot N$ for $1982 < s < 1984$,

$I(s) = 0.2426 \cdot N$ for $1984 < s < 1987$,

Solution As in the Background discussion, the number of expected AIDS cases is given by the integral

$$NA(1994, 1995; 1987) = \int_{1980}^{1987} I(s)[F(1995 - s) - F(1994 - s)]\, ds$$

The integrand has exponential terms like

$$F(1995 - s) = 1 - e^{-\frac{1}{9}(1995 - s)^{2.286}}$$

We can not integrate such functions using the FTC as we do not know an antiderivative of an exponential of s raised to a non-integer exponent. The integral must be approximated by numerical integration. We shall use the Trapezoid Rule (see Section 5.5) that gives

$$\int_a^b g(s)\, ds \approx \sum_{n=1,N} w_n\, g(s_n)$$

where the weight constants are $w_n = 1$ for $1 \le n < N$ and $w_0 = w_N = 0.5$. The evaluation points $\{s_n\}$ are a partition of the integration interval $[a,b]$. For this example the function

$$g(s) = I(s)[F(1995 - s) - F(1994 - s)]$$

and the partition must include the times at which the incidence function changes. Consequently, we use a uniform partition of $N = 7$: $s_n = 1980 + n$, $n = 0, 1, 2, \ldots, 7$. The integrand involves differences in the incubation delay function

$$\Delta F = F(1995 - s_n) - F(1994 - s_n)$$

whose values are given in Table 1.

Table 1. Delay onset Probabilities.

year s	F(1995 - s)	F(1994 - s)	ΔF
1980	0.959834	0.935796	0.024039
1981	0.935796	0.901510	0.034285
1982	0.901510	0.854886	0.046624
1983	0.854886	0.794451	0.060435
1984	0.794451	0.719823	0.074627
1985	0.719823	0.632121	0.087703
1986	0.632121	0.534178	0.097943
1987	0.534178	0.430493	0.103685

The steps and data of the numerical integration are indicated in Table 2. The numerical integration predicts that approximately 11,555 AIDS cases would be reported in the 1994-95 year due to the HIV infections that occurred in the 1980-87 period. This is based on the above model and data, which may prove to be inaccurate when more observation data is available. It illustrates an estimation method that requires an integration that can only be done numerically.

Table 2. Trapezoid Rule Evaluation
Step size is h = 1

s	I(s)	Prob = ΔF	Weight W	I(s)*ΔF*W
1980	190.8	0.024039	0.5	2.293302
1981	13976.1	0.034285	1	479.177167
1982	20431.5	0.046624	1	952.604548
1983	20431.5	0.060435	1	1234.782253
1984	20431.5	0.074627	1	1524.742728
1985	38573.4	0.087703	1	3382.999778
1986	38573.4	0.097943	1	3777.995298
1987	38573.4	0.103685	0.05	199.973922
			Sum of terms	11554.568996

EXERCISE SET 7.4

1 Determine the constant k that makes the function f(x) a probability density function over the interval I. Then, determine the probability that the random variable x is in the indicated interval.

   a) $f(x) = k$, over $I = [0,5]$; $P[1 < x < 2]$

   b) $f(x) = kx$ for $x \in I = [0,5]$; $P[1 < x < 2]$

   c) $f(x) = k(x - 1)$ for $x \in I = [0,1]$; $P[0.75 < x < 1]$

   d) $f(x) = kx(10 - x)$ for $x \in I = [0,10]$; $P[0 < x < 5]$

   e) $f(x) = ke^{-x}$ for $x \in I = [0,10]$; $P[0 < x < 5]$

   f) $f(x) = k(x - 0.5)(x - 1.5)^2(x - 2.5)$ on $I = [0.5, 2.5]$; $P[1 < x < 2]$
   This is the density function sketched in Figure 7.17.

2 Determine the mean values of each of the distributions in Exercise 1.

3 In a dart game, the probability of a well tossed dart hitting the center disk of a target depends on the size of the disk and the distance, x, that the thrower is from the target. Assume that the density function is proportional to the area of the disk times a D(x) that decreases with increasing distance x. Assume the area of the small center target is 10 cm² and $D(x) = 1 - 0.01x^2$.

   a) What is the form of the density function f(x) if the proportionality constant is k?

   b) What must be the value of k to make $\int_0^{10} f(x)\,dx = 1$?

482 Chapter 7 Applications Involving Integration

  c) What is the probability that a person standing between 3 and 5 meters from the dart board will hit the center disk?

  d) At what distance should the "toss-line" be drawn so that the probability of hitting the target center is 0.5?

4 If the probability of a dart hitting the "bulls-eye" in the previous exercise is given by the dart board target of area 10 cm^2 and $D(x) = e^{-2x}$

  a) What is the form of the density function f(x) if the proportionality constant is k?

  b) What must be the value of k to make $\int_0^\infty f(x)\,dx = 1$?

  c) What is the probability that a person standing between 3 and 5 meters from the dart board will hit the bulls-eye?

  d) At what distance should the "toss-line" be drawn to give a 10% chance of hitting the bulls-eye?

  The **variance** of a probability density f(x) is defined as $V = \int_I (x - \mu)^2 f(x)\,dx$ where μ is the mean of the density function and the integral is taken over the entire range of x-values. Use this formula to work the following exercises.

5 a) What are the mean and variance of the density $f(x) = 2 - 2x$ on [0,1]?

  b) What are the mean and variance of a uniform density over the interval [a,b]?

  c) Compute the variance for the three density functions of Examples 1-3 of this section.

6 Compute the variance of each density in Exercise 1.

7 Assume that a personal computer has a failure hazard rate of λ = 0.15 years. What is the probability that it will experience a failure in the first 180 days of service?

8 Assume that a long life light bulb has a constant hazard rate of λ = 0.5 years for failure. If a light bulb has been burning continuously on a marquee for 15 years what is the probability of this occurring? What is the probability that it will burn another 15 years?

9 If a person that comes into contact with a flu virus will develop the flu in t-days with an exponential distribution that has constant hazard rate of λ = 0.1 day^{-1} what is the time T for which the person has a 50% chance of showing flu symptoms after exposure?

10 A Chemist measures the time of contact of protein in a solution with a lipid surface using a florescence technique. Assume that the time between successive contacts of protein molecules with the lipid is exponentially distributed with a "hazard rate" λ = 0.5 sec^{-1}.

  a) What is the mean time between hits on the lipid?

  b) What is the probability that a protein molecule contacts the lipid within $5 \cdot 10^{-3}$ sec after a molecule hits the lipid?

  c) What is the time period after a collision for which there is a 95% probability that another molecule will hit the lipid?

11 For what value k is the function $f(x) = k \cdot (0.1x - x^2/360)$ $0 \le x \le 36$ a density function? What is the mean value of a random variable with this density function? What is the variance of this distribution?

12 Use the Trapezoid Rule to approximate the integral and estimate the expected number of AIDS cases in the year 1996-1997 due to the HIV infectives of the period 1980 to 1987 using the model and data of Example 5?

Section 7.5 Chemistry, Physics and Hydrology.

Chemistry

Example 1 **Enzyme Kinetics: The Integrated Michaelis-Menten equation.**

Background

Enzymes are proteins that significantly increase the rate of a biochemical reaction but are not altered due to the reaction. The simplest enzyme reaction mechanism involves an enzyme E and substrate S that combine to form an enzyme-substrate complex ES. The ES complex either reverts to enzyme and substrate or converts the substrate to product P freeing up the enzyme. The schematic model of this reaction is

$$E + S \rightleftharpoons ES \longrightarrow E + P \qquad \text{Model Enzyme Reaction}$$

Biochemists normally measure the rate of product formation, which is the negative of the rate that substrate is depleted by the reaction. This rate is called the velocity of the reaction and is denoted by v, or v(t). The equation normally given to describe the velocity of this model is the classical Michaelis-Menten equation. It is expressed in one of two forms, one employing the concentration of product P and the initial substrate concentration S_0, and the other simply in terms of the substrate concentration S:

$$v = P' = V_{max}(S_0 - P)/(K_m + S_0 - P)$$

or **Michaelis-Menten Equation**

$$v = -S' = V_{max}S/(K_m + S)$$

The independent variable is t (for time) and v, P, and S are all functions of t, e.g., v(t), P(t), and S(t). These equations involves two constants that are used to characterize an enzyme and its interaction with the substrate. In the numerator is the term V_{max}, which is the maximum possible velocity. v(t) approaches V_{max} as S or S_0 approaches ∞. In the denominator is the "Michaelis" constant K_m, which has the same concentration units as the substrate and indicates the concentration of substrate at which $v = 0.5V_{max}$. The two equations are the same because of the principal of conservation of mass, which in this case says that the sum of the free substrate S and product P must equal the total amount S_0 of substrate present at the start of the experiment:

$$S_0 = S + P \quad \text{or} \quad S = S_0 - P \qquad \text{Conservation of Mass}$$

Chemist often write these symbols in square brackets, like [S], to indicate they are measured in concentration units. To simplify notation we will not use this notation.

Problem The Michaelis-Menten equation is a differential equation since the velocity v is a derivative. Determine the integrated form of the Michaelis-Menten equation.

Solution The procedure used to "integrate" this equation is called "separation of variables". The derivative is expressed as a ratio of differentials, $P' = dP/dt$ or $S' = dS/dt$, and all the t-terms are placed on one side of the equation with all S or P terms on the other side. We consider the first form of the equation. Multiplying by dt gives

$$dP = V_{max}\{(S_0 - P)/(K_m + S_0 - P)\} \cdot dt$$

Then, dividing by the rational term in brackets on the right side gives:

$$dP/\{(S_0 - P)/(K_m + S_0 - P)\} = V_{max} \cdot dt$$

or,

$$[(K_m + S_0 - P)/(S_0 - P)] \, dP = V_{max} \, dt$$

Writing the fraction on the left as the sum of two terms, i.e., $(A + B)/C = A/C + B/C$, gives

$$[K_m /(S_0 - P) + 1] \, dP = V_{max} \, dt$$

The 1 appears because the $S_0 - P$ terms cancel in the second fraction. This equation is "integrated" taking the integral of the terms on the left of the = sign with respect to the variable P and taking the integral of the right side with respect to t:

$$\int K_m /(S_0 - P) + 1 \, dP = \int V_{max} \, dt$$

Applying the Natural Log and Power antiderivative Rules gives

$$-K_m \, ln(S_0 - P) + P = V_{max} \cdot t + C$$

where C is an integration constant. Note that $P < S_0$ so that the difference $S_0 - P$ is positive. Assuming that at time $t = 0$ the reaction has yet to proceed, so that there is no product, gives the **initial condition** at $t = 0$, $P(0) = 0$. Substituting these into the above equation allows us to solve for C:

$$C = -K_m \, ln(S_0)$$

Hence, solving for the $V_{max} t$ term gives

$$V_{max} \cdot t = P - K_m \, ln(S_0 - P) + K_m \, ln(S_0)$$

or, combining the two ln terms,

Product Form of the Integrated Michaelis-Menten Equation

$$V_{max} \cdot t = P + K_m \, ln(S_0/(S_0 - P))$$

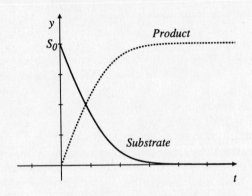

This equation can be expressed in terms of substrate rather than product concentrations by using the conservation of mass equations. Substituting $S = S_0 - P$ gives the equivalent equation:

$$V_{max} \cdot t = S_0 - S + K_m \, ln(S_0/S)$$

Substrate Form of the Integrated Michaelis-Menten Equation

"Integrated" Michaelis-Menten equations cannot be solved explicitly for either S or P as functions of t. The graph of S and P versus t is illustrated above for a case where $K_M = 2$ and $V_{max} = 4$.

☑

Physics

Example 2   **Work: the integral of force over distance.**

Background   The concept of "work" as used in Physics is rather special. **Work** is the total effort that is expended when a force is applied to move an object. If the force is constant the corresponding **work** is the product of the amount of force being applied and the distance through which it is applied. Let W, F and D denote the work, force, and distance. The basic relationship for a constant force is

$$W = F \cdot D$$ **Work due to a constant force.**

When the force is not constant, and it varies as a function F(x) of the position x along a line, then the "work" done by exerting the force F(x) over an x-interval [a,b] is given by an integral:

$$W = \int_a^b F(x)\, dx \qquad \text{Work due to a variable force.}$$

To see why the total work is given by such an integral we apply the basic *method of calculus*. The total amount of work performed is approximated by partitioning the interval [a,b] into subintervals, approximating the work over each subinterval by replacing the function F(x) by a constant, and adding these approximations over all the subintervals. The resulting sum has the form of a Riemann Sum. As the number of subintervals increases to ∞, the limit of the approximation provides the actual amount of work associated with the variable force function; but the limit of the Riemann Sum is by definition a definite integral.

More specifically, assume that a continuous force function F(x) is applied over an interval [a,b]. Let $\{x_i\}$ be a partition of [a,b]. The force F(x) is approximated on the i-th subinterval $I_i = [x_{i-1}, x_i]$ by the constant $F(x_i^*)$, where x_i^* is some x-value in the interval I_i. Denoting the length of I_i by Δx_i, the approximate work resulting from applying this constant force over the interval I_i is given by

$$W_i = F(x_i^*) \cdot \Delta x_i$$

The approximate total work done by applying these constant forces over the subintervals is

$$W \approx \Sigma_{i=1,N}\, W_i = \Sigma_{i=1,N}\, F(x_i^*) \cdot \Delta x_i$$

This summation has the form of a Riemann Sum. Its limit, as $N \to \infty$, gives the integral determination of the total work, as state above.

$$W = \lim_{N \to \infty} \Sigma_{i=1,N}\, W_i = \lim_{N \to \infty} \Sigma_{i=1,N}\, F(x_i^*) \cdot \Delta x_i = \int_a^b F(x)\, dx$$

The SI or metric unit for force is a Newton, N, whose units are kilograms times meters divided by seconds squared; $kg \cdot m/s^2$. The distance unit is in meters, m, giving the units of work as Newton-meters, which is called a joule, J. You may also find reference to work in the British units, with force in pounds, lb, and distance in feet, ft, giving work in units of foot-pounds, ft-lb.

Problem A   A group of students need to push a stalled car to a garage. Assume the car weighs 2400 lbs and the garage is 320 ft away. Assume that the force needed to push the car is initially one fourth the weight of the car and that due to a downhill slope of the road it decreases with the distance pushed being given by

$$F(x) = 600(x + 1)^{-2}, \text{ when the car has been pushed x feet.}$$

Determine the total work required of the students to push the car to the garage.

Solution   The work is given by

$$W = \int_0^{320} 600(x+1)^{-2}\, dx = -600(x+1)^{-1} \Big|_0^{320}$$

$$= -600[1/321 - 1] \approx 598.1 \text{ ft-lb}$$

Problem B   A weight-lifting machine in a gym employs cams to vary the force required to lift a weight. Assume that the force required to lift y kg of weight when the weight is a distance x meters from the resting position is

$$F(x) = y \cdot (4x)^{1/3}$$

Compare the work exerted lifting a resting 100kg weight 25cm to that required to lift the same resting weight 50cm.

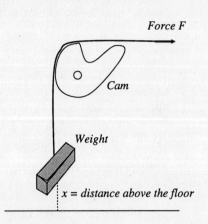

Solution   The two values to be compared are calculate as the integrals of F over [0, 0.25] and [0, 0.50].

$$W_{25} = \int_0^{0.25} 100(4x)^{1/3} \, dx$$

$$= 100(3/4)(1/4)(4x)^{4/3} \big|_0^{0.25} = 300/16 = 18.75 \text{ J}$$

$$W_{50} = \int_0^{0.5} 100(4x)^{1/3} \, dx$$

$$= 100(3/4)(1/4)(4x)^{4/3} \big|_0^{0.5} = (300/16)2^{4/3} \approx 47.247 \text{ J}$$

Consequently, as $W_{50} - W_{25} \approx 1.52 \, W_{25}$, the work required to lift the weight an additional 25cm is slightly more than one and a half times that required to lift it the initial 25cm.

Example 3   Hooke's Law for stretching and compressing springs.

Background   Hooke's Law states that the force required to compress or stretch an "ideal" spring is proportional to the distance the spring length changes from a "natural" un-compressed length. The proportionality constant, traditionally denoted by k, is called the *spring constant* and is determined by the physical characteristics and composition of the spring. If the *natural length* of a spring is L_0, then a constant force F will stretch or compress the spring to a new length L given by

$$L = L_0 + k \cdot F$$

Hooke's Law is often stated as in terms of the change in spring length, $x = L - L_0$, as

$$F = k \cdot x$$ **HOOKE'S LAW**

If F is positive the spring is stretched and if F is negative the spring is compressed.

Problem   Assume a force of 50 N compress a spring from its natural length of 10cm to 8cm.

(a) What is the spring constant?

(b) What is the work required to stretch the spring to 25cm from its original length?

(c) What is the work required to change this spring's length from 6cm to 15 cm?

(d) What will be the spring's length after 0.2J of work is expended to compress the spring from 8cm?

Solution (a) The distance the spring is compressed is x = 10 - 8 = 2cm = 0.02m. The constant force that stretched the spring is F = 50 kg·m/s^2. Solving Hooke's law, 50 = k · 0.02 for the spring constant k gives

$$k = 50/0.02 = 2500 \text{ kg/s}^2$$

(b) If the spring is stretched from its natural length 10 cm to 25 cm, the distance through which the force is applied is x = 0.15m. The force required to stretch the spring will vary with the distance, being given by F(x) = 2500x, thus the corresponding work required to stretch the spring is

$$W = \int_0^{0.15} F(x)\, dx = \int_0^{0.15} 2500\, x\, dx = 1250\, x^2 \Big|_0^{0.15} = 28.125 \text{ J}$$

(c) The work involved in changing the springs length from L_a to L_b is given by

$$W = \int_{L_a - L_0}^{L_b - L_0} kx\, dx$$

The distances for the given problem, in meters, are $L_0 = 0.1$, $L_a = 0.06$, and $L_b = 0.15$. Thus the work is given by

$$W = \int_{-0.04}^{0.05} 2500\, x\, dx = 1250((0.05)^2 - (-0.04)^2) = 1.125 \text{ J}$$

(d) Applying the same principle used in part (c), the given work, 0.2J is the integral of the force kx from -0.02 = 0.08 - 0.1, as the starting length is 8 cm, to L - 0.1, where L is the unknown new length. The integral is evaluate, set equal to 0.2 J and the resulting equation is solved for L:

$$W = 0.2 = \int_{-0.02}^{L - 0.1} 2500\, x\, dx = 1250\, [(L - 0.1)^2 - (-0.02)^2]$$

Solving this equation for L gives

$$L = 0.1 \pm \text{SQRT}(0.2/1250 + 0.02^2)$$

There are two possible answers:

$L_1 \approx 12.37$ cm corresponding to the + sign,

or

$L_2 \approx 7.63$ cm corresponding to the - sign.

The correct solution is L_2 which represents a further compressed spring, as specified in the question. The other solution L_1 represents the spring's state if it is stretched applying the same amount of work. Remember, this is an ideal spring and the energy released allowing it to return to its resting state can be utilized in stretching it an equal distance past the resting length, then the given work will stretch it the remaining distance to reach L_1.

Example 4 **Distance traveled: the integral of velocity over time traveled**.

Problem Determine the frequent flyer miles a traveler earns on a flight for which the plane takes off at 160 MPH, accelerates over a one hour period to a cruising speed of 600 MPH, at which it cruises for 8 hours, and then it decelerates for one hour, landing at 160 MPH. Assume the planes velocity is the function sketched in 24, which is given by

$$v(t) = \begin{cases} 160 + 220[(2t-1)^{1/3} + 1] \text{ MPH} & \text{for } 0 < t < 1 \text{ hr,} \\ 600 \text{ MPH} & \text{for } 1 < t < 5 \text{ hr,} \\ v(10-t) \text{ MPH} & \text{for } 5 < t < 10 \text{ hr.} \end{cases}$$

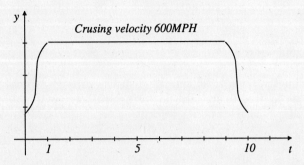

Crusing velocity 600MPH

Solution The basic principle required to answer this question is:

The distance D an object moves in a time interval [a,b] is the integral of its instantaneous velocity over the time interval:

$$D = \int v(t)\, dt$$

As the velocity function v(t) is symmetric about the time t = 5, the total flight distance D is twice the integral over the first half of the flight time. Thus

$$D = 2 \int_0^5 v(t)\, dt$$

This integral is evaluated by splitting it at the points where the function's definition changes:

$$\begin{aligned} D/2 &= \int_0^1 v(t)\, dt + \int_1^5 v(t)\, dt \\ &= \int_0^1 160 + 220[(2t-1)^{1/3} + 1]\, dt + \int_1^5 600\, dt \\ &= 380t + (165/2)(2t-1)^{4/3} \Big|_0^1 + 2400 \\ &= 380 + 165/2 - 165/2 + 2400 = 2780 \end{aligned}$$

Thus the total flight distance was D = 5,560 miles.

☑

Hydrology

Example 5 Reservoir Storage Capacity.

Background

Hydrologists measure the rates of inflow, I, and outflow, Q, for storage reservoirs. (Yes, the letter Q, not the letter O, is used here and in the study of heat flow to represent the quantity that flows out.) The difference I - Q is the rate of change in the amount of water stored in the reservoir. In the Hydrology literature the equation:

The Rate of Change in Storage

$$S' = I - Q$$

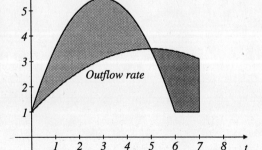

is called the *hydrologic continuity equation*. S' indicates the rate of change in the storage. The cumulative storage over a period [0,T] is

The Net Storage $\qquad S = \int_0^T I(t) - Q(t)\, dt$

The fluid pressure acting on the strip S_i is approximately $p_i = w \cdot h_i$. Consequently, the total force acting on the i-th strip is approximately

$$F_i = w \cdot h_i^* \cdot L(h_i^*) \Delta h_i$$

Adding these forces for each strip gives the approximate force on the membrane

$$F \approx \Sigma_{i=1,N} F_i = \Sigma_{i=1,N} w \cdot h_i^* \cdot L(h_i^*) \Delta h_i$$

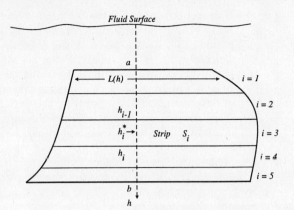

This sum has the form of a Riemann Sum, and hence in the limit as $N \to \infty$ it gives the total force as a definite integral, with respect to the depth variable h. We refer to this force as a *horizontal force* as it is the force that acts, in opposition directions, on each side of the membrane.

Horizontal Fluid Force on a Vertical Membrane

$$HF = \int_a^b w \cdot h \cdot L(h) \, dh$$

If the fluid is compressible, the density will actually vary with the depth and be a function w(h).

Problem What is the total fluid force exerted on a circular cylindrical drum that is floating up-right at a depth of 20 meters in water, whose density is 1000kg/m^3, if the drum is 50 cm high with radius 10 cm?

Solution The total force is the sum of three forces. The force on the top, the bottom and the side of the can. The vertical forces on the top and bottom are easily computed as these are horizontal surfaces. The bottom surface is at a depth h = 20, the drum's top surface is at h = 19.5. Thus, as the top and bottom surface areas are $\pi \cdot 0.1^2$ m^2, the vertical force on the can's

top is: $\quad VF_{top} = 1000 \cdot 19.5 \cdot \pi \cdot 0.1^2 \approx 612.6$ kg

bottom is: $\quad VF_{bottom} = 1000 \cdot 20 \cdot \pi \cdot 0.1^2 \approx 628.4$ kg

The horizontal force acting on the side of the can is the same force that would be exerted on the vertical membrane formed by slitting the can vertically and opening it flat to form a rectangle. The rectangle would have width equal to the circumference of the can, thus $L(h) = 2\pi r = 0.2\pi$ for $19.5 \leq h \leq 20$. Applying the HF integral formula with a = 19.5 and b = 20 gives :

$$HF = \int_{19.5}^{20} 1000 \cdot h \cdot 0.2\pi \, dh = 100\pi h^2 \big|_{19.5}^{20} = 1975\pi \approx 6204.6 \text{ kg}$$

The total force acting on the can is the sum of the three calculated forces, approximately 7445.6 kg.

EXERCISE SET 7.5

Chemistry

1   A **first order chemical reaction** proceeds at a rate proportional to the concentration of the reactants. If reactants A and B combine to form a chemical complex C, at a first order rate k, the schematic model is A + B --> C . The corresponding Dynamical Model consists of the velocity equation describing the rate of C formation, it is a differential equation: $v = C' = k \cdot A \cdot B$.

a) If the concentrations of the reactants A and B are monitored and described as functions of time, A(t) and B(t), then the concentration of C at time T is the initial concentration C_0 plus the increase over the time period [0,T]. This is given by the definite integral $C(T) = k\int_0^T A(t)B(t)\, dt$.
Calculate C(5) if A = 10, and B(t) = 3 + sin(t/4π).

b) If the concentration of B decreases at the same rate as the concentration of product increases, then B' = -C' or B' = -k · A · B. Express this equation as an integral equation by expressing B' as the ratio of differentials dB/dt, separating the variables and integrating each side of the equation.

c) If the concentration of A is 50 μM throughout an experiment and the initial concentration at t = 0 of B is 20μM, evaluate the integrals in part b) and solve for the concentration B at time t = T.

2 A reaction in which two molecules of the same chemical combine to form a second chemical is modeled by A + A ---> B. The rate of change in the concentration [A] of chemical A is then described by $d[A]/dt = -k[A]^2$. By "separating the variables" and integrating, determine the function [A] as a function of t and the initial amount of the chemical.

3 For a **second order chemical reaction** in which A and B react to form a complex X the concentration of A and B changes as X is formed. The corresponding Dynamical Model is the velocity equation for the rate of X formation:

$$dx/dt = k(a - x)(b - x)$$

where a and b are the initial concentrations of A and B and x is the concentration of X. a) Use separation of variables and the partial fraction method to integrate this rate equation when a = 10 and b = 3, assuming x = 0 at time t = 0. Can you solve the solution explicitly for x? b) Determine the general solution for arbitrary constants a and b and initial concentration x_0 at t = 0.

4 An **autocatalytic reaction** involves a situation where the product acts to promote the reaction at a faster rate. The rate of an autocatalytic reaction A → X can be described by the Dynamical Model equation v = dx/dt = kx(a - x).
a) If a = 100 and k = 0.3, and x(0) = 0, what is the amount of X formed over the interval [0,5] ?
b) If at t = 5 the reaction in part a) is continued with the addition of 100 more units of A and 50 units of X, what will be the amount of A left at time t = 10?

5 **"Concentration work"**. This problem deals with the "work" involved when the concentration of a substance in a fluid is increased. Fluid properties are described by the same parameters used to model gases. An "ideal" gas satisfies the equation

$$pV = NkT \quad \text{or} \quad p = NkT/V$$

where p is the pressure on the gas, V is its volume, N is the number of gas particles, T is its temperature in degrees Kelvin, and k is the Boltzman's constant, $k \approx 1.38 \times 10^{-23}$ J K^{-1}. An alternative formulation in terms of the gas concentration C = N/V is p = CkT. In Chemistry, the concentration of a dilute solute in a solution is modeled by the properties of an "ideal" gas; that is, the number of solute molecules and solute concentration are assumed to also satisfy the above equations. If a solution is concentrated by a change in its volume, as when a piston compresses the volume in a cylinder, then the "concentration work" is given by the integral of the pressure, p = NkT/V, with respect to volume, from the initial volume V_1 to the altered volume V_2:

$$W_{conc} = -\int_{V_1}^{V_2} p\, dV = -\int_{V_1}^{V_2} NkT/V\, dV$$

a) If N, k and T are constant, then the "concentration work" is a function of the volumes V_1 and V_2. In this case what is W_{conc}?

b) Express the "concentration work" as a function of the concentrations C_1 and C_2 of the solute at the volumes V_1 and V_2 when N, k and T are constant. Remember that concentration is amount per unit volume. The concentration of a substance in Molar units, which is Moles/l, is obtained from the volume and number of molecules N by the equation $C = (N/N_A)/V$ where the constant N_A is Avogadro's number, $N_A \approx 6.022 \times 10^{23}$ molecules mole^{-1}.

c) What is the work required to concentrate 2 moles of solute by a factor of 10, i.e., when $C_2 = 10\, C_1$, at temperature T = 310°K ?

Section 7.5   CHEMISTRY, PHYSICS AND HYDROLOGY

Problem   Calculate the cumulative storage of a lake over the interval [0,7] days if it rained for 4 days and the Inflow and Outflow functions are given by

$$\text{Inflow} \quad I = \begin{cases} 1 - 0.5t(t-6) \; 10^3 m^3 d^{-1} & \text{for } 0 \leq t \leq 6 \\ 1 & \text{for } 6 \leq t \leq 7 \end{cases}$$

and

$$\text{Outflow} \quad Q = 1 - 0.1t(t-10) \; 10^3 m^3 d^{-1} \quad \text{for } 0 \leq t \leq 7$$

Solution   The net storage is the integral over 7 days, which is split into two integrals, one over the first 6 days and one for day 7. The first integral is

$$\int_0^6 [1 - 0.5t(t-6)] - [1 - 0.1t(t-10)] \, dt$$
$$= \int_0^6 2t - (1/5)t^2 \, dt = t^2 - (1/15)t^3 \Big|_0^6 = 36/5$$

The second integral is

$$\int_6^7 1 - [1 - 0.1t(t-10)] \, dt = \int_6^7 t^2/10 - t \, dt = t^3/30 - t^2/2 \Big|_6^7 = -34/15$$

The net accumulation of water in the lake is the Storage

$$S = \int_0^7 I(t) - Q(t) \, dt = 36/5 - 34/15 = 74/15 = 4.9\underline{3} \; 10^3 \, m^3$$

This is the signed area between the curves I and Q as sketched in 25. The total global volume of all surface water is about $1.3 \cdot 10^{12} \, m^3$.

☑

Example 6   **Fluid infiltration of porous soils: The Horton Equation.**

Background   In modeling the variation of soil moisture due to rainfall, Hydrologist use an empirical formula called the *Horton Equation* to describe the depth infiltration rate or *fluid capacity* of various soil types. This equation has the form

$$f = f_c + (f_0 - f_c)e^{-\alpha t} \quad \text{in. h}^{-1}$$

Using the hydrologist term of *capacity* for the mathematical term rate, the parameters of the Horton Equation are

     f_0 = initial infiltration capacity of the soil;
     f_c = the equilibrium final capacity of the soil (after a long period of time);
     α = a parameter that reflects the rate of water diffusion in the soil.

Problem   A three hour storm deposited 4 in. of water over soil with characteristics $\alpha = 0.15$/hr. If the soil had initial infiltration capacity $f_0 = 0.5$/hr and equilibrium capacity $f_c = 0.2$, what was the total infiltration and what was the runoff from this storm?

Solution   The **total infiltration volume**, V, is the integral of the capacity of infiltration, f, as given by the Horton Equation, over the period of precipitation:

$$V = \int_0^T f(t) \, dt = \int_0^3 0.2 + 0.3e^{-0.15t} \, dt$$

$$= \{0.2t - 0.3e^{-0.15t}/0.15\} \Big|_0^3$$

$$= 0.6 - 2e^{-0.45} + 2 \approx 1.32 \text{ in.}$$

The **runoff** is the difference between the depth of water deposited and the depth infiltrated:

$$\text{runoff} = 4 - V = 4 - 1.32 = 2.68 \text{ in.}$$

☑

Example 7 Fluid Pressure and Fluid Force.

Background

A fluid is a substance that will deform or change its shape if it is not constrained by a container. A basic principle concerning fluid pressure states that a fluid exerts an equal force in all directions. If a fluid is in an open container the upward fluid force at the surface, the interface with the atmosphere, must counter-balance the atmospheric air pressure that presses down on the fluid surface. Below the surface, at any point the upward acting fluid force must counter-balance the downward acting force due to the gravitational pull on the fluid above the point. This concept leads to a formula for the fluid pressure or force that is experienced by an object located below the fluid surface. We shall consider the force exerted by the fluid on a flat horizontal membrane which is parallel to the fluid surface and the horizontal force exerted by the fluid on a membrane perpendicular to the fluid surface.

The fluid force exerted on a point at depth h (meters) is determined by the density of the fluid. The same force is exerted in every direction and results in a pressure

$$p = w \cdot h \qquad \textbf{Fluid Pressure}$$

where w (kg/m^3) is the weight-density of the fluid and h is the height in meters of the fluid surface above the point. Note that the unit for pressure is mass/area, kg/m^2. The total downward force exerted on a section of a horizontal membrane is the product of the pressure at each point and the membrane's surface area, A. We shall refer to this as a *vertical force* as it is the downward force on the top of the membrane and is also the upward force acting on the underside of the membrane:

Vertical Fluid Force on a Horizontal Membrane

$$VF = \text{pressure} \cdot (\text{surface area}) = p \cdot A = w \cdot h \cdot A,$$

The fluid pressure on a **vertical membrane** varies with the depth of the membrane. The total force on a vertical membrane is determined using the *fundamental method of calculus*. Since the pressure is constant at a fixed depth, the membrane is divided into narrow horizontal strips. For each strip, the fluid pressure acting on the strip is approximated by a constant, chosen as the pressure at some point within the strip. The approximate fluid force acting on the strip is then this constant pressure times the area of the strip. Adding these approximate forces for each strip gives an approximation of the total force on the membrane. This approximation has the form of a Riemann Sum. Taking the limit as the number of strips increases to infinity then gives the exact force on the membrane as the limit of Riemann Sums, i.e., as an integral.

Specifically, assume that a membrane's top and bottom edges are parallel to the surface, at depths a and b, respectively and that its width at a depth h is given by a function L(h), see 26. A partition $\{h_i\}_{i=0,N}$ of the interval [a,b] determines N membrane strips, where strip S_i is the portion of the membrane between depths h_{i-1} and h_i. If we arbitrarily choose a number h_i^* in the interval $[h_{i-1}, h_i]$, then the width of strip S_i is approximately $L(h_i^*)$ and its area is approximately

$$A_i = L(h_i^*) \cdot \Delta h_i$$

d) The concentration of Hydrogen ions, H^+, in gastric fluid has an average pH of about 2 while the concentration of H^+ in blood plasma is about pH 7. Thus the stomach lining concentrates the H^+ by a ratio factor of 10^5. How much concentration work is done in concentrating one mole of ions?

e) Often the temperature of a solution changes as the volume is changed. What is the equation for concentration work when $T = T_0 - \alpha V^{1/2}$ where α is a constant and $T_0 - \alpha$ is the temperature when the solution volume is one liter. State the integral equation and the integrated form.

Physics

6   Assume that the velocity of a dog chasing a frisbee is $s(t) = 10t - t^2$ m/sec at time t. a) what is the distance the dog travels in the first 5 seconds? b) When will the dog stop running? c) If the dog catches the frisbee when his speed is zero, how far did the frisbee travel? d) Assuming the dog immediately returns to the owner, as indicated by a negative velocity, how far away from the owner will the dog be 4 seconds after catching the frisbee?

7   A ball is attached to one end of a rubber line, the other is attached to a fixed peg. The line is stretched and the ball released. It travels with a velocity $v(t) = -e^{-2t}\cos(t)$; when v is negative the ball moves in the direction from its release point towards the peg. Its speed is $s(t) = |v(t)|$. How far does the ball travel over the period $[0,\pi]$?

8   The speed of a hit baseball is $s(t) = 30t(10 - t^3)$ ft/s. a) How far has the ball traveled in time T? b) How much reaction time does an outfielder have to catch the hit ball 270 ft from home plate?

9   If the force in kilo-Newtons required to push a stone across a yard is $F(x) = 3x + 1/(x + 1)$, when the stone is x meters from the starting point, what is the work expended pushing the stone 30 m from the starting position?

SPRING PROBLEMS

10 A spring has a spring constant k = 20 and natural length 30 cm.
    a) What is the work required to compress the spring to a length of 10 cm from its natural length.
    b) How much force is required to stretch the spring from its natural length to a length of 50 cm?
    c) What is the length of the spring if a force of 10 N is applied to it when its length is 20 cm?
    d) What is the work expended in stretching the spring from 35 cm to 45 cm in length?

11 A force of 100 N stretches a garage door spring from its natural length of 1 m to 1.2 m.
    a) What is the spring constant?
    b) How much work is done in stretching the spring from its natural position to the length of 1.5m?
    c) What would be the length of the spring if 5J of work is expended in stretching the spring from its natural length?

12 A force of 1000 N compresses an automobile spring from its natural length of 0.3 m to 0.2 m.
    a) What is the spring constant for this spring?
    b) How much work is done in further compressing the spring from 0.2 to 0.1 m?
    c) What would be the length of the spring if 150J of work is expended in compressing it from its natural length?

Hydrology

13 Following a storm the flow rate over the spillway of a dam is given by $V'(t) = e^{-Kt}$, where the constant K is related to the size of the spillway and the volume of water introduced to the reservoir by the storm. Express the outflow volume as an integral and integrate to give V(t). What interpretation could you give to the constant of integration?

14 Let the flow rates of a fluid into and out of a chamber be $Q_{in}(t) = t^2 + 3$ and $Q_{out}(t) = 4t + 3$.
    a) Sketch the graph of $Q_{in}(t)$ and $Q_{out}(t)$ over [0,6].
    b) Compute the volume, V(t) in the chamber at time t assuming V(0) = 10.
    c) What is the minimum volume of fluid in the chamber and when does it occur?

494   Chapter 7 Applications Involving Integration

15. The initial infiltration capacity of a watershed is estimated as 4.5 in./hr and its equilibrium capacity is 0.4in./hr. Its time constant is $\alpha = 0.35$ hr^{-1}. a) Use the Horton Equation of Example 6 to determine the values of f at time t = 10min, 30 min, 1 hr, 3 hr, and 5 hr and plot these values on an infiltration graph of f verses time.
b) Integrate the Horton equation to give the volume of infiltration over the 5 hr period.

16. Calculate the infiltration volume of a watershed over a six hour interval, [0,6], using the Horton infiltration rate f when the initial infiltration capacity is 2.1 in./hr. and the time constant $\alpha = 0.05$ hr^{-1} if the equilibrium final capacity of the watershed is 0.2 in/hr. See Example 6

17. The Philip infiltration capacity (rate) equation has the form $f = 0.5At^{-1/2} + B$ (in./hr.) where the constants A and B relate to soil type and water movement factors. What is the cumulative infiltration using this function.

18. Hydrologists use the term "Instantaneous Unit Hydrograph" (IUH) to denote the function u(t) that gives the rate of runoff for a particular watershed in response to a "unit impulse" of rain. A "unit impulse" is one unit of rain that is assumed to fall all at once. The runoff resulting from rain over an interval [0,T] is given by a special type of integral that involves the rainfall function I(t) and a shifted IUH function u(T - t). Using an integration variable s instead of t, this is the "convolution integral"

$$Q(T) = \int_0^T I(s) \cdot u(T - s)\, ds$$

a) Assume that $u(t) = 2 - 2t$ for $0 \leq t \leq 1$ and $u(t) = 0$ if $t > 1$. Compute the runoff Q(T) associated with a constant drizzle, I(t) = 0.1, at times i) T = 0.5 ii) T = 1.5 iii) T = 5.

b) Repeat part a) for a rain fall that is given by I(t) = 3/(t + 1).

c) Repeat parts a) and b) when the IUH function is $u(t) = (3/4)(t - 1)^2 + 1/4$. Hint: to evaluate the integral in part b) use the fact that $(T - s - 1)^2 = ((s + 1) - T)^2 = (s + 1)^2 - 2T(s + 1) + T^2$.

FLUID PRESSURE EXERCISES (See Example 7.)

19. Calculate the total fluid-force on the four sides of a full drink container that has a rectangular base of dimensions 4cm by 6 cm and height of 10 cm, assuming the density of the juice is 1.2 g/cm^3.

20. What is the fluid-pressure on the end of a trough full of water (density 1g/cm^3) that has a triangular shaped cross-section of width 10 cm at the top and has a height of 20cm? First, sketch the end and indicate the height variable h and the width length L(h).

21. A circular light cover of radius 20cm is on the side wall of a swimming pool being centered 2 m below the water surface. Calculate the force acting on the cover. Carefully sketch the side of the pool, and the light cover to see that the pressure integral is for h in the interval [1.8, 2.2].

22. A cylindrical storage tank that is 10 m high with radius 2m is filled with corn syrup that has a density of 2.8 g/cm^2. a) What is the vertical pressure on the bottom of the tank? b) What is the horizontal pressure on the side of the tank?

23. Compute the fluid force on a cubic box with edge length 10 cm if it is submerged to a depth of 3 m (at its center) in salt water of density 1.1 g/cm^2.

24. If a 300m high dam is u-shaped, having its lower edge given by the curve $y = 0.3x^2$ for $-1 \leq x < 1$, x and y in units of km, what is the total force acting on the dam when the reservoir it forms is a) full of water? b) The water level is at the 200m height?

Section 7.6 Physiology and Biology

Example 1   **Metabolic Rates and Basal Metabolism.**

Background

An animal or person's metabolic rate reflects internal biochemical activity. The **basal metabolic rate**, BMR, is the base rate at which energy is consumed when the individual is not subject to physical stress. Measured over several days, a **metabolic rate**, MR(t), is often periodic, corresponding to cycles of rest and activity. The MR is measured in various ways, as heat produced (kcal/hr), as oxygen consumed per unit body weight per minute ($cm^3 \, O_2 \, / \, g \cdot min$), or as carbon dioxide expelled per unit time ($cm^3 \, CO_2/hr$). An individual's total **Basal Metabolism**, BM, over a period of time $[t_1, t_2]$ is the integral of their Metabolic Rate over this time period:

$$BM = \int_{t_1}^{t_2} MR(t) \, dt$$

For a North American or European adult human the average Basal Metabolism is about 2000 kcal/day, this represents the daily energy required from food sources.

Problem   Calculate the daily Basal Metabolism of a mouse if its Metabolic Rate is

$$MR(t) = 0.3 - 0.15\cos(\pi t/12) \text{ kcal/hr.}$$

Solution   The mouse's Basal Metabolism is the integral of its MR over the interval [0,24]:

$$BM = \int_0^{24} 0.3 - 0.15\cos(\pi t/12) \, dt$$

$$= 0.3t - (1.8/\pi)\sin(\pi t/12) \Big|_0^{24}$$

$$= 7.2 - (1.8/\pi)\{\sin(2\pi) - \sin(0)\} = 7.2 \text{ kcal/day.}$$

Example 2   **Cardiac Output.**

Background

Cardiac Output, CO, is the average volume of blood pumped by the heart per minute. One classical method of measuring CO is called the *Dye dilution method*. It consists of injecting and monitoring a marker that can be detected in the blood. Early experiments utilized ink dyes, hence the name; now tritium labeled indicators are often utilized as the "dye". A dye injected into a vein on the right side of the heart then circulates with the blood through the heart, to the lungs, back to the heart, and into the arterial circulation system. The concentration of the dye is measured at an artery giving a function c(t) from which the CO is estimated. Eventually the dye recirculates and the measurement becomes complicated to interpret, consequently only the data over the interval [0,T] is utilized where T is the recirculation time. If D is the amount injected then all of this dye must pass the observation point during the interval [0,T]. The amount of dye passing the observation point at time t is the product of the dye concentration and the blood flow rate, $a(t) = c(t) \cdot f(t)$. The total amount of dye is then

$$D = \int_0^T c(t) \cdot f(t) \, dt$$

The dye dilution method utilizes an approximation that results from replacing f(t) by its average value, the cardiac output CO, which can then be factored out of the integral (This is similar to the generalized Integral Mean Value theorem.):

$$D \approx \int_0^T c(t) \cdot CO \, dt = CO \int_0^T c(t) \, dt$$

Hence, as T is usually given in seconds, the formula found in physiology text is

$$CO = 60[D / \int_0^T c(t)\, dt]$$ **Cardiac Output**

where the factor 60 converts the T units of seconds to the minute unit of Cardiac Output.

Problem Determine the Cardiac Output for an individual if Clinical measurements following a 5mg injection of dye at time $t = 0$ indicate dye only after 3 s and the measurement for the interval [3,18] were fit to a polynomial. There was no observed dye for the interval [18,20] and then dye reappeared. The resulting equation for c(t) was

$$c(t) = \begin{cases} 0 & \text{if } 0 < t < 3; \\ 10^{-2}(t^3 - 40t^2 + 453t - 1026)\,\text{mg/l} & \text{if } 3 < t < 18; \\ 0 & \text{if } 18 < t < 20. \end{cases}$$

Solution Using the value of $T = 18$, the person's Cardiac Output is given by

$$CO = 60 \cdot 5 / \int_3^{18} 10^{-2}(t^3 - 40t^2 + 453t - 1026)\, dt$$

$$= 30{,}000 / \int_3^{18} t^3 - 40t^2 + 453t - 1026\, dt$$

The integral is evaluated as

$$\int_3^{18} (t^3 - 40t^2 + 453t - 1026)\, dt = t^4/4 - (40/3)t^3 + (453/2)t^2 - 1026t \,\big|_3^{18}$$

The evaluation of this polynomial at the limits of integration can be more easily performed on a calculator if the polynomial is factored as:

$$= (((t/4 - 40/3)t + 453/2)t - 1026)t \,\big|_3^{18} = 3402 - (-1379.25) = 4781.25$$

The Cardiac Output is therefore

$$CO = 30{,}000 / 4781.25 = 6.275\ \text{l/min}$$

which is about normal for an adult human.

Example 3 Oxygen debt due to exercise.

Background

Exercise, in general using muscles, requires the release of energy, principally derived from ATP, creatine phosphate, and glycogen stored in the muscle cells. This energy consumption is another form of the Metabolic Rate discussed in Example 1, here we shall call it the **rate of energy consumption**, E'. The stored energy is regained in the cells by an "aerobic" process that utilizes oxygen to metabolize energy sources in the blood. The **rate of oxidative metabolism**, OM', is limited by the lung's capacity to deliver oxygen and consequently during strenuous exercise "lags" the rate of energy consumption. The **oxygen debt**, OD, resulting from exercise over a time interval [a,b] is the integral of the difference in the rate oxygen is used and the rate oxygen is replenished in the muscle tissue:

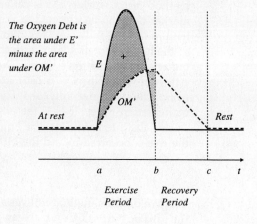

The Oxygen Debt is the area under E' minus the area under OM'

$$OD_{a,b} = \int_a^b E'(t) - OM'(t)\, dt$$

This is the signed area between the OM' and E' curves in the following figure. If the exercise terminates, eventually, at some time t = c, all of the oxygen is restored. The **recovery time**, c is thus the time at which

$$OD_{a,c} = \int_a^c E'(t) - OM'(t)\, dt = 0$$

Splitting the oxygen debt integral over [a,c] into two integrals over [a,b] and [b,c], we see that this equation is equivalent to the identity $OD_{a,b} = -OD_{b,c}$:

$$\int_a^b E'(t) - OM'(t)\, dt = -\int_b^c E'(t) - OM'(t)\, dt.$$

The **recovery period** is the difference c - b between the end of exercise at time t = b, and the return to the base oxygen levels at time t = c.

Problem Assume a person rests for 2 minutes and then vigorously exercises for two minutes and has an energy consumption rate

$$E'(t) = 5 - 4(t-3)^2 \text{ if } 2 < t < 4 \quad \text{and base level} \quad E'(t) = 1, \text{ if } t \leq 2 \text{ or } t \geq 4$$

Assume that the oxidative rate normally matches the energy rate E', is a quadratic function during exercise, and then decreases linearly to the recovery time c, at which the oxygen debt is recouped:

$$OM'(t) = \begin{cases} 1 & \text{if } t \leq 2 \text{ or } t \geq c; \\ 3 - (t-4)^2/2, & \text{if } 2 < t < 4; \\ \text{a linear function} & \text{if } 4 \leq t \leq c. \end{cases}$$

What is the recovery period following this exercise?

Solution We must first determine the recovery time c. The linear portion of OM' passes through the point (4,3) and the unknown point (c,1). It is described by the point-slope equation

$$y = 3 + [(3-1)/(4-c)](t-4).$$

The value of c is found by integrating and solving for c in the equation

$$\int_2^4 E'(t) - OM'(t)\, dt = \int_4^c E'(t) - OM'(t)\, dt.$$

Substituting the formulas for the functions OM' and E' on the intervals [2,4]

$$OD_{2,4} = \int_2^4 [5 - 4(t-3)^2] \cdot [3 - (t-4)^2/2]\, dt$$

$$= \int_2^4 [5 - 4(t-3)^2]\, [3 - (t-4)^2/2]\, dt$$

$$= \int_2^4 -26 + 20t - (7/2)t^2\, dt = -26t + 10t^2 - (7/6)t^3 \big|_2^4 = 8/3$$

On the interval [4,c], the formulas give

$$OD_{4,c} = \int_4^c 1 - [3 + [2/(4-c)](t-4)]\, dt$$

$$= -2t - (t-4)^2/(4-c) \big|_4^c = 8 - 2c + (c-4) = 4 - c$$

Setting $OD_{2,4} = -OD_{4,c}$, we solve for c = 20/3. The recovery period is c - 4 = 8/3 min, or 2 min. 40 sec.

There are various types of Mathematical Models used to describe biological processes and populations. Some focus on detailed aspects of a biological process, such as the physiological models in the previous Examples, while others consider more general and aggregate descriptions of "populations", for instance describing the magnitude or size of a cell culture or a population of animals. Models that describe how a population or organism changes over time in mass or quantity are referred to as *growth* and *decay* models. These models are actually used in virtually every discipline in which the change in quantities is studied.

Example 4 Growth and Decay Models.

Background

In population biology **growth and decay models** are dynamical models that specify the rate of change of the population with respect to time. When the change occurs in discrete steps, such as annual hatching of chicks, the model is given by "difference equations". When the change is continuous, or can be viewed as continuous on average, such models are given by "differential equations". In either situation, the rate of change in the populations will depend on biological reproductive and mortality characteristics of the population, intrinsic model parameters reflecting environment factors and external influences, and on the current state or size of the population. The basic continuous growth model has the form

$$X'(t) = R - D \qquad \textbf{GROWTH/DECAY MODEL}$$

where $X(t)$ is the size of the population at time t, R is the reproductive rate and D is the death rate for the population The derivative $X'(t)$ is the net instantaneous rate of change in the population size. The model describes the population's *growth* if X' is positive and its *decay* if X' is negative. If R and D are known functions of time only then the state of the population is given by the integral-function

$$X(t) = X(0) + \int_0^t R(s) - D(s) \, ds$$

However, for most models the reproduction and death rates also depend on the current size of the population. If the time dependent factors can be separated and factored out of the reproduction and death rates so that

$$R = R(X) \cdot g(t) \qquad \text{and} \qquad D = D(X) \cdot g(t)$$

where $g(t)$ is an environmental factor dependent only on time (such as seasonal temperature) then the Growth/Decay Model can be solved by the method of "separation of variables". We set $X' = dX/dt$ and separating the variables X and t in the differential equation:

$$dX/dt = R - D = R(X) \cdot g(t) - D(X) \cdot g(t) = [R(X) - D(X)] \, g(t)$$

$$dX \,/\, [R(X) - D(X)] = g(t) \, dt$$

This equation is then "integrated", in a special way since the left side is integrated with respect to X and the right side with respect to t. Formally, this gives the integral equation:

$$\int [R(X) - D(X)]^{-1} \, dX = \int g(t) dt$$

The integral on the right is simply taken over the interval [0,T]. The left integral is evaluated from $X = X_0$, the initial population size at time $t = 0$, to $X(T)$, the population size at time $t = T$.

Problem

The size of a liver is modeled by a *Logistic* growth rate modulated by a periodic factor due to the daily administration of a drug. Assume $R(X) = 4X$, $D(X) = X^2$, and $g(t) = 0.5 \cos(2\pi t)$ and the liver's initial size is $X_0 = 1$ kg at $t = 0$. Determine the size of the liver as a function of time.

Solution The dynamic model is the equation

$$X' = [4X - X^2] \cdot 0.5 \cos(2\pi t); \quad X(0) = 1$$

This equation is of the "separable" form and is converted to the integral form

$$\int 1/(4X - X^2) \, dX = \int 0.5 \cos(2\pi t) \, dt$$

The integral on the right is

$$\int 0.5 \cos(2\pi t) \, dt = (0.25/\pi) \sin(2\pi t) + C$$

The integration constant C will be determined from the initial condition. The integral on the left is evaluated by the method of partial fractions (see Section 6.3). To express the integrand as a sum of fractions, set

$$1/(4X - X^2) = 1/[X(4 - X)] = A/X + B/(4 - X) = \{A(4 - X) + BX\} / \{X(4 - X)\}$$

The numerators are equated giving the system of two equations for A and B.

$$1 = [-A + B]X + 4A \implies \begin{array}{l} -A + B = 0 \\ 4A = 1 \end{array} \quad \begin{array}{l} \text{equating the coefficients of X} \\ \text{equating the constant terms.} \end{array}$$

The solution is $A = 1/4$ and $B = 1/4$. Consequently, assuming $X < 4$,

$$\int 1/(4X - X^2) \, dX = \int 1/4X \, dX + \int (1/4)/(4 - X) \, dX = (1/4)[ln(X) - ln(4 - X)] = 1/4 \, ln(X/(4 - X))$$

The population size is found by equating the two antiderivatives and solving for X:

$$(1/4)ln(X/(4 - X)) = (1/4) \sin(2\pi t)/\pi + C$$

$$X/(4 - X) = e^{\sin(2\pi t)/\pi + 4C}$$

This equation has the form $X/(4 - X) = B$, where B is the exponential term. Solving for X gives

$$X = 4B/(B + 1) = 4/[1 + 1/B]$$

The term 1/B is the exponential with a negative exponent, thus the liver size is

$$X(t) = 4/[1 + e^{-4C - \sin(2\pi t)/\pi}]$$

This can be simplified by setting $k = e^{-4C}$, then the solution has the form similar to the Logistic Equation (see Example 2 of Section 6.3):

$$X(t) = 4/[1 + k \, e^{-\sin(2\pi t)/\pi}]$$

The constant k is determined from the initial condition $X(0) = 1$ kg by setting $t = 0$:

$$1 = 4/[1 + k] \implies k = 3$$

The particular solution is thus

$$X(t) = 4/[1 + 3e^{-\sin(2\pi t)/\pi}]$$

Example 5 Ruminant Digestion; the Passage of food through a sheep's stomach.

Background Sheep, as ruminants, have a complicated stomach with a storage compartment called a "rumen". Ruminants can ingest low nutrient content food, which repeatedly is chewed, passed to the rumen for digestion and re-chewed as "cud". Quantitative models of this process give the rate, r(t), of passage of food through the digestive tract. Animal scientists establish such models by measuring the amount of "recovered" feed that passes through the animal at time t after eating as a percentage of the total amount of the feed. Since it is difficult to actually measure what an animal eats, the total feed is measured as the total amount of material that eventually passes through the digestive tract. The amount that passes through over an interval [0,t] is the integral of the digestive rate r(t) over the interval [0,t]. Thus, the **Percent Recover** of food at time t is given by

$$PR(t) = \{ \int_0^t r(s) \, ds \,/\, \int_0^\infty r(s) \, ds \} \cdot 100\%$$

Problem A two compartment model of digesta passage for sheep has a lag of T hours, before which no feed is digested, and is given by

$$r(t) = 0 \text{ for } 0 < t < T$$

$$r(t) = e^{-\alpha(t - T)} - e^{-\beta(t - T)} \text{ for } t \geq T$$

The constants α and β are related to biological rates of passage from the rumen and stomach and $\alpha < \beta$. Determine the percentage of material that has been recovered two days after ingestion if

$$T = 20 \quad \alpha = 0.0001 \quad \text{and} \quad \beta = 0.05$$

Solution Since r(t) = 0 on [0,20], $\int_0^{20} r(t) \, dt = 0$. Thus the integrals in the PR formula are evaluated from t = 20 rather than from t = 0. The upper limit for the numerator integral is 48 since 2 days corresponds to t = 48 hr:

$$\int_{20}^{48} e^{-0.0001(t - 20)} - e^{-0.05(t - 20)} \, dt$$

$$= -10^4 e^{-0.0001(t - 20)} + 20 e^{-0.05(t - 20)} \Big|_{20}^{48}$$

$$= -10^4(e^{-0.0028} - 1) + 20(e^{-1.4} - 1) \approx 27.96 - 15.07 = 12.89$$

The integral in the denominator is an improper integral, as discussed in Section 5.4, since it is over an infinite interval. It is evaluated as a limit:

$$\int_{20}^\infty e^{-0.0001(t - 20)} - e^{-0.05(t - 20)} \, dt = \lim_{s \to \infty} \int_{20}^s e^{-0.0001(t - 20)} - e^{-0.05(t - 20)} \, dt$$

$$\lim_{s \to \infty} -10^4(e^{-0.0001(s - 20)} - 1) + 20(e^{-0.05(s - 20)} - 1) = 9{,}980$$

as each exponential term approaches zero. The percentage of passage at the end of day two is quite small, less than 1%,

$$PR(48) = 100 \cdot 12.89 \,/\, 9{,}980 \approx 0.13\%$$

EXERCISE SET 7.5

Medical applications

1. Compute the basal metabolism over the interval [2,4] of an animal whose $MR(t) = -0.5\, t\, \cos(\pi t/12) + 10$ kcal/hr.

2. Creatinine is released into the blood during exercise. In an experiment a dog walked on a treadmill and its circulating creatinine varied with the speed of the treadmill. It was determined as the integral of $y = 4s^{1/5}$ over the speed interval $[0, M(w)]$ where $M(w)$ is the maximum speed for a dog of weight w. If $M(w) = 5(1 + \sin((\pi/10)(w - 10)))$ what is the equation for circulating creatinine as a function of weight?

3. Following the administration of a drug a patient's systolic blood pressure changes at a rate $r(t) = 0.5\cos(t/4)\sin(t/4)$ mm Hg/hr. What would be the patient's blood pressure in two hours if it was 96 mm Hg when the drug was administered? When will it first return to its pre-drug level?

4. As in Example 3, compute the recovery time C following exercise corresponding to the following functions: The energy consumption function is
   $E'(t) = \{5 - (t - 4)^2$ if $2 < t < 6$, and $E'(t) = 1$ otherwise,
   and the oxidative metabolic rate is
   $$OM'(t) = \begin{cases} 1 & \text{if } t < 2 \text{ or } t \geq C; \\ 1 + (t - 2)^3/64 & \text{if } 2 < t < 6; \\ \text{linear} & \text{if } 6 < t < C. \end{cases}$$

5. a) A drug is eliminated via urine excretion t hr after administration at the rate $y' = -3t^{1/2}$ mg/hr. Assume that 10 g of drug was administered at time $t = 0$. What will be the amount of drug in the patient at time $t > 0$? At what time will the residual amount of drug in the person be 10^{-3} g? When will the patient be drug free? What if "drug free" means below a detectable amount of 10^{-9}g?
   b) Repeat part a) with the rate function $y' = -3t^{-1/2} \cdot y$.

6. The rate of elimination of a drug can be used to determine the total amount of a drug available to a body. Another method is to measure the assimilation rate, the rate at which the drug enters the blood system. This type of curve has been used in television commercials to "compare" different aspirin-type drugs. A typical rate of assimilation function has the form
   $$f(t) = kt(t - b)^2 \quad 0 < t < b$$
   The total amount of drug assimilated is given by
   $$A = \int_a^b f(t)\, dt$$

Drug	k	b
d_1	0.15	3
d_2	0.01	6
d_3	0.001	9

   Assume that equal amounts of three drugs, d_1, d_2, and d_3, are administered and found to be assimilated with the parameter values given in the table.
   a) Sketch on the same graph the assimilation rate functions, f_1, f_2, and f_3 corresponding to the values of k and b for the three drugs.
   b) Which drug is assimilated at a faster initial rate and which drug is assimilated over the longest period of time?
   c) Which drug provides the greatest amount A and which drug provides the least?

Biology.

7. Solve the basic Growth/Decay Model (Example 4) for the indicated reproductive and death rates if $X(0) = 2$.
   a) $R(t) = 3t$ $D(t) = t^{1/2}$
   b) $R(t) = 3X$ $D(t) = 1$
   c) $R(t) = 3X$ $D(t) = X^{1/2}$
   d) $R(t) = 3X$ $D(t) = X^2$
   e) $R(t) = 3$ $D(t) = X^2$
   f) $R(t) = X\sin(\pi t/24)$ $D(t) = \sin(\pi t/24)$

8. A model of limited growth is given by $X' = R(X) = -3(X + 1)ln((X + 1)/8)$.
   a) Sketch the graph of $y = R(X)$ for $0 \leq x \leq 15$. When is the growth rate positive? What is the "carrying capacity" of this model, that is, what is the size M such that $x' > 0$ when $X < M$ and $x' < 0$ if $X > M$?
   b) Write the integral form of this model, integrate and solve for X as a function of t. Hint: use the substitution $u = X + 1$ to simplify the X integral.

9. An animal gives off radiant "heat" above its normal body temperature depending on its activities. Assume that over a six-hour period the radiated heat given off by the animal at time t is

$$-H'(t) = h(t) = 64 - (t - 2)^2(t - 4)^2, \text{ kcal/hr} \quad t \in [0,6]$$

   a) Sketch a graph of the function, $h(t)$. Indicate the local maximum and minimum values.
   b) Estimate the total heat loss over the time intervals [0,3] and [2,4].
   c) Compute the actual total heat loss over the intervals [0,3] and [2,4] by expanding $h(t)$ and integrating over the appropriate interval.

10. A population of size $N(t)$ has an instantaneous rate of change $R(t) = -t(t - 100)$. If $N(0) = 50$, what is the equation for $N(t)$? If $N(0) = 50$, what is $N(50)$?

11. A population $N(t)$ has an instantaneous growth rate $R = 3N(t)$. a) What is the equation for $N(t)$? b) If the population size is $3.2 * 10^4$ at $t_0 = 0$, what will be its size at $t_1 = 10$? c) If the population size is $5 * 10^4$ at $t_0 = 2$, what will be its size at $t_1 = 10$?

12. Assume a population, $N(t)$, grows at a rate $N'(t) = 3t \cdot N(t)$. a) Write an "integral" equation for this model. Evaluate the integrals and solve for $N(t)$. b) If $N(0) = 3.2 \, 10^4$ what is $N(10)$? c) If $N(2) = 0.5 \, 10^6$ what is $N(10)$?

13. Assume that the growth rate of a plant is $g'(x) = 15t(365 - t)^2 \, 10^{-5}$ cm/day, at time t in days. The maximum period of growth is [0,T], where T is a function of rainfall, x inches/yr, given by $T = 0.365x(200 - x)$ days.
   a) Express the total growth $G(x)$ of the plant as an integral over time with variable limits of integration.
   b) Evaluate the integral form $G(x)$ and determine $G(30)$ if $G(5) = 2$.
   c) What is the instantaneous rate of change in the total growth with respect to variations in the rainfall, x?

14. Solve for $X(t)$ the equation equating the antiderivatives in the Growth/Decay Example 4 when the initial liver size is $X_0 = 5$ kg. (The absolute value term is now different.)

Section 7.7 Toxicology and Environment Models

Example 1 **Uptake of pollutants by freshwater fish.**

Background

The amount of a pollutant that an animal or organism internalizes, called pollution uptake, depends on the concentration of pollutant in the animals environment, the physical mechanisms by which the pollutant enters the body, and the internal physiology that leads to retention and elimination. Fish uptake pollutants via two principal pathways: ingestion, eating polluted feed, and through respiration, breathing by filtering oxygen and other substances through their gills. Two input or **uptake models** for these pathways are:

Uptake by ingestion: $I_E = C_f(0.25 m^{0.8} + 2\, dm/dt) f_1$ g/day

Uptake by respiration, $I_B = (10^3\, m^{0.8}) \cdot C_w \cdot f_2$ g/day

where the model components and parameters are:

m = mass of the fish, in units g
dm/dt = growth rate of the fish, in units g/day
f_1 = fraction of ingested pollutant that is absorbed,
f_2 = fraction of pollutant entering the gills that is absorbed,
C_f = the mass concentration of pollutant in food, g/g
C_w = concentration of pollutant in water, g/cm^3 or g/g
t = the time in days
$I_E(t)$ = rate of uptake by ingestion at time t, g/day
$I_B(t)$ = rate of uptake by breathing at time t, g/day

The term $m^{0.8}$ reflects the metabolic rate of the fish, while the numerical factors 0.25 and 10^3 reflect the consumption rate of food and oxygen per unit body mass. For mature fish the growth rate dm/dt can be set to zero while for fingerlings this term is significant.

Problem Calculate the total amount of methylmercury uptake by a mature 1.2kg trout over a 25 day period in which the mercury concentration in its food increased linearly from 0 to $5 \cdot 10^{-6}$ g/g at day 25 and the water mercury concentration is given by the double exponential decay function

$$C_w = 0.35 \cdot 10^{-6}(e^{-0.002t} - e^{-0.05t})\ \text{g/g}$$

Assume that the fractional uptake parameters are $f_1 = 0.4$ and $f_2 = 0.07$.

Solution The total amount of mercury entering the fish is the integral

$$A = \int_0^{25} I_E(t) + I_B(t)\, dt$$

The formulas for the uptake rates are as indicated above, and the food concentration of mercury is given by the linear function

$$C_f = 5 \cdot 10^{-6} (t/25) = 2 \cdot 10^{-7} t$$

Assuming $dm/dt = 0$, since the trout is "mature", gives

$$I_E(t) = 2 \cdot 10^{-7} t\, 0.25 \cdot 1200^{0.8} \cdot 0.4 \approx 5.8 \cdot 10^{-6} t$$

$$I_B = (10^3 \cdot 1200^{0.8}) \cdot 0.35 \cdot 10^{-6}(e^{-0.004t} - e^{-0.05t}) \cdot 0.07$$

$$= 7.2 \cdot 10^{-3}(e^{-0.004t} - e^{-0.05t})$$

The total uptake is thus

$$A = \int_0^{25} 5.8 \cdot 10^{-6} t + 7.2 \cdot 10^{-3}(e^{-0.004t} - e^{-0.05t}) \, dt$$

$$= 2.9 \cdot 10^{-6} t^2 + 7.2 \cdot 10^{-3}(-250 e^{-0.004t} + 20 e^{-0.05t}) \big|_0^{25}$$

$$= 2.9 \cdot 10^{-6} \cdot 625 + 7.2 \cdot 10^{-3}(-250(e^{-0.1} - 1) + 20(e^{-1.25} - 1)) = 70.3 \cdot 10^{-3} \text{g}$$

Observe that 97.5% of the uptake is due to the I_B component, i.e., through the gills, even though the fraction of pollutant retained was only 0.07.

Example 2 Lake contamination and purification.

Background

In a well mixed lake the presence of a contaminate is almost uniform. If the lake remains constant in size the daily water inflow to the lake must match the outflow. If a lake of volume V m^3 has an outflow rate of Q m^3/d and the concentration of a contaminate is indicated by x(t) g/m^3, then the rate of change in the total amount of pollutant in the lake is given by the equation

$$D_t\{x(t) \cdot V\} = I(t) - Q \cdot x(t)$$

where I(t) (g/m^3/d) is the rate new contaminate enters the lake and $-Q \cdot x(t)$ is the amount of pollutant that leaves the lake with the outflow.

Problem

If the volume of lake Erie is $V = 458 \cdot 10^9$ m^3 and its outflow rate is about $Q = 4.79 \cdot 10^8$ m^3/d, how long would it take to reduce the level of a pollutant in the lake to 5% of its current level assuming there is no additional input of the pollutant?

Solution

The volume of the lake is constant and the input $I(t) = 0$ by assumption. The pollutant level in the lake is found by solving the dynamical model equation

$$D_t\{x(t) \cdot V\} = V \cdot dx/dt = -Qx$$

Separating the variables, this is equivalent to the integral equation

$$\int 1/x \, dx = \int -Q/V \, dt$$

The solution is $ln(x) = -(Q/V)t + C$, where C is the integration constant. Solving for x gives the exponential decay equation, with $x(0) = e^C$:

$$x(t) = x(0) e^{-(Q/V)t}$$

The problem is to find $t = t_{0.05}$ such that the pollutant in the lake at time $t_{0.05}$ is 5% of the amount in the lake at time $t = 0$. Solving

$$x(t)/x(0) = e^{-(Q/V)t} = 0.05$$

for $t = t_{0.5}$ gives

$$t_{0.05} = -(V/Q)ln(0.05) \approx (458 \cdot 10^9/4.79 \cdot 10^8)(2.995) \approx 2864 \text{ days}$$

Thus it would take about 7.8 years for the pollution level to decrease by 95%.

Example 3 The Dose Commitment resulting from toxic exposure.

Background The "dose" of a toxin that an individual or organ incurs at time t due to toxic exposure over an earlier time interval [0,T] is determined by two functions. One is the **uptake** or "input" function I(t) and the other is the **retention** function, R(t).

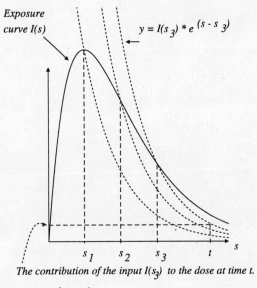

The contribution of the input I(s_3) to the dose at time t.

$I(s_3) * e^{(t-s_3)}$

The retention function indicates the fraction of toxin that remains in the body or organ at a time t after it first enters the body. If an amount I(s) of toxin enters the body at time s, then at a later time t this toxin has been in the body t - s units of time. At time t the fraction of this toxin remaining in the body is thus R(t - s). Thus, the amount of toxin entering at time s and remaining at time t is

$$d(s,t) = I(s) \cdot R(t-s)$$

This is illustrated for an exponential decay retention function in 28. Three decay curves are sketched, corresponding to input times s_1, s_2, and s_3. These are each the same basic decay curve $R(s) = e^{-ks}$ shifted horizontally s_i units to the right and scaled by the input at s_i, the term $I(s_i)$

$$y(s) = I(s_i)\exp(-k(s - s_i))$$

At time t, the contribution to the individual's internal dose from an input at time s_i is then the y-coordinate on this decay curve when s = t,

$$d(s_i,t) = y(t) = I(s_i)\exp(-k(t - s_i))$$

Replacing s_i with an un-subscripted variable s gives the function d(s,t) for the exponential decay retention function: $d(s,t) = I(s)e^{(-k(t-s))}$.

The total internal dose at time t, D(t), is the residual amount of toxin in the body from all exposures prior to time t. This is the "sum" of the terms d(s,t) for prior times s from s = 0 to s = t. A "sum" corresponding to all values over an interval is given a definite integral, the integral of d(s,t) with respect to s over the s-interval [0,t]. This type of integral has a special form that occurs in many mathematical problems, it is called a **convolution integral**:

$$D(t) = \int_0^t I(s) \cdot R(t-s) \, ds \qquad \text{DOSE AT TIME t}$$

The total **dose commitment**, DC, indicates the amount of internal Dose that an individual experiences from exposure to and retention of a toxin from a specified source over all future time. While no one lives for ever, in theoretical models this future interval is simply taken as the infinite interval [0,∞).

$$DC = \int_0^\infty D(t) \, dt. \qquad \text{DOSE COMMITMENT}$$

Combining the two integral formulas gives an equation involving the sequential evaluation of two integrals. Such integrals are called "iterated integrals" and these are studied in Chapter 9.

$$DC = \int_0^\infty \{\int_0^t I(s)R(t-s)\,ds\}\,dt \qquad \text{DOSE COMMITMENT}$$

In toxicological studies one is concerned with the "risk" of some damage resulting from a toxic exposure. The **risk** is the product of the total Dose Commitment and the "potency" of the toxin, typically denoted by β. The "potency" relates the physiological damage the toxin can cause.

$$\text{Toxic Risk} = \beta \cdot DC$$

Problem Determine the Dose, and the Dose Commitment resulting from the following situations:

(a) **Chronic exposure**: $I(t) = A$ for all $t > 0$, and
first order elimination: given by the retention function $R(t) = e^{-kt}$.

(b) **Constant exposure over a finite interval**: $I(t) = A$ if $0 < t < T$ and $I(t) = 0$ if $t > T$, with first order elimination: $R(t) = e^{-kt}$.

Solutions (a) The convolution integral corresponding to Chronic exposure is

$$D(t) = \int_0^t A e^{-k(t-s)}\,ds = (A/k)e^{-k(t-s)}\Big|_{s=0}^{s=t} = A/k - (A/k)e^{-kt} = (1 - e^{-kt})A/k$$

The Dose commitment is then the improper integral

$$DC = \int_0^\infty A/k - (A/k)e^{-kt}\,dt = \lim_{r \to \infty} \int_0^r A/k - (A/k)e^{-kt}\,dt$$

$$= \lim_{r \to \infty} Ar/k + (A/k^2)e^{-kr} - A/k = \infty$$

The exponential term goes to zero and the term Ar/k goes to infinity. The input functions, retention functions, and consequent Dose functions are illustrated in the followint three graphs.

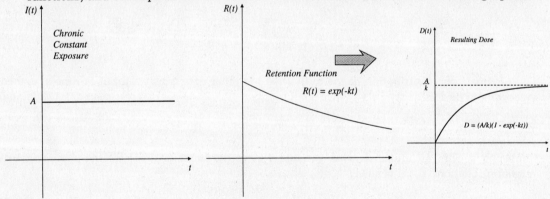

(b) For a constant exposure over a finite interval the input, retention and Dose functions are illustrated in the net three graphs. These conditions are identical to the chronic exposure conditions for $t < T$.

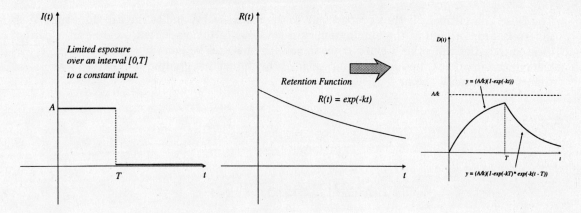

After t exceeds the period of exposure, the integrals must be evaluated by splitting them at T. Using the integration of part (a) the Dose is given by

$$D(t) = A/k - (A/k)e^{-kt} \text{ for } 0 < t < T$$

For t > T the integral is split at T:

$$D(t) = \int_0^t I(s)R(t-s)\,ds = \int_0^T I(s)R(t-s)\,ds + \int_T^t I(s)R(t-s)\,ds$$

The second integral is always zero since $I(s) = 0$ for $s > T$. Thus

$$D(t) = \int_0^T I(s)R(t-s)\,ds$$

$$= \int_0^T A e^{-k(t-s)}\,ds = (A/k)e^{-k(t-s)}\Big|_{s=0}^{s=T}$$

$$= (A/k)(e^{-k(t-T)} - e^{-kt}) \quad \text{for } t > T$$

The Dose Commitment is also calculated as a split integral

$$DC = \int_0^\infty D(t)\,dt = \int_0^T D(t)\,dt + \int_T^\infty D(t)\,dt$$

The first integral on the right is

$$\int_0^T D(t)\,dt = \int_0^T (1 - e^{-kt})A/k\,dt = \{T + (1/k)(e^{-kT} - 1)\}A/k = AT/k + (A/k^2)(e^{-kT} - 1)$$

The second integral is improper and is evaluated as a limit:

$$\int_T^\infty D(t)\,dt = \lim_{r \to \infty} \int_T^r D(t)\,dt$$

$$= \lim_{r \to \infty} \int_T^r (A/k)(e^{-k(t-T)} - e^{-kt})\,dt$$

$$= \lim_{r \to \infty} (-A/k^2)(e^{-k(t-T)} - e^{-kt})\Big|_T^r$$

$$= \lim_{r \to \infty} (-A/k^2)(e^{-k(r-T)} - 1 - e^{-kr} + e^{-kT})$$

$$= (-A/k^2)(-1 + e^{-kT}) = (A/k^2)(1 - e^{-kT})$$

since $e^{-kr} \to 0$ as $r \to \infty$. The Dose Commitment is the sum of these two integrals

$$DC = AT/k + (A/k^2)(e^{-kT} - 1) + (A/k^2)(1 - e^{-kT}) = AT/k$$

The Dose Commitment is thus proportional to the toxin level A and to the exposure time T. It is inversely proportional to the elimination rate k. When a toxic accident occurs and a patient is treated in a Detoxification Center there is nothing that can be done to change A or T. Detoxification efforts then focus on increasing k by inducing a greater elimination rate than would normally occur.

EXERCISE SET 7.7

1. An individual suffering from toxic poisoning is monitored. The toxin is excreted by kidney filtration at a measured rate of $3(t + 5)^{-1/2}$ mg/hr, t hours after admission to a detoxification center. Assume that the individual ingested 100 mg of toxin 2 hr prior to admission to the center. a) What will be the amount of toxin in the patient 4 hr after admission? b) At what time will the person's toxic residual be 0.1 mg? c) When will the person be free of toxin?

2. An air pollution index gives the average pollutant concentration above a threshold level C_T:
   Index = $(1/24)\int_0^{24} C(t) - C_T \, dt$, where C(t) is the concentration at time t.
   a) What is the Index value corresponding to a particulate matter for which the threshold value is C_T = 5 ppm, and the measured levels over the interval [0,24] are given by $C(t) = 3\sin(\pi t) + 6$ ppm?
   b) What is the Index value corresponding to a pollutant with $C_T = 20$ and $C(t) = 50te^{-0.1t}$?

3. If the volume of a polluted lake is $V = 4 \cdot 10^5 m^3$ and its outflow rate is about $Q = 8 \cdot 10^3 \, m^3/d$,
   a) How long would it take to reduce the level of a pollutant in the lake to 5% of its current level assuming there is no additional pollutant input?
   b) If the pollutant concentration in the lake is 2 ppb and the volume inflow to the lake equals the outflow, but the inflow contains pollution at a level of 3ppb, when will the lake concentration reach 2.5 ppb?
   c) What would be the answer to part b) if the inflow rate equalled the outflow rate Q plus an evaporation rate $E = 5 \cdot 10^2 m^3/d$?

4. In reference to Example 1, assume that the fish pollution model parameters are:
   $I_E = C_f(0.25m^{0.8} + 2 \, dm/dt)f_1$ g/day $I_B = (10^3 \, m^{0.8}) \cdot C_w \cdot f_2$ g/day
   with model components m = 1500g, $f_1 = 0.3$, and $f_2 = 0.01$.
   What is the total pollutant uptake of the fish over a one year (365 d) period if it is subjected to constant pollutant levels of $C_f = 0.0001$ g/g, $C_w = 0.5 \cdot 10^{-7}$ g/g when i) dm/dt = 0 g/day? ii) dm/dt = 0.02 g/day?

5. If the exposure levels in Exercise 4 are $C_f = 0.005$ g/g and $C_w = 0.4 \cdot 10^{-6}(e^{-0.001t} - e^{-0.05t})$ g/g what will be the fishes exposure over a 100 day period?

6. Determine the Dose Commitment of an individual subjected to exposure to a toxin that has first order elimination, i.e, the retention function $R(t) = e^{-kt}$, under the following situations:
   a) "Chronic" exposure to $I(t) = 5 \cdot 10^{-9}$ for all t > 0.
   b) A limited exposure, I(t) = 0.003 for 0 < t < 120 days, and I(t) = 0 if t > 120.
   c) A periodic exposure at work, where I(t) = 0.006 for $0 \le t < 5$, I(t) = 0 for 5 < t < 7 and I(t) is periodic with period T = 7, for all future time.
   d) A seasonally varying exposure level $I(t) = 0.001(1 + \sin((\pi/365)t))$.

7. In many instances a more accurate elimination model is given by a "two compartment" model whose retention function is $R(t) = e^{-0.1t} - e^{-t}$. Repeat Exercise 6 for this retention function.

Chapter 8 *Differential Equations*

You have already seen differential equations, they are simply equations that involve derivatives. To solve a differential equation is to find a function $y = f(x)$ that satisfies the equation, i.e., such that when it and its derivatives (as required) are substituted into the differential equation, the resulting expression will be true for all values of x. Actually, in many models the independent variable will be t instead of x, especially for dynamical system models in which a system is changing with time.

Differential equations form the basis for continuous models, and hence for a major discipline of applied mathematics and modeling. Most universities offer at least four courses studying differential equations. Thus, in this Chapter only a brief introduction to the study of differential equations is presented. However, the material in this Chapter should provide a sufficiently strong foundation in differential equations to allow the reader to become literate enough to read the models and analysis that appear in other discipline journals and in general interest science magazines.

Section 8.1   First-order differential equations. 510

Classifying differential equations
Separable equations.
Linear first-order equations.
Initial Value Problems.

Section 8.2 Stability and Equilibriums of Dynamical Systems 520

Introduction to Dynamical Systems.
Equilibriums.
Graphing Dynamical Systems.
Stability of Equilibriums.

Section 8.3   Higher-order differential equations 525

Introduction: n^{th}-order linear differential equations.
Second-order differential equations.
Systems of differential equations.

Section 8.1 First-Order Differential Equations

Introduction.

The dynamics of continuously varying systems are modeled by equations expressing the rate of change in the system. Such models utilize derivatives and the equations are called **differential equations**. Throughout this text we have introduced such models, and to illustrate the various integration techniques, we have solved the models by converting the differential equations into equations involving integrals that we could integrate directly. This procedure is only possible when the differential equations are *separable*. Most differential equations however, are not separable and this approach does not work. Alternative solution methods are introduced to solve some fundamental types of differential equations. However, in practice many differential equations can not be solved by known methods and their solutions must be approximated by numerical methods. In this Chapter we will only be able to introduce a few methods for solving simple differential equations. Most universities offer several courses devoted to the solution and analysis of more complicated differential equations.

Classifying Differential Equations.

Differential equations are classified first by their *order*, which is simply the highest order derivative in the equation, and then by their form, whether they are *linear* or *nonlinear*, and how the independent variable appears in the equation. The terminology is directly analogous to that introduced in Chapter 2 to describe Difference Equations. We will introduce formal function notation for the generic *standard form* of an nth-order differential equation and then focus on the simplest first-order equations. In a subsequent Section we will only be able to solve 2nd-order and higher-order equations when they are linear and have constant coefficients.

The **standard form** of an nth-order differential equation is

$$y^{(n)}(t) = F(t, y, y', y'', \ldots, y^{(n-1)})$$

The function $y(t)$ is symbolic, can be replaced by any function name. The differentiation is assumed to be with respect to the variable t, which may or may not be explicitly stated and of course may be replaced by another independent variable.

When $n = 1$ the equations are called **first-order** differential equations. Examples are given by

$$y' = 3y \qquad f'(x) = 1 - x \cdot f(x) \qquad u' = u/t \qquad y' = y^2 - t^2$$

When $n = 2$ the equations are called **second-order** differential equations. This is sometimes abbreviated to simply **2nd-order**. These equations may involve the first derivative as well as the functions. Examples are

$$y''(t) = 2y(t) - y'(t) \qquad g''(x) = 4g(x) \qquad y'' = \sin(t)y \qquad y'' = y^2 - 1/y'$$

When $n > 2$ the derivatives are usually written with the parenthetical exponent notation.

$$y^{(4)} = 3y'' - 2y \quad \text{is a } 4^{th}\text{-order differential equation.}$$

The second set of adjectives used to describe differential equations are derived from the form of the function F appearing in the above general definition. If it is linear in the function y and the

Section 8.1 First-Order Differential Equations

derivatives of order less than n, i.e., each term only occurs to the power one and none are multiplied together or otherwise combined into a single term, then the differential equation is said to be **linear**. Otherwise, the equation is said to be **nonlinear**.

Linear equations:	Nonlinear equations:
$y' = 3 - 2y$	$y' = 2y^2$
$y' = y + \sin(t)$	$y' = \sin(y)$
$y' = t^2 y$	$y' = t/y$
$w' = -xw + e^x$	$y'' = y' \cdot y$
$y'' = 3y' - y$	$v'' = v'/v$
$v'' = v'/t + \ln(t)v$	$R''(x) = [1 + R(x)]^{1/2}$

For the remainder of this section we will only discuss first-order differential equations.

Separable Equations.

The equation $y' = F(t,y)$ is **separable** when the function $F(t,y)$ can expressed as a product

$$F(t,y) = g(y) \cdot h(t)$$

In this case the differential equation can be solved by integration. The method, called *separation of variables*, consists of expressing y' as a ratio of differentials and then rearranging the differential equation so that only y-functions are on the left and t-functions on the right. Then, the two sides are integrated as ordinary integrals.

$$y' = f(t,y) \quad \Rrightarrow \quad dy/dt = g(y) \cdot h(t) \quad \Rrightarrow \quad 1/g(y)\, dy = h(t)\, dt$$

These equations are then integrated:

$$\int 1/g(y)\, dy = \int h(t)\, dt$$

When the integrals are evaluated, this results in an equation of the form

$$I(y) = H(t) + c \quad \text{an integration constant}$$

The solution y is then found by solving this last equation for y.

Example 1 Solving a separable differential equation.

Problem Determine the solution of $y' = y^2 + ty^2$.

Solution To see that the equation is separable, the right side is expressed as $y^2 \cdot (1 + t)$. Therefore the equation can be separated as

$$dy/dt = y^2(1 + t) \quad \text{or} \quad 1/y^2\, dy = (1 + t)\, dt$$

Integrating,

$$\int 1/y^2 \, dy = \int 1 + t \, dt$$

or

$$-y^{-1} = t + t^2/2 + C$$

Solving for y gives

$$y = -1/(t + t^2/2 + C) = -(t + t^2/2 + C)^{-1}$$

Often the separation of variables method results in integrals that are very difficult to integrate. Even when we can evaluate the integrals, it is then sometimes impossible to solve the final equation explicitly for the solution y(t). Especially when the original differential equation is nonlinear. The linear case is always easy to solve.

Linear first-order equations.

The standard form of a linear first-order differential equation can be written as[†]

$$y'(t) = a(t) \, y(t) + b(t) \qquad \text{Linear First-Order D.E.}$$

where a(t) and b(t) are functions of t only. The term *homogeneous* is used to indicate when the function b(t) = 0, i.e., does not exist in the equation:

$$y'(t) = a(t) \, y(t) \qquad \text{Homogeneous Linear First-Order D.E.}$$

The homogeneous equation. The homogeneous equation is clearly separable! Its solution is straight forward using separation of variables:

$$dy/dt = a(t) \, y(t) \quad \Longrightarrow \quad 1/y \, dy = a(t) \, dt$$

Integrating, if A(t) is an antiderivative of a(t), gives

$$ln(y) = \int 1/y \, dy = \int a(t) \, dt = A(t) + c$$

Solving for y,

$$y(t) = e^{\int a(t) \, dt} = e^{A(t) + c} = C e^{A(t)}$$

where the integration constant c is removed from the exponent to give $C = e^c$. This equation can be applied any time we encounter a homogeneous linear first-order differential equation. What a mouth full! We will dispense with all of the adjectives whenever possible. You should become very quick to form the solutions of such equations.

The *general* Solution of $y'(t) = a(t) \, y(t)$ is

$$y(t) = C e^{A(t)} \text{ where } A(t) = \int a(t) \, dt \text{ and C is a constant.}$$

> When the *general* formula is constructed, normally the integration constant is <u>not included</u> in the function A(t).

[†] Differential Equation will be abbreviated as D.E.

Example 2 *General* solutions of homogeneous first-order D.E.s

Problem Determine the *general* solution of the given differential equations.

a) $y' = 3y$ b) $y' = t^3 y$ c) $r' = \sin(\theta) r$ d) $g'(x) = g(x)/x$

Solutions In each case the solution just requires the integration of the coefficient. The independent variable is assumed to be t if not indicated, and the function argument variable if one is indicated.

a) $a(t) = 3$ and thus $A(t) = 3t$. The *general* solution is $y = Ce^{3t}$

b) $a(t) = t^3$ and hence $A(t) = t^4/4$, giving the *general* solution $y = C \exp(t^4/4)$

c) The independent variable is θ. $a(\theta) = \sin(\theta)$ so $A(\theta) = -\cos(\theta)$ and the *general* solution is $r = Ce^{-\cos(\theta)}$

d) The independent variable is x. $a(x) = 1/x$, therefore $A(x) = ln(x)$ and the *general* solution is

$$g(x) = Ce^{ln(x)} = C \cdot x$$

☑

The solution with the arbitrary constant C is called a *general* solution because <u>every</u> particular solution of the differential equation has this form for some numerical value of the constant C. Solutions can not have any other form. We consider the determination of specific solutions at the end of this section when we consider *initial value* problems.

The nonhomogeneous equation. The solution of the non-homogeneous equation $y' = a(t)y + b(t)$ is similarly found via integration. However, its solution requires two integrations and is formed as a sum of two functions. One is a particular solution of the equation and the other involves a constant and is the *general* solution of the associated homogeneous equation, formed by setting $b(t) = 0$. To state this as a formula we add subscripts h and p to functions $y(t)$ to denote the solution of the associated homogeneous equation and a particular solution of the given non-homogeneous equation.

The *general* solution of $y' = a(t)y + b(t)$ has the form

$$y(t) = y_h(t) + y_p(t)$$

where

$$y_h(t) = Ce^{A(t)} \quad \text{with} \quad A(t) = \int a(t)\, dt \quad \text{and} \quad C \text{ is a constant;}$$

$y_p(t)$ is **any particular solution** of the non-homogeneous equation:

$$y_p' = a(t)y_p + b(t)$$

Consequently, if we can "observe" any particular solution y_p then adding this to the associated function y_h gives the *general* solution. We know that a function $y_p(t)$ is a solution if the equation resulting from substituting y_p and y_p' into the given differential equation results in a true mathematical statement.

Example 3   Determining the *general* solution of a non-homogeneous D.E.

Problem   Verify that the indicate function y_p is a solution of the given differential equation and then state the *general* solution of the equation.

a) D.E. $y' = 3y - 5$; $y_p(t) = 5/3$ a constant function.

b) D.E. $y' = y + \sin(t)$; $y_p(t) = -0.5[\sin(t) + \cos(t)]$

Solution   a) To verify the particular solution, first take the derivative: $y_p' = d/dt\ 3/5 = 0$. Substitute the derivative and the function into the D.E. and simplify:

$$0 = 3 \cdot 3/5 - 5$$

As the right side of the equation indeed is zero, the equation is true and thus the given y_p is a solution. The solution of the associated homogeneous equation, $y' = 3y$, is $y_h = Ce^{3t}$. Consequently, the *general* solution of the non-homogeneous equation $y' = 3y - 5$ is

$$y = y_p + y_h = 3/5 + Ce^{3t}$$

b) The first step is to differentiate the given function y_p:

$$y_p' = -0.5[\cos(t) - \sin(t)]$$

Then, this derivative and y_p are substituted into the given D.E. and we check to see if this results in a true equation. If it does then y_p is a particular solution.

$$y_p' = y_p + \sin(t)$$

becomes

$$-0.5[\cos(t) - \sin(t)] = -0.5[\sin(t) + \cos(t)] + \sin(t)$$

The right side of this equation simplifies to the left side, so it is always true; hence, y_p is a particular solution. The solution of the associated homogeneous equation $y' = y$ is just $y_h = Ce^t$. Thus, the *general* solution of the given D.E. is

$$y = y_p + y_h = -0.5[\sin(t) + \cos(t)] + Ce^t$$

☑

In the previous example the particular solutions were given. An obvious question is "How do you find particular solutions?" There is a formula that gives a particular solution for linear first-order D.E.s. The solution y_p is given by an integral involving b(t), the non-homogeneous term, and the integral A(t) of the y-coefficient a(t).

A particular solution of $y' = a(t)y + b(t)$ is

$$y_p(t) = e^{A(t)} \int^t b(s)\, e^{-A(s)}\, ds \quad \text{where} \quad A(t) = \int a(t)\, dt$$

The integration variable was arbitrarily changed to s to emphasize that the first function $e^{A(t)}$ <u>can not cancel the term</u> $e^{-A(s)}$ inside the integral. The s-integral is however evaluated at s = t after integrating. An example illustrates this more clearly. Following the example a *Rationale* for this formula is given.

Example 4 Solving a non-homogeneous D.E.

Problem Determine the *general* solution of $y' = 2y/t + t^2 - 1$.

Solution The D.E. is recognized as being linear and non-homogeneous with $a(t) = 1/t$ and $b(t) = t^2 - 1$.

The first step is to evaluate $A(t) = \int 2/t \, dt = 2\ln(t)$. Then the *general* solution of the associated homogeneous equation is

$$y_h = Ce^{2\ln(t)} = Ct^2$$

A particular solution of the given equation is found by evaluating

$$\begin{aligned}
y_p &= e^{A(t)} \int^t b(s) \, e^{-A(s)} \, ds \\
&= e^{2\ln(t)} \int^t [s^2 - 1] e^{-2\ln(s)} \, ds \\
&= t^2 \int^t [s^2 - 1] s^{-2} \, ds = t^2 \int^t 1 - s^{-2} \, ds \\
&= t^2 \cdot [s + s^{-1}]\big|^t = t^2 \cdot [t + t^{-1}] = t^3 + t
\end{aligned}$$

The *general* solution of $y' = 2y/t + t^2 - 1$ is the sum of y_p and y_h :

$$y = t^3 + t + Ct^2$$

☑

Notice that the formula for the particular solution always requires the integration of the product $b(s) \cdot e^{A(s)}$. Consequently, evaluating this integral frequently requires the use of *Integration by Parts*. This technique was presented in Chapter 6, Section 6.2. You may need to review this method to work the exercises.

Rationale for the y_p formula.

The equation for y_p is derived from the homogeneous solution y_h by replacing the arbitrary integration constant C by a function, $c(t)$. Then, substituting the resulting function and its derivatives into the given non-homogeneous differential equation results in a different second homogeneous D.E. for $c(t)$. Solving this by separation of variables leads to the formula given above. To see how this works, consider the function y_p defined by

$$y_p \equiv c(t) \cdot e^{A(t)} \quad \text{where} \quad A(t) = \int a(t) \, dt$$

Its derivative is, by the Product Rule,

$$y_p' = D_t c(t) \cdot e^{A(t)} + c(t) \cdot D_t e^{A(t)} = c'(t) \cdot e^{A(t)} + c(t) \cdot e^{A(t)} \cdot a(t)$$

Substituting these into $y' = a(t)y + b(t)$ gives:

$$c'(t) \cdot e^{A(t)} + c(t) \cdot e^{A(t)} \cdot a(t) = a(t) \cdot c(t) \, e^{A(t)} + b(t)$$

Canceling the product term $a(t) \, c(t) \, e^{A(t)}$ from both sides of this equation results in the equation

$$c'(t) \cdot e^{A(t)} = b(t)$$

Solving this for c(t) by separation of variables, we divide by the exponential term, which is equivalent to multiplying by it with a negative sign in the exponent, and convert c' to dc/dt, then integrate

$$dc = b(t) e^{-A(t)} dt \quad \Longrightarrow \quad c(t) = \int b(t) e^{-A(t)} dt$$

Thus the particular solution

$$y_p = c(t) \cdot e^{A(t)} = e^{A(t)} \cdot c(t) = e^{A(t)} \cdot \int b(t) e^{-A(t)} dt$$

To emphasize that the first function $e^{A(t)}$ can not be simply passed inside the integral we have elected to use a different integration variable in the above formula for y_p.

Initial value problems.

The *general* solutions of both the homogeneous and non-homogeneous D.E.s involve a transformed integration constant. When the constant is replaced by a numerical value the solution becomes a particular solution. There are thus as many different particular solutions as there are real numbers. If we sketch the solutions for different numerical values of C the curves all are "similar". For different values of C they never intersect. The curves at the right are solutions $y = Ce^{-\cos(t)}$ of the D.E. $y' = \sin(t)y$. The dotted curves are for $C = 0.1$, increasing by 0.1 to 0.9 (uppermost) and then the solid curves are for $C = 1$, increasing by 1 to 10. In each case the y-intercept is

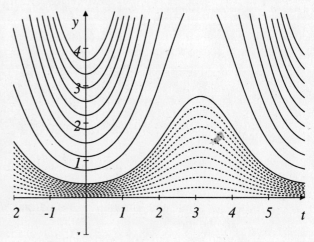

$$y(0) = C \cdot e^{-1} \approx 0.37\ C$$

A common problem, which is given the name an **initial value** problem, is to determine the constant C so that the particular solution passes through a given y-intercept.

Initial Value Problem
$$y' = F(t,y) \quad y(0) = y_0 \quad \text{a given constant.}$$

Algebraically, this is done by solving the differential equation and then solving $y(0) = y_0$. When the differential equation is non-linear the dependence of solutions on C may not be obvious. However for the linear cases it is fairly direct. The following formulas indicate the form of the solution but we recommend that you do not memorize them as it is simple enough to just work through the solutions for each problem.

The solution of $y' = a(t)y + b(t)$; $y(0) = y_0$ Non-homogeneous Initial Value Problem
is $y = Cy_h + y_p$ with $C = [y_0 - y_p(0)] / y_h(0)$
where y_h and y_p are as above.

In the homogeneous case, when $b(t) = 0$, the initial value problems solution can be expressed as the exponential function $e^{A(t)}$ scaled by the factor y_0 with a sort of "shift" in the exponent.

The solution of $y' = a(t)y$; $y(0) = y_0$ is   Homogeneous Initial Value Problem
$$y = y_0 e^{A(t) - A(0)} \quad \text{where } A(t) = \int a(t)\, dt$$

Section 8.1 FIRST-ORDER DIFFERENTIAL EQUATIONS 517

Example 5 Initial Value Problems.

Problem Determine the solution of the I.V. problem:

a) $y' = t^3 y$ $y(0) = 4$ b) $g'(x) = 3\, g(x) - \cos(2x)$ $y(0) = 5$

Solution a) The *general* solution of $y' = t^3 y$ is $y = C \exp(t^4/4)$ as $A(t) = t^4/4$. Setting

$$y(0) = 4 \text{ gives } C \cdot e^0 = 4 \text{ or simply } C = 4$$

Thus, the I.V. problem solution is

$$y = 4 \exp(t^4/4)$$

b) To determine the solution of the initial value problem we must first find a particular solution of $g' = 3 \cdot g(x) - \cos(2x)$. The solution of the associated homogeneous part is $g_h = C\, e^{3x}$. A particular solution of the non-homogeneous equation is given by

$$g_p = e^{3x} \int -\cos(2s)\, e^{-3s}\, ds$$

Don't forget the negative sign in the exponent!

To evaluate this integral requires the application of Integration by Parts twice and then an algebraic rearrangement. Let

$$I(s) = \int -\cos(2s)\, e^{-3s}\, ds$$

Then by the I.P. formula, $I = \int u\, v' = uv - \int v\, u'$, with $u = e^{-3s}$ and $v' = -\cos(2s)$ gives

$$I = e^{-3s} \cdot (-\sin(2s)/2) - \int (-\sin(2s)/2) \cdot (-3e^{-3s})\, ds$$

Apply the I.P. formula again. This time use $u = 3e^{-3s}$ and $v' = \sin(2s)/2$:

$$I = e^{-3s} \cdot (-\sin(2s)/2) - \{ -\cos(2s)/4 \cdot 3e^{-3s} - \int -\cos(2s)/4 \cdot (-9e^{-3s})\, ds \}$$

Which simplifies to

$$I = (-1/2)\, e^{-3s}\sin(2s) + (3/4) \cos(2s) e^{-3s} - (9/4) \int -\cos(2s) e^{-3s}\, ds$$

But, the last integral term is simply -9/4 times the integral I. Solving this equation for I gives

$$I = (4/13)\, [\, (-1/2)\, e^{-3s}\sin(2s) + (3/4) \cos(2s) e^{-3s}\,]$$

Replacing the variable s by x and multiplying by e^{3x} gives the particular solution:

$$g_p = e^{A(x)} \cdot I(x) = e^{3x} \cdot [\, (-2/13)\, e^{-3x}\sin(2x) + (3/13) \cos(2x) e^{-3x}\,]$$

The exponential terms e^{3x} and e^{-3x} cancel upon multiplication to give

$$g_p(x) = (-2/13)\sin(2x) + (3/13) \cos(2x)$$

Now, to answer the posed problem. We must find C so that

$$g(0) = g_h(0) + g_p(0) = 5$$

Substituting into this the values

$g_h(0) = C \cdot e^0 = C$ and $g_p(0) = (-2/13) \sin(0) + (3/13) \cos(0)] = 3/13$

We find $C + 3/13 = 5$ or $C = 62/13 \approx 4.77$. The solution to the I.V. problem is thus

$$g(x) = (-2/13)\sin(2x) + (3/13) \cos(2x) + (62/13) e^{3x}$$

☑

A related type of problem called a **boundary value** problem, replaces the time t = 0 with a specified time t_0. The problem is then stated as

$$y' = F(t,y) ; y(t_0) = y_0$$

These problems are solved analogously to the method applied for the initial value problems, with the *general* solution being evaluated at t_0 instead of at zero to determine the appropriate constant's value. More general conditions can be imposed to specify the numerical value of a solution or its derivative at a specific t-value. One common condition is to specify that $y'(t_0) = y_1$, a given value. In this case you must differentiate the *general solution*, or use the differential equation to evaluate y'. The equation $y' = y_1$ is then solved for the constant C.

Exercise Set 8.1

1 The following equations are separable: determine the *general* solution of the given equation.

    a) $y' = 2t - e^t$ b) $y' = 2ty$ c) $y' = \cos(2t)$ d) $y' = \tan(y) 5t$

    e) $y' = ye3^t$ f) $y' = te^{3y}$ g) $y' = t - ty$ h) $yy' = \sin(t)$

    i) $(t + 1)y' = y$ j) $t \ln(t) y' = y \ln(y)$ k) $y' = t \sin(t)/y$ l) $y' = t^3/y^3$

2 Find the *general* solution of the following linear equations.

    a) $y' = -0.2y$ b) $y' = 3y$ c) $y' = 3y + 2$ d) $y' = 3ty$ e) $y' = 3ty + 2$

    f) $y' = 5y/t$ g) $y' = 5y/t - 2$ h) $y' = \sin(t)y$ i) $y' = \sin(t)y - e^{-\cos(t)}$

    j) $y' = -2y + \sin(3t)$ k) $y' - 3y = t^2$ l) $y' = -2ty + t^3$

3 Find the specific solution of the given equation satisfying the indicated condition.

    a) $y' = 7y$, $y(0) = 1$ b) $y' = 7y$, $y(2) = 1$ c) $y' = 3y$, $y'(5) = 1$

    d) $y' = \sin(t)y^2$, $y(0) = 2$ e) $y' = y/(t + 1)$, $y(0) = 2$ f) $y' = t^2/y^3$, $y(1) = 16$

    g) $y' = 5y$, $y(2) + y'(2) = 3$ h) $y' = t/y$, $y(1) = 2$ i) $y' = t/y$, $y'(1) = 2$

4 Determine the specific solution y satisfying the indicated condition.

    a) $y' = 3y + 2$, $y(0) = 3$ b) $y' = 3y + 2$, $y(1) = 3$ c) $y' = 3y + t$, $y(0) = 3$

    d) $y' = 3y + t$, $y(1) = 2$ e) $y' = y/t + 1$, $y(1) = 2$ f) $y' = -y + \cos(t)$, $y'(0) = 1$

    g) $y' = (y/t) + t$, $y(1) = 0$ h) $y' = (y/t) + t$, $y(1) = 3$ i) $y' = (-y/t) + e^t$, $y(1) = 1$

5. Newton's Law of cooling states that the rate of change in temperature of a body is proportional to the difference between the temperature T, and the surrounding temperature, T_0.
   a) State Newton's Law as a differential equation
   b) Solve this differential equation.
   c) How long will it take a body of temperature $T_0 = 37$ degrees C immersed in a 15 degrees C bath to reach a temperature of 26 degrees C if the proportionality constant is -k = -2 degrees/hour?

6. A wet porous material with adequate ventilation will loose its moisture at a rate proportional to its moisture content. If a wet blanket that is being used to protect a seal in transit looses 30 percent of its moisture in one hour, when will only 20 percent of its original moisture remain?

7. A tank contains 1000 liters of a brine containing 10 kilos of salt. The tank water is diluted by adding fresh water at a rate of 25 liters/min., which mixes instantaneously with the water in the tank. The tank over-spills at a rate equal to the amount of water added so that its volume remains constant.
   a) Set up a differential equation to describe the amount of salt in the tank.
   b) How much salt will be in the tank after 15, 30, and 60 minutes?
   c) When will the tank contain 2 kilos of salt?

8. One published model of the global human population used a growth equation of the form $N'(t) = \tau N(t)/(D - t)$ with "initial" population $N(0) = k D^{-\tau}$.
   a) Solve this equation (k, τ, D constants.)
   b) If $k = 1.79 \times 10^{11}$, $\tau = 0.99$, and $D = 2026.87$, how close is this estimate to the 1975 world population of 3.97 billion people? (Use t = 1975.)
   c) Using the parameters (b), why was this model titled "Doomsday: Friday, 15 November 2026"?

9. In the study of epidemics, a usual assumption is that the rate of new infections is proportional to the number of contacts between infectives, I, and susceptibles, S.
   a) Express this as a differential equation (assume that no one gets over the disease).
   b) In a finite population of size N, the number of infectives plus susceptible equals N. Use this to describe the change in the number of susceptible as a function of S (and the constant N).
   c) Solve the equation derived in part (b).

10. The equation $N(t) = 450/[1 + \exp(5.041 - 1.022t)]$ was found to fit experimental data for the growth of the bacterial Paramecium aurelia. Determine the differential equation describing this growth function.

11. The yield of a healthy crop is predicted by $dw/dt = \tau(t)[1 - w(t)/W]w(t)$ where W equals the dry weight at maximum yield, w(t) equals the dry weight at time t, and τ equals the coefficient of growth as a function of time.
    a) Solve for w(t) if τ equals a constant.   b) Solve for w(t) if τ equals t.
    c) Solve for w(t) if τ equals t(T - t), where T is a fixed constant.

12. In predicting epidemics of leaf rust in plants, crop scientists have used the equation $dx/dt = kx(1 - x)$ where x is the rusted fraction of the leaf area and k relates the susceptibility of the plant to the blight fungi. k has been found to be sensitive to temperature, T, with $k = k_{max} - c(T - 20)$, c and k_{max} constants. The function for the mean daily maximum temperature is $T = a + b \sin(\pi t/120)$. Solve for x(t) when a = 15, b = 5 and c = 1.

13. An alternative way to determine a particular solution utilizes the fact that the only way a function sin(kt) or cos(kt) can appear as b(t) in the derivative is for the function y to contain both these trig functions. In particular when a(t) = a, a constant, then a particular solution y_p of $y' = ay + \sin(kt)$ will have the form
$$y_p(t) = \alpha \sin(kt) + \beta \cos(kt)$$
To determine α and β, y_p and y_p' are substituted into the differential equation and the right side is simplified. Then, the coefficients of sin(kt) on each side of the equation are equated to give an equation involving the coefficients α and β. Next the coefficients of cos(kt) are similarly equated resulting in a second equation for α and β. Solving these two simultaneously gives the desired particular solution. Apply this method to solve the given D.E.
   a) $y' = 3y - \cos(2t)$ b) $y' = -2y + \sin(5t)$ c) $y' = -3y + 2\sin(\pi t) - \cos(\pi t)$

Section 8.2 Stability and Equilibriums of Dynamical Systems.

Introduction to Dynamical System.

In this section we will consider a simply dynamical system that is described by a first-order differential equation

$$y' = F(t,y) \text{ with initial state } y(0) = y_0$$

The state of the system that is being modeled is described by the trajectory of a solution as t increases. These trajectories are curves in the t-y coordinate system. Our objective is to describe qualitative aspects of these solution curves utilizing information provided by the differential equation. That is, from the properties of the function $F(t,y)$. Normally, in the study of dynamical systems this function will also contain parameters that represent numerical conditions affecting the system. A whole branch of the study of dynamical systems focuses on the way solution trajectories change as the model parameters change.

Two important concepts that are used to describe systems and to characterize their intrinsic behavior are the concepts of *equilibriums* and their *stability*. In simple words, a system is at an *equilibrium* if it does not change. The concept of *stability* refers to how the system will evolve if for some reason it is started at a state that is nearby an *equilibrium state*. If all nearby solutions approach the *equilibrium state* as $t \to \infty$, then the *equilibrium* is said to be **stable.** If, on the other hand, some nearby solution approaches another state as $t \to \infty$ then it is **not stable**. We will not enter into the theory to the extent of rigorously defining what we mean by "nearby". What we will do, is to show that a dynamical system can be linearized, i.e., its differential equation can be approximated by a homogeneous linear D.E. The behavior of solutions to these equations are easily found and we can quickly determine if these linearized solutions approach zero or ∞. The two alternatives correspond to being *linearly stable* or *linearly unstable*.

If you think of a dynamical system, perhaps an automobile engine, or an ecosystem, or a hormonal feedback mechanism, you realize that the dynamics of the system depend on the internal state of the system and on external influences. The system evolves or changes as time increases, this is described by the differential equation. However, the system does not depend explicitly on time, that is, on time measured from some starting point. A system with a clock, a microwave oven, of course does depend on time explicitly. In the following we shall exclude this time-dependent type of system and only consider time independent systems, which mathematicians call *autonomous systems*. These are systems modeled by a differential equation of the form

$$y' = f(y) \text{ with no explicit t-variable.}$$

The reason for this is simple. There is a vast theory that applies to such equations. The *non-autonomous* equations however are very difficult to analyze. We also assume f is continuous.

Equilibriums.

A dynamical system governed by the equation $y' = f(y)$ is said to have an **equilibrium at** y_E if whenever a solution $y(t) = y_E$, its derivative $y'(t) = 0$. Thus, the

The Equilibrium states of $y' = f(y)$ are the roots of $f(y) = 0$.

Example 1 **Equilibriums of Growth Models.**

Problem   What are the equilibrium of the Exponential and Logistic Growth models?

Solution To answer this you need to know the differential equation of each model. The

EXPONENTIAL GROWTH MODEL: $y' = ky$

has a single equilibrium, $y_E = 0$ since this is the only root of $f(y) = ky = 0$. The

LOGISTIC GROWTH MODEL: $y' = r(1 - y/M)y$

is a nonlinear model. Setting $f(y) = r(1 - y/M)y = 0$ gives two roots.

$$y_{E1} = 0 \quad \text{and} \quad y_{E2} = M$$

In the logistic model the population y remains static, it does not change, when it reaches the carrying capacity M.

Graphing Dynamical Systems.

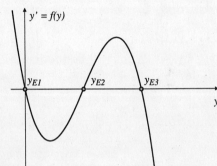

To graphically identify the equilibrium states of a system we introduce a graph of the differential equation, not of the solution $y(t)$. This is a graph with the horizontal axis denoted by y and the vertical axis denoted by y'. The function $y' = f(y)$ is plotted; we will refer to this graph as the **DE-graph**. The equilibrium states y_E are the y-values where the graph of f intersects the horizontal y-axis. There are three equilibriums illustrated in the graph at the right.

The DE-graph can be viewed as a parametric curve using the time variable t as the parameter. The equilibrium points divide the curve into segments and different solutions $y(t)$ will *generate* these segments as t varies. This is done by visualizing a point on the graph of $f(y)$ as $P_t = (y(t), y'(t))$. As t changes the corresponding points P_t will move along a segment of the DE-graph. How fast, in t-units, and in which direction the points P_t move will naturally depend on the values of $f(y)$ since this gives the rate of change in y as a function of time; remember $y' = D_t y = f(y)$.

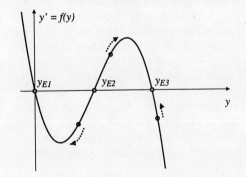

If $y'(t) > 0$ then as t increases $y(t)$ increases and the point P_t will move to the right.

If $y'(t) < 0$ then $y(t)$ decreases as t increases and the point P_t will move to the left.

If $y'(t) = 0$ then P_t is at an equilibrium point and does not change as t increases.

The speed at which the point P_t moves is governed by the values of y' and $y'' = D_y f(y) \cdot y'$. The speed is

$$s_t = |y'(t)| \cdot \text{SQRT}\{ 1 + [D_y f(y(t))]^2 \}$$

Thus if the point P_t is above the y-axis it moves to the right and if it is below the y-axis it moves to the left. The further a point is from the horizontal y-axis the faster it travels. As points approach the horizontal y-axis they slow down. In fact, they can never reach the y-axis in a finite amount of time.

"What is the limit of P_t as t approaches infinity?"

The answer depends on which segment of the DE-graph the points lie. Consider the solution of the initial value problem $y' = f(y)$, $y(0) = y_0$. Let $y'_0 \equiv f(y(0)) = f(y_0)$. Then, at time $t = 0$ the corresponding point on the DE-graph is

$$P_0 = (y(0), y'(0)) = (y_0, y'_0))$$

For this solution $y(t)$, $\lim_{t \to \infty} P_t$ depends on the relationship of y_0 to the systems equilibriums and whether y'_0 is positive or negative.

If $y'_0 > 0$ then the point P_0 is above the y-axis and $y(t)$ will increase as t increases. Consequently, P_t will approach either the first equilibrium point to the right of y_0 or infinity if there is no equilibrium to the right of y_0.

If $y'_0 < 0$ then the point P_0 is below the y-axis and $y(t)$ will decrease as t increases. Consequently, P_t will approach either the first equilibrium point to the left of y_0 or negative infinity if there is no equilibrium to the left of y_0.

Example 2 Limits of solutions of the Logistic model.

Problem A population's size is described by the solutions of the Logistic Equation $y' = 0.8(1 - y/6)y$. What will be the limiting size of the population? Illustrate the population's dynamics if its initial size, at $t = 0$, is a) $y_0 = 1$ b) $y_0 = 4$ and c) $y_0 = 8$. Use the DE-graph to sketch these solutions without solving for them explicitly.

Solution The sketch of the DE-graph is quite simple as the function $f(y)$ is a quadratic in factored form. It has two equilibriums as it passes through the y-axis at the origin, $y_{E1} = 0$, and at $y_{E2} = 6$. Assuming that the initial size y_0 is positive, we see from the graph that

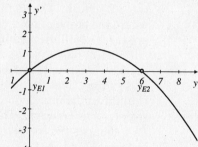

$$\lim_{t \to \infty} y(t) = y_{E2} = 6$$

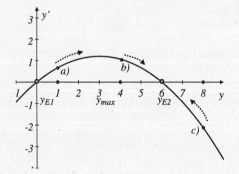

If the solution begins above 6, as in case c), then it decreases toward 6. If it begins below 6, as in cases a) and b), it will increase toward 6. Using the DE-graph we can infer even more about how the solution curves will look when plotted on a t-y axis system.

In case a) $y_0 = 1$ is less than the value $y_{max} = 3$ at which the function $f(y) = 0.2(1 - y/6)y$ has its maximum. Consequently, the solution to this Initial Value problem, which is denoted by $y_a(t)$ will increase at an increasing rate as the derivative $y_a'(t) = f(y_a(t))$ becomes greater as the point corresponding P_t moves along the DE-curve, approaching the maximum point on this curve.

After the point P_t passes through this apex it moves downward, toward the equilibrium point at (6,0). As it moves downward, the derivative y_a' becomes smaller and the $y_a(t)$ curve begins to level off, to approach the equilibrium asymptotically. This increasing and then decreasing of the derivative of y_a results in a sigmoidal or s-shaped curve, as illustrated in the graph below.

In case b) $y_0 = 4$ and the solution starts past the peak of the DE-curve. Consequently the graph of the solution $y_b(t)$ is not s-shaped. It simply approaches the line $y = 6$ asymptotically, looking very much like an exponential decay curve.

In case c) $y_0 = 8$ and the solution $y_c(t)$ starts above the equilibrium. In this case it simply "decays" toward $y = 6$, approaching this horizontal line from above.

The equations of these various solution curves are given by the Logistic function y(t) that is found by solving the differential equation. Logistic equations are solved by separation of variables as discussed in Section 8.1. You should however be able to sketch a graph of solutions like this without solving. You will not be able to indicate the scale on the t-axis in this case.

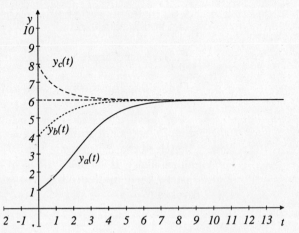

Stability of Equilibriums.

An equilibrium y_E is **stable** if there is an open interval I containing y_E such that for all $y_0 \in I$ the solution of the initial value problem $y' = f(y)$, $y(0) = y_0$ approach y_E as t goes to infinity.

If $y(0) \in I$ then $\lim_{t \to \infty} y(t) = y_E$. **A Stable Equilibrium**

If this is not true, if no matter how small an interval one takes there is always some solution that either diverges to infinity or approaches another equilibrium point, then y_E is said to be **unstable**. This concept of stability is sometimes referred to as being **locally stable**. When the interval I is the entire real line, I = $(-\infty, \infty)$, then a stable equilibrium is said to be **globally stable.**

Employing the graphical approach we can see that an equilibrium is stable when the DE-graph crossed the y-axis from above to below at the equilibrium point. We can indicate this condition algebraically using the derivative of f(y), with respect to y. If the derivative is negative then indeed the DE-curve will cross the y-axis from above to below. If the derivative is positive, then it will cross from below to above and the equilibrium will be unstable. When the derivative is zero additional information is required to indicate whether or not the DE-curve crossed the y-axis and in which direction. These observations are stated as a theorem.

STABILITY THEOREM.

> An equilibrium y_E of $y' = f(y)$ is **stable** if $D_y f(y_E) < 0$.
>
> An equilibrium y_E of $y' = f(y)$ is **unstable** if $D_y f(y_E) > 0$.

Example 3 **Determining the stability of equilibriums.**

Problem Classify the stability of the equilibriums of $y' = y(y - 2)(4 - y)^2$.

Solution The function f(y) = y(y - 2)(4 - y) has roots $y_{E1} = 0$, $y_{E2} = 2$ and $y_{E3} = 4$. The derivative of f(y) is found by applying the product rule twice.

$$f'(y) = (y - 2)(4 - y)^2 + y(4 - y)^2 - 2y(2 - y)(4 - y)$$

This derivative is evaluating at the equilibrium values and the Stability Theorem is applied.

$f'(0) = -32$ and hence by the Theorem $y_{E1} = 0$ is a stable equilibrium.

$f'(2) = 8$ and hence $y_{E2} = 2$ is an unstable equilibrium.

$f'(4) = 0$ The theorem does not indicate any conclusion in this case!

The system in fact is not stable at $y_{E3} = 4$. This can be seen graphically from a sketch of the DE-graph. The DE-curve touches the y-axis at 4 but remains above it on both sides of 4. Consequently, a solution slightly below 4 will approach 4 as $t \to \infty$. But, a solution slightly above 4 will diverge to ∞ as $t \to \infty$.

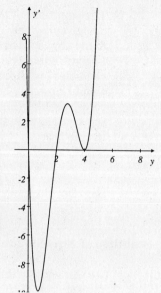

Exercise Set 8.2

1. Sketch the DE-graph for the given equation. Identify all equilibriums on the graph and decide graphically if they are stable or not.

   a) $y' = 2 - y$ b) $y' = -3y + 2$ c) $y' = y(y - 5)$

   d) $y' = 4 - y^2$ e) $y' = y(y - 2)(y + 3)$ f) $y' = y(y - 2)^2$

   g) $y' = 8 - y^3$ h) $y' = y^2(y + 4)$ i) $y' = y(y - 2)^3$

2. Sketch the solution curves for the given equation and indicated initial y-values without actually solving the equation. Use the DE-graph as a guide.

   a) $y' = 2y - 3$ $y_0 = 1$ and 3
   b) $y' = y(2y - 3)$ $y_0 = 0.25, 1.25$ and 3
   c) $y' = 0.1y(y - 2)(y - 5)$ $y_0 = 0.5, 1.25, 2.25, 4.5$ and 6
   d) $y' = (y + 2)(y - 1)^2$ $y_0 = -3, -1, 0,$ and 2

3. Determine the equilibrium states of the given equation and apply the Stability Theorem if it applies to determine their stability.

   a) $y' = ln(y)$ b) $y' = ln(4/y)$ c) $y' = y^3 - 2y$

   d) $y' = \sin(\pi y)$ e) $y' = (1 - y)^2$ f) $y' = \sin^2(\pi y)$

4. For what values of k is zero a stable equilibrium of $y' = ky$?

5. For what positive values of r is $y_E = 4$ a stable equilibrium of $y' = y^{r-1}(y - 4)^r$?

6. If $y' = 3y(y - 6) \, ln(y - 4)$ evaluate $\lim_{t \to \infty} y(t)$ if $y(0)$ equals a) 4.5   b) 7

Section 8.3 Higher-Order Differential Equations.

Introduction.

In this section we will only consider linear differential equations. These are equations that can be written in the form

$$a_n y^{(n)} + a_{n-1} y^{(n-1)} + \ldots + a_2 y'' + a_1 y' + a_0 y = b \qquad n^{th}\text{-order L.D.E.}$$

The coefficients $a_0, a_1, a_2, a_3, \ldots$ and the term b may be constants or functions of the independent variable only. We shall actually focus on the case where they are constants. If $b = 0$ then the equation is *homogeneous* and this is the case we will consider first. An important aspect of linear differential equations is that multiples of their solutions can be added to form other solutions. If $y_1(t)$ and $y_2(t)$ are two solutions of the above equation then $y(t) \equiv Ay_1(t) + By_2(t)$ is a solution for any numbers A and B.

The nth-order equation is a generalization of the first-order equation $y' = ay + b$ that we considered in Section 8.1. Recall that for a constant coefficient a the solution of the homogeneous equation $y' = ay$ or $y' + (-a)y = 0$ was simply $y(t) = Ce^{-at}$. The solution of an nth-order constant coefficient homogeneous differential equation is actually just the sum of n-terms having this same exponential form. It will involve n constants, which we subscript as C_i for $i = 1$ to n and n coefficients in the exponent. Traditionally these are denoted by λ_i. The form of the solution is then

$$y(t) = \Sigma_{i=1,n} C_i \exp(\lambda_i t)$$

The coefficients $\lambda_1, \lambda_2, \ldots$ are called *eigenvalues* (a German term) and are found as the n roots of an nth degree polynomial. When there are repeated roots there must be a modification of this formula with one of the C_i being multiplied by t. When the roots are complex numbers, not real, another modification is introduced in which two of the exponential terms are replaced by sine and cosine functions. This case is not a rarity. In fact, it occurs whenever the system being described by the differential equation has oscillations. For instance a violin string vibrating or a daily fluctuation in temperature. We will have to consider this case, but, you will not need to actually do complex arithmetic.

Many systems have more than one dependent variable. When there are two or more state variables or functions that depend on the independent dynamic variable t and possibly on each other the dynamical model of the system will be given by a system of two or more differential equations. As we shall see, a system of n first-order equations can be expressed as a single n^{th}-order differential equation. This will allow us to consider some simple systems using our previous knowledge of solutions of higher-order equations. This relationship actually works both ways. In advanced courses the theory for higher order equations is actually related to systems of first-order equations because using vector and matrix notation these are easier to analyze. In Section 10.4 we consider the equilibrium and stability analysis of systems of differential equations.

Second-Order Differential Equations.

The second order equation can be written with the coefficient $a_2 = 1$. (If its not, simply divide all coefficients by a_2 and then relabel them.) Consider first the homogeneous equation:

$$y'' + a_1 y' + a_0 y = 0$$

Assuming that a solution is given by $y = ce^{\lambda t}$ we will differentiate this function twice, substitute into the differential equation, and obtain a quadratic equation for the exponent λ. Each root of this equation will provide a term in the *general* solution. To establish the formula for the λ's, let

$$y(t) = Ce^{\lambda t} \quad \text{then} \quad y'(t) = \lambda Ce^{\lambda t} \quad \text{and} \quad y''(t) = \lambda^2 Ce^{\lambda t}$$

Substituting these into the general second-order differential equation gives

$$\lambda^2 Ce^{\lambda t} + a_1 \cdot \lambda Ce^{\lambda t} + a_0 \cdot Ce^{\lambda t} = 0$$

Factoring the sum gives

$$[\lambda^2 + a_1 \cdot \lambda + a_0] Ce^{\lambda t} = 0$$

The exponential $e^{\lambda t}$ can never equal zero. If the constant $C = 0$ then our solution would be the trivial zero function. The only other way this equation can be true is for the term in square brackets to be zero. This gives the key equation for λ which is called either a *characteristic* equation or an *auxiliary* equation.

$$\lambda^2 + a_1 \cdot \lambda + a_0 = 0 \qquad \textbf{Characteristic Equation}$$

Denote the two roots of this equation by λ_1 and λ_2. Each root yields an exponential solution. Using the linearity of the equations the sum of the two solutions is also a solution. In fact, every possible solution is a linear combination of these two solutions. There are actually three forms of the *general* solution corresponding to the three possible types of roots of a quadratic: (i) when they are real and not equal, (ii) when the are the same, repeated roots, and (iii) when the roots are *complex* numbers.

If λ_1 and λ_2 are real and not equal then

$$y(t) = C_1 e^{\lambda_1 t} + C_2 e^{\lambda_2 t} \qquad \textbf{General Solution (i)}$$

Particular solutions correspond to specific numerical values of the C coefficients. Since there are two C's two conditions must be specified to determine these uniquely.

An Initial Value Problem specifies $y(0) = y_0$ and $y'(0) = y_1$

A two-point Boundary Value Problem specifies the value of the function at two t-values
$$y(t_0) = y_0 \quad \text{and} \quad y(t_1) = y_1$$

Example 1   **Solving linear 2nd-order differential equations**.

Problem   Determine the *general* solution of $y'' + 4y' + 3y = 0$ and find the particular solution with initial values $y(0) = 2$ and $y'(0) = -1$.

solution   The first step is to write the *characteristic* equation. This is the quadratic equation formed by replacing y'' by λ^2 and y' by λ in the differential equation.

$$\lambda^2 + 4\lambda + 3 = 0$$

The quadratic factors as $(\lambda + 1)(\lambda + 3)$. Thus its two roots are $\lambda_1 = -3$ and $\lambda_2 = -1$. The *general* solution is then

$$y(t) = C_1 e^{-3t} + C_2 e^{-t}$$

The particular solution is found by first differentiating the *general* solution:

$$y'(t) = (-3)C_1 e^{-3t} + (-1)C_2 e^{-t}$$

Now y and y' are evaluated at $t = 0$. As $e^0 = 1$,

$$y(0) = C_1 + C_2 \quad \text{and} \quad y'(0) = -3C_1 - C_2$$

Equating these expressions to the given initial values gives two equations in the unknown C_1 and C_2. These are solved.

$$2 = C_1 + C_2 \quad \text{and} \quad -1 = -3C_1 - C_2 \quad \Longrightarrow \quad C_1 = -1/2 \quad \text{and} \quad C_2 = 5/2$$

The solution of the initial value problem is then

$$y(t) = -1/2\, e^{-3t} + 5/2\, e^{-t}$$

When the roots of the *characteristic* equation are the same, the two exponential functions in solution (i) are identical and hence can be factored. Then the two constants act as only one constant, $C_3 = C_1 + C_2$. In this case a second solution is found by multiplying one of the terms by the independent variable t. This gives the second form of the *general* solution.

If λ_1 and λ_2 are equal then

$$y(t) = [C_1 + C_2 \cdot t] e^{\lambda_1 t} \qquad \textbf{\textit{General} Solution (ii)}$$

Example 2   **Solving an equation with repeated roots.**

Problem   Determine the solution of the initial value problem $y'' - 4y' + 4 = 0$, $y(0) = 2$ and $y'(0) = 5$.

Solution   First form the *characteristic* equation $\lambda^2 - 4\lambda + 4 = 0$. The roots are both $\lambda = 2$. The *general* solution is the of form (ii)

$$y(t) = [C_1 + C_2\, t]\, e^{2t}$$

To match the given initial values we first differentiate the *general* solution. Applying the product rule and simplifying gives

$$y'(t) = [2C_1 + C_2 (1 + 2t)]\, e^{2t}$$

Then setting $t = 0$ in these functions and equating them to the given values results in two equations for the C's.

$$C_1 + C_2 = 2 \quad \text{and} \quad 2C_1 + C_2 = 5 \quad \Longrightarrow \quad C_1 = 3 \quad \text{and } C_2 = -1$$

The solution is then

$$y(t) = [3 - t]\, e^{2t}$$

The third case arises when the roots of the *characteristic* equation are *complex* numbers. If this occurs the two roots will have the form

$$\lambda_1 = \alpha + \beta i \quad \text{and} \quad \lambda_2 = \alpha - \beta i$$

In engineering courses these roots are used in form (i) of the *general* solution as engineers seem comfortable working with *complex* arithmetic and functions. Indeed, in mathematics there is a whole branch of function theory that considers *complex* analysis of such functions. In this limited introduction we will simply slide over the details and state the resulting formula.

If λ_1 and λ_2 are *complex* roots: $\lambda = \alpha \pm \beta i$ then

$$y(t) = [C_1 \cos(\beta t) + C_2 \sin(\beta t)]\, e^{\alpha t} \qquad \textbf{\textit{General} Solution (iii)}$$

Example 3 A solution when the eigenvalues are complex.

Problem Determine the solution of the boundary value problem $y'' + 4y = 0$, $y(0) = 2$ and $y(\pi/4) = 3$.

Solution The *characteristic* equation for this problem does not have a λ term as there is no y' term in the differential equation. It is just $\lambda^2 + 4 = 0$. The roots of $\lambda^2 = -4$ are purely imaginary with $\alpha = 0$ and $\beta = 2$: $\lambda = \pm 2i$. Thus, in the *general* solution the exponential term $e^{\alpha t} = e^0 = 1$ and does not appear as a factor.

$$y(t) = C_1 \cos(2t) + C_2 \sin(2t)$$

The boundary conditions are

At $t = 0$: $\quad 2 = y(0) = C_1 \cos(0) + C_2 \sin(0) = C_1$

At $t = \pi/4$: $\quad 3 = y(\pi/4) = C_1 \cos(\pi/2) + C_2 \sin(\pi/2) = C_2$

Thus, the particular solution is

$$y(t) = 2 \cos(2t) + 3 \sin(3t)$$

The *general* solution of an nth-order linear constant coefficient differential equation is very similar to that given above for the 2nd-order equation. It of course will have n-constants and will be a sum of either exponential terms, exponential terms times polynomials (corresponding to multiple roots) and/or trigonometric functions (when *complex* roots occur).

Systems of differential equations.

A system of two or more linked first-order equations can be converted to a higher-order equation by identifying the different functions with derivatives. Let us demonstrate this method with an example.

Example 4 Solving a 1st-order system.

Problem Determine the second-order differential equation that is equivalent to the system of two differential equations in the functions $u(t)$ and $v(t)$:

$$u' = 3u - 2v \quad \text{and} \quad v' = -u + v$$

Solution The method is to differentiate one equation, thereby introducing a second derivative. Then, the other function term and its derivative are algebraically eliminated from the second-order differential equation. First, take the derivative of the u' equation

$$u'' = 3u' - 2v'$$

Next, solve the first equation for v and substitute into the v' equation:

$$v = [u' - 3u]/2 \quad \Longrightarrow \quad v' = -u + [u' - 3u]/2 = u'/2 - 5u/2$$

Now, substitute this equation for v' into the above u'' equation:

$$u'' = 3u' - 2[u'/2 - 5u/2]$$

Rearrange this equation to the standard form:

$$u'' - 2u' - 5u = 0$$

The *characteristic* equation for this differential equation is then

$$\lambda^2 - 2\lambda - 5 = 0$$

Its roots are given by the quadratic formula: $\lambda = [2 \pm \text{SQRT}(4 + 20)]/2$

$$\lambda_1 = 1 + \sqrt{6} \quad \text{and} \quad \lambda_2 = 1 - \sqrt{6}$$

Thus the *general* solution for u(t) is

$$u(t) = C_1 e^{(1 + \sqrt{6})t} + C_2 e^{(1 - \sqrt{6})t}$$

The corresponding function v(t) can be found from the equation $v = [u' - 3u]/2$. We first differentiate u(t):

$$u'(t) = (1 + \sqrt{6})C_1 e^{(1 + \sqrt{6})t} + (1 - \sqrt{6})C_2 e^{(1 - \sqrt{6})t}$$

Substituting these into the v equation and simplifying gives

$$v(t) = [-1 + \sqrt{6}/2] C_1 e^{(1 + \sqrt{6})t} - [1 + \sqrt{6}/2] C_2 e^{(1 - \sqrt{6})t}$$

☑

Exercise Set 8.3

1. Determine the *general* solution of the following equations.

   a) $y'' + 3y' - 4y = 0$ b) $y'' - y' - 2y = 0$ c) $y'' + y' - 2y = 0$

   d) $y'' - y' + 2y = 0$ e) $y'' - 4y = 0$ f) $y'' + 4y = 0$

   g) $2y'' - y'' + y = 0$ h) $y'' - y' = 0$ i) $y'' - 3y' - 3y = 0$

   j) $y'' + 2y = 0$ k) $y'' + y' - 6y = 0$ l) $y'' - 10y' + 25y = 0$

2. Solve the following equations subject to the initial conditions $y(0) = 1$, $y'(0) = 2$.

   a) $y'' - 2y' - 3y = 0$ b) $y'' - 2y' + y = 0$ c) $y'' + 5y = 0$

   d) $y'' - 5y = 0$ e) $2y'' - 3y' + y = 0$ f) $2y'' + 4y' - 7y = 0$

3. Solve the following boundary-value problems.

   a) $y'' + 9y = 0$ $y(0) = 1, y(2) = 2$ b) $y'' - 9y = 0$ $y(0) = 0, y(2) = 1$

   c) $y'' - 2y' + y = 0$ $y(0) = 1, y(2) = -1$ d) $y'' + 2y' + 2y = 0$ $y(0) = 1, y(2) = -1$

   e) $y'' + 9y = 0$ $y(0) = 0, y(\pi/6) = 2$ f) $y'' + y = 0$ $y(0) = 3, y(\pi/4) = 0$

   g) $y'' + 25y = 0$ $y(0) = 1/2, y(\pi/10) = 2$ h) $y'' + 4y = 0$ $y(0) = 1, y(\pi/12) = 1$

4. Find a second-order differential equation having both of the indicated functions as solutions.

   a) e^{-t}, e^{-2t} b) e^{3t}, e^{-3t} c) e^{2t}, e^{-t} d) e^{2t}, te^{2t}

   e) $\cos(3t), \sin(3t)$ f) $e^t \cos(3t), e^t \sin(3t)$ g) e^{at}, e^{bt} h) te^{-kt}, e^{-kt}

5 The diffusion of a compound between two compartments separated by a membrane of thickness depends on the concentrations of the substance in each compartment and the diffusion constant D, which is characteristic of the substance and membrane. If A is the area of the membrane, V is the volume of compartment 1, y_1 and y_2 are the concentrations in compartments 1 and 2, respectively.
Then from Fick's Law it follows that $dy_1/dt = -(y_1 - y_2)DA/V$ assume that $y_1(0) = 0$ and $y_2(0) = 100$.
   a) Solve for y_1 when y_2 is held constant, i.e., $y_2' = 0$.
   b) Solve the problem if $y_2(t) = k[1 + \sin^2(t)]$. (What is k?)
   c) If $dy_2/dt = (y_1 - y_2)DA/V$ solve for both y_1 and y_2.
   d) Theoretically will y_1 ever equal y_2?

6 Solve the system of equations.

   a) $u' = -2u + v$ $v' = +2u - v$ b) $u' = +2u - 2v$ $v' = +2u - 2v$

   c) $u' = -3u + v$ $v' = 3u + v$ d) $u' = -3u + v$ $v' = 3u - 2v$

   e) $u' = -3u + v$ $v' = 3u - 3v$ f) $u' = -u + 3v$ $v' = +u - 4v$

   g) $u' = u + v$ $v' = -u + v$ h) $u' = u + 2v$ $v' = -2u + v$

7 Extend the method used to solve 2nd-order equations to solve the third-order equation:

   a) $y''' + 5y' = 0$ b) $y''' + 3y'' + 2y' = 0$ c) $y''' - 8y = 0$

Chapter 9   *Multiple Integration.*

9.1   **Iterated Integration.** 531

9.2   **The Double Integral.** 537

9.3   **Double Integrals Over General Regions: Determining the limits of integration.** 544

9.4   **Areas and Volumes as Multiple Integrals.** 556

9.5   **Applications Involving Multiple Integration:** 564

Section 9.1 *Iterated Integration.*

Iterated integrals are the generalization of a definite integral to functions of two or more variables. "They are said to be iterated because they are nested integrals that are evaluated sequentially, considering only one variable at a time, and using the value of this *partial integration* with respect to one variable as the function to be integrated in the next integration. In this section we present the "mechanics" of evaluating "iterated integrals" as a prelude to the study of "multiple integration" in the following sections. We begin by considering "double integrals" which involve the integration of functions of two variables. The procedure will be quite simple - it will consist of evaluating two ordinary integrals sequentially. The methods used for two dimensions is easily extended to higher dimensions. Applying the same procedure, an iterated integrals of functions of three variables, called a "triple" iterated integral, can be evaluated by performing three integrations in a row.

Assume that the function F(x,y) is an algebraic combination of elementary terms involving x, or y, or both x and y. These terms may be "rational" terms such as

$$x + 3y \quad \text{or} \quad 3xy - y^4 \quad \text{or} \quad \{x - y\} / \{x + y\},$$

or terms involving elementary trigonometric, exponential, or logarithm functions such as

$$xe^y \quad \text{or} \quad \sin(x) - y \quad \text{or} \quad ln(x/y)$$

An "iterated integral of F(x,y)" is like a definite integral and has specific limits of integration. However, since the integrand F(x,y) is a function of two variables, an iterated integral of F involves two integrals, each with its own limits of integration. An iterated integral is expressed as one set of symbols with two integral signs and two corresponding differential symbols that indicate the "order" of integration. The two basic forms are:

             "dy dx Form"   **"dx dy Form"**

$$\int_a^b \int_{g(x)}^{f(x)} F(x,y) \, dy \, dx \quad \text{and} \quad \int_c^d \int_{l(y)}^{r(y)} F(x,y) \, dx \, dy$$

The integral symbols dx and dy are paired with the differential symbols $\int$ in an ordered fashion. The innermost differential and integral sign are linked and the outer symbols are linked. Thus an iterated integral is really just two "nested" ordinary integrals. For a "dy dx Form" the two corresponding integrals have limits of the form:

        the innermost integral   *the outer integral*

$$\int_{g(x)}^{f(x)} \underline{\qquad} dy \quad \text{and} \quad \int_a^b \underline{\qquad} dx$$

Iterated integrals are evaluated by evaluating the innermost integral first and then the outer integral. Thus the "dy dx Form" indicates an integration with respect to y first and then with respect to x, while the "dx dy Form" indicates the order of integration is first with respect to x and then with respect to y.

The limits of integration for the first integral may be functions. In the "dy dx Form" the integration limits for y are indicated by f(x) and g(x) and in the "dx dy Form" the integration limits for x are denoted by l(y) and r(y). The outermost integral must have constant limits.

DEFINITION OF ITERATED INTEGRALS.

If $f(x)$ and $g(x)$ are continuous functions for $x \in [a,b]$ and $F(x,y)$ is a continuous function of x and y for $a \leq x \leq b$ and $g(x) \leq y \leq f(x)$ then the **iterated integral**

$$\int_a^b \int_{g(x)}^{f(x)} F(x,y) \, dy \, dx = \int_a^b \left(\int_{y=g(x)}^{y=f(x)} F(x,y) \, dy \right) dx$$

If $l(y)$ and $r(y)$ are continuous functions for $y \in [c,d]$ and $F(x,y)$ is a continuous function of x and y for $c \leq y \leq d$ and $l(y) \leq x \leq r(y)$ then the **iterated integral**

$$\int_c^d \int_{l(y)}^{r(y)} F(x,y) \, dx \, dy = \int_c^d \left(\int_{x=l(y)}^{x=r(y)} F(x,y) \, dx \right) dy$$

Each form of the iterated integral is evaluated by performing two integrations. The first integration is evaluated by treating the second variable as a constant. A "partial antiderivative" is determined using the Integral Rules for single variable functions using the variable of integration. The integral is then the difference in the value of this "partial antiderivative" at the upper integration limit minus its value at the lower integration limit. The resulting expression will be a function of the second variable. It is then integrated over the outer limits of integration as an ordinary integral. This procedure is much simpler in practice than it appears when described in words.

Example 1 **Evaluating iterated integrals**.

Problem A   Evaluate the iterated integral $\int_2^3 \int_4^5 y + x \, dx \, dy$.

Solution   The first integration is with respect to x, since the dx term precedes the dy term. Treating y as a constant, a "partial" antiderivative of $F(x,y) = y + x$ with respect to the x-variable is

$$\int y + x \, dx = yx + (1/2)x^2$$

The y term is integrated to give yx just as a constant 4 would integrate to 4x, with the y treated as the constant 4. Notice that a general "constant of integration" is not included in the antiderivative. In this example the integration limits for the first integral are both constants. The iterated integral is then evaluated as

$$\int_2^3 \int_4^5 y + x \, dx \, dy = \int_2^3 \left(yx + (1/2)x^2 \big|_{x=4}^{x=5} \right) dy$$

$$= \int_2^3 \left([5y + (1/2)5^2] - [4y + (1/2)4^2] \right) dy$$

Section 9.1   ITERATED INTEGRATION

$$= \int_2^3 y + 9/2 \ dy \ = (1/2)y^2 + (9/2)y|_2^3$$

$$= [(1/2)3^2 + (9/2)3] - [(1/2)2^2 + (9/2)2] = 7$$

Problem B   Evaluate the integral $\int_0^1 \int_x^{3x} 1 + y + x \ dy \ dx$.

Solution   The first integral to be evaluated is the innermost integral $\int_x^{3x} 1 + y + x \ dy$.

An antiderivative of $1 + y + x$, with respect to y, is

$$\int 1 + y + x \ dy = y + y^2/2 + yx.$$

Therefore, the first integral is

$$\int_x^{3x} 1 + y + x \ dy = y + y^2/2 + yx \Big|_{y=x}^{y=3x}$$

$$= [3x + (3x)^2/2 + 3x \times x] - [x + x^2/2 + x \times x]$$

$$= 3x + (9/2)x^2 + 3x^2 - x - x^2/2 - x^2$$

$$= 3x + 6x^2$$

Using this the iterated integral is then evaluated as

$$\int_0^1 \left(\int_x^{3x} 1 + y + x \ dy \right) dx = \int_0^1 3x + 6x^2 \ dx$$

$$= (3/2)x^2 + 2x^3 \Big|_0^1 = 3/2 + 2 = 7/2$$

Problem C   Evaluate the iterated integral $\int_2^3 \int_1^y xy \ dx \ dy$.

Solution   The first integral to be evaluated is with respect to the variable x, since the dx term precedes the dy term. A partial antiderivative of xy with respect to x is

$$\int xy \ dx = (1/2)x^2 y.$$

The y factor is treated as a constant multiple. Therefore, the iterated integral is evaluated as:

$$\int_2^3 \int_1^y xy \ dx \ dy = \int_2^3 \left((1/2)x^2 y \Big|_{x=1}^{x=y} \right) dy$$

$$= \int_2^3 \left((1/2)y^2 y - (1/2)1^2 y \right) dy$$

$$= \int_2^3 (1/2)y^3 - (1/2)y \ dy \ = (1/8)y^4 - (1/4)y^2 \Big|_2^3$$

$$= [(1/8)3^4 - (1/4)3^2] - [(1/8)2^4 - (1/4)2^2] = 55/8$$

Problem D   Evaluate the integral $\int_0^{\sqrt{\pi/2}} \int_0^{r^2} r \sin(\theta) \ d\theta \ dr$.

Solution   In this problem the variables are r and θ. The first integration is with respect to θ. An antiderivative of $r \sin(\theta)$ with respect to θ, treating r as a constant, is

$$\int r \sin(\theta) \ d\theta = -r \cos(\theta).$$

Thus the iterated integral is

$$\int_0^{\sqrt{\pi/2}} \int_0^{r^2} r \sin(\theta) \ d\theta \ dr = \int_0^{\sqrt{\pi/2}} \left(-r \cos(\theta) \Big|_{\theta=0}^{\theta=r^2} \right) dr$$

$$= \int_0^{SQRT(\pi/2)} -r\cos(r^2) + r\cos(0)\, dr$$

$$= \int_0^{SQRT(\pi/2)} -r\cos(r^2) + r\, dr \qquad \text{As } \cos(0) = 1.$$

$$= [-(1/2)\sin(r^2) + (1/2)r^2]\,|_0^{SQRT(\pi/2)}$$

$$= [-(1/2)\sin((SQRT(\pi/2))^2) + (1/2)(SQRT(\pi/2))^2]$$
$$\quad - [-(1/2)\sin(0^2) + (1/2)0^2]$$

$$= -(1/2)\sin(\pi/2) + \pi/4 = \pi/4 - 1/2$$

☑

When evaluating iterated integrals we have indicated the limits at which the "partial" antiderivative should be evaluated by equations to help you identify the variable that is being evaluated. This is a good practice that may help avoid mistakes.

Higher dimensional iterated integrals.

Iterated integrals involving three or more variables are evaluated by extending the method of successive integrations to as many variables and thus integrals as given. When considering a function of three variables, for instance H(x,y,z), a "triple" integrated integral can be formed. There are six different orders in which such integrals can be formed, corresponding to the six possible ways in which the variables x, y, and z can be ordered. In these integrals the limits of integration can be constants or functions of the variables that have yet to be integrated. In such cases it is again very helpful to write the limits of integration as equations. The integral symbols with limits of integration and the differential symbols are paired, like left and right parentheses, with the innermost integration being evaluated first. For three variables there are six different orders in which iterated integrals can occur. For four variables this jumps to 24 different forms. Instead of indicating the most general forms we illustrate the process of evaluating such integrals by the following example.

Example 2   **Evaluating triple iterated integrals.**

Example A   Evaluate the "triple" iterated integral with constant limits of integration

$$I = \int_0^4 \int_3^8 \int_1^2 x + y + z\, dx\, dy\, dz$$

Solution   This integral is evaluated as three iterated integrals, first with respect to x, then with respect to y and finally with respect to z. Treating y and z as constants, a "partial antiderivative" of the integrand is

$$\int x + y + z\, dx = x^2/2 + yx + zx$$

Thus,

$$I = \int_0^4 \int_3^8 \left(\int_1^2 x + y + z\, dx\right) dy\, dz$$

$$= \int_0^4 \int_3^8 (x^2/2 + yx + zx)|_{x=1}^{x=2}\, dy\, dz$$

$$= \int_0^4 \int_3^8 (2^2/2 + 2y + 2z) - (1^2/2 + y + z)\, dy\, dz$$

$$= \int_0^4 \int_3^8 1 + y + z\, dy\, dz$$

This "double" integral is then evaluated by "partial" integrating with respect to y and then integrating with respect to z. Continuing the integration from above:

$$I = \int_0^4 \left(\int_3^8 1 + y + z \, dy \right) dz$$

$$= \int_0^4 \left. (y + y^2/2 + zy) \right|_{y=3}^{y=8} dz$$

$$= \int_0^4 (8 + 8^2/2 + 8z) - (3 + 3^2/2 + 3z) \, dz$$

$$= \int_0^4 32.5 + 5z \, dz = \left. 32.5z + 5z^2/2 \right|_0^4 = 170$$

Problem B Evaluate the "triple" integral $I = \int_0^2 \int_0^{5x} \int_0^{x-2y} 1 + zy - yx \, dz \, dy \, dx$.

Solution The integral is evaluated by first integrating with respect to z, treating y and x as constants, and evaluating this "partial antiderivative" at the upper z-limit minus its value at the lower z-limit. The resulting expression will be a function of y and x but <u>not z</u>. It is then integrated with respect to y, evaluating the "partial antiderivative" for $0 \le y \le 5x$. At this stage the resulting integrand is a function of x only and is integrated over the interval [0,2].

A "partial" antiderivative with respect to z of $H(x,y,z) = 1 + zy - yx$ is

$$\int 1 + zy - yx \, dz = z + (1/2)z^2 y - yxz$$

The first integration is thus

$$I = \int_0^2 \int_0^{5x} \left(\int_0^{x-2y} 1 + zy - yx \, dz \right) dy \, dx$$

$$= \int_0^2 \int_0^{5x} \left(\left. z + (1/2)z^2 y - yxz \right|_{z=0}^{z=x-2y} \right) dy \, dx$$

$$= \int_0^2 \int_0^{5x} x - 2y + (1/2)(x - 2y)^2 y - yx(x - 2y) \, dy \, dx$$

As each term is zero at $z = 0$.

$$= \int_0^2 \int_0^{5x} x - (2 + (1/2)x^2)y + 2y^3 \, dy \, dx$$

The integrand resulting after the z-integration is rearranged by grouping terms with the same power of y so that the second integration, with respect to y, treating x as a constant can be more easily evaluated. Continuing from above:

$$I = \int_0^2 \left(\left. xy - (2 + (1/2)x^2)y^2/2 + y^4/2 \right|_{y=0}^{y=5x} \right) dx$$

$$= \int_0^2 x \times 5x - (2 + (1/2)x^2)(5x)^2/2 + (5x)^4/2 \, dx$$

As each term is zero when $y = 0$.

$$= \int_0^2 5x^2 - (2 + (1/2)x^2)25x^2/2 + 625x^4/2 \, dx$$

$$= \int_0^2 -20x^2 + (1225/4)x^4 \, dx$$

Finally, integrating with respect to x, the value of the iterated integral is found:

$$I = \left. (-20/3)x^3 + (1225/20)x^5 \right|_0^2$$

$$= (-20/3)2^3 + (1225/20)2^5 = 2090.416666...$$

After each "partial integration" the resulting integrand was algebraically simplified as much as possible and rearranged, gathering terms with like powers of the next variable to be integrated.

Exercise Set 9.1

1. Evaluate the given iterated integral.

   a) $\int_0^1 \int_3^5 1 \, dx \, dy$
   b) $\int_0^1 \int_3^5 1 \, dy \, dx$
   c) $\int_0^1 \int_0^3 x \, dx \, dy$

   d) $\int_0^1 \int_0^3 x \, dy \, dx$
   e) $\int_0^1 \int_x^2 y - x \, dy \, dx$
   f) $\int_0^1 \int_y^2 y - x \, dx \, dy$

   g) $\int_0^1 \int_3^5 xy^2 \, dx \, dy$
   h) $\int_0^1 \int_3^5 xy^2 \, dy \, dx$
   i) $\int_0^1 \int_0^x 2xy - y \, dy \, dx$

   j) $\int_0^1 \int_0^y 2xy - y \, dx \, dy$
   k) $\int_{-1}^2 \int_0^{x^2} xe^y \, dy \, dx$
   l) $\int_1^2 \int_x^{x^2 - x} y/x \, dy \, dx$

2. Evaluate the given iterated integral.

   a) $\int_{-\pi}^{\pi} \int_\theta^{\theta^2} r \, dr \, d\theta$
   b) $\int_{-2}^{1/2} \int_0^{r^2} r \, d\theta \, dr$
   c) $\int_{-\pi}^{\pi} \int_\theta^{\theta^2} r - \theta \, dr \, d\theta$

   d) $\int_0^2 \int_{-r}^r r - \theta \, d\theta \, dr$
   e) $\int_0^1 \int_0^{r^2} r \cos(\theta) \, d\theta \, dr$
   f) $\int_0^1 \int_0^\theta r \cos(\theta^3) \, dr \, d\theta$

   g) $\int_0^1 \int_0^r r \sin(\theta) \, d\theta \, dr$
   h) $\int_0^1 \int_0^\theta r \sin(\theta) \, dr \, d\theta$

3. Evaluate the given integral.

   a) $\int_0^2 \int_1^3 s + t \, ds \, dt$
   b) $\int_0^2 \int_3^0 s + t \, ds \, dt$
   c) $\int_0^2 \int_1^s s + t \, dt \, ds$

   d) $\int_0^2 \int_{-t}^0 s + t \, ds \, dt$
   e) $\int_0^2 \int_1^s se^{-t} \, dt \, ds$
   f) $\int_0^2 \int_1^s te^{-s} \, dt \, ds$

   g) $\int_1^2 \int_1^s t/s \, dt \, ds$
   h) $\int_1^2 \int_1^s s/t \, dt \, ds$

4. Evaluate the given "triple" integral.

   a) $\int_0^1 \int_0^2 \int_1^3 1 \, dx \, dy \, dz$
   b) $\int_1^3 \int_0^2 \int_0^1 1 \, dz \, dy \, dx$
   c) $\int_0^1 \int_0^2 \int_1^3 x \, dx \, dy \, dz$
   d) $\int_0^1 \int_0^2 \int_1^3 x \, dy \, dx \, dz$
   e) $\int_0^1 \int_0^2 \int_1^3 y \, dx \, dy \, dz$
   f) $\int_0^1 \int_0^2 \int_1^3 y \, dz \, dy \, dx$
   g) $\int_0^1 \int_0^2 \int_1^3 z \, dx \, dy \, dz$
   h) $\int_0^1 \int_0^2 \int_3^4 x + 2y + z \, dx \, dy \, dz$
   i) $\int_0^1 \int_0^2 \int_3^4 x + 2y + z \, dz \, dy \, dx$
   j) $\int_0^1 \int_0^2 \int_3^4 x + 2y + z \, dy \, dz \, dx$
   k) $\int_0^1 \int_0^{2z} \int_z^y x \, dx \, dy \, dz$
   l) $\int_0^1 \int_0^{2z} \int_z^y x + y + z \, dx \, dy \, dz$
   m) $\int_0^1 \int_{-x}^x \int_5^y x \, dz \, dy \, dx$
   n) $\int_0^1 \int_0^{2x} \int_x^y x + y + z \, dz \, dy \, dx$

Section 9.2 The Double Integral

Introduction.

In this section the concept of integrating a function of two variables over a given region R is introduced. We focus on the integration of a function F(x,y) of the two variables x and y over a region R in the x-y plane. This type of integral is called a **double integral** and is defined, like the ordinary definite integral was defined in Chapter 5, as all integrals are, as the limit of a specific type of sum called a Riemann Sum. In this section it is shown that double integrals can be expressed as simple iterated integrals when the region R is rectangular. In the next section more general regions are considered and again it will turn out that the double integral can be expressed as iterated integrals, with function limits of integration. In Sections 9.4 and 9.5 applications of the double integral are introduced.

To start, assume that the region R is finite and is bounded by a set of curves that are the graphs of equations in x and y. The function F(x,y) is assumed to be defined and continuous on the region R. The concept of "continuity" for functions of two variables is a generalization of the concept defined for a single variable function; intuitively, a continuous function's graph will not have "jumps". This is the simple interpretation of the limit condition required for F(x,y) to be continuous at a point (x_0, y_0), that

$$\lim_{(x,y) \to (x_0, y_0)} F(x,y) = F(x_0, y_0)$$

An integral over a multi-dimensional region R is called a **multiple integral**; in the case we are considering since F(x,y) is a function of two variables the term **double integral** is used. A double integral is denoted by two integral signs, without limits of integration, followed by the name of the region, R, the function F(x,y) and then a "differential" symbol, which may be either "dA" or a pair of symbols "dx dy" or "dy dx":

$$\iint_R F(x,y)\, dA \qquad \text{or} \qquad \iint_R F(x,y)\, dx\, dy$$

The double integral of F(x,y) over the region R

The "differential" symbols "dA" and "dx dy" both represent the concept of an area increment. Both symbols are frequently used and represent the same concept.

Region R is the union of subregions.

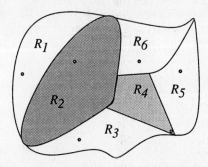

• *Evaluation Points* (x_n^*, y_n^*)

Like the definition of an ordinary single-variable definite integral the mathematical definition of a double integral is given by a limit. The limit definition is general and difficult to apply to evaluate specific integrals. However, when the region R is "nice" the double integral can be evaluated by rearranging the limit as two "nested" limits. It follows that the double integral can be expressed as an iterated integral and thus it can be evaluated using the Integral Rules of single variable calculus. To give the limit definition of the Double Integral we must first introduce some notation and terms.

A region R is said to be **partitioned**, or divided, into a set of subregions $\{R_1, R_2, \ldots, R_N\}$, as illustrated on the next page. We assume that R is the union of the subregions and the subregions do not intersect, except at their boundaries, i.e.,

$$R = \cup\, R_n \quad \text{and} \quad \text{interior}(R_i) \cap \text{interior}(R_j) = \phi \quad \text{the empty set, if } i \neq j$$

The **area** of the n-th subregion R_n is denoted by A_n.

An **evaluation** point in R_n is denoted by a subscripted point with asterisks, (x_n^*, y_n^*).

The **partial sum,** or **Double Riemann Sum,** associated with a partition $\{R_n\}_{n=1,N}$ and a corresponding set of evaluation points $\{(x_n^*, y_n^*)\}_{n=1,N}$ has the form:

Partial Sum or **Double Riemann Sum**

$$S = \Sigma_{n=1,N} \; F(x_n^*, y_n^*) A_n$$

The double integral is defined as the limit of such Riemann Sums. In turn, each such sum having this form will provide an approximation of the double integral.

DEFINITION OF THE DOUBLE INTEGRAL.

If $F(x,y)$ is continuous on a region R of the x-y plane, then the **double integral of F(x,y) over R** is

$$\iint_R F(x,y) \, dA = \lim_{\substack{N \to \infty \text{ and } A_n \to 0}} \Sigma_{n=1,N} \; F(x_n^*, y_n^*) A_n,$$

where the limit is considered over all possible Riemann Sums corresponding to partitions $\{R_n\}_{n=1,N}$ of R and sets of evaluation points $\{(x_n^*, y_n^*)\}_{n=1,N}$.

The limit used in the definition of the double integral involves three limit concepts simultaneously:

1) The limit as the number N of subregions becomes infinite.
2) The limit as the area of each subregion approaches zero.
3) The limit for all possible choices for the evaluation points (x_n^*, y_n^*).

The limit involved in the definition of the double integral is very complicated. An important observation is that if it exist, all partition strategies lead to the same limit. If the double integral exists, then the limit of the Riemann sums for any choice of partitions and evaluation points will lead to the value of the integral. The only restriction is that the areas of the subregions go to zero as the number of subregions increases to infinity.

> This means that we have a great freedom to make the partitions and
> choose the evaluation points to suit our convenience, that is, to make
> the summations as simple as possible to evaluate.

The limit definition of the double integral is used in theoretical situations to establish mathematical theorems and properties of multiple integrals and to develop numerical approximation formulas. It is also used in applications, when modeling physical or biological processes, to show that a specific quantitiy can be evaluated by a double integral. This is done by first showing that the process can be approximated by a Double Riemann type sum, like that in the definition of the double integral. Then, it is argued that the process being modeled is actually described by the limit of such approximating sums, a limit that by definition gives the integral over a region R of a specific function.

The limit definition of a double integral is very rarely used to evaluate a specific integral. Instead, a two-dimensional version of the Fundamental Theorem of Calculus is employed. This consists of representing a double integral $\iint_R F(x,y)\, dA$ as an "iterated integral" and applying the Integration Rules for single variable calculus to integrate the integral.

The main challenge in representing a double integral as an iterated integral is to establish the appropriate limits of integration. The limits of integration are determined from the boundary curves of the region R. **The function F(x,y) does not affect the limits of integration, they are determined solely by the region R.**

In this section we consider the simplest type of region, a rectangular region whose boundaries are constant and hence give rise to iterated integrals with constant limits of integration. Double integrals over more general regions are considered in the next section. As we shall see there, double integrals over regions that have two constant sides and the other two sides consist of the graphs of functions can also be evaluated as iterated integrals. If a region R cannot be split into sub-regions of this type, then we will not be able to evaluate double integrals over it as iterated integrals.

REPRESENTING A DOUBLE INTEGRAL OVER A RECTANGLE AS AN ITERATED INTEGRAL.

The general process of converting a double integral to an iterated integral consists of making a specific choice for the partitions of the region R into subregions. The choice is made so that the Double Riemann sum representing the double integral can be rearranged as two "nested" sums over single variables. Each of the "nested" sums is in turn seen to have the form of a simple Riemann sum, as introduced in Chapter 4 to define the Definite Integral of a single variable function. Thus, in the limit each single variable sum gives a definite integral in the corresponding single variable. The net result, like the Fundamental Theorem of Calculus, is that double integrals can be evaluated as an iterated integrals.

Double integrals over rectangular regions.

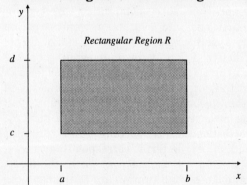

Consider a rectangular region R as illustrated at the left. The set-notation description of this region is

$$R = \{(x,y) \mid a \leq x \leq b \text{ and } c \leq y \leq d\}$$

To evaluate a double integral over R as a limit, the first step is to systematically partition R into subregions. It is easy to partition a rectangle into rectangular subregions. This is done by constructing a grid system. We start partitioning the intervals on the x and y axes that correspond to the boundary of the rectangle R.

Let $\{x_i\}_{i=0,N}$ be a partition of the x-interval [a,b] into N subintervals:

$$x_0 = a < x_1 < x_2 < \ldots < x_N = b$$

Similarly, let $\{y_j\}_{j=0,M}$ be a partition of the y-interval [c,d] into M subintervals:

$$y_0 = c < y_1 < y_2 < \ldots < y_M = d$$

The rectangular grid of the region R is formed by the intersection with R of the

       vertical lines: $x = x_i$, for $i = 0, 1, \ldots, N$

and

horizontal lines: $y = y_j$, for $j = 0, 1, ..., M$

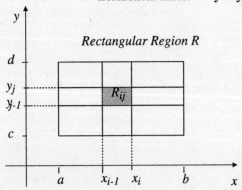
Rectangular Region R

The grid divides R into rectangular subregions that are numbered by rows and columns, relative to the lower left corner of R.

The sub-rectangle in the i-th row up and j-th column to the right is denoted by $R_{i,j}$. This rectangle is bounded by four lines, as illustrated at the left,

by the vertical lines $x = x_i$, on the right, and
$x = x_{i-1}$, on the left;

by the horizontal lines $y = y_j$, above, and
$y = y_{j-1}$, below.

The area of the sub-rectangle $R_{i,j}$ is the product of its width and height:

$$A_{i,j} = \Delta x_i \cdot \Delta y_j \quad \text{or} \quad A_{i,j} = \Delta y_j \cdot \Delta x_i,$$

where its width Δx_i is the length of the i-th subinterval of [a,b] and its height Δy_j is the length of the j-th subinterval of [c,d]. These lengths are the differences

$$\Delta x_i = x_i - x_{i-1} \quad \text{and} \quad \Delta y_j = y_j - y_{j-1}$$

To choose the evaluation point for the rectangle $R_{i,j}$ a number x_i^* is chosen in the i-th x-interval $[x_{i-1}, x_i]$ and a second number y_j^* is selected in the j-th y-interval $[y_{j-1}, y_j]$. The representative function value in the rectangle $R_{i,j}$ is the value of F at this point: $F(x_i^*, y_j^*)$.

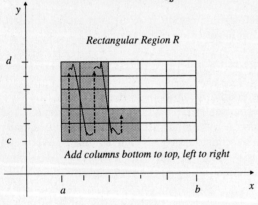
Rectangular Region R
Add columns bottom to top, left to right

The Riemann Sum corresponding to this type of rectangular partitions and evaluation points can be expressed very systematically as "nested sums". This can be done in two different forms depending on the order in which the terms are added. The first form is illustrated at the left. The nested sum consists of adding the terms corresponding to each subregion. Starting in the lower left corner with the rectangle $R_{1,1}$, the terms corresponding to rectangles in the first column are added, i.e. the terms corresponding to $R_{1,2}$, $R_{1,3}$, ... up to the top rectangle $R_{1,M}$. Then, the terms corresponding to the second column are added from the bottom to the top, i.e., corresponding to $R_{2,1}$, $R_{2,2}$, ... , $R_{2,M}$, in order.

This process is repeated, adding the terms in each column to the right until the last column has been added. This adding process is indicated by a Riemann Sum of the form:

$$S = \Sigma_{i=1,N} \Sigma_{j=1,M} \; F(x_i^*, y_j^*) \cdot (\Delta y_j \cdot \Delta x_i)$$

The index "j" is associated with the y-values, so that the first summation, $\Sigma_{j=1,M}$, indicates a sum up a column, which column is indicated by the i-value, that is held fixed.

The index "i" is associated with the x-variables so the second sum, $\Sigma_{i=1,N}$, indicates the addition of all the column-sums, starting with the left column.

In expanded form this sum is:

$$S = \sum_{i=1,N} \sum_{j=1,M} F(x_i^*,y_j^*) \cdot (\Delta y_j \cdot \Delta x_i) = \sum_{i=1,N} \left(\sum_{j=1,M} F(x_i^*,y_j^*) \cdot \Delta y_j \cdot \Delta x_i \right)$$

$$= \underbrace{\sum_{j=1,M} F(x_1^*,y_j^*) \cdot \Delta y_j \cdot \Delta x_1}_{\substack{\text{sum over 1st column} \\ i = 1}} + \underbrace{\sum_{j=1,M} F(x_2^*,y_j^*) \cdot \Delta y_j \cdot \Delta x_2}_{\substack{\text{sum over 2nd column} \\ i = 2}}$$

$$+ \underbrace{\sum_{j=1,M} F(x_3^*,y_j^*) \cdot \Delta y_j \cdot \Delta x_3}_{\substack{\text{sum over 3rd column} \\ i = 3}} + \ldots + \underbrace{\sum_{j=1,M} F(x_N^*,y_j^*) \cdot \Delta y_j \cdot \Delta x_N}_{\substack{\text{sum over the right-most N-th column} \\ i = N}}$$

Notice that in expanded form each term itself has the form of a Riemann Sum for a function of one variable. Such Riemann Sums in the limit become definite integrals. For instance, the limit of the sum over the 3rd column, as $M \to \infty$ and $\Delta y_j \to 0$ is

$$\lim_{M \to \infty, \Delta y \to 0} \sum_{j=1,M} F(x_3^*,y_j^*) \cdot \Delta y_j \cdot \Delta x_3 = \int_c^d F(x_3^*,y) \, \Delta x_3 \, dy$$

The limits of integration, c and d, are the endpoints of the y-interval partitioned by the y_j-values.

The double integral is evaluated as the limit of the Riemann Sum S as both N and M approach ∞ and both Δx and Δy approach zero. Taking the limits of M and Δy first gives a sum of integrals with respect to y. Then taking the limits of N and Δx gives a second iterated integral with respect to x.

$$\iint_R F(x,y) \, dA = \lim_{N \to \infty \, M \to \infty \, \Delta x \to 0 \, \Delta y \to 0} \sum_{i=1,N} \sum_{j=1,M} F(x_i^*,y_j^*) \cdot \Delta y_j \cdot \Delta x_i$$

$$= \lim_{N \to \infty \, \Delta x \to 0} \sum_{i=1,N} \left[\lim_{M \to \infty \, \Delta y \to 0} \sum_{j=1,M} \left(F(x_i^*,y_j^*) \cdot \Delta y_j \right) \Delta x_i \right]$$

$$= \lim_{N \to \infty \, \Delta x \to 0} \sum_{i=1,N} \left[\left(\int_c^d F(x_i^*,y) \, dy \right) \Delta x_i \right]$$

$$= \int_a^b \int_c^d F(x,y) \, dy \, dx$$

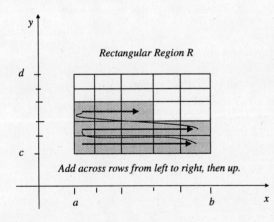

Add across rows from left to right, then up.

Another way to systematically add a term for each rectangle is to add across rows. Start in the bottom left again, but this time add across each row from left to right and at the end of a row returning to the left of the next row up. Adding across a row consists of a sum over the "i" index, from the left side, i = 1, to the right side, i = N. Adding the subtotals for individual rows is indicated by a second summation over the "j" index, to indicate the sum moving up from the bottom row, j = 1, to the top row j = M. The total sum over the entire set of sub-rectangles is the same value S describe above. It is given this time by the nested Riemann Sums with the order of the summations reversed from the above formula:

$$S = \sum_{j=1,M} \sum_{i=1,N} F(x_i^*,y_j^*) \cdot (\Delta x_i \cdot \Delta y_j)$$

$$= \underbrace{\sum_{i=1,N} F(x_i^*,y_1^*) \cdot \Delta x_i \cdot \Delta y_1}_{\substack{\text{sum across the 1st row} \\ j = i}} + \underbrace{\sum_{i=1,N} F(x_i^*,y_2^*) \cdot \Delta x_i \cdot \Delta y_2}_{\substack{\text{sum across the 2nd row} \\ j = 2}}$$

$$+ \underbrace{\sum_{i=1,N} F(x_i^*,y_3^*) \cdot \Delta x_i \cdot \Delta y_3}_{\substack{\text{sum across the 3rd row} \\ j = 3}} + \ldots + \underbrace{\sum_{i=1,N} F(x_i^*,y_M^*) \cdot \Delta x_i \cdot \Delta y_M}_{\substack{\text{sum across the top, M-th row} \\ j = M}}$$

Using this order of summation for the Riemann Sum S, the double integral can again be evaluated by taking the limits as M and N $\to \infty$ and Δx and $\Delta y \to 0$. The double integral is then given by

$$\iint_R F(x,y) \, dA = \lim_{M \to \infty \, N \to \infty \, \Delta x \to 0 \, \Delta y \to 0} \Sigma_{j=1,M} \Sigma_{i=1,N} \, F(x_i^*, y_j^*) \bullet (\Delta x_i \bullet \Delta y_j)$$

$$= \lim_{M \to \infty \, N \to \infty \, \Delta y \to 0 \, \Delta x \to 0} \Sigma_{j=1,M} \Sigma_{i=1,N} \left(F(x_i^*, y_j^*) \bullet \Delta x_i \right) \bullet \Delta y_j$$

$$= \lim_{M \to \infty \, \Delta y \to 0} \Sigma_{j=1,M} \left[\lim_{N \to \infty \, \Delta x \to 0} \Sigma_{i=1,N} \left(F(x_i^*, y_j^*) \bullet \Delta x_i \right) \bullet \Delta y_j \right]$$

$$= \lim_{M \to \infty \, \Delta y \to 0} \Sigma_{j=1,M} \left[\left(\int_a^b F(x, y_j^*) \, dx \right) \Delta y_j \right]$$

$$= \int_c^d \int_a^b F(x,y) \, dx \, dy$$

The double integral of F(x,y) over a rectangular region

$$R = \{(x,y) \mid a \leq x \leq b \text{ and } c \leq y \leq d\}$$

can be evaluated as an iterated integral in

the **dx dy form**:

$$\iint_R F(x,y) \, dA = \int_c^d \int_a^b F(x,y) \, dx \, dy$$

or in

the **dy dx form**:

$$\iint_R F(x,y) \, dA = \int_a^b \int_c^d F(x,y) \, dy \, dx$$

Example 1   **Evaluating Double Integrals over rectangular regions.**

Problem   Evaluate the double integral of $F(x,y) = x + y^2$ over the indicated regions:
   (a) the rectangle determined by the inequalities $-1 \leq x \leq 5$ and $2 \leq y \leq 7$;
   (b) the rectangle with opposite vertices (1,2) and (3,5);
   (c) the rectangle determined by the inequalities $|x| \leq 2$ and $|y - 3| \leq 1$.

Solutions (a) The double integral is evaluated as the iterated integral of the dx dy form:

$$\iint_R x + y^2 \, dA = \int_2^7 \int_{-1}^5 x + y^2 \, dx \, dy$$

$$= \int_2^7 x^2/2 + y^2 x \big|_{-1}^5 \, dy = \int_2^7 \{25/2 + 5y^2\} - \{1/2 - y^2\} \, dy$$

$$= \int_2^7 12 + 6y^2 \, dy = 12y + 2y^3 \big|_2^7 = \{94 + 2 \cdot 343\} - \{24 + 16\} = 740$$

(b) The rectangle R is $\{(x,y) \mid 1 \leq x \leq 3 \text{ and } 2 \leq y \leq 5\}$. Therefore, using the "dy dx form" the double integral is evaluated as

$$\iint_R x + y^2 \, dA = \int_1^3 \int_2^5 x + y^2 \, dy \, dx$$

$$= \int_1^3 xy + y^3/3 \big|_2^5 \, dx = \int_1^3 3x + 39 \, dx = 3x^2/2 + 39x \big|_1^3 = 90$$

(c) The rectangle R is given in set notation by $\{(x,y) \mid -2 \le x \le 2 \text{ and } -1 \le y - 3 \le 1\}$ or

$$R = \{(x,y) \mid -2 \le x \le 2 \text{ and } 2 \le y \le 4\}$$

Thus,

$$\iint_R x + y^2 \, dA = \int_{-2}^{2} \int_{2}^{4} x + y^2 \, dy \, dx$$

$$= \int_{-2}^{2} xy + y^3/3 \, \big|_{2}^{4} \, dx = \int_{-2}^{2} 2x + 56/3 \, dx$$

$$= x^2 + (56/3)x \, \big|_{-2}^{2} = 224/3 = 74.6$$

Exercise Set 9.2

In exercises 1 and 2:
(i) Sketch the indicated rectangular region R on an x-y axis system.
(ii) Describe the region R in set notation.
(iii) Express the double integral of F(x,y) over R as an iterated integral in both the "dy dx form" and the "dx dy form".
(iv) Evaluate $I = \iint_R x^3 - y \, dA$.

1. R is the rectangular shaped region whose opposite vertices are:

   a) (0,0) and (3,2)   b) (0,0) and (2,2)

   c) (0,0) and (-1,4)   d) (2,1) and (3,3)

   e) (2,1) and (-3,3)   f) (2,1) and (-1,2)

   g) (-1,2) and (2,-1) h) (8,2) and (1,4)

2. R is the region satisfying the given inequalities:

   a) $|x| \le 5$ and $|y| \le 2$ b) $|x| \le 3$ and $|y - 7| \le 3$

   c) $|x - 3| \le 5$ and $|y - 10| \le 2$ d) $|x - 4| \le 3$ and $|y - 7| \le 3$

   e) $|x + 3| \le 5$ and $|y - 1| \le 2$ f) $|x + 4| \le 3$ and $|y + 7| \le 3$

3. Evaluate the integral of $F(x,y) = xy^2$ over the indicated regions:

   a) the rectangle determined by the inequalities $-1 \le x \le 5$ and $2 \le y \le 7$;

   b) the rectangle with opposite vertices (1,2) and (3,5);

   c) the rectangle determined by the inequalities $|x| \le 2$ and $|y - 3| \le 1$.

4. Evaluate the integral of $F(x,y) = r \sin(\theta)$ over the indicated regions:

   a) the rectangle determined by the inequalities $0 \le r \le 4$ and $-\pi \le \theta \le \pi$;

   b) the rectangle with opposite vertices $(0,\pi)$ and $(1,\pi/2)$;

   c) the rectangle determined by the inequalities $0 \le r \le 2$ and $|\theta - \pi/3| \le \pi/6$.

5. If $\iint_R x - y \, dA = -1$, and R is a rectangle whose height is twice its width, with one vertex at (1,2) what are the coordinates of the vertex of R that is opposite this given vertex? Are there two possible answer?

Section 9.3 Double Integrals Over General Regions: Determining limits of integration.

Establishing iterated integrals over non-rectangular regions.

When the region R is not a simple rectangle the double integral $\iint_R F(x,y)\, dA$ can often be evaluated as an iterated integral with variable limits of integration. The limits of the iterated integrals will depend on the boundary of the region R and not on the actual function being integrated. To establish the limits of integration we initially shall consider regions that have two sides "parallel" to a coordinate axes. The other two sides may be straight lines or curves that are the graphs of functions. Regions with these types of boundaries are illustrated in Figure 9.10. The term "parallel" is used in quotation marks since one or both of the "parallel" sides may only be a single point where the "curved" sides intersect.

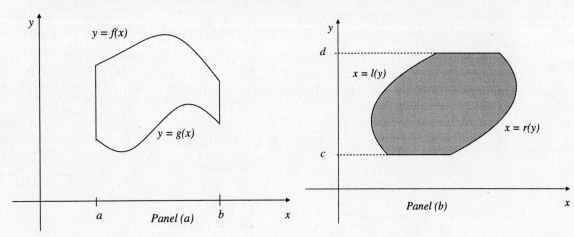

The region R is depicted in panel (a) has two "parallel" sides formed by the vertical lines x = a and x = b. R is bounded above by the graph of a function y = f(x) and below by the graph of a function y = g(x). It is important to note which function is at the top of the region and which is at the bottom. In this case the region R can be expressed in set notation as

$$R = \{(x,y)\mid a \leq x \leq b \text{ and } g(x) \leq y \leq f(x)\}$$

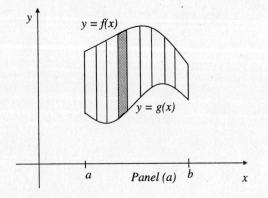

The double integral over this region R of F(x,y) can be evaluated as an iterated integral of the "dy dx form". The corresponding iterated integral over this region is evaluated by performing a "partial" integration with respect to the y-variable over the interval [g(x),f(x)], followed by integrating with respect to x over the interval [a,b]. The limits of integration for the y-integral are simply the boundary functions for the y-variable. These integration limits arise when the region is partitioned by first dividing it into strips parallel to the "parallel" sides, as illustreated at the left. The y-integral can be viewed as integrating over individual strips, its integration limits are the values of g and f at some point x between the x-values that form the strip.

Section 9.3   DOUBLE INTEGRALS OVER GENERAL REGIONS   545

$$\int_{g(x)}^{f(x)} F(x,y)\, dy$$

Then, this is integrated from x = a to x = b, giving the iterated integral:

$$\iint_R F(x,y)\, dA = \int_a^b \int_{g(x)}^{f(x)} F(x,y)\, dy\, dx \qquad \textbf{dy dx Form}$$

A similar approach gives the iterated integrals corresponding to the region R depicted in panel (b) on the previous page. This region has two "parallel" sides formed by the horizontal lines y = c and y = d. The right and left side boundaries of this region are given by functions of the y-variable.

The function giving the right boundary of the region is denoted by x = r(y) and the left boundary by x = l(y). Using the function notation r(y) and l(y) to helps identify the right and left sides. Again, it is important to distinguish the order of the boundary functions. This region R can be expressed in set notation as

$$R = \{(x,y)\mid\ c \leq y \leq d \quad \text{and} \quad l(y) \leq x \leq r(y)\ \}$$

The double integral of F(x,y) over this region R is evaluated as an iterated integral of the "dx dy form":

$$\iint_R F(x,y)\, dA = \int_c^d \int_{l(y)}^{r(y)} F(x,y)\, dx\, dy \qquad \textbf{dx dy Form}$$

Note: In each case the variable with constant bounds is taken as the second variable of integration.

Example 1   **Determining limits of integration when R is specified by a graph.**

Problem   Establish the limits of the iterated integrals that evaluate the double integral of F(x,y) over the regions indicated in the following graphs

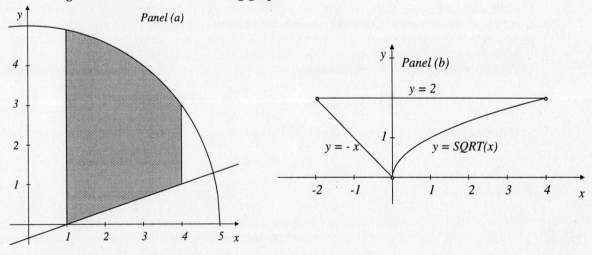

Solutions **Panel (a).**   The illustrated region R has two sides "parallel" to the y-axis. The left side of R is a segment of the line x = 1 and the right side is a segment of the line x = 4. Consequently, the second or outer integral will be with respect to x and the corresponding iterated integral will be of the "dy dx form". The upper boundary of R appears to be a portion of a circle centered at the origin of radius 5; thus the equation of the upper limit of integration is

$$f(x) = \text{SQRT}(25 - x^2)$$

The lower boundary is a line through the points (1,0) and (4,1); thus the lower limit of integration for y is given by the equation of a line through these points

$$y = g(x) = (x - 1)/3$$

The set description of the region R is thus

$$\{(x,y) \mid 1 \le x \le 4 \text{ and } (x-1)/3 \le y \le \text{SQRT}(25 - x^2)\}$$

The corresponding iterated integral over R is

$$\int_1^4 \int_{(x-1)/3}^{\text{SQRT}(25 - x^{\wedge}2)} F(x,y)\, dy\, dx$$

Panel (b). The region R indicated in panel (b) is bounded above by the horizontal line y = 2. The lower boundary is just a single point, the origin. This point constitutes the other "parallel" line and is the intersection of the line y = 0 with the region R. Since R has two sides "parallel" to the x-axis the iterated integral will be given in the "dx dy form". The limits of integration for the second y-integral are the constants determining the lower and upper boundaries: y = 0 and y = 4.

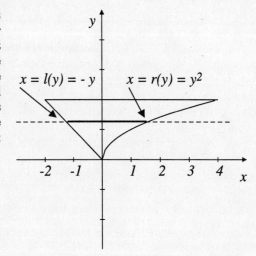

The appropriate left and right limits for first integral with respect to x can be found by a graphical approach. A horizontal line is drawn that cuts through the region R. Choosing any y-value in the interval [0,4], this line will intersect the region R in a segment, as illustrated at the right.

The left endpoint of this segment has x-coordinate given by the left boundary of the region, which is the curve x = l(y) = -y. The right endpoint of the segment has x-coordinate given by the right boundary curve, x = r(y) = y^2.

The region is described in set notation by

$$R = \{(x,y) \mid 0 \le y \le 4 \text{ and } -y \le x \le y^2\}$$

The corresponding iterated integral of F(x,y) over R is

$$\int_0^2 \int_{-y}^{y^{\wedge}2} F(x,y)\, dx\, dy$$

☑

Double integrals that can be evaluated two different ways.

If a region R has "parallel" sides parallel to both coordinate axes then the double integral over the region can be directly evaluated by iterated integrals of both forms. This occurs when R is a rectangle but can also occur when R is a right triangle with sides parallel to each axis and in some cases when R is determined by the intersection of two curves, which can be viewed either as the upper and lower boundaries or as the left and right boundaries.

Example 2 Two iterated integrals over the same region.

Problem Let the region R be the finite region bounded by the intersection of the curves $y = x^3$ and $y = \text{SQRT}(x)$ in the first quadrant. Express $\iint_R F(x,y)\, dA$ as an iterated integral in both the "dy dx form" and the "dx dy form".

Solution The region R is sketched at the right. Its "parallel" boundaries are both points. The two points are found by solving simultaneously the boundary equations $y = x^3$ and $y = \text{SQRT}(x)$.

$$x^3 = \text{SQRT}(x) \implies x = 1 \text{ or } x = 0$$

The most obvious set notation description of R uses the boundary functions as given. The upper boundary is $f(x) = x^{1/2}$ and the lower boundary is $g(x) = x^3$. Using the set description

$$R = \{(x,y) \mid 0 \leq x \leq 1 \text{ and } x^3 \leq y \leq \text{SQRT}(x)\}$$

the corresponding iterated integral is

$$\int_0^1 \int_{x^3}^{x^{1/2}} F(x,y)\, dy\, dx$$

On the other hand, the Region R can be expressed by solving each boundary curve for x as a function of y. The left boundary is then $x = l(y) = y^2$ and the right boundary is given by $x = r(y) = y^{1/3}$. Employing the set description

$$R = \{(x,y) \mid 0 \leq y \leq 1 \text{ and } y^2 \leq x \leq y^{1/3}\}$$

the corresponding iterated integral is

$$\int_0^1 \int_{y^2}^{y^{1/3}} F(x,y)\, dx\, dy$$

Example 3 Integrating over a triangular region.

Problem Express the double integral of $F(x,y)$ over the finite region R bounded by the lines $y = 3x$, $y = 0$ and $x = 2$ as an iterated integral of each form.

Solution First sketch the region as in panel (a). To establish a "dy dx form" iterated integral, first determine the constant limits for the x variable. While these limits may seem easy to determine visually from graph, they can also be determined algebraically from the intersection points of the boundary curves. The lower x-limit, $x = 0$, is actually the x-coordinate at the intersection of $y = 0$ and $y = 3x$. The upper x-limit is $x = 2$. The two functions $y = 3x$ and $y = 0$ provide the upper and lower limits for the y-integration.

For more complicated regions the following graphical approach may be used to determine which is the upper curve and which is the lower curve. The technique is to sketch a

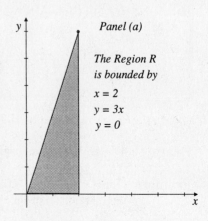

Panel (a)

The Region R is bounded by

$x = 2$
$y = 3x$
$y = 0$

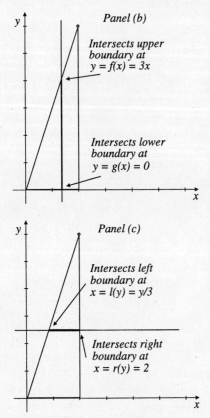

vertical line, corresponding to a constant x-value in the interval [0,2]. The line intersects the region R in the thick segment as illustrated in panel (b). Then the y-coordinates of this segment vary between the lower intersection coordinate, y = g(x) = 0, and the upper intersection coordinate, y = f(x) = 3x. The "dy dx form" of the iterated integral over R is

$$\int_0^2 \int_0^{3x} F(x,y) \, dy \, dx$$

To determine the limits for the "dx dy form" of the iterated integral, first establish the constant limits for y. These are the lower limit y = 0 and the upper limit y = 6 that is found algebraically by solving simultaneously the two boundary curves y = 3x and x = 2. To establish the left and right limits for x, we sketch a horizontal line through the region at height y. As sketched in panel (c), this line intersects the region in a segment. The x-limits of integration are given by solving for the x-coordinates of the left and right endpoints of this segment. These are obtained by solving the boundary equations for x as functions of y. The left endpoint of the segment is determined by the line x = l(y) = y/3 and the right endpoint is given by the vertical line x = r(y) = 2. The iterated integral is

$$\int_0^6 \int_{y/3}^2 F(x,y) \, dx \, dy$$

Integrating over regions that must be divided into subregions of the "parallel" form.

Frequently we need to evaluate double integrals over regions that do not have one of the two "parallel" forms considered above. This can be accomplished using iterated integrals if the region R can be split into subregions that do have one of the two "parallel" forms. This is possible because of a general Rule that states the double integral over a region that is the union of two non-overlapping subregions is the sum of the integrals over the subregions.

MULTIPLE INTEGRAL OVER A SPLIT REGION

If the region R is the union of two regions R_1 and R_2 that intersect only on their boundaries,

$$R = R_1 \cup R_2 \quad \text{and} \quad (\text{interior of } R_1) \cap (\text{interior of } R_2) = \emptyset$$

then

$$\iint_R F(x,y) \, dA = \iint_{R_1} F(x,y) \, dA + \iint_{R_2} F(x,y) \, dA$$

Example 4 **Splitting a Region into subregions for integration.**

Problem How would you integrate a function over the region R that is inside the polygonal line connecting in order the points (0,0), (0,6), (2,6), (2,4), (6,5), (6,0) and (0,0)?

Solution The region R is first sketched, as illustrated. The simplest approach is to split the region R into two regions by introducing the vertical line x = 2. This creates the rectangle region

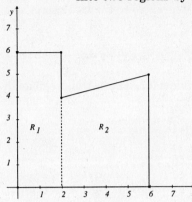

$$R_1 = \{(x,y) \mid 0 \le x \le 2 \text{ and } 0 \le y \le 6 \}$$

and the region R_2 as illustrated. To describe R_2 we need an equation for the line segment from the point (2,4) to the point (6,5). The point slope equation of this line is

$$y = 4 + 0.5(x - 2)$$

The set description of R_2 is thus

$$R_2 = \{(x,y) \mid 2 \le x \le 6 \text{ and } 0 \le y \le 4 + 0.5(x - 2) \}$$

The integral over R is then evaluated as the sum of the integrals over R_1 and R_2.

$$\iint_R F(x,y) \, dA = \iint_{R1} F(x,y) \, dA + \iint_{R2} F(x,y) \, dA$$

$$= \int_0^2 \int_0^6 F(x,y) \, dy \, dx + \int_2^6 \int_0^{4 + 0.5(x-2)} F(x,y) \, dy \, dx$$

☑

Example 5   Two ways to split a region for integration.

Problem Determine the two forms of iterated integrals over the region bounded above by the lines y = 3 - 1.5x and y = 3 + 1.5x and bounded below by the lower half of the circle $x^2 + y^2 = 4$.

Solution The region is illustrated in panel (a), below. This region does not have the simple form that we require for specifying the limits of integration since its boundary does not have any "parallel" sides. In such cases the region must be divided into subregions that do have "parallel" sides. There are two easy ways to do this, as illustrated in panels (b) and (c). In panel (b) the subregions R_1 and R_2 have sides "parallel" to the y-axis, given by the lines x = -2, x = 0, and x = 2.

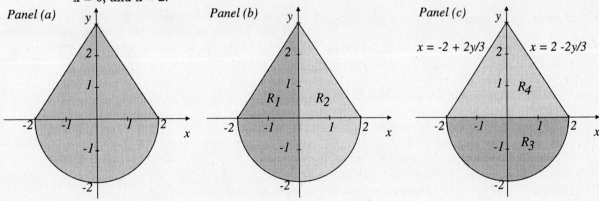

These regions are described in set notation by:

$$R_1 = \{(x,y) \mid -2 \le x \le 0 \text{ and } -\sqrt{4-x^2} \le y \le 3 + 1.5x \},$$

$$R_2 = \{(x,y) \mid 0 \le x \le 2 \text{ and } -\sqrt{4-x^2} \le y \le 3 - 1.5x \}$$

Thus, the "dy dx form" of the integral over R is the sum of the integrals over the two subregions R_1 and R_2.

$$\iint_R F(x,y)\, dA = \int_{-2}^{0}\int_{-\sqrt{4-x^2}}^{3+1.5x} F(x,y)\, dy\, dx + \int_{0}^{2}\int_{-\sqrt{4-x^2}}^{3-1.5x} F(x,y)\, dy\, dx$$

Alternatively, the region R can be partitioned by the x-axis as in panel (c). The two subregions in this case are given by:

$$R_3 = \{(x,y)\mid -2 \leq y \leq 0 \text{ and } -\sqrt{4-y^2} \leq x \leq \sqrt{4-y^2}\},$$

$$R_4 = \{(x,y)\mid 0 \leq y \leq 3 \text{ and } 2 + (2/3)y \leq x \leq 2 - (2/3)y\}$$

The corresponding integral over R in the "dx dy form" is the sum of the integrals over the two subregions:

$$\iint_R F(x,y)\, dA = \int_{-2}^{0}\int_{-\sqrt{4-y^2}}^{\sqrt{4-y^2}} F(x,y)\, dx\, dy + \int_{0}^{3}\int_{2+(2/3)y}^{2-(2/3)y} F(x,y)\, dx\, dy$$

☑

Determining the limits of integration.

The most challenging part of multiple integration problems is often to establish the appropriate limits of integration corresponding to a given region. The main difficulty is that the region R is not always indicated in the clear form you require. Three basic ways to describe a region are frequently encountered.

(i) To specify R explicitly in a sentence or by a set of inequalities.

(ii) To describe R in a sentence that indirectly provides the functions and constants needed to describe the boundary of R. The boundary functions must be determined and the extent of the x or y-variables determined algebraically.

(iii) To illustrate R on an x-y coordinate system. This type of representation is frequently the most illuminating, but again requires the ability to determine the equations of the boundary curves from the graph. (You must know the equations from prior experience!)

Example 6 **Expressing a region R in set notation.**

An example is given to illustrate each of the three common methods used to express a region. The problem is to express each region in set notation. From these, appropriate limits for an iterated integral over the region can be read off directly as the bounds for the x and y variables.

Problem A A type (i) description, in which R is explicitly indicated in a sentence, is

"The region R is bounded above the curve $y = x + 1/2$ and bounded below by the curve $y = \sin(x)$, for x between 0 and π".

Solution This sentence translates directly to a set description of R with the boundary functions $f(x) = x + 1/2$ and $g(x) = \sin(x)$:

$$R = \{(x,y)\mid 0 \leq x \leq \pi \text{ and } \sin(x) \leq y \leq x + 1/2\}$$

Problem B

A type (ii) description of a region is:

"The region R is the intersection of two unit circles, one centered at the origin and the other centered at the point (1,0)."

Solution This description must be read with a practical interpretation of the wording. Technically, the two circles intersect at two points, so the desired region R is actually the intersection of the circular regions inside the two unit circles. Another way to describe R is "the region interior to the two given circles." To describe this region the equations of the circles must be determined. It is assumed that you know the equation of a circle centered at (x_0, y_0) of radius r is

$$(x - x_0)^2 + (y - y_0)^2 = r^2.$$

The unit circle centered at the origin is $x^2 + y^2 = 1$.

The unit circle centered at (1,0) is $(x-1)^2 + y^2 = 1$.

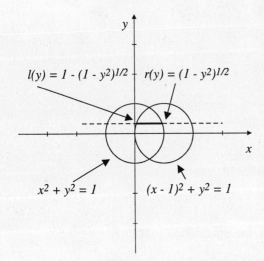

These circles and the region R of their intersection are illustrated at the left. The points of intersection of these circles are obtained by solving the first equation for y^2, $y^2 = 1 - x^2$, and substituting this for y^2 into the second equation:

$$(x-1)^2 + y^2 = 1 \implies (x-1)^2 + 1 - x^2 = 1$$

$$x^2 - 2x + 1 + 1 - x^2 = 1 \text{ or } x = 1/2$$

Then, substituting $x = 1/2$ into either circle equation gives the two intersection points:

$$(x_1, y_1) = (1/2, -\text{SQRT}(3)/2)$$

and

$$(x_2, y_2) = (1/2, +\text{SQRT}(3)/2)$$

The "parallel" sides are the intersection points that are considered to be parallel to the x-axis.

The y-coordinates of these points give the limits for y in the region R. The lower limit is the y-coordinate of the lower intersection point, $c = -\text{SQRT}(3)/2$, and the upper limit is the y-coordinate of the upper intersection point, $d = \text{SQRT}(3)/2$.

To establish the appropriate left and right boundary functions we sketch a horizontal line for any fixed y-value in the interval $[-\text{SQRT}(3)/2, \text{SQRT}(3)/2]$. This line will intersect the region R in a segment, as illustrated by the thick segment in the graph. The left endpoint of this segment is on the left side of the circle centered at (0,1). Its x-coordinate is found by solving the equation of this circle for x. NOTE: The negative square root is used to yield the left side of the circle:

$$x = l(y) = 1 - \text{SQRT}(1 - y^2)$$

The right endpoint of the thick segment is on the circle centered at (0,0). Solving the equation of this circle for x, taking the positive root this time to give the right side of the circle, yields the right boundary function:

$$x = r(y) = \text{SQRT}(1 - y^2)$$

Thus the set description of the region R is

$$R = \{(x,y) \mid -\text{SQRT}(3)/2 \leq y \leq \text{SQRT}(3)/2 \text{ and } 1 - \text{SQRT}(1 - y^2) \leq x \leq \text{SQRT}(1 - y^2) \}$$

Problem C A common way to describe a region is simply to give a graph illustrating it. This type of representation is frequently the most illuminating. As you may have observed in the previous two examples, sketching the region is usually done (at least mentally) when establishing the boundaries of a region. A graphical representation must be clear enough that the boundary curves can be determined. This requires clear indications of the scale and specific points on the boundary. It also requires the reader to perceive geometric shapes like circles and straight lines. To extract the boundary functions you must assume the curves are of specific standard forms and determine the specific equations whose graphs form the boundaries from clearly identifiable points that they pass through. An example of a type (iii) description is:

Determine two set-notation descriptions of the region R which is shaded in the graph at the right.

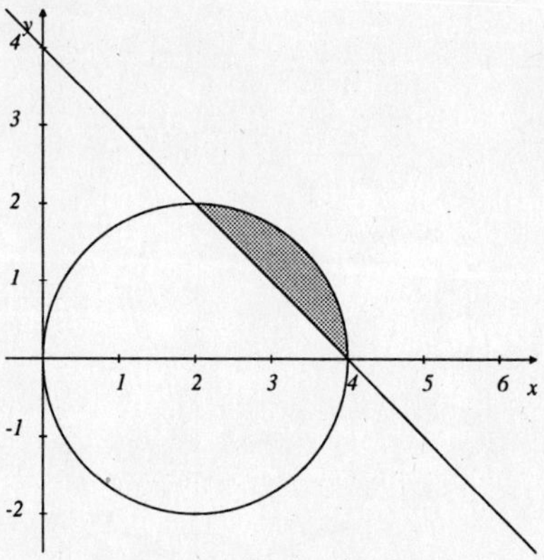

Solution Looking at the graph, we must make assumptions about the boundary curves. They appear to be parts of a line and of a circle. The line passes through the two points (0,4) and (4,0) and the circle is centered at (0,2) with radius 2. The equations of the line and circle are

$$y = 4 - x \quad \text{or} \quad x = 4 - y$$

and

$$x^2 + (y-2)^2 = 4$$

The points of intersection of the two boundary curves are found by solving these equations simultaneously. Setting $x = 4 - y$ in the equation of the circle gives

$$(4 - y)^2 + (y - 2)^2 = 4$$

Expanding the squares and gathering terms, this gives the quadratic equation

$$2y^2 - 12y + 16 = 0 \quad \text{or} \quad 2(y - 2)(y - 4) = 0$$

Using the two roots, $y = 2$ and $y = 4$, gives the intersects points $(x_1,y_1) = (0,4)$ and $(x_2,y_2) = (2,2)$. The set descriptions of the region R are thus

$$R = \{(x,y) \mid 0 \leq x \leq 2 \text{ and } 4 - x \leq y \leq 2 + \text{SQRT}(4-x^2) \}$$

and, solving for x as a function of y,

$$R = \{(x,y) \mid 2 \leq y \leq 4 \text{ and } 4 + y \leq x \leq \text{SQRT}(4 - (y-2)^2) \}$$

☑

Changing the order of integration.

Thus far in this Section we have concentrated on the region R and establishing iterated integrals to evaluate double integrals. Sometimes the integrand F(x,y) will influence the choice of order for integration. A classic example involves the integral of $F(x,y) = \exp(x^2)$. Since an antiderivative of $\exp(x^2)$ can not be expressed using only a finite number of elementary functions, the integral

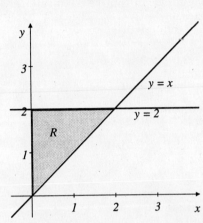

$$\iint_R \exp(x^2)\, dA = \int_0^2 \int_y^2 \exp(x^2)\, dx\, dy$$

cannot be evaluated directly. However, by recognizing that the integration region R is the triangular region with vertices (0,0), (0,2), and (2,2), bounded by the y-axis, the line y = 2 and the line y = x, the order of integration can be reversed, with an appropriate change of limits. The equivalent integral is

$$\iint_R \exp(x^2)\, dA = \int_0^2 \int_0^x \exp(x^2)\, dy\, dx$$

The inner integral is now with respect to y and can be evaluated as

$$\int_0^2 \int_0^x \exp(x^2)\, dy\, dx = \int_0^2 y \cdot \exp(x^2)\Big|_{y=0}^{y=x}\, dx = \int_0^2 x \cdot \exp(x^2)\, dx$$

This integral can now be evaluated!

$$\int_0^2 x \cdot \exp(x^2)\, dx = (1/2)\exp(x^2)\Big|_0^2 = (1/2)e^4 - 1/2 \approx 26.8$$

The above is a rather dramatic example that actually occurs frequently in statistics and physic problems.

Example 7 Changing the order of integration.

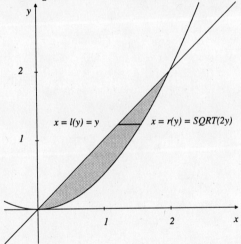

Problem Determine the "dx dy form" of the integral

$$\int_0^2 \int_{0.5x^2}^{x} F(x,y)\, dy\, dx$$

Solution First, sketch the region R. From the limits of integration we infer that: the upper boundary of R is the line y = x and the lower boundary is $y = 0.5x^2$. These two curves intersect at x = 0 and x = 2. To change the order of integration we must determine the boundaries of R as functions of y. Solving the left boundary for x gives x = l(y) = y. Similarly, solving the right boundary equation for x gives x = r(y) = SQRT(2y). As y varies between 0 and 2 the desired integral is

$$\int_0^2 \int_y^{SQRT(2y)} F(x,y)\, dx\, dy$$

Exercise Set 9.3

For problems 1 to 4, (i) Sketch the indicated region.
(ii) Describe the region two ways in set notation.
(iii) Can a double integral over the region be evaluated as an iterated integral? If so, indicate the limits of integration. If not, indicate why and indicate if the region can be "split" into sub-regions over which appropriate iterated integrals may be evaluated. In this case indicate the limits of integration for the sub-regions.

1. The triangular shaped region with vertices
   a) (0,0), (2,3) and (0,3)   b) (0,0), (2,3) and (2,0)   c) (2,0), (2,3) and (4,3)
   d) (-1,0), (1,0) and (0,4)   e) (0,0), (2,3) and (2,-3)   f) (0,0), (2,3) and (4,1)

2. The region interior to the closed polygonal curve connecting the given points.
   a) (0,0), (2,0), (2,3) (-1,3)   b) (0,0), (2,-3), (2,3) (1,1)   c) (0,0), (2,3), (0,3) (-1,1)
   d) (0,0), (2,-4), (2,3) (-2,3) e) (0,0), (2,3), (3,0) (1,-1)   f) (0,0), (0,-4), (2,3) (-2,2)

3. The finite region between the given curves.
   a) $y = x$ and $y = x^2$ b) $y = 2x$ and $y = x^2$ c) $y = x$ and $y = -x^2+2$
   d) $y = -x$ and $y = x^2-2$ e) $y = x^2$ and $y = -x^2+2$ f) $y = 2x^2$ and $y = x^3$
   g) $y = x^2$ and $y = x^{1/3}$ h) $y = x^2$ and $x = y^2$ i) $y = x/2$ and $x = y^2$
   j) $y = x$ and $y = x^3$

4. a) The region interior to the circle $x^2 + y^2 = 4$ and above the line $y = 0.5x$.
   b) The region interior to the ellipse $4x^2 + 9y^2 = 36$ with x positive.
   c) The region interior to the two circles $x^2 + y^2 = 9$ and $(x - 3)^2 + y^2 = 9$.
   d) The region interior to the two circles $x^2 + y^2 = 16$ and $(x + 4)^2 + y^2 = 16$.

5. Give the two set notation descriptions of the indicated region.

(A)
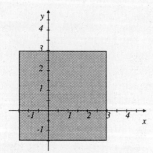

(B)
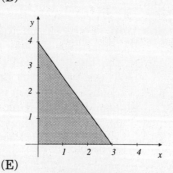

(C)
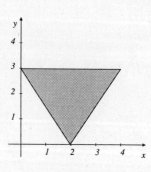

(D)
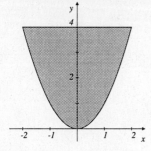

(E)
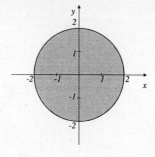

(F)
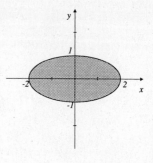

6. Give the two set notation descriptions of the indicated region.

(A)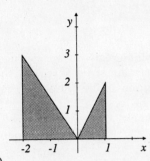

(B)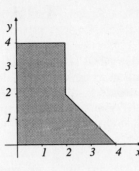

(C)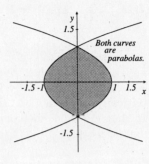

Both curves are parabolas.

(D)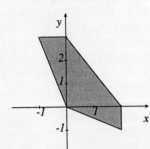

(E)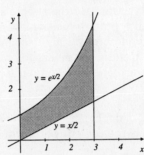

$y = e^{x/2}$

$y = x/2$

(F)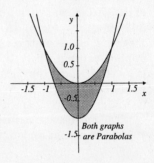

Both graphs are Parabolas

7. For each region in exercise 5 evaluate $\iint_R x - y \, dA$.

8. For each region in exercise 6 evaluate $\iint_R x + y \, dA$.

9. (i) Sketch the region of integration for the given iterated integral. (ii) Determine the limits for equivalent integral with the order of integration reversed. If necessary, split the region into appropriate subregions.

a) $\int_1^4 \int_2^5 F(x,y) \, dy \, dx$

b) $\int_1^4 \int_2^5 F(x,y) \, dx \, dy$

c) $\int_1^4 \int_{x/4}^1 F(x,y) \, dy \, dx$

d) $\int_1^{2.5} \int_{2x}^5 F(x,y) \, dy \, dx$

e) $\int_0^4 \int_{-y}^y F(x,y) \, dx \, dy$

f) $\int_1^5 \int_2^{3x} F(x,y) \, dy \, dx$

g) $\int_1^4 \int_{-y}^y F(x,y) \, dx \, dy$

h) $\int_1^5 \int_{1/x}^{3x} F(x,y) \, dy \, dx$

i) $\int_0^4 \int_{-y^2}^{2y} F(x,y) \, dx \, dy$

j) $\int_0^5 \int_{1-x}^{3+x} F(x,y) \, dy \, dx$

k) $\int_1^5 \int_{-\sqrt{25-y^2}}^{1+2y} F(x,y) \, dx \, dy$

l) $\int_0^3 \int_{\sqrt{9-x^2}}^{9-x^2} F(x,y) \, dy \, dx$

10. Sketch the region below the cubic curve $y^3 = x$ and above the quadratic $6y = x^2 - 5x - 12$. Establish the limits of integration to evaluate a double integral of $F(x,y)$ over this region.

Section 9.4 Areas and Volumes as Multiple Integrals.

In the preceding sections a double integral $\iint_R F(x,y) \, dA$ over a plane region R was defined to be a numerical value, given by a complicated limit. Fortunately, for certain types of regions these values can be found by integrating iterated integrals. The integration limits were seen to be critically dependent on the shape of the region and did not depend on the function being integrated. In this and the following section the role of the integrand $F(x,y)$ and various interpretations of the double integral $\iint_R F(x,y) \, dA$ are introduced.

The interpretation of an integral depends on prior interpretations of the function and the units of the variables. When $F(x,y)$ is simply assumed to be a real valued function without any associated meaning, $\iint_R F(x,y) \, dA$ evaluates as a pure dimensionless number. As such, it has no implied meaning, no geometric significance, and no physical or biological interpretation. However, in an applied setting, where the variables x and y are given units of measure and the integrand $F(x,y)$ is given a specific physical or conceptual connotation, then $\iint_R F(x,y) \, dA$ will also have units and it can be interpreted correspondingly.

The most common interpretations of double integrals are "the area of a plane region R" and "the volume of a solid region, whose base is the region R and whose height at (x,y) is $F(x,y)$".

Interpretation of $F(x,y)$ and units of x and y	Interpretation of $\iint_R F(x,y) \, dA$
$F(x,y) = 1$ x and y have linear units, e.g., cm, km, miles, light years	$\iint_R F(x,y) \, dA = $ **Area of R**. Its units are linear units squared.
$z = F(x,y)$ is the height or elevation of a surface over the region R. All coordinates in linear units.	$\iint_R F(x,y) \, dA = $ **Volume of the solid body with base R and height given by $F(x,y)$**. The integral's unit is the product of the x, y, and z units.

Other interpretations of double integrals are presented in the following section.

Area As a Double Integral.

The area of a region R can be evaluated by the integral over R of the constant function $F(x,y) = 1$.

$$\text{Area} = \iint_R 1 \, dA \qquad \qquad \textbf{Area of the region R.}$$

This interpretation of the integral comes directly from the "additive" property of areas:

and the definition of the integral as a limit of Riemann Sums:

$$\iint_R F(x,y) \, dA = \lim_{N \to \infty, \, A\_n \to 0} \sum_{n=1,N} F(x_n^*, y_n^*) A_n$$

where A_n is the area of the subregion R_n, R is the union of the subregions R_n, and the subregions are disjoint. When $F(x,y) = 1$ for all (x,y), each Riemann Sum simply gives the total area of R for every choice of partitions.

$$\iint_R 1 \, dA = \lim_{N \to \infty, \, A\_n \to 0} \sum_{n=1,N} 1 \times A_n = \lim_{N \to \infty, \, A\_n \to 0} \sum_{n=1,N} A_n = \text{Area of R}.$$

Section 9.4   AREAS AND VOLUMES AS MULTIPLE INTEGRALS   557

ADDITIVE PROPERTY FOR AREAS

> If the region R is the union of two regions R_1 and R_2 that intersect only at their boundaries,
>
> $$R = R_1 \cup R_2 \quad \text{and} \quad (\text{interior of } R_1) \cap (\text{interior of } R_2) = \emptyset$$
>
> Then
>
> $$\text{Area}\{R\} = \text{Area}\{R_1\} + \text{Area}\{R_2\}$$

To give an area unit of measure to this integral the coordinate variables must be uniformily scaled. Frequently, specific units for the variables are not indicated. You may be asked to find the area of a region R bounded by a set of curves, given in simple function notation like $y = x^2 - 1$. In such cases a <u>generic</u> measurement unit is assumed and the unit of the integral is normally not indicated, but is (generic unit)^2. In some applications the x and y variables may have different units, in which case the unit for the area integral is the product of the two axis units,

$$\text{area unit} = (\text{x-unit}) \times (\text{y-unit})$$

For instance if the unit on the x-axis is cm while the y-axis has a unit of m, the area unit will be cm × m = 100 cm^2 or 0.01 m^2.

Example 1   Computing an area as a double integral.

Problem   A fable tells the story of a peasant who saved the Pharaoh from a lion and was rewarded with the amount of land that he could encircle by plowing in one day. The peasant plowed a straight line 250 rods long, then he made a right angle and plowed a straight line 400 rods in length. At this point he plowed a parabolic arc back to the starting point. How much land was he granted?

Solution   The first step is to sketch the region. Assume that the peasant started at the origin and plowed first along the positive x-axis, to the point (250,0). Then assume he plowed in the direction of the positive y-axis, to the point (250,400). At the point we must make an assumption about the arc that was described simply as "parabolic". This description is not really clear since two points, the origin and the point (250,400), do not uniquely determine a parabola. It takes three points! To get an answer we must guess or "assume" the form of the arc. One simple parabola passing through the two points is the graph of the parabola opening downward with its apex at the point (250,400). This curve is the graph of an equation of the form

$$y = -C(x - 250)^2 + 400.$$

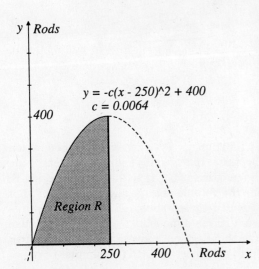

The constant C is chosen so that this curve passes through the origin. Setting x = 0 and y = 0 gives C = 4/625 = 0.0064. The area of the region that the peasant enclosed is given by the iterated integral of F(x,y) = 1 over the region R. To save writing its decimal value repeatedly, we leave the constant C in the boundary equation and substitute its value at the end of the calculations.

$$\iint_R 1 \, dA = \int_0^{250} \int_0^{-C(x-250)^2 + 400} 1 \, dy \, dx$$

$$= \int_0^{250} y \big|_0^{-C(x-250)^2 + 400} \, dx$$

$$= \int_0^{250} -C(x-250)^2 + 400 \, dx$$

$$= -(C/3)(x-250)^3 + 400x \big|_0^{250}$$

$$= [-(C/3)(250-250)^3 + 400 \times 250] - [-(C/3)(0-250)^3 + 400 \times 0]$$

$$= 400 \times 250 + (C/3)(-250)^3$$

$$= 100{,}000 - (C/3) \times 15{,}625{,}000 = 6{,}666.\underline{6} \text{ square rods.}$$

A rod was an ancient unit of measure that was carried into this century as a surveyor's unit: 1 rod = 5.029 m = 5.5 yards. It is still used in horse racing, where one furlong is 40 rods. The total area calculated is thus about 1.686 km² or 168.6 hectares in SI or metric measures or about 416.6 acres in English units.

Calculating Volumes as Double Integrals.

The graph of a continuous function z = F(x,y) is a surface in three dimensions. If $F(x,y) \geq 0$ for $(x,y) \in R$, then this surface is above the x-y plane over the region R. In analogy to the "area under a curve y = f(x)", the "volume under the surface z = F(x,y)" over a region R is the volume of the solid three-dimensional body that has the region R in the x-y plane as a "base" and a "top" consisting of the portion of the surface z = F(x,y) above R. Like a cylinder, the sides of this solid are parallel to the z-axis along the boundary of the region R. We refer to this solid and its volume simply as "the solid under F over R" or "the volume under F over R". When the region is clearly identified, the phrase "over R" may be omitted.

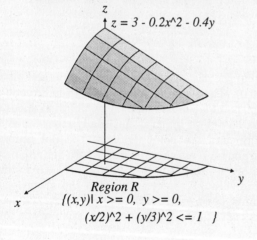
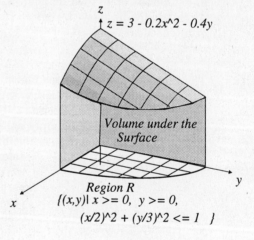

The volume under a given surface, i.e., the graph of a function z = F(x,y), over a specified region R, is evaluated as the integral of F(x,y) over R. Therefor, the primary skills required to determine volumes are the ability to establish the limits of integration corresponding to the specific region R, and then to evaluate the definite integrals. To determine the volume of a specific solid you will not need to sketch or work with three-dimensional surfaces. To sketch such solid regions requires practice and perhaps a little artistic skill to represent three-dimensional objects on a two-dimensional sheet of paper. While the material of this section does not require you to sketch such functions, it is sometimes helpful and we encourage you to attempt to sketch the simpler surfaces and volumes.

Rectangular Solid

Volume = $L * W * H$

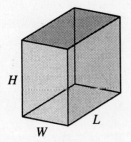

To understand why integrating F(x,y) over the region provides the desired volume let us consider some simple situations. Assume the region R is rectangular and consider a positive function F(x,y). If the function F(x,y) is actually a constant, for instance F(x,y) = 5, then the solid region under the graph of F(x,y) over R would be a rectangular shaped solid. Its volume could be directly calculated as the product of its width, length, and height:

$$v = w \times l \times h \qquad \text{or} \qquad v = A \times h$$

where its base width = w, base length = l, and height = h. The term A = w × l is the area of the rectangular base.

If F(x,y) is not constant over R then the solid under z = F(x,y) is similar to the rectangular solid just considered, except that its "top" is not flat.

We apply the **Fundamental Method of Integral Calculus** to determine the volume under the graph of a general positive function F over a region R. The desired volume, V, is found by a sequence of steps.

Step I First the region R is partitioned into subregions, $R_{i,j}$.

Step II Then, for each subregion, the volume $v_{i,j}$ under the surface z = F(x,y) over subregion $R_{i,j}$, is approximated by approximating the function by a constant over the subregion.

Step III An approximation of the volume V is given by the sum of these approximate subregion volumes. The sum over all subregions has the form of a Riemann Sum.

Step IV Then, the limit of the approximating sums of Step III is taken as the subregion areas $A_{i,j}$ approach zero and the number of sub-regions becomes infinite. In the limit the error of the approximation of V by the sum goes to zero, giving the desired volume V. But, this limit is by definition the multiple integral $\iint_R F(x,y) \, dA$. Thus the volume V is given by the integral $\iint_R F(x,y) \, dA$.

The details of this application of The Fundamental Method of Calculus are as follows:

Step I First, we partition R into sub-rectangles, following the method and notation used in Section 9.2, when the double integral of F(x,y) over R was developed. To determine the volume under z = F(x,y) over one of these sub-rectangles is conceptually and practically identical to the problem that we began with. The only difference is that the sub-rectangle $R_{i,j}$ will have a base that is smaller in area. Recall that the area of $R_{i,j}$ is

$$A_{i,j} = \Delta x_i \times \Delta y_j.$$

Step II To approximate the volume $v_{i,j}$ under F over $R_{i,j}$ the function F is replaced by a constant value over $R_{i,j}$. The constant is chosen as the value of the function F at some point (x_i^*, y_j^*) in $R_{i,j}$. The point (x_i^*, y_j^*) is arbitrary, it can be chosen at our convenience to make actual numerical calculations easier or to simplify formulas. Frequently it is chosen as a corner point of the rectangle $R_{i,j}$. Let

$$z_{i,j}^* = F(x_i^*, y_j^*).$$

Then, over the sub-rectangle $R_{i,j}$, the volume $v_{i,j}$ under the surface z = F(x,y) is approximated by the volume under the horizontal surface $z = z_{i,j}^*$, which is simply the volume of a rectangular solid.

Approximate Volume Element

$$v_{i,j} \approx z_{i,j}^* \times A_{i,j} = z_{i,j}^* \times \Delta x_i \times \Delta y_j.$$

Step III The desired volume V is then approximated by the sum of the approximate volumes:

$$V \approx \Sigma_{n=1,N} v_{i,j} \approx \Sigma_{n=1,N} z_{i,j}^* \times A_n$$

or, substituting the function values for $z_{i,j}^*$,

Approximate Total Volume

$$V \approx \Sigma_{n=1,N} F(x_n^*, y_n^*) \times A_n$$

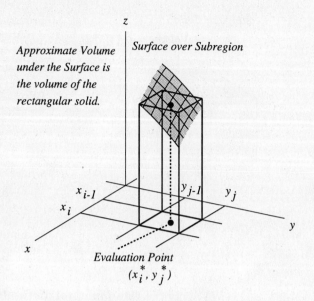

Approximate Volume under the Surface is the volume of the rectangular solid.

Surface over Subregion

Evaluation Point (x_i^*, y_j^*)

This approximation can be used when the function F(x,y) is not actually known, but its value is known by making measurements at specific grid points (x_i^*, y_j^*). In most experimental work this is the formula used to estimate volumes.

Step IV The exact volume is found as the limit of the approximations,

$$V = \lim_{N \to \infty, A\_n \to 0} \Sigma_{n=1,N} F(x_n^*, y_n^*) \times A_n.$$

The limit of this Riemann Sums is by definition equal to the integral of F over R.

The Volume under the surface z = F(x,y) over R is

$$V = \iint_R F(x,y)\, dA$$

Example 2   **Volume of a solid wedge.**

Problem   Determine the volume of the solid wedge that is the solid bounded above by the surface z = y over the rectangular region R = {(x,y) | $0 \le x \le 10$ and $0 \le y \le 3$ } and below by the x-y plane.

Solution   The volume V is the area under the surface z = F(x,y) = y. It is given by the double integral $\iint_R y\, dA$. This is evaluated as the iterated integral

$$\int_0^3 \int_0^{10} y\, dx\, dy = \int_0^3 yx \big|_0^{10} dy = \int_0^3 10y\, dy = 5y^2 \big|_0^3 = 45 \text{ (units}^3\text{)}.$$

In the above discussion the region R was assumed to be a rectangle. This assumption was made to simplify the discussion of subregions and the calculation of their areas easier. The same result follows for any finite region whose boundary is a continuous curve in the x-y plane. Furthermore, if the function F(x,y) is not strictly positive, then the double integral can be interpreted as the "net" volume between the surface z = F(x,y) and the x-y plane. The "net" volume may be positive or negative; it will be negative if the surface is below the x-y plane, i.e., if z = F(x,y) is negative.

Example 3 The Volume of a Diamond.

Background A regular diamond-shaped solid with 8 faces has 6 corners or vertices and is characterized by the lengths of its principal axes connecting opposite vertices. To describe a diamond mathematically, it is viewed as a solid centered at the origin of an x-y-z coordinate system with its principal axes aligned with the coordinate axes, as in the figure at the right. If the length of the diamond's axes are 2a, 2b, and 2c, their endpoints give the six vertices. These are the points (0,0,c) and (0,0,-c) on the z-axis; (0,b,0) and (0,-b,0) on the y-axis, and (a,0,0) and (-a,0,0) on the x-axis. The surface of the diamond consists of the eight planes passing through three adjacent vertices.

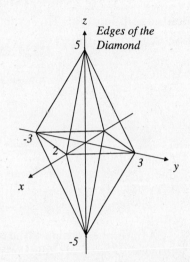

Edges of the Diamond

Problem Determine the volume of a regular diamond with principal axes of lengths 4, 6, and 10.

Solution **Modeling Component**. To determine the volume of the diamond we use its symmetry. Visualizing the diamond as sketched above, it consist of eight similar subregions, consisting of the part of the diamond that is in each octant of the x-y-z space. Thus its volume is eight times the volume of one of these. We consider the subregion that is in the "First Octant", where x, y, and z are all positive. In this octant the volume of the diamond is the volume under the surface that is described by the plane connecting the three points (2,0,0), (0,3,0), and (0,0,5). The equation of this plane is

$$(x/2) + (y/3) + (z/5) = 1$$

or

$$z = -(5/2)x - (5/3)y + 5.$$

Thus, the volume of the diamond in the first octant will be the double integral of the corresponding function

$$F(x,y) = -(5/2)x - (5/3)y + 5$$

over a region R in the x-y plane. The region R is triangular with vertices (0,0,0), (2,0,0) and (0,3,0). Ignoring the z-coordinate that is zero in the x-y plane, this region R is

$$\{(x,y) \mid 0 \le x \le 2 \text{ and } 0 \le y \le 3 - (3/2)x \}.$$

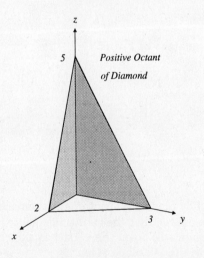

Positive Octant of Diamond

The upper y-limit is determined by the equation of the boundary line y = 3 - (3/2)x. This is the equation of the line through (2,0) and (0,3) in an x-y coordinate system. You can determine this equation using the two-point equation for lines. Alternatively, use the observation that this boundary is the intersection of the plane z = F(x,y) given above with the x-y plane. The equation of the x-y plane is z = 0. The intersection of the two planes is the line found by equating the z-equations of the planes and solving for y. Setting

$$0 = -(5/2)x - (5/3)y + 5$$

gives the upper boundary equation y = 3 - (3/2)x.

Calculus application: The octant volume of the diamond is then given by

$$V = \iint_R F(x,y)\, dA = \int_0^2 \int_0^{3-(3/2)x} -(5/2)x - (5/3)y + 5 \, dy\, dx$$

$$= \int_0^2 -(5/2)xy - (5/6)y^2 + 5y \Big|_{y=0}^{y=3-(3/2)x} dx$$

$$= \int_0^2 15/2 - (15/2)x + (15/8)x^2 \, dx$$

$$= (15/2)x - (15/4)x^2 + (5/8)x^3 \Big|_0^2 = 5 \text{ units}^3.$$

The total volume of the diamond is eight times the octant volume, 40 units^3.

☑

A natural extension of the concept of an "area between two curves" is the "volume between two surfaces". Given two functions F(x,y) and G(x,y), that are continuous on a region R, and satisfy F(x,y) ≥ G(x,y) for all points (x,y) ∈ R the

Volume between the graphs of F and G is

$$V = \iint_R F(x,y) - G(x,y) \, dA$$

If the inequality F ≥ G is not satisfied at some points in R then the number V may be positive or negative, in which case it is interpreted as the "net" volume between the surfaces. Generally, when asked to calculate the volume of a solid bounded by two surfaces the appropriate formula that gives the true physical volume between the surfaces no matter which is "above" the other is

$$V = \iint_R |F(x,y) - G(x,y)| \, dA$$

Physical Volume of the solid bounded by the graphs of F and G

Example 4   **Volume of a Parabolic trough.**

Problem   Determine the volume of a "Parabolic" trough of length 25cm that is 6cm wide at its top and 5cm deep.

Solution   The trough can be visualized as a surface that is formed by placing a sheet of metal on the x-axis and bending it upward at the edges parallel to the x-axis so that the resulting surface has a parabolic cross-sectional shape. The bottom of the trough will be on the x-axis, as sketched on the next page. The trough lies above a rectangle R in the x-y plane given by

$$R = \{(x,y) \mid 0 \leq x \leq 25 \text{ and } -3 \leq y \leq 3\}.$$

One end of the trough is in the y-z plane and is described by a parabola $z = Ay^2$, with x = 0. The constant A is determined by the condition that at y = 3, z = 5. Thus A = 5/9. The other end of the trough is at x = 25 and is again described by the same parabola curve $z = (5/9)y^2$. For any x between 0 and 25 the trough is described by the same parabolic equation,

$$z = (5/9)y^2.$$

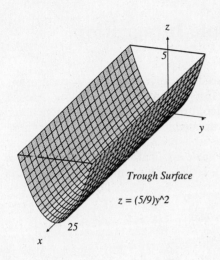

Trough Surface
$z = (5/9)y^2$

The volume of the trough is the volume of the region between the trough and the horizontal plane that forms the top of the trough, the plane z = 5. Employing the formula for the volume between two curves, with

$$F(x,y) = 5 \quad \text{and} \quad G(x,y) = (5/9)y^2.$$

This volume is

$$V = \iint_R F(x,y) - G(x,y) \, dA = \int_0^{25} \int_{-3}^{3} 5 - (5/9)y^2 \, dy \, dx$$

$$= \int_0^{25} 5y - (5/27)y^3 \Big|_{-3}^{3} \, dx$$

$$= \int_0^{25} 20 \, dx = 20x \Big|_0^{25} = 500 \text{ cm}^3.$$

Exercise Set 9.4

1. Determine the area of the indicated region. Sketch the region.

   a) The region enclosed by the parabola $y = x^2$ and the line $y = -x + 2$.

   b) The triangular region with vertices (1,1), (4,1), and (4,4).

   c) The triangular region with vertices (1,1), (4,2), and (4,4).

   d) The finite region determined by the parabolas $y = -x^2 + 2$ and $y = x^2 - 2$.

   e) The region bounded by the parabola $y = -x^2$ and the line $y = -5x$.

   f) The region bounded by the curve $y = e^{2x}$, the x-axis and the lines $x = -1$ and $x = 2$.

   g) The finite region between the curve $y = x(x - 1)^2$ and the x-axis.

   h) The finite regions bounded by $y = x^3$ and $y = x^{1/3}$.

2. Determine the volume of the indicated solid region. Sketch the solid.

   a) The solid under the plane $z = 0.5x$ over the rectangle $0 \le x \le 5$ and $0 \le y \le 10$.

   b) The solid under the plane $z = x + y$ above the region bounded by $x^2 + y^2 = 1$, $x = 0$, and $y = 0$.

   c) The solid below $z = xy$ above the rectangle with opposite vertices (0,0) and (1,5).

   d) The solid above the x-y plane and below the surface $z = 100 + x - y^2$ for $x \le 0$.

   e) The solid under the surface $z = 1/xy$ over the rectangle with opposite vertices (2,3) and (5,7).

   f) The solid under the surface $z = y \cos(xy)$ over the rectangle $0 \le x \le \pi/4$, $0 \le y \le 2$.

   g) The solid below the surface $z = 2e^{-xy} - 1$ and above the x-y plane with $y \ge 0$ and $1 \le x \le e$.

Section 9.5 Applications Involving Multiple Integration.

In this section several different "applications" are introduced that involve multiple integration. In each example the double integral is introduced as the continuous analog of a summing process, to "add up" the total amount of a quantity over a two dimensional region or in a three dimensional volume. Several of the examples require a modeling component in which the "application" terminology is introduced and the mathematical problem is established. The calculus component then consists of evaluating specific double integrals. The two basic applications introduced involve computing Total Amounts and Averages of variable quantities. The basic interpretations of the double integral are:

Interpretation of F(x,y) and units of x and y	Interpretation of $\iint_R F(x,y)\, dA$
$z = F(x,y) =$ density at point (x,y). x and y have linear units, and z has units of Amount/(linear units)^2.	$I = \iint_R F(x,y)\, dA = $ **Total Amount over R**. If Amount is mass, then I is total mass. If Amount is individuals, e.g. number of spiders, then I is the total population of the region R.
F(x,y) a quantity, the amount of something measurable, defined for (x,y) ∈ R.	$\text{Ave}\{F\} = \iint_R F(x,y)\, dA\, /\, \iint_R 1\, dA$ The **Average** of F has the same unit as F(x,y).
F(x,y) any quantity and a weight specified by a non-negative function, $w(x,y) \geq 0$.	$\text{WAve}\{F\} = \iint_R w(x,y) \times F(x,y)\, dA / \iint_R w(x,y)\, dA$ is the **Weighted Average** of F(x,y) over the region R with weight w(x,y)

Densities and total amounts.

A "density" is a measure of amount per unit area. A mass density may be in units g/cm^2, a population density may be in units animals/km^2, a construction cost rate may have units dollars/ft^2. The basic concept that gives the total amount of a substance of constant density over an area is

$$\text{Total} = \text{density} \times \text{area}.$$

This is valid only if the density is a constant. If the density varies over a region R. i.e, is given by a function

$$D = D(x,y),$$

then the total amount over R is computed in exactly the same way that the volume over R was computed. The process is analogous, with the density value replacing the "height" value use to compute a volume. The net result is that the total amount is given by the double integral of the density D over R.

Total Amount of Substance having density D(x,y) over a region R.

$$\text{Total Amount} = \iint_R D(x,y)\, dA$$

If the density is a "probability density" then the integral over a region R indicates the probability that an observation (x,y) will be in the region R.

Example 1 Determining the Total Population from a Population Density.

Problem Assume that the population of a city is maximum at its center and decrease as the driving distance to the city center increases. One density model with this property is given by the density function

$$D(x,y) = D_0 / \{|x| + |y| + 1\}.$$

The constant D_0 is the maximum density at the city center, which is the origin (0,0). The units of D are persons/area. If x and y are given in km then D has units people/km^2. What is the total population of a square shaped city that is 20 km on edge and has a core density of 40,000 people per km^2?

Solution **Modeling component.** The total population is the same as the total volume under the population density surface, i.e. the

$$\text{Total Population} = \iint_R D(x,y) \, dA.$$

The density function in this example involves absolute values. Its denominator is one plus a measure of the "street" or "taxi cab" distance from the center of the city, the origin (0,0), to a point (x,y). To avoid integrating a function with absolute values and having to deal with the different cases for evaluating the absolute values, we reduce the problem to an easier one. Observe that the density function is symmetric about the origin. By the symmetry of D(x,y) the total population is the same in each quadrant sector of the city. Therefor, we only need to calculate the population in the first quadrant, where $|x| = x \geq 0$ and $|y| = y \geq 0$. The total population will be four times this quadrant population. Since the city is a square with side lengths 20 km the first quadrant is the region

$$R = \{(x,y) | \ 0 \leq x \leq 10 \text{ and } 0 \leq y \leq 10\}$$

Calculus application. The quadrant population is the integral over R of the density function D. In units of 1,000 people, this is:

The Quadrant Population = $\int_0^{10} \int_0^{10} 40/\{x + y + 1\} \, dy \, dx$

$$= \int_0^{10} 40 \, ln(x + y + 1)|_{y=0}^{y=10} \, dx$$

$$= 40 \int_0^{10} ln(x + 11) - ln(x + 1) \, dx$$

To evaluate this integral we need to know an antiderivative of $ln(x + c)$ for c a constant. This can be obtained from an integration table, or by integration by parts, using $u = ln(x + c)$ and $dv = dx$, we find the derivative $du = dx/(x + c)$ and choose the antiderivative of 1 as $v = x + c$. The resulting formula is

$$\int ln(x + c) \, dx = (x + c)ln(x + c) - \int (x + c)/(x + c) \, dx = (x + c)ln(x + c) - x.$$

This formula is used twice, with $c = 11$ and $c = 1$, to evaluate the above integral:

The Quadrant Population = $40\left(\{[(x + 11)ln(x + 11) - x] - [(x + 1)ln(x + 1) - x]\}|_0^{10}\right)$

$$= 40\left(\{[21ln(21) - 10] - [11ln(11) - 1]\} - \{[11ln(11) - 0] - [1ln(1) - 0]\}\right)$$

$$= 40\left(21ln(21) - 22ln(11) - 9\right) \approx 87.251 \ (000) \text{ people}.$$

The total city population is four times this number, approximately 349 thousand people.

Example 2 Amount of solute in a solution with variable concentration.

Background A rectangular channel contains solution of substance S. For instance, S could be salt, a specific protein or a suspended inorganic compound. In this example assume that S is protein molecules being manufactured in a biotechnology firm, by extraction from a hyperdomo-cell line that has been engineered to produce the protein. The channel carries the protein in solution from a breeding tank that maintains a constant concentration of S entering the channel. At the other end of the channel the solute is removed by an osmotic process that keeps the concentration of solute at zero. In a steady state the concentration of S in the channel is a linear function of the distance along the length of the channel.

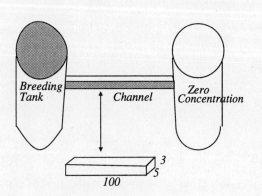

Problem If the concentration of S at the breeding tank is 50 Moles/l and the channel is 5mm wide, 3mm deep and 10cm long, how much solute is in the channel at steady state?

Solution **Modeling component.**

The channel is modeled as a box over the region

$$R = \{(x,y) \mid 0 \le x \le 100 \text{ and } 0 \le y \le 5 \ ; \ x \text{ and } y \text{ in mm}\}.$$

The concentration function $C(x,y)$ is chosen to be linear in x, and satisfy the boundary conditions:

$C(0,y) = 50$ Breeder input level $C(100,y) = 0$ Outlet level.

The concentration function is then

$$C(x,y) = 50(1 - x/100) \text{ Moles}/l.$$

The basic relationship between concentration and amount is

$$\text{Amount} = \text{Concentration} \times \text{Volume}.$$

The volume of the channel over any mm^2 subregion of the base R is $3mm^3$, (height 3mm times area $1mm^2$). Thus, the amount of solute above a square mm of the channel is 3 times the concentration. However, we must multiply by a conversion factor to convert the given concentration in units of Moles/liter to the units Moles/mm^3 since the volume unit being used is mm^3. One liter equals 10^6 cubic millimeters as

$$l = (10cm)^3 = (10^2 \text{ mm})^3 = 10^6 \text{ mm}^3.$$

Thus the amount of S above a unit area of R, at the point (x,y) is

$$A(x,y) = 50(1 - x/100)\text{Moles}/l \times 3mm^3$$

$$= 150(1 - x/100)(\text{Moles}/l) \times mm^3 / (10^6 mm^3/l)$$

Canceling the mm^3 in the numerator and denominator, writing $150 = 1.5 \times 10^2$, $1/100$ as 0.01, and $1/10^6$ as 10^{-6}, the density of solute at (x,y) is given by

$$D(x,y) = A(x,y)/mm^2 = 1.5(1 - 0.01x)10^{-4} \text{ Moles}/mm^2.$$

Calculus application.

The total amount of S is then the integral of D(x,y) over R.

$$\text{Total Amount of Solute} = \iint_R D(x,y) \, dA = \int_0^{100} \int_0^5 1.5(1 - 0.01x) 10^{-4} \, dy \, dx$$

$$= \int_0^{100} 1.5(1 - 0.01x) 10^{-4} \times y \Big|_0^5 \, dx$$

$$= \int_0^{100} 7.5(1 - 0.01x) 10^{-4} \, dx = 7.5(x - 0.005x^2) 10^{-4} \Big|_0^{100}$$

$$= 7.5(100 - 0.005 \times 100^2) 10^{-4} = 375 \times 10^{-4} = 3.75 \, 10^{-2} \text{Moles}.$$

To give the number of molecules of S in the channel, we multiply the number of Moles by Avogadro's number, $A^o \approx 6.025 \times 10^{23}$. This is the number of molecules per Mole. Thus the total number of solute molecules in the channel at steady-state is approximately

$$3.75 \times 10^{-2} \times 6.025 \times 10^{23} \approx 2.259 \times 10^{22} \text{ molecules}.$$

☑

Average value of a function F(x,y).

Recall that an average of a set of numbers involves two sum, one to add up the numbers and one to count how many numbers were added.

$$\text{Average of } \{a_1, a_2, \ldots, a_n\} = \Sigma_{i=1,n} a_i / \Sigma_{i=1,n} 1 = \Sigma_{i=1,n} a_i / n$$

The average of F(x,y) over a region R is similarly defined, with the Summations Σ replaced by the double integral $\iint_R \underline{\quad} \, dA$, which in effect "adds" the functions values over the entire region R. In the denominator, the integral of the constant 1 corresponds to "counting" how many terms were "added", which for a region is simply the area of R. Thus, the average of a function over a region is obtained as the integral over the region divided by the area of the region:

$$\text{Ave}\{F\} = \iint_R F(x,y) \, dA \Big/ \iint_R 1 \, dA \qquad \textbf{Average of F(x,y)}$$

Example 3   Computing an Average function value.

Problem   Compute the average value of $F(x,y) = xy^2$ over the triangular region with vertices (0,0), (2,0), and (0,3).

Solution   The Ave{F} is by definition $\text{Ave}\{F\} = \iint_R xy^2 \, dA \Big/ \iint_R 1 \, dA$. The integral in the denominator is simply the area of the triangle, which is 3. As the line passing though the two vertices (0,2) and (3,0) is $y = 3 - 1.5x$, the double integral of F is evaluated as

$$\iint_R xy^2 \, dA = \int_0^2 \int_0^{3 - 3x/2} xy^2 \, dy \, dx.$$

$$= \int_0^2 xy^3/3 \Big|_0^{3 - 3x/2} \, dx = \int_0^2 x(3 - 3x/2)^3 \, dx$$

$$= \int_0^2 -(9/8)x^4 + (27/4)x^3 - (27/2)x^2 + 9x \, dx$$

$$= -(9/40)x^5 + (27/16)x^4 - (9/2)x^3 + (9/2)x^2 \Big|_0^2 = 9/5$$

The average function value is thus $\text{Avg}\{F\} = (9/5)/3 = 3/5$.

☑

Example 4 Computing the Average value over a circular region.

Problem Compute the average value of the function $F(x,y) = 3 + x \exp(x^2 + y^2)$ over the interior of the circle or radius 2 centered at the origin.

Solution To compute the Ave{F} we first evaluate the double integral of $F(x,y)$ over the interior of the circle

$$x^2 + y^2 = 4.$$

There are two ways to express the double integral as an iterated integral. If we choose to represent the region by solving the boundary equation for y as a function of x, then

$$R = \{(x,y) \mid -2 \leq x \leq 2 \text{ and } -\text{SQRT}(4 - x^2) \leq y \leq \text{SQRT}(4 - x^2)\}.$$

The corresponding iterated integral is

$$I = \int_{-2}^{2} \int_{-\text{SQRT}(4-x^2)}^{\text{SQRT}(4-x^2)} 3 + xe^{(x^2 + y^2)} \, dy \, dx.$$

This integral <u>can not be evaluated</u> because there is <u>no</u> antiderivative of e^{y^2} in terms of elementary functions. But, if we solve the boundary equation for x as a function of y, the circular region R is given by

$$R = \{(x,y) \mid -2 \leq y \leq 2 \text{ and } -\text{SQRT}(4 - y^2) \leq x \leq \text{SQRT}(4 - y^2)\}.$$

For this representation, the corresponding iterated integral is

$$I = \int_{-2}^{2} \int_{-\text{SQRT}(4-y^2)}^{\text{SQRT}(4-y^2)} 3 + xe^{(x^2 + y^2)} \, dx \, dy.$$

This integral can be evaluate, because of the factor x times the exponential. Treating y as a constant,

$$D_x e^{(x^2 + y^2)} = 2xe^{(x^2 + y^2)}.$$

Therefore, the above integral is evaluated as

$$I = \int_{-2}^{2} 3x + 0.5 e^{(x^2 + y^2)} \Big|_{x = -\text{SQRT}(4-y^2)}^{x = +\text{SQRT}(4-y^2)} \, dy$$

When x is evaluated at each limit the exponent reduces to the constant 4:

$$x^2 + y^2 = (4 - y^2) + y^2 = 4$$

as the y^2 terms cancel. Therefore, after evaluating the antiderivative at the limits for x, the integral becomes

$$I = \int_{-2}^{2} [3 \, \text{SQRT}(4 - y^2) + 0.5e^4] - [3 \times -\text{SQRT}(4 - y^2) + 0.5e^4] \, dy$$

$$= 3 \int_{-2}^{2} 2 \, \text{SQRT}(4 - y^2) \, dy.$$

The resulting integral is again one that we can not evaluate directly. However, there is a way out of this dilemma. We are seeking the Ave{f} which equal the integral I divided by the area of R. The integral I is found to be 3 times an integral that is exactly equal to the area of the region R. To see this we evaluate the area of R as an iterated integral which after the first integration results in the same undetermined integral:

Section 9.5 APPLICATIONS INVOLVING MULTIPLE INTEGRATION

$$A = \int_{-2}^{2} \int_{-\text{SQRT}(4-y^2)}^{+\text{SQRT}(4-y^2)} 1 \, dx \, dy$$

$$= \int_{-2}^{2} \Big|_{-\text{SQRT}(4-y^2)}^{+\text{SQRT}(4-y^2)} \, dy$$

$$= \int_{-2}^{2} 2\,\text{SQRT}(4-y^2) \, dy$$

Consequently, using these integral forms we find the average I/A equal 3:

$$\text{Ave}\{F\} = 3 \int_{-2}^{2} 2\,\text{SQRT}(4-y^2) \, dy \,\Big/\, \int_{-2}^{2} 2\,\text{SQRT}(4-y^2) \, dy = 3.$$

☑

Weighted Averages.

An important generalization of the concept of an Average is that of a "weighted" average. A "weight" function is any non-negative function w(x,y). A simple average gives equal value to all the numbers being averaged, while a "weighted" average gives unequal value to the terms being averaged. In statistics, if equal data values are grouped, the weight indicates the number of data with the same value. The "weight" function specifies the weight to be given as a function of the domain coordinate (x,y). If w(x,y) > 1 then the product w(x,y)×F(x,y) gives a greater value than F(x,y). The weighted average WA{f} is obtained by "adding" the weighted values of F and dividing by the "Total amount of Weights."

Assume that a function F(x,y) and a non-negative weight function w(x,y) are defined over a region R. The "weighted average" of F over R, with weight w is the ratios of two integrals over the same region R:

The Average of F(x,y) over R with weight W(x,y) is

$$\text{Ave}\{F;w\} = \text{WAve}\{F\} = \iint_R F(x,y) \times w(x,y) \, dA \,\Big/\, \iint_R w(x,y) \, dA$$

Example 5 **Computing weighted averages.**

Problem What is the weighted average of the function F(x,y) = 1 + x over the triangular region with vertices (0,0), (2,0) and (2,4) with the weight $w(x,y) = e^{-x-y}$?

Solution The weighted average is calculated by evaluating the integrals of w(x,y)F(x,y) and w(x,y) over R as iterated integrals. The integral of the weight function is simply

$$\iint_R w(x,y) \, dA = \int_0^2 \int_0^{2x} e^{-x-y} \, dy \, dx$$

$$= \int_0^2 -e^{-x-y}\Big|_{y=0}^{y=2x} \, dx$$

$$= \int_0^2 -e^{-3x} + e^{-x} \, dx$$

$$= e^{-3x}/3 - e^{-x}\Big|_0^2$$

$$= e^{-6}/3 - e^{-2} + 2/3.$$

The integral of the weighted function requires slightly more work to evaluate:

$$\iint_R F(x,y) \times w(x,y) \, dA = \int_0^2 \int_0^{2x} (1+x)e^{-x-y} \, dy \, dx$$

$$= \int_0^2 -(1+x)e^{-x-y}\Big|_{y=0}^{y=2x} \, dx$$

$$= \int_0^2 -(1+x)e^{-x-2x} + (1+x)e^{-x} \, dx$$

$$= \int_0^2 -e^{-3x} - xe^{-3x} + e^{-x} + xe^{-x} \, dx$$

Integration by parts is required to integrate the two terms involving products. Leaving the intermediate steps to the reader to complete the basic antiderivatives are:

$$\int -xe^{-3x} \, dx = xe^{-3x}/3 + e^{-3x}/9$$

$$\int xe^{-x} \, dx = -xe^{-x} - e^{-x}$$

Combining similar terms, this integral becomes

$$\iint_R F(x,y) \times w(x,y) \, dA = (-2/9 + x/3)e^{-3x} - (2+x)e^{-x} \Big|_0^2$$

$$= (-2/9 + 2/3)e^{-6} - (2+2)e^{-2} + 2/9 + 2$$

$$= (4/9)e^{-6} - 4e^{-2} + 20/9$$

Thus the weighted average is

$$\text{WAve}\{F\} = [(4/9)e^{-6} - 4e^{-2} + 20/9] / [e^{-6}/3 - e^{-2} + 2/3] \approx 3.16$$

☑

The following example illustrates a very common type of application in which a weighted integral is used to "integrate" the interaction between people and their environment, or a contaminate in the environment. This is a very important type of calculation that is required whenever toxicological or health impact assessments are performed.

Example 6 **Risk Assessment. Computing an expected community risk due to a toxin.**

Background In toxicological Risk Assessment, the risk associated with a specific toxin is indicated by a "risk factor", r, which is the "probability of a harmful effect" per exposure of an individual to a unit amount of toxin. These risk factors are generally estimated by experiments with different species and from limited accidental exposure histories.

The "linear risk model" assumes that the "risk" to an individual exposed to an amount A of the toxin is $A \times PR$, where PR is the Probability of Risk per unit exposure.

Individual Risk = Probability of Risk × amount of exposure.

$$IR = PR \times A \qquad \text{Individual Risk}$$

The "risk" to a general population is calculated as the sum of the risks to each member. If every individual is exposed to the same amount of toxin, then this is simply the product of the Population size and the Individual Risk:

Population Risk = IR × size of the population.

$$PR = IR \times N_P. \qquad \text{Population Risk}$$

Two main factors must be accounted for in assessing the consequent Population Risk due to a toxin release. One is the actual physical distribution of the toxin, resulting in variable exposure levels over a geographic region R. Assume the resulting toxin distribution is described by

Section 9.5 — APPLICATIONS INVOLVING MULTIPLE INTEGRATION

$T(x,y)$ = level of toxin at position (x,y).

The other principal factor influencing the Population Risk is the distribution of the "population at risk" over the region R. Assume that

$P(x,y)$ = Population density at (x,y).

In this situation the risk will vary among the population but an "Average Population Risk" is computed. To account for the variability in the exposure level and the population size this is computed as a weighted average. The population density is treated as a "weight".

$$\text{Total Population Risk} = \iint_R T(x,y) \times P(x,y) \, dA$$

$$\text{Average Population Risk} = \iint_R T(x,y) \times P(x,y) \, dA \,/\, \iint_R P(x,y) \, dA$$

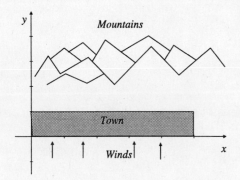

Problem

A western community is exposed to an air-borne toxin. Assume the community occupies a rectangular shaped region 5 km long and 1 km wide. The town is on the slopes of the Rockies with a desert to its west. The wind carrying a toxin blows off the desert and deposits the toxin as it moves up the slopes. Assume that the town area is modeled by the region

$$R = \{(x,y) \mid 0 \le x \le 5 \text{ and } 0 \le y \le 1\}$$

with East being in the positive y-direction. Assume the amount of the toxin deposited is

$$T(x,y) = 0.1 e^{5y} \text{ ppm (parts per million)}.$$

Assume that the population density is

$$P(x,y) = 600x(5 - x)y.$$

Thus the population likes to live in the hills and near the center of the town. Assume the risk factor for this toxin is "one in a million":

$$PR = 10^{-6} \text{ tumors per exposure to 1 ppm}.$$

What is the average population risk to this exposure situation?

Solution The Average risk is the ratio of the total risk to the total number of persons in the town.

$$\text{Total Population} = \int_0^5 \int_0^1 P(x,y) \, dy \, dx = \int_0^5 \int_0^1 600x(5-x)y \, dy \, dx$$

$$= \int_0^5 300x(5-x)y^2 \Big|_{y=0}^{y=1} dx = \int_0^5 300x(5-x) \, dx$$

$$= 750x^2 - 100x^3 \Big|_0^5 = 6{,}250 \text{ persons}.$$

The total population risk, TPR, is the integral over R of the product of the risk factor, the population density, and the toxin deposition level:

$$TPR = \int_0^5 \int_0^1 PR \times T(x,y) \times P(x,y) \, dy \, dx$$

$$= \int_0^5 \int_0^1 10^{-6} \times 600x(5-x)y \times 0.1 e^{5y} \, dy \, dx$$

$$= 6 \times 10^{-5} \int_0^5 \int_0^1 (5x - x^2) y e^{5y} \, dy \, dx$$

$$= 6\times 10^{-5} \int_0^5 (5x - x^2)(y/5 - 1/25)e^{5y}\big|_{y=0}^{y=1} dx$$

as integration by parts gives $\int y e^{5y} dy = (y/5 - 1/25)e^y$, thus the above integral becomes

$$\text{TPR} = [(24/25)e^5 - 4/25]6\times 10^{-5} \int_0^5 (5x - x^2) dx$$

$$= [(24/25)e^5 - 4/25]6\times 10^{-5} \; (5/2)x^2 - (1/3)x^3\big|_0^5$$

$$= [(24/25)e^5 - 4/25]6\times 10^{-5} \times 125/6 = 5(24e^5 - 1)10^{-5} \approx 0.178 \text{ tumors}$$

The Average Population Risk, APR, is thus

$$\text{APR} = 0.178 / 6{,}250 \approx 2.85 \; 10^{-5} \text{ tumors per person.}$$

This risk would be stated as a "probable risk of 28.5 tumors per million population."

Exercise Set 9.5

1. Compute the total mass of a rectangular steel plate that has density given by $D(x,y) = 1 + 0.5\sin(x - y)$ g/m^2 for $0 \leq x \leq 2\pi$ m and $0 \leq y \leq \pi$ m.

2. What is the total mass of a rectangular steel plate whose density is given by $D(x,y) = 0.8(x + 2y + 1)$ g/cm^2 for $0 \leq x \leq 5$cm and $0 \leq y \leq 3$cm?

3. The density of soft wood on a 2km^2 tract of land was estimated from satellite images and fit to the model equation $D(x,y) = 0.5(x + y/2 + 0.1) \; 10^3$ bft/ km^2 where $1 \leq x \leq 3$ km and $0 \leq y \leq 1$ km. What is the total board foot bft of soft wood on the tract?

4. If the mass density of a mold in a square growth dish is $D = 2 - |x| - |y|$ μg/cm^2 for $|x| \leq 1$ cm and $|y| \leq 1$cm, determine the total mass of mold in the dish.

5. In Example 2 the problem was solved by first converting the concentration to an amount above a mm^2 area of the base of the channel. Alternatively, the problem could be solved by simply integrating the concentration function over the three-dimensional volume of the trough. Even though the concentration does not vary with the height in this example we can treat the function C as a function of the height variable z, i.e. use $C(x,y,z) = 50(1 - 0.01x)$ Moles/l. Then the total amount of solute would be given by a triple integral

$$A = \iiint_S C(x,y,z) \, dV$$

where the integral is over the solid region $S = \{(x,y,z) \mid 0 < x < 100, 0 < y < 5, \text{ and } 0 < z < 3\}$.
(a) Express this triple integral as an iterated integral over the three variables and evaluate the total amount A.
(b) What would be the total amount of solute in the channel if due to gravity the concentration of solute were $C(x,y,z) = 50(1 - 0.01x)(1 - z/3)$?

6. Assume the probability density function for two measurements to have the values x and y, for a rectangular region $0 \leq x \leq 1$ and $0 \leq y \leq 2$ is given by $D(x,y) = 6e^{-(2x + 3y)}$.

   a) What is the total probability over this region?

   b) What is the probability that x and y are in the rectangle and their sum is less than 1?

7. The mass density of a square plate is given by $D(x,y) = \sin(\pi x)\sin(\pi y)$ for $0 \leq x \leq 1$ and $0 \leq y \leq 1$. What is the total mass of the plate?

8. The "centroid" of a region R is by definition a point $(\bar{x},\bar{y})$ that is the mass center of the region when the region R has a constant density. The coordinates $\bar{x}$ and $\bar{y}$ are defined by

$$\bar{x} = \iint_R x \, dA \bigg/ \iint_R 1 \, dA$$

$$\bar{y} = \iint_R y \, dA \bigg/ \iint_R 1 \, dA$$

   a) Compute the centroid of the rectangular region $0 \leq x \leq 2$ and $2 \leq y \leq 4$.

   b) Compute the centroid of the triangular region with vertices (0,0),(3,0) and (3,4).

   c) Compute the centroid of the triangular region with vertices (0,0),(3,1) and (4,6).

   d) Compute the centroid of the region inside the triangle with vertices (-4,0),(4,0) and (0,6) but outside the triangle with vertices (-4,0),(4,0) and (0,3). Sketch the region first.

9. Compute the average value of the function $z = F(x,y)$ over the indicated region.

   a) $F(x,y) = x - y$ over the square with opposite vertices (0,0) and (-2,-2).

   b) $F(x,y) = (x - 1)^2 y$ over the triangle with vertices (1,1), (4,-2), and (4,5).

   c) $F(x,y) = x \sin(y)$ over the region bounded by $x = 0$, $y = 0$ and $y = 2 - x/3$.

10. Compute the average value of the function $F(x,y) = 2y \exp(-(x^2 + y^2))$ over the interior of the circle $x^2 + y^2 = 9$.

11. What is the average value of $z = x^2$ over the trapezoid with vertices (-8,0), (8,0), (5,4), and (-5,4)?

12. The temperature of a sheet of steel is given by $T(x,y) = 340 e^{-2x}$ °C for $0 \leq x \leq 10$ and $0 \leq y \leq 2$. What is the average temperature of the sheet?

13. For the function $F(x,y) = x + y$ and the triangle R with vertices (0,0),(5,0), and (5,4):

   a) Compute the average of F over R.

   b) Compute the average of F over R with the weight function $w(x,y) = xy$.

   c) Compute the average of F over R with the weight function $w(x,y) = e^{-x}$.

14. A rectangular region is exposed to a toxic substance. Assume the population distribution $P(x,y)$ and the exposure levels $T(x,y)$ are given by

$$P(x,y) = 300 + 200\cos(\pi x) + 50y \text{ persons/km}^2$$

$$T(x,y) = 0.2 e^{-0.1y} \text{ mg/km}^2$$

over the region $R = \{(x,y) | -1 \leq x \leq 1 \text{ and } -2 \leq y \leq 2\}$. Assume the x and y units are km. If the toxicity of the substance is 1 tumor per μg exposure, what is the expected Average Population Risk due to this exposure? See Example 6.

Chapter 10 *Partial Derivatives*.

10.1 Introduction to Partial Derivatives. 575

10.2 Tangent Planes and Slopes of Surfaces. 583

10.3 Linear Approximations and the Differential of F(x,y). 594

10.4 Linearization of Multivariable Functions; Stability Analysis of Dynamical Systems. 607

An Introductory Discussion.

The graph of a multivariable function $z = f(x,y)$ is a surface in three-dimensional space. When the function is sufficiently "nice" its graph can be describe locally by a tangent plane, which is determined by the function's "partial derivatives". A "partial derivative" is a generalization of the ordinary derivative concept to functions of two or more variables.

> A *partial derivative* is the instantaneous rate of change of a multivariable function due to an incremental increase in only one of its independent variables, while all the other variables are held constant.

Naturally, a partial derivative is defined as a limit, a limit of a difference quotient, the same type of limit used to define ordinary derivatives. Consequently, it should not be surprising that partial derivatives can be evaluated using the basic Derivative Rules and properties established in Chapter 3 for ordinary derivatives. These Rules are applied to establish the rates of change in f(x,y) as either x increases and y is held fixed, or visa-versa as y increases and x is held constant. This means that "partial derivatives" of every function can be formally determined without evaluating limits. In Section 10.1 simple partial derivatives are defined and the fundamental rules for computing partial derivatives are introduced. Then, higher order partial derivatives are introduced. This section ends with a brief introduction to *partial differential equations*, which are refered to by mathematicians as PDEs, and are simply an equations that involve partial derivatives.

In Section 10.2 geometric applications of partial derivatives are introduce. Partial derivatives are used to describe and characterize surfaces that are graphs of multivariable functions. Recall that for single variable functions the derivative provided the slope of the tangent line to the function's graph. For multivariable functions, the partial derivative provides the slopes needed to describe *tangent planes* to the surface that is the graph of the function.

Another important use of partial derivatives is to provide approximations of functions. In Section 10.3 the concept of a linear approximation is introduced with the notion of the "differential" of a multivariable function. A special application of this process, introduced in Section 10.4, is the process of *Linearizing a dynamical model* about *equilibrium* points. This technique is used to study the *stability* of such models to perturbations of the primary variables. Stability analysis is a method used to analyze dynamical systems ranging from simple mechanical mechanisms, like a pendulum, to complex ecosystem models. The presentation in this section is intended to introduce the simplest aspects of this important application.

Section 10.1 Introduction to Partial Derivatives.

Partial Derivatives.

The partial derivative of a function $F(x,y)$ is a function that describes the rate of change in the function's value as one of the independent variables increases. Since there are two independent variables, x and y, there are correspondingly two partial derivatives. These are denoted using the dell symbol ∂ to denote a partial derivative. The partial derivative with respect to x is denoted by $\partial/\partial x\, F(x,y)$ and is defined by a limit process identical to that used to define $d/dx\, f(x) = f'(x)$ for a single variable function $f(x)$. Consequently, all of the Derivative Rules and Formulas that were developed for ordinary derivatives are applicable to evaluate partial derivatives. The key observation is that the variable that is not being differentiated is treated as a constant. Thus when we evaluate $\partial/\partial y\, F(x,y)$ the x-variable is treated like a constant and the derivative rules are applied to the y-variable.

DEFINITION OF FIRST PARTIAL DERIVATIVES.

The **first partial derivative of $z = F(x,y)$ with respect to x** is

$$\partial/\partial x\, F(x,y) = \lim_{h \to 0} F(x+h, y) - F(x,y) / h$$

provided the limit exists. This partial derivative is also denoted by

$$\partial z/\partial x, \qquad z_x \qquad \text{and} \qquad F_x(x,y) \quad \text{or simply } F_x.$$

The **first partial derivative of $F(x,y)$ with respect to y** is

$$\partial/\partial y\, F(x,y) = \lim_{h \to 0} F(x, y+h) - F(x,y) / h,$$

provided the limit exists. This partial derivative is also denoted by

$$\partial z/\partial y, \qquad z_y, \qquad \text{and} \qquad F_y(x,y) \quad \text{or simply } F_y.$$

All of the Algebraic Rules established for ordinary derivatives apply by extension to partial derivatives. For instance, the partial derivative of a linear combination of differentiable functions is evaluated as the linear combination of the partial derivatives of each function:

$$\partial/\partial x\, [A \cdot F(x,y) \pm B \cdot G(x,y)] = A \cdot \partial/\partial x\, [F(x,y)] \pm B \cdot \partial/\partial x\, [G(x,y)]$$

for constants A and B. Frequently partial derivatives are simply referred to as "partials".

Example 1 Computing Partial Derivatives.

Problem For the given function compute the partial derivatives $F_x(x,y)$ and $F_y(x,y)$.

Solutions a) If $F(x,y) = x^2 + 5x + y^3$, the partials of F are

$$F_x(x,y) = \partial/\partial x(x^2) + \partial/\partial x(5x) + \partial/\partial x(y^3) = 2x + 5$$

since y is treated as a constant $\partial/\partial x \, y^3 = 0$; similarly,

$$F_y(x,y) = \partial/\partial y \, (x^2 + 5x) + \partial/\partial y \, y^3 = 0.3y^2$$

since x is treated as a constant $\partial/\partial y \, (x^2 + 5x) = 0$.

b) If $F(x,y) = 1 + x^3 y^2$, then the first partial derivatives of F are:

$$F_x(x,y) = 3x^2 y^2 \text{ since the factor } y^2 \text{ is treated as a constant;}$$

$$F_y(x,y) = 2x^3 y \quad \text{since the factor } x^3 \text{ is treated like a constant.}$$

c) If $F(x,y) = \sin(xy)$, each partial derivative is taken applying the Chain Rule or the General Sine Rule: let $u = xy$, then $\partial u/\partial x = y$ and $\partial u/\partial y = x$, thus

$$F_x(x,y) = \partial/\partial x \, \sin(xy) = D_u \sin(u) \times \partial/\partial x \, u$$

$$= \cos(u) \times y = y \cos(xy)$$

Similarly, $F_y(x,y) = \partial/\partial y \, \sin(xy) = x \cos(xy)$

d) If $F(x,y) = x/y + 2y$, then

$$F_x(x,y) = 1/y \quad \text{and} \quad F_y(x,y) = -x/y^2 + 2$$

e) If $F(x,y) = 3x + 2y + e^{xy}$ the partial derivative of the first two terms of F are evaluated by the Power Rule, while the General Exponential Rule is applied to evaluate the partial derivative of the third term e^{xy}:

$$F_x(x,y) = 3 + y e^{xy} \quad \text{and} \quad F_y(x,y) = 2 + x e^{xy}$$

f) If $F(x,y) = (3x - 2xy + y^3)^5$ then the General Power Rule is used to evaluate

$$F_x(x,y) = 5(3x - 2xy + y^3)^4 \partial/\partial x(3x - 2xy + y^3) = 5(3 - 2y)(3x - 2xy + y^3)^4$$

$$F_y(x,y) = 5(3x - 2xy + y^3)^4 \partial/\partial y(3x - 2xy + y^3) = 5(-2x + 3y^2)(3x - 2xy + y^3)^4$$

Second Order Partial Derivatives.

The first partial derivatives $F_x(x,y)$ and $F_y(x,y)$ are themselves functions of the two variables x and y. Consequently, just as higher derivatives can be computed for single variable functions, higher order partial derivatives can be computed. However, there are more second order partial derivatives. In general, if the function F is a function of N independent-variables it will have N first partial derivatives, each of which can be differentiated again in N different ways. Consequently there are N^2 possible ways to compute "second" partial derivatives for such a function. For a function of two variables, N = 2, this means there are four "second order" partial derivatives.

DEFINITION OF SECOND PARTIAL DERIVATIVES.

The four **second partial derivatives** of $z = F(x,y)$ are :

The second partial derivative of F with respect to x:

$$\partial^2/\partial x^2 \, F(x,y) = F_{xx}(x,y) \equiv \partial/\partial x \, F_x(x,y) = \partial/\partial x \left(\partial/\partial x \, F(x,y) \right)$$

The second partial derivative of F with respect to y:

$$\partial^2/\partial y^2 \, F(x,y) = F_{yy}(x,y) \equiv \partial/\partial y \, F_y(x,y) = \partial/\partial y \left(\partial/\partial y \, F(x,y) \right)$$

The second (mixed) partial derivative of F with respect to x and then y:

$$\partial^2/\partial y \partial x \, F(x,y) = F_{xy}(x,y) \equiv \partial/\partial y \, F_x(x,y) = \partial/\partial y \left(\partial/\partial x \, F(x,y) \right)$$

The second (mixed) partial derivative of F with respect to y and then x:

$$\partial^2/\partial x \partial y \, F(x,y) = F_{yx}(x,y) \equiv \partial/\partial x \, F_y(x,y) = \partial/\partial y \left(\partial/\partial y \, F(x,y) \right)$$

The derivatives F_{xy} and F_{yx} are called **mixed derivatives**. Notice that the notation for mixed partial derivatives indicates the order in which the partial derivatives should be taken. However, the order of the notation is reversed between the ∂^2 form of the second derivative and the subscripted form of the derivative:

Caution   In the subscripted form F_{xy}, the derivatives are computed in the order that the subscripts appear, <u>from left to right</u>, F_{xy} denotes the derivative first with respect to x and then with respect to y;

whereas, in the $\partial^2 F/\partial\_\partial\_$ form, the order of taking derivatives is from <u>right to left</u>, $\partial^2/\partial x \partial y \, F$ is evaluated by differentiating first with respect to y and then with respect to x.

Example 2 Computing second order partial derivatives.

Problem Determine all second order partial derivatives of the given function.

(a) $F(x,y) = x^3 + 2xy - y^3$ (b) $F(x,y) = x \ln(y) - ye^x$ (c) $F(x,y) = x \tan(y)$

Solutions (a) First, compute the first order partials:

$$F_x(x,y) = 3x^2 + 2y \quad \text{and} \quad F_y(x,y) = 2x - 3y^2$$

Then, differentiating these gives the second partials:

$$F_{xx}(x,y) = \partial/\partial x \, (3x^2 + 2y) = 6x \quad F_{xy}(x,y) = \partial/\partial y \, (3x^2 + 2y) = 2$$

$$F_{yy}(x,y) = \partial/\partial y \, (2x - 3y^2) = -6y \quad F_{yx}(x,y) = \partial/\partial x \, (2x - 3y^2) = 2$$

(b) Again, the first step is to compute the first order partials:

$$F_x(x,y) = \ln(y) - ye^x \qquad F_y(x,y) = x/y - e^x$$

Then, the second partials are

$$F_{xx}(x,y) = \partial/\partial x \, (\ln(y) - ye^x) = ye^x$$

$$F_{xy}(x,y) = \partial/\partial y \, (\ln(y) - ye^x) = 1/y - e^x$$

$$F_{yy}(x,y) = \partial/\partial y \, (x/y - e^x) = -x/y^2$$

$$F_{yx}(x,y) = \partial/\partial x \, (x/y - e^x) = 1/y - e^x$$

(c) The first partial derivatives of $F(x,y) = x \tan(y)$ are

$$F_x(x,y) = \tan(y) \quad \text{and} \quad F_y(x,y) = x \sec^2(y)$$

Then, the second partial derivatives are

$$F_{xx}(x,y) = \partial/\partial x \, \tan(y) = 0$$

$$F_{xy}(x,y) = \partial/\partial y \, \tan(y) = \sec^2(y)$$

$$F_{yy}(x,y) = \partial/\partial y \, \sec^2(y) = 2\sec^2(y)\tan(y)$$

$$F_{yx}(x,y) = \partial/\partial x \, [x \sec^2(y)] = \sec^2(y)$$

In the above example you may have noticed that the mixed partials F_{xy} and F_{yx} are equal for each function. Is this a coincidence? No! Computing a mixed partial derivative involves taking two limits sequentially. If the function $F(x,y)$ is nice enough, at and near the point where the derivatives are being evaluated, the order in which the limits are taken can be reversed, resulting in the other mixed partial derivative. A sufficient condition for this to occur is that the mixed partial derivatives are continuous in a neighborhood of the point (x,y). This will occur as long as the function and its partial derivatives do not have a singularity caused by division by zero or attempting to take an even root of a negative term or a log or $\ln$ of a non-positive term.

EQUALITY OF MIXED PARTIAL DERIVATIVES.

> If the partial derivatives F_{xy} and F_{yx} are continuous in a circular region about a point (x,y) then
>
> $$F_{xy}(x,y) = F_{yx}(x,y).$$

Higher Order Partial Derivatives.

Higher order partial derivatives are defined by repeated partial differentiation. For instance, the third partial derivative of F with respect to x is denoted with subscript notation by $F_{xxx}(x,y)$ and is evaluated as the partial with respect to x of the second partial with respect to x,

$$F_{xxx} = \partial^3/\partial^3 x \, F(x,y) = \partial/\partial x \, F_{xx}(x,y)$$

There are many higher order partial derivatives of each order, as many as there are ways to choose which variable to differential with respect to. One of the eight possible third partial derivatives of F is the partial $F_{xxy}(x,y)$, which is denoted in the ∂ notation as

$$F_{xxy} = \partial^3/\partial y \partial^2 x \, F(x,y) = \partial/\partial y \, F_{xx}(x,y)$$

This notation may be compared to the mixed partial $F_{xyy}(x,y)$, which is defined as

$$F_{xyy} = \partial^3/\partial y^2 \partial x \, F(x,y) = \partial/\partial y \, F_{xy}(x,y),$$

and to the mixed partial $F_{yxy}(x,y)$, which is defined as

$$F_{yxy} = \partial^3/\partial y \partial x \partial y \, F(x,y) = \partial/\partial y \, F_{yx}(x,y)$$

Example 3 **Computing higher order partial derivatives.**

Problem Compute the third order partial derivative $F_{xxy}(x,y)$ of $F(x,y) = x^2 y - \sin(xy)$.

Solution The appropriate first and second order partial derivatives must be evaluated. Reading the order of the subscripts, from left to right, of F_{xxy} indicates the necessary partial derivatives. From left to right the subscripts are x, x, and then y, which indicates to differentiate first with respect to x, then by x again, and finally by y. This will give F_x, then F_{xx} and finally, taking the partial with respect to y, F_{xxy}.

$F_x(x,y) = 2xy - y \cos(xy)$

$F_{xx}(x,y) = \partial/\partial x \, (2xy - y \cos(xy)) = 2y + y^2 \sin(xy)$

$F_{xxy}(x,y) = \partial/\partial y \, (2y + y^2 \sin(xy)) = 2 + 2y \sin(xy) + y^2 x \cos(xy)$

The last differentiation required the Product Rule and the General Rule for differentiating sin(u).

Partial Differential Equations.

For many functions the various partial derivatives are related. They may be equal to each other or the relationship may be more complicated. When we write an equation that involves partial derivatives it is called a **partial differential equation**, which is often shortened to **PDE**. Examples of such equations are given by

$$F_{xx}(x,y) = 0, \qquad F_{xy}(x,y) = F_{yx}(x,y), \quad \text{and} \quad F_{xx}(x,y) + F_{yy}(x,y) = 0$$

In dynamical models the variable t, for time, is often a major independent variable. If the function's variables are not indicated, the system variables are inferred from the symbols used to represent the partial derivatives. Thus, in the PDE

$$U_t = 4U_{xx}$$

the function is U and the implied variables are t and x, hence, U must be a function U(t,x).

A particular function F(x,y) is said to be a **solution** of a specific PDE if the given PDE is a valid equation, i.e., is true for all x and y in a specific domain when the appropriate partial derivatives of the function F are evaluated and substituted into the PDE. If the resulting equation is not valid, the function F is **not a solution** of the PDE.

There are normally many functions that are formally solutions to the same partial differential equation. This is because the solution of a PDE involves integration and hence the introduction of "integration constants". However, the integration involved is "partial integration" and consequently the "integration constants" may actually be functions of the other independent variables. In real problems the solutions are generally restricted by imposing boundary conditions that they must satisfy.

Solving PDEs is generally more difficult than solving ordinary Differential Equations. In fact, many equations that arise in direct applications, such as modeling the flame of a candle, have yet to be solved theoretically. The study of PDEs is thus an open and inviting area of mathematics. We will not actually solve any PDEs in this section, but we can verify that particular functions are, or are not, solutions of specific PDEs. This is done by evaluating the partial derivatives of the function that are involved in the PDE, substituting these into the PDE, and ascertaining if the resulting equation is, or is not, valid.

Example 4   **Confirming solutions to PDEs.**

Problem A   Show that $z = y^3 x + 3 ln(y)$ is a solution of the PDE $z_{xx} = 0$.

Solution   Taking the partial derivative of z with respect to x, twice, gives

$$z_x = y^3 \text{ and } z_{xx} = 0$$

Thus, for $z = y^3 x + 3 ln(y)$ the equation $z_{xx} = 0$ is true for all x and y.

Problem B   Show that the function $F(x,y) = e^{3x}\cos(3y)$ is a solution to the **Laplace Equation**

$$\partial^2/\partial^2 x \, F(x,y) + \partial^2/\partial^2 y \, F(x,y) = 0$$

Solution   Compute the second partial derivatives as follows:

$$F_x(x,y) = 3e^{3x}\cos(3y) \qquad \text{and} \qquad F_{xx}(x,y) = 9e^{3x}\cos(3y)$$

$$F_y(x,y) = -3e^{3x}\sin(3y) \quad \text{and} \quad F_{yy}(x,y) = -9e^{3x}\cos(3y)$$

Substituting into the Laplace PDE the second partials just computed gives

$$9e^{3x}\cos(3y) - 9e^{3x}\cos(3y) = 0$$

This equation is true! Thus $F(x,y) = e^{3x}\cos(3y)$ is a solution of Laplace's Equation $F_{xx} + F_{yy} = 0$.

Problem C   Show that the function $F(x,y) = x^3 \cos(5y)$ is <u>not</u> a solution of the **Laplace Equation**

$$\partial^2/\partial^2 x\, F(x,y) + \partial^2/\partial^2 y\, F(x,y) = 0$$

Solution   Compute the second partial derivatives as follows:

$$F_x(x,y) = 3x^2 \cos(5y) \quad \text{and} \quad F_{xx}(x,y) = 6x \cos(5y)$$

$$F_y(x,y) = -5x^3 \sin(5y) \quad \text{and} \quad F_{yy}(x,y) = -25x^3 \cos(5y)$$

Substituting into the PDE the second partials just computed gives

$$6x \cos(5y) - 25x^3 \cos(5y) = 0$$

Which is not true for all x and y in the domain of F! This is the equation

$$[6x - 25x^3] \cos(5y) = 0$$

which implies either $6x - 25x^3 = 0$ or $\cos(5y) = 0$. Both equations are true for particular x and y-values but are not true for all x and all y. Thus the given function F is not a solution of Laplace's Equation.

Exercise Set 10.1

1. Compute F_x and F_y for the indicated functions.

   a) $F(x,y) = 3x - y^4$ b) $F(x,y) = yx - y + x$ c) $F(x,y) = 2x^3 - y^4$

   d) $F(x,y) = y/x$ e) $F(x,y) = 5x - x^2y^4$ f) $F(x,y) = (x - y)/x$

   g) $F(x,y) = (x + y)/(x - y)$ h) $F(x,y) = (x + y)^{-2}$ i) $F(x,y) = 3x - e^{xy}$

   j) $F(x,y) = x^2 e^{xy}$ k) $F(x,y) = x^2 \ln(y)$ l) $F(x,y) = \ln(xy)$

   m) $F(x,y) = 2x^2 - 3xy + \ln(y^2)$ n) $F(x,y) = -7x^2 \log(y^2)$ o) $F(x,y) = x^y$

   p) $F(x,y) = x^{1/y}$ q) $F(x,y) = y \cos(x)$ r) $F(x,y) = xy - \sin(xy^2)$

2. Evaluate $F_x(x_0, y_0)$ and $F_y(x_0, y_0)$ at the indicated point (x_0, y_0).

   a) $F(x,y) = x^2 + 3y^2$; $(x_0, y_0) = (1, -2)$ b) $F(x,y) = 3xy - x^{1/2}$; $(x_0, y_0) = (4, 1)$

   c) $F(x,y) = x^2 y^{-2}$; $(x_0, y_0) = (1, 3)$ d) $F(x,y) = 3xe^y$; $(x_0, y_0) = (2, 0)$

   e) $F(x,y) = \sin(xy)$; $(x_0, y_0) = (0.5, \pi)$ f) $F(x,y) = \sin(xy)$; $(x_0, y_0) = (3, 1)$

   g) $F(x,y) = y^2 - xy$; $(x_0, y_0) = (-1, 2)$ h) $F(x,y) = xy^2 - yx^2$; $(x_0, y_0) = (2, 1)$

3. Compute the four second partials: F_{xx}, F_{yy}, F_{xy} and F_{yx} and verify that the mixed partials are equal.

   a) $F(x,y) = x + y^2 - 3$ b) $F(x,y) = x^3 y^2$ c) $F(x,y) = y e^{3x}$

   d) $F(x,y) = y e^{y/x}$ e) $F(x,y) = xy + y^2 - 3ln(x/y)$ f) $F(x,y) = e^{3xy}$

   g) $F(x,y) = y \cos(x)$ h) $F(x,y) = \sin(xy)$ i) $F(x,y) = \cos(x) + y^2 - 3$

   j) $F(x,y) = \sin(xy) e^{3x}$ k) $F(x,y) = \sin(x) + 2x^2 y - e^{yx} + 2y^2 - 1$ l) $F(x,y) = ln(x+y) + e^{xy}$

4. Compute the partial derivatives F_x, F_y, F_z, F_{xz} and F_{zy} of the given function F.

   a) $F(x,y,z) = xyz$ b) $F(x,y,z) = xy + 2xz + 3yz$ c) $F(x,y,z) = 3x^2 - 2xy + 3yz^3$

   d) $F(x,y,z) = 2x e^{zy}$ e) $F(x,y,z) = xy - xz$ f) $F(x,y,z) = \cos(xy) - \sin(xz)$

   g) $F(x,y,z) = x/y - xz^{2.5}$ h) $F(x,y,z) = x^{y^{\wedge}z}$

5. At what point(s) (x_0, y_0) are both first partial derivatives of $F(x,y) = x^2 - 6xy + 4y^3$ equal to zero?

6. Compute the first partial derivative of the function with respect to each variable.

   a) $F(x,y,z) = 3xyz$ b) $F(x,y,z) = ln(x) e^{y/z}$ c) $F(x,y,z) = (x^y)^z$

   d) $G(u,v,t) = u/v - e^{-3t}$ e) $G(u,v,t) = (u^2 - v^2 + 2t^2) 1/2$ f) $G(u,v,t) = u ln(t/3v)$

   g) $H(r,\theta,\phi) = r^2 \sin(\theta) \cos(\phi)$ h) $H(r,\theta,\phi) = r \tan(\theta - \phi^2)$ i) $H(r,\theta,\phi) = \phi e^{2r\theta}$

   j) $z = (x - u)^2 + (y - 3v)^{1/3}$ k) $w = PV + \phi V v^2 / 2g$ l) $v = 4\pi r^3 \sin(\theta/s)$

7. A formula that relates the surface area, A, of a human to the person's weight, w, and height, h, is $A = cw^p h^q$, where c, p, and q are parameters. What is the rate of change in Area due to a unit change in height, or in weight? What is the rate of change in A due to a unit change in the parameter p? What is the sum of the changes in A due to a unit change in all the variables and parameters individually?

8. In business, the first partial derivatives are called "marginals". What is the marginal profit for increasing cost if the profit is a function $P = 200 - 3x/y + (x - y)^2$, where x = cost and y = demand if x = 10 and y = 6?

9. Show that $u(x,y) = \cos(2x) e^{-3y}$ is a solution of the PDE $u_{xx} + (4/9) u_{yy} = 0$.

10. Show that $g(x,y) = ln(x^2 + y^2)$ satisfies Laplace's Equation: $g_{xx} + g_{yy} = 0$.

11. Show that $g(x,y) = 3x^4 + 6x^2 y^2 + 2y^4$ is a solution of the PDE: $x g_x(x,y) + y g_y(x,y) = 4g(x,y)$.

12. For the given function $z = F(x,y)$ determine the point(s) (x_0, y_0) at which both $\partial/\partial x\, F(x_0, y_0) = 0$ and $\partial/\partial y\, F(x_0, y_0) = 0$.

   a) $z = x^2 + y^2$ b) $z = xy$ c) $z = x^2 - y^2$ d) $z = 3x^3 - 4x + 2y - y^2$

   e) $z = xy^2 - yx^2 - 3y$ f) $z = xy^2 - 3x^2 y - 9x$ g) $z = x \sin(\pi y) + 0.5x$ h) $z = x^3 - 3xy + 2y$

13. For each function considered in Exercise 12, compute the three second order partial derivatives. At the points (x_0, y_0) where both first partials are zero, determine the sign (+, -, or 0) of the quantity
$$D = [F_{xy}(x,y)]^2 - F_{xx}(x,y) \cdot F_{yy}(x,y)$$

Section 10.2   Tangent Planes and Slopes of Surfaces.

Introduction.

Partial derivatives are used in the same way that ordinary derivatives are use, to describe, analyze, and approximate functions. The basic geometric applications of partial derivatives are to describe the first and second order characteristics of a surface corresponding to the graph of a multivariable function. In this section we consider the first order approximation of a three-dimensional surface that is the graph of $z = F(x,y)$. At a point (x_0, y_0, z_0) on the surface $z = F(x,y)$ the approximating linear surface is in fact a plane, called a tangent plane. The linear function whose graph is the tangent plane is a first degree polynomial, called a Taylor Polynomial. In Section 10.3 the first order Taylor Polynomial is used to locally approximate the values of the function $F(x,y)$ when the point (x,y) is near the point (x_0, y_0).

A plane is characterized by the fact that its rate of change in directions parallel to the x or y axis is constant. On a tangent plane these rates of change are provided by the partial derivatives of F at the point (x_0, y_0). An even more useful fact is that the rate of change on the tangent plane in any direction can be derived from these same derivatives. In this section the Tangent Plane is introduced and some "graphing techniques" are introduced to help illustrate three-dimensional graphs on two-dimensional paper.

The Tangent Plane.

A tangent plane is a generalization of the concept of a tangent line. Before discussing tangent planes we shall briefly review the fundamental principles of a tangent line to the graph of a single variable function.

For a single variable function $g(x)$, the derivative $g'(x_0)$ is the instantaneous rate of change in the function's values $g(x)$, as x increases from the value x_0. By definition, $g'(x_0)$ is the slope of the tangent line at $(x_0, g(x_0))$. The tangent line is always given in the point-slope equation of a line: $y = y_0 + g'(x_0)(x - x_0)$. Observe that the slope of the tangent line is the change in the y-coordinate on the tangent line due to a unit increase in x; as illustrated in Figure below, $g'(x_0)$ is the change in the y-coordinate on the tangent line from x_0 to $x_0 + 1$.

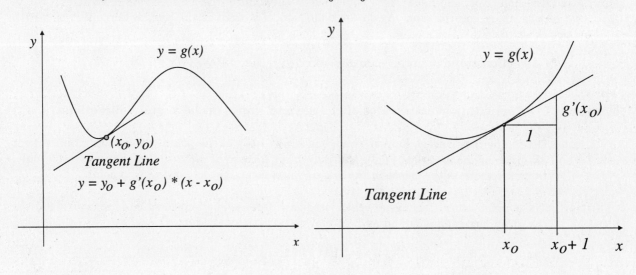

The graph of a continuous function of two variables, $z = F(x,y)$, is a surface in the x-y-z coordinate system. The surface is said to be "smooth" if at each point $(x, y, F(x,y))$ on the surface there is a tangent plane. The tangent plane in three dimensions corresponds to a tangent line in two dimensions. Just as a tangent line was determined by its slope, a tangent plane can be characterized by its two slopes, its "slope in the x-direction" and its "slope in the y-direction".

To explore these concepts we introduce "difference notation" to describe the change in a multivariable function. The capital Greek letter delta, Δ, is again used to denote a difference, however, in a multi-variable situation a subscript is added to distinguish which independent variable is changing. Initially, we consider the difference Δz in the z-values on a surface induced by changes in only one of the independent variables. Recall that an increment of an independent variable may be described either as a "difference", Δx or Δy, or as a "differential", dx or dy; the terms are interchangeable for independent variables.

The change in a surface $z = F(x,y)$ due to an increment in the x-variable, Δx, is the difference:

$$\Delta z_x = z(x + \Delta x, y) - z(x, y) \quad \text{or} \quad \Delta F_x = F(x + \Delta x, y) - F(x, y)$$

The average rate of change of z (in the x-direction) between (x, y) and $(x + \Delta x, y)$, is the difference quotient $\Delta z_x / \Delta x$.

The change in a surface $z = F(x,y)$ due to an increment in the y-variable, Δy, is the difference:

$$\Delta z_y = z(x, y + \Delta y) - z(x, y) \quad \text{or} \quad \Delta F_y = F(x, y + \Delta y) - F(x, y)$$

The average rate of change of z (in the y-direction) between (x, y) and $(x, y + \Delta y)$, is the difference quotient $\Delta z_y / \Delta y$.

The *x-direction slope* is the rate of change in z at a specific point (x, y) as x increases and y is held fixed, which is defined as the limit of the average rate of change in the x-direction as the increment Δx goes to zero. By definition, this limit gives the partial derivative of z with respect to x:

"x-direction slope" $= \lim_{\Delta x \to 0} \Delta z_x / \Delta x = \lim_{\Delta x \to 0} \{z(x + \Delta x, y) - z(x, y)\}/\Delta x = \partial z/\partial x$.

Similarly, the *y-direction slope* is the rate of change in z at a particular point (x, y) as y increases and x is held fixed. This is defined as the limit of the average rate of change in the y-direction as Δy goes to zero. Again, by definition, this limit is the partial derivative of z with respect to y:

"y-direction slope" $= \lim_{\Delta y \to 0} \Delta z_y / \Delta y = \lim_{\Delta y \to 0} \{z(x, y + \Delta y) - z(x, y)\}/\Delta y = \partial z/\partial y$.

Example 1 Computing average rates of change and slopes in the x- and y-directions.

Problem Determine the Average Rate of Change in $z = x^2 - 2xy$ at the point $(1, 3)$ due to an increment $\Delta x = 0.1$ in the x-variable only and due to an increment $\Delta y = -0.2$ in the y-variable only. Compare these to the slopes in the x and y-directions, respectively, at this point.

Solution The average rates of change are calculated as difference quotients evaluated at $(x, y) = 1, 3)$ for the given increments Δx and Δy.

$$\Delta z_x/\Delta x = \{z(1 + 0.1, 3) - z(1, 3)\} / 0.1 = \{[1.1^2 - 2 \times 1.1 \times 3] - [1^2 - 6]\}/ 0.1 = -3.9$$

$$\Delta z_y/\Delta y = \{z(1, 3 - 0.2) - z(1, 3)\} / -0.2 = \{[1^2 - 2 \times 1 \times (3 - 0.2)] - [1^1 - 6]\}/ -0.2 = 2$$

The slopes are calculated as partial derivatives, again evaluated at the reference point.

Slope in the x-direction is $\partial z/\partial x \,|_{(x,y) = (1,3)} = 2x - 2y|_{(1,3)} = -4$

Slope in the y-direction is $\partial z/\partial y \,|_{(x,y) = (1,3)} = -2x|_{(1,3)} = -2$

The average rate of change in the x-direction differs from the slope in the x-direction because the function z is nonlinear in x. As the increment Δx is made smaller we expect this difference to approach zero. However, as z is essentially linear in the y-variable, the slope in the y-direction and the average rate of change for the increment $\Delta y = -0.2$ are identical. This will be true for any increment Δy.

What distinguishes *plane* surfaces from *curved* surfaces is that a plane has the same "slopes" at every point, whereas, the "slopes" of a curved surface will vary from one point to the next. The "slopes" of a plane are the coefficients of the x and y variables when its equation is expressed in the **function form**:

$$z = Ax + By + C \quad \textbf{Function Equation of a Plane}$$

To see this, consider a plane given by the above general function form. The difference Δz induced by a difference Δx in the x-variable is the constant $A \times \Delta x$, for all points (x,y):

$$\Delta z_x = [A(x + \Delta x) + By + C] - [Ax + By + C] = A \times \Delta x$$

Hence, the x-direction slope of the plane is

$$\lim_{\Delta x \to 0} \Delta z_x/\Delta x = \lim_{\Delta x \to 0} \{A \times \Delta x\}/\Delta x = A$$

Similarly, it follows that the y-direction slope of this plane is the coefficient B:

$$\lim_{\Delta y \to 0} \Delta z_y/\Delta y = \lim_{\Delta y \to 0} \{B \times \Delta y\}/\Delta y = B$$

The "point-slope" equation of a plane is a generalization of the point-slope equation of a line. It immediately indicates a point (x_0, y_0, z_0) on the plane and the slopes in the x and y directions.

DEFINITION OF THE POINT - SLOPE EQUATION OF A PLANE.

The **point-slope** equation of the plane surface containing the point (x_0, y_0, z_0) with "x-direction slope" = A and "y-direction slope" = B is

$$z = z_0 + A(x - x_0) + B(y - y_0)$$

A tangent plane is defined at a point on a surface $z = F(x, y)$ as the plane that passes through the given point and has the same "x and y-slopes" as the graph of F has at that point. Using the reference point (x_0, y_0) in the domain of F, the associated tangent plane is tangent to the graph of $z = F(x, y)$ at the point (x_0, y_0, z_0). The slopes of the surface at this point are given by the partial derivatives evaluated at (x_0, y_0):

the slope in the x-direction: $\partial/\partial x \, F(x_0, y_0)$

the slope in the y-direction: $\partial/\partial y \, F(x_0, y_0)$

The tangent plane is defined using these slopes and the point-slope equation of a plane.

DEFINITION OF THE TANGENT PLANE.

> The **tangent plane to the surface $z = F(x,y)$ at the point (x_0, y_0, z_0)**, where $z_0 = F(x_0, y_0)$, is
>
> $$z = z_0 + \partial/\partial x F(x_0, y_0)(x - x_0) + \partial/\partial y F(x_0, y_0)(y - y_0)$$

Example 2 **Determining the equations of tangent planes at specific points.**

Problem What are the equations of the tangent planes to the surface $z = x^2 + 3y^2$ at the points $P_1 = (-1, 2, 13)$, $P_2 = (2, 1, 7)$, and $P_3 = (0, 0, 0)$?

Solution The surface is the graph of the function $F(x,y) = x^2 + 3y^2$. To determine the tangent planes, first compute the partial derivatives of $z = F(x,y)$.

$$z_x = F_x(x,y) = 2x \quad \text{and} \quad z_y = F_y(x,y) = 6y$$

At the point P_1 the domain point is $(-1, 2)$. As $F_x(-1, 2) = -2$ and $F_y(-1, 2) = 12$, the tangent plane at P_1 is

$$z = 13 - 2(x + 1) + 12(y - 2)$$

At the point $P_2 = (2, 1, 7)$, the corresponding partial derivatives are $z_x = 4$ and $z_y = 6$. Thus the equation of the tangent plane at $(2, 1, 7)$ is

$$z = 7 + 4(x - 2) + 6(y - 1)$$

The domain point corresponding to $P_3 = (0, 0, 0)$ is the origin, $(0, 0)$, at which the partial derivatives are both zero. Thus the tangent plane at $(0, 0, 0)$ is just the x-y plane, $z = 0$. ☑

Example 3 **Comparing surface z-values and tangent plane z-values.**

Problem Let S be the surface $z = 3xe^{-2y}$. To compare the z-values on this surface and a tangent plane to this surface:
(i) Determine the tangent plane, denoted P, to the surface S at the point $(4, 0, 12)$.
(ii) Determine the difference between the z-value on the surface S and the corresponding z-value on the tangent plane P at the positions:
(a) $(x,y) = (5,1)$ (b) $(x,y) = (4,1)$ and (c) $(x,y) = (4.1, -0.01)$.

Solution (i) The plane P is found by first computing the partial derivatives:

$$z_x = 3e^{-2y} \quad \text{and} \quad z_y = -6xe^{-2y}$$

Evaluated at $x = 4$ and $y = 0$, these partial derivatives provide the slopes $A = z_x = 3$ and $B = z_y = -24$. Therefore, the equation of the tangent plane at $(4, 0, 12)$ is

$$z = 12 + 3(x - 4) - 24y$$

(ii) The difference in the z-values on the surface and on the plane is the error, E, incurred when using a tangent plane to approximate a surface.

$$E = \text{[surface z-value]} - \text{[tangent plane z-value]}$$

$$= 3xe^{-2y} - [12 + 3(x - 4) - 24y]$$

(a) At $(x,y) = (5,1)$ the difference $E = 15e^{-2} - 39 \approx -36.97$.

(b) At $(x,y) = (4,1)$, the difference $E = 12e^{-2} + 12 \approx 13.62$

(c) At $(x,y) = (4.1,-0.01)$, the difference $E = 12.3e^{0.02} - 12.54 \approx 0.008$.

Observe that the difference is relatively large at the first two points whose (x,y) coordinates are over one unit away from $(4,0)$. However, the difference is fairly small at the third point that is a little over 0.1 units away from $(4,0)$.

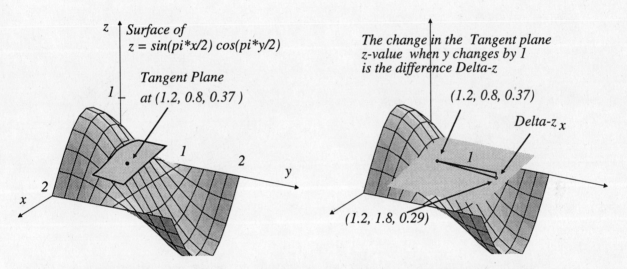

An important and useful observation is that the x and y-slopes of a tangent plane can be calculated as the change or differences Δz_x and Δz_y <u>on the tangent plane itself</u> corresponding to unit increments Δx and Δy, respectively. This is easily verified algebraically for the plane $z = z_0 + A(x - x_0) + B(y - y_0)$. The difference Δz_x when $\Delta x = 1$ is the coefficient A:

$$\Delta z_x = [z_0 + A(x + 1 - x_0) + B(y - y_0)] - [z_0 + A(x - x_0) + B(y - y_0)] = A$$

Similarly, the "y-direction slope" is $B = \Delta z_y$ when $\Delta y = 1$. These differences are illustrated in the above Figure, where both slopes illustrated are negative.

A main objective in science is to determine functional relationships from experimental data. To describe "first-order" characteristics of an experimental system a first step is to estimate the partial derivatives of the system under specific conditions. This is commonly done by performing two sets of experiments, the first at prescribed reference conditions and the second set with a unit change in one of the primary variables. The observed change in the measured output of the system is then interpreted as the partial derivative. It is called the sensitivity of the system with respect to the variable that was altered.

Sketching surfaces and tangent planes.

Level curves and the intersection of surfaces.

To sketch a three-dimensional surface on a two-dimensional piece of paper requires the use of "perspective" and imagination. In the following we consider and compare two surfaces, the graph of the function F(x,y) and a tangent plane to the surface at a specific point.

One approach used to illustrate a surface and its curvature is to "reduce the dimensionality" by considering lines in the surface. In general the intersection of two three-dimensional surfaces determines a space curve, which is referred to as the curve where one surface "cuts" through the other surface. When one of the surfaces is a plane, the "cut" curve can then be expressed as a two-dimensional curve in the plane. It is usually much easier to visualize a curved surface by sketching the cut curves associated the surface and several planes. The planes are usually chosen to be parallel to one of the coordinate planes and regularly spaced, in which case the cut curves resemble the "ribs" of a wooden boat.

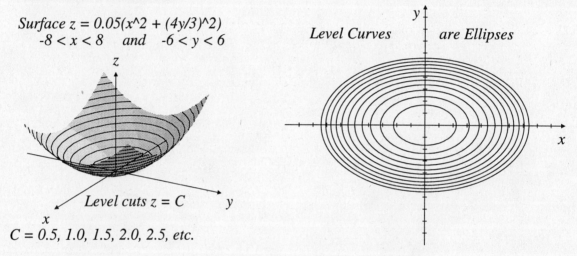

Surface $z = 0.05(x^2 + (4y/3)^2)$
$-8 < x < 8$ and $-6 < y < 6$

Level cuts $z = C$
$C = 0.5, 1.0, 1.5, 2.0, 2.5,$ etc.

Level Curves are Ellipses

When the "cut" planes are all horizontal, i.e., parallel to the x-y plane with equations z = C for a constant C, the resulting intersection curves are called "level curves"; these are the curves that appear in contour maps. A topographical map is formed by considering the surface cuts corresponding to an evenly spaced range of C values, e.g., C = 100, 110, 120, ... etc., and projecting these three-dimensional level curves onto the x-y plane. In the left panel of Figure 10.3 a portion of the parabolic surface

$$z = 0.05[x^2 + (4/3)y^2]$$

is sketched with level curves at increments of 0.5. The right panel illustrates the corresponding projections of these curves onto the x-y plane.

The curvature of a surface can be illustrated in a three-dimensional coordinate system by considering the "cut" curves generated by a series of planes perpendicular to the x-y plane. The simplest planes perpendicular to the x-y plane are those parallel to one of the other coordinate planes.

Planes parallel to the x-z plane have equations of the form y = constant;

planes parallel to the y-z plane are the planes where x = constant.

The curvature of the surface is most easily visualized if "cut" curves for regularly spaced planes are sketched. The following computer-generated graphs utilize this technique.

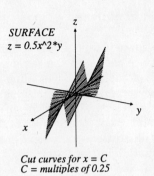

Cut curves for $x = C$
C = multiples of 0.25

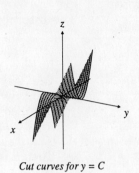

Cut curves for $y = C$

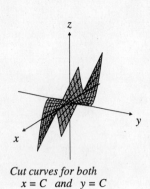

Cut curves for both
$x = C$ and $y = C$

Lines in three-dimensions.

A line as the intersection of two planes.

When the surface being considered is actually a plane, the "cut" curves formed by intersecting it with other planes are straight lines.

Recall that two planes either

(i) do not intersect, in which case they are parallel,

(ii) coincide, i.e., are the same plane, or

(iii) intersect in a straight line.

Line of intersection of two planes.

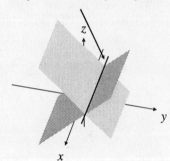

The equations of lines in three-space are more complicated than line equations in two-space. Just as a point in two dimensions is the intersection of two lines, a line in three-dimensions is determined by the intersection of two planes. One way to indicate a three-dimensional line is to indicate two planes whose intersection forms the line. A line L is specified by two simultaneous equations of the form:

Line in three-space as the intersection of two planes.

$$L = \{(x,y,z) \mid z = z_0 + A_0 x + B_0 y \text{ and } z = z_1 + A_1 x + B_1 y\}$$

This type of representation is not unique, since there are infinitely many planes containing a line. (Think of the planes rotating about the line.) If one of the planes is parallel to the y-z or x-z coordinate planes then this reduces to a pair of equations of the form:

$$L: z = c + my \text{ \textbf{and} } x = x_0 \quad \text{or} \quad L: z = c + mx \text{ \textbf{and} } y = y_0.$$

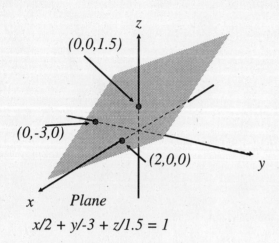

The intercept equation of a plane.

A plane in three-space is uniquely determined by three points (that are not all on the same line). One way to indicate a plane is by labeling the points where it intersects the coordinate axes, provided the plane is not parallel to one of the axes. Assume the intercept points are $(a,0,0)$, $(0,b,0)$ and $(0,0,c)$ as illustrated in the next graph. Then, the plane's equation in the "intercept" form is :

Intercept Equation of a Plane

$$x/a + y/b + z/c = 1$$

Lines in tangent planes; a generalization of the point-slope equation.

The plane most often considered in calculus problems is a tangent plane. When the tangent point is not near the origin, or if the axis-intercepts differ greatly in magnitude, sketching the intercepts may not be practical or provide adequate information about the plane in a region near the tangent point.

Another approach to illustrating a tangent plane is to sketch two lines in the tangent plane that pass through the tangent point and are parallel to the x and y axis, respectively. To illustrate a tangent plane at (x_0, y_0, z_0) and the corresponding partial derivatives of $z = F(x,y)$ the two "cut" lines in the tangent plane determined by the planes $y = y_0$ and $x = x_0$ are sketched. This can be done in three-dimensions or in a two-dimensional cut plane.

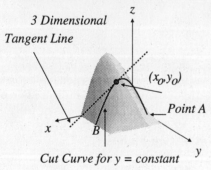
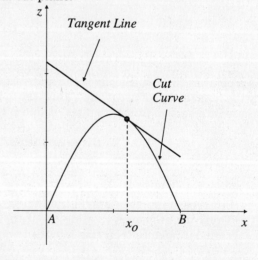

The plane $y = y_0$ cuts the surface $z = F(x,y)$ along a curve $z = F(x, y_0)$. In the above figure this plane is depicted on the left in a three-dimensional x-y-z graph, which we abbreviate as 3-D, and on the right as a two-dimensional plane having x and z coordinate axes. Observe the orientation of the points A, x_0, and B in the two graphs. In the two-dimensional graph the z-surface appears as a curve. This curve is the intersection of the $z = F(x,y)$ surface by the plane $y = y_0$. The indicated tangent line is described in the $y = y_0$ plane by the standard point-slope equation

$$z = z_0 + [\partial/\partial x\, F(x_0, y_0)]\, (x - x_0)$$

with $z_0 = F(x_0, y_0)$ and the slope indicated by the derivative evaluated at x_0, which is actually a partial derivative since y is assumed to be fixed and equal to y_0. The above equation considered in the x-y-z space is actually the equation of a plane, with zero slope in the y-direction, hence the absence of a y-term. The equation we are seeking is one to describe the tangent line in three-dimensions that is the intersection of this plane with the "cut" plane $y = y_0$. This intersection is described by two simultaneous equations.

Equation of the line in the tangent plane parallel to the x-axis.

$$y = y_0 \quad \text{and} \quad z = z_0 + [\partial/\partial x \, F(x_0, y_0)] \, (x - x_0)$$

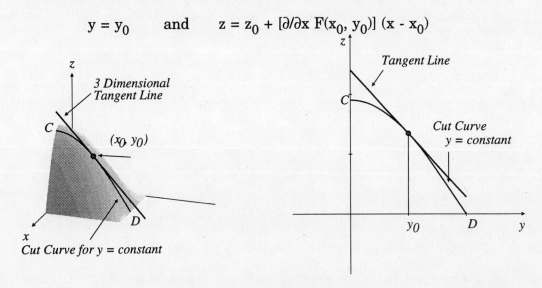

Next, consider the plane $x = x_0$; the corresponding "cut" curve where it intersects the surface $z = F(x,y)$ is illustrated in the left figure above. The resulting curve $z = F(x_0, y)$ is depicted in a 3-D system on the left and on the right in a 2-D y-z plane corresponding to the plane $x = x_0$. The point-slope equation of the tangent line in the y-z plane is

$$z = z_0 + [\partial/\partial y \, F(x_0, y_0)] \, (y - y_0)$$

with $z_0 = F(x_0, y_0)$. This line passes through the point (y_0, z_0) and has slope given by the partial derivative $F_y(x_0, y_0)$. In the 3-D system, this same tangent line is described by two equations.

Equations of the line in the Tangent plane parallel to the y-axis.

$$x = x_0 \quad \text{and} \quad z = z_0 + [\partial/\partial y \, F(x_0, y_0)](y - y_0)$$

Recall that in three dimensions, two intersecting lines uniquely determine a plane that contains the two lines. Hence the two lines in the tangent plane given by the above equations will in fact characterize, or uniquely determine, the tangent plane. The slopes of the tangent lines give the "slopes" of the tangent plane in the respective directions. Including these lines will sometimes help identify and illustrate the tangent plane in a sketch. This is illustrated in Figure at the right.

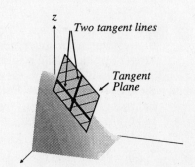

Level lines for constant z-values are indicated on tangent plane.

Many applications and systems involve more than two independent variables. For a higher dimensional function, like $z = F(x,y,u,v,t)$, the "linear surface" that corresponds to a "tangent plane" is called a "hyper-plane"; it would have an equation of the form $z = Ax + By + Cu + Dv + Et$. Such "linear surfaces" are of dimension one less than the space, e.g., like a line in a plane, a plane in 3-space, or a three-dimensional linear surface in four-dimensional space. The equations and properties of higher dimensional "hyper-planes" and, as we discuss in the following sections, linear approximations and differentials, are directly analogous to those for a two-variable function $F(x,y)$, with the pattern or form given for x and y extended to the additional independent variables. We do not discuss higher dimensional surfaces here, but you will see them when you study Statistics, where they are the primary model for statistical analysis of many variables.

Exercise Set 10.2

1. a) Determine the Average Rate of Change in $z = x^2 + xy$ at the point (1,4) due to an increment $\Delta x = 0.1$ in the x-variable only and due to an increment $\Delta y = -0.2$ in the y-variable only.

   b) What are the slopes of the z surface in the x and y directions at the point (1,4,5)?

   c) Compare the slopes in the x and y-directions in part b) to the Average Rates of Change found in part a).

2. Repeat Exercise 1 with a) $z = -3x + 2y$ and b) $z = 20x/y$.

3. What are the equations of the tangent planes to the surface $z = x^2 + 3y$ at the points $P_1 = (-1,2,7)$, $P_2 = (2,1,7)$, and $P_3 = (0,0,0)$?

4. Determine the equation of the tangent plane to the surface $z = F(x,y)$, at the point (x_0, y_0, z_0) when:

   a) $z = x^2 - xy$; $(x_0, y_0) = (1,-2)$ b) $z = xy$; $(x_0, y_0) = (2,1)$

   c) $z = xe^{-y}$; $(x_0, y_0) = (3, ln(2))$ d) $z = x^2 + y^2 - 9$; $(x_0, y_0) = (0,3)$

   e) $z = 4x^2 - 2x + y^2$; $(x_0, y_0) = (2,-2)$ f) $z = x^2 + y^2 + xy$; $(x_0, y_0) = (1,-2)$

   g) $z = 4x^2 - e^{xy}$; $(x_0, y_0) = (2,-2)$ h) $z = xy/(xy + 1)$; $(x_0, y_0) = (0,3)$

5. Let S be the surface $z = 3xy^2$. a) Determine the tangent plane to the surface S at the point (1,2,12). b) Determine the difference between the z-value on the surface S and the corresponding z-value on the tangent plane at the positions: (a) $(x,y) = (1,1)$ (b) $(x,y) = (2,2)$ and (c) $(x,y) = (1.1,1.9)$.

6. Sketch the graph of $z = F(x,y)$, indicating the cut curves for $x = x_0$ and $y = y_0$ and any others that may help you to visualize the graph. (Try the intersection with the coordinate planes $x = 0$ and $y = 0$.) Sketch the tangent plane at the point with x_0 and y_0 as indicated.

   a) $F(x,y) = 2x + y^2$ $(x_0, y_0) = (2,1)$ b) $F(x,y) = (x - 5)^2$ $(x_0, y_0) = (2,3)$

   c) $F(x,y) = 25 - x^2 - y^2$ $(x_0, y_0) = (1,4)$ d) $F(x,y) = (x - 5)^2 + 2y$ $(x_0, y_0) = (2,3)$

7. Express the equation of the indicated plane in the Equation Form, in the Intercept Form, and in the Point-Slope Form.

   a) The plane with slope 2 in the x-direction, slope -3 in the y-direction, and passing through the point (1,1,1).

   b) The plane with slope -3 in the x-direction, slope 5 in the y-direction, and passing through the point (2,-1,4).

   c) The plane that contains the three points (1,2,4), (-1,2,3) and (-2,4,3).

   d) The plane that contains the three points (1,3,2), (-1,-2,3) and (-2,0,-3).

   e) The plane that is parallel to the plane $z = 1 + 3x + 0.5y$ and contains the point (-1,2,3).

   f) The plane that is parallel to the plane $z = 1 - 3(x - 4) + 5(y + 2)$ and contains the point (1,2,3).

   g) The tangent plane to $z = y + x^2$ at (1,2,3).

   h) The tangent plane to $z = y \sin(\pi x) + y$ at (1,2,2).

8. Consider the surface $z = 2x + y^2$.

   a) Determine the equations of the cut curves where this surface intersects the planes $y = c$ for $c = 0$, 1, 2, and 3.

   b) Sketch the projections of these cut curves onto the x-z plane.

   c) Repeat parts a) and b) with the role of x and y interchanged.

   d) Use the curves in parts a) to c) to sketch the corresponding cut curves in an x-y-z system.

9. Determine the equation of the lines parallel to the x- and y-axis in the tangent plane to

   a) $z = 3y + 0.01x^2$ at the point (10,1,4).

   b) $z = x/y$ at the point (4,2,2).

   c) $z = x^{1/2}y$ at the point (4,3,6).

10. For each function in Exercises 1 and 2, sketch the level curves corresponding to $z = 0$, 1, 2, 3, and 4. Indicate the equation of each level curve.

Section 10.3 Linear Approximations and the Differential of F(x, y).

"Linear approximations" of nonlinear functions $F(x, y)$ are used to calculate numerical approximations, to "simplify" formulas, and to simplify difficult mathematical problems. The process of creating a linear problem by replacing a nonlinear function by its first-order approximation is called "linearizing". This technique is introduced here and considered in greater detail in Section 10.4.

The first-order approximation of a function $F(x, y)$ is a polynomial in x and y of degree 1 that has a special form, a form referred to as a Taylor Polynomial. It is actually just the point-slope equation of a tangent plane to the surface $z = F(x, y)$. When this first-order approximation is used repeatedly to approximate the function's values for (x, y) near a particular reference point (x_0, y_0), it is considered as a function of the increments or differentials dx and dy of change in the independent x and y-variables from point (x_0, y_0). In this situation it is referred to as the "differential" of the function F. The differential is denoted dF and is a generalization of the differential df established for single variable functions $f(x)$. Among the applications that utilize the differential is an important type referred to as "sensitivity analysis". This section finishes with a brief look at problems of this type.

First-order Taylor Polynomials.

A tangent plane to $z = F(x, y)$ is the graph of a first-order polynomial in x and y. As introduced in the previous Section, the equation of the tangent plane has a special form, with constants derived from the values of the function F and its partial derivatives at the reference point. It is a polynomial of degree one in a class of multivariable polynomials called "Taylor Polynomials". In Chapter 12, Section 12.1, higher order Taylor Polynomials and nonlinear approximations will be introduced.

The first-order Taylor Polynomial of F at (x_0, y_0) is

$$P_1(x, y) = F(x_0, y_0) + \{F_x(x_0, y_0)(x - x_0) + F_y(x_0, y_0)(y - y_0)\}$$

Example 1   **Determining a first-order Taylor Polynomial.**

Problem   What is the first-order Taylor Polynomial for $z = x^3 + xy - y^4$ at (1,2)?

Solution   At (1,2) the corresponding $z = -13$. The partial derivatives of z are evaluated at (1,2):

$$z_x = 3x^2 + y, \quad \text{at } (1,2) \; z_x = 5$$
$$z_y = 4y^3 + x, \quad \text{at } (1,2) \; z_y = 33$$

These provide the coefficients of the Taylor Polynomial P_1:

$$P_1(x,y) = -13 + 5(x - 1) + 33(y - 2)$$

$P_1(x,y)$ could be simplified to the form $P_1(x,y) = -84 + 5x + 33y$. However, this is usually not done. In most applications the polynomial will be established at a specific reference point because we need to evaluate it at points near the reference point. Keeping it in the "shifted-form" will make the calculations easier and actually reduce round-off errors in numerical evaluations.

Linear approximations.

A tangent plane approximates a function's graph "locally". A Taylor expansion of a function, $z = P_1(x,y)$, provides a first-order linear approximation of a function near the reference point (x_0, y_0). The approximation $F(x,y) \approx P_1(x,y)$ can be used to numerically estimate a function's value, to analyze models involving the function, or to simplify a mathematical problem involving the function.

DEFINITION OF A LINEAR APPROXIMATION.

> The **linear approximation** of a function $F(x,y)$ at a point (x_0, y_0) is given by the **first order Taylor expansion of $F(x,y)$ at (x_0, y_0)**:
>
> $$F(x,y) \approx P_1(x,y)$$
> or
> $$F(x,y) \approx F(x_0, y_0) + F_x(x_0, y_0)(x - x_0) + F_y(x_0, y_0)(y - y_0)$$

Example 2 **Determining a Linear Approximation.**

Problem What is the linear approximation of $F(x,y) = xe^{xy} + \cos(y)$ about the point $(x_0, y_0) = (2,0)$?

Solution The function's value at $(2,0)$ is $F(2,0) = 2 \cdot e^0 + \cos(0) = 3$.

The first partial derivatives of $F(x,y)$ are computed and evaluated at $(2,0)$.

$$\partial/\partial x \, F(x,y) = e^{xy} + yxe^{xy} \quad \text{and thus} \quad F_x(2,0) = 1$$

$$\partial/\partial y \, F(x,y) = x^2 e^{xy} - \sin(y) \quad \text{and thus} \quad F_y(2,0) = 4$$

Using these values as coefficients, the Taylor approximation of $F(x,y)$ about $(2,0)$ is

$$F(x,y) \approx 3 + (x - 2) + 4(y - 0) = 1 + x + 4y$$

The differential of a multivariable function.

Recall that a differential of an independent variable is simply a number. Considering a function $z = F(x,y)$, each independent variable, x and y, has a corresponding differential, denoted dx and dy. Differentials may be positive, negative, or even zero, but usually are "relatively" small. A differential is often thought of as an increment. If a reference point (x_0, y_0) is given then the differentials dx and dy determine a second point with coordinates

$$x = x_0 + dx \quad \text{and} \quad y = y_0 + dy$$

Solving for the differentials, we see that these differentials are identical to differences, which are denoted with "delta" notation:

$$dx = x - x_0 = \Delta x \quad \text{and} \quad dy = y - y_0 = \Delta y$$

The differential of a function F is another function denoted by dF. The differential dF is a function of twice as many variables, of all the independent variables and of all their differentials. In most applications, the dependence of dF on these variables is not explicitly indicated. A differential of a function z = F(x,y) may be denoted by either dF or dz.

DEFINITION OF THE (TOTAL) DIFFERENTIAL OF A FUNCTION.

> The **(Total) differential of a function z = F(x,y)** is
>
> $$dF = dz = F_x(x,y) \times dx + F_y(x,y) \times dy$$
>
> where dx and dy are differentials of x and y.

Differentials are used both in formula and in numerical approximations. If a function F has more independent variables, then the differential is just the sum over **all the variables** of the product of the partial derivative of F with respect to a variable times the differential of the same variable.

Example 3 **Computing Differentials**.

Problem Determine the total differential of the function $z = xe^{xy}$.

Solution First compute the partial derivatives:

$$\partial z/\partial x = (x + y)e^{xy} \quad \text{and} \quad \partial z/\partial y = x^2 e^{xy}$$

The differential of z is therefore

$$dz = (x + y)e^{xy} \, dx + x^2 e^{xy} \, dy$$

Example 4 **An Application in Chemistry involving Differentials**.

Problem In chemistry the term *enthalpy* is used to describe the thermodynamics of a chemical reaction. The *enthalpy*, denoted by H, is related to the pressure, P, the volume, V, and the energy, E, of the chemical reaction by the simple equation

$$H = E + PV$$

What is the differential of the *enthalpy*?

Solution In this problem there are three independent variables, E, P, and V. As

$$\partial H/\partial E = 1, \quad \partial H/\partial P = V, \quad \text{and} \quad \partial H/\partial V = P,$$

the differential of H is

$$dH = dE + P \, dV + V \, dP$$

The terms dE, dP, and dV are differentials of the respective variables. In chemistry, these differentials are the changes in experimental conditions and the differential dH is the expected first-order change in the observed *enthalpy*.

Approximating the difference ΔF by the differential dF.

Let us consider the z-values on the graph of a function $z = F(x, y)$ and its tangent plane at a point near (x_0, y_0), say the point (x, y). The differences in the coordinate values from (x_0, y_0) to (x, y) may be thought of as differentials, dx and dy:

$$dx = x - x_0 \quad \text{and} \quad dy = y - y_0$$

The difference in the function's values between (x_0, y_0) and (x, y) is the difference:

$$\Delta F = F(x_0 + dx, y_0 + dy) - F(x_0, y_0) = F(x, y) - F(x_0, y_0)$$

Solving for $F(x, y)$ gives

$$F(x, y) = F(x_0, y_0) + \Delta F \qquad \textbf{Difference Formula}$$

The difference ΔF is the change in z-values on the surface $z = F(x, y)$. Theoretically, if we know $F(x_0, y_0)$ and ΔF this equation gives the value of $F(x, y)$. In many applications the function F is not known; it is the object of experimental observations to establish the function. If we do not know the function, it is likely that the difference ΔF cannot be calculated, hence, $F(x, y)$ cannot be evaluated by the Difference Formula. However, $F(x, y)$ can be approximated by approximating ΔF by the differential dF. (The coefficients of the differential can be estimated by performing a series of experiments or observations varying only one variable at a time.) Replacing ΔF by dF in the Difference Equation gives the Differential Approximation formula:

Differential Approximation

$$F(x, y) \approx F(x_0, y_0) + dF = F(x_0, y_0) + F_x(x_0, y_0) \times dx + F_y(x_0, y_0) \times dy$$

A standard problem in calculus texts that is solved by a differential approximation is to estimate the value of a numerical expression that is difficult to evaluate; for instance, to estimate $(SQRT(4.5) \times 2^{3.1})^3$ without using a calculator. The solution is to introduce a function of several variables that gives the desired numerical value when evaluated at a specific point. The function is chosen so that it and its partial derivatives are easily evaluated at a nearby reference point. The desired value is approximated by the Differential Approximation of the function based at the reference point.

Example 5 **Numerical Approximations.**

Problem Approximate the number $(8.1^{1/3} + 1.98)^{1/2}$.

Solution Represent the number as a function value. Introduce the function

$$F(x,y) = (x^{1/3} + y)^{1/2}$$

The desired number is $z = F(8.1, 1.98)$. The function is easily evaluated at $(x_0, y_0) = (8, 2)$.

$$F(x_0, y_0) = F(8,2) = (8^{1/3} + 2)^{1/2} = 2$$

Compute the partial derivatives of F and evaluate these at the reference point.

$$F_x(x,y) = (1/6)x^{-2/3}(x^{1/3} + y)^{-1/2} \qquad F_x(8,2) = (1/6)(1/4)(1/2) = 1/48$$

$$F_y(x,y) = (1/2)(x^{1/3} + y)^{-1/2} \qquad F_y(8,2) = 1/4$$

The differentials of x and y are given by the differences in the x and y coordinates between the reference point and the point where the approximation is being given:

$$dx = 8.1 - 8 = 0.1 \quad \text{and} \quad dy = 1.98 - 2 = -0.02$$

The order is important! The differential

$$dx = (\text{x-value at desired point}) - (\text{reference x-value})$$

Similarly, for dy. The desired value is then estimated by the Differential Approximation

$$F(8.1, 1.98) \approx F(8,2) + dF = F(8,2) + F_x(8,2) \times 0.1 + F_y(8,2) \times (-0.02)$$

$$(8.1^{1/3} + 1.98)^{1/2} \approx 2 + (1/48)(1/10) + (1/4)(-1/50) = 473/240 \approx 1.971$$

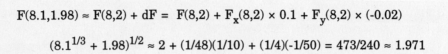

Error estimation.

Background and terminology.

When a function is evaluated with uncertain data the resulting value is also uncertain. For instance, if a function $F(x,y)$ is to be evaluated at a data point (x_1, y_1) but the values of x_1 and y_1 are measured as the values x_0 and y_0 with errors ε_1 and ε_2, respectively, then the resulting function value $F(x_0, y_0)$ will differ from the desired function value $F(x_1, y_1)$ by an error of

$$E = |F(x_1, y_1) - F(x_0, y_0)| \quad \textbf{Absolute Error}$$

If the true values x_1 and y_1 are not known the error E cannot be calculated. A bound for the error E can be derived from the first-order approximation of F:

$$F(x_1, y_1) \approx F(x_0, y_0) + F_x(x_0, y_0)\Delta x + F_y(x_0, y_0)\Delta y$$

This can be solved to give an estimate of the absolute error

$$E = |F(x_1, y_1) - F(x_0, y_0)| \approx |F_x(x_0, y_0)\Delta x + F_y(x_0, y_0)\Delta y| \quad \textbf{Error Estimate}$$

Treating the approximation $\approx$ as an equality and using the general inequality for the absolute value of a sum, $|A + B| \leq |A| + |B|$, this approximation leads to an error bound

$$E \approx |F_x(x_0, y_0)\Delta x + F_y(x_0, y_0)\Delta y| \leq |F_x(x_0, y_0)\Delta x| + |F_y(x_0, y_0)\Delta y| = EB \quad \textbf{Error Bound}$$

The relationships between the true x and y values and the measured values are:

$$x_1 = x_0 + \varepsilon_1 \quad \text{and thus} \quad \Delta x = x_1 - x_0 = \varepsilon_1$$

$$y_1 = y_0 + \varepsilon_2 \quad \text{and thus} \quad \Delta y = y_1 - y_0 = \varepsilon_2$$

As the true values x_1 and y_1 are not known, neither are the errors ε_1 and ε_2! The problem in practice is to estimate E using estimates of the measurement errors ε_1 and ε_2. In many instances **absolute error bounds** are available of the form:

$$|x_1 - x_0| \leq E_x \quad \text{and} \quad |y_1 - y_0| \leq E_y, \quad \textbf{Sample error bounds}$$

where E_x and E_y are known positive constants. Replacing Δx by its bound E_x and Δy by E_y in the Error Bound equation gives a first-order bound for the error E subject to constant measurement errors:

Constant Error bound

$$E = |F(x_1, y_1) - F(x_0, y_0)| \leq |F_x(x_0, y_0)|E_x + |F_y(x_0, y_0)|E_y = CEB$$

Section 10.3 — Linear Approximations and the Differential of F(x,y)

Another type of measurement error bounds are **relative error bounds** of the form

$$|x_1 - x_0|/|x_1| \le E_x \quad \text{and} \quad |y_1 - y_0|/|y_1| \le E_y$$

The relative error bounds are the errors divided by the true values. Again, since the true values x_1 and y_1 are not known, the relative error is generally not known. Instead, a "sample estimate" of the relative error is used. This is obtained by dividing by the **sample** or **measured** estimates x_0 and y_0 of the true x and y-values:

Relative Sample Error Bounds
$$|x_1 - x_0|/|x_0| \le E_x \quad \text{and} \quad |y_1 - y_0|/|y_0| \le E_y$$

This type of error bound can be rearranged, multiplying by the denominators, to give "proportional" error bounds, in which the absolute errors are bounded by the product of the error rate and the measured amount:

Proportional Sample Error Bounds
$$|x_1 - x_0| \le |x_0| \times E_x \quad \text{and} \quad |y_1 - y_0| \le |y_0| \times E_y.$$

Using the proportional error bounds to replace Δx and Δy in the error bound formula gives the

Proportional Error bound
$$|F(x_1,y_1) - F(x_0,y_0)| \le |F_x(x_0,y_0)| \times |x_0| \times E_x + |F_y(x_0,y_0)| \times |y_0| \times E_y = PEB$$

Example 6 Error estimation: computing the volume of a bacteria.

Problem Assume that the measured radius and length of a cylindrical rod-shaped bacteria are

$$\text{radius}: x_0 = 2\mu m \quad \text{and} \quad \text{length } y_0 = 5\mu m$$

What are the first-order error bounds for the resulting estimate of the bacteria volume,

$$v_0 = 40\pi \ \mu m^3,$$

(a) if the measurement error bounds are constants $E_x = E_y = 0.5\mu m$, and
(b) if the measurement error bounds are proportional, at 5% of the measured values?

Solution For the volume function $v = F(x,y) = \pi x^2 y$, the partial derivatives are $\partial v/\partial x = 2\pi xy$ and $\partial v/\partial y = \pi x^2$. As all terms are positive the absolute value signs can be omitted. The first-order error bound becomes

$$\text{Error Bound} = EB = 2\pi x_0 y_0 \Delta x + \pi x_0^2 \Delta y$$

(a) Substituting the measurement values $x_0 = 2\mu m$ and $y_0 = 5\mu m$. and the constant error bounds into the constant error bound equation gives

$$CEB = 2\pi \cdot 2 \cdot 5 \cdot 0.5 + \pi \cdot 4 \cdot 0.5 = 12\pi \ \mu m^3$$

(b) For 5% proportional measurement errors are

$$x_0 \times E_x = 2 \times 0.05 = 0.1 \quad \text{and} \quad y_0 \times E_y = 5 \times 0.05 = 0.25$$

The Proportional Error Bound is thus

$$PEB = 2\pi \cdot 2 \cdot 5 \cdot 0.1 + \pi \cdot 4 \cdot 0.25 = 3\pi \mu m^3$$

Sensitivity analysis.

An important application of mathematical modeling is to establish the way a dynamical system, like a computer, an ecosystem, an electrical grid network, or an organ like the heart, will react to changes or perturbations in its environment, components or design, or the external inputs to the system. The differential approximation formula provides a first-order estimate of the change in a function resulting from changes in the primary variables and forms the basic tool for analyzing dynamic models. This type of analysis is called "sensitivity analysis" and is a part of the mathematical field of study called "Control Theory". A completely parallel set of analysis is employed in business application where the term "sensitivity" is replaced by the term "marginal"; "marginal analysis" is a core component of applied economics.

Dynamical Systems: Background and terminology.

A dynamical system is any process, social, economical, physical or biological that involves change over time. Even the simplest dynamical systems are usually influenced by many conditions, bio-physical constraints, inputs from outside the system, and relationships that must be satisfied. A mathematical model of a dynamical system attempts to represent all of the things that affect the system by variables or parameters and similarly to represent the observable behavior of the system by other variables. The dynamics of the system is usually modeled by difference or differential equations that describe the way the system components are related and vary with time or position.

In system analysis and the area of system control additional terminology is introduced to refer to the Model variables and parameters. The term **state variables** refers to the independent variables of a System Model and the behavior of the system is described by a **response function**. A response function R may be given explicitly, may be determined as a solution of differential equations, or may be known only as a data set quantifying observed system performance. Real systems have natural limitations on their operating conditions. If the pressure is too high pipes may burst, if the temperature is too low plants may die, etc. A combination of state variables that is feasible for the system is represented by a point, S, in the domain of the response function, which is sometimes called an **operating point** or **operating state**. Think of S as the conditions for an experiment or the settings on a machine.

Sensitivity analysis is concerned with predicting the change in the system response due to incremental changes in the state variables.

The objective of sensitivity analysis is to know in advance what will be the system's operational response to changes in the operating state, in particular, to identify the "sensitivity" or rate of response to changes in state variables.

Incremental changes in state variables are specified by differences or differentials of the variables. For instance, the operating points of a system with three state variables, u, v, and w, would be denoted by an ordered triple

$$S = (u,v,w) \quad\quad \textbf{Operating State}$$

and increments, or perturbations, of the state variables would be indicated by the corresponding ordered triple of differentials

$$\Delta S = (du, dv, dw) \quad\quad \textbf{Change in State}$$

The response function for this system would be denoted either as

$$R(u,v,w) \quad \text{or simply} \quad R(S) \quad\quad \textbf{Response Function}$$

Section 10.3   Linear Approximations and the Differential of F(x,y)

When an operating point S is perturbed by an increment ΔS the consequent change in the response function R is the difference:

$$\Delta R = R(S + \Delta S) - R(S) \qquad \textbf{Change in Response}$$

Sensitivity analysis is concerned with knowing this difference in advance, before the system is changed and therefore before the perturbed response $R(S + \Delta S)$ can be known. The basic approach is to approximate the difference ΔR by the differential dR. The critical components of the differential dR are the rates of change in R due to changes in specific state variables, which are the partial derivatives of R. In Control Theory terminology these partial derivatives are called the "sensitivities" of the system:

The sensitivity of R to a particular state variable
is the partial derivative of R with respect to that variable.

The "**total sensitivity**", or simply "**the sensitivity**" of R
is the differential of R, dR, corresponding to a state differential ΔS.

The total sensitivity dR is the first-order approximation of the change ΔR resulting from a perturbation ΔS of the operating point S.

The differential dR is the sum, over each state variable, of the sensitivity of R to the state variable multiplied by the differential of the state variable. Sensitivity is measured at a reference operating point. For example, using the notation introduced above, for a specific reference operating point $S_0 = (u_0, v_0, w_0)$,

the (total) sensitivity of R at (u_0, v_0, w_0) is

$$dR = R_u(u_0,v_0,w_0)\, du + R_v(u_0,v_0,w_0)\, dv + R_w(u_0,v_0,w_0)\, dw$$

In applications the units and magnitudes of the state variables and the response variable are sometimes very different. A common way of avoiding problems associated with this is to re-scale the equations. Two important ways to scale variables are (1) to use "relative" values or (2) to use "percentage" values. These concepts can be combined to give "relative percentage" values.

"relative"   means to divide by the indicated variable.

"percentage"   means to multiply by 100%, i.e, multiply by 100 and add a % sign at the end as a unit.

"relative percent"   means to do both, divide by the variable, multiply by 100, and add a % sign as a unit.

Applied to the sensitivity analysis of a response R at a state S, these terms are used as follows:

For a state variable u	Evaluated at a point u_0 with an increment du
The **relative change in a variable u** is the differential divided by the operating value of u.	du / u_0
The **relative percentage change in u** is 100% times the relative change in u.	$100 du / u_0$ %

A **P% change in a variable u** is the percent P times the operating value u, times 0.01.	$du = P \times u_0 \times 0.01$
For a Response function R	Evaluated at a state S and increment ΔS
The **relative sensitivity of R to a state variable u** is the sensitivity divided by the value of R.	$[\partial R/\partial u] / R$
The **relative (total) sensitivity of R**, or relative approximate change in R, is the differential dR divided by R.	dR / R
The **relative percentage sensitivity of R**, or approximate percentage change in R, is 100% times the relative sensitivity.	$dR \times 100\% / R$

Example 7   **Sensitivity analysis of a Manufacturing Revenue Model.**

Problem   The revenue of a company is a function of manufacturing capacity, u, annual sales, v, and taxes, w. Assume the revenue function is

$$R = R(u,v,w) = [(v/2u) - w]^{0.7}$$

(a)   Determine the relative sensitivity of the revenue at the point $S = (2,16,3)$.
(b)   What is the relative % change in revenue if taxes increase by 10% and sales decrease by 5% with no change in capacity?

Solution   (a) The relative sensitivity is given by dR/R. To calculate the differential, we first calculate the partial derivatives:

$$R_u = [(v/2u) - w]^{-0.3} \times (-v/2u^2)$$

$$R_v = [(v/2u) - w]^{-0.3} \times (1/2u)$$

$$R_w = -[(v/2u) - w]^{-0.3}$$

The relative total sensitivity is

$$dR/R = [[(v/2u) - w]^{-0.3}(-v/2u^2) / R]\, du + [[(v/2u) - w]^{-0.3}(1/2u) / R]\, dv + [-[(v/2u) - w]^{-0.3} / R]\, dw$$

Substituting for R the function value $[(v/2u) - w]^{0.7}$, each term is seen to have the denominator

$$[(v/2u) - w]^{0.3}[(v/2u) - w]^{0.7} = [(v/2u) - w]$$

Upon rearranging the relative sensitivity dR/R becomes:

$dR/R = [(-v/2u^2) / [(v/2u) - w]] \, du + [(1/2u) / [(v/2u) - w]] \, dv + [-1 / [(v/2u) - w]] \, dw$

$= [-(v/u)/(v - 2uw)] \, du + [1/(v - 2uw)] \, dv + [-2u/(v - 2uw)] \, dw$

Evaluated at $S = (2,16,3)$, i.e., setting $u = 2$, $v = 16$, and $w = 2$, the relative sensitivity is

$$dR/R = -2du + (1/4)dv - dw$$

Thus, at operating point $(2,16,3)$, the relative response sensitivity is least sensitive to changes in v (because of the 1/4 coefficient), and is four times more sensitive to changes in w and eight times more sensitive to changes in u than it is to changes in v. The sign of a differential's coefficient indicates whether the response will increase, for a positive sign, or decrease, for a negative sign, when the differential is positive.

(b) A 10% change in sales is indicated by the differential $dw = 0.1w$. A 5% decrease in sales is indicated by $dv = -0.05v$, and no change in capacity is indicated by $du = 0$. These are indicated by the differential $\Delta S = (0, -0.05v, 0.1w)$. The first-order estimate of the relative percent change in revenue is the relative percentage sensitivity evaluated at ΔS.

$100\% \, dR/R = 100\{[-(v/u)/(v - 2uw)\times 0] + [(1/(v - 2uw))\times(-0.05v)] + [-2u/(v - 2uw)\times 0.1w]\}\%$

$= 100\{-0.05v/(v - 2uw) + -0.2uw/(v - 2uw)\}\%$

$= -(5v + 2uv)/(v - 2uv) \, \%$

At the operating point $(2,16,3)$ used in part (a), this would be -36%, or a 36% decrease in revenue.

☑

Example 8 Parameter sensitivity of the Logistic Model.

Problem The Logistic model describing limited growth of a population involves two biological parameters, an intrinsic growth rate, r, and a carrying capacity, M. If the "size" of the population is denoted by X, one form of the Logistic Growth rate is

$$R = rX(1 - X/M)$$

(a) What is the sensitivity of R to each of the parameters r and M?
(b) To which parameter is R more sensitive?
(c) In environmental impact assessments of a proposed development or modification of the environment, a normal question is to determine the impact on an indigenous population. If the population is assumed to be governed by a Logistic Model, what is the total relative sensitivity of R to changes in the parameters and population size?

Solution (a) The sensitivity to each parameter is given by the partial derivative:

Sensitivity to $r = \partial R/\partial r = X(1 - X/M)$

Sensitivity to $M = \partial R/\partial M = rX^2/M$

(b) To which parameter is the growth rate more sensitive? The answer depends on the population size X and the relative magnitude of the state parameters r and M. The ration of the two partial derivatives is

$$R_r / R_M = (M/X - 1)/r$$

If the natural growth rate $r < (M/X - 1)$ the ratio R_r/R_M is greater than 1 and the Logistic Growth rate R is more sensitive to changes in r than to changes in M. The reverse holds if the inequality is reversed. Thus when the population size X is close to the carrying capacity M, variations in the biological growth rate are less critical than changes in M.

(c) The total sensitivity of R to perturbations in both the parameters and the population size is

$$dR = X(1 - X/m) \, dr + rX^2/M \, dM + r(1 - 2X/M) \, dX$$

The next example works in the other direction, in the differential cannot be evaluated by taking partial derivatives because the function is not known. Instead, the differential is approximated by a difference obtained from numerical observations of a system.

Example 9   Estimating sensitivity.

Background   In complex real systems the Response function R may not be known as a mathematical function. It may simply be a measurable characteristic of the system, such as total production. In such cases the sensitivity of the system must be estimated by estimating the sensitivity to each state variable, i.e., the partial derivatives of R with respect to the state variables. In this situation a practical approach is to estimate the sensitivities by an observed operating sensitivity:

the "operating sensitivity" of R to a state variable is the observed change in system response to a **unit change** in that variable only.

In most applications an exact unit change in a state variable is not possible. In this case, the sensitivity or partial derivative is estimated by a difference quotient $\Delta R/\Delta u$ of the observed change divided by the actual increment of the state variable.

Problem   Estimate the total sensitivity of a system $S(u,v,w)$ at the point $S_0 = (1,2,3)$ if the response to the system was measured at this state and three nearby states as:

$R(1,2,3) = 5$ $R(1.5,2,3) = 6$ $R(1,3,3) = 4.5$ $R(1,2,6) = 7$

Solution   The differences in the response values are used to estimate the partial derivatives.

$\partial R/\partial u \approx \Delta R/\Delta u = [R(1.5,2,3) - R(1,2,3)] / [1.5 - 1] = 2$

$\partial R/\partial v \approx \Delta R/\Delta v = [R(1,3,3) - R(1,2,3)] / [3 - 2] = -0.5$

$\partial R/\partial w \approx \Delta R/\Delta w = [R(1,2,6) - R(1,2,3)] / [6 - 3] = 2/3$

The estimated total sensitivity of the system at $(1,2,3)$ is

$$dR \approx 2du - 0.5dv + (1/3)dw$$

Exercise Set 10.3

1. Determine a first-order estimate of $F(x_1, y_1)$ using the indicated reference point (x_0, y_0).

   a) $F(x,y) = x^2 + y$; $(x_0, y_0) = (1,2)$; $(x_1, y_1) = (1.5, 3)$

   b) $F(x,y) = x^2 - 2y$; $(x_0, y_0) = (1,2)$; $(x_1, y_1) = (0.5, 2.5)$

   c) $F(x,y) = x - y^3$; $(x_0, y_0) = (2,1)$; $(x_1, y_1) = (2.5, 2)$

   d) $F(x,y) = x^2 - 2y + xy$; $(x_0, y_0) = (1,2)$; $(x_1, y_1) = (0.5, 2.5)$

   e) $F(x,y) = yx^2 - xy^2$; $(x_0, y_0) = (1,1)$; $(x_1, y_1) = (0.5, 1.1)$

   f) $F(x,y) = x^{1/2} - 2y$; $(x_0, y_0) = (1,2)$; $(x_1, y_1) = (1.5, 2.1)$

   g) $F(x,y) = x/y$; $(x_0, y_0) = (6, 24)$; $(x_1, y_1) = (5, 25)$

   h) $F(x,y) = SQRT(y/x)$; $(x_0, y_0) = (6, 24)$; $(x_1, y_1) = (5, 25)$

   i) $F(x,y) = x \cos(y)$; $(x_0, y_0) = (2, 0)$; $(x_1, y_1) = (2.1, -0.2)$

   j) $F(x,y) = x^2 e^{3y}$; $(x_0, y_0) = (1, 0)$; $(x_1, y_1) = (1.2, 0.3)$

2. Determine the first-order Taylor Expansion $P_1(x,y)$ for the given function at the point $(x_0, y_0) = (1, 0)$. Use this to approximate $F(1.2, -0.5)$. What is the error in using this approximation?

   a) $F(x,y) = x^2 - xy$
   b) $F(x,y) = \cos(xy)$
   c) $F(x,y) = xe^{-y}$

   d) $F(x,y) = (x^2 + y^2 - 9)$
   e) $F(x,y) = 4x^2 - 2x + y^2$
   f) $F(x,y) = (x^2 + y^2)^{1/2}$

   g) $F(x,y) = 4x^2 - e^{xy}$
   h) $F(x,y) = xy/(xy + 1)$
   i) $F(x,y) = y \ln(x)$

3. Determine first-order Error Bounds for the compute areas with the indicated measurement error bounds.

   a) The area of a right triangle is computed. The measured base was 0.7 cm and height was 0.3 cm. The measurements are each subject to a 0.03 cm error.

   b) The same problem as a), but with a 5% relative error in all measurements.

   c) The surface area of a box is computed. The box has a square base measured to be 0.5 cm on an edge, and a height measured to be 5 cm. The error in each measurement is not more than 0.05 cm.

   d) The same as c) but with a maximum 10% relative error in all measurements.

4. Approximate the given value using differentials:

   a) $3.9^{0.5} - SQRT(26)$ b) $9.2^{1/2} + e^{0.31}$ c) $\ln(3)\tan(0.2)$

5. The cross-sectional area of a blood vessel is given by $A = \pi(R^2 - r^2)$ where r and R denote the inner and outer radius, respectively, of the vessel.

   a) What is the approximate rate of distortion in the area A if the inner vessel radius increases at a rate of 10^{-4} cm/sec, and the outer radius remains fixed?

   b) At what rate must the outer radius expand to maintain a constant area A if the inner radius changes as in a)?

6. In studying the ionic flow in a nerve axon, the concentration of a solute is given by

   $$c(x,t) = Ax_0/(\alpha^2 + 4Dt)^{1/2} \exp[-(x - ut - x_0)^2/x_0^2 + 4Dt],$$

   where A, α, and x_0 are constants, D is a diffusion coefficient and u is an indicator of the electron potential gradient. Compute the total differential of c(x,t).

7. A manufacturing process involves four state variables: x, y, u, and v. Assume that the response function is
   $$R = x/u + (y/v)^2 - 3xv$$

   a) What is the sensitivity of R to each state variable?

   b) What is the relative sensitivity of R to each state variable?

   c) What is the total sensitivity of R at the operating point (0,1,2,-3)?

   d) What is the estimated relative percentage change in the response R to a 5% relative increase in x and y and a 7% decrease in u and v, at the point (0,1,2,-3)?

8. In chemistry the Michaelis-Menten equation describes the rate of an enzyme reaction. Its form is

   $$v = V_{max}ES/\{K_M + S\}$$

   Considering the parameters V_{max} and K_M as state variables, what is the total sensitivity of the velocity v at a fixed level of substrate S and enzyme E? If an experiment increased V_{max} by 10% and decreased K_M by 5%, what would be the relative sensitivity of v when S = 500, E = 5 at the operating states (V_{max}, K_M) equal to

   a) (10,500)   b) (10,25)   c) (10, 5000)?

Section 10.4 Linearizing Multivariable functions and Stability Analysis of Dynamical Systems.

The method of Linearization.

Linearization is the process of reducing a nonlinear problem to a linear problem.

Some mathematical problems that are "difficult" to solve because they involve nonlinear functions can be reduced to "tractable" problems that are more easily solved by replacing a nonlinear function by its linear approximation. Examples of such problems include solving non-linear algebraic equations, integrating complicated functions, and solving differential equations that involve even simple non-linear functions. The mathematics involved in solving such problems can be very difficult and explicit solutions may even be impossible. On the other hand, solving comparable problems with linear functions can usually be accomplished with basic Rules.

As introduced in Section 10.3, the linear approximation of a function $F(x, y)$ at a point (x_0, y_0) is given by

$$F(x, y) \approx F(x_0, y_0) + F_x(x_0, y_0)(x - x_0) + F_y(x_0, y_0)(y - y_0)$$

Therefore, the first step in "linearizing" a function $F(x, y)$ is to choose the reference point (x_0, y_0). The first consideration in choosing this reference point is to pick a point at which the function and its partial derivatives can be easily evaluated. A second criteria is to chosen (x_0, y_0) so that the resulting linear problem is easier to solve. Recall that a linear function is said to be homogeneous if its constant term is zero, like $z = 3x + y$, and is non-homogeneous if it does have a constant term, like $z = 2 + 3x + y$. Generally, even linear problems are more easily solved if they are homogeneous. For this reason, the most common choice for reference points is a "zero-point" or "root" of the function $F(x, y)$, i.e., a point (x_0, y_0) such that $F(x_0, y_0) = 0$.

Example 1   **Linearizing nonlinear problems.**

Problem   For each nonlinear problem construct a corresponding linear problem by replacing the nonlinear function by its homogeneous linear approximation:

(a) Solve for x the equation: $x + y = 5 ln(x - 2y)$.

(b) Evaluate the integral: $\int_0^1 \int_0^1 \exp(-(x^2 + y^2))\, dy\, dx$.

Solutions   In problem (a) the nonlinear term is $5 ln(x - 2y)$. Since $ln(1) = 0$ we choose an evaluation point (x_0, y_0) satisfying $x_0 - 2y_0 = 1$. Any point on the line $y = 0.5(x - 1)$ would satisfy this condition. An arbitrary choice would be $x = 3$ and $y = 1$. To form the linear approximation of $F(x, y) = 5 ln(x - 2y)$ at $(3,1)$ we first compute the partial derivatives :

$$\partial/\partial x \; 5 ln(x - 2y) = 5/(x - 2y) \quad \text{thus} \quad F_x(3,1) = 5$$

$$\partial/\partial y \; 5 ln(x - 2y) = -10/(x - 2y) \quad \text{thus} \; F_y(3,1) = -10$$

These partial derivatives are the coefficients for the linear approximation:

$$5 ln(x - 2y) \approx 5(x - 3) - 10(y - 1).$$

The linearized problem becomes:

Solve for x the equation: $x + y = 5(x - 3) - 10(y - 1)$.

The solution of this problem is x = 0.25(5 + 11y); if y = 1 this linear problem gives x = 4. The solution of the original nonlinear problem when y = 1 is approximately x = 6.34.

Problem (b) cannot be solved explicitly. To linearize the problem, the exponential term is replaced by a linear approximation. As the exponential is never zero the reference point (x_0, y_0) cannot be chosen to give a homogeneous function. In this case we choose a point (x_0, y_0) that is within the region of integration at which the partial derivatives are easily evaluated.

$$\partial/\partial x \exp{-(x^2 + y^2)} = -2x \exp{-(x^2 + y^2)}$$

$$\partial/\partial y \exp{-(x^2 + y^2)} = -2y \exp{-(x^2 + y^2)}$$

The simplest choice for (x_0, y_0) is the origin (0,0), but at this point both partials are zero. Instead, we choose $(x_0, y_0) = (1,1)$. Then both partial derivatives evaluate to $-2e^{-2}$ and the linear approximation is

$$\exp{-(x^2 + y^2)} \approx e^{-2} - 2e^{-2}(x - 1) - 2e^{-2}(y - 1) = e^{-2}[5 - 2x - 2y]$$

The corresponding linearized problem is then to evaluate

$$\int_0^1 \int_0^1 e^{-2}[5 - 2x - 2y] \, dy \, dx$$

This integral evaluates using the basic Power Rule to $3e^{-2} \approx 0.406$, whereas numerical integration of the nonlinear problem evaluates to ≈ 0.558.

☑

The next example involves four independent variables. The same process is followed to linearize the model equation. The nonlinear function is approximated by its linear expansion at a reference point.

Example 2 **Linearizing a model of an Infectious Disease**.

Background The "Black Death" in 14th century Europe (the Plague that was primarily transmitted by fleas on rats) was spread at a rate, V, that was a function of a transmission coefficient r, a diffusion coefficient D (this is a measure of the distance people traveled), a mortality rate A, and the population density S. Epidemiologist established a model for this disease spread given by the equation

$$V = 2(rDS)^{1/2}[1 - A/(rS)]^{1/2}$$

Problem Establish a linear model of the velocity V about the point (r_0, D_0, S_0, A_0) corresponding to the following estimates of these parameters in the year 1347 :

the transmission rate $r_0 = 0.4$ the diffusion rate $D_0 = 100$

the population density $S_0 = 50$ and the mortality rate $A_0 = 15$

Solution First, the model is simplified from the standard epidemiological form by simply multiplying the two square root terms, canceling the rs in the second term and factoring:

$$V = 2D^{1/2}(rS - A)^{1/2}$$

The partial derivatives of V are computed and evaluated at (r_0, D_0, S_0, A_0) as follows:

$\partial V/\partial r = S \cdot D^{1/2}(rS - A)^{-1/2}$

At $(0.4, 10^4, 50, 15)$: $V_r = 50 \cdot 10 (0.4 \cdot 50 - 15)^{-1/2} = 5 \cdot 100 \cdot 5^{-1/2} = 5^{1/2} \cdot 100 \approx 223.6$

$\partial V/\partial D = D^{-1/2}(rS - A)^{1/2}$

At $(0.4, 10^4, 50, 15)$: $V_D = 0.1(0.4 \cdot 50 - 15)^{1/2} = 5^{1/2} \cdot 0.1 \approx 0.22$

$\partial V/\partial S = rD^{1/2}(rS - A)^{-1/2}$

At $(0.4, 10^4, 50, 15)$: $V_S = 0.4 \cdot 10 \cdot 5^{1/2} \approx 8.9$

$\partial V/\partial A = -D^{1/2}(rS - A)^{-1/2}$

At $(0.4, 10^4, 50, 15)$: $V_A = -10 \cdot 5^{-1/2} \approx -4.5$

The linear model is then given by

$$V = V_r(r - r_0) + V_D(D - D_0) + V_S(S - S_0) + V_A(A - A_0)$$

or

$$V = 223.6(r - 0.4) + 0.22(D - 100) + 8.9(S - 50) - 4.5(A - 15)$$

The reference point in this example does not correspond to a zero velocity. The velocity is zero if $D = 0$, which corresponds to no communications between villages, or when $rS = A$. Assuming D_0 is not zero, the second case occurs when $r_0 = 0.3$ instead of the 0.4 value used above. The difficulty in this case is that three of the four partial derivatives become infinite at this point and the other is identically zero.

Linearization is a process that can reduce a difficult nonlinear problem to a solvable associated linear problem. The solution of the linear problem may be used to approximate the solution of the original nonlinear problem. Generally, the linear problem's solution will only provide a good approximation of the desired solution of the nonlinear problem for values of (x, y) near the reference point (x_0, y_0). Usually, the linearized problem's solution will be a poor approximation of the solution of the nonlinear problem at points (x, y) that are far from the reference point.

However, for many types of problems analysis of the associated linear problem's solutions will indicate important qualitative characteristics of the nonlinear problem's solutions near the reference point. A special class of such problems involves the character of solutions to systems of differential equations, in particular, determining the character of the solutions over a long time period. The principal question concerns whether a solution that is initially (at $t = 0$) near an "equilibrium" point will eventually (as $t \to \infty$) approach the equilibrium. This type of analysis is referred to as Stability Analysis. The remainder of this section introduces this area of mathematical analysis, which is the starting point for the study of Dynamical Systems and "Chaos".

Stability Analysis of Dynamical Systems

In this subsection the basic concepts involved in the "stability analysis" of equilibrium points of Dynamical Systems are introduced. The study of Dynamical Systems is itself an active field of modern mathematics. Only the most basic concepts and terminology are introduce; detailed developments of the theory and the geometrical nature of solutions are not presented in this very brief introduction. The intention here is to give an important example of Linearization and to illustrate the use of partial derivatives to classify the qualitative aspects of solutions of Dynamical Systems.

Dynamical Systems: Terminology.

Consider a two-dimensional dynamical system involving two "state" variables, X and Y. The dynamics of the system are described by a system of differential equations indicating the time derivative of each variable. We assume that the system is not influenced by external forces or by the measurement of time, which is usually impose on the system for our recording purposes. The dynamics of such systems at any time t will only depend on the state of each variable. In particular it will not depend explicitly on the time variable t. Thus the basic dynamical system equations have the form

$$D_t X = F(X, Y) \quad \text{and} \quad D_t Y = G(X, Y) \qquad \textbf{Dynamical Model}$$

A solution of such systems consist of a pair of functions, X(t) and Y(t). A **general solution** will involve two arbitrary constants, corresponding to integration constants, since two integrals must be evaluated to solve such systems. A **specific solution** is one in which the arbitrary constants are replaced by numerical values so that the solution satisfies specific conditions. For an **initial value problem**, these conditions specify the values of X and Y at t = 0:

$$X(0) = X_0 \quad \text{and} \quad Y(0) = Y_0 \qquad \textbf{Initial conditions}$$

If the functions in the Dynamical Model, F and G, are nonlinear, the Dynamical System can be very difficult or impossible to solve. When F and G are linear, the solutions of such systems are easily determined. Characteristics of nonlinear systems are inferred by considering associated linear systems.

A solution specifies a parametric curve in an X-Y **phase plane**. Such curves are called **phase curves** and a graph that illustrates these curves is called a **phase portrait**. A solution trajectory or path is a parametric curve specified by

$$\{ (X(t), Y(t)) \mid -\infty < t < \infty \}.$$

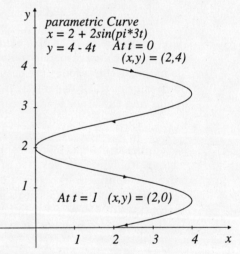

Recall that a parametric curve has a direction, corresponding to how the point (X(t), Y(t)) varies as t increases; this is sometimes indicated by arrow-heads on the curve.

Equilibrium Points

The first step in the study of Dynamical Systems is to identify "equilibrium points" of the system. Equilibrium points are points at which the system remains static, i.e., at which its time derivatives are zero.

> The Equilibrium, or Stationary states of a dynamical system are the solutions (X_E, Y_E) of
>
> $$D_t X = 0 \quad \text{and} \quad D_t Y = 0.$$
>
> For the above dynamical system these are the simultaneous roots of the equations:
>
> $$F(X, Y) = 0 \text{ and } G(X, Y) = 0.$$

To classify an equilibrium point the nature of the trajectories of solution curves that start near the point is considered. An equilibrium point is said to be *stable* if it is a limit point, in the sense that solutions that pass sufficiently close to the point will eventually either approach it as $t \to \infty$ or will always stay close to the point. If this is not the case, and solutions that are very close to the point eventually approach another point or become infinite, the equilibrium is said to be *unstable*.

Equilibrium points are further classified by the geometric nature of trajectories near them. Solutions may approach a point along a simple curved trajectory or in a spiraling fashion. They may not actually approach the point but may follow closed "orbits" and move around a closed loop containing the equilibrium point. These are called "qualitative" aspects of the solutions.

Example 3 Equilibrium of a pendulum system.

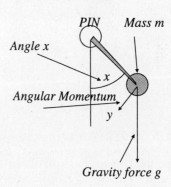

Model Pendulum

Background   A pendulum only subject to gravity swings or rotates about a pin. Its angle of deviation from the downward direction is x and its angular momentum is y. If the length of the pendulum is a, its mass is m, and the force of gravity is g then its movement is governed by its initial position x_0, initial angular velocity y_0, and the retardation force of air resistance and friction at the pin. These forces are assumed to be a constant c times the pendulum's velocity. The pendulum's movement is described by a system of two differential equations:

$D_t x = (1/ma)y$

$D_t y = -mg \sin(x) - cy$

Problem   What are the equilibrium states of a pendulum system and what are their characteristics?

Solution   Intuitively, the pendulum has two equilibrium states: (1) at rest hanging straight down and (2) at rest pointing straight up. These states are called "equilibriums" because if the pendulum is in one of these positions with no momentum it will not move (due to gravity) unless it is perturbed by an outside force.

In the first equilibrium state, hanging downward, a slight movement of the pendulum results in it swinging back and forth due to the forces of gravity, and due to friction it eventually comes to rest at the downward position; the less friction on its pin the longer it will take to come to rest. Because it returns to the equilibrium, the equilibrium is said to be "stable".

However, if the pendulum in the second equilibrium position, pointing straight upward, is perturbed just slightly, it will begin to swing downward and then will oscillate about the downward position, which it will eventually approach. Because the perturbed system does not return to the upright equilibrium, this state is said to be "unstable".

Mathematically, the equilibrium states are found by solving simultaneously the two equations

$$D_t x = (1/ma)y = 0 \quad \text{and} \quad D_t y = -mg \sin(x) - cy = 0.$$

The only solution of the first equation is for zero momentum: $y = 0$. The second equation, with $y = 0$, then has an infinite number of solutions corresponding to $\sin(x) = 0$: $x = n\pi$, for every integer n. When n is even the pendulum is pointing downward, in the stable equilibrium state. When n is odd the pendulum is point upward in the unstable state. Why one equilibrium state is stable and the other is not stable can be established by the stability analysis method that is developed in the remainder of this section.

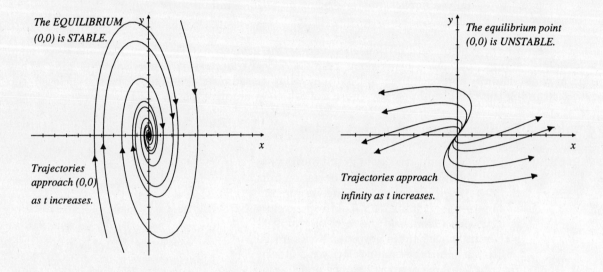

Stable and Unstable Equilibrium

If a system is slightly perturbed from an equilibrium state and as time increases the system returns to the same equilibrium, or always stays near it, then the equilibrium is called a **stable equilibrium**, otherwise, the equilibrium is **unstable**.

Illustrations of stable and unstable equilibriums are depicted in the above figure. We will not discuss these graphs further as it would require too great a digression from our applications of partial derivatives. Since most nonlinear systems cannot be solved, the stability of their equilibrium points is difficult to determine. However, linear dynamical systems can be easily solved, and the stability of their solutions is easily analyzed. The only equilibrium of a homogeneous linear system is the origin, $(0,0)$.

The stability of an equilibrium point of a nonlinear dynamical system is analyzed by **linearizing the system about the equilibrium**. If for the linearized system the origin is stable, or unstable, then the original equilibrium point is said to be **linearly stable**, or **unstable** respectively.

The following general method outlines the steps involved in analyzing the stability of a nonlinear dynamical system. It is not essential at this introductory level that you understand all the details of this method, in particular the conclusions of Step IV, but reading through this outline should help you understand the method better. The final results that you will use to work exercises are given in the Table of Solutions for the corresponding Linear System: $x' = Ax + By$ and $y' = Cx + Dy$. The Solution Table has two parts that correspond to two situations, when $B \cdot C = 0$ and when $B \cdot C \neq 0$.

Method to analyze the Linear Stability of a Dynamical System

$$D_t X = F(X, Y) \quad \text{and} \quad D_t Y = G(X, Y)$$

Step I Identify the equilibrium states (X_E, Y_E), which are the simultaneous roots of the equations $F(X, Y) = 0$ and $G(X, Y) = 0$.

Step II Linearize the system functions F and G about an equilibrium. Evaluate the differentials of F and G at an equilibrium (X_E, Y_E). Since F and G both vanish at the equilibrium point, the linear approximations of F and G are given by:

$$F(X, Y) \approx F_X(X_E, Y_E) \cdot (X - X_E) + F_Y(X_E, Y_E) \cdot (Y - Y_E)$$

$$G(X, Y) \approx G_X(X_E, Y_E) \cdot (X - X_E) + G_y(X_E, Y_E) \cdot (Y - Y_E)$$

Step III For each equilibrium point create a linear system by (1) replacing F and G in the nonlinear system by their linear approximations and (2) introducing deviation variables, denoted by lower case letters:

$$x = (X - X_E) \quad \text{and} \quad y = (Y - Y_E).$$

As X_E and Y_E are constants, the derivatives of x and y equal to those of X and Y:

$$D_t x = D_t X - D_t X_E = D_t X \quad \text{and} \quad D_t y = D_t Y - D_t Y_E = D_t Y.$$

Thus, the desired linear system has the form:

$$D_t x = F_X(X_E, Y_E) x + F_Y(X_E, Y_E) y$$

$$D_t y = G_X(X_E, Y_E) x + G_y(X_E, Y_E) y$$

Linearized System

This linear system can be written in a simpler form:

$$x' = Ax + By, \quad y' = Cx + Dy$$

Where the four coefficient constants are:

$$A = F_X(X_E, Y_E) \quad B = F_Y(X_E, Y_E)$$

$$C = G_X(X_E, Y_E) \quad D = G_y(X_E, Y_E)$$

Step IV Analyze the solutions of the linear system. The nature of the solutions of the linear system and the "linear stability" of the associated equilibrium are indicated in the **Table of Solutions of the Linear System**, found at the end of this section. The solutions and their stability are completely determined by the roots of the associated **characteristic equation**:

$$\lambda^2 - (A + D)\lambda + (AD - BC) = 0.$$

You can determine the stability of an equilibrium by simply solving this equation for λ, using the numbers A, B, C, and D from a specific problem, and referring to the Table of Solutions. The quadratic equation has three possible types of roots:

(i) two different real roots,
(ii) one repeated real root, or

(iii)   two "complex" roots involving of i = SQRT(-1).

The roots of the characteristic equation depend on the relative values of the parameters A, B, C, and D. Mathematicians traditionally denote the two roots of the characteristic equation by λ_1 and λ_2. These numbers are also called **eigenvalues**, a German term that has no direct translation.

If both eigenvalues λ (or their real part) are negative the equilibrium is STABLE.

Otherwise, it is UNSTABLE.

The Table of Solutions has several cases indicated to cover all of the possibilities.[†] The general solution will involve two arbitrary constants, corresponding to integration constants, and two related constants. In the event that the product B • C is zero, the system is not strongly coupled, as one of the equations then only involves a single variable, e.g., if B = 0 then the first equation is simply x' = Ax. In this case at least one of the constants must be zero. The table is split into two parts to handle all cases.

By linearizing, the equilibrium point (X_E, Y_E) of the nonlinear system is transformed to the only equilibrium point of the linear system, the origin (0,0). The linear system will be stable if its solution approaches the point (0,0) as time t goes to infinity. Notice in the Table of Solutions that all solutions, with one exception, involve an exponential terms of the form $e^{\lambda t}$ or $e^{\alpha t}$. As

$$\lim_{t \to \infty} e^{\lambda t} = \text{zero if } \lambda < 0 \text{ and is } +\infty \text{ if } \lambda > 0$$

it follows that the linear stability analysis hinges on the sign of λ or α. The only exception is the one case in which the solutions are simply the sum of two trigonometric functions, with no exponential term. In this case, the solutions are periodic with constant amplitudes. As $t \to \infty$ the solutions do not approach (0,0) but simply orbit around it, and consequently also never leave the vicinity of the origin. This gives the one special type of stability in which the equilibrium is not approached.

Example 4   **Stability analysis of a Host-Parasite or Predator-Prey System.**

Background   In Population Biology or Ecology, the dynamical interaction between two species is modeled by differential equations of the form:

$$dH/dt = F(H, P) \quad \text{and} \quad dP/dt = G(H, P) \qquad \textbf{Dynamic Equations}$$

where F(H, P) and G(H, P) are the time-rates of change of two populations whose sizes are denoted by H and P. A. J. Lotka studied this type of model in 1925 and considered H to represent a "Host" or "Prey" species and P to represent a "Parasite" or "Predator" species. The term "predator-prey" is more commonly encountered in modern ecology but we will utilize the "Host-Parasite" terminology to give two different variables, H and P. Generally, the interaction functions F(H, P) and G(H, P) are nonlinear functions that specify the natural growth and decay of the species subject to the affect on their growth due to interactions. A Host-Parasite model is defined as one in which

---

[†]   The representation of the constants is not unique. Basically, there are four constant terms and it is possible to pick two as arbitrary and to solve for the other two. The reasons for the second portion of the Table of Solutions is to account for an alternative solution when B = 0.

interactions between the two species lead to a decrease in the Host population and increase in the Parasite population.

Equilibrium population levels, H_E and P_E, are the simultaneous solutions of

$$F(H, P) = 0 \quad \text{and} \quad G(H, P) = 0.$$

The local stability of an equilibrium is determined from the linearized system expressed in terms of the deviation variables

$$h(t) = H(t) - H_E \quad \text{and} \quad p(t) = P(t) - P_E.$$

The coefficients of the linear system are partial derivatives of $F(H, P)$ and $G(H, P)$ evaluated at the equilibrium point. The linearized system is:

$$D_t h(t) = \partial/\partial H \; F(H_E, P_E) \bullet h(t) + \partial/\partial P \; F(H_E, P_E) \bullet p(t)$$

$$D_t p(t) = \partial/\partial H \; G(H_E, P_E) \bullet h(t) + \partial/\partial P \; G(H_E, P_E) \bullet p(t)$$

Problem Consider an example in which the Host population grows according to a Logistic Growth Model in the absence of Parasites and the Parasite population decays exponentially in the absence of Host. The interaction of the two populations is assumed to occur at a rate proportional to the product of their sizes and to result in a proportional rate of decrease in the Host and increase in the Parasite population levels. For a specific example we assume that :

the intrinsic growth rate of the Host is $r = 2$;
the carrying capacity of the Host is $M = 10^4$;
the natural decay rate of the Parasite is $k = 0.5$;
the "cost" to the Host of interactions with Parasites is $c = 0.1$ per interaction;
and the "benefit" to the Parasites of interactions with the Host is $b = 0.005$ per interaction.

What are the non-zero equilibrium population levels for this model and are they stable?

Solution With the specified parameters the Host-Parasite model is given by the nonlinear dynamical system

$$D_t H = F(H, P) = 2H(1 - H/10^4) - 0.1HP$$

$$D_t P = G(H, P) = -0.5P + 0.005HP$$

Solving the equations $F(H, P) = 0$, $G(H, P) = 0$ for the equilibrium values:

$$2(1 - H/10^4)H - 0.1HP = 0 \quad \text{and} \quad -0.5P + 0.005HP = 0$$

can be factored as

$$H(2 - 2 \bullet 10^{-4}H - 0.1P) = 0 \quad \text{and} \quad P(-0.5 + 0.005H) = 0.$$

Setting the factors to zero gives the two possible equilibriums: the trivial case $H = 0$ and $P = 0$, which corresponds to no populations, and one non-zero equilibrium, that occurs when

$$-0.5 + 0.005H = 0, \text{ and } 2 - 2 \bullet 10^{-4}H - 0.1P = 0$$

The first equation gives $H = 100$ and using this in the second gives $P = 19.8$. Thus

$$(H_E, P_E) = (100, 19.8)$$

To analyze the stability of the non-zero equilibrium, first compute the partial derivatives of F and G and evaluate them at $(H_E, P_E) = (100, 19.8)$:

$$A = \partial/\partial H \; F(H_E, P_E) = 2 - 4 \cdot 10^{-4} H_E - 0.1 P_E = -0.02$$

$$B = \partial/\partial P \; F(H_E, P_E) = -0.1 H_E = -10$$

$$C = \partial/\partial H \; G(H_E, P_E) = 0.005 P_E = 0.099$$

$$D = \partial/\partial P \; G(H_E, P_E) = -0.5 + 0.005 H_E = 0$$

The corresponding linear system is therefore

$$h'(t) = -0.02 h(t) - 10 p(t) \quad \text{and} \quad p'(t) = 0.099 \, h(t)$$

The characteristic equation for this system is

$$\lambda^2 - (A + D)\lambda + (AD - BC) = \lambda^2 + 0.02\lambda + 0.99 = 0$$

Its roots, using the quadratic formula, are complex

$$\lambda = \alpha \pm \beta i \text{ with } \alpha = -0.01 \text{ and } \beta = \text{SQRT}(0.9899)$$

Without concerning ourselves with the precise nature of the solution, since $B \cdot C \neq 0$, from the last row of the first Table Of Solutions we see that because α is negative we know that the linear system is **stable at the origin** and thus the nonlinear system is **linearly stable at (100, 19.8)**.

Using the form of solution given in the center column of the Table of Solutions of Linear Systems, the solution of the linearized system is

$$h(t) = [c_1 \sin(\beta t) + c_2 \cos(\beta t)] e^{-0.01 t},$$

$$p(t) = [(0.003 \, c_1 + \beta/10 \, c_2) \sin(\beta t) - (0.003 c_1 + \beta/10 \, c_2) \cos(\beta t)] e^{-0.01 t},$$

where $\beta = \text{SQRT}(0.9899) \approx 0.9949$ radians, and c_1 and c_2 represent arbitrary constants. These solutions will oscillate because of the trigonometric functions but as $t \to +\infty$, both $h(t)$ and $p(t)$ will approach zero; this is because the real part of the roots $\alpha = -10$ is negative and consequently the exponential term has a negative exponent. Thus, as

$$\lim_{t \to \infty} h(t) = 0 \quad \text{and} \quad \lim_{t \to \infty} p(t) = 0$$

the equilibrium (100, 19.8) is a stable equilibrium.

Section 10.4 — Linearizing Multivariable Functions

Table of Solutions of the Linear System
$$x' = Ax + By, \quad y' = Cx + Dy$$

If the Roots λ_1 and λ_2 of $\lambda^2 - (A+D)\lambda + (AD-BC) = 0$ are :	The General Solutions to the linear system are:	Linear Stability of equilibrium (X_E, Y_E)
If $B \cdot C \neq 0$, one of the following cases holds:		
Real and distinct roots: λ_1 and λ_2 given by the quadratic formula $\lambda = 0.5[(A+D) \pm \{(A+D)^2 - 4(AD-BC)\}^{1/2}]$	$x(t) = c_1 \exp(\lambda_1 t) + c_2 \exp(\lambda_2 t)$ $y(t) = c_3 \exp(\lambda_1 t) + c_4 \exp(\lambda_2 t)$ Where c_1 and c_2 are constants and $c_3 = (\lambda_1 - A)c_1/B$ $c_4 = (\lambda_2 - A)c_2/B$	If both roots are negative, $\lambda_1 < 0$ and $\lambda_2 < 0$, then the equilibrium is **STABLE**. If either root is positive the equilibrium is **UNSTABLE**.
Real and equal roots: $\lambda_1 = \lambda_2$ $\lambda_1 = \lambda_2 = (A+D)/2$	$x(t) = [c_1 + c_2 t]\exp(\lambda_1 t)$ $y(t) = [c_3 + c_4 t]\exp(\lambda_1 t)$ Where c_1 and c_2 are constants and $c_3 = [(\lambda_1 - A)c_1 + c_2]/B$ $c_4 = (\lambda_2 - A)c_2/B$	If $\lambda_1 < 0$, the equilibrium is **STABLE**. If $\lambda_1 \geq 0$, the equilibrium is **UNSTABLE**.
Complex roots: $\lambda = \alpha \pm \beta i$ $i^2 = -1$ $\alpha = (A+D)/2$ $\beta = 0.5\{-(A+D)^2 + 4(AD-BC)\}^{1/2}$	$x(t) = [c_1 \sin(\beta t) + c_2 \cos(\beta t)]e^{\alpha t}$ $y(t) = [c_3 \sin(\beta t) + c_4 \cos(\beta t)]e^{\alpha t}$ Where c_1 and c_2 are constants and $c_3 = [(\alpha - A)c_1 - \beta c_2]/B$ $c_4 = [(\alpha - A)c_2 + \beta c_1]/B$	If $\alpha \leq 0$ the equilibrium is **STABLE**. If $\alpha > 0$ the equilibrium is **UNSTABLE**.

Table of Solutions of the Linear System
$$x' = Ax + By, \quad y' = Cx + Dy$$

If the Roots λ_1 and λ_2 of $\lambda^2 - (A+D)\lambda + (AD - BC) = 0$ are:	The General Solutions to the linear system are:	Linear Stability of equilibrium (X_E, Y_E)
If $B \cdot C = 0$, the eigenvalues are $\lambda_1 = A$ and $\lambda_2 = D$.		
If $A \neq D$ the roots are distinct.	$y_1 = c_1 e^{At} + c_2 e^{Dt}$ $y_2 = c_3 e^{At} + c_4 e^{Dt}$ with constants c_1 and c_4 and $c_3 = c_1 \cdot C/(A - D)$ $c_2 = c_4 \cdot B/(A - D)$ Note that $c_2 = 0$ if $B = 0$, and $\quad c_3 = 0$ if $C = 0$, One of which must hold!	The equilibrium is **STABLE** if both $A < 0$ and $D < 0$. If $A \geq 0$ or $D \geq 0$ the equilibrium is **UNSTABLE**.
If $A = D$ the eigenvalues are repeated: $\lambda_1 = \lambda_2 = A$	$x(t) = [c_1 + c_2 t]e^{At}$ $y(t) = [c_3 + c_4 t]e^{At}$ with constants c_1 and c_3 and four equations that must hold: $c_4 = c_1 \cdot C \quad B \cdot c_4 = 0$ $c_2 = c_3 \cdot B \quad C \cdot c_2 = 0$ Note that $\quad c_2 = 0$ if $B = 0$, and $\quad c_4 = 0$ if $C = 0$. At least one of these must hold as $B \cdot C = 0$ implies that $B = 0$ or $C = 0$.	If $A < 0$ the equilibrium is **STABLE**. If $A \geq 0$, the equilibrium is **UNSTABLE**.

Exercise Set 10.4

1. Construct a linear problem by replacing the nonlinear part of the given problem by its homogeneous linear approximation.
   a) Solve for x the equation $xy = 5ln(x - 2y)$ expanding about $(x_0, y_0) = (5,2)$.
   b) Solve for x the equation $x - 2y = 5(x^2 - 2xy)$ expanding about $(x_0, y_0) = (2,1)$.
   c) Estimate the value of the integral $\int_0^1 \int_0^1 exp\text{-}(x^2 + y - 2y^2) \, dy \, dx$ by replacing the integrand by its linear expansion about $(x_0, y_0) = (0,0)$ and for $(x_0, y_0) = (1,1)$.

2. a) Evaluate the integral $\int_1^2 \int_1^{0.5} x \, exp(-x^2 + y) \, dy \, dx$. Hint: write the exponential term as a product of two exponentials.
   b) Approximate the integral $\int_1^2 \int_1^{0.5} x \, exp(-x^2 + y) \, dy \, dx$ by linearizing the integrand about the point $(x_0, y_0) = (1,1)$.
   c) Compare the results of parts a) and b).

3. A model of the reactions velocity between an enzyme and a larger protein that is rod-like, i.e., has a circular cylinder shape, is $v = \pi b [1 - ln(4b/l)]$, where l is the length of the reactive part of the molecule which is of radius b. Construct a linear approximation of this model about the point $b_0 = 20$ nm and $l_0 = 80$.

4. In Example 3 dynamics of a pendulum were modeled by the following system of differential equations: $D_t x = (1/ma)y$ and $D_t y = -mg \, sin(x) - cy$. The equilibriums for this system were found to correspond to angles x that were multiples of π. Express the linearized system of equations corresponding to a) $y_0 = 0$ and $x_0 = 0$ and b) $y_0 = 0$ and $x_0 = \pi$.

For the Host-Parasite Models in Exercises 5-8:
i)   Identify all non-zero equilibrium points.
ii) Determine the linearized system for each equilibrium point.
iii) Determine the linear stability of each equilibrium.

5. A model in which the Host grows exponentially while the Parasite decays exponentially:

   $H'(t) = F(H, P) = 10H - H \cdot P \quad P'(t) = G(H, P) = -2P + 0.5HP$

6. A model in which the Host has logistic growth and the Parasite decays exponentially:

   $H'(t) = F(H, P) = 2(1 - H/50)H - 2HP \quad P'(t) = G(H, P) = -5P + 0.25HP$

7. A model in which the Host experience a Gompertz type limit growth and the Parasite decays exponentially:

   $H'(t) = F(H, P) = 2Hln(50/H) - 2HP \quad P'(t) = G(H, P) = -5P + 0.25HP$

8. A model in which both the Host and the Parasite experience limited, logistic type, growth:

   $H'(t) = F(H, P) = 2(1 - H/50)H - HP \quad P'(t) = G(H, P) = (1 - P/100)P + 0.5HP$

9. A model of two species that compete for the same resources has the form:

   $$D_t X = F(X, Y) \quad D_t Y = G(X, Y)$$

   where F(X, Y) and G(X, Y) represents the natural growth of each species including the effect of their competition. Competition is assumed to decrease the resources available to each species, thus the usual form of these functions is

$F(X, Y) = f(X) - Q(X, Y)$ and $G(X, Y) = g(Y) - kQ(X, Y)$.

Here $f(X)$ is the growth rate of species X in the absence of species Y; $g(Y)$ denotes the natural growth of species Y in the absence of species X. $Q(X, Y)$ denotes the effects of competition.

For the following "competition models" determine the equilibrium populations (X_E, Y_E) and the associated linearized system. Use the Table of Solutions to Linear systems to determine the linear stability of each equilibrium.

a) $X'(t) = 3X - 2XY;$ $\qquad Y'(t) = 2Y - 0.5XY^2$

b) $X'(t) = 3X - 0.5X^2 - 2XY;$ $\qquad Y'(t) = Y - 0.5XY$

c) $X'(t) = 10X - 0.5X^2 - XY;$ $\qquad Y'(t) = 2Y - 0.5Y^2 - 0.5XY$

d) $X'(t) = 10 ln(5/X) - XY;$ $\qquad Y'(t) = 20Y - Y^2 - 0.5XY$

e) $X'(t) = X - 2XY;$ $\qquad Y'(t) = 2Y - 0.5XY^2$

10. Determine the stationary points for the given systems and determine their linear stability.

a) $X' = 2X + Y^2 \qquad Y' = X^2 - 25$ $\qquad$ b) $X' = X - Y^2 \qquad Y' = X^2 - 125Y$

c) $X' = 25 - X^2 - Y^2 \qquad Y' = (X - 5)^2 + 25Y$ $\qquad$ d) $X' = X^2 - Y^2 \qquad Y' = X^2 + Y^2 - 100$

e) $X' = Xe^{-Y} \qquad Y' = X^2 + Y^2 - 9$ $\qquad$ f) $X' = 8X - Y^3 \qquad Y' = ln(X)(9 - Y^2)$

g) $X' = X^2 - XY \qquad Y' = sin(\pi(X+Y)) \qquad$ Group the equilibrium by stability type.

Chapter 11 *Directional Derivatives*.

11.1 **Vectors and Vector valued-functions: Parametric Curves.** 622

11.2 **Derivatives of Vector Functions: Tangent Vectors.** 630

11.3 **Directional Derivatives.** 635

11.4 **Level Curves and Curves of Steepest Ascent or Descent** 644

In this Chapter we consider derivatives and rates of change involving higher dimensional functions of two types. The first type consist of **multi-valued**, or **vector-valued**, functions of a single variable, these are functions that describe parametric curves in two or higher dimensional space, like $\underline{z}(t) = (f(t), g(t))$, this function's domain is one-dimensional while its range is two-dimensional. The graphs of these functions can be considered as parametric curves, and they are actually formed from two or more ordinary single variable functions. To differentiate such functions we simply differentiate the component functions separately. But, combining the individual component derivatives provides additional information about the parametric curve in the form of a "tangent vector" that indicates not only the "direction" associated with the curve, but also the "speed" that the curve is traversed by a particle whose position at time t is indicated by the point $\underline{z}(t)$. These concepts are introduced in Section 11.2. To prepare for this, in Section 11.1 the basic vocabulary and algebraic properties of "vectors" are introduced so that we can discuss parametric curves using vectors.

The second type of function that is considered is called a **multivariable** *function*. These are single-valued functions of several variables, like $z = F(x,y)$, whose domain is two-dimensional and range is one-dimensional. The graphs of multivariable functions are surfaces in higher-dimensional space. The graph of $z = F(x,y)$ is a surface in 3-dimensional space, while the graph of $z = F(x, y, u, v)$ is a surface in 5-dimensional space. To mathematically describe and analyze such surfaces we will use partial derivatives, just as the ordinary derivative was used to analyze a curve $y = f(x)$. Recall that a partial derivative, $F_x(x_0, y_0)$ or $F_y(x_0, y_0)$, indicates the rate of change in the function when only one of the independent variables, x or y, increases from (x_0, y_0). This is the rate of change in the z-variable on the surface $z = F(x, y)$ as the domain points (x, y) move on a line past the derivative point (x_0, y_0), the line must be parallel to the x or y coordinate axis, respectively. The direction of movement along the line is important; it must correspond to an increase in the partial derivative variable.

A point moving on a line can only move in one of two directions, however, when we consider points moving in two-dimensions there are an infinite number of directions that a point can moved. These directions correspond to different *oriented lines* through a reference point. If L is a line in the x-y plane through the reference point (x_0, y_0), we can assign an *increasing direction* to L and consider the change in $F(x, y)$ as the point (x, y) moves along the line in this direction past the point (x_0, y_0). The rate of change in these function values at the point (x_0, y_0) is also a derivative, called a *directional derivative*. Directional derivatives are calculated using a vector-valued functions called *gradients*. These concepts are introduced in Section 11.3 and then are utilized in Section 11.4 to describe *level curves* and *curves of steepest assent* or *descent* on the surface formed by the graph of $F(x,y)$. These concepts are utilized in Chapter 12 where maximum and minimum problems for multivariable functions are considered.

Section 11.1 Vectors and Vector-valued functions: Parametric Curves

To specify a direction in higher-dimensional space the concept of a *vector* is utilized. In this section the vocabulary of vectors is introduced along with the basic aspects of vector-algebra. Additionally, we introduce the vector form of a parametric curve. The following development is primarily done in two-dimensions, but can be easily extended to higher dimensional situations.

Vectors

A **vector** is a quantity that has both "length" and "direction". Vectors can be pictured graphically as arrows (Figure below); a vector has a direction indicated by placing an arrow-head at its terminal end and a dot at its beginning point. A vector thus indicates the direction of the terminal point from the initial point, and its length is the distance between the points. A **position vector** is a vector that starts at the origin. A position vector and its terminating point are both denoted by the same notation, in two dimensions as a ordered pair, like (3,-5) or (x, y), and similarly in higher dimensions, e.g., as (2,-1,0.1) or (r_1, r_2, r_3) three-dimensions. However, an alternative notation is to refer to a vector by a single name, indicated in this text by an underlined variable, e.g., $\underline{v}$. The ordered pair representation of a position vector $\underline{v}$ typically utilizes the vector name subscripted to refer to each coordinate of its terminal point.

Position Vector $\underline{v}$ with components v_1 and v_2

$\underline{v} = (v_1, v_2)$

The **position vector** $\underline{v} = (v_1, v_2)$ is the vector from the point (0,0) to the point (v_1, v_2). Notice that the same notation is used to represent the terminal point and the position vector.

The **components of the vector $\underline{v}$** are the terms v_1 and v_2. v_1 is called the **first component of $\underline{v}$**, and v_2 is called the **second component** of $\underline{v}$. If the axes of the coordinate system are x and y, v_1 is called the x-component and v_2 is the y-component of $\underline{v}$.

The **length of the vector $\underline{v}$**, also called the **norm** of the vector, is denoted by $\|\underline{v}\|$. The length of a position vector is the Euclidean distance from the origin (0,0) to the point (v_1, v_2):

$$\|\underline{v}\| = \text{SQRT}\{v_1^2 + v_2^2\}$$

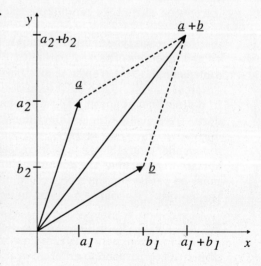

The algebra of vectors can be visualized geometrically. To add two vectors, say $\underline{a}$ and $\underline{b}$, we picture a vector that is not a position vector, that has the same length and direction as the vector $\underline{b}$ but originates from the terminal point of the vector $\underline{a}$. This vector can be constructed by shifting the position vector $\underline{b}$ parallel to the coordinate axes, so that its initial point coincides with the terminal point of $\underline{a}$. When this is done, the terminal point of the shifted vector can be represented as a new position vector that is by definition the vector sum $\underline{a} + \underline{b}$. As depicted in the above figure, if the roles of $\underline{a}$ and $\underline{b}$ were reversed, and the vector $\underline{a}$ were shifted to originate from the terminal point of $\underline{b}$, the resulting terminal point would be the same physical point. The two position vectors $\underline{a}$ and $\underline{b}$ and their shifted counterparts form a parallelogram. The sum $\underline{a} + \underline{b}$ is the position vector that is the diagonal of this parallelogram. The terminal point being the same for both orders of shifting the vectors corresponds to the commutative property of addition:

Section 11.1   Vectors, Vector Functions, and Parametric Curves

$$\underline{a} + \underline{b} = \underline{b} + \underline{a}$$

Algebraically, vector addition is achieved by adding the respective vector components. If

$$\underline{a} = (a_1, a_2) \quad \text{and} \quad \underline{b} = (b_1, b_2)$$

the components of the sum $\underline{a} + \underline{b}$ is defined to be

$$\underline{a} + \underline{b} = (a_1, a_2) + (b_1, b_2) \equiv (a_1 + b_1, a_2 + b_2).$$

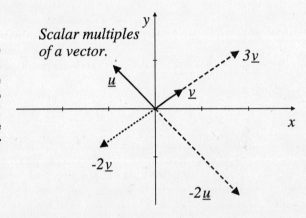

Scalar multiples of a vector.

"Scalar multiplication" refers to multiplying a vector by a number, which is called a "scalar" because it indicates the scale or relative length of the multiplied vector to the length of the vector prior to multiplication. Algebraically, scalar multiplication is done "term wise", multiplying each vector component by the scalar.

If $\underline{v} = (v_1, v_2)$ and k is a constant, then
$$k\underline{v} \equiv (kv_1, kv_2)$$

The vector $k\underline{v}$ will have the same direction as $\underline{v}$ if k is positive and the opposite direction if k is negative. The constant k is referred to as a "scalar" since multiplying by k "scales" the vector to a different length; the length of the vector $k\underline{v}$ is $|k|$ times the length of $\underline{v}$:

$$\|k\underline{v}\| = |k| \times \|\underline{v}\|$$

Two vectors are **parallel** if they have either the same direction or the opposite direction. This can occur only if one of the vectors is a multiple of the other:

$$\underline{u} \text{ and } \underline{v} \text{ are \textbf{parallel}} \quad \Longleftrightarrow \quad \underline{u} = \lambda \underline{v} \text{ for some constant } \lambda.$$

If $\lambda > 0$ then $\underline{u}$ and $\underline{v}$ have the same direction
if $\lambda < 0$ then $\underline{u}$ and $\underline{v}$ have the opposite direction.

For instance, $\underline{u} = (3,6)$ is parallel to both $\underline{v} = (1,2)$ and $\underline{w} = (-9,-18)$. As $\underline{u} = 3\underline{v}$ the vectors $\underline{u}$ and $\underline{v}$ have the same direction. As $\underline{u} = -1/3\underline{w}$ the vectors $\underline{u}$ and $\underline{w}$ are opposite in direction.

The **zero vector** is the vector with zero for each component: $\underline{z} = (0,0)$; it is the only vector with length zero and the zero vector has no direction.

A **unit vector** is a vector whose length is one.

Given any vector $\underline{v}$, there is a unit vector $\underline{u}$ that has the same direction as $\underline{v}$. This unit vector is obtained by multiplying the vector $\underline{v}$ by the reciprocal of its length, which is simply referred to as dividing by its length:

$$\underline{u} = (1/\|\underline{v}\|)\,\underline{v} = \underline{v}\,/\,\|\underline{v}\| \quad \textbf{Unit vector } \underline{u} \textbf{ in direction } \underline{v}.$$

One form of vector multiplication is called the **dot product**. The dot product of two vectors is a number, not another vector. If $\underline{w} = (w_1, w_2)$ and $\underline{v} = (v_1, v_2)$ the **dot product of $\underline{w}$ and $\underline{v}$** is the sum of the products of their respective components:

$$\underline{w} \bullet \underline{v} \equiv w_1 \times v_1 + w_2 \times v_2 \qquad \textbf{The Dot Product.}$$

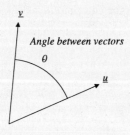

Angle between vectors

In higher dimensions the dot product is defined similarly, e.g., in three dimensions the dot product of $\underline{a} = (a_1, a_2, a_3)$ and $\underline{b} = (b_1, b_2, b_3)$ is

$$\underline{a} \bullet \underline{b} = a_1 \times b_1 + a_2 \times b_2 + a_3 \times b_3.$$

The dot product has an interesting and useful alternative formulation. Even in higher dimensional spaces, any two vectors $\underline{u}$ and $\underline{v}$, considered as position vectors, lie in a plane. Let the angle between the vectors be denoted by θ, as illustrated at the left. It can be shown, using the geometric definition of the sine and cosine functions, that the dot product of $\underline{u}$ and $\underline{v}$ is equal to the cosine of the angle θ between the vectors multiplied by the lengths of the vectors:

$$\underline{u} \bullet \underline{v} = \|\underline{u}\| \times \|\underline{v}\| \times \cos(\theta) \qquad \textbf{Dot Product Angle formula}.$$

One important aspect of this formula is that it gives a quick way to check if two vectors are parallel or perpendicular. Two vectors are **parallel** if the angle between them is either $\theta = 0$ or $\theta = \pi$. When they point in the same direction $\theta = 0$, in which case $\cos(\theta) = 1$. When they point in opposite directions $\theta = \pi$, in which case $\cos(\theta) = -1$. The dot product can be used to test if two vectors are parallel:

$\underline{u}$ and $\underline{v}$ are parallel with the same direction $\iff$ $\underline{u} \bullet \underline{v} = \|\underline{u}\| \times \|\underline{v}\|$.

$\underline{u}$ and $\underline{v}$ are parallel with the opposite direction $\iff$ $\underline{u} \bullet \underline{v} = -\|\underline{u}\| \times \|\underline{v}\|$.

For instance, the dot product of the parallel vectors $\underline{u} = (3,6)$ and $\underline{v} = (1,2)$ is

$$\underline{u} \bullet \underline{v} = (3,6) \bullet (1,2) = 3 \times 1 + 6 \times 2 = 15$$

Their lengths are $\|(3,6)\| = \text{SQRT}(45)$ and $\|(1,2)\| = \text{SQRT}(5)$, thus,

$$\|\underline{u}\| \times \|\underline{v}\| = \text{SQRT}(45)\text{SQRT}(5) = \text{SQRT}(225) = 15 = \underline{u} \bullet \underline{v}$$

Two vectors $\underline{u}$ and $\underline{v}$ are **perpendicular (or orthogonal)** if the angle between them is a right angle, either $\theta = \pi/2$ or $\theta = -\pi/2$. In either case $\cos(\theta) = 0$ and hence their dot product is zero. This provides a test for determining if two vectors are perpendicular:

$$\underline{u} \text{ and } \underline{v} \text{ are perpendicular} \iff \underline{u} \bullet \underline{v} = 0 \qquad \textbf{Perpendicular Test}.$$

For any vector $\underline{v} = (v_1, v_2)$ a perpendicular vector is obtained by interchanging the components and changing the sign of one component.

A vector perpendicular to (v_1, v_2) is $(-v_2, v_1)$.

The dot product can be used to determine the angle between two vectors by solving the Dot Product Angle formula for θ:

$$\cos(\theta) = \underline{u} \bullet \underline{v} / (\|\underline{u}\| \times \|\underline{v}\|)$$

or

$$\theta = \cos^{-1}\left(\underline{u} \bullet \underline{v} / (\|\underline{u}\| \times \|\underline{v}\|)\right) \qquad \textbf{Angle between } \underline{u} \textbf{ and } \underline{v}$$

Section 11.1   VECTORS, VECTOR FUNCTIONS, AND PARAMETRIC CURVES

Example 1   **Dot products and orthogonal vectors.**

Problem Evaluate the dot product of the two vectors and indicate if they are orthogonal or parallel. If they are not parallel, what is the angle between the vectors? Sketch the vectors to illustrate the angle.

a) $\underline{u} = (2,5)$ and $\underline{v} = (3,-1)$ b) $\underline{u} = (1,-1)$ and $\underline{v} = (2,2)$
c) $\underline{u} = (3,2)$ and $\underline{v} = (-4,-8/3)$.

Solutions a) The vectors $(2,5)$ and $(3,-1)$, plotted in the figure at the right, are not perpendicular since their dot product is non-zero:

$$(2,5) \bullet (3,-1) = 2 \times 3 + 5 \times (-1) = 1$$

The vectors are not parallel. As $\|(2,5)\| = \text{SQRT}(29)$ and $\|(3,-1)\| = \text{SQRT}(10)$, the angle between the vectors is obtained (using the INV COS or ARC COS features of a scientific calculator) as

$$\theta = \cos^{-1}(1/\text{SQRT}(290)) = 1.512 \text{ radians or about } 86.6°.$$

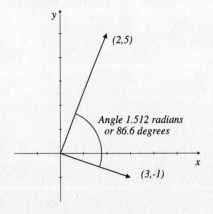

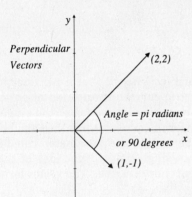

b) The vectors $(1,-1)$ and $(2,2)$ (plotted at the left) are orthogonal, or perpendicular, since their dot product is zero:

$$(1,-1) \bullet (2,2) = 1 \times 2 + (-1) \times 2 = 0$$

The positive angle between the vectors is $\theta = \pi/2$ radians.

c) The vectors $(3,2)$ and $(-4,-8/3)$ are sketched below, their dot product is

$$(3,2) \bullet (-4,-8/3) = -12 - 16/3 = -52/3$$

We could verify that these vectors are parallel with opposite direction by showing that $\|(3,2)\| \times \|(-4,-8/3)\| = 52/3$. Alternatively, we can simply proceed to evaluate the angle between the vectors:

$$\theta = \cos^{-1}\left(-52/3 \, / \, \{\|(3,2)\| \times \|(-4,-8/3)\|\}\right)$$

$$= \cos^{-1}\left(-52/3 \, / \, \{\text{SQRT}(13) \times \text{SQRT}(16 + 64/9)\}\right)$$

$$= \cos^{-1}\left(-52/3 \, / \, \{\text{SQRT}(13) \times \text{SQRT}(208/9)\}\right)$$

$$= \cos^{-1}(-1) = \pi$$

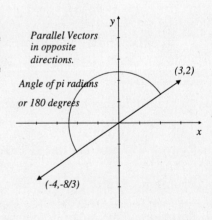

The fraction reduced to -1 as the denominator reduced to 52/3 since

$$\text{SQRT}(208/9) = \text{SQRT}(16 \times 13/9) = (4/3)\text{SQRT}(13).$$

Example 2 Determining vectors with specific properties.

Problem For the vector $\underline{v} = (2,5)$ compute i) a unit vector $\underline{u}$ in the direction of $\underline{v}$; ii) a vector $\underline{w}$ twice as long as $\underline{v}$ but in the opposite direction; iii) the dot product $\underline{v} \bullet \underline{r}$ where $\underline{r} = (1,2)$; iv) the angle between $\underline{v}$ and the vector $\underline{r} = (1,2)$; and v) determine a vector orthogonal to $\underline{v}$ with length 2.

Solution i) The unit vector in the direction $\underline{v}$ is $\underline{u} = \underline{v} / \|\underline{v}\| = (2/\text{SQRT}(29), 5/\text{SQRT}(29))$, as the length of $\underline{v}$ is

$$\|\underline{v}\| = \text{SQRT}(2^2 + 5^2) = \text{SQRT}(29) \approx 5.38.$$

ii) The vector $\underline{w} = -2\underline{v} = (-4,-10)$.

iii) $\underline{v} \bullet \underline{r} = (2,5) \bullet (1,2) = 2 \cdot 1 + 5 \cdot 2 = 12$.

iv) The angle between $\underline{v}$ and $\underline{r}$ is

$$\theta = \cos^{-1}(\underline{v} \bullet \underline{r} / [\|\underline{v}\| \cdot \|\underline{r}\|]) = \cos^{-1}(12/[\text{SQRT}(29) \cdot \text{SQRT}(5)])$$

$$= \cos^{-1}(12/\text{SQRT}(145)) \approx 0.0831 \text{ radians or about } 4.76°.$$

v) A vector orthogonal to $\underline{v}$ with the same length as $\underline{v}$ is the vector $(-5,2)$. A unit vector perpendicular to $\underline{v}$ is thus $(-5/\text{SQRT}(29), 2/\text{SQRT}(29))$.

☑

Parametric curves.

Vector notation is very useful when describing parametric curves. Recall that a parametric curve is more general than the graph of a function. A parametric curve has an orientation or direction. The direction of a parametric curve is derived from the positive direction of the single parameter used to "generate" the curve. The most frequent choice for this parameter is the variable t, however, like an integration variable, the symbol used to represent the parameter can be changed. In an x-y plane a parametric curve is specified by two single-valued functions of the parameter t, which are often subscripted to indicate the component they describe. The vector form of a two-dimensional parametric curve is

$$\underline{r}(t) = (r_1(t), r_2(t)) \qquad \textbf{Parametric curve}$$

where $r_1(t)$ and $r_2(t)$ are real valued functions and

$$\text{Domain}(\underline{r}) = \text{Domain}(r_1) \cap \text{Domain}(r_2)$$

To discuss different parametric curves we shall employ different symbols to represent the curves and the component functions, e.g., $\underline{p}(t) = (p_1(t), p_2(t))$ or $\underline{l}(s) = (l_1(s), l_2(s))$, etc.

Example 3 A nonlinear parametric curve.

Problem Sketch and describe the orientation of the curve $\underline{p}(t) = (t^2, e^{-t})$.

Solution This curve describes portions of two exponential curves. The x-y equations of these are obtained by solving the x-component equation $x = f(t)$ for t and substituting the resulting function of x, $t = f^{-1}(x)$, into the y-component equation for t.

In this example we first set $x = t^2$, from the first component equation. Then, there are two possible solutions for t in terms of x, but only if $x \geq 0$:

$$t = \text{SQRT}(x) \qquad \text{and} \qquad t = -\text{SQRT}(x)$$

Substituting these, one at a time, into the second component equation $y = e^{-t}$ gives the two curves.

If $t = \text{SQRT}(x) \geq 0$, then $y = e^{-\text{SQRT}(x)}$
and
if $t = -\text{SQRT}(x) \leq 0$, then $y = e^{\text{SQRT}(x)}$.

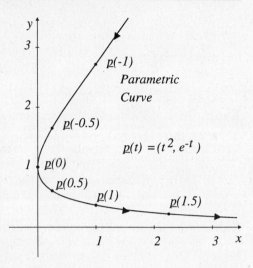

The parametric curve $\underline{p}$ consists of the portion of each curve with positive x-values. As sketched in the above figure, the orientation depends on the value of t. If t is negative the parametric curve describes the upper exponential curve, $y = e^{\text{SQRT}(x)}$, with an orientation decreasing from right to left. When t is negative and increases towards zero, the point $\underline{p}(t)$ moves along the upper curve from right to left. The points corresponding to $t = -1$ and $t = -0.5$ are $\underline{p}(-1) = (1, e)$ and $\underline{p}(-0.5) = (2.5, e^{0.5})$, which are indicated on the graph. As t approaches zero the points $\underline{p}(t)$ approach $\underline{p}(0) = (0,1)$. Then, as t increases through positive values the point $\underline{p}(t)$ moves along the lower exponential curve $y = e^{-\text{SQRT}(x)}$ from left to right. Notice that the curve $\underline{p}$ is not the graph of $y = f(x)$.

Parametric equation of a line.

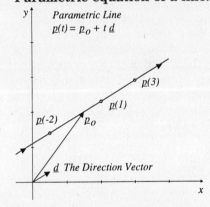

The simplest parametric curve is a straight line. A parametric or vector equation of a line is determined by a point on the line $\underline{p}_0 = (x_0, y_0)$ and a **direction** specified by a **direction vector** $\underline{d} = (d_1, d_2)$, as illustrated at the left.

Parametric Equation of a Line through $\underline{p}$ with direction $\underline{d}$

$$\underline{p}(t) = \underline{p}_0 + t\underline{d}$$

The direction vector $\underline{d}$ indicates both the orientation of the line and the slope of the line.

The **slope** of a line $\underline{p}(t) = \underline{p}_0 + t\underline{d}$, with direction $\underline{d} = (d_1, d_2)$ is

$$m = d_2/d_1.$$ **Slope of a line.**

The equations for ordinary x-y lines are unique. If the constants and coefficients of two equations are different, like $y = 1 + 2x$ and $y = 3 - 5x$, then the lines they describe are different. For parametric equations this is not the case. For instance, the parametric lines

$$L_1: \underline{p}(t) = (2,3) + t(3,-1) \quad \text{and} \quad L_2: \underline{p}(t) = (2,3) + t(6,-2)$$

describe the same physical line, the line through the point $(2,3)$ with slope $m = -1/3$. The difference between the two parametric lines is that the line L_2 is traversed twice as fast as the line L_1. At $t = 0$ the position $\underline{p}(0) = (2,3)$ for both lines. However, for other values of t the corresponding points on the two lines are different. At $t = 1$ the corresponding point on line L_1 is $\underline{p}(1) = (5,2)$ whereas the point on line L_2 is $\underline{p}(1) = (8,-1)$. Similarly, a point on the line will corresponds to different parameter-values for the two equations: for instance the point $(5,2) = \underline{p}(1)$, on L_1 and $(5,2) = \underline{p}(0.5)$ on line L_2. Another line that has the same slope, passes through the same point $(2,3)$ but has opposite direction to the lines L_1 and L_2 is the line

$$L_3: \underline{p}(t) = (2,3) + t(-0.3, 0.1)$$

The same x-y line is also the path of the parametric line

$$L_4: \underline{p}(t) = (5,2) + t(-6,2)$$

Example 4 Parametric equations of lines.

Problems (a) Determine the parametric vector equation of the line L_1 that passes through (1,5) at $t = 0$ and passes through the point (7,1) at $t = 10$.

(b) Determine the parametric equation of the line that is perpendicular to the line in part (a) and passes through the point (1,5) at $t = -3$.

Solutions (a) The vector equation of the line has the form

$$L_1: \underline{p}(t) = (1,5) + t(d_1, d_2)$$

The components d_1 and d_2 are found by solving the equation $\underline{p}(10) = (7,1)$ component-wise.

$$\underline{p}(10) = (1,5) + 10(d_1, d_2) = (1 + 10d_1, 5 + d_2) = (7,1)$$

implies that $1 + 10d_1 = 7$, or $d_1 = 0.6$, and that $5 + 10d_2 = 1$, or $d_2 = -0.4$. The line L_1 has the equation $\underline{p}(t) = (1,5) + t(0.6, -0.4)$.

(b) The line that is perpendicular to the line L_1 of part (a) will have an equation of the form L_2: $\underline{P}(t) = \underline{P} + t\underline{D}$, where the direction vector $\underline{D}$ is perpendicular to the direction vector $\underline{d} = (0.6, -0.4)$. One choice for this vector is obtained by interchanging the components and changing the sign of one component: $\underline{D} = (0.4, 0.6)$. Using this as the direction vector, the initial point $\underline{P}$ corresponding to $t = 0$, must be chosen so that $\underline{P}(-3) = (1,5)$. This equation is expressed using the vector $\underline{P} = (P_1, P_2)$ and is then solved component-wise for P_1 and P_2. Setting,

$$\underline{P}(-3) = (P_1, P_2) + -3(0.4, 0.6) = (P_1 - 1.2, P_2 - 1.8) = (1,5)$$

implies that $P_1 = 2.2$ and $P_2 = 6.8$. Thus the equation for L_2 is

$$\underline{P}(t) = (2.2, 6.8) + t(0.4, 0.6)$$

This solution is not unique! The vector equation of the line L_2 involves four parameters, P_1, P_2, D_1, and D_2. The statement of Problem (b) provides only three bits of information: the time associated with the point (1,5) gives two values, but the information on the slope only provides one value, the ratio $m = D_2/D_1$. As the slope of the line L_1 is $m_1 = -2/3$, the slope of L_2 must be $m_2 = -1/(-2/3) = 3/2$. Another choice for the direction vector would be $\underline{D} = (2,3)$. For this choice, a similar calculation gives the point $\underline{P} = (7,14)$, and an alternative equation for the line L_2 is

$$\underline{P}(t) = (7,14) + t(2,3)$$

☑

Exercise Set 11.1

1. For the given vector $\underline{v}$ determine
   i) a unit vector $\underline{u}$ in the direction of $\underline{v}$; ii) a vector $\underline{w}$ twice as long as $\underline{v}$ but in the opposite direction;
   iii) the dot product $\underline{v} \cdot \underline{r}$ where $\underline{r} = (1,2)$; iv) the angle between $\underline{v}$ and the vector $\underline{r} = (1,2)$;
   v) a vector orthogonal to $\underline{v}$ with length 2.

   a) $\underline{v} = (0,2)$ b) $\underline{v} = (-3,0)$ c) $\underline{v} = (1,2)$ d) $\underline{v} = (-1,-1)$ e) $\underline{v} = (1,5)$

   f) $\underline{v} = (1,1)$ g) $\underline{v} = (1,-1)$ h) $\underline{v} = (-1,2)$ i) $\underline{v} = (-2,-2)$ j) $\underline{v} = (2,5)$

   k) $\underline{v} = (2,-5)$ l) $\underline{v} = (-2,1)$ m) $\underline{v} = (1,4)$ n) $\underline{v} = (-4,3)$ o) $\underline{v} = (-3,4)$

2. i) Give a parametric equation and the x-y equation of the line through the point (x_0, y_0) with the direction given by $\underline{d}$.
   ii) What is the point on this line one unit away from (x_0, y_0) in the direction $\underline{d}$?

   a) $(x_0, y_0) = (1,1)$ and $\underline{d} = (0,-1)$ b) $(x_0, y_0) = (1,0)$ and $\underline{d} = (0,1)$

   c) $(x_0, y_0) = (1,2)$ and $\underline{d} = (1,1)$ d) $(x_0, y_0) = (3,-4)$ and $\underline{d} = (4,3)$

   e) $(x_0, y_0) = (0,3)$ and $\underline{d} = (-1,-1)$ f) $(x_0, y_0) = (2,3)$ and $\underline{d} = (8,6)$

In Exercises 3 - 6: For the indicated parametric curve $\underline{r}(t)$,
   i) Sketch the position vectors $\underline{r}(t)$ for $t = 0, 1, -1$ (if the vector exists).
   ii) Determine the x-y equation of the curve.

3. a) $\underline{r}(t) = (t, t^2 + 1)$ b) $\underline{r}(t) = (t^2, t + 1)$ c) $\underline{r}(t) = (1 + 3t, t^2 - 2)$

   d) $\underline{r}(t) = (t^2, 1/t)$ e) $\underline{r}(t) = (t^2, t/(t+1))$ f) $\underline{r}(t) = (1 - t^3, t^2)$

4. a) $\underline{r}(t) = (t^{1/3}, t - 9)$ b) $\underline{r}(t) = (e^t, e^{2t})$ c) $\underline{r}(t) = (e^{3t}, e^{2t})$

   d) $\underline{r}(t) = (t^{1/2}, t^3 - 9)$ e) $\underline{r}(t) = (e^t, e^{-2t})$ f) $\underline{r}(t) = (e^{-3t}, e^{2t})$

5. a) $\underline{r}(t) = (t, e^{-2t})$ b) $\underline{r}(t) = (t^2, e^{-2t})$ c) $\underline{r}(t) = (2t, e^{3t})$

   d) $\underline{r}(t) = (-\ln(t), \ln(t))$ e) $\underline{r}(t) = (\ln(t), t)$ f) $\underline{r}(t) = (\ln(t), t^2)$

   g) $\underline{r}(t) = (-\ln(t), t \cdot \ln(t))$ h) $\underline{r}(t) = (t, \ln(t^2))$ i) $\underline{r}(t) = (1/t, \ln(t^2))$

6. a) $\underline{r}(t) = (\sin(t), \cos(t))$ b) $\underline{r}(t) = (\cos(\pi t), \sin(\pi t))$ c) $\underline{r}(t) = (3\cos(2t), 3\sin(2t))$

   d) $\underline{r}(t) = (\sin(t), 2\cos(t))$ e) $\underline{r}(t) = (3\cos(\pi t), \sin(\pi t))$ f) $\underline{r}(t) = (3\cos(2t), 2\sin(2t))$

   g) $\underline{r}(t) = (t, \cos(\pi t))$ h) $\underline{r}(t) = (t^2, \cos(\pi t))$ i) $\underline{r}(t) = (t, \cos^2(\pi t))$

7. Determine two parametric equations for the graph of a circle centered at the origin of radius 4. Choose them so that the trajectory they describe passes through $(4,0)$ at $t = 0$ but in the opposite direction.

8. Determine two parametric equations for the parabola $y = 3x^2$ that have the same direction.

9. Determine a parametric equation for the hyperbola $x^2 - (y/9)^2 = 1$, that passes through the point $(1,0)$ at $t = 2$.

Section 11.2 Derivatives of Vector Functions: Tangent Vectors

Derivatives of Vector functions.

In this section we consider smooth parametric curves that have tangents at almost every point on the curve. Points at which a tangent line does not exist are called **singular points**. As parametric curves have a direction, this induces a corresponding direction on their tangent lines. Furthermore, if a parametric curve is viewed as a trajectory of a moving particle that traces out the curve as time increases, at each point on the curve we should be able to identify the speed that the particle is traveling. Both of these concepts, the direction of travel and the speed of travel, are indicated by a "tangent vector".

The tangent vector to a curve $\underline{r}(t)$ is defined as the limit of a vector-difference quotient, $\Delta \underline{r}(t)/\Delta t$, as $\Delta t \to 0$. However, it turns out that the tangent vector is determined by differentiating the component functions $x = r_1(t)$ and $y = r_2(t)$ with respect to t. To see how this happens, consider first the numerator of the vector-difference quotient corresponding to an increment $h \equiv \Delta t$. This is a vector difference

$$\Delta \underline{r}(t) = \underline{r}(t + h) - \underline{r}(t)$$

that is evaluated component-wise, and results in a vector. Dividing this difference by h is also done component-wise and again results in a vector. Thus the difference quotient is a vector that can be expressed as

$$\Delta \underline{r}(t)/\Delta t = \{\underline{r}(t + h) - \underline{r}(t)\} / h$$

$$= (r_1(t + h) - r_1(t) , r_2(t + h) - r_2(t)) / h$$

$$= (r_1(t + h) - r_1(t) / h , r_2(t + h) - r_2(t) / h)$$

The **tangent vector at the point $\underline{r}(t)$** is defined as the limit of the vector $\Delta \underline{r}(t)/h$, as $h \to 0$. This limit can be evaluated component-wise if the component limits exist:

$$D_t \, \underline{r}(t) = \lim_{h \to 0} \{\underline{r}(t + h) - \underline{r}(t)\} / h$$

$$= \lim_{h \to 0} (\{r_1(t + h) - r_1(t)\} / h , \{r_2(t + h) - r_2(t)\} / h)$$

$$= (\lim_{h \to 0} \{r_1(t + h) - r_1(t)\} / h , \lim_{h \to 0} \{r_2(t + h) - r_2(t)\} / h)$$

If each component function of $\underline{r}$ is differentiable, then taking each component limit in the above equation gives a vector with the derivatives $r_1{'}$ and $r_2{'}$ as its components. If either r_1 or r_2 are not differentiable, then the derivative of $\underline{r}$ will not exist.

Definition of a Tangent Vector.

If $\underline{r}(t) = (r_1(t), r_2(t))$ where $r_1(t)$ and $r_2(t)$ are differentiable functions, then the derivative of $\underline{r}$ is the **tangent vector $\underline{r}'(t)$** obtained by differentiating the components of $\underline{r}$:

$$D_t \underline{r}(t) = \underline{r}'(t) = (r_1'(t), r_2'(t))$$

Derivatives of vector functions are obtained by ordinary differentiation, therefor, all the algebraic differentiation Rules established in Chapter 3 also apply to derivatives of vector-valued functions. However, the Dot product rule corresponds to the ordinary product rule.

Vector Differentiation Rules

Scalar Multiple Rule	$D_t\, c\, \underline{r}(t) = c\, D_t \underline{r}(t)$ c a constant
Sum Rule	$D_t(\underline{u} + \underline{v}) = D_t \underline{u} + D_t \underline{v}$
Difference Rule	$D_t(\underline{u} - \underline{v}) = D_t \underline{u} - D_t \underline{v}$
Dot-Product Rule	$D_t(\underline{u} \cdot \underline{v}) = [D_t \underline{u}] \cdot \underline{v} + \underline{u} \cdot [D_t \underline{v}]$
Chain Rule	$D_s \underline{r}(t) = D_t \underline{r}\, ds/dt$ where $t = t(s)$

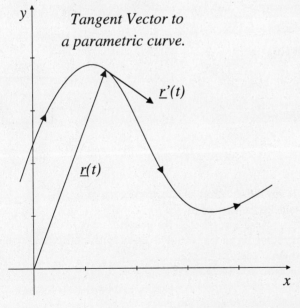

Tangent Vector to a parametric curve.

Tangent vectors and lines.

Normally, the tangent vector $\underline{r}'(t)$ is not sketched as a position vector, but as a vector emanating from the point $\underline{r}(t)$ on the parametric curve, as illustrated at the left. This provides a visual indication of the curve's direction and, by the length of the tangent vector, the speed at which the curve is being traversed.

At each point on a parametric curve where a tangent vector is defined there is a tangent line. The tangent line can be defined as a parametric line using the tangent point and the tangent vector as a direction. At the point $\underline{r}(t_0) = (x_0, y_0)$ on the curve $\underline{r}(t)$, that is the particle position at time $t = t_0$, the tangent vector is $\underline{r}'(t_0)$ and the tangent line equation is :

TL: $\underline{l}(s) = \underline{r}(t_0) + s\underline{r}'(t_0)$ **Tangent line to $\underline{r}(t)$ at $\underline{r}(t_0)$**

Notice that we have introduced a different parameter symbol, s rather than t, for the tangent line. This is to avoid confusion when referring to the curve $\underline{r}(t)$ and a tangent line $\underline{l}(s)$ in the same sentence. Observe that the curve $\underline{r}(t)$ passes through the point (x_0, y_0) at time $t = t_0$, whereas the tangent line passes through this tangent point at $s = 0$.

The "speed" at which a parametric curve is traversed is conceptualized by thinking of the parameter as time and the component units as distances. At any point r(t), or simply at time t, the **speed of the trajectory is the length of the tangent vector at r(t)**:

$$\text{Speed at the position } \underline{r}(t) \text{ is } SP(t) = \text{SQRT}(r_1'(t)^2 + r_2'(t)^2) \quad \text{Speed}$$

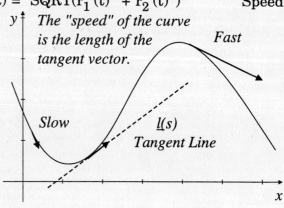

If the units of t or the component coordinates x and y are other than time or distance units then the "speed" might be interpreted differently. The unit of the speed SP(t) is (the x-y unit)/(t-unit) when x and y have the same distance units. As sketched at the right, a short tangent vector corresponds to a slow moving trajectory and a long tangent to a fast moving trajectory.

Example 1 **Tangent Vectors**.

Problem Determine the tangent vectors to the given curve and the speed of the trajectory traversing the curve when t = 2.

(a) The parabola: $\underline{r}(t) = (t, t^2 - 4t + 5)$.
(b) The curve $y = [10 ln(x)]^{-3}$ as the trajectory of $\underline{r}(t) = (e^{-0.1t}, t^{-3})$.
(c) The expanding wave curve $\underline{r}(t) = (t^2, t \sin(\pi t))$.

Solution (a) The tangent vector is $\underline{r}'(t) = (1, 2t - 4)$ so that $\underline{r}'(2) = (1, 0)$, a unit vector in the direction of the positive x-axis. Hence SP(2) = 1.

(b) For this curve $\underline{r}'(t) = (-0.1 e^{-0.1t}, -3t^{-4})$. Thus $\underline{r}'(2) = (-0.1 e^{-0.2}, -3 \cdot 2^{-4}) \approx (-0.082, -0.1875)$. At the position $\underline{r}(2)$ the parametric trajectory is moving so that both x and y are decreasing. The trajectory's speed is

$$SP(2) = \|(-0.1 e^{-0.2}, -3 \cdot 2^{-4})\| = \text{SQRT}(0.01 e^{-0.4} + 9 \cdot 2^{-8}) \approx 0.205$$

(c) The tangent vector is $\underline{r}'(t) = (2t, \sin(\pi t) + \pi t \cos(\pi t))$. Hence, $\underline{r}'(2) = (4, 2\pi)$ and $SP(2) = \text{SQRT}(16 + 4\pi^2) \approx 7.448$.

Example 2 **Tangents to a parametric ellipse.**

Problem The parametric curve $\underline{r}(t) = (4\sin(t), \cos(t))$ is an ellipse that is the trajectory of a particle traversing the ellipse in a counter-clockwise direction.
(a) Verify that the ellipse is the graph of $(x/4)^2 + y^2 = 1$ and sketch its graph.
(b) Determine the tangent vectors, the trajectory speed and the tangent lines to the ellipse at the points $\underline{r}(\pi/3)$, $\underline{r}(\pi/2)$ and $\underline{r}(\pi)$. Sketch the tangent vectors.
(c) What are the maximum and minimum speeds of this trajectory?

Solution (a) That the curve $\underline{r}(t)$ traverses the ellipse whose x-y equation is

$$(x/4)^2 + y^2 = 1$$

This is found by substituting $x = 4\sin(t)$ and $y = \cos(t)$ into the ellipse equation, and after canceling the 4's observing that this results in Pythagorean's identity, and hence is valid for all t:

$$(x/4)^2 + y^2 = (4\sin(t)/4)^2 + (\cos(t))^2$$
$$= \sin^2(t) + \cos^2(t) = 1$$

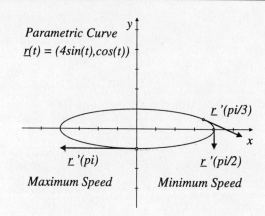

(b) The point $\underline{r}(0) = (0,1)$ is at the top of the skjetched ellipse. The other three points are:

$\underline{r}(\pi/3) = (2\ \text{SQRT}(3), 1/2)$ $\underline{r}(\pi/2) = (4,0)$

$\underline{r}(\pi) = (0,-1)$

As sketched on the graph, notice that $\underline{r}(\pi/2)$ is the right apex and $\underline{r}(\pi)$ is the bottom apex of the ellipse. The tangent vectors are found by differentiating the components of $\underline{r}(t)$:

$$\underline{r}'(t) = (4\cos(t), -\sin(t))$$

The tangent vector at $\underline{r}(\pi/3)$ is $\underline{r}'(\pi/3) = (2, -\text{SQRT}(3)/2)$. Thus the tangent line at this point is

$$\underline{l}(s) = (2\ \text{SQRT}(3), 1/2) + s(2, -\text{SQRT}(3)/2)$$

This is the line with x-y slope of $m = [-\text{SQRT}(3)/2]/2 = -\text{SQRT}(3)/4$. As $\underline{r}'(\pi/2) = (0,-1)$, the tangent line to the ellipse at $(4,0)$ is

$$\underline{l}(s) = (4,0) + s(0,1)$$

This is the vertical line $x = 4$, with an orientation pointing in the downward direction. At the bottom apex, $\underline{r}'(\pi) = (-4,0)$ and the tangent line is given in parametric form by

$$\underline{l}(s) = (0,-1) + s(-4,0)$$

This is the horizontal line $y = -1$ with the orientation point toward the left.

The trajectory speed, SP, which varies as $\underline{r}(t)$ moves around the ellipse is

$$SP(t) = \|\underline{r}'(t)\| = \|(4\cos(t), -\sin(t))\| = \text{SQRT}(16\cos^2(t) + \sin^2(t))$$

At the three points considered above, the speed is given by:

when $t = \pi/3$ $SP = \|\underline{r}'(\pi/3)\| = \|(2, -\text{SQRT}(3)/2)\| = \text{SQRT}(5.25) \approx 2.29$

when $t = \pi/2$ $SP = \|\underline{r}'(\pi/2)\| = \|(0,-1)\| = 1$

when $t = \pi$ $SP = \|\underline{r}'(\pi)\| = \|(-4,0)\| = 4$

(c) The maximum and minimum speed of the trajectory are found by identifying the critical points of the speed function. As the speed is defined for all t, these are simply the points at which $SP'(t) = d/dt\ \|\underline{r}'(t)\| = 0$. The derivative of the speed function is

$$D_t \|\underline{r}'(t)\| = D_t\ (16\cos^2(t) + \sin^2(t))^{1/2} = -15\sin(t)\cos(t) / \{(16\cos^2(t) + \sin^2(t))^{1/2}\}$$

This derivative is zero when the numerator $\sin(t)\cos(t) = 0$; this occurs when t is an integer multiple of π or an odd integer multiple of $\pi/2$. Thus, for each n, the points $t = n\pi$ and $t = n\pi + \pi/2$ are critical points. Applying the First Derivative Test for extrema, as the denominator of the above derivative is always positive we see that the maximum speed, $SP = 4$, occurs at the points $t = n\pi$ and the minimum speed, $SP = 1$, occurs at the points $t = \pi/2 + n\pi$.

Exercise Set 11.2

1. Determine the tangent vectors to the given curve and the speed of the trajectory traversing the curve at the point corresponding to t = 2.
   (a) The parabola: $\underline{r}(t) = (t, t^2 + 3t + 4)$
   (b) The linear curve $\underline{r}(t) = (e^{-t}, 2e^{-t})$
   (c) The spiral curve $\underline{r}(t) = (t \cos(\pi t), t \sin(\pi t))$
   (d) The logarithmic curve $\underline{r}(t) = (e^t, t)$
   (e) The quadratic curve $\underline{r}(t) = (e^{2t}, e^{4t} + 3)$
   (f) The circle $\underline{r}(t) = (5 \cos(3t), 5 \sin(3t))$

2. For each function $\underline{r}(t)$ compute the tangent vectors $\underline{r}'(t)$, i) evaluate the speed of the trajectory $\underline{r}(t)$ at t = 0 and at t = 10.
   a) $\underline{r}(t) = (t, \cos(\pi t))$
   b) $\underline{r}(t) = (t^2, t/(t+1))$
   c) $\underline{r}(t) = (1 + 3t, t^2 - 2)$
   d) $\underline{r}(t) = (t^{1/3}, t - 9)$
   e) $\underline{r}(t) = (e^{3t}, e^{2t})$
   f) $\underline{r}(t) = (ln(t), t^2)$
   g) $\underline{r}(t) = (t^{1/2}, t^3 - 9)$
   h) $\underline{r}(t) = (2t, e^{-2t})$
   i) $\underline{r}(t) = (-ln(t), ln(t)t)$

3. Observe that for any differentiable function f(t), by the Chain Rule $D_t \text{SQRT}(f(t)) = f'(t)/2\text{SQRT}(f(t))$. Consequently, critical points for SQRT(f(t)) are the critical points for f(t) plus any points at which f(t) = 0. Use this observation to determine the critical points at which the speed SP(t) of the trajectory $\underline{r}(t)$ may have a maximum or minimum value.

   a) $\underline{r}(t) = (t, t(t - 3))$
   b) $\underline{r}(t) = (t^2, t^{-1})$
   c) $\underline{r}(t) = (t, \cos(3t))$
   d) $\underline{r}(t) = (t^2 - 1, t + 3)$
   e) $\underline{r}(t) = (ln(t), t^2)$
   f) $\underline{r}(t) = (2 - t^2, t^{3/2})$

Section 11.3 Directional Derivatives

Introduction.

"Directional derivatives" indicate the rates of change of multivariable functions corresponding to different directions of change in the domain space. This concept applies with the same notation for any number of dimensions, but we will initially consider the two-dimensional function $z = F(x, y)$. Assume that F is defined in a region about a reference point (x_0, y_0). Then, if the function F is not constant the value of $F(x, y)$ will vary as (x, y) moves in any direction from (x_0, y_0). The question that we pose is this:

What is the rate of change in $z = F(x, y)$ if (x, y) moves from (x_0, y_0) along a directed line?

The rate is a derivative, and since it is associated with a specific direction it is called a "directional derivative".

Example 1. A hiker's problem.

Problem Some hikers consider possible paths to travel by referring to the adjacent contour map. Assume the map indicates level curves for the altitude function $A = F(x, y)$ at map coordinates (x, y) and the hikers are at the point whose map coordinates are (x_0, y_0).
(a) What would be their change in altitude if they traveled eastward?
(b) A more interesting question is "What would be the change in the hikers' altitude if they walked in the compass direction 30°, where North is 0°?"

Solution The hikers are at an altitude of $A_0 = F(x_0, y_0)$. To travel eastward is to walk parallel to the x-axis. The rate of change in this direction is provided by the partial derivative of F with respect to x. If the hikers proceed eastward their altitude would initially be changing at a rate of $F_x(x_0, y_0)$. If the units for the x and y variables are equal to one (giant) step, then to a first order approximation their altitude after one step would be

New height ≈ Old height + Rate of change × Distance covered.

$$A_1 \approx A_0 + F_x(x_0, y_0) \times 1.$$

To answer question (b) you might estimate how the altitude would change by looking at the contour map and sketching a line through the point (x_0, y_0) in the desired direction. As illustrated in the figure, you would expect their altitude to increase as they followed this line since the line crosses level curves corresponding to increasing altitude values. But what is the actual rate of change, just as they leave the point (x_0, y_0)? How big a step up will their first step be? The contour map can not reveal this precise information since it indicates only some of the function values. To answer these questions we must know the actual function $F(x, y)$ and introduce the concept of a "directional derivative". We revisit this problem in Example 4 where a specific altitude function is used to give a numerical answer.

Directional Derivatives.

DIRECTIONAL DERIVATIVE

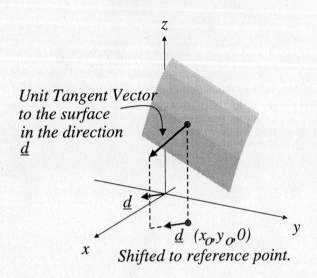

Unit Tangent Vector to the surface in the direction $\underline{d}$

Shifted to reference point.

The directional derivative of a function $F(x, y)$ at a reference point (x_0, y_0) is the initial rate of change in the values $z = F(x, y)$ as (x, y) changes in a specific direction from the reference point. A direction will be specified by a vector $\underline{d}$ that in turn prescribes a "direction line" L through the reference point.

Restricting the domain points (x, y) to the line L in effect reduces the computation of the directional derivative to a one-dimensional problem. Then, as always, the derivative is found as the limit of an average rate of change in the function over an interval, ... the limit as the length of the interval approaches zero. The restriction to the line L will allow us to evaluate the directional derivative using our existing knowledge of single variable differentiation.

The "reduction in dimension" is accomplished by considering only the change in function values as (x, y) varies along the line L. This in effect restricts us to considering the change in F along the curve which is the intersection of the surface $z = F(x, y)$ with the plane through the line L and perpendicular to the x-y plane. The directional derivative of F at the point (x_0, y_0) will be the rate of change in the z-coordinate that moves along this restricted curve just as it passes the point $(x_0, y_0, F(x_0, y_0))$. The direction of movement is important, for like the Hiker's situation in Example 1, if traveling in one direction along this path result in an increase in the z-coordinate, traveling in the opposite direction will result in a decrease in the z-coordinate. The direction on this surface curve is taken to correspond to the direction of the line L. To make this association and to reduce the problem to one involving a single variable, we use a parametric equation of the line.

For an evaluation point (x_0, y_0) and a direction given by $\underline{d} = (d_1, d_2)$, the "directed line" through (x_0, y_0) with direction $\underline{d}$ is the parametric line

$$L: \quad \underline{p}(t) = (x_0, y_0) + t\underline{d}$$

The reference point is $\underline{p}(0) = (x_0, y_0)$ and any point (x, y) on L in the direction of $\underline{d}$ must have the form

$$(x, y) = (x_0 + td_1, y_0 + td_2), \quad \text{for some } t > 0$$

Restricting (x, y) to the line L means that the function values $z = F(x, y)$ can be considered as a function of the parameter t, setting

$$F(x, y) = F(\underline{p}(t)) \text{ where } (x, y) = \underline{p}(t)$$

The change in the function values from the reference point $\underline{p}(0)$ and a point $\underline{p}(t)$ in the direction $\underline{d}$ is the difference

$$\Delta F = F(\underline{p}(t)) - F(\underline{p}(0)) \quad \quad \textbf{Change in z-values.}$$

The average rate of change in F between the reference point and the point on L is the change in the function's values divided by the distance between the points $\underline{p}(t)$ and $\underline{p}(0)$ on the line. This distance, which is the length of the vector $\underline{p}(t) - \underline{p}(0)$, depends on t:

$$h = h(t) = \| \underline{p}(t) - \underline{p}(0) \|$$ **Distance moved in the x-y plane.**

The average rate of change in F is thus the difference quotient

$$\Delta F/h = \{F(\underline{p}(t)) - F(\underline{p}(0))\} / \| \underline{p}(t) - \underline{p}(0) \|$$ **Average Rate of Change in F.**

To work with this difference quotient, it is viewed as the ratio of two difference quotients, one indicating the change in F due to a change in t, and the other indicating the change in distance along the line L due to a change in t:

$$\Delta F/h = [\Delta F(\underline{p}(t)) / \Delta t] / [\|\Delta \underline{p}\| / \Delta t]$$

since the Δt terms cancel. The derivative, or "instantaneous rate of change of F in the direction $\underline{d}$", is the limit of the average rate of change as the point $\underline{p}(t)$ approaches the reference point $\underline{p}(0)$. Unlike the limit involved in the derivative of a single variable function, to maintain the direction of movement in this case a one sided limit must be used, the limit as the parameter t approaches zero through positive values only, $t \to 0+$.

The **Instantaneous Rate of Change of F** in the direction $\underline{d}$ at (x_0, y_0) is

$$\lim_{t \to 0+} F(\underline{p}(t)) - F(\underline{p}(0)) / \| \underline{p}(t) - \underline{p}(0) \|$$

Notice that $\underline{p}(t) - \underline{p}(0) = t\underline{d}$ is a vector with the same direction as $\underline{d}$ since $t > 0$, so that:

$$h = \| \underline{p}(t) - \underline{p}(0) \| = \| (x_0, y_0) + t\underline{d} - (x_0, y_0) \| = \| t\underline{d} \| = |t| \|\underline{d}\| = t\, \text{SQRT}(d_1^2 + d_2^2)$$

Using this identity to evaluate the denominator in the above limit leads to the following definition of a "directional derivative".

DEFINITION OF THE DIRECTIONAL DERIVATIVE $D_{\underline{d}} F(x,y)$.

The **directional derivative** of $F(x,y)$ at the point (x_0, y_0) in the direction $\underline{d} = (d_1, d_2)$ is

$$D_{\underline{d}} F(x_0, y_0) = \lim_{t \to 0+} \{F(x_0 + td_1, y_0 + td_2) - F(x_0, y_0)\} / \{t\, \text{SQRT}(d_1^2 + d_2^2)\}$$

Alternatively, if $\underline{u} = (u_1, u_2)$ is a unit vector in the direction $\underline{d}$, then

$$D_{\underline{d}} F(x_0, y_0) = D_{\underline{u}} F(x_0, y_0) = \lim_{t \to 0+} \{F(x_0 + tu_1, y_0 + tu_2) - F(x_0, y_0)\} / t$$

Example 2. **Evaluating a directional derivative as a limit.**

Problem   Compute the directional derivative of $F(x, y) = x^2 + y$ at $(1,2)$ in the direction $\underline{d} = (3,4)$.

Solution   At the evaluation point $F(1,2) = 3$. The derivative $D_{(3,4)} F(1,2) = 2$ is evaluated as:

$$D_{\underline{d}} F(1,2) = \lim_{t \to 0+} \{F(1 + 3t, 2 + 4t) - F(1,2)\} / \{t\, \text{SQRT}(3^2 + 4^2)\}$$

$$= \lim_{t \to 0+} \{[(1 + 3t)^2 + (2 + 4t)] - 3\} / 5t$$

$$= \lim_{t \to 0+} \{10t + 9t^2\} / 5t$$

$$= \lim_{t \to 0+} 2 + (9/5)t = 2$$

Partial derivatives are directional derivatives resulting from specific choices for the direction vector. A unit vector in the direction of the positive x-axis is $\underline{d} = (1,0)$, while the unit vector $\underline{d} = (0,1)$ points in the direction of the positive y-axis. The partial derivatives of F are actually directional derivatives in these two directions:

$$\partial/\partial x\, F(x,y) = D_{(1,0)}F(x,y) \quad \text{and} \quad \partial/\partial y\, F(x,y) = D_{(0,1)}F(x,y).$$

Just as the partial derivatives F_x and F_y are the slopes of the tangent plane to the surface $z = F(x,y)$ at (x_0, y_0), the slopes in the (positive) "x-direction" and in the (positive) "y-direction", respectively, other directional derivatives can be visualized as "slopes" on the tangent plane in the "direction $\underline{d}$". Recall that the equation of the tangent plane is

$$P:\; z = z_0 + F_x(x_0, y_0)(x - x_0) + F_y(x_0, y_0)(y - y_0)$$

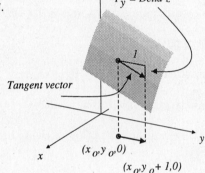
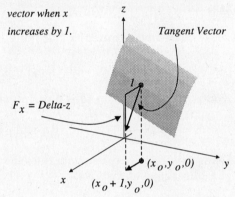

The partial derivative in the "y-direction" is the change in the z-coordinate of the tangent vector when y increases by 1.

The partial derivative in the "x-direction" is the change in the z-coordinate of the tangent vector when x increases by 1.

Furthermore, recall that $F_x(x_0, y_0)$ and $F_y(x_0, y_0)$ are equal to the changes in the z-coordinates on the tangent plane from (x_0, y_0) to $(x_0 + 1, y_0)$ and to $(x_0, y_0 + 1)$, respectively. In a similar fashion, the directional derivative $D_{\underline{d}}F(x_0, y_0)$ is the change in the z-coordinate on the tangent plane from (x_0, y_0) to a point one unit away, in the direction $\underline{d}$. Denoting the unit vector in the direction $\underline{d}$ as $\underline{u} = (u_1, u_2)$, the (x, y) point one unit in the direction $\underline{d}$ from (x_0, y_0) is

$$(x_1, y_1) = (x_0 + u_1, y_0 + u_2).$$

On the tangent plane P the z-coordinate at the point (x_1, y_1) is

$$z_1 = z_0 + F_x(x_0, y_0) \times u_1 + F_y(x_0, y_0) \times u_2$$

since $x_1 - x_0 = u_1$ and $y_1 - y_0 = u_2$. The directional derivative $D_{\underline{d}}F(x_0, y_0)$ is then the difference $z_1 - z_0$. This observation provides an extremely useful alternative way to calculate directional derivatives:

$$D_{\underline{d}}F(x_0, y_0) = D_{\underline{u}}F(x_0, y_0) = F_x(x_0, y_0) \times u_1 + F_y(x_0, y_0) \times u_2$$

where $\underline{u}$ is a unit vector in the direction $\underline{d}$.

This equation expresses the directional derivative as a sum of products, which is precisely the form of a dot product. The right side of the equation is the dot product of the vector $\underline{u}$ with the vector $(F_x(x_0, y_0), F_y(x_0, y_0))$. This vector of partial derivatives is very important, and is called **the gradient of F** or simply **the gradient vector**. The gradient is denoted using the symbol ∇, and ∇F is read as "the gradient of F".

$$\nabla F(x, y) \equiv (F_x(x, y), F_y(x, y)) \qquad \textbf{The gradient of F.}$$

Evaluated at a particular point, $\nabla F(x_0, y_0)$ is a vector. When the point is not specified, but arbitrary, $\nabla F(x, y)$ is a vector-valued multi-variable function. Its domain is multi-dimensional, the set of points (x, y) for which both partial derivatives $F_x(x, y)$ and $F_y(x, y)$ are defined, and its range is a set of vectors. Using the gradient vector the directional derivative of F at an arbitrary point (x, y) in a direction $\underline{d}$ is given by:

Gradient form of Directional Derivative

$$D_{\underline{d}} F(x, y) = \nabla F(x, y) \cdot \underline{d} / \|\underline{d}\|$$

Note: if $\underline{d}$ is a unit vector then the denominator $\|\underline{d}\| = 1$. Using the gradient allows us to evaluate directional derivatives as linear combinations of ordinary partial derivatives.

Example 3. **Computing Gradients and directional derivatives.**

Problem   Compute the gradient and use the gradient formula to evaluate $D_{\underline{d}}F(x, y)$ at the point $(0,2)$.

a) $F(x, y) = x^2 - 2xy + y^3$, $\underline{d} = (1,3)$ b) $F(x, y) = x + e^{xy} - x/y$ $\underline{d} = (2,-5)$

Solutions a) First compute the partial derivatives of F to obtain the gradient function:

$$\nabla F(x, y) = (2x - 2y, 2x - 3y^2)$$

Using this function form of the gradient, the directional derivative of F in the direction $\underline{d} = (1,3)$ at an arbitrary point (x, y) is

$$D_{(1,3)}F(x, y) = \nabla F(x, y) \cdot (1,3) / \|(1,3)\|$$

$$= (2x - 2y, 2x - 3y^2) \cdot (1,3) / \text{SQRT}(10)$$

$$= \{(2x - 2y) \times 1 + (2x - 3y^2) \times 3\} / \text{SQRT}(10)$$

$$= \{8x - 2y - 9y^2\} / \text{SQRT}(10)$$

The directional derivative at $(0,2)$ is obtained by setting $x = 0$ and $y = 2$:

$$D_{(1,3)}F(0,2) = -40 / \text{SQRT}(10) \approx -12.65$$

b) In this solution we focus on the point $(0,2)$. Evaluating the gradient at the point $(0,2)$ and then taking the dot product with the normalized direction vector gives the specific directional derivative at the point $(0,2)$. The gradient of $F(x, y) = x + e^{xy} - x/y$ is

$$\nabla F(x, y) = (1 + ye^{xy} - 1/y, xe^{xy} + x/y^2)$$

Evaluated at the reference point $(0,2)$ the gradient is

$$\nabla F(0,2) = (1 + 2 - 1/2, 0) = (2.5, 0)$$

The desired directional derivative is

$$D_{(2,-5)}F(0,2) = \nabla F(0,2) \cdot (2,-5) / \|(2,-5)\|$$

$$= (2.5, 0) \cdot (2,-5) / \text{SQRT}(2^2 + (-5)^2) = 5/\text{SQRT}(29) \approx 0.93$$

Example 4. **The Hiker's Problem revisited.**

Problem   Consider a more specific version of the "Hiker's Problem" introduced in Example 1. Assume that i) the hikers are at the position $(x_0, y_0) = (2,-1)$ on a section map with North being the direction of the positive y-axis and East being the direction of the positive x-axis, ii) the x and y variables are measured in meters, iii) the terrain for the map-region is given by the altitude function

$$A = F(x,y) = 1500 + 1000x^2 \exp(-x^2 - 4y^2) \text{ m}$$

The Hikers' Problem asks: "What would be the rate of change in their altitude if the hikers walked in the compass direction 30° from the grid point (2,-1)? Would they step up or down? If they stepped 1m forward, how "big" would this step be?"

Solution The "rate of change" in the altitude as you proceed in a direction of 30° is given by a directional derivative. The first step is to determine the direction as a vector. First we convert the compass angle 30° to an angle of inclination relative to the horizontal axis. This is an "inclination angle" of 90° - 30° = 60°. Hikers refer to angles by degrees but mathematicians refer to slopes as pure numbers, indicating the ratio of the change in y to the change in x. The slope corresponding to 60° is tan(60°) = SQRT(3). We can choose as a direction vector any vector $\underline{d} = (d_1, d_2)$ such that its slope $m = d_2/d_1 = $ SQRT(3). A simple choice is $\underline{d} = (1, \text{SQRT}(3))$. The length of $\underline{d}$ is 2:

$$\| (1, \text{SQRT}(3)) \| = \text{SQRT}(1^2 + \text{SQRT}(3)^2) = \text{SQRT}(4) = 2$$

The next step is to compute the gradient of F:

$$\nabla F(x,y) = (\ 10^3[2x - 2x^3]\exp(-x^2 - 4y^2)\ ,\ -8 \cdot 10^3 y x^2 \exp(-x^2 - 4y^2))$$

At the point (2,-1) the gradient is

$$\nabla F(2,-1) = (-12e^{-8} \times 10^3\ ,\ 32e^{-8} \times 10^3) = 10^3 e^{-8}(-12, 32)$$

Therefore the directional derivative is

$$D_{(1, \text{SQRT}(3))} F(2,-1) = \nabla F(2,-1) \bullet (1, \text{SQRT}(3)) / \| (1, \text{SQRT}(3)) \|$$

$$= 10^3 e^{-8}(-12, 32) \bullet (1, \text{SQRT}(3)) / 2$$

$$= 10^3 e^{-8}[-6 + 16\ \text{SQRT}(3)] \approx 7.28$$

Since the directional derivative is positive, they would be stepping upward. Their first step one meter forward would be very large. They would have to be scaling a steep slope since their approximate change in altitude would be 7.28 meters.

Maximum, minimum and zero directional derivatives.

Directional derivatives at the same surface point $(x_0, y_0, F(x_0, y_0))$ will vary as the direction $\underline{d}$ changes. To determine the maximum and minimum possible values for these direction derivatives a different form of the Gradient equation is utilized. The dot-product equation for the directional derivative can be expressed by replacing the component-product form of the dot-product by the angle formulation:

$$\nabla F(x,y) \bullet \underline{d} / \| \underline{d} \| = \| \nabla F(x,y) \| \times \| \underline{d} \| \times \cos(\theta) / \| \underline{d} \| = \| \nabla F(x,y) \| \cos(\theta)$$

where θ is the angle between the direction $\underline{d}$ and the gradient vector $\nabla F(x, y)$. This gives an angle formula for the directional derivative:

$$D_{\underline{d}}F(x, y) = \| \nabla F(x, y) \| \cos(\theta) \quad \text{Angle Formula for } D_{\underline{d}}F(x, y)$$

As $\cos(\theta)$ varies between -1 and 1, from the angle formula we see that the directional derivative can only assume values between $-\| \nabla F(x, y) \|$ and $+\| \nabla F(x, y) \|$:

$$- \| \nabla F(x, y) \| \leq D_{\underline{d}}F(x, y) \leq \| \nabla F(x, y) \|$$

The gradient vector not only provides the maximum and minimum values for the derivative, it also provides the directions that correspond to these extreme values.

The **minimum directional derivative** of $F(x,y)$ is

$$D_{\underline{d}}F(x,y) = -\|\nabla F(x,y)\|$$

which occurs when the direction of $\underline{d}$ is opposite to $\nabla F(x,y)$, in which case $\underline{d}$ is said to be the **direction of greatest descent, or rate of decrease.**

The maximum directional derivative of $F(x,y)$ is

$$D_{\underline{d}}F(x,y) = \|\nabla F(x,y)\|$$

which occurs when $\underline{d}$ is in the same direction as $\nabla F(x,y)$, which is called the **direction of greatest ascent, or rate of increase**.

$$\underline{d} = -\nabla F(x, y) = (-F_x(x, y), -F_y(x, y)) \quad \textbf{Direction of greatest descent.}$$

$$\underline{d} = \nabla F(x, y) = (F_x(x, y), F_y(x, y)) \quad \textbf{Direction of greatest ascent.}$$

As the direction $\underline{d}$ varies from $-\nabla F$ to ∇F the derivative $D_{\underline{d}}F$ varies from a positive value to a negative value. If the derivative varies continuously, for some direction it must equal zero. The direction $\underline{d}$ for which $D_{\underline{d}}F = \nabla F(x, y) \bullet \underline{d}/\|\underline{d}\| = 0$ must be perpendicular to the gradient. There are two vector-directions perpendicular to the gradient, with corresponding angles to the gradient of $\theta = \pi/2$ or $-\pi/2$. As the directional derivative in either of these directions is zero, the value of $F(x,y)$ will not change when (x, y) is varied in these directions, i.e., these are the directions in which the function $F(x, y)$ remains constant. The simplest choices for these "constant directions" are the vectors formed by interchanging the components of the Gradient, and changing the sign of one component.

$$\underline{d} = (-F_y(x, y), F_x(x, y))$$

and **Directions of constant F(x, y).**

$$\underline{d} = (F_y(x, y), -F_x(x, y))$$

Example 5. Determining the directions of greatest ascent, greatest descent, and constant function values.

Problem For the function $F(x,y) = x^2 - 2xy + 5y^3$ at the point $(1,-2)$, determine the greatest and least values for the directional derivative $D_{\underline{d}}F(1,-2)$, the directions in which these occur, and the directions in which the function remains constant.

Solution First compute the gradient of F and evaluate it and its length at $(1,-2)$.

$$\nabla F(x,y) = (2x - 2y, -2x + 15y^2)$$

$$\nabla F(1,-2) = (6, 58)$$

$$\|\nabla F(1,-2)\| = \|(6,58)\| = \text{SQRT}(6^2 + 58^2) = \text{SQRT}(3400) = 10\,\text{SQRT}(34) \approx 58.31$$

The direction of greatest ascent is $\underline{d} = \nabla F(1,-2) = (6,58)$. The maximum possible directional derivative of F at $(1,-2)$ is

$$M = D_{(6,58)}F(1,-2) = \|\nabla F(1,-2)\| = 10\,\text{SQRT}(34)$$

The direction of greatest descent is $\underline{d} = -\nabla F(1,-2) = (-6,-58)$. As the norm of this vector is the same as the norm of $(6,58)$ the minimum possible value for the directional derivative of F at $(1,-2)$ is

$$-M = D_{(-6,-58)}F(1,-2) = -\|\nabla F(1,-2)\| = -10\,\text{SQRT}(34)$$

Two vectors perpendicular to $\nabla F(1,-2)$ are obtained by interchanging the components of ∇F and changing the sign of one component. This gives the directions $\underline{d} = (-58, 6)$ and $\underline{d} = (58,-6)$ for which $D_{\underline{d}}F(1,-2) = 0$.

Exercise set 11.3

1. Use the limit definition to evaluate the directional derivative of $F(x,y) = 2x - 5y$ at an arbitrary point (x,y) with direction $\underline{d} = (3,4)$.

2. Use the Limit Definition to compute the derivative of the given function $F(x,y)$ at the origin in the direction $\underline{d} = (1,2)$.

   a) $F(x,y) = 4 + 3x - 2y$ b) $F(x,y) = 1 + x - y^2$ c) $F(x,y) = \text{SQRT}(x^2+y^2)$

3. Using the Gradient Form to compute $D_{\underline{d}}F(x,y)$ for each of the two given vectors $\underline{d}$, and evaluate these at the point $(x_0, y_0) = (-2, 3)$.

   a) $F(x,y) = x^2 - 2xy + 3y^2$ $\underline{d} = (2,1)$ and $\underline{d} = (-2,1)$

   b) $F(x,y) = x^3 - 2xy$ $\underline{d} = (1,1)$ and $\underline{d} = (-1,1)$

   c) $F(x,y) = x - 3xy^2 + 2y^3$ $\underline{d} = (2,1)$ and $\underline{d} = (-2,-1)$

Section 11.3   DIRECTIONAL DERIVATIVES   643

   d) $F(x,y) = (x-y)^2$ $\underline{d} = (2,-1)$ and $\underline{d} = (3,2)$

   e) $F(x,y) = y\cos(3\pi/x)$ $\underline{d} = (1,1)$ and $\underline{d} = (-1,1)$

   f) $F(x,y) = (x+y)^{-1}$ $\underline{d} = (1,0)$ and $\underline{d} = (-2,0)$

   g) $F(x,y) = xye^{-2x}$ $\underline{d} = (ln(2),1)$ and $\underline{d} = (2,1)$

4. For each function $F(x,y)$ in Exercise 3, at the point $(x_0, y_0) = (-2,3)$ determine:
   i) the maximum and minimum directional derivatives and <u>unit</u> direction vectors that give these derivatives;
   ii) a direction in which the directional derivative is zero.

5. Consider the surface determined by the equation

$$z = (x^2+y^2)(2-x^2-y^2) + 3.$$

   a) Evaluate the gradient vector of this surface.

   b) Determine the tangent plane to the surface at the point $(1,-1,3)$.

   c) In what directions from the point $(1,-1)$ will the slope of the surface be i) maximum, ii) minimum, and iii) zero?

   d) For what values of (x,y) will the gradient vanish? What will be the equation of the tangent planes at these points?

6. A hiker is walking a path and her position at time t is $\underline{r}(t) = (t\cos(\pi t), t\sin(\pi t))$. At any given time the direction of the path is $\underline{r}'(t)$. Assume the altitude of the terrain is described by the surface

$$z = (x+y)^2 e^{-0.001(x+y)} + 10^4.$$

   a) What will be the position and altitude of the hiker at times $t_0 = 0$, $t_1 = 0.5$, $t_2 = 1$ and $t_3 = 3.5$?

   b) What will be the rate of change in the hiker's altitude at the times t_0, t_1, t_2 and t_3?

7. The net profit associated with a product is a function of the price P and the amount spent on advertising A: $NP = P(9 - 6A - A^2)$. If the current price is $P = 1$ and the advertising cost is $A = 2$,
   i) What will be the rate of change in profit if the advertising is decreased at twice the rate the price is decreased?

   ii) What would be the ratio of change in price and advertising that would yield the greatest rate of increase in profit?

   iii) Determine the equation of the curve that would yield the quickest ascent to reach the maximum profit. What would this be and how is it seen from the curve?

8. In what unit direction $\underline{u} = (u_1, u_2)$ from the position $(x_0, y_0) = (2,1)$ does the surface $z = y^2x^2 - 2x + y^2$ have a 6% "grade"? (6% is the maximum grade, or rate of increas, for most railway tracks.) (Hint: construct two equations for the components of $\underline{u}$ from the given information.)

Section 11.4 Level Curves and Curves of Steepest Ascent or Descent.

Introduction.

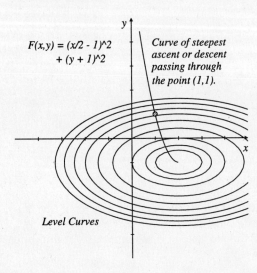

In previous sections Level curves were described as curves in the domain space on which a function's values are constant. In this section we explore how the equations for such curves can be derived using the gradient of the function. We also introduce another family of curves, called the curves of "steepest ascent (or descent)", which are referred to simply as "Steepest curves". This is a family of curves that indicate the path in the domain along which the function's values change most rapidly. If we consider a point (x, y) moving along one of these curves in one direction the function's values increase as fast as possible, and if the direction is reversed the function values will decrease as quickly as possible. A family of level curves and a single Steepest Curve are indicated at the left for the function $F(x, y) = (x/2 - 1)^2 + (y + 1)^2$.

Steepest curves and Level curves are related, in fact they are "orthogonal" or perpendicular to each other. At a point of intersection of a Level curve and a Steepest curve, the tangent lines to the two curves are perpendicular. This will become apparent in the following discussion when these curves are described using directional derivatives.

To conceptualize the concepts of Steepest and Level curves it is often useful to think of the analogy in which a surface formed by the graph of $z = F(x, y)$ is thought of as the geographical terrain of a region. In this context, a Level curve is the easiest hiking path because it goes neither up nor down but winds around a hill at a given level. Similarly, the Steepest curve representing the curve of steepest decent is the path that a stream or river would follow, always descending as quickly as possible. The Steepest curve considered as the curve of steepest ascent is the path you would follow moving upstream, towards the peak of the continental divide. Since the direction in which such a curve is followed distinguishes whether it is a curve of ascent or descent, it would seem natural to describe such curves in parametric form so that the direction can also be indicated. As the gradient $\nabla F(x, y)$ provides the directions for maximum and minimum directional derivatives, as well as the direction for zero derivatives, it should not be surprising that the gradient is used to describe these special curves.

The method that we shall use to describe these curves is indirect. Instead of explicitly giving the curves in the function form $y = f(x)$ or in parametric form as $\underline{r}(t) = (r_1(t), r_2(t))$ they are indicated by specifying their tangents at each point. This is done in the form of a differential equation, either giving $f'(x)$ equals a function of x or giving the tangent vector $\underline{r}'(t) = (r_1'(t), r_2'(t))$. To find the explicit equation of the curve the given differential equation(s) must be solved, generally by integration, subject to the conditions that the solutions pass through a fixed reference point (x_0, y_0), i.e., that $y_0 = f(x_0)$ or $\underline{r}(0) = (x_0, y_0)$.

Level curves.

Level curves were described as the projections onto the x-y plane of "cut" curves, where the surface $z = F(x, y)$ intersects horizontal planes of the form z = constant. You are familiar with such curves; for instance, level lines on contour maps, as isobars on weather maps showing lines of constant barometric pressure, or as lines on a geographic map connecting points that have the same temperature, called isotherms, or the same mean temperature during the coldest months, called isocrymes. (The prefix "iso" is from the Greek *isos*, which means equal.) We will denote a Level curve by a subscripted capital Greek letter Gamma; the subscript is used to indicate the constant value c:

$$\Gamma_c \equiv \{ (x, y) \mid F(x, y) = c \} \quad \textbf{Level Curve for F(x, y) = c.}$$

The following discussion introduces a method of determining Level curves from the gradient of the function. Consider a point (x_0, y_0) on a Level curve Γ_c where $c = F(x_0, y_0)$. Assume that this Level curve is a parametric curve $\underline{r}(t)$ and $(x_0, y_0) = \underline{r}(0)$, for some unknown vector function $\underline{r}(t)$, which we must determine. The Level curve is

$$\Gamma_c = \{(x, y) \mid (x, y) = \underline{r}(t) \text{ for some real value } t\}$$

We assume that the curve Γ_c is smooth enough to have a tangent at each point. Then at each point (x, y) the tangent vector to Γ_c is $\underline{r}'(t) = (r_1'(t), r_2'(t))$.

A basic assumption is that if a point $(x, y) = \underline{r}(t)$ moves along the curve Γ_c, it must move in the direction of the tangent vector to the curve, the direction $\underline{r}'(t)$. Since value of $F(x, y)$ does not change on a Level curve, Γ_c, the directional derivative as it travels along the curve must be zero, i.e., the derivative $D_{\underline{r}'}F(x, y) = 0$. Recalling that this derivative is the dot-product of the direction $\underline{r}'$ and the gradient ∇F, this requires that at each point the tangent vector $\underline{r}'$ must be perpendicular to ∇F:

$$D_{\underline{r}'}F(x, y) = \nabla F(x, y) \bullet \underline{r}' / \|\underline{r}'\| = 0 \quad \Longrightarrow \quad \nabla F(x, y) \bullet \underline{r}' = 0$$

The tangent vector $\underline{r}'(t)$ must therefor be parallel to any other vector perpendicular to ∇F; in particular, it must be parallel to the vector $(-F_y(x, y), F_x(x, y))$. Recall that if two vectors are parallel, then one is a multiple of the other. This means that their x and y components must both be related by the same multiple. Thus, there must exist a constant λ for which the following two equations must hold

$$F_x(x, y) = \lambda \, r_2'(x, y) \quad \text{and} \quad F_y(x, y) = -\lambda \, r_1'(x, y)$$

The sign of λ will indicate the direction of the curve relative to the direction of the gradient vector and $|\lambda|$ is the relative length of the tangent vector to the length of the gradient vector. In the present discussion the function F is given and we are seeking the function $\underline{r}(t)$ that gives the Level curve Γ_c through (x_0, y_0). To find $\underline{r}$ it will simplify the arithmetic if we assume that the constant $\lambda = 1$. Then the component functions of $\underline{r}$ are determined as the solutions of the two initial value problems:

$$r_1'(t) = -F_y(x, y) \; ; \; r_1(0) = x_0$$

$$r_2'(t) = F_x(x, y) \; ; \; r_2(0) = y_0$$

This is a system of differential equations whose solution gives the component functions r_1 and r_2. The Level curve is then expressed as a parametric curve, giving its coordinate functions

$$x = r_1(t) \quad \text{and} \quad y = r_2(t)$$

The difficulty arises when we realize that the point (x, y) cannot be expressed as a function of t unless we know the functions r_1 and r_2 already, or are given the partial derivatives F_x and F_y as functions of t! An alternative to solving the system for these component functions is to assume that the Level curve is (at least locally) also the graph of a function $y = f(x)$. In this case the derivative $y' = D_x f(x)$ is the slope of the tangent line to the Level curve. This tangent line, however, has the same slope as the tangent vector to the Level curve, when it is considered as a parametric curve. The slope of the tangent vector $\underline{r}'(t) = (r_1'(t), r_2'(t))$ is $r_2'(t)/r_1'(t)$. Equating these two slopes, $y' = r_2'/r_1'$, replacing the derivatives r_1' and r_2' by the equivalent partial derivatives of F as given above, we find the Level curve through (x_0, y_0) is the graph of the solution of the following

Differential Equation Describing a Level Curve.

$$y' = -\{F_x(x, y) / F_y(x, y)\} \quad ; \quad y = y_0 \text{ when } x = x_0.$$

Example 1 Determining Level Curves.

Problem Determine the Level curves for the function

$$F(x, y) = 1500 + 1000x^2 \exp(-x^2 - 4y^2)$$

What is the easiest path through the point $(x_0, y_0) = (2, -1)$ considered in that example?

Solution The function F was considered in the Hikers Example 1, 4 of Section 11.3. First, determine the gradient vector:

$$\nabla F(x, y) = (10^3[2x - 2x^3]\exp(-x^2 - 4y^2), -8 \times 10^3 y x^2 \exp(-x^2 - 4y^2))$$

Using these functions the slope of each Level curve is given by the ratio of the partial derivatives:

$$y' = -[10^3[2x - 2x^3]\exp(-x^2 - 4y^2)] / [-8 \times 10^3 y x^2 \exp(-x^2 - 4y^2)]$$

which is defined if $y \neq 0$ and $x \neq 0$. This simplifies considerably, to give

$$dy/dx = [x - x^3] / [4yx^2] = [1/x - x] / 4y$$

This differential equation is separable and is solved by separating the y and x terms treating the derivative as a ratio of differentials dy and dx :

$$4y \, dy = [1/x - x] \, dx$$

Then integrating each side of the equation gives

$$\int 4y \, dy = \int 1/x - x \, dx$$

$$2y^2 = \ln|x| - x^2/2 + K$$

where K is an integration constant. The corresponding function forms

$$y = \pm \text{SQRT}(2\ln|x| - x^2 + K/2)$$

are only defined when the expression $\ln|x| - x^2/2 + K$ in the square root is zero or positive. For each K there are two values $x_{1\text{-}k}$ and $x_{2\text{-}k}$ such that the function is only defined when $x_{1\text{-}k} \leq |x| \leq x_{2\text{-}k}$. For instance, when K = 5 the corresponding level curve is only defined for $x_{1\text{-}5} \approx 0.00637377 \leq |x| \leq x_{2\text{-}5} \approx 3.53948$.

Sketches of Level curves corresponding to five values of the integration constant K are indicated in the following figure. When K is sufficiently small, these curves look like slightly distorted ellipses that are concentric around the points (1,0) and (-1,0). For larger values of K the graphs of the level curves look more like ellipses flattened against the y-axis. Note that the graphs have a gap at the y-axis since they are not defined for x = 0. On each Level curve the function F(x, y) is constant. The following table lists the constant function values on the five illustrated curves.

Section 11.4 LEVEL CURVES AND CURVES OF STEEPEST ASCENT OR DESCENT

Constant Function values on Level Curves $y = \pm \sqrt{2\ln\|x\| - x^2 + K/2}$ of $F(x,y) = 1000(1.5 + x^2 \exp(-x^2 - 4y^4))$		
K	Sample point on the curve	Constant $F(x,y)$
0.6	(1, 0.2236)	1801.1942
1	(1, 0.5)	1635.335
2	(1, 0.866)	1518.3156
3.3	(2, 1.0)	1500.3354
5	(1, 1.5)	1500.0454

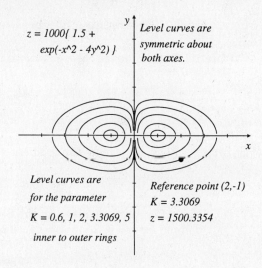

$z = 1000\{1.5 + \exp(-x^2 - 4y^2)\}$

Level curves are symmetric about both axes.

Level curves are for the parameter $K = 0.6, 1, 2, 3.3069, 5$ inner to outer rings

Reference point (2,-1)
$K = 3.3069$
$z = 1500.3354$

The integration constant K corresponding to the Level curve passing through the point (2,-1) is found by setting x = 2 and y = -1 in the above integral equation and solving for K:

$$2(-1)^2 = \ln(2) - 2^2/2 + K \quad \text{gives a calculator value of} \quad K \approx 3.3069$$

Thus the Level curve through the point (2,-1) is the graph of

$$\Gamma_c: 2y^2 - \ln|x| + x^2/2 - 3.3069 = 0$$

On this curve

$$F(x,y) = F(2,-1) = 1500 + 10^3 \times 4 \times e^{-8}$$

Thus $c \approx 1500.3355$.

The graph of the Function F(x, y) is an almost level surface when x or y are large, since the exponential term is then very small. It has two small "hill like" protrusions centered at the points (± 1, 0). The graph illustrates the three-dimensional graph $z = 3x^2 \exp(-x^2 - 4y^2)$, which is a function derived from F by subtracting 1500 and multiplying by 0.003 to scale the z-axis.

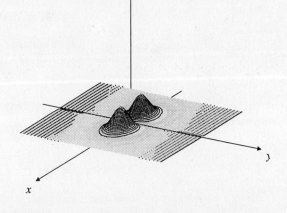

☑

Steepest curves.

The curves of steepest descent and ascent are determined by the same procedure used to determine Level curves. The only difference is that the tangent vector to the steepest ascent curve is in the direction of $\nabla F(x,y)$, while the tangent vector to the curve of steepest descent is in the direction $-\nabla F(x,y)$. In either case the slope of the tangent line, considering y locally as a function of x, is given by

Differential Equation Describing a Steepest curve

$$y' = F_y(x,y) / F_x(x,y)$$

The Steepest curve is the graph of the solution of this equation with $y = y_0$ when $x = x_0$.

Example 2 Determining Steepest curves.

Problem Determine the Steepest curves for the same function F(x,y) considered in the previous example:

$$F(x, y) = 1500 + 1000x^2 \exp(-x^2 - 4y^2)$$

What is the path taken by a river flowing through the point $(x_0, y_0) = (2, -1)$?

Solution First, determine the gradient vector for F:

$$\nabla F(x, y) = (\ 10^3[2x - 2x^3]\exp(-x^2 - 4y^2)\ ,\ -8\cdot 10^3 yx^2 \exp(-x^2 - 4y^2))$$

The differential equation for the Steepest curve is found as the solution of $y' = F_y/F_x$, which is not defined for x = 0, 1, or -1 because of division by zero. Thus, we need to solve

$$y' = [-8\cdot 10^3 yx^2 \exp(-x^2 - 4y^2)] / [10^3[2x - 2x^3]\exp(-x^2 - 4y^2)]$$

which, upon simplifying, is just

$$y' = 4yx/[x^2 - 1]$$

This equation is solved by separating the variables and integrating. Writing $y' = dy/dx$ and treating dx as a differential that can be transferred to the right side of the equation:

$$dy/4y = x/[x^2 - 1]\ dx, \quad \text{for } y \neq 0$$

Introducing an integration of both sides of this equation gives

$$\int 1/4y\ dy = \int x/[x^2 - 1]\ dx$$

$$(1/4)\ln|y| = (1/2)\ln|x^2 - 1| + K$$

where K is a constant of integration. Solving explicitly for y gives

$$y = \pm C\ (x^2 - 1)^2$$

where $C = e^{4K}$ is a modified integration constant, which would always be positive. Notice that the integral of 1/y is given as the natural log of the absolute value of y. This is necessary to represent all the solutions and leads to the ± sign in front of the constant C; in particular the negative sign must be used to represent the curve through (2,-1). To determine C for the curve passing through (2,-1) we set x = 2 and y = -1: solving

$$-1 = -C\ (2^2 - 1)^2 \quad \text{gives } C = 1/9$$

Thus the Steepest curve through (2,-1), which is the path a river flowing through this point would follow, is

$$y = -(1/9)(x^2 - 1)^2.$$

The x-axis is also a Steepest curve, y = 0, given by the differential equation $y' = 0$. The family of Steepest curves is symmetric about both axes. The curves with the plus sign are above the x-axis in the left panel of the following figure, while the curves with the negative sign are below the x-axis. The sketch indicates eight Steepest curves, taking the plus and minus signs for C = 0.01, 1/9 (this passes through the point (2,-1)), 1, and 3. The C value is the y-intercept of each curve. The directions of increasing function values are indicated by the arrows. Observe that the Steepest curves all approach one of the critical points (1,0) or (0,1). It follows that these points must be where the function z = F(x, y) has maximum values.

(This topic is discussed in Chapter 12, Section 12.2.)

There are three x-values at which these Steepest curves are not defined because the partial derivative $F_x = 0$ and hence the derivative $dy/dx = F_y/F_x$ could not be utilized. These are $x = 0, 1$, and -1. When $x = 0$, the function $F(0, y) = 1500$, a constant and hence this is actually a Level curve of the function. When $x = \pm 1$ the function $F(x, y)$ decreases as $|y|$ increases, e.g.,

$$F(1, y) = 1500 + \exp(1 - 4y^2)$$

Hence, these vertical lines are Steepest curves with the direction of increasing function values pointing towards the critical points $(1,0)$ and $(-1,0)$. These are indicated in the right panel of the graph along with the four Level curves for F.

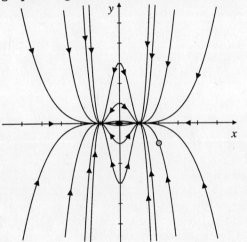 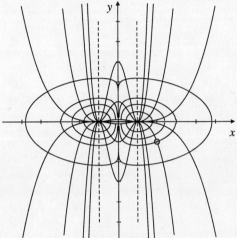

The curves at the left are the steepest curves for $z = 1500 + x \exp(x^2 - 4y^2)$. The illustrated curves are the graphs of $y = \pm C(x^2 - 1)^2$ for four values of the constant C, from upper left to right, 0.01, 0.1111, 1.0 and 3.0. At the right are the level curves, as sketched in the previous graph.

☑

Example 3 Curves orthogonal to an ellipse.

Problem The family of curves having the equations $(x/4)^2 + (y/9)^2 = c^2$ for different values of the constant c are ellipses. There is a corresponding family of curves that have the property that they are orthogonal to these ellipses. This is illustrated in the next figure. When one of these orthogonal curves intersects an ellipse, its tangent is perpendicular to the ellipse's tangent.

The problem is to determine the equations of this family of orthogonal curves.

Solution The orthogonal curves can be determined as a family of Steepest curves. To do this we first determine a function F such that the ellipses are Level curves of F. Then the family of orthogonal curves are the corresponding Steepest curves for the function F. Clearly, the family of ellipses are the Level curves of the function

$$F(x, y) = (x/4)^2 + (y/9)^2$$

To compute the Steepest curves for this function we first determine the gradient:

$$\nabla F(x, y) = (x/2, (2/9)y)$$

The Steepest curves are the integral curves specified (for $x \neq 0$) by

$$y' = F_y/F_x = 4y/9x, \quad \text{for } x \neq 0$$

This is a separable equation whose solution, when $y \neq 0$, is obtained by separating the variables and integrating:

$$9/y \, dy = 4/x \, dx$$

$$\int 9/y \, dy = \int 4/x \, dx$$

$9 \ln|y| = K 4 \ln |x|$ + integration constant.

We solve for y by dividing by 9, taking the exponential of both sides and introducing a positive constant $k = e^{K/9}$. This gives

$$y = \pm kx^{4/9}, \quad x \neq 0$$

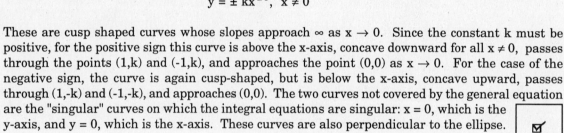

Ellipses
$(x/4)^2 + (y/9)^2 = C^2$
$C = 1, 1/2, 1/3,$ and $1/4$

Orthogonal Curves
$y = +/- kx^{4/9}$
$k = 3$
$k = 2$
$k = 1$
$k = 1$
$k = 2$
$k = 3$

These are cusp shaped curves whose slopes approach ∞ as $x \to 0$. Since the constant k must be positive, for the positive sign this curve is above the x-axis, concave downward for all $x \neq 0$, passes through the points (1,k) and (-1,k), and approaches the point (0,0) as $x \to 0$. For the case of the negative sign, the curve is again cusp-shaped, but is below the x-axis, concave upward, passes through (1,-k) and (-1,-k), and approaches (0,0). The two curves not covered by the general equation are the "singular" curves on which the integral equations are singular: $x = 0$, which is the y-axis, and $y = 0$, which is the x-axis. These curves are also perpendicular to the ellipse.

Exercise Set 11.4

1. For the given functions: (i) Determine the equations of the Level curves and the Steepest curves. (ii) Sketch three Level curves and three Steepest curves on the same x-y graph.

   a) $z = F(x, y) = -1 + 3(x - 1) - 2(y - 2)$
   b) $z = F(x, y) = -1 + 2x^2 - 2(y - 2)$
   c) $z = F(x, y) = xy$
   d) $z = F(x, y) = x/y$
   e) $z = F(x, y) = x^2 + 2y$
   f) $z = F(x, y) = x^2 y^3$

2. For the given function: (i) Determine the equations of the Level curves and the Steepest curves. (ii) Is there a Level curve and the Steepest curve passing through the origin? (iii) Sketch the curves of part (ii) on the same graph.

   a) $z = F(x, y) = x \cdot e^y$
   b) $z = F(x, y) = e^{x-y}$
   c) $z = F(x, y) = xy \cdot e^y$
   d) $z = F(x, y) = x \cdot e^{2x-y}$
   e) $z = F(x, y) = \exp(x^2 + y^2)$
   f) $z = F(x, y) = \exp(x^2 - y^2)$
   g) $z = F(x, y) = x \cdot \exp(x^2 + y^2)$
   h) $z = F(x, y) = y^2 \exp(x^2 - 3y^2)$
   i) $z = F(x, y) = x + \cos(y)$
   j) $z = F(x, y) = \cos(xy)$
   k) $z = F(x, y) = x \cos(y)$
   l) $z = F(x, y) = xy \cos(xy) - \sin(xy)$

3. For each function in Exercises 1 and 2, determine the equations of the Level Curve and the Steepest curve passing through the point (3,1).

4. You have been assigned to plan the route for a commuter "light rail" train track between two towns. Assume the elevations are given by the surface $z = F(x,y)$, where the units of x, y and z are kilometers, and that the towns are located at the points $\underline{A} = (2,1)$ and $\underline{B} = (0,0)$. You are constrained that the "grade" never exceeds 6%, the maximum grade that trains can ascend or descend safely.

   i) What factors would affect the path you plan? (Ignore environmental aspects and focus on the Level curves for the surface.)

   ii) How would you maximize the amount of level track while minimizing the total amount of track on the route?

   iii) If the towns are located on a flat plane, say $F(x,y) = [2x - 3y]10^{-2}$, determine a possible route for the track. Sketch the route on an x-y plane and indicate the equations of the curves.

   iv) a) If the terrain is more variable, given by $F(x,y) = 0.5 + 0.1x^2$ what are the Level curves?
   b) In this case the town B is located at the bottom of a valley that extends in the direction of the y-axis and the town A is 400m higher than B. What are the curves of steepest ascent and descent through the two towns?
   c) At any point (x,y) what is the unit vector $\underline{u}(x,y)$ that corresponds to a grade of $S = 0.06$?
   d) Establish the differential equation for the "grade" curves $y = f(x)$ that have constant slope $S = 0.06$.
   e) Although you can not solve the differential equation, you can find bounds for the length of the track required as follows.

   A lower bound is obtained by assuming the track follows the line L_1: $y = [u_2(0,0)/u_1(0,0)]x$ from the town B until $y = 1$ and then follow the Level curve to town A.

   An upper bound for the total length is obtained by assuming the tract follows the line L_2: $y = 1 + [u_2(2,1)/u_1(2,1)](x - 2)$ from the town A until $y = 0$ and then follows the Level curve to town B.

   Sketch these two extreme paths on a graph to visualize why their lengths provide bounds for the lengths of the solution.

Chapter 12 *Applications of Higher-Order Partial Derivatives:*

12.1 Taylor Polynomials In Two-Dimensions: Nonlinear Approximations of F(x,y) 653
   Polynomials in Two-Variables.
   Taylor Polynomials and Approximations.

12.2 Maximum and Minimum of Multivariable functions 662
   Maximum, minimum, and critical points.
   A Second Partial Derivative Test for Extreme Values.
   Optimization: Determining Maximum and Minimum Values of z = F(x, y).

12.3 Optimization with Constraints. 673
   The Method of Lagrange Multipliers.
   Why the Lagrange Method works.

Introduction.
Higher order partial derivatives, like higher order ordinary derivatives, arise by repeatedly taking partial derivatives. Higher order partial derivatives provide information about the non-linear characteristics of a function. In Section 12.1 higher order partial derivatives are used to form multivariable polynomials that can be used to approximate a multivariable function near a specific reference point. These polynomials have a special form and are called "Taylor Polynomials". They of course generalize the single variable Taylor Polynomials considered in Chapter 4, Section 4.9.

Optimization problems involving two independent variables usually consist of determining the maximum or minimum of a function $z = F(x, y)$. This is accomplished by employing a "Second Partial Derivative Test", which is introduced in Section 12.2. However, in applications involving multivariable functions there are often constraints that must be imposed on the independent variables, such as physical limitations or known relationships. To determine the extreme values of a function in this case is then a "constrained optimization problem". Such problems are solved by the method of "Lagrange Multipliers", which is presented in Section 12.3. Although this method is developed using directional derivatives, it is simple to apply without actually introducing directional derivatives.

Section 12.1 Taylor Polynomials In Two-Dimensions; Nonlinear approximations of F(x, y).

Polynomials in two variables.

Polynomials in two variables, say x and y, are linear combinations of terms of the form $x^n y^m$, where n and m are positive integers or zero. For example

$$P(x, y) = 1 + 2x^2 y - 3xy^4 \quad \text{or} \quad P(x, y) = x - y + xy$$

The "degree" or "order" of a term $x^n y^m$ is the sum of the exponents: $n + m$. The **order** of a polynomial is the highest order of its individual terms. Thus the polynomial on the left above has order 5 and the polynomial on the right has order 2.

An alternative form of a polynomial, referred to as a "shifted" form, involves powers of the differences $x - x_0$ and $y - y_0$, where (x_0, y_0) is a specific reference point. When "shifted" to a reference point (x_0, y_0), the general polynomial term is a number times a product of the form

$$(x - x_0)^n (y - y_0)^m \qquad \textbf{General Polynomial Term of degree n + m}$$

For instance,

$$P(x, y) = 1 - (x - 1)(y - 5)^2 \quad \text{and} \quad P(x, y) = (x - 7) - 3(x - 7)(y + 3) + (y + 3)^3$$

are third order polynomials in shifted form, with reference points (1,5) and (7,-3), respectively. By expanding the powers and gathering like terms, shifted polynomials can always be expressed in the standard form, involving only powers of x and y.

A general N-th order polynomial $P(x, y)$ is a linear combination of power terms involving x and y of order N or less. Thus $P(x, y)$ can be expressed as a sum, which is a finite series, of the form:

$$P(x, y) = \Sigma_{n = 0, N} \text{[Terms of order n]}$$

$$= \text{[a constant]} + \text{[1st order terms]} + \text{[2nd order terms]}$$

$$+ \text{[3rd order terms]} \quad \ldots \quad + \text{[N-th order terms]}$$

The general bracketed expression [n-th order terms] represents a linear combination of all the mixed x-y-polynomial terms of order n. If we are interested in shifted polynomials, these terms will be expanded about a point (x_0, y_0). The numerical coefficient of each term will be denoted by a subscripted parameter of the form $c_{i,j}$.

$c_{i,j}$ will denote the coefficient of the i-j-th polynomial term, $x^i y^j$ or $(x - x_0)^i (y - y_0)^j$, which is of order i+j.

For instance, one 5-th order term is $c_{3,2}(x - x_0)^3 (y - y_0)^2$. For each integer n there are exactly (n + 1) terms of order n, since this is the number of pairs of non-negative exponents, i and j, that add to n: these are the pairs

$$0,n \quad 1,n-1 \quad 2,n-2 \quad \ldots \quad n-1,1, \quad n,0$$

The expression [n-th order terms] thus refers to a sum of terms of the form

$$c_{i,j}(x - x_0)^i(y - y_0)^j$$

with $i + j = n$. These terms can be systematically listed and summed by varying the exponents of $(x - x_0)$ and $(y - y_0)$ in tandem. As the two exponents must add to n, $i + j = n$, we can always express the exponent j in terms of i as $j = n - i$. Thus, all such terms can be described by letting the index i range from zero to n, $i = 0, 1, 2, ..., n$. Thus the expression [n-th order terms] represents a sum of the form

$$[\text{n-th order terms}] = \Sigma_{i = 0, n} \, c_{i, n - i}(x - x_0)^i(y - y_0)^{n - i} \qquad \textbf{n-th Order Terms}$$

The general polynomial P of degree N can thus be expressed as a double summation of the form

$$P(x, y) = \Sigma_{n = 0, N}[\, \Sigma_{i = 0, n} \, c_{i, n - i}(x - x_0)^i(y - y_0)^{n - i} \,]$$

N-th Order Polynomial shifted to (x_0, y_0)

The expanded form of this general shifted polynomial is

$P(x, y) = c_{0,0}$ \hfill $n = 0$

$\qquad + c_{1,0}(x - x_0) + c_{0,1}(y - y_0)$ \hfill $n = 1$

$\qquad + c_{2,0}(x - x_0)^2 + c_{1,1}(x - x_0)(y - y_0) + c_{0,2}(y - y_0)^2$ \hfill $n = 2$

$\qquad + c_{3,0}(x - x_0)^3 + c_{2,1}(x - x_0)^2(y - y_0) + c_{1,2}(x - x_0)(y - y_0)^2 + c_{0,3}(y - y_0)^3$ \hfill $n = 3$

....

$\qquad + c_{4,0}(x - x_0)^4 + ...$ \hfill until $n = N$

Taylor Polynomials and Approximations.

A "Taylor Polynomial" is a particular form of polynomial that represents a function $F(x, y)$ near a reference point (x_0, y_0). The form of a "Taylor Polynomial" is specified by defining the coefficients of shifted power terms using the values of partial derivatives of F evaluated at the reference point. The rational for defining the "Taylor Polynomial" is to make it match the function F as closely as possible at the reference point. To do this the Taylor Polynomial is constructed by specifying its coefficients $c_{i,j}$ so that it and its partial derivatives have exactly the same values as the function F and its derivatives at the reference point. Shifted polynomials are a natural choice because, as we shall see, they and their derivatives are easy to evaluate at the reference point. An added benefit in applications is to avoid numerical round-off errors. If (x, y) is near the reference point (x_0, y_0) then the terms $(x - x_0)$ and $(y - y_0)$ will be small. When these terms are less than one then their powers will all be even smaller.

The Taylor Polynomial P_N is referred to as **the Taylor expansion of F about (x_0, y_0)**. The **N-th Order Taylor Approximation** of F at (x_0, y_0) consist of representing $F(x, y)$ by the polynomial $P_N(x, y)$:

$$F(x, y) \approx P_N(x, y) \qquad \textbf{N-th Order Taylor Approximation}$$

The general form for Taylor Polynomials can be obtained by determining the coefficients $c_{i, n - i}$ of the general polynomial. This is done by equating the values of the same order partial derivatives of $F(x, y)$ and $P_N(x, y)$ at (x_0, y_0). This process results in a set of equations that give the $c_{i,j}$-coefficients. The algebra becomes simple since writing P_N in the shifted form makes the evaluation

Section 12.1 — Taylor Polynomials In Two-Dimensions

Definition of Taylor Polynomials.

> The **N-th Order Taylor Polynomial expansion of the function F about a point (x_0, y_0)** is the N-th degree polynomial P_N that has the same value as the function F and its partial derivatives of order less than or equal to N at (x_0, y_0). This requires that for each choice of the indices n and i, with $0 \leq i \leq n \leq N$,
>
> $$\partial^n/\partial^{n-i}x\, \partial^i y\, P_N(x_0, y_0) = \partial^n/\partial^{n-i}x\, \partial^i y\, F(x_0, y_0)$$

of P_N and all of its partial derivatives at (x_0, y_0) easy. Setting $(x, y) = (x_0, y_0)$ makes each term involving a product of $(x - x_0)$ or $(y - y_0)$ zero, and thus the only non-zero term is the constant term. The constant term in a partial derivative $\partial^n/\partial x^i \partial y^j\, P_N(x, y)$ is simply $i!\, j!\, c_{i,j}$. We shall establish the coefficients of the Taylor Polynomials for N = 1 and N = 2, and then simply state the general form of the coefficients for arbitrary N.

CASE N = 1 The first order Taylor Polynomial is a linear function of the form

$$P_1(x, y) = c_{0,0} + c_{1,0}(x - x_0) + c_{0,1}(y - y_0).$$

Matching the function $F(x, y)$ and the polynomial $P_1(x, y)$ at (x_0, y_0) we set $P_N(x_0, y_0) = F(x_0, y_0)$ which gives the first constant $c_{0,0}$:

$$P_1(x_0, y_0) = c_{0,0} = F(x_0, y_0).$$

Matching the first partial derivatives, we set

$$\partial/\partial x\, P_1(x_0, y_0) = \partial/\partial x\, F(x_0, y_0) \quad \text{or} \quad c_{1,0} = F_x(x_0, y_0).$$

and

$$\partial/\partial y\, P_1(x_0, y_0) = \partial/\partial y\, F(x_0, y_0) \quad \text{or} \quad c_{0,1} = F_y(x_0, y_0)$$

With these coefficients, we form the

First-order Taylor Polynomial

$$P_1(x, y) = F(x_0, y_0) + F_x(x_0, y_0)(x - x_0) + F_y(x_0, y_0)(y - y_0)$$

This is the same first order function introduced in Chapter 10, Section 10.2, that describes the tangent plane to the graph of $z = F(x, y)$ at the point $(x_0, y_0, F(x_0, y_0))$.

CASE N = 2 The second order Taylor Polynomial has the form

$$P_2(x, y) = c_{0,0} + c_{1,0}(x - x_0) + c_{0,1}(y - y_0) + c_{2,0}(x - x_0)^2 + c_{1,1}(x - x_0)(y - y_0) + c_{0,2}(y - y_0)^2$$

The coefficients $c_{0,0}$, $c_{1,0}$, and $c_{0,1}$ are the same as those derived above for the first order Taylor Polynomial. This is because the first partial derivatives of the second order terms still involve a term $(x - x_0)$ or $(y - y_0)$ and hence are zero when evaluated at (x_0, y_0). Matching the second partial derivatives results in only three equations since the mixed partial derivatives are equal:

$$\partial^2/\partial^2 x \, P_2(x_0, y_0) = \partial^2/\partial^2 x \, F(x_0, y_0) \quad \text{or} \quad 2 \cdot c_{2,0} = \partial^2/\partial^2 x \, F(x_0, y_0)$$

$$\partial^2/\partial^2 y \, P_2(x_0, y_0) = \partial^2/\partial^2 y \, F(x_0, y_0) \quad \text{or} \quad 2 \cdot c_{0,2} = \partial^2/\partial^2 y \, F(x_0, y_0)$$

$$\partial^2/\partial x \partial y \, P_2(x_0, y_0) = \partial^2/\partial x \partial y \, F(x_0, y_0) \quad \text{or} \quad c_{1,1} = \partial^2/\partial x \partial y \, F(x_0, y_0)$$

In the first two equations the coefficient 2 is introduced when the second-degree term is differentiated, e.g.,

$$\partial/\partial x \, P_2(x, y) = c_{1,0} + 2 \cdot c_{2,0}(x - x_0) + c_{1,1}(y - y_0)$$

and then

$$\partial^2/\partial^2 x \, P_2(x, y) = 2 \cdot c_{2,0}.$$

Solving for the second order c-coefficients, and using subscript notation for the partial derivatives of F gives:

The second-order Taylor Polynomial

$$P_2(x, y) = F(x_0, y_0) + F_x(x_0, y_0)(x - x_0) + F_y(x_0, y_0)(y - y_0)$$

$$+ (1/2)[F_{xx}(x_0, y_0)(x - x_0)^2 + 2 \cdot F_{xy}(x_0, y_0)(x - x_0)(y - y_0) + F_{yy}(x_0, y_0)(y - y_0)^2]$$

Notice that the factor 1/2 multiplies all the second order coefficients. This 1/2 term is present to cancel the 2 that is introduced when the $(x - x_0)^2$ or $(y - y_0)^2$ terms are differentiated twice. On the other hand, the multiple of 2 introduced to the mixed-product term is because there are two equivalent mixed second order partial derivatives, F_{xy} and F_{yx}, these are equal and thus we write 2(one of them) rather than $F_{xy} + F_{yx}$. In the formula we used F_{xy} but you this term could be F_{yx}.

Example 1 **Second order Taylor Polynomials.**

Problem   Determine the Taylor Polynomial $P_2(x, y)$ for the indicated function and reference point.

   1 $F(x, y) = y e^{5(x - 3)}$ at $(x_0, y_0) = (3, 1)$ 2 $F(x, y) = \sin(x/2 - y)$ at $(x_0, y_0) = (\pi, 0)$.

Solutions a) The first step is to compute the partial derivatives of F and evaluate them at the reference point:

$$F(x, y) = y e^{5(x - 3)} \qquad F(3,1) = 1$$

$$F_x(x, y) = 5y e^{5(x - 3)} \qquad F_x(3,1) = 5$$

$$F_y(x, y) = e^{5(x - 3)} \qquad F_y(3,1) = 1$$

$$F_{xx}(x, y) = 25y e^{5(x - 3)} \qquad F_{xx}(3,1) = 25$$

$$F_{xy}(x, y) = 5 e^{5(x - 3)} \qquad F_{xy}(3,1) = 5$$

$$F_{yy}(x, y) = 0 \qquad F_{yy}(3,1) = 0$$

Substituting the values for the partial derivatives into the formula for the second order Taylor Polynomial gives:

$$P_2(x,y) = 1 + 5(x-3) + 1(y-1) + (1/2)[25(x-3)^2 + 2\times 5(x-3)(y-1) + 0(y-1)^2]$$

or just

$$P_2(x,y) = 1 + 5(x-3) + (y-1) + (25/2)(x-3)^2 + 5(x-3)(y-1).$$

Normally, such polynomials are left in shifted form and are not expanded and simplified.

b) The same procedure used for problem a) is repeated in this case. The partial derivatives are:

$$F(x,y) = \sin(x/2 - y) \qquad F(\pi,0) = 1$$

$$F_x(x,y) = (1/2)\cos(x/2 - y) \qquad F_x(\pi,0) = 0$$

$$F_y(x,y) = -\cos(x/2 - y) \qquad F_y(\pi,0) = 0$$

$$F_{xx}(x,y) = -(1/4)\sin(x/2 - y) \qquad F_{xx}(\pi,0) = -1/4$$

$$F_{xy}(x,y) = (1/2)\sin(x/2 - y) \qquad F_{xy}(\pi,0) = 1/2$$

$$F_{yy}(x,y) = -\sin(x/2 - y) \qquad F_{yy}(\pi,0) = -1$$

Consequently, as both linear coefficients are zero, the second order Taylor Polynomial expansion of $\sin(x/2 - y)$ about $(\pi,0)$ is:

$$P_2(x,y) = 1 + (1/2)[-1/4\,(x-\pi)^2 + 2(1/2)(x-\pi)(y-0) - (y-0)^2]$$

or

$$P_2(x,y) = 1 + -1/8\,(x-\pi)^2 + 1/4\,(x-\pi)y - 1/2\,y^2$$

☑

When the reference point is the origin, the expansion is about $(x_0, y_0) = (0,0)$, and the Taylor expansion of F is referred to as a "**Maclaurin** expansion". The "**Maclaurin**" polynomials are special cases of Taylor Polynomials and have a simpler form since the differences $(x - x_0)$ and $(y - y_0)$ are in this case simply x and y.

The first and second order Maclaurin Polynomials have the same form as the Taylor Polynomials with the simplified powers of x and y:

First Order Maclaurin Expansion: $P_1(x,y) = F(0,0) + F_x(0,0)\,x + F_y(0,0)\,y$

Second Order Maclaurin Expansion:

$$P_2(x,y) = F(0,0) + F_x(0,0)\,x + F_y(0,0)\,y + 0.5\{F_{xx}(0,0)\,x^2 + 2F_{xy}(0,0)\,xy + F_{yy}(0,0)\,y^2\}$$

GENERAL CASE, ARBITRARY ORDER N. The number of terms in a Taylor Polynomial increases as the order increases. The general N-th order Taylor Polynomial has $[N^2 + 3N]/2$ terms. Since there are $(i + 1)$-terms of order i, this is the sum of the number of terms of order i for $i = 0$ to $i = N$:

$$\Sigma_{i = 0,N} (i + 1) = N[N + 1]/2 + N = [N^2 + 3N]/2 \text{ terms}$$

Applying this formula, we see that the general 5-th degree Taylor Polynomial $P_5(x, y)$ would have 20 terms, while $P_{10}(x, y)$ would have 75 terms. However, as we shall see in the following examples, many of these terms may not actually occur as their coefficients may be zero, because the corresponding partial derivatives are zero at (x_0, y_0). Continuing the process used for the cases N = 1 and N = 2, it can be shown that the general term of a Taylor Polynomial has the form

$$c_{i,n-i}(x - x_0)^i(y - y_0)^{n-i}, \text{ for } 0 \leq i \leq n \leq N.$$

The form of the polynomial is thus

$$P_N(x, y) = \Sigma_{n = 0,N} \Sigma_{i = 0,n} c_{i,n-i}(x - x_0)^i(y - y_0)^{n-i}$$

The coefficient $c_{i,n-i}$ is determined by equating the n-th order partial derivatives:

$$\partial^n/\partial x^i \partial y^{n-i} P_N(x_0, y_0) = \partial^n/\partial x^i \partial y^{n-i} F(x_0, y_0).$$

Because of the shifted form of the polynomial, all terms with powers of $(x - x_0)$ greater than i or powers of $(y - y_0)$ greater than $(n - i)$ still retain powers of these factors after differentiating. Hence, they have value zero when (x_0, y_0) is substituted for (x, y). Furthermore, all terms with powers less than these values are reduced to zero by the differentiating. The net result is that the only term remaining in the derivative of P_N is the derivative of the term with x-exponent i and y-exponent $(n - i)$, and this derivative is the constant, $c_{i,n-i}\, i!\,(n-i)!$. Thus $c_{i,n-i}$ is found by solving

$$\partial^n/\partial x^i \partial y^{n-i} F(x_0, y_0) = \partial^n/\partial x^i \partial y^{n-i} c_{i,n-i}(x - x_0)^i(y - y_0)^{n-i} = c_{i,n-i} \cdot i! \cdot (n-i)!$$

Thus equating

$$c_{i,n-i} \cdot i! \cdot (n-i)! = \partial^n/\partial^{n-i}x \partial^i y\, F(x_0, y_0)$$

we solve for

$$c_{i,n-i} = \partial^n/\partial^{n-i}x \partial^i y\, F(x_0, y_0) / [i! \cdot (n-i)!]$$

The formula usually given in reference books for this coefficient involves a rearrangement of the factorial terms. The standard formula is

$$c_{i,n-i} = \partial^n/\partial^{n-i}x \partial^i y\, F(x_0, y_0) \cdot C(n,i) / n! \quad \textbf{General Taylor Coefficient}.$$

where the capital C function is defined by

$$C(n,i) = n! / [(n-i)!\, i!]$$

represents the number of combinations, or ways, i-items can be chosen from a group of n-items. $C(n,i)$ is read as "the (number of) combinations of n things taken i at a time". This happens to be exactly the number of equivalent partial derivatives of the same total order. For instance, there are $C(5,3) = 5!/2!3! = 120/(2 \times 6) = 10$ different 5-order partial derivatives that involve taking the derivative with respect to x 3-times and with respect to y 2-times, e.g., F_{xyxyx} and F_{yxxxy} are two of the ten possible derivatives.

N-th-Order Taylor Polynomial

$$P_N(x, y) = \sum_{n=0,N} \left[\sum_{i=0,n} \partial^n/\partial^{n-i}x \partial^i y\, F(x_0, y_0) \cdot [C(n,i)/n!]\, (x - x_0)^i (y - y_0)^{n-i} \right]$$

Example 2 **Higher-order Taylor Polynomials.**

Problem Determine the 6-th order Taylor Polynomial expansion $P_6(x, y)$ for $F(x, y) = y \sin(\pi x)$ at $(x_0, y_0) = (0.5, 3)$.

Solutions The first step is to compute the partial derivatives of F and evaluate them at (0.5,3). As $F_y(x, y) = \sin(\pi x)$ is a function of x only, all higher partial derivatives with respect to y are zero. In particular all of the mixed partials involving two or more derivatives with respect to y are zero. We therefore only need to evaluate the partials with respect to x and the mixed partials with only one derivative with respect to y.

$$F_x(x, y) = \pi y \cos(\pi x) \qquad F_{xy}(x, y) = \pi \cos(\pi x)$$

$$F_{xx}(x, y) = -\pi^2 y \sin(\pi x) \qquad F_{xxy}(x, y) = -\pi^2 \sin(\pi x)$$

$$F_{xxx}(x, y) = -\pi^3 y \cos(\pi x) \qquad F_{xxxy}(x, y) = -\pi^3 \cos(\pi x)$$

$$F_{xxxx}(x, y) = \pi^4 y \sin(\pi x) \qquad F_{xxxxy}(x, y) = \pi^4 \sin(\pi x)$$

$$F_{xxxxx}(x, y) = \pi^5 y \cos(\pi x) \qquad F_{xxxxxy}(x, y) = \pi^5 \cos(\pi x)$$

$$F_{xxxxxx}(x, y) = -\pi^6 y \sin(\pi x)$$

The partials are easily evaluated at $x_0 = 0.5$ and $y_0 = 3$ as $\sin(0.5\pi) = 1$ and $\cos(0.5\pi) = 0$.

$$F(0.5, 3) = 3 \qquad F_y(0.5, 3) = 1$$

$$F_x(0.5, 3) = 0 \qquad F_{xy}(0.5, 3) = 0$$

$$F_{xx}(0.5, 3) = -3\pi^2 \qquad F_{xxy}(0.5, 3) = -\pi^2$$

$$F_{xxx}(0.5, 3) = 0 \qquad F_{xxxy}(0.5, 3) = 0$$

$$F_{xxxx}(0.5, 3) = 3\pi^4 \qquad F_{xxxxy}(0.5, 3) = \pi^4$$

$$F_{xxxxx}(0.5, 3) = 0 \qquad F_{xxxxxy}(0.5, 3) = 0$$

$$F_{xxxxxx}(0.5, 3) = -3\pi^6$$

These are used to evaluate the Taylor coefficients $c_{0,0}$, $c_{0,1}$, and for n = 1, 2, ...,6, $c_{n,0}$ and $c_{n-1,1}$. We need to know the corresponding number of combinations, which are given by

$$C(n,0) = n!/[n!\cdot 0!] = 1 \quad \text{and} \quad C(n-1, 1) = n!/[(n-1)!\cdot 1!] = n.$$

(Remember 0! = 1 by definition.) Hence, the only non-zero Taylor coefficients are

$$c_{0,0} = 1\cdot 3/0! = 3 \qquad c_{0,1} = 1\cdot 1/1! = 1 \qquad c_{2,0} = -3\pi^2\cdot 1/2! = -(3/2)\pi^2 \qquad c_{2,1} = -\pi^2\cdot 3/3! = -\pi^2/2$$

$$c_{4,0} = 3\pi^4\cdot 1/4! = \pi^4/8 \qquad c_{4,1} = \pi^4\cdot 5/5! = \pi^4/24 \qquad c_{6,0} = -3\pi^6\cdot 1/6! = -\pi^6/240$$

The 6-th order Taylor expansion of $y \sin(\pi x)$ about $(0.5, 3)$ is thus a polynomial that is the sum (or difference) of 7 power-terms:

$$P_6(x, y) = 3 + (y - 3) - (3/2)\pi^2(x - 0.5)^2 - (1/2)\pi^2(x - 0.5)^2(y - 3) + (1/8)\pi^4(x - 0.5)^4$$
$$+ (1/24)\pi^4(x - 0.5)^4(y - 3) - (1/240)\pi^6(x - 0.5)^6.$$

Observe that because of the division by n!, where n is the order of the term, the coefficients remain relatively small.
For instance the last coefficient is $c_{6,0} = -\pi^6/240 \approx -4.006$.

The work involved in computing higher order Taylor Polynomials increases exponentially as the order increases. But, the accuracy of using an approximation $F(x, y) \approx P_N(x, y)$ also increases as the order increases. This is seen in the following example.

Example 3 **Approximating numerical values with P_N.**

Problem In the previous example the 6-th order Taylor polynomial was constructed for the function $F(x, y) = y \sin(\pi x)$ at the point $(0.5, 3)$. Using the coefficients generated in that example, compare the error in the approximation of $F(0.6, 2.9) = 2.9 \sin(0.6\pi)$ using the Taylor polynomials of order N = 0, 1, 2, ..., 6.

Solution The lower-order Taylor polynomials are simply the 6-th order polynomial given above, ignoring the higher-order terms. The lower-order Taylor Polynomials for $y \sin(\pi x)$ are:

$P_0(x, y) = 3$ has only the constant term;

$P_1(x, y) = 3 + (y - 3)$ adds the 1st order terms;

$P_2(x, y) = 3 + (y - 3) - (3/2)\pi^2(x - 0.5)^2$ includes the 2nd order terms;

$P_3(x, y) = 3 + (y - 3) - (3/2)\pi^2(x - 0.5)^2 - (1/2)\pi^2(x - 0.5)^2(y - 3)$
adds the one non-zero 3rd order term;

$P_4(x, y) = 3 + (y - 3) - (3/2)\pi^2(x - 0.5)^2 - (1/2)\pi^2(x - 0.5)^2(y - 3) + (1/8)\pi^4(x - 0.5)^4$ etc.;

$P_5(x, y) = 3 + (y - 3) - (3/2)\pi^2(x - 0.5)^2 - (1/2)\pi^2(x - 0.5)^2(y - 3) + (1/8)\pi^4(x - 0.5)^4$
$+ (1/24)\pi^4(x - 0.5)^4(y - 3)$

The value of each of these polynomials and P_6, evaluated at the point $(0.6, 2.9)$ using a pocket calculator as indicated in teh Table .
The resulting absolute error

$E = |2.9 \sin(0.6\pi) - P_N(0.6, 2.9)|$

is calculated using the decimal value

$F(0.6, 2.9) = 2.9 \sin(0.6\pi) = 2.7580638$

It is clear that the accuracy of this approximation improves dramatically as the degree of the Taylor Polynomial increases.

N	$P_N(0.6, 2.9)$	|Error|
0	3	0.2419362
1	2.9	0.1419362
2	2.751955933	0.0061079
3	2.756890735	0.0011731
4	2.758108349	0.0000445
5	2.758067762	0.0000039
6	2.758063756	0.0000001

Exercise Set 12.1

1. Determine the first and second order Taylor Polynomials for the given function at the indicated reference point.

   a) $F(x,y) = 3x + y^2$ at $(x_0, y_0) = (0,0)$
   b) $F(x,y) = x^3 + xy$ at $(x_0, y_0) = (0,0)$
   c) $F(x,y) = 3x + y^2$ at $(x_0, y_0) = (1,2)$
   d) $F(x,y) = xy$ at $(x_0, y_0) = (1,-1)$
   e) $F(x,y) = 2xy^3$ at $(x_0, y_0) = (2,-1)$
   f) $F(x,y) = 1 + xy - x^2y^2$ at $(x_0, y_0) = (0,2)$
   g) $F(x,y) = y\ln(x)$ at $(x_0, y_0) = (1,1)$
   h) $F(x,y) = e^{x-y}$ at $(x_0, y_0) = (0,0)$
   i) $F(x,y) = y^x$ at $(x_0, y_0) = (1,2)$
   j) $F(x,y) = e^x - xy$ at $(x_0, y_0) = (0,1)$
   k) $F(x,y) = \cos(3x + y^2)$ at $(x_0, y_0) = (0,0)$
   l) $F(x,y) = \sin(x) + xy$ at $(x_0, y_0) = (0,0)$

2. Determine the second order Taylor approximation of $F(x,y)$ about the point (x_0, y_0). Use a calculator to find the error in this approximation at the point (x_1, y_1): $E = |P_2(x_1, y_1) - F(x_1, y_1)|$.

   a) $F(x,y) = 3x^2 - 2xy + y^2$; $(x_0, y_0) = (1,2)$, $(x_1, y_1) = (1.1, 1.9)$
   b) $F(x,y) = \sin(\pi xy)$; $(x_0, y_0) = (1, 0.5)$, $(x_1, y_1) = (1.05, 0.53)$
   c) $F(x,y) = \exp{-(x^2 + y^2)}$; $(x_0, y_0) = (0,0)$ $(x_1, y_1) = (0.2, 0.1)$
   d) $F(x,y) = x\ln(y)$; $(x_0, y_0) = (2,1)$. $(x_1, y_1) = (2.1, 0.8)$

3. Determine the Taylor coefficient $c_{3,2}$ of the power term $(x-2)^3(y-5)^2$ in the Taylor expansion of $F(x,y)$ about the point $(2,5)$.

   a) $F(x,y) = x^8 - y^8$
   b) $F(x,y) = x^4 e^{-y}$
   c) $F(x,y) = x^5 y$
   d) $F(x,y) = x^4 y^3$
   e) $F(x,y) = x^{4/3} y^3$
   f) $F(x,y) = \ln(x-1)/y$

4. Determine the 3rd order Taylor Polynomial expansion of $F(x,y)$ about the point $(1,2)$.

   a) $F(x,y) = x^2(y-2)$
   b) $F(x,y) = y/x$
   c) $F(x,y) = x^2 e^{(y-2)}$
   d) $F(x,y) = y \sin(\pi x)$
   e) $F(x,y) = \cos(\pi y)\sin(\pi x)$
   f) $F(x,y) = e^{xy}$

5. For each function in Exercise 4 compare the error in the approximation $F(x,y) \approx P_N(x,y)$ for $N = 0, 1, 2$, and 3 at the indicated point.

   a) $(1.1, 2.2)$
   b) $(1.5, 2.1)$
   c) $(2, 3)$

6. Show that the 6-th order Taylor Polynomial expansion of $F(x,y) = \sin(x)\cdot\cos(y)$ at $(x_0, y_0) = (0,0)$ is

$$P_6(x,y) = x^5/120 + x^3 y^2/12 - x^3/6 + xy^4/24 - xy^2/2 + x$$

Section 12.2 Maximum and Minimum of Multivariable Functions.

Maximum, minimum, and critical points.

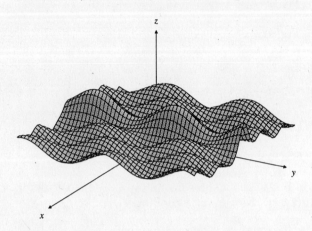

Most multivariable functions have more complex local behavior than single variable functions. If you visualize a surface $z = F(x, y)$, as illustrated in the figure at the left, as an old, relatively smooth, mountain range, you will see that the critical points that characterize the terrain or surface are the local peaks, basins and passes between adjacent hills. The principal problem of this subsection is to determine the local extrema, the maximum and minimum values of a continuous function $z = F(x, y)$. The domain points at which extrema occur are called maximum or minimum points.

DEFINITION OF LOCAL MAXIMUM AND MINIMUM.

Maximum. A function $z = F(x,y)$ has a **local maximum at a point (x_0, y_0)** if, for all points (x,y) within a sufficiently small circle centered at (x_0, y_0),

$$F(x,y) \leq F(x_0, y_0).$$

In this case, **the value $z_0 = F(x_0, y_0)$ is a local maximum value of $F(x,y)$.**

Minimum. A function $z = F(x,y)$ has a **local minimum at a point (x_0, y_0)** if, for all points (x,y) within a sufficiently small circle centered at (x_0, y_0),

$$F(x,y) \geq F(x_0, y_0).$$

In this case, **the value $z_0 = F(x_0, y_0)$ is a local minimum value of $F(x,y)$.**

The procedure for determining extrema of a function $F(x, y)$ is to first identify the critical domain points at which an extrema might occur. Then, a test is applied to these points to decide whether or not they correspond to local maximum or minimum values. A key observation that can be applied when the surface $z = F(x, y)$ is smooth enough to have tangent planes at each point, is:

If $z = F(x, y)$ has an extrema at a point (x_0, y_0) then the tangent plane at (x_0, y_0, z_0) will be horizontal and hence both partial derivatives will be zero at this point:

$$F_x(x_0, y_0) = 0 \quad \text{and} \quad F_y(x_0, y_0) = 0.$$

This follows intuitively by holding one of the variables constant and considering F as a function of the other variable. A local extreme value $z_0 = F(x_0, y_0)$ must also be a (local) extreme value of the

Section 12.2   MAXIMUM AND MINIMUM OF MULTIVARIABLE FUNCTIONS

one-dimensional functions $z = F(x, y_0)$ and $z = F(x_0, y)$, which are obtained by fixing one of the variables, $y = y_0$ and $x = x_0$, respectively. As the derivative, which in this case is actually a partial derivative, must be zero at extrema of one-dimensional differentiable functions, both first partial derivatives of F must vanish at (x_0, y_0). Therefore, the tangent plane must be the horizontal plane $z = z_0$. This implies that the first-order approximation of a function near a maximum or minimum point is a constant, and for this reason the function is said to be "stationary" at the point (x_0, y_0).

DEFINITION OF STATIONARY AND CRITICAL POINTS.

> A function $F(x,y)$ is **stationary at a point (x_0, y_0)** if
>
> $$\partial/\partial x \, F(x_0, y_0) = 0 \quad \text{and} \quad \partial/\partial y \, F(x_0, y_0) = 0.$$
>
> In this case, (x_0, y_0) is called a **stationary point**.
>
> The **critical points of a function $F(x,y)$ on a region D** are:
>
> i) all stationary points for F in D;
> ii) all points $(x,y) \in D$ for which either $F_x(x,y)$ or $F_y(x,y)$ does not exist;
> iii) points on the boundary of the region D.

If $F(x, y)$ is continuous then the only points at which extrema values can occur are critical points.

At the first type of critical point, a stationary point, the function's graph varies smoothly and has a horizontal tangent plane.

At the second type of critical point the function's graph changes shape abruptly, as at least one of its partial derivatives does not exist. Functions can have extreme values at such points.

The third type of critical point, a point on the boundary of the domain D, corresponds to an end point of an interval [a,b] in the single variable situation. Recall that we had to check the value of the function at both endpoints, a and b, when determining the "global" extrema of a function over the interval [a,b]. In two dimensions, a point is on the boundary of D if every circle centered at the point includes points outside of D. If a problem seeks only extreme values at points "interior" to a region, which is equivalent to a single variable problem on an open interval (a,b), it is not necessary to check for extrema on the boundary. If the region D is not specified we assume it is the natural domain of the function F.

Example 1   **Finding Critical Points**.

Problem   Determine the critical points of the function $F(x, y) = x^3 + 3xy - y^4$.

Solution   First, compute the partial derivatives:

$$F_x(x, y) = 3x^2 + 3y \quad \text{and} \quad F_y(x, y) = 3x - 4y^3.$$

Stationary points are identified by solving simultaneously the equations $F_x = 0$ and $F_y = 0$. $F_x = 0$ gives $y = -(x^2)$. Substituting x^2 for y in the equation $F_y = 0$ gives

$$F_y = 3x - 4x^6 = 0 \quad \text{or} \quad x(3 - 4x^5) = 0 \quad ; \text{hence } x = 0 \text{ or } x = (3/4)^{1/5}.$$

There are thus two stationary critical points:

$$(0,0) \quad \text{and} \quad ((3/4)^{1/5}, -(3/4)^{2/5}) \approx (0.944, -0.891).$$

Since F and both partial derivatives are defined for all (x, y) these are the only critical points.

☑

Example 2 Critical points that are not stationary points.

Problem Determine the critical points of the function $F(x, y) = x^{1/2}(y - 2)^{2/3}$.

Solution The natural domain of this function is $\{(x, y) \mid x \geq 0\}$. Hence all points on the y-axis are boundary points for the domain of F. The partial derivatives of F are

$$F_x(x, y) = 0.5 x^{-1/2}(y - 2)^{2/3} \quad \text{and} \quad F_y(x, y) = (2/3) x^{1/2}(y - 2)^{-1/3}.$$

On the y-axis the partial F_x is singular and not defined. The partial F_y is singular and is not defined on the horizontal line $y = 2$. Thus F may have a maximum or minimum value on each of these lines. Since $F(x, y) > 0$ if (x, y) is not on one of these lines and $F = 0$ on the lines we see that at each point on these lines the function assumes its minimum value.

☑

A Second Partial Derivative Test for Extreme Values.

For the remainder of this section we shall assume that the function F has continuous second order partial derivatives in a domain D. Then the only critical points interior to D will be stationary points. We will establish a multi-variable test for extrema that is a generalization of the "Second Derivative Test" which was given in Chapter 4, Section 4.3, for critical points of a single variable function.

Review It may prove helpful at this point to quickly review aspects of the single variable "Second Derivative Test". Recall that the second order characteristics of a function g(x) at a critical point are those of its quadratic approximation. The quadratic approximation of g(x) at x_0 is provided by the **second order Taylor Polynomial expanded about x_0**:

$$g(x) \approx p_2(x) = g(x_0) + g'(x_0)(x - x_0) + [g''(x_0)/2](x - x_0)^2.$$

At a stationary point x_0 the derivative $g'(x_0) = 0$ and thus the polynomial $p_2(x)$ is the equation of a shifted parabola with its apex at $x = x_0$:

$$g(x) \approx p_2(x) = g(x_0) + [g''(x_0)/2](x - x_0)^2$$

If $g''(x_0) > 0$, the parabola p_2 opens upward and thus for x near x_0,

$$g(x) \approx p_2(x) \geq g(x_0), \quad \text{i.e.,} \quad g(x_0) \text{ is a local minimum value.}$$

If $g''(x_0) < 0$, the parabola opens downward and $g(x_0)$ is a maximum value.

If $g''(x_0) = 0$, the quadratic degenerates to a constant $p_2(x) = g(x_0)$ and no conclusion is drawn because higher order terms are necessary to determine the true nature of the function. (For instance, $g(x) = x^n$, where $n > 2$ at $x_0 = 0$.)

A similar situation occurs in the multivariable situation. At a stationary critical point the second order Taylor Polynomial $z = P_2(x, y)$ provides a "quadratic" approximation of $F(x, y)$. The graph of P_2 is a second order "tangent surface" to the graph $z = F(x, y)$ at the point $(x_0, y_0, F(x_0, y_0))$. Consequently, the character of the function $F(x, y)$ when (x, y) is near the point (x_0, y_0) is similar to that of its Tangent Quadratic P_2. In particular, it can be shown that:

If the Taylor Polynomial P_2 for $F(x, y)$ has a maximum or a minimum at a stationary point (x_0, y_0) then the function F will also have a (local) maximum or minimum, respectively, at this point.

When (x_0, y_0) is a stationary point of F, the first order terms in the Taylor expansion of F vanish, leaving only the constant and quadratic terms. To simplify matters, we introduce letters to represent the three second order partial derivatives. Set:

$$A = F_{xx}(x_0, y_0) \qquad B = F_{xy}(x_0, y_0) \qquad C = F_{yy}(x_0, y_0)$$

Then, the quadratic approximation of $F(x, y)$ is given by its Taylor Polynomial:

$$F(x, y) \approx P_2(x, y) = F(x_0, y_0) + (1/2)[A(x - x_0)^2 + 2B(x - x_0)(y - y_0) + C(y - y_0)^2]$$

To establish a "Second Partial Derivative Test" we need to know the nature of the basic quadratic surface

$$z = z_0 + A(x - x_0)^2 + 2B(x - x_0)(y - y_0) + C(y - y_0)^2$$

This is the shifted form of the basic quadratic centered at the origin, whose equation is:

$$z = Ax^2 + 2Bxy + Cy^2$$

For the moment let us consider the case when $B = 0$, as this simplifies the discussion. There are three basic types of quadratic surfaces, as illustrated in Figure 12.2. The equations of these surfaces are :

(a) $z = x^2 + y^2$;   a circular cone shaped surface opening upward. It has a **minimum** at the origin.

(b) $z = -x^2 - y^2$;   a circular cone shaped surface opening downward. It has a **maximum** at the origin.

(c) $z = x^2 - y^2$;   a "saddle" shaped surface that is like a parabola opening upward along the x-axis, when $y = 0$, and is like a parabola opening downward along the y-axis, when $x = 0$. The surface has **neither a maximum nor a minimum** at the origin.

The graphs of equations (a) and (b) are surfaces that are called **circular paraboloids**. The term "circular" refers to the fact that level curves of these surfaces are all circles. The "-oid" ending is added to the shape of the intersection of the surface with a plane perpendicular to the x-y plane, for instance the axis planes. Thus a paraboloid has a parabola shape when x is constant or when y is constant. The graphs of these three functions is indicated on the next page.

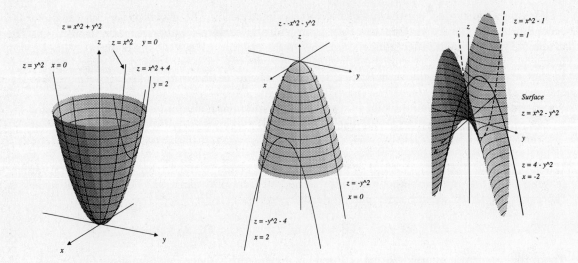

The graph of equation (c) is in the right panel. It is said to have a **saddle** shape and the point (0,0) is called a **saddle point.** This type of curve is a **hyperbolic paraboloid**, a graph that has hyperbola shaped level curves (when z is constant) and parabola shaped cut curves when x is a constant or y is a constant. For instance, the cut curves for the planes x = k, k a constant, are downward opening parabolas $z = k^2 - y^2$. However, for y held fixed, y = k, the cut curves are the upward opening parabolas $z = x^2 + k^2$. There is no ambiguity in this case, the origin **is nether a maximum nor a minimum** point on this surface.

When the coefficients A and C of the general quadratic function are not 1 or -1, as was the case in equations (a), (b) and (c) above, the corresponding surface still has a shape similar to the graph of one of these basic functions. For non-unit coefficients the graph's level curves then become ellipses in the first two cases and the surfaces are called **elliptic paraboloids**. These are not surfaces of revolution. In the third case the parabolic cut curves simply open wider or narrower; the fundamental shapes remain the same.

The three basic surfaces cited above contain no mixed xy terms as we assumed that B = 0. If B ≠ 0, the graph of a quadratic will be congruent to a surface that is the graph of a quadratic with no xy-term that is found geometrically by a rotation of the axis, or equivalently a rotation of the surface, to make the surface symmetric about axis planes. Without going into the details of this rotation, the critical step involves the algebraic sign of a term D, called the "discriminant" of the quadratic form, which is given by

$$D = B^2 - AC.$$

If D is not zero then the surface can be "rotated" into one of the above three basic forms. The character of the surface, in particular whether or not it has an extrema and the type of extrema, can be determined from the sign of D and the sign of the coefficients A and C.

If D is positive, the rotated surface has the shape of the surface (c) $z = x^2 - y^2$, for which D = 1, and there is no extreme value.

If D is negative, the rotated surface will have a shape similar to that of surfaces (a) and (b), for which D = -1, and there is an extreme value. To determine which type requires a further check.

If A or C is positive then a minimum occurs, as in surface (a).
If A or C is negative then a maximum occurs, as in surface (b).

If D = 0, then more elaborate analysis is needed because the surface is actually a higher order surface. This situation parallels the undetermined single-variable situation of the second derivative vanishing at a critical point.

Comparing the value of D, and the coefficients A or C to the values in the three basic surfaces discussed above is the basis of the following important test.

THE SECOND PARTIAL DERIVATIVE TEST.

Let (x_0, y_0) be a stationary point of a function $F(x,y)$ and set

$$A = F_{xx}(x_0, y_0) \qquad B = F_{xy}(x_0, y_0) \qquad C = F_{yy}(x_0, y_0)$$

and

$$D = B^2 - AC.$$

i) If **D > 0** the point (x_0, y_0) is a saddle point and the graph of $F(x,y)$ **does not have a maximum or a minimum at (x_0, y_0)**.

ii) If **D < 0** (x_0, y_0) is a local extreme point; then

   a) $F(x,y)$ has a (local) **maximum** at (x_0, y_0)
      if $A < 0$ or $C < 0$;

   b) $F(x,y)$ has a (local) **minimum** at (x_0, y_0)
      if $A > 0$ or $C > 0$.

iii) If **D = 0** **no conclusion is drawn**. The function $F(x,y)$ may or may not have an extreme value at (x_0, y_0).

NOTE   The Second Partial Derivative Test can only be applied to stationary points.

Example 3   **Applying the Second Partial Derivative Test.**

Problem   Determine if the function $F(x, y) = \sin(x) + y^2$ has a maximum or a minimum at the critical points (a) $(x_0, y_0) = (\pi/2, 0)$ and (b) $(x_0, y_0) = (3\pi/2, 0)$.

Solution   (a) First verify that the point $(x_0, y_0) = (\pi/2, 0)$ is a stationary point of the function. To do this calculate the first partial derivatives of $F(x, y)$ and evaluate these at (x_0, y_0).

$$F_x(x, y) = \cos(x) \qquad \text{and} \qquad F_y(x, y) = 2y$$

$$F_x(\pi/2, 0) = \cos(\pi/2) = 0 \qquad \text{and} \qquad F_y(\pi/2, 0) = 2 \cdot 0 = 0$$

Since the first derivatives vanish, the Second Partial Derivative Test can be applied. Compute the second partial derivatives and evaluate these at (x_0, y_0). Set

$$A = F_{xx}(\pi/2, 0) = -\sin(\pi/2) = -1 \qquad B = F_{xy}(\pi/2, 0) = 0 \qquad C = F_{yy}(\pi/2, 0) = 2.$$

Therefore, the discriminate

$$D = B^2 - AC = 2 > 0$$

By case i) of the Second Partial Derivative Test, the point $(\pi/2, 0)$ is a saddle point. The function **does not have a maximum or minimum at $(\pi/2, 0)$**.

(b) The point $(3\pi/2, 0)$ is also a stationary point for the function, since

$$F_x(3\pi/2, 0) = \cos(3\pi/2) = 0 \qquad \text{and} \qquad F_y(3\pi/2, 0) = 2 \cdot 0 = 0$$

Evaluating the second partial derivatives at (x_0, y_0) gives

$$A = -\sin(3\pi/2) = 1 \qquad B = 0 \qquad \text{and} \qquad C = 2$$

Therefore, $D = -2$ is negative and case ii) of the Test is applicable. The next step is to check the sign of either A or C. Note that they must have the same sign in this case so either will suffice. As $A > 0$, the Test concludes that the point $(3\pi/2, 0)$ is a local minimum point for the function. The local minimum value is $F(3\pi/2, 0) = -1$.

Optimization: Determining Maximum and minimum values of $z = F(x, y)$.

"Optimization" problems are concerned with determining the extrema of functions, usually functions used to model or describe quantities whose values are functions of controllable independent variables. The goal is to know from the model equations what choices for the input variables will give a desired optimal output value. This is the practical setting for traditional maximum or minimum problems. The first step in solving an "optimization" problem is to identify critical points of the specified function. If the function $F(x, y)$ is differentiable, the stationary points are found by simultaneously solving the equations $F_x(x, y) = 0$ and $F_y(x, y) = 0$. This can be difficult if these equations are non-linear. At these points the Second Partial Derivative Test is applied.

Non-stationary critical points, at which a partial derivative is not defined, must be analyzed separately. Also, if the natural domain, or the "acceptable region for (x, y) values" has a finite boundary, an analysis of the function's values on the boundary must be undertaken to see if the function attains an extrema on the boundary. In Section 12.3 this topic is explored as a "constrained optimization problem".

Example 4   **Determining the characteristics of a polynomial.**

Problem   Determine the character of the graph of $F(x, y) = x^2 y - 5x - 4y$.

Solution   As a polynomial, the values of F approach infinity as x and y become large. The function is positive for large $|x|$ and positive y and for small x and large -y. The character of the graph for finite values of x and y is determined by identifying its critical points and classifying its shape at these. The stationary critical points are found by calculating the first partials, and then solving simultaneously the equations

$$F_x(x, y) = 2xy - 5 = 0 \qquad \text{and} \qquad F_y(x, y) = x^2 - 4 = 0$$

The $F_y = 0$ equation only involves x so it is easily solved for $x = 2$ or $x = -2$. Substituting these into the $F_x = 0$ equation gives the corresponding y-values, $y = 5/2x$: $y = 5/4$ and $y = -5/4$. Since both partials are defined for all x and y the only critical points are the stationary points

Section 12.2   MAXIMUM AND MINIMUM OF MULTIVARIABLE FUNCTIONS

$$(x_0, y_0) = (2, 1.25) \quad \text{and} \quad (x_1, y_1) = (-2, -1.25)$$

The second partial derivatives of F are:

$$F_{xx}(x, y) = 2y \qquad F_{xy}(x, y) = 2x \qquad F_{yy}(x, y) = 0$$

At (2,1.25)   A = 2.5   B = 4   C = 0

At (-2,-1.25)   A = -2.5   B = -4   C = 0

Because $C = F_{yy}(x, y) = 0$, $D = B^2 - AC = B^2 > 0$ at both stationary points. Thus both points are saddle points.

☑

Example 5   Pricing competitive items to maximize profit.

Background   A retailer stocks two different brands of the same item. Because of advertising, more consumers tend select brand A over brand B when they are priced the same. However, if there is a differential in price this trend changes. The retailer actually has different costs and hence profit for the two brands. Let x be the selling price of item A and x_c its unit cost. Similarly let y and y_c be the selling price and unit cost of item B. The retailer wishes to price the items to maximize the profit function

$$P(x, y) = (x - x_c) \cdot SA(x, y) + (y - y_c) \cdot SB(x, y)$$

where the functions SA and SB indicate the average number of sales of products A and B per week. In a stable market, the total sales will remain constant, i.e., SA + SB = N will not change. However, this may not be the case if "sale" prices or price increases alter the total sales.

Problem   Determine the optimal pricing policy for two competing products, A and B, if the wholesale cost of A is $20 and of B is $30 and the expected number of sales, pricing A at $x and B at $y, are

$$SA(x, y) = 100 + 990(y - x) \quad \text{and} \quad SB(x, y) = 500 + 1210(x - y)$$

Solution   The profit function is

$$P(x, y) = (x - 20)(100 + 990(y - x)) + (y - 30)(500 + 1210(x - y))$$

which simplifies to

$$P(x, y) = -10[99x^2 + 20x(82 - 11y) + 121y^2 - 1700y + 1700]$$

The partial derivatives of the profit function are

$$P_x(x, y) = -20[99x - 82 + 11y] \quad \text{and} \quad P_y(x, y) = 20[110x - 121y + 850]$$

Solving $P_x = 0$ and $P_y = 0$ simultaneously, gives the critical point y = 50 and x = 520/11, which is approximately 47.27. That this corresponds to a maximum for P is determined by applying the Second Partial Derivative Test. The second partials of P are constants:

$$A = P_{xx}(x, y) = -1980, \quad B = P_{xy}(x, y) = -220, \quad C = P_{yy}(x, y) = -2420$$

Hence, the discriminate $D = (-220)^2 - (-1980)(-2420) = -4743200$ is negative. Since A is negative it follows that P has a maximum value at this point. As this is the only critical point it appears that the optimal selling price for product A is $47.27 and for product B is $50.00.

However, there is a problem. If we calculate the number of units sold at these prices, using the function $SB(47.27, 50) = 500 + 1210(47.27 - 50) = -2803.3$, a negative number. What has given an apparent maximum value for the profit does not correspond to a real situation. The model equation must be flawed.

☑

Example 6 Determining Local Extreme values.

Problem Determine the local extreme values of the function $F(x, y) = xye^{-0.005x^2 - 0.02y^2}$.

Solution First, to identify the critical points the partial derivatives of F must be equated to zero and solved simultaneously. To simplify this process we introduce a new variable that is equal to the exponential term in F. Let

$$U = e^{-0.005x^2 - 0.02y^2}$$

The partial derivatives of U are

$$\partial/\partial x\, U = -0.01xU \quad \text{and} \quad \partial/\partial y\, U = -0.04yU$$

Expressing the function F as the product

$$F(x, y) = xyU$$

the partial derivatives of F are obtained by the Product Rule:

$$\partial/\partial x\, F(x, y) = \partial/\partial x(xy) \cdot U + (xy) \cdot \partial/\partial x\, U$$

$$= yU - 0.01x^2 yU = y(1 - 0.01x^2)U$$

$$\partial/\partial y\, F(x, y) = \partial/\partial y(xy) \cdot U + (xy) \cdot \partial/\partial y\, U$$

$$= xU - 0.04xy^2 U = x(1 - 0.04y^2)U$$

Expressing the derivatives in a factored form makes it easier to identify the roots of the equations $F_x = 0$ and $F_y = 0$. As $U(x, y) \neq 0$,

$$F_x = 0 \implies y = 0 \text{ or } 1 - 0.01x^2 = 0$$

$$F_y = 0 \implies x = 0 \text{ or } 1 - 0.04y^2 = 0$$

Consequently, there are five stationary points given by the simultaneous solutions of these equations:

$$(0,0) \quad (10,5) \quad (10,-5) \quad (-10,5) \quad (-10,-5)$$

To apply the Second Partial Derivative Test to these points we need the second partial derivatives:

$$F_{xx}(x, y) = \partial/\partial x[y(1 - 0.01x^2)U]$$

$$= [y(1 - 0.01x^2)] \cdot \partial/\partial x U + yU \cdot \partial/\partial x (1 - 0.01x^2)$$

$$= y(1 - 0.01x^2) \cdot -0.01xU + yU(-0.02x)$$

$$= -0.01xy(3 - 0.01x^2)U$$

Similar differentiation and rearrangement gives:

$$F_{yy}(x, y) = -0.04xy(3 - 0.04y^2)U$$

$$F_{xy}(x, y) = (1 - 0.01x^2)(1 - 0.04y^2)U$$

Notice that at the stationary points other than $(0,0)$ the mixed partial F_{xy} is zero and the value of the term $U = e^{-1}$. The following table lists the values of the derivatives and the discriminate at the stationary points.

Stationary points	$A = F_{xx}$	$B = F_{xy}$	$C = F_{yy}$	$D = B^2 - AC$	Conclusion
$(0,0)$	0	1	-0	1	A saddle point. No maximum or minimum.
$(10,5)$ and $(-10,-5)$	$-e^{-1}$	0	$-4e^{-1}$	$-4e^{-2}$	A maximum occurs since $D < 0$ and $A < 0$.
$(10,-5)$ and $(-10,5)$	e^{-1}	0	$4e^{-1}$	$-4e^{-2}$	A minimum occurs since $D < 0$ and $A > 0$.

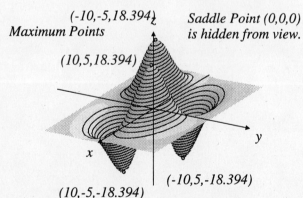

Surface $z = x*y*exp(-0.005x^2 - 0.02y^2)$
Level Curves at z = integer values.

$(-10,-5,18.394)$
Maximum Points
$(10,5,18.394)$

Saddle Point $(0,0,0)$ is hidden from view.

$(-10,5,-18.394)$
$(10,-5,-18.394)$
Mimimum Points

The same maximum value occurs at the points $(10,5)$ and $(-10,-5)$:

$$F(10,5) = F(-10,-5) = 50e^{-1} \approx 18.394.$$

The minimum value occurs at the critical points in the second and fourth quadrants :

$$F(-10,5) = F(10,-5) = -50e^{-1} \approx -18.394.$$

The function F has no singular points where it or its derivatives are not defined. Furthermore, as x or y become infinite the exponential term approaches zero. Therefore, these local maximum and minimum are actually "global" extrema.

Exercise Set 12.2

Determine all stationary points of the given function. Use the Second Partial Derivative Test to classify the character of each stationary point.

1. $F(x, y) = x^2 - 2x + y^2$
2. $F(x, y) = 3y - x^2 - y^2$
3. $F(x, y) = x^2 + 4x - y^2 + 3y + 5$
4. $F(x, y) = x^3 - y^2 - 3y + 1$
5. $F(x, y) = x \cos(y)$
6. $F(x, y) = x^2 y + 2x - 9y$
7. $F(x, y) = x^2 y^2 - 4x + y$
8. $F(x, y) = xy^2 - 4xy + 2y^2$
9. $F(x, y) = xy(8 - x - y)$
10. $F(x, y) = (y - 3)^2(x^2 - 4x) + y^2 - 8y$
11. $F(x, y) = \text{SQRT}(x^2 + y^2 + 1)$
12. $F(x, y) = x^3 + y^3 - 3xy$
13. $F(x, y) = (4 - x - y)xy$
14. $F(x, y) = x^{1/3} y$
15. $F(x, y) = x^{1/3} y - 8y$
16. $F(x, y) = 1/y - 1/x - 4x + y$
17. $F(x, y) = xy + 8/x + 1/y$
18. $F(x, y) = x^3 - y^3 - 4xy$
19. $F(x, y) = x^3 + 12xy + y^3$
20. $F(x, y) = x^3 - 4y^3 - 4xy$
21. $F(x, y) = x^4 + 24xy + 2y^3$
22. $F(x, y) = x^{1/3} - 4y^{2/3} - 9x + y/3$

Section 12.3 Optimization with Constraints

Problems in which the maximum or minimum values of a "performance function" $F(x, y)$ is to be found are referred to as "optimization" problems. In many applications the objective is to determine the input or "control" variables, the x and y variables that yield optimal performance of a system. However, in most applications an optimization problem is subject to constraints on the admissible values of x and y. Constraints are often imposed to limit the variables to realistic values and can assume many forms.

We consider two types of constraints. One type involves restricting the points (x, y) to be in a specific region D and the other type restricts the admissible points (x, y) to a particular curve in the x-y plane. Examples of the first type would be to require (x, y) lie in the half plane $x > 0$, or to lie within a circle or radius 3 centered at (1,-2), i.e., to satisfy the inequality

$$(x - 1)^2 + (y + 2)^2 \le 9.$$

Regional constraints are usually described algebraically by inequalities that provide the equations for the finite boundary of the region. The boundary curve is found by replacing the inequalities by equal signs. The second type of constraint specifies a relationship between admissible x and y values by an equation, which can be rearranged into the generic form

$$g(x, y) = 0. \qquad \textbf{Constraint Equation}$$

In this form, the constraint curve is seen to be a level curve of $z = g(x, y)$, the level curve for $z = 0$. Examples of such constraints are:

If $x = 2y$ then the function $g(x, y) = x - 2y$;

If the points (x, y) lie on a circle of radius 5 then $g(x, y) = x^2 + y^2 - 25$;

If $x = \sin(y)$ then $g(x, y) = x - \sin(y)$.

First, let us consider the solution of the Constrained Optimization, restricted to a curve. This problem is posed as:

> Determine the extreme values of $z = F(x, y)$,
> subject to the constraint $g(x, y) = 0$.

Restricting the points (x, y) to a curve $g(x, y) = 0$ effectively reduces the two-dimensional optimization of $F(x, y)$ to a single variable problem. If the curve $g(x, y) = 0$ can be solved for y as a function of x, $y = f(x)$, then the constrained optimization problem is reduced to optimizing $z = F(x, f(x))$ as x varies over some interval.

Example 1 **Reducing the constrained optimization of $F(x, y)$ to a single variable problem.**

Problem Determine the minimum z-coordinate on the plane

$$2(x - 1) + 3(y + 2) + 5(z - 2) = 0$$

if (x, y) is restricted to the parabola that is the graph of $g(x, y) = y - 1 + x^2 = 0$.

Solution As the g-curve is simply the graph of $y = 1 - x^2$, the term $1 - x^2$ is substituted for y into the function

$$z = F(x, y) = -(2/5)(x - 1) - (3/5)(y + 2)$$

to give

$$z = -(2/5)(x - 1) - (3/5)(3 - x^2)$$

The optimum z-value is then found by minimizing the quadratic. Considering z as a function of x only: $z' = -2/5 + 6x/5$ and $z'' = 6/5$. Hence the minimum z-value occurs at $x = 1/3$. The corresponding minimum value $z_{min} = -7/5$ is also the minimum of the surface values and occurs at the optimal point $(1/3, 8/9)$. ☑

In many applications the curve $g(x, y) = 0$ is not the graph of a function. Then solving for y and substituting will not be possible. Additionally, the g-curve may actually consist of portions of graphs of several equations, for instance, describing different portions of the boundary of a region D.

A fundamental principle states that:

> the optimal value of a function $F(x, y)$ on a domain D will occur either at the critical points of F interior to D or on the boundary of D.

Thus to solve an optimization problem constrained to a region D you must:

(i) determine the unconstrained optimization interior to D, and

(ii) determine the optimization on the boundary curve of D, i.e., solve a constrained optimization of the second type.

This in effect says that the solution of regional constraint problems must include the solutions of constrained optimization problems on boundary curves.

Example 2 Optimizing over a region D.

Problem Determine the maximum and minimum values of the function $F(x, y) = 10 - (x + y)^2$ over the triangular region D with vertices $(0,0)$, $(2,3)$ and $(2,-1)$.

Solution The region D is illustrated in the inset in the graph on the next page, the boundary curves for D are:

C_1: $y = 3x/2$, $0 \leq x \leq 2$

C_2: $x = 2$, $-3 \leq y \leq 1$

C_3: $y = -x/2$, $0 \leq x \leq 2$

We begin by considering the possible extrema of F without imposing a constraint. The first partial derivatives of F are the same:

$$F_x = F_y = -2(x + y)$$

Consequently, each point on the line $y = x$ is a stationary critical point. As the second order partials are all the same constant, $F_{xx} = F_{xy} = F_{yy} = -2$, the discriminant $D = 0$. In this case the Second Partial Derivative Test is inconclusive: it cannot be used to classify these stationary points! In this case we shall use the inequality definition of a maximum and simple algebra to show that each point on the line $y = x$ is actually a maximum point. Assume (x_0, y_0) is a point on the line.

Then $x_0 + y_0 = 0$. If (x, y) is any point not on this line, $x + y \neq 0$. Hence

$$(x + y)^2 > 0 = (x_0 + y_0)^2$$

Subtracting from the constant 10 reverses the inequality and gives the function inequality

$$F(x, y) = 10 - (x + y)^2 < 10 - (x_0 + y_0)^2 = F(x_0, y_0)$$

As the line $y = x$ intersects the region D we conclude that the maximum of F on D is $F(x, y) = 10$, which occurs for $y = x$ and $0 \leq x \leq 2$.

Next, consider the boundary of D. Since all the stationary points of F corresponded to the function's maximum, the points that give the minimum of F over D must occur on the boundary. (Like single variable functions, a continuous function must attain its maximum and its minimum values on a closed set of domain points, i.e., region D that includes its boundary curves.) For each boundary curve we reduce the problem to a one-dimensional optimization problem. The smallest of the minimum function values on the three boundary lines will provide the minimum of F on D.

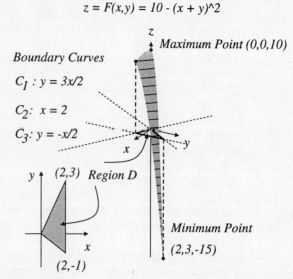

Figure 12.

On the curve C_1, $y = 3x/2$ and the function F reduces to the single variable function:

$$z = 10 - (5x/2)^2 \text{ for } 0 \leq x \leq 2.$$

As $z' = -25x/2$ and $z'' = -25/2$ this function has a maximum at $x = 0$. Its minimum value must occur at a boundary. Hence the minimum of F on C_1 occurs at the point $(2,3)$ and is given by $z_1 = -15$.

On the curve C_2, $x = 2$ and the function F reduces to a different single variable function:

$$z = 10 - (2 + y)^2 \text{ for } -1 \leq y \leq 3.$$

As a function of y, $z' = -2(2 + y)$ and $z'' = -2$, so this function has its maximum at $y = 2$, which is consistent with the global extrema found above when $x = y$. The minimum values of f on this line segment must occur at the boundaries. At $y = -1$, we find $z = 9$ and $z = -15$ at $y = 3$. The minimum of F on C_2 is thus $z_2 = -15$, at the point $(2,3)$.

On the curve C_3, $y = -x/2$ and the function F reduces to yet another single variable function:

$$z = 10 - (2x/3)^2 \text{ for } 0 \leq x \leq 2.$$

As $z' = -2x/9$ and $z'' = -2/9$, this function has a maximum at $x = 0$ and must have its minimum at the other boundary. The minimum of F on C_3 is then $z_3 = 9$.

The minimum of F on D is the least of the minimum values z_1, z_2, and z_3. Thus the minimum of F on D is -15, which occurs at the point $(2,3)$.

The Method of Lagrange Multipliers

The above examples illustrated the algebraic reduction of a multivariable optimization problem to a linear problem. This is possible when the constraint curve $g(x, y) = 0$ can be solved as a function of one variable. We now introduce a method that is more general and can be applied to a wider variety of problems. This method consists of introducing a new variable, called a "multiplier", that is traditionally denoted by the lower case Greek lambda, λ. In this method, the optimization problem is transformed into an algebra problem that involves the simultaneous solution of three equations in the three unknowns, x, y, and the parameter λ. The optimal function values occur at the points determined by the roots x and y. The value of λ is not utilized in computing the optimal function value but is usually essential to find the corresponding x and y. It is called the "Lagrange multiplier method" after Joseph Lagrange (1736-1813), who pioneered the branch of mathematics called "calculus of variations" and developed this method.

We shall state the method and illustrate its application to problems. We finish this section with a discussion of why the Lagrange Method involves the parameter λ and why it leads to an extreme value. This is a geometric development that can be visualized in terms of reaching the crest of a hiking trail. The same mathematical arguments apply in higher dimensional space, and allow the method to expand to deal with functions of more than two variables.

THE METHOD OF LAGRANGE MULTIPLIERS.

> The optimum values of a differentiable function $F(x,y)$, subject to the constraint $g(x,y) = 0$, occur at the points (x,y) that are the roots of the system of **Lagrange Equations**:
>
> $$F_x(x,y) = \lambda \, g_x(x,y)$$
>
> $$F_y(x,y) = \lambda \, g_y(x,y)$$
>
> $$g(x,y) = 0$$
>
> for a constant **Lagrange Parameter**, λ.

Example 3   Optimization using Lagrange Multipliers.

Problem   Find the maximum and minimum z-coordinates on the plane

$$2(x + 1) + 3(y + 2) + (z - 2) = 0$$

when x and y lie on the unit circle $x^2 + y^2 = 1$.

Solution   The function F and constraint function g are

$$z = F(x, y) = 6 - 2x - 3y \quad \text{and} \quad g(x, y) = x^2 + y^2 - 1$$

Their partial derivatives are

$$F_x(x, y) = -2, \quad F_y(x, y) = -3, \quad g_x(x, y) = 2x, \quad \text{and} \quad g_y(x, y) = 2y$$

The system of Lagrange Equations for this problem is:

$$F_x(x,y) = -2 = \lambda \cdot 2x$$

$$F_y(x,y) = -3 = \lambda \cdot 2y$$

$$g(x,y) = x^2 + y^2 - 1 = 0$$

From the first two equations we find $\lambda = -1/x$ and $\lambda = -3/2y$. Equating the right side of these two equations gives $x = 2y/3$. Substituting for x in the last equation gives

$$(4/9)y^2 + y^2 - 1 = 0 \quad \text{or} \quad y^2 = 9/13, \quad y = \pm 3/\text{SQRT}(13)$$

Therefore there are two possible extreme points

$$(x_0, y_0) = (2/\text{SQRT}(13), -3/\text{SQRT}(13)) \quad \text{and} \quad (x_1, y_1) = (-2/\text{SQRT}(13), -3/\text{SQRT}(13))$$

Evaluating $F(x, y)$ at these points gives the corresponding z-coordinates on the plane:

$$z_0 = F(x_0, y_0) = 6 - 4/\text{SQRT}(13) - 9/\text{SQRT}(13) = 6 - 13/\text{SQRT}(13) = 6 - \text{SQRT}(13) \approx 2.4$$

Similarly, we find $z_1 = 6 + \text{SQRT}(13) \approx 9.6$. We conclude that the point (x_0, y_0) corresponds to the minimum value z_0, while (x_1, y_1) corresponds to the maximum z-value.

☑

Example 4 Constrained extrema on a cone.

Problem Determine the extreme points on the conic surface $z = x^2 + y^2$ subject to the constraint that (x, y) is on the parabola $x = 3y^2 - 1$.

Solution In this problem the surface and constraint functions are:

$$F(x,y) = x^2 + y^2 \quad \text{and} \quad g(x,y) = -1 - x + 3y^2$$

The partial derivatives of these functions are straightforward and are used to form the corresponding Lagrange System of equations:

$$2x = \lambda(-1); \quad 2y = \lambda 6y; \quad -1 - x + 3y^2 = 0$$

To solve this system, first notice that the second equation can only be valid if $\lambda = 1/3$ or if $y = 0$.

    If $y = 0$,   then the third equation gives $x = -1$ and thus one optimal point is $(x_0, y_0) = (-1, 0)$; the corresponding z-value is $z_0 = 1$. Note that we do not need to know λ in this case to establish the critical point. The value of λ must be 2, to satisfy the first equation.

    If $\lambda = 1/3$,   from the first equation we find $x = -1/6$ and then from the last equation $y^2 = 5/18$. Thus two additional possible extreme points are

$$(x_1, y_1) = (-1/6, \text{SQRT}(5/18)) \quad \text{and} \quad (x_2, y_2) = (-1/6, -\text{SQRT}(5/18))$$

Next, calculate the function's value at these three points.

$z_0 = F(-1,0) = 1$, and $z_1 = F(-1/6, \sqrt{5/18}) = z_2 = F(-1/6, -\sqrt{5/8})) = 11/36$.

The value maximum function value is thus 1, which occurs at $(-1,0)$, while the minimum is $11/36$ occurs at both (x_1, y_1) and (x_2, y_2).

☑

Why the Lagrange Method Works.

To motivate the Lagrange solution to the constrained optimization problem we use the "hiker's trail" analogy to visualize the problem. To do this, think of the graph of $z = F(x, y)$ as a ground surface, with $F(x, y)$ being the altitude of the terrain at map coordinates (x, y). This terrain can be depicted on a topographical map that indicates level curves of the function F drawn on an x-y plane. Next, think of the graph of the relation $g(x, y) = 0$ as a curve on this topographical map that describes a "hiking trail". The constrained optimization problem is then to determine the highest and lowest positions on the trail.

The argument in essence consists of identifying the direction one hikes along the trail as a tangent vector, and at each point on the trail identifying the steepest uphill direction of the terrain. The positions in which the trail goes at a right angle to the uphill direction, i.e., the trail's tangent vector is perpendicular to the vector that points uphill, is a point where the a hiker is traveling horizontally. Since at such a point the hiker moves neither upward nor downward, it must be a local extreme point, like the situation in one dimension when the derivative is zero.

To development this argument mathematically, we utilize the results concerning tangent vectors, level curves, and directional derivatives introduced in Chapter 11, Sections 11.1-11.3.

The Lagrange Method is deduced from three observations that show how the constrained problem is really a one dimensional problem whose solution is obtained via multivariable techniques. The key is to think of the constraint as a single dimension curve specified by a parameter; the hiker's position on the trail at time t is, in theory, a simple way to parameterize the curve g. The argument is made assuming that such a parametric description of the curve $g(x, y) = 0$ exists, but does not actually require that a specific parametric function is known. The key that ties the following observations together is the fact that at a highest (or lowest) point on the trail the direction of the trail must be perpendicular to the direction that points uphill. This allows us to relate the partial derivatives of g and F.

<u>Observation I.</u>

First, we introduce a parametric representation of the "trail", which is the curve $g(x, y) = 0$ (see the graph on the following page). The exact form of this parameterization is not important! We will simply assume there is some vector-valued function $\underline{r}(t)$ that describes the curve $g(x, y) = 0$, i.e., that every point (x, y) on this curve can be represented as

$$\underline{r}(t) = (x(t), y(t))$$

In Chapter 11, Section 11.1, we introduced the tangent vector to such curves and established that at any point $\underline{r}(t)$ the tangent vector is given by

$$\underline{T} = \underline{r}'(t).$$

This notation is illustrated in Figure 12.5, where the constraint curve is

$$g(x, y) = x^2 - 3xy + (y/2)^2 + y^3 = 0.$$

Section 12.3 — Optimization with Constraints

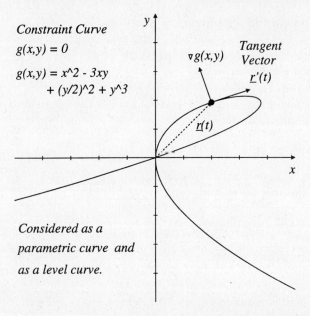

Constraint Curve
$g(x,y) = 0$

$g(x,y) = x^2 - 3xy + (y/2)^2 + y^3$

Considered as a parametric curve and as a level curve.

At the point $\underline{r}(t)$ two vectors have been sketched, a tangent vector $\underline{r}'$ and the gradient vector $\nabla g(x, y)$.

Observation II.

Next, we observe that the curve $g(x, y) = 0$ is in fact a level curve of the surface that is the graph of $z = g(x, y)$; the level curve with $z = 0$. The importance of this observation is the fact, established in Chapter 11, Section 11.3, that

the directional derivative of $z = g(x, y)$ in the direction of the level curve is zero.

Thus, at any point $(x, y) = \underline{r}(t)$ on the constraint curve, the directional derivative in the direction $\underline{T}$ is zero:

$$D_{\underline{T}} g(x, y) = \nabla g(x, y) \cdot \underline{T} / \|\underline{T}\| = 0$$

As $\underline{T} = \underline{r}'(t)$, this means that the dot product of $\underline{r}'(t)$ with the gradient of g equals zero, $\nabla g(x, y) \cdot \underline{r}'(t) = 0$, which implies that $\underline{r}'$ is perpendicular to ∇g.

Observation III.

The third observation is that if a point $(x, y) = \underline{r}(t)$ corresponds to an extreme value of $F(x, y)$ then the directional derivative of $F(x, y)$ in the direction of the path g must be zero. The trail is level at this point. This means that

$$D_{\underline{r}'(t)} F(x, y) = \nabla F(x, y) \cdot \underline{r}'(t) / \|\underline{r}'(t)\| = 0.$$

But this also indicates that $\nabla F(x, y) \cdot \underline{r}'(t) = 0$, which means the gradient vector ∇F is perpendicular to the tangent vector $\underline{r}'$.

Combining the last two observations we see that $\nabla F(x, y)$ and $\nabla g(x, y)$ are perpendicular to the same vector $\underline{r}'(t)$ (which we do not need to know!) Consequently, the two vectors $\nabla F(x, y)$ and ∇g must be parallel. This means that there must be a constant, which is denoted by λ, such that

$$\nabla F(x, y) = \lambda \, \nabla g(x, y)$$

The constant λ is the **Lagrange Multiplier**. The x and y-component equations of this vector equation provide the first two equations of the Lagrange System. The third Lagrange equation is provided by the original constraint: $g(x, y) = 0$. The Lagrange Multiplier has no numerical significance, it does not in any way tell us something about the points (x, y) where extrema for F occur, nor does it tell us anything about the actual maximum or minimum values of F. However, in specific problems we often must determine the numerical value of λ to be able to solve for the numerical values of the optimal points (x, y).

Example 5 Parallel Gradient vectors for a constrained optimization problem.

Problem Establish the greatest and least values of the product xy on the ellipse $(x/2)^2 + (y/3)^2 = 1$. What are the tangent vectors at a point where the maximum occurs?

Solution The function to be optimized is $F(x, y) = xy$ and $g(x, y) = (x/2)^2 + (y/3)^2 - 1 = 0$ is the constraint equation. The Lagrange Equations for this problem are

$$y = \lambda x/2 \qquad x = \lambda(2/9)y \qquad \text{and} \qquad (x/2)^2 + (y/3)^2 - 1 = 0$$

The optimum points are found algebraically by deducing from the first two equations that $\lambda = 3$. Then, substituting $y = 3x/2$ in the constraint equation we find the optimal points occur when

$$x = \pm \text{SQRT}(2) \qquad \text{and} \qquad y = \pm 3/\text{SQRT}(2)$$

From these equations it follows that the maximum and minimum values of xy constrained to the ellipse are 3 and -3.

The geometric description of this problem is illustrated in Figure 12.6, where in the left panel the constraint curve is the ellipse and the other curves are level curves for $F(x, y)$, many of which intersect the ellipse. Notice that the level curves in general intersect the ellipse twice, except for four curves that appear to touches the ellipse only once. These four curves are the graphs of xy = 3 and xy = -3 and they touch the ellipse at the four optimum points found by the Lagrange Method. At these points the level curves are actually tangent to the ellipse. The two points on xy = 3 are the constrained maximum points while the two points on xy = -3 give the constrained minimum values of F.

In the right panel of Figure 12.6 only the positive quadrant is sketched to show two sets of gradient vectors. The thick (shorter) vectors are the gradients of the constraint function, $\nabla g = (x/2, 2y/9)$, plotted at points of intersection of the ellipse with the function F's level curves, Γ_c for c = 0.25, 3, and 4. The narrower vectors emanating from the same points are the gradients of F: $\nabla F = (y, x)$. Observe that at most points the two types of gradients are not parallel. But, at the point where the ellipse is tangent to the level curve the two gradient vectors are parallel, in this case in the same direction. This occurs only at the point where the constraint and level curves are tangent. At the point

$$(x_0, y_0) = (\text{SQRT}(2), 3/\text{SQRT}(2))$$

the two gradient vectors are

$$\nabla g(x_0, y_0) = (1/\text{SQRT}(2), \text{SQRT}(2)/3)$$

$$\nabla F(x_0, y_0) = (3/\text{SQRT}(2), \text{SQRT}(2))$$

Thus $\nabla F(x_0, y_0) = 3 \cdot \nabla g(x_0, y_0)$. The number 3 is the value of λ.

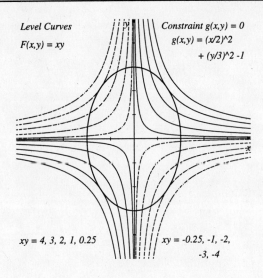
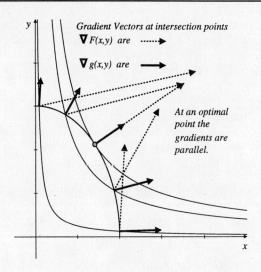

We end this discussion with the observation that the development of the Lagrange Method using a vector or parametric curve is very general, it actually made no reference to the spatial-dimensions of the vector functions. Indeed, from this development we see that the same type of Lagrange System arises if the optimization problem is higher dimensional. For instance, if F is a function of four variables, x, y, u, and v, the optimization of $F(x,y,u,v)$ subject to a constraint $g(x,y,u,v) = 0$, is found at the roots of the corresponding four-dimensional system consisting of five equations:

Lagrange System of Equations

$$F_x(x,y,u,v) = \lambda\, g_x(x,y,u,v), \qquad F_y(x,y,u,v) = \lambda\, g_y(x,y,u,v)$$

$$F_u(x,y,u,v) = \lambda\, g_u(x,y,u,v) \qquad F_v(x,y,u,v) = \lambda\, g_v(x,y,u,v)$$

$$\lambda \text{ a constant}, \quad \text{and} \quad g(x,y,u,v) = 0$$

The difficulty of such problems is solving five equations in five unknowns. If F and g are not linear this can be quite a challenge. Many business problems, like scheduling airplanes or routing parcel deliveries, are constrained optimization problems where the dimensions may be large, like the number of airports, and consequently the mathematical optimization problem is theoretically simple but realistically impractical to solve.

Exercise Set 12.3

1. Find three positive numbers x, y, and z whose sum is 50 and

    a) whose product is a maximum;

    b) such that xy^2z^2 is a maximum;

    c) such that $x^2 + y^2 + z$ is a minimum.

2. What are the maximum and minimum temperatures on a circular plate described by the region satisfying $x^2 + y^2 \leq 1$ if the temperature at (x, y) is $T = x^2 - x + 2y^2$?

3. What are the extrema of the function $F(x, y) = x^2 - 3x + xy + y^2 - 5$ on the rectangle $D: 0 \leq x \leq 2$ and $0 \leq y \leq 5$?

4. What are the extrema of $F(x, y) = 6xy - 4x^3 - 3y^2$ on the unit square $0 \leq x \leq 1$ and $0 \leq y \leq 1$.

5. What are the extrema of $F(x, y) = 2x^2 + y^2 - 6x - 4y - 1$ on the triangle-shaped region with vertices $(0,0)$, $(5,0)$ and $(5,10)$?

6. Determine the maximum and minimum value of the functions

   i) xy ii) $x + 2y$ iii) $x^2 - y$ iv) e^{xy}

   on the following regions:

   a) The line segment $y = -x + 5$ in the first quadrant.

   b) The circle $x^2 + y^2 = 4$.

   c) The curve $y = 2/x$.

   d) The parabola $x - y^2 = 4$.

Use the Lagrange Multiplier Method to solve the following constraint problems.

7. Find the maximum and minimum values of $z = F(x, y)$ subject to $g(x, y) = 0$.

   a) $F(x, y) = 2x^2 + y^2 - 2y$; $g(x, y) = 2x - y + 1$

   b) $F(x, y) = x^2 + y^2$; $g(x, y) = -x^2 - y^2$

   c) $F(x, y) = x^2 + 2y^2$; $g(x, y) = x^2 + y^2 - 4$

   d) $F(x, y) = x^2 + 2y^2$; $g(x, y) = x - 2y + 3$

   e) $F(x, y) = x^3 + y^3$; $g(x, y) = x^2 + y^2 - 2$

   f) $F(x, y) = x + y$; $g(x, y) = x^2 + y^2 - 1$

   g) $F(x, y) = x^2 e^{-y}$; $g(x, y) = y - x^2$

   h) $F(x, y) = \sin(\pi xy)$; $g(x, y) = 3 - x$

8. Express the given problem as a constrained optimization problem and solve it using the Lagrange Method.
   a) Determining the minimum distance between the origin and the line $y = 2x + 3$.
   b) Determine the rectangle with the largest area that can be inscribed in a circle of radius 4.
   c) What is the largest isosceles triangle that can be inscribed in a circle of radius 9?
   d) What is the largest rectangle that can be inscribed in the ellipse $x^2 + 4y^2 = 4$?

9. A builder has 4,800 m² of blocks and wishes to construct a walled compound 2 m high to hold waste materials. For sorting purposes an additional interior wall that divides the compound in half must be also constructed. What is the maximum volume of the compound that the builder can build?

10. In economics the Cobb-Douglas production function gives the number N of units that can be manufactured using x-units of labor and y-units of capital (equipment and materials). The form of this function is

$$N(x, y) = 16x^{\alpha}y^{\beta}$$

Assume that for a particular product $\alpha = 0.25$ and $\beta = 0.75$. If a labor unit cost $50 and a capital unit cost $200, the total cost of a project is $C(x, y) = 50x + 200y$. What is the maximum number of units that can be built if the total budget is $4 million? How would the resources be distributed to achieve this maximum production?

11. What would be the maximum number of units produced in Exercise 11 if $\alpha = 0.6$ and $\beta = 0.2$?

12. A chicken farmer must decide on the optimal mix of two types of feed. The constraints on the choice is that the feed must provide sufficient calories and the farmer wishes to minimize the total cost. Assume feed A cost $0.50/kg and feed B cost $0.75/kg. If the calories supplied by a mixture of x kg feed A and y kg feed B results in $C(x, y) = 50xy - x^{1/2}$ calories, what is the optimal mix of feed to provide 10,000 calories of food?

13. Apply the Lagrange method to determine the maximum of the product $P = xyz$ subject to the constraint: $3x + 4y + 3z = 24$.

14. Apply the Lagrange Multiplier method to determine the minimum of the quadratic "objective" function $F(x, y, z, w) = x^2 + y^2 + z^2 + w^2$ subject to the linear constraint $3x + 2y - 4z + w = 15$.

15. If the objective function is a function of more than two variables, then more constraints may be required to reduce the optimization problem to a one-dimensional problem. The Lagrange method generalizes to incorporate two or more constraint functions. Assume that the objective function that we seek the optimum of is F subject to two constraints that can be expressed in terms of functions g and h as g = 0 and h = 0. One way to establish the appropriate equations is to introduce two parameters, λ and the Greek letter μ, and consider the function $T = F - \lambda g - \mu h$. The Lagrange equations are then the system of equations obtained by setting the partial derivatives of T equal to zero. One equation is formed for each independent variable, and two more are formed by the partial derivatives with respect to λ and μ. For instance, if one of the variables is x, then one of these equations is

$$T_x = F_x - \partial/\partial x \, \lambda g - \partial/\partial x \, \mu h = F_x - \lambda g_x - \mu h_x = 0, \text{ or } F_x = \lambda g_x + \mu h_x$$

Since the functions F, g, and h do not depend on the introduced parameter λ, one of the auxiliary equations is $T_\lambda = F_\lambda - \partial/\partial \lambda \, \lambda g - \partial/\partial \lambda \, \mu h = g = 0$, i.e., the first constraint equation.

a) Write the system of Lagrange equations for a function $F(x, y, z)$ of three variables with two constraints $g(x, y, z) = 0$ and $h(x, y, z) = 0$.

b) Determine the maximum of the function $F(x, y, z) = 10 + 2x + 3y + z^2$ on the curve that is the intersection of the sphere $x^2 + y^2 + z^2 = 16$ and the plane $2x + y + z = 2$.

16. A container is to be designed as a circular cylinder topped by a cylindrical cone. What is the maximum volume of such a container if its surface area is 25π m^2 and $r + x + y = 30$? Hint: If the cylinder part of the container has radius r and height x, and the cone has height y, then the surface area is $A = 2\pi rx + \pi r^2 + 4\pi ry/3$ and the volume is $V = \pi r^2 x + \pi r^2 y/3$.

Answers to Selected Exercises

Graphs are grouped at the end of the exercise set. When a graph is provide it is indicated by **GRAPH HERE** in the the solution sets.

CHAPTER 1 ANSWERS

Exercise Set 1.1

1.1.1 a) [0,8]   b) (0,8)   c) $\{x \mid -1 \leq x - 3 \leq 1\} = \{x \mid 2 \leq x \leq 4\} = [2,4]$
d) $\{x \mid |x + 2| > 3\} = \{x \mid x + 2 < -3 \text{ or } x + 2\} = \{x \mid x < -5 \text{ or } x > 1\} = (-\infty,-5) \cup (1,+\infty)$
e) $\{x \mid x > 2 \text{ and } x \leq 8\} = (2,8]$ f) $\{x \mid x > 0 \text{ or } x \leq -2\} = (-\infty,-2] \cup (0,\infty)$
g) $\{x \mid x^2 > 4\} = \{x \mid SQRT(x^2) > \sqrt{4}\} = \{x \mid |x| > 2\} = \{x \mid x < -2 \text{ or } x > 2\} = (-\infty,-2) \cup (2,+\infty)$
h) $\{x \mid x^2 - 4 \leq 5\} = \{x \mid x^2 \leq 9\} = \{x \mid SQRT(x^2) \leq 9\} = \{x \mid |x| \leq 3\} = \{x \mid -3 \leq x \leq 3\} = [-3,3]$

1.1.2 a) $D_f = (-\infty,+\infty)$ b) $D_g = (-\infty,3) \cup (3,+\infty)$ c) $D_h = (-\infty,0) \cup (0,+\infty)$ d) $D_f = [4,+\infty)$
e) $D_f = (-\infty,0) \cup (0,+\infty)$ f) $D_g = (-\infty,-2) \cup (-2,2) \cup (2,+\infty)$ g) $D_h = (-\infty,0) \cup [1/2,+\infty)$ h) $(-\infty,-2] \cup [2,+\infty)$ i) $[-2,2]$

1.1.3 a) $R_f = [-1,+\infty)$ b) $R_g = (-\infty,1) \cup (1,+\infty)$ c) $R_h = (-\infty,2) \cup (2,+\infty)$ d) $R_f = [0,+\infty)$ e) $R_f = (0,+\infty)$ f) $R_g = (-\infty,+\infty)$
g) $R_h = [0,\sqrt{2}) \cup (\sqrt{2},+\infty)$ h) $[0,+\infty)$ i) $[0,+\infty)$

1.1.4 a) no   b) yes   c) no   d) yes   e) no   f) yes   g) yes   h) no

1.1.5 a) $D_f = \{2,3,5,6\}$ $D_g = \{2,3,4,5\}$ b) **GRAPH HERE**   **GRAPH HERE**
c) $f + g = \{(2,6), (3,4), (5,8)\}$; $D_{f+g} = \{2,3,5\}$ $fg = \{(2,9),(3,0),(5,12)\}$; $D_{fg} = \{2,3,5\}$ $f/g = \{(2,1),(5,1/3)\}$; $D_{f/g} = \{2,5\}$
$f \circ g = \{(2,4),(4,3),(5,5)\}$ $D_{f \circ g} = \{2,4,5\}$ $gf = \{(2,0),(3,2),(5,3),(6,6)\}$; $D_{gf} = \{2,3,5,6\}$
d) $f^{-1} = \{(3,2),(4,3),(2,5),(5,6)\}$ $g^{-1} = \{(3,2),(0,3),(2,4),(6,5)\}$

1.1.6 a) $(f + g)(x) = (1/x) + x^2 - 3$ $D_{f+g} = (-\infty,0) \cup (0,+\infty)$ $(g \circ f)(x) = x^2 - 3 - (1/x)$ $D_{g-f} = (-\infty,0) \cup (0,+\infty)$ $(fg)(x) = (x^2 - 3)/x$
$D_{fg} = (-\infty,0) \cup (0,+\infty)$ $(f/g)(x) = 1/[x(x^2 - 3)]$ $D_{(f/g)} = \{x \mid x \neq -\sqrt{3}, 0, \sqrt{3}\}$ $(f \circ g)(x) = 1/(x^2 - 3)$ $D_{f \circ g} = \{x \mid x \neq -\sqrt{3}, \sqrt{3}\}$
$(g \circ f)(x) = (1/x^2) - 3$ $D_{g \circ f} = \{x \mid x \neq 0\}$
b) $(f + g)(x) = 2x - 1 + \sqrt{x}$ $D_{f+g} = [0,+\infty)$ $(g-f)(x) = \sqrt{x} - 2x + 1$ $D_{g-f} = [0,+\infty)$ $(fg)(x) = (2x - 1)\sqrt{x}$ $D_{fg} = [0,+\infty)$
$(f/g)(x) = (2x - 1)/\sqrt{x}$ $D_{f/g} = (0,+\infty)$ $(f \circ g)(x) = 2\sqrt{x} - 1$ $D_{f \circ g} = [0,+\infty)$ $(g \circ f)(x) = SQRT(2x - 1)$ $D_{g \circ f} = [1/2,+\infty)$
c) $(f + g)(x) = x^2 - 2x + 1/(x + 1)$ $D_{f+g} = (-\infty,-1) \cup (-1,+\infty)$ $(g - f)(x) = 1/(x+1) - x^2 + 2x$ $D_{g-f} = (-\infty,-1) \cup (-1,+\infty)$
$(fg)(x) = (x^2 - 2x)/(x + 1)$ $D_{fg} = (-\infty,-1) \cup (-1,+\infty)$ $(f/g)(x) = (x + 1)(x^2 - 2x)$ $D_{f/g} = (-\infty,-1) \cup (-1,+\infty)$
$(f \circ g)(x) = (-2x - 1)/(x + 1)^2$ $D_{f \circ g} = (-\infty,-1) \cup (-1,+\infty)$ $(g \circ f)(x) = 1/(x - 1)^2$ $D_{g \circ f} = (-\infty,1) \cup (1,+\infty)$

1.1.7 a) $f(x - 3) + 2$ b) $f(x - 1) - 3$ c) $f(x + 2) - 1$ d) $f(x + 4) + 5$

1.1.8 a) graph of f(x) shifted up 2 units   b) graph of f(x) shifted to the right 3 units
c) graph of f(x) shifted to the right 5 units and down 2 units.   d) graph of f(x) shifted to the left 2 units and up 3 units
e) graph of f(x) shifted down 2 units
f) graph of f(x) shifted to the right 2 units, stretched by a factor of two away from the x-axis and then reflected through the x-axis.

1.1.9 a) graph of $f(x) = x^2$ shifted up 3 units   b) graph of $f(x) = x^2$ shifted to the left 3 units
c) graph of $f(x) = \sqrt{x}$ shifted to the right 5 units   d) graph of $f(x) = x^2$ shifted to the right 4 units and down 2 units
e) graph of $f(x) = x - x^2$ shifted to the left 2 units   f) graph of $f(x) = x - \sqrt{x}$ shifted to the right 1 unit and up 4 units

1.1.10 a) $g(x) = f(x - 1) + 2$ b) $g(x) = f(x - 3) - 1$ c) $g(x) = f(x + 3) + 1$ d) $g(x) = f(x - 1.5) + 2.5$ e) $g(x) = f(x + 1.5) + 2.5$
f) $g(x) = f(x + 4) - .5$ g) $g(x) = f(x - 2.2) + 2.8$ h) $g(x) = f(x + 2) - 2.5$ i) $g(x) = f(x + 3.5) - 2$

1.1.11 a) $f(x) = g(x + 1) - 2$ b) $f(x) = g(x + 3) + 1$ c) $f(x) = g(x - 3) - 1$ d) $f(x) = g(x + 1.5) - 2.5$ e) $f(x) = g(x - 1.5) - 2.5$
f) $f(x) = g(x - 4) + .5$ g) $f(x) = g(x + 2.2) - 2.8$ h) $f(x) = g(x - 2) + 2.5$ i) $f(x) = g(x - 3.5) + 2$

1.1.12 a) **GRAPH HERE**   b) **GRAPH HERE**   c) **GRAPH HERE**   d) **GRAPH HERE**

Selected Answers

Graphing Solutions to Chapter 1 Section 1

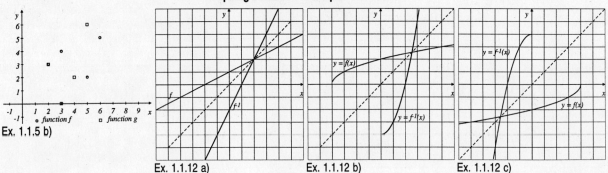

Ex. 1.1.5 b)   Ex. 1.1.12 a)   Ex. 1.1.12 b)   Ex. 1.1.12 c)

Exercise Set 1.2

1.2.1 Use $0 < a < b$ implies $a^2 < b^2$

1.2.2 Use i) $x_1 = -2, x_2 = 1$ and ii) $x_1 = -2, x_2 = 3$

1.2.3 a) $f(-x) = (-x)^2 = x^2 = f(x)$ b) $f(-x) = (-x)^3 = -x^3 = -f(x)$

1.2.4 a) GRAPH HERE   b) GRAPH HERE   c) GRAPH HERE

1.2.5 a) $y = 6 + 2x$; 1st   b) $y = -6 + x + x^2$; 2nd   c) $y = -1 + (x/2)$; 1st   d) $y = 5 - 2x^2 + x^5$; 5th
       e) $y = (11/2)x - 5x^2$; 2nd   f) $y = x + x^2 - 3x^3$; 3rd

1.2.6 a) $y = 5 + 5(x - 2) = 5x - 5$; y intercept -5 b) $y = 3 - 1(x + 2) = -x + 1$; y intercept 1   c) $y = 1 + 0(x - 6) = 1$; y intercept 1
       d) $y = 2 - (1/2)(x - 2) = (-1/2)x + 3$; y intercept 3   e)   f) $y = 3.4 - 2.7(x - 0.1) = -2.7x + 3.67$; y intercept 3.67

1.2.7 a) $y = -9 + 6(x - 1)$ b) $y = 3 - 3(x - 1)$ c) $y = 10/3 - (1/6)(x - 1)$ d) $y = 9 + 3(x - 1)$
       e) $y = (33/2) + 5(x - 1)$ f) $y = 5.57 + 0.7(x - 1)$ g) $y = (-9/8) + (3/8)(x - 1)$ h) $y = 0.6 - 0.05(x - 1)$

1.2.8 Suppose $y = mx + b$, $z = ny + c$, then $z = n(mx + b) + c$ or $z = nmx + nb + c = px + d$ for $p = nm$ and $d = nb + c$. Hence, z is linearly related to x.

1.2.9 $m_1 < m_4 < m_3 < m_2$

1.2.10 a) $y = (x - 3)^2 + 1$ b) $y = (x + 1)^2$ c) $y = (x - 2)^2 - 1$ d) $y = -(x - 0)^2 + 3$ e) $y = -1/2(x + 1/4)^2 + 1/32$
       f) $y = (x + 5/2)^2 - 33/4$ g) $y = (x - 1/2)^2 - 1/4$ h) $y = -(x - 7/2)^2 + 9/4$

1.2.11 a) $3^{1/3}$ b) $(x - 1)^{1/3}$ c) $(x^2 - 2x)^{1/3}$ d) $(x^2 + 2xh + h^2 - 1)^{1/3}$

1.2.12 $y = -2(x - 2)^2 - 3$

1.2.13 a) GRAPH HERE b) GRAPH HERE c) GRAPH HERE c) GRAPH HERE

1.2.14 a) shifting the graph 2 units to the right and stretching by a factor of 3
       b) shifting the graph 3 units to the left, reflecting through the x-axis and then shifting up 1 unit.
       c) shifting the graph 3 units to the left, stretching by a factor of 2 and reflecting through the x-axis
       d) shifting the graph 4 units to the left, compressing by a factor of 1/2 and shifting down 2 units

1.2.15 Matches: b - A; c - C; e - B; g - D

1.2.16 $r(t) = -0.01(t - 20)^2 + 4$

1.2.17 a) $N = S(n) + I(n)$ b) 62   c) $\Delta(n) = kS(n)$

Graphing Solutions to Chapter 1 Section 2

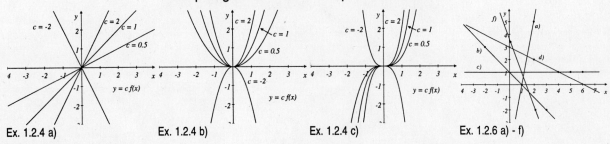

Ex. 1.2.4 a)   Ex. 1.2.4 b)   Ex. 1.2.4 c)   Ex. 1.2.6 a) - f)

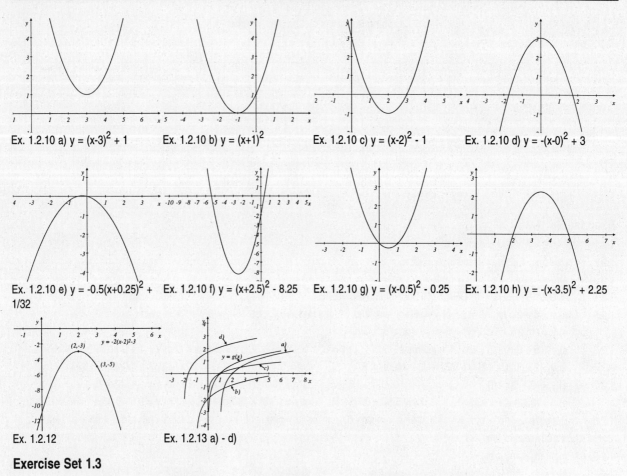

Ex. 1.2.10 a) $y = (x-3)^2 + 1$ Ex. 1.2.10 b) $y = (x+1)^2$ Ex. 1.2.10 c) $y = (x-2)^2 - 1$ Ex. 1.2.10 d) $y = -(x-0)^2 + 3$

Ex. 1.2.10 e) $y = -0.5(x+0.25)^2 + 1/32$ Ex. 1.2.10 f) $y = (x+2.5)^2 - 8.25$ Ex. 1.2.10 g) $y = (x-0.5)^2 - 0.25$ Ex. 1.2.10 h) $y = -(x-3.5)^2 + 2.25$

Ex. 1.2.12   Ex. 1.2.13 a) - d)

Exercise Set 1.3

1.3.1   a) 2.2973967   b) -1.4422495   c) -1.4422495   d) 8.8249778
      e) $(-2)^{3.25} = (-2)^{13/4}$ undefined (even root of a negative number)   f) $(-2)^{0.64} = (-2)^{16/25} = 1.5583292$
      g) $(-4)^{0.8} = (-4)^{4/5} = 3.0314331$ h) $(-5)^{-1.1} = 1/(-5)^{11/10}$ undefined (even root of a negative number)

1.3.2   a) GRAPH HERE   b) GRAPH HERE   c) GRAPH HERE   d) GRAPH HERE   e) GRAPH HERE   f) GRAPH HERE
      g) GRAPH HERE   h) GRAPH HERE

1.3.3   a) GRAPH HERE   b) GRAPH HERE   c) GRAPH HERE   d) GRAPH HERE   e) GRAPH HERE   f) GRAPH HERE
g) GRAPH HERE   h) GRAPH HERE   i) GRAPH HERE   j) GRAPH HERE

1.3.4   a) GRAPH HERE   b) GRAPH HERE   c) GRAPH HERE   d) GRAPH HERE   e) GRAPH HERE   f) GRAPH HERE
      g) GRAPH HERE   h) GRAPH HERE   i) GRAPH HERE

1.3.5   a) C   b) D   c) A   d) B

1.3.6   a)-l) GRAPH HERE

1.3.7   a) - h) GRAPH HERE

Graphing Solutions to Chapter 1 Section 3

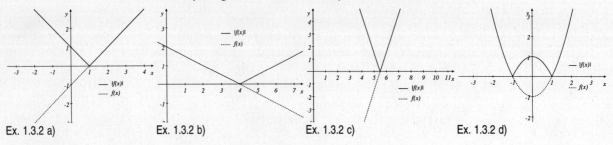

Ex. 1.3.2 a)   Ex. 1.3.2 b)   Ex. 1.3.2 c)   Ex. 1.3.2 d)

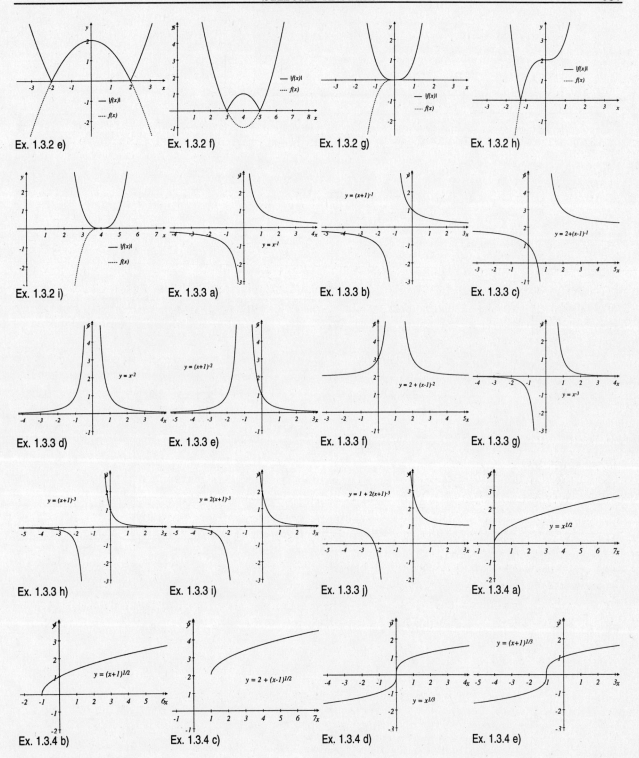

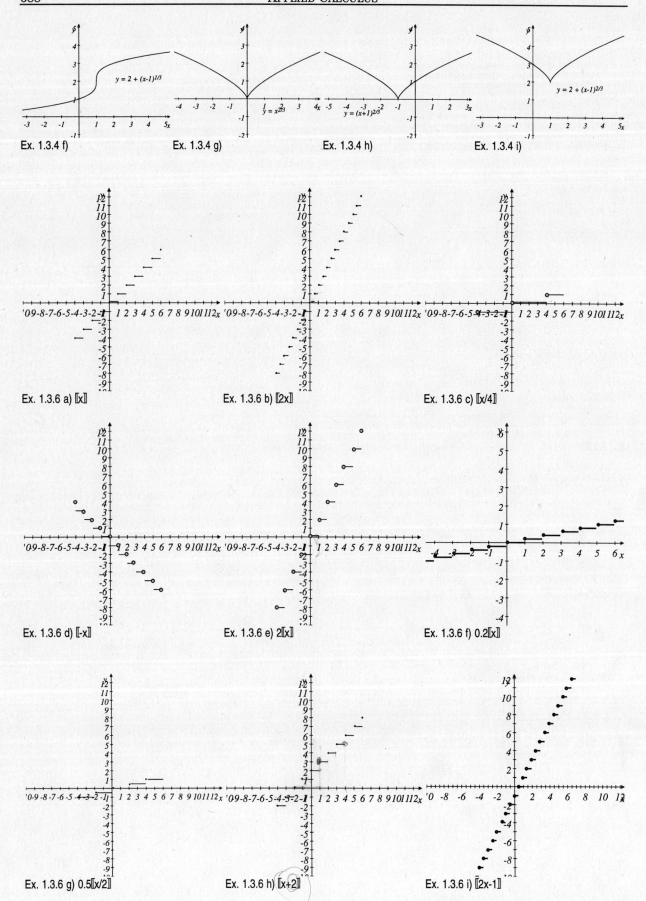

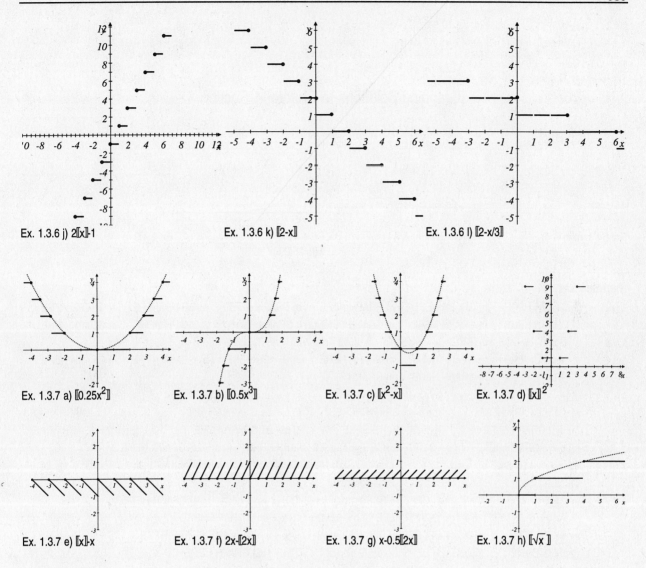

Ex. 1.3.6 j) 2⟦x⟧-1   Ex. 1.3.6 k) ⟦2-x⟧   Ex. 1.3.6 l) ⟦2-x/3⟧

Ex. 1.3.7 a) ⟦0.25x²⟧   Ex. 1.3.7 b) ⟦0.5x³⟧   Ex. 1.3.7 c) ⟦x²-x⟧   Ex. 1.3.7 d) ⟦x⟧²

Ex. 1.3.7 e) ⟦x⟧-x   Ex. 1.3.7 f) 2x-⟦2x⟧   Ex. 1.3.7 g) x-0.5⟦2x⟧   Ex. 1.3.7 h) ⟦√x⟧

Exercise Set 1.4

1.4.1   a) 2^5 b) 2^a if $a > 0$, 2^{-a} if $a < 0$, $2^a = 2^{-a}$ if $a = 0$
         c) since $e < \sqrt{10}$ then $e^x < \sqrt{10}^x$ for $x > 0$ and $e^x > \sqrt{10}^x$ for $x < 0$. For $x = 0$,
         $e^0 = \sqrt{10}^0$ d) 3^2 e) 3^5 f) 3   g) 3   h) log(5)   i) 4   j) a^x
         k) $a^x > \log_a(x)$ for $x > c$, $a_x < \log_a(x)$ $0 < x < c$ where c is the solution to $a^x = \log_a(x)$

1.4.2   a) $3000(2)^{144}$ b) $3000(1.6)^n$ c) $3000(1.6)^{144}$

1.4.3   a) GRAPH HERE   b) GRAPH HERE   c) GRAPH HERE   d) GRAPH HERE   e) GRAPH HERE   f) GRAPH HERE

1.4.4 $A(10) = 400e^{-(ln(2))(10)/15} = 400(1/2)^{2/3}$, $A(15) = 200$ $A(20) = 400e^{-(ln(2))(20)/15} = 400(1/2)^{4/3}$

1.4.5   a) $K = ln(5/2)/3$ b) $3ln(2)/ln(5/2) \approx 2.269$ c) $25e^{2ln(5/2)/3} = 25(5/2)^{2/3}$ d) GRAPH HERE

1.4.6   a) $t_{1/4} = -ln(3/4)/k$ b) $7e^{-12ln(3/4)/5} = 7(4/3)^{12/5} \approx 13.962g$

1.4.7 $t = -50ln(25/37) \approx 19.6$ min; $t = -5ln(25/37) \approx 1.96$ min

1.4.8   a) $A(t) = A_0 e^{-ln(2)t/8}$; $(A(t) - A(t+1))/A(t) = 1 - e^{-ln(2)/8} \approx 8.3\%$ b) $t = 8ln(4)/ln(2) = 16$ days   c) $5 - 5e^{-ln(2)/8} \approx 0.415$ units.

1.4.9   a) GRAPH HERE   b) GRAPH HERE   c) GRAPH HERE   d) GRAPH HERE   e) GRAPH HERE   f) GRAPH HERE

1.4.10 a) $-ln(3.5)/2$ b) $10^{0.1}/2 \approx 0.629$ c) $ln(7)/3$ d) $(e^7 + 1)/2 \approx -0.389$ e) $log_5(10) = log(10)/log(5) = 1/log(5)$
        f) $1/3 ln(1/7) = -ln(7)/3$ g) 9/5   h) $(6 + ln(1/2))/3$ i) $3(10^{-7})$ j) 0, -2   k) $x = -10\,ln(2)$ l) 384

1.4.11 $\varepsilon \approx 0.25$ 1.4.12 $\varepsilon = (1 - log(1.6))/0.36 \approx 2.21$; using a 0.18 molar solution

1.4.13 a) $\varepsilon = log(10.5)/3 \approx 0.34040$ b) $50(10.5)^{5/3} \approx 2,517.34$

1.4.14 a) $10^{2.8123946} = 649.224$ b) $10^{-4.155634} \approx 6.99 \times 10^{-5}$ c) $10^{2.8123734} = 649.19232$ d) $10^{.7420952} = 5.5219845$

1.4.15 a) Since $I_0 = k10^0 = k$, then $I = I_0 10^L$ or $(I/I_0) = 10^L$. solving for L, $L = \log(I/I_0)$ b) $L = 10\log(3) + 30 \approx 34.771213$ c) 10^5 d) 10^{10}

1.4.16 a) $R = \log(I/I_0)$ b) 398

Graphing Solutions to Chapter 1 Section 4

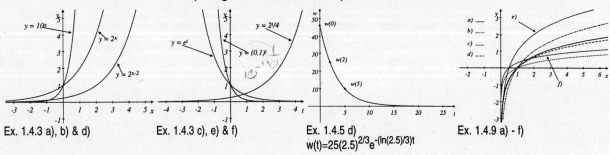

Ex. 1.4.3 a), b) & d)   Ex. 1.4.3 c), e) & f)   Ex. 1.4.5 d) $w(t)=25(2.5)^{2/3}e^{-(\ln(2.5)/3)t}$ Ex. 1.4.9 a) - f)

Exercise Set 1.5

1.5.2 a) GRAPH HERE b) GRAPH HERE c) GRAPH HERE d) GRAPH HERE e) GRAPH HERE f) GRAPH HERE
       g) GRAPH HERE h) GRAPH HERE h) GRAPH HERE i) GRAPH HERE j) GRAPH HERE f) GRAPH HERE

1.5.3 a) 1 b) 0 c) .8415 d) .5403 e) -0.4161 f) .9093 g) .7071 h) .7071 i) 1 j) 2 k) 1.1547 l) 1.7321

1.5.4 a) GRAPH HERE b) GRAPH HERE c) GRAPH HERE d) GRAPH HERE e) GRAPH HERE f) GRAPH HERE
       g) GRAPH HERE h) GRAPH HERE h) GRAPH HERE

1.5.5 a) $p = 2\pi$ GRAPH HERE b) $p = 2\pi$ GRAPH HERE c) $p = 2\pi$ GRAPH HERE d) $p = 2\pi$ GRAPH HERE
       e) $p = \pi$ GRAPH HERE f) $p = 4\pi$ GRAPH HERE g) $p = 2\pi$ GRAPH HERE h) $p = 4\pi$ GRAPH HERE
       i) $p = 2$ GRAPH HERE j) $p = 4$ GRAPH HERE k) $p = 2\pi$ GRAPH HERE l) $p = 2\pi$ GRAPH HERE
       m) $p = 2$ GRAPH HERE n) $p = 2$ GRAPH HERE o) $p = 6$ GRAPH HERE p) $p = 2/3$ GRAPH HERE

1.5.6 a) $\cos(x^2 + 1)$; .5403 b) $(\cos(x))^2 + 1$; 2 c) $\sin(x - \pi)$; 0 d) $(\sin(x))^2 + 1$; 2 e) $(\cos(x - \pi))^2 + 1$; 1 f) $\sin(x^2 + 1)$; .8415
       g) $(\cos(x))^2 + 1$; 2 h) $\cos(\sin(x))$; 1 i) $\sin(\cos(x))$; 0

1.5.7 a) $p = \pi$ $D = \{x \mid x \neq (2n + 1)\pi/2\}$ GRAPH HERE b) $p = \pi/2$ $D = \{x \mid x \neq (2n + 1)\pi/4\}$ GRAPH HERE
       c) $p = 2\pi$ $P\{x \mid x \neq (2n + 1)\pi\}$ GRAPH HERE d) $p = 3\pi$ $D = \{x \mid x \neq 3n\pi\}$ GRAPH HERE
       e) $p = 2\pi$ $D = \{x \mid x \neq (2n + 1)\pi/2\}$ GRAPH HERE f) $p = 1$ $D = \{x \mid x \neq (2n + 1)/2\}$ GRAPH HERE
       g) $p = 2$ $D = \{x \mid x \neq 2n\}$ GRAPH HERE h) $p = 2$ $D = \{x \mid x \neq (2n + 1)/2\}$ GRAPH HERE
       i) $p = 2\pi$ $D = \{x \mid x \neq (2n + 1)\pi/2\}$ GRAPH HERE j) $p = 1$ $D = \{x \mid x \neq n\}$ GRAPH HERE

1.5.8 a) $y = \sin(\pi x) + 2$ b) $y = 2\sin(\pi(x - 1/2))$ c) $y = 1/2 \sin(\pi x) + 3$ d) $y = \sin(\pi x/2) + 1/2$
       e) $y = 3\sin(2\pi x) - 1$ f) $y = 1/4 \sin(\pi x/3) - 2$

1.5.9 a) $y = \cos(\pi(x - 1/2)) + 2$ b) $y = 2\cos(\pi(x - 1))$ c) $y = 1/2 \cos(\pi(x - 1/2)) + 3$ d) $y = \cos(\pi/2 (x - 1)) + 1/2$
       e) $y = 3\cos(2\pi(x - 1/4)) - 1$ f) $y = 1/4 \cos(\pi/3 (x - 3/2)) - 2$

1.5.10 a) GRAPH HERE b) GRAPH HERE c) GRAPH HERE d) GRAPH HERE e) GRAPH HERE

1.5.11 a) GRAPH HERE   1.5.12 a) GRAPH HERE

1.5.13 a) 5 units of time b) $v(t) = a \sin(\pi/3)t$ if $0 \leq t < 3$ and $v(t) = 0$ if $3 \leq t < 5$ c) GRAPH HERE

1.5.14 a) $\text{SQRT}(4 - x^2)/2$ b) $x/\text{SQRT}(x^2 + 9)$ c) $1/\text{SART}(1 - 4x^2)$ d) $x/\text{SQRT}(16 - x^2)$ e) $1 / \text{SQRT}(x^2 + 2)$ f) $2 / \text{SQRT}(4 - x^2)$

1.5.15 a) GRAPH HERE b) 7.5×10^{13} cps c) GRAPH HERE d) $(3/7)(10)^{15}$ cps to $(3/4)(10)^{15}$ cps e) GRAPH HERE

1.5.16 a) $P = 1/70$; $BP(t) = 20 \sin(140\pi(t - 1/140)) + 100$ b) $P = 1/80$; $BP(t) = 20 \sin(160\pi(t - 1/160)) + 100$
        c) $P = 1/80$; $BP(t) = 20 \sin(160\pi(t - 1/160)) + 120$

1.5.18 a) Fundamental tone: $y = 2\sin(t)$, second overtone: $y = 2\sin(3t)$ First overtone: $y = 2\sin(2t)$, third overtone: $y = 2\sin(4t)$

Graphing Solutions to Chapter 1 Section 5

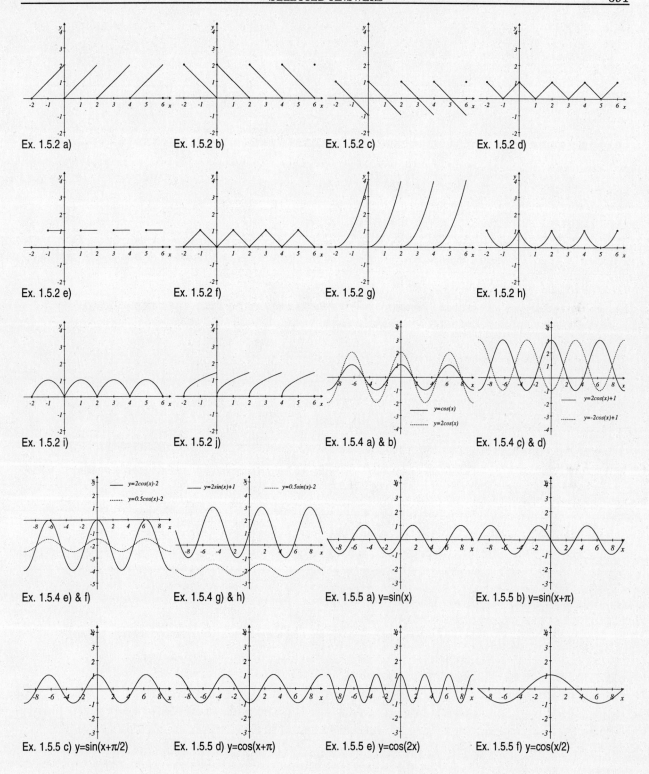

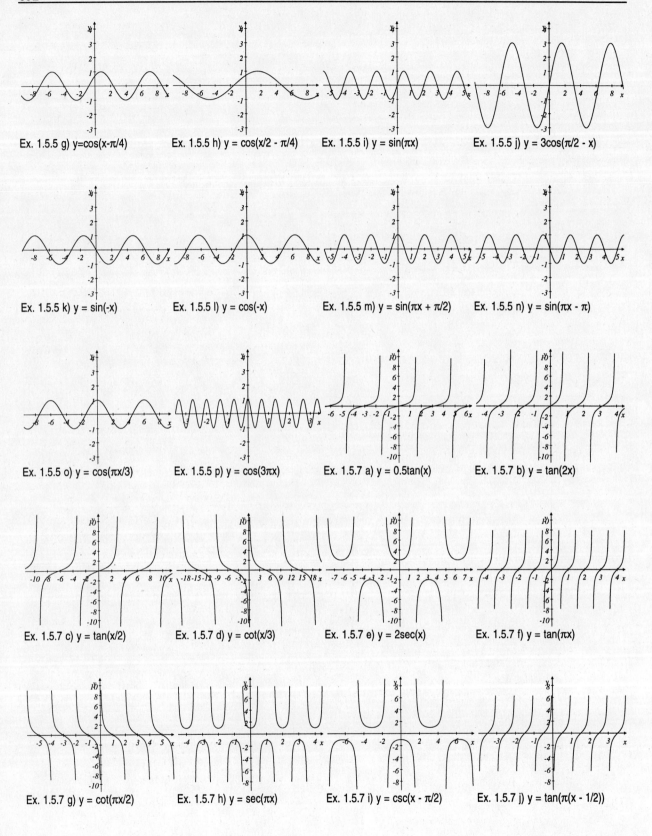

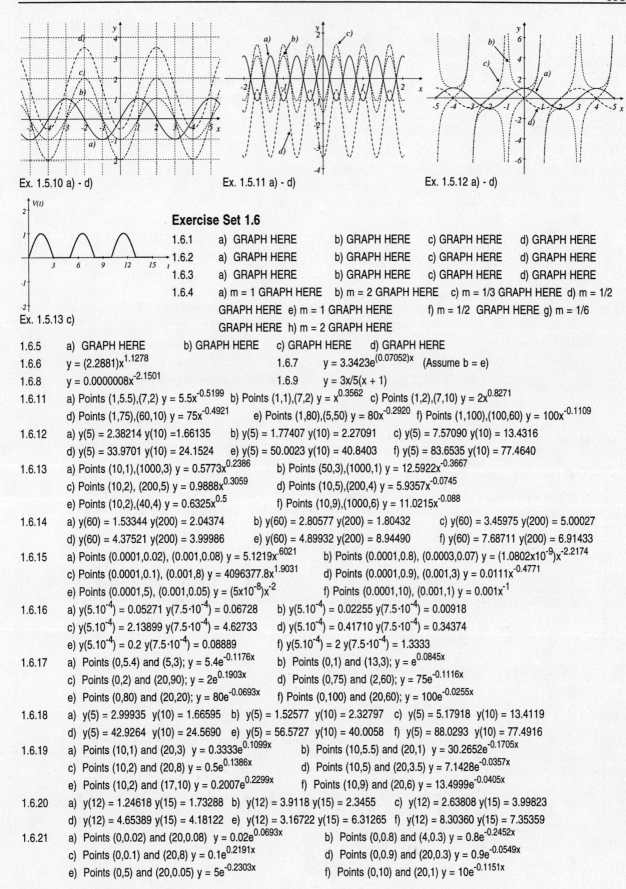

Ex. 1.5.10 a) - d)

Ex. 1.5.11 a) - d)

Ex. 1.5.12 a) - d)

Ex. 1.5.13 c)

Exercise Set 1.6

1.6.1   a) GRAPH HERE   b) GRAPH HERE   c) GRAPH HERE   d) GRAPH HERE

1.6.2   a) GRAPH HERE   b) GRAPH HERE   c) GRAPH HERE   d) GRAPH HERE

1.6.3   a) GRAPH HERE   b) GRAPH HERE   c) GRAPH HERE   d) GRAPH HERE

1.6.4   a) m = 1 GRAPH HERE   b) m = 2 GRAPH HERE   c) m = 1/3 GRAPH HERE   d) m = 1/2 GRAPH HERE   e) m = 1 GRAPH HERE   f) m = 1/2 GRAPH HERE   g) m = 1/6 GRAPH HERE   h) m = 2 GRAPH HERE

1.6.5   a) GRAPH HERE   b) GRAPH HERE   c) GRAPH HERE   d) GRAPH HERE

1.6.6 $y = (2.2881)x^{1.1278}$ 1.6.7 $y = 3.3423e^{(0.07052)x}$ (Assume b = e)

1.6.8 $y = 0.0000008x^{-2.1501}$ 1.6.9 $y = 3x/5(x+1)$

1.6.11   a) Points (1,5.5),(7,2) $y = 5.5x^{-0.5199}$ b) Points (1,1),(7,2) $y = x^{0.3562}$ c) Points (1,2),(7,10) $y = 2x^{0.8271}$

        d) Points (1,75),(60,10) $y = 75x^{-0.4921}$ e) Points (1,80),(5,50) $y = 80x^{-0.2920}$ f) Points (1,100),(100,60) $y = 100x^{-0.1109}$

1.6.12   a) y(5) = 2.38214 y(10) =1.66135   b) y(5) = 1.77407 y(10) = 2.27091   c) y(5) = 7.57090 y(10) = 13.4316

        d) y(5) = 33.9701 y(10) = 24.1524   e) y(5) = 50.0023 y(10) = 40.8403   f) y(5) = 83.6535 y(10) = 77.4640

1.6.13   a) Points (10,1),(1000,3) $y = 0.5773x^{0.2386}$ b) Points (50,3),(1000,1) $y = 12.5922x^{-0.3667}$

        c) Points (10,2), (200,5) $y = 0.9888x^{0.3059}$ d) Points (10,5),(200,4) $y = 5.9357x^{-0.0745}$

        e) Points (10,2),(40,4) $y = 0.6325x^{0.5}$ f) Points (10,9),(1000,6) $y = 11.0215x^{-0.088}$

1.6.14   a) y(60) = 1.53344 y(200) = 2.04374   b) y(60) = 2.80577 y(200) = 1.80432   c) y(60) = 3.45975 y(200) = 5.00027

        d) y(60) = 4.37521 y(200) = 3.99986   e) y(60) = 4.89932 y(200) = 8.94490   f) y(60) = 7.68711 y(200) = 6.91433

1.6.15   a) Points (0.0001,0.02), (0.001,0.08) $y = 5.1219x^{.6021}$ b) Points (0.0001,0.8), (0.0003,0.07) $y = (1.0802 \times 10^{-9})x^{-2.2174}$

        c) Points (0.0001,0.1), (0.001,8) $y = 4096377.8x^{1.9031}$ d) Points (0.0001,0.9), (0.001,3) $y = 0.0111x^{-0.4771}$

        e) Points (0.0001,5), (0.001,0.05) $y = (5 \times 10^{-8})x^{-2}$ f) Points (0.0001,10), (0.001,1) $y = 0.001x^{-1}$

1.6.16   a) $y(5 \cdot 10^{-4}) = 0.05271$ $y(7.5 \cdot 10^{-4}) = 0.06728$ b) $y(5 \cdot 10^{-4}) = 0.02255$ $y(7.5 \cdot 10^{-4}) = 0.00918$

        c) $y(5 \cdot 10^{-4}) = 2.13899$ $y(7.5 \cdot 10^{-4}) = 4.62733$ d) $y(5 \cdot 10^{-4}) = 0.41710$ $y(7.5 \cdot 10^{-4}) = 0.34374$

        e) $y(5 \cdot 10^{-4}) = 0.2$ $y(7.5 \cdot 10^{-4}) = 0.08889$ f) $y(5 \cdot 10^{-4}) = 2$ $y(7.5 \cdot 10^{-4}) = 1.3333$

1.6.17   a) Points (0,5.4) and (5,3); $y = 5.4e^{-0.1176x}$ b) Points (0,1) and (13,3); $y = e^{0.0845x}$

        c) Points (0,2) and (20,90); $y = 2e^{0.1903x}$ d) Points (0,75) and (2,60); $y = 75e^{-0.1116x}$

        e) Points (0,80) and (20,20); $y = 80e^{-0.0693x}$ f) Points (0,100) and (20,60); $y = 100e^{-0.0255x}$

1.6.18   a) y(5) = 2.99935 y(10) = 1.66595   b) y(5) = 1.52577 y(10) = 2.32797   c) y(5) = 5.17918 y(10) = 13.4119

        d) y(5) = 42.9264 y(10) = 24.5690   e) y(5) = 56.5727 y(10) = 40.0058   f) y(5) = 88.0293 y(10) = 77.4916

1.6.19   a) Points (10,1) and (20,3) $y = 0.3333e^{0.1099x}$ b) Points (10,5.5) and (20,1) $y = 30.2652e^{-0.1705x}$

        c) Points (10,2) and (20,8) $y = 0.5e^{0.1386x}$ d) Points (10,5) and (20,3.5) $y = 7.1428e^{-0.0357x}$

        e) Points (10,2) and (17,10) $y = 0.2007e^{0.2299x}$ f) Points (10,9) and (20,6) $y = 13.4999e^{-0.0405x}$

1.6.20   a) y(12) = 1.24618 y(15) = 1.73288   b) y(12) = 3.9118 y(15) = 2.3455   c) y(12) = 2.63808 y(15) = 3.99823

        d) y(12) = 4.65389 y(15) = 4.18122   e) y(12) = 3.16722 y(15) = 6.31265   f) y(12) = 8.30360 y(15) = 7.35359

1.6.21   a) Points (0,0.02) and (20,0.08) $y = 0.02e^{0.0693x}$ b) Points (0,0.8) and (4,0.3) $y = 0.8e^{-0.2452x}$

        c) Points (0,0.1) and (20,8) $y = 0.1e^{0.2191x}$ d) Points (0,0.9) and (20,0.3) $y = 0.9e^{-0.0549x}$

        e) Points (0,5) and (20,0.05) $y = 5e^{-0.2303x}$ f) Points (0,10) and (20,1) $y = 10e^{-0.1151x}$

1.6.22   a) $y(8) = 0.034817$ $y(15) = 0.056556$ b) $y(8) = 0.112506$ $y(15) = 0.020218$ c) $y(8) = 0.577073$ $y(15) = 2.67490$
         d) $y(8) = 0.580096$ $y(15) = 0.395003$ e) $y(8) = 0.79218$ $y(15) = 0.158015$ f) $y(8) = 3.98200$ $y(15) = 1.77905$

Exercise Set 1.7

1.7.1   a) $(0, +\infty)$ b) $(-\infty, +\infty)$ c) $\{x \mid x \geq 2, x \neq (2n+1)/2, n \geq 2, n \in I\}$
        d) $(-\infty, 2) \cup (2, +\infty)$ e) $\{x \mid x \geq 0, x \neq (2n+1)\pi/2, n \geq 0, n \in I\}$ f) $[2, +\infty)$

1.7.2   a) -5   b) $2/e$ c) $\sin(-1) \approx -0.8415$ d) $2e$ 1 f) 24

1.7.3   a) 0   b) $+\infty$ c) 2   d) 1   e) 0   f) 3

1.7.4   a) $\lim_{x \to 1-} |x| = 1,$; $\lim_{x \to 1+} (1/x) = 1$; continuous on $(-\infty, +\infty)$
        b) $\lim_{x \to 1-} |x - 1| = 0$; $\lim_{x \to 1+} (1 - x) = 0$, continuous on $(-\infty, +\infty)$
        c) $\lim_{x \to 1-} \text{SQRT}(1-x) = 0$; $\lim_{x \to 1+} \sqrt{x} - 1 = 0$; continous on $(-\infty, +\infty)$
        d) $\lim_{x \to 0-} 2x-1 = -1$; $\lim_{x \to 0+} 1 - x = 1$; continuous on $(-\infty, 0) \cup (0, +\infty)$
        e) $\lim_{x \to 1-} 2x = 2$; $\lim_{x \to 1+} \sqrt{x} = 1$; $\lim_{x \to 1-} \sqrt{x} = 2$; $\lim_{x \to 4-} \sqrt{x} = 2$; $\lim_{x \to 4+} 4 - x^2 = -12$; cont. on $(-\infty, 1) \cup (1, 4) \cup (4, \infty)$
        f) $\lim_{x \to 1-} 2x - 1 = 1$; $\lim_{x \to 1+} \sqrt{x} = 1$; $\lim_{x \to 4-} \sqrt{x} = 2$; $\lim_{x \to 4+} x^2 - 4x = 0$; continuous on $(-\infty, 4) \cup (4, +\infty)$
        g) $\lim_{x \to -2-} (-3) = -3$; $\lim_{x \to -2+} [x] = -2$; $\lim_{x \to 4-} [x] = 3$; $\lim_{x \to 4+} 5 - 0.5x = 3$; cont. everywhere except $x = -2, -1, 0, 1, 2, 3$
        h) $\lim_{x \to 1-} \sin(\pi x) = 0$; $\lim_{x \to 1+} x^3 = 1$; $\lim_{x \to 2-} x^3 = 8$; $\lim_{x \to 2+} 4\sqrt{2x} = 8$; continuous on $(-\infty, 1) \cup (1, \infty)$

1.7.5   a) 14   b) -25/4   c) 0   d) 1/2   e) $\lim_{x \to 3-} f(x) = -\infty$; $\lim_{x \to 3+} f(x) = \infty$ f) 0

1.7.6   a) Yes, (Apply the Intermediate Value Thm to $f(x) = x^2 - 2$ on $[-1,2]$)   b) Yes (Apply IVT to $f(x) = e^x - 10x$ on $[-1,2]$)
        c) Yes (Apply IVT to $f(x) = xe^x - 2$ on $[-1,2]$ d) Yes (Apply IVT to $f(x) = \cos(x^2) - x$ on $[-1,2]$)
        e) Solution exists on $[1/2, 2]$. Apply IVT.   f) No ($e^x \geq x + 1$ and $x + 1 > \ln(x + 2)$. Hence $e^x >$
        $\ln(x + 2)$. To see this graph $f(x) = e^x$, $g(x) = x + 1$, $h(x) = \ln(x + 2)$ on the same axes)   g) $f(x) = x - \tan(\pi x)$ is discontinuous at
        $x = (2n + 1)/2, n \in I$. Cannot apply IVT to $[-1,2]$. Use $[-1/4, 1/4]$ or $[1, 1.4]$ etc. to show existence of solutions.

Chapter 2 ANSWERS

Exercise Set 2.1

2.1.1   a) 27   b) 396   c) 715   d) 1287   e) 35   f) 2.079   g) $t + t^2 + t^3 + t^4$
        h) $(t_1 - t_0) + (t_2 - t_1) + (t_3 - t_2) + (t_4 - t_3) + (t_5 - t_4) = t_5 - t_0$

2.1.2   a) $(1/2) + (1/3) + ... + (1/21)$ b) $16 + 25 + 36 + 49 + 64$ c) $\dot{x}_1 \Delta(x_1) + \dot{x}_2 \Delta(x_2) + ... + \dot{x}_N \Delta(x_N)$
        d) $(1/3) + 1 + 3 + 9 + 27 + 81 + 243$ e) $2 + 2 + 2 + 2 + 2 + 2 + 2 + 2$ f) $\Delta x + 2\Delta x + 3\Delta x + ... N\Delta x$

→ 2.1.3   a) $\Sigma_{i=1,5} i^2$ b) $\Sigma_{i=1,10} (1/2)^i$ c) $\Sigma_{i=1,5} (2i + 3)$ d) $\Sigma_{i=1,\infty} (i a_i + b_i)$ e) $\Sigma_{i=1,4} (ai + i + 2)$ f) $\Sigma_{i=1,8} 2 - (-1)^i$ g) $\Sigma_{i=5,8} (i - (i-1))$ h) $\Sigma_{i=1,8} 4(-1)^{i+1}$ i) $\Sigma_{i=-1,3} 6(3)^i$ j) $\Sigma_{i=1,6} 8(1/2)^i$

→ 2.1.4   a) 210   b) 465   c) 204   d) 1296   e) 1365   f) 1860   g) 2522   h) 108   i) 139   j) 80   k) 2045   l) 73576

→ 2.1.5   a) $3(2^{10} - 1)$ b) $6(1 - (1/2)^{10})$ c) $(3/2)(2^{11} - 1)$ d) $4(1 - (1/2)^{11})$ e) $2(1 - (1/2)^{10})$ f) $2(2^{10} - 1)$
        g) $(10/9)(1 - (0.1)^{11})$ h) $4((0.25)^9 - 1)/3$ i) $10(1 + 0.1^{11})/11$ j) $4(1 - (0.25)^9)$ k) $20((0.1)^{25} - 1)/9$ l) $(1 - (0.2)^{100})/4$

2.1.6   a) $3(2^7 - 1)$ b) $10(1 - (0.1)^7)/3$ c) $5((0.6)^7 - 1)$ d) $2(3) + 2(4) + 2(5) = 24$ e) 271   f) $185(10)^{-5}$

2.1.7   a) $S(10) = 10$ b) $S(n) = n(n + 1)/2 + 3n$ c) $S(10) = 2(2^{10} - 1)$ $S(100) = 2(2^{100} - 1)$ d) $S(n) = hn(n + 3)/2$
        e) $S(n) = (n + 3)/2$ f) if $n_1 < n_2$, then $S(n_1) < S(n_2)$ since $\Sigma_{i=1,n_1} a_i < \Sigma_{i=1,n_1} a_i + \Sigma_{i=n_1,n_2} a_i = \Sigma_{i=1,n_2} a_i$

2.1.8   I, II, III, IV, V are not convergent. VI is convergent if $-1 < r < 1$.

2.1.9   a) 385 diverges   b) $1 - (1/2)^{10}$ converges to 1 since $(1/2)^n \to 0$.   c) $S(10) \approx 2.93$ diverges   d) 1 diverges.
        e) $S(10) = ((1/2)^{10} - 1)/3$ converges to $-(1/3)$
        f) $S(10) = 1 + 0 - (1/3) + 0 + (1/5) + 0 - (1/7) + 0 + (1/9) + 0 \approx 0.83492$ converges to $\pi/4$

2.1.10   GRAPH HERE The rectangle is divided into symmetrical halves. The area of each half is seen to be the sum of the area of the columns, $S = 1 + 2 + 3 + ... + N$. Hence $S = 1/2 N(N + 1)$.

2.1.11 $2^{31} - 1 = 214748367 \approx 2.1 \times 10^9$ 2.1.12 $20(1 - (1/2)^7) = 635/32 \approx 19.84$ gms.

2.1.13 $P(10) = 467$; $P(N) = 102 + N(N + 1)(2N + 1)/6 - 2N$ 2.1.14 $\Sigma_{i=n,N} (a_i - a_{i-1}) = a_N - a_{n-1}$; for $n = 1$ $a_N - a_0$

2.1.15   a) $20 - 9.8(1 - (1/2)^4) = 10.8125$ b) Amount left after N days = $20 - 9.8(1 - (1/2)^N)$. As $n \to +\infty$, $(1/2)^N \to 0$ so there will always be at least 0.2 mg in the system.

2.1.16 $\sin(x) = \Sigma_{n=0,\infty} (-1)^n x^{2n+1}/(2n + 1)!$ $\cos(x) = \Sigma_{n=0,\infty} (-1)^n x^{2n}/(2n)!$
        $e^x = \Sigma_{n=0,\infty} x^n/n!$ $1/1+x = \Sigma_{n=0,\infty} (-1)^n x^n$ $\ln(1 + x) = \Sigma_{n=1,\infty} (-1)^{n-1} x^n/n$.

Selected Answers

2.1.17   a) $1 + x - x^2/2! - x^3/3! + x^4/4! + x^5/5! - x^6/6! - x^7/7! + ...$ b) $-1 + x + x^2/2! - x^3/3! - x^4/4! + x^5/5! - x^6/6! - x^7/7! + ...$
    c) $x^2 - x^4/3! + x^6/5! - x^8/7! + x^{10}/9! + ...$ d) $3 + 3x + 3x^2/2! + 3x^3/3! + 3x^4/4! + ...$ e) $e + x^2 + x^3/2! + x^4/3! + x^5/4! + ...$
    f) $x - x^2 + x^3 - x^4 + x^5 - x^6 + ...$ for $|x| , 1$ g) $1 + x^2/2 - 2x^3/3 + 3x^4/4 - 4x^5/5 + ...$ for $|x| < 1$
    h) $1 - x^2/3! + x^4/5! - x^6/7! + x^8/9! - ...$ i) $1 - x/2 + x^2/3 - x^3/4 + x^{4/5} - ...$

2.1.18   a) $1 + 0 + (0)^2/2! + (0)^3/3! + ... = 1$ b) $1 + (1)^2/2! + (1)^3/3! + (1)^4/4! + ... = e$ c) $1 - x + x^2/2! - x^3/3! + x^4/4! + ...$
    d) $1 + 3t + 9t^2/2! + 27t^3/3! + 81t^4/4! + ...$ e) $1 + x^3 + x^6/2! + x^9/3! + x^{12}/4! + ...$

2.1.19   a) $1 + (t + x) + (t + x)^2/2! + (t + x)^3/3! + (t + x)^4/4! + ...$

Exercise Set 2.2

2.2.1   a) 128   b) 128   c) 42   d) 95/16   e)

2.2.2   a) $x_{n+1} = 2x_n$ b) $x_{n+1} = x_n + 2x_{n-1}$ c) $x_{n+1} = 3x_n - 2$ d) $x_{n+1} = 0.5x_n + 3$ e) f) $x_{n+1} = x_n + x_{n-1}$

2.2.3   a) $189/32 = 5.90625$ b) -2   c) 571,580,604,870   d) 5.5   e) 96   f) 4.40454

2.2.4   a) $X_{n+1} = 3X_n$; Order = 1   b) $y_{n+1} = 0.1y_{n-2}$; Order = 3   c) $X_{n+1} = X_n + 3X_{n-1}$; Order = 2
    d) $y_{n+1} = y_n + 0.1y_{n-3}$; Order = 4   e) $z_{n+1} = 1/2\, z_n + 1/2\, z_{n-1}$; Order = 2   f) $x_{n+1} = x_{n-2} - x_{n-4}$; Order = 5
    g) $H_{n+1} = -H_n + H_{n-1}$; Order = 2   h) $X_{n+1} = 3 - 2X_n$; Order = 1
    i) $Y_{n+1} = -3Y_n + Y_{n-1} + 2$; Order = 2   j) $X_{n+1} = 2X_{n+1} + 1$; Order = 0

2.2.5   a) $\Delta M_n = 0.1 M_n$ or $M_{n+1} = 1.1 M_n$ b) $\Delta H_n = 11$ or $H_{n+1} = H_n + 11$
    c) $\Delta P_n = 0.5 P_n (1000 - P_n)$ or $P_{n+1} = 501 P_n - 0.5 P_n^2$
    d) $\Delta S_n = 0.05(S_n + S_{n-1} + S_{n-2} + ... + S_{n-11})/12$ or $S_{n+1} = (12.05/12)S_n + 0.5(S_{n-1} + S_{n-2} + ... + S_{n-11})/12$

2.2.6   a) $\Delta X_n = -0.9 X_n + 2$ b) $\Delta X_n = -1.5$ c) $\Delta X_n = X_n^2 - 2x_n$
    d) $\Delta X_n = -0.9\, X_n + 2$ e) $\Delta X_n = -X_n + 2X_{n-1}$ f) $\Delta X_n = X_{n-1} - 2X_n$

2.2.7   a) No   b) Yes   c) Yes   d) Yes   e) No   f) Yes   g) No   h) No

2.2.8   a) Yes   b) No   c) Yes   d) Yes   e) No   f) NO   g) Yes   h) No

2.2.9   a) No   b) Yes   c) No   d) No   e) No   f) No   g) No   h) No

2.2.10   a) Yes   b) No   c) Yes   d) No   e) No   f) Yes   g) No   h) No

2.2.11 $\Delta Y_n = -0.22 Y_n + 25$ $Y_0 = 100$; $Y_5 = 109.7$

2.2.12   a) $\Delta T_n = 0.1(T - T_n)/T$ and   b) $\Delta T_n = 0.1(T - T_n)/T_n$

2.2.13 $\Delta R_n = 0.05 R_n$ $R_n = 1.05^n R_0$, $R_{10} = 8{,}470{,}252$

2.2.14 $\Delta R_n = -0.01 R_n + 2000$ $R_n = 0.99^n R_0 + 200{,}000(1 - 0.99^n)$ $R_{10} = 4{,}721{,}910$

2.2.15 $\Delta D_n = -0.05 D_n$ $D_0 = 136$ Solution: $D_n = D_0 (0.95)^n$. Set $D_n = 30$ and solve for n: $n = \ln(30/136)/\ln(0.95) = 29.4669$ so the drug should be replaced about every month.

Exercise Set 2.3

2.3.1   a) $x_{E1} = 0$ $x_{E2} = 1$ $x_{E3} = 4$ b) $x_0 = 3$, then $x_1 = 5$ c) $x_7 \approx 1.3$
    d) The graph of Δx_n appears to be a maximum at $x \approx 2.5$.

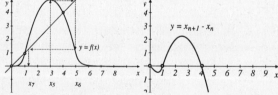

2.3.2   a) There are six equilibriums where the curve intersects $y = x$. These appear to be at the integer values $X_{En} = n$ for $n = 0, 1, ..., 5$.

    b) x_1 will be the largest at $x_0 = 4.5$, then $x_1 \approx 5.5$. There is no positive x_0 that gives a "smallest" value of x_1.

    c) If $x_5 = 3$ then there are three candidates for x_4, 2.25, 3, or 3.75 the x-coordinates where the line $y = 3$ intersects the $f(x)$ curve.

    d) The graph of $y = f(x) - 0.5$ is the graph of $f(x)$ shifted downward by 0.5.

    e) The graph of $y = \Delta x = f(x) - x$ sketched as a dotted curve. There are six x-values at which the graph of Δx intersects the line $y = 0.5$. These are indicated by the arrows pointing toward the x-axis.

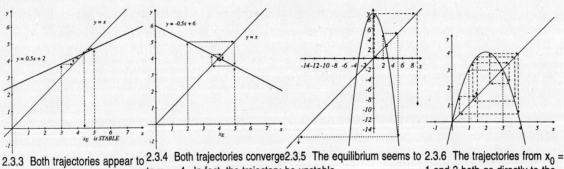

2.3.3 Both trajectories appear to converge to $x_E = 4.5$. Hence it is a stable equilibrium.

2.3.4 Both trajectories converge to $x_E = 4$. In fact, the trajectory starting at $x_0 = 5$ is identical to the trajectory starting at $x_0 = 3$ since then $x_1 = 5$.

2.3.5 The equilibrium seems to be unstable.

2.3.6 The trajectories from $x_0 = 1$ and 3 both go directly to the equilibrium $x_E = 3$. The dashed trajectory from $x = 1.5$ does not directly approach 3. The equilibrium does not seem stable.

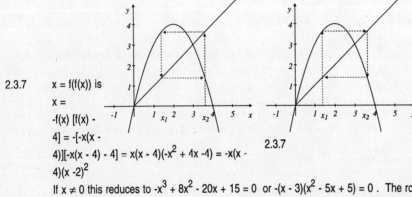

2.3.7 $x = f(f(x))$ is
$x = -f(x)[f(x) - 4] = -[-x(x - 4)][-x(x - 4) - 4] = x(x - 4)(-x^2 + 4x - 4) = -x(x - 4)(x - 2)^2$

2.3.7

If $x \ne 0$ this reduces to $-x^3 + 8x^2 - 20x + 15 = 0$ or $-(x - 3)(x^2 - 5x + 5) = 0$. The roots are the equilibrium $x_E = 3$ and the two cyclic roots $\qquad x_1 = 0.5(5 - \sqrt{5}) \approx 1.38$ and $x_2 = 0.5(5 + \sqrt{5}) \approx 3.618$

2.3.9 Solve the equation $x = 0.5(x + A/x) : x^2 = A$ or $x = \sqrt{A}$. The solution sequence for $A = 9$ and $x_0 = 1$ is {1, 5, 17/5, 3.02352, 3.00009, 3.0000, ...} while the solution sequence for $x_0 = -2$ is {-2, -13/4, -3.0961, -3.00000, ...}.

2.3.11 Try the equation $x_{n+1} = 2\sin(x_n)$ The first 30 itterates of this with $x_0 = 1.5$ and with 12 digit arithmetic are listed with $[n, x_n]$:
[[0,3/2], [1,1.99498997320], [2,1.82274182615], [3,1.93685852823], [4,1.86748816441], [5,1.91261777721], [6,1.88429134356], [7,1.90252314113], [8,1.89096274232], [9,1.89836611235], [10,1.89365408633], [11,1.89666517523], [12,1.89474587563], [13,1.89597124359], [14,1.89518971889], [15,1.89568849455], [16,1.89537030537], [17,1.89557334559], [18,1.89544380542], [19,1.89552646140], [20,1.89547372458], [21,1.89550737364], [22,1.89548590426], [23,1.89549960279], [24,1.89549086255], [25,1.89549643923], [26,1.89549288106], [27,1.89549515134], [28,1.89549370280], [29,1.89549462703], [30,1.89549403733]]
Notice that it takes 29 iterations to reach the number 1.895494

Exercise Set 2.4

2.4.1 a) $X_n = 2(-1)^n$ $X_{10} = 2$ b) $X_n = 2$ $X_{10} = 2$ c) $X_n = 2^n$ $X_{10} = 1024$ d) $X_n = 5 \cdot 3^n$ $X_{10} = 295{,}245$
 e) $X_n = 0.5(0.1)^n$ $X_{10} = 5 \cdot 10^{-11}$ f) $X_n = 2(1/4)^n$ $X_{10} = 2^{-19} \approx 1.9073486 \cdot 10^{-6}$

2.4.2 a) $X_n = 2$ $X_{10} = 2$ b) $X_n = 2^{n+1}$ $X_{10} = 2048$ c) $X_n = 0$ $X_{10} = 0$
 d) $X_n = 5 \cdot 4^n$ $X_{10} = 5242880$ e) $X_n = 0.5(1.1)^n$ $X_{10} = 1.296871230$ f) $X_n = 2(3/4)^n$ $X_{10} = 0.112627029$
 g) $X_n = (1/2)^n$ $X_{10} = 9.765625 \cdot 10^{-4}$ h) $X_n = 2(-0.3)^n$ $X_{10} = 1.18098 \cdot 10^{-5}$

2.4.3 $P_n = P_0(1 + I)^n$; $P_{25} = 500(1.04)^{25} \approx \$1{,}332.92$. Solve $500(1.04)^n = 100{,}000$ for $n = \ln(100000/500)/\ln(1.04) \approx 135.08978$ yrs. It would take 136 years of compound reinvestment.

2.4.4 The interest for one year of monthly compounding is $(1 + I/12)^{12} - 1$. $I = 0.24$ Annual interest rate is $(1.02)^{12} - 1 = 26.824\%$.

2.4.5 $\Delta P_n = (P_n + Ir)$ where $I = P_n \cdot r$. Thus $\Delta P_n = (P_n + P_n r)r$ or $P_{n+1} = (1 + r + r^2)P_n$. For $r = 10\%$, $P_{n+1} = 1.11 P_n$. For \$10,000 the "in advance" interest is \$1,100 compared to "not in advance" interest of \$1,000. For monthly compounding the difference in rates is: "in advanced" $(1 + 0.1/12 + (0.1/12)^2)^{12} - 1$ compared to $(1 + 0.1/12)^{12} - 1$ or in decimals 10.5626399% for "in advanced" and 10.4713067% for "not in advanced".

2.4.6 a) $X_n = 2 + n$ $X_5 = 7$ b) $X_n = (-2)^n + 1$ $X_5 = -31$ c) $X_n = (9/5)6^n - 4/5$ $X_5 = 13,996$
       d) $X_n = (107/21)(-20)^n - 2/21$ $X_5 = -16304762$ e) $X_n = (11/6)(1/4)^n - 4/3$ $X_5 = -2727/2048 \approx -1.331542969$
       f) $X_n = (8/3)(1/4)^n - 2/3$ $X_5 = -85/128 = -0.6640625$

2.4.7 a) $X_n = 2 + n$ $X_5 = 7$ b) $X_n = 2^n + 1$ $X_5 = 33$ c) $X_n = (-1/2)3^n + 1/2$ $X_5 = -121$ d) $X_n = 6 \cdot 4^n - 1$ $X_5 = 6143$
       e) $X_n = 2.5(1.1)^n - 2$ $X_5 = -81051/40000 = 2.026275$ f) $X_n = 8(3/4)^n - 6$ $X_5 = -525/128 = -4.1015625$

2.4.8 a) $-4/3$ b) $-2/3$ c) -2 d) ∞ e) 0.25 f) $5/12$ g) ∞ h) ∞

2.4.9 $A = \$6,401.19$ 7.3.10 $P_0 = \$3,395,685.70$

2.4.11 I_n is the number of infected individuals. $\Delta I_n = 0.2 I_n$ Sol. $I_n = I_0(1.2)^n$

2.4.12 $\Delta A_n = (R/100) A_n - 150$ Sol. $A_n = (A_0 - 15000/R)(1 + R/100)^n + 15000/R$ The minimum size is $A_0 > 15000/R$ so that it reproduces more than the predation. and $A_0 = 2000$ it takes 16.886 years for the colony to double in size.

2.4.13 a) 376.32 app.   b) $r = 2.6826$ c) $r = 0.998$ $Y_0 = -10010.2408$

2.4.14 a) $\Delta D_n = 1.2 D_n/2 - 0.2 D_n$ or $D_{n+1} = 1.4 D_n$ b) $D_{n+1} = 1.4 D_n - H$ c) $D_{10} = 76,304.34$ when $D_0 = 3000$ and $H = 150$. Since the population has a reproduction rate greater than 1, the population can only reach a steady state if the harvest equals the reproduction $H = D_0 * 0.4$. For $D_0 = 3000$ the harvest must be $H = 1200$. The only way a steady state could be reached is for the birth rate/doe to be less than twice the population death rate.

Exercise Set 2.5

2.5.1 a) $C_1 + C_2(-3)^n$ b) $C_1 2^n + C_2(-2)^n$ c) $C_1(-4)^n + C_2(-2)^n$ d) $C_1(\sqrt{3/2})^n + C_2(-\sqrt{3/2})^n$ e) $(C_1 + C_2 n)(-3)^n$
       f) $(C_1 + C_2 n) 2^n$ g) $C_1(5/2 + \sqrt{21}/2)^n + C_2(5/2 - \sqrt{21}/2)^n$ h) $C_1 + C_2(-1)^n$ i) $C_1 + C_2(-1/4)^n$
       j) $C_1 + C_2 2^n$ k) $C_1[(1+\sqrt{5})/2]^n + C_2[(1-\sqrt{5})/2]^n$
       l) $C_1[(3+\sqrt{13})/2]^n + C_2[(3-\sqrt{13})/2]^n$ m) $\sqrt{3}^n (C_1 \cos(n \text{Arccos}(-1/3^{1/2})) + C_2 \sin(n \text{Arccos}(-1/3^{1/2})))$
       n) $5^{-n}(C_1 \cos(n \pi/2) + C_2 \sin(n \pi/2))$ o) $(C_1[(-5+\sqrt{41})/2]^n + C_2[(-5-\sqrt{41})/2]^n$
       p) $2^n(C_1 \cos(n \text{Arccos}(0.25)) + C_2 \sin(n \text{Arccos}(0.25)))$ q) $10^n(C_1 \cos(n \text{Arccos}(-0.6)) + C_2 \sin(n \text{Arccos}(-0.6)))$
       r) $(C_1 + C_2 n)(-0.1)^n$

2.5.2 a) $16 \cdot 5^n/3 - 10 \cdot 8^n/3$ b) $100 \cdot 8^n/387 - 100 \cdot 5^n/387$ 2.5.3 a) $3^n(1 + 7n/3)$ b) $25 \cdot 3^n n/324$

2.5.4 a) $3^n(\cos(n \text{Arccos}(5/6)) - 5/\sqrt{11} \sin(n \text{Arccos}(5/6)))$ b) $3^n(50/\sqrt{11}) \sin(n \text{Arccos}(5/6))$

2.5.5 $f_n = C_1((1+\sqrt{5})/2)^n + C_2((1-\sqrt{5})/2)^n$ $C_1 = (5+\sqrt{5})/10$ $C_2 = (5-\sqrt{5})/10$

2.5.6 a) $x_{n+2} - 7 x_{n+1} + 10 x_n = 0$ b) $y_{n+2} - 6 y_{n+1} + 9 y_n = 0$ c) $x_{n+2} - 0.6 x_{n+1} + 0.05 x_n = 0$
       d) $y_{n+2} - 8 y_{n+1} + 16 y_n = 0$ e) $w_{n+2} + w_n = 0$ f) $z_{n+2} - 4 \cos(1) z_{n+1} + 4 z_n = 0$
       g) $w_{n+2} + 4 w_n = 0$ h) $z_{n+2} - 0.5 z_{n+1} + 0.25 z_n = 0$

2.5.7 For equations a), b), f), g)

2.5.8 ??   2.5.9 a) $b = -10$ b) $b = 9/4$ c) $b = 9$

2.5.10 For $k \in (-\sqrt{2}, \sqrt{2})$ complex roots with $b < 1$, hence stable steady state $y = 0$. If $\sqrt{2} < |k|$ 3/2 repeated real roots with $|\lambda| = |\pm \sqrt{2}/2| < 1$. If $|k| > 3/2$ then roots are real and distinct and both are less than one so zero is the steady-state.

2.5.11 Model equation $y_{n+2} - 2 y_{n+1} + F y_n = 0$ $0 < F < 1$.

2.5.12 a) $C_1 2^n + C_2(-3)^n - 1$ b) $C_1(-4)^n + C_2 + 2n$ c) $C_1 + C_2 n + 2 n^2$ d) $C_1 + C_2(-2)^n + 2n/3$
       e) $C_1(1+\sqrt{5})^n + C_2(1-\sqrt{5})^n - 3/5$ f) $3^n(C_1 \cos(2 n \pi/3) + C_2 \sin(2 n \pi/3)) + 2/13$
       g) $C_1((1+\sqrt{17})/2)^n + C_2((1-\sqrt{17})/2)^n - 3/4$ h) $(-5)^n(C_1 + C_2 n) + 1/18$

2.5.13 a) $11/5 \cdot 2^n + 4/5 \cdot (-3)^n - 1$ b) $3/5 (-4)^n + 7/5 + 2n$ c) $1 - 3n + 2n^2$ d) $16/9 + 2/9 (-2)^n + 2n/3$
       e) $(8+\sqrt{5})/10 (1+\sqrt{5})^n + (8-\sqrt{5})/10 (1-\sqrt{5})^n - 3/5$ f) $3^n(24/13 \cos(2 n \pi/3) + 68/(39\sqrt{3}) \sin(2 n \pi/3)) + 2/13$
       g) $(51 - 11\sqrt{17})/136 ((1+\sqrt{17})/2)^n + (51 - 11\sqrt{17})/136 ((1-\sqrt{17})/2)^n - 3/4$ h) $(-5)^n (17/18 - 17 n/15) + 1/18$

2.5.14 a) $x_{n+2} - 7 x_{n+1} + 10 x_n = 12$ b) $y_{n+2} - 6 y_{n+1} + 9 y_n = -0.8$ c) $x_{n+2} - 0.6 x_{n+1} + 0.05 x_n = -0.9$
       d) $y_{n+2} - 8 y_{n+1} + 16 y_n = -9$ e) $w_{n+2} + w_n = -1$ f) $z_{n+2} - 4 \cos(1) z_{n+1} + 4 z_n = 7(5 - 4\cos(1))$
       g) $w_{n+2} + 4 w_n = -5$ h) $z_{n+2} - 0.5 z_{n+1} + 0.25 z_n = -0.75$

2.5.15 a) $x_{n+2} - 4 x_{n+1} + 3 x_n = 5$; $x_n = C_1 + C_2 3^n - 5n/2$ b) $C_1 = -1$ $C_2 = 2$ and then $x_{20} = 2 \cdot 3^{20} - 51$

2.5.16 $\alpha_2 = 1/2$ $\alpha_3 = 5/8$ $\alpha_{20} = 518815/524288 \approx 0.989561$

2.5.17 a) $x_{n+2} - 2 x_{n+1} - 3 x_n = -4$ b) $x_{n+2} - x_{n+1} + 0.25 x_n = -0.5$ c) $x_{n+2} - 7 x_{n+1} + 10 x_n = -4$
       d) $x_{n+2} + 4.8 x_{n+1} - 1 = 0.48$ e) $x_{n+2} + 4 x_n = 25$ f) $x_{n+2} - 1/4 x_{n+1} + 1/16 x_n = 39/16$

Chapter 3 ANSWERS

Exercise Set 3.1

3.1.1   a) 2   b) 2   c) -4/3   d) -4   e) -800   f) 800   g) $5(e^{-2} - 1) \approx 1.1070$ h) $10\, ln\,(1.1) \approx 0.0953$
      i) $5(e^{5.2} - e^5) \approx 164.2954$ j) $10\, ln\,(1.05) \approx 0.4879$

3.1.2   a) 2   b) $2t_0 + 0.1$ c) $-1/t(t + 0.1)$ d) $3x_0^2 + .3x_0 + 0.01$ e) $1/(SQRT(z_0 + 0.1) + SQRT(z_0))$
      f) $10.2^t(2^{0.1} - 1) \approx 2^t(.7177)$ g) $10e^x(e^{0.1} - 1) \approx 1.0517 e^x$ h) $10e^{-x}(e^{-0.1} - 1) \approx -0.9516 e^{-x}$

3.1.3   a) 4   b) $(\sqrt{3} - 1)/2 \cong .3660$ c) 0   d) 1/2   e) 3   f) -1   g) $(e^3 - e)/2 \approx 8.6836$ h) $(e^6 - e^2)/2 \approx 198.0199$

3.1.4   3/2 km   2.1.5   b) 3.6 cm   c) 0.1333 cm/day; 0.9333 cm/week

3.1.6   The graph is constructed by connecting the points in the following table with straight line segments.

t	0	10	20	30	40	50	60	70	80	90	100	110	120
c(t)	50	52	67	87	105	123	139	153	161	161	151	116	76

3.1.7   a) $y = x + 3$ b) $y = 0.001x + 3$ c) $y = 0.000001x + 3$

3.1.8 $\Delta t = 1$ a) $y = 3t - 2$ b) $y = (\sqrt{2} - 1)t + 2 - \sqrt{2}$ c) $y = 0$ d) $y = \log_3(2)(t - 1)$ e) $y = 2t$ f) $y = 2$
      g) $y = e^2(t - 1) + e(2 - t)$ h) $y = e^4(t - 1) + e^2(2 - t)$

      $\Delta t = -1/2$ a) $y = 3/2\,t - 1/2$ b) $y = (2 - \sqrt{2})t + \sqrt{2} - 1$ c) $y = -2t + 2$ d) $y = 2\log_3(2)(t - 1) = \log_3(4)(t - 1) = \log_3(4)t - \log_3(4)$
      e) $y = (4 - 2\sqrt{2})t + (2\sqrt{2} - 2)$ f) $y = 3/2\,t + 1/2$ g) $y = 2(e - \sqrt{e})(t - 1) + (2\sqrt{e} - e)$ h) $y = 2(e^2 - e)t + (2e - e^2)$

      $\Delta t = 1/2$ a) $y = (5/2)t - 3/2$ b) $y = (\sqrt{6} - 2)t + (3 - \sqrt{6})$ c) $y = y = -2t + 2$ d) $y = 2\log_3(3/2)(t - 1)$
      e) $y = 4(\sqrt{2} - 1)t + 6 - 4\sqrt{2}$ f) $y = 1/2\,t + 3/2$ g) $y = 2e(\sqrt{e} - 1)t + 3e - 2e\sqrt{e}$ h) $y = 2e^2(e - 1)t + 3e^2 - 2e^3$

3.1.9   a) $y = (2/\pi)x$ b) $y = (4/\pi)x$ c) $y = 2x + 1$ d) $y = (3\sqrt{3}/2)(x - 1)$ e) $y = (8/\pi)(\sqrt{2} - 1)x + 2$ f) $y = (4/\pi)(\sqrt{2} - 1)x + 1$

3.1.10   a) $-2.8\pi(\Delta c/\Delta x)$ b) HINT: Let $c(x) = k$ then $\Delta c/\Delta x = 0$ if $\Delta x \neq 0$ c) see exercises

3.1.11   a) $H(1) = (11,0)$ $F(1) = (1,9)$ b) $H(3) = (13,0)$ $F(3) = (241/72, 7)$
      c) $H(1) = (11,0)$ $F(1) = (11/10,9)$ $H(3) = (13,0)$ $F(3) = (1313/360,7)$

3.1.12   140 mg/l   2.1.13 $T = ln\,4/5$; average rate of change is $5/ln\,4 \approx 3.61$

3.1.14   10 jumps; shortest horizontal jump $\approx .16$ unit from $(5 - \sqrt{10}, 0)$ to $(2,1)$ longest horizontal jump 1 unit from $(4,9)$ to $(5,10)$

Exercise Set 3.2

3.2.1   a) $f'(1) = 2$ $f'(2) = 2$ $f'(0) = 2$ b) $g'(1) = -5$ $g'(2) = -5$ $g'(3) = -5$

3.2.2   a) $f'(0) = 1$ $f'(1) = 3$ $f'(2) = 5$

3.2.3   a) $f'(0) = 1/2$ b) $f'(1) = 1/2\sqrt{2} = \sqrt{2}/4$ c) $f'(100) = 1/2\sqrt{101} = \sqrt{101}/202$

3.2.4   a) $y' = -5$ b) $y' = 6x$ c) $y' = -2/(2x + 5)^2$

3.2.5   a) $y = 3x - 1$ b) $y = -1$ c) $y = t - 2$ d) $y = 0$ e) $y = 2t + 1$ f) $y = 4\pi t - 4\pi$ g) $y = -t + 1$ h) $y = 1$

3.2.6   a) $f'(-2/3) = 3$ b) $f'(2) = 4$; $f'(-2) = -4$ c) $M'(2) = 12$ d) $g'(3) = 5$; $g'(-2) = -5$
      e) $g'(0) = 2$; $g'(2) = -2$ f) $f'(2) = 0$

3.2.7   a) $D^+_x f(1) = D^-_x f(1) = 2$ b) $D^+_x f(3) = D^-_x f(3) = 0$ c) $D^+_x f(0) = 3$; $D^-_x f(0) = -3$
      d) $D^+_x f(0) = +\infty$; $D^-_x f(0)$ does not exist   e) $D^+_x f(1/3) = 3$; $D^-_x f(1/3) = -3$ f) $D^+_x f(0) = +\infty$; $D^-_x f(0) = -\infty$
      g) $D^+_x f(1) = -2$; $D^-_x f(1) = 2$ h) $D^+_x f(0) = 1/2$; $D^-_x f(0) = 2$ i) $D^+_x f(2) = 2$; $D^-_x f(2) = 1$
      j) $D^+_x f(2) = -1/2$; $D^-_x f(2) = -1/4$

3.2.8 $y = |x + 2| = x + 2$ if $x \geq -2$ thus $D^+_x f(-2) = 1$
      $y = |x + 2| = -(x + 2)$ if $x < -2$ thus $D^-_x f(-2) = -1$ Since $D^+_x f(-2) \neq D^-_x f(-2)$, f is not differentiable at $x = -2$

3.2.9 $y = [3x]$ is 1 if $1/3 \leq x < 2/3$ and is 2 if $2/3 \leq x < 1$
      $D^+_x f(2/3) = \lim_{h \to 0+} [f(2/3 + h) - f(2/3)]/h = 0$ since if $0 < h < 1/3$ then $2/3 < 2/3 + h < 1$
      $D^-_x f(2/3) = \lim_{h \to 0-} [f(2/3 + h) - f(2/3)]/h = +\infty$ since if $-1/3 < h < 0$ then $1/3 < 2/3 + h < 2/3$

2.2.10 a) $f'(2) = 3$; $\Delta f(2) = 3h$ b) $f'(2) = 4$; $\Delta f(2) = h(4 + h)$

h	$\Delta f(2)$	$\Delta f(2)/h$	Error
-0.1	-0.3	3	0
0.1	0.3	3	0
0.01	0.03	3	0
0.001	0.003	3	0

h	$\Delta f(2)$	$\Delta f(2)/h$	Error
-0.1	-0.39	3.9	0.1
0.1	0.41	4.1	0.1
0.01	0.401	4.01	0.01
0.001	0.4001	4.001	0.001

c) $f'(2) = 0$; $\Delta f(2) = h^2$

h	$\Delta f(2)$	$\Delta f(2)/h$	Error
-0.1	0.01	-0.1	0.1
0.1	0.01	0.1	0.1
0.01	0.001	0.01	0.01
0.001	0.000001	0.001	0.001

2.2.11

	h = 0.1	h = -0.05
(a)	0.4559	0.6915
(b)	1.0033	1.0926
(c)	-2.9552	-1.9950
(d)	2.5892	1.3675
(e)	25.8925	13.6754
(f)	0.7177	0.5858
(g)	1.0517	0.7869
(h)	-0.9516	-1.2954
(i)	21.1241	15.8061
(j)	0.9516	1.3863
(k)	0.4139	0.6021
(l)	0.0995	0.1026

3.2.12 a) Let $h = \Delta t$

$\Delta ln(a)/\Delta t = [ln(a + h) - ln\ a]/h$

$= 1/h\ ln[(a + h)/a]$

$= 1/h\ ln(1 + h/a) = ln(1 + h/a)^{1/h}$

b) NOTE: $\varepsilon = ha \to 0$ as $h \to 0$)

$\lim_{h \to 0} ln[(1 + h/a)^{1/h}] = \lim_{\varepsilon \to 0} ln[(1 + \varepsilon)^{1/a\varepsilon}]$

$= ln[(\lim_{\varepsilon \to 0} (1 + \varepsilon)^{1/\varepsilon})^{1/a}] = ln[e^{1/a}]$

$= 1/a\ ln\ e = 1/a \cdot 1 = 1/a$

c) $f(t) = ln(t)$

3.2.13 a) The Average = $\{[f(x + h) - f(x)]/h + f(x - h) - f(x)]/(-h)\}/2$

$= \{(f(x + h) - f(x) + f(x) - f(x - h))/h\}/2$

$= (f(x + h) - f(x - h))/2h$

b) $f'(2) = -1$ $\{f(2 + h) \cdot f(2 - h)\}/2h = \{(h^2 - h + 1) - (h^2 + h + 1)\}/2h = -2h/2h = -1$

Hence Central Difference Approximation is exactly $f'(2)$ $\therefore$ No error.

c) Central Difference Approximation = $\{SQRT(3.5(2 + h) + 7) - SQRT(3.5(2 - h) + 7)\}/2h$

h	C.D.A.	C.D.A. Error	Error $\Delta f/h$
0.5	0.46827	0.0009198	0.013768013
0.1	0.4677437	0.0000366	0.002887191
0.01	0.4677075	0.0000004	0.000291952

Exercise Set 3.3

3.3.1 a) 0   b) 7   c) $2x$ d) $2.6x^{1.6}$ e) $12x^{11}$ f) $-5x^{-6}$ g) $-x^{-2}$ h) $-4.6x^{-5.6}$ i) $3/2\ x^{1/2}$
j) $2/3\ x^{-1/3}$ k) $1/8\ x^{-7/8}$ l) $0.6x^{-0.4}$

3.3.2 a) $6x - 2$ b) $20x^3 - 10x^4$ c) $9x^2 - 70x^9$ d) $3/2\ x^2 - 4x$ e) $3/2\ x^{1/2}$ f) $(1/2)x^{-1/2}$ g) $1/4\ x^{-3/4} + 1$
h) $1/2\ x^{-1/2} + 1$ i) $3x^{-1/4} - 2x$ j) $1.1x^{0.1} - 6x$ k) $-2/3x^{-5/3} + 2/3\ x^{-1/3}$ l) $5/2\ x^{3/2} - 2/5\ x^{-3/5}$

3.3.3 a) $9x^2 - 4x + 3$ b) $2x + 5x^{-6}$ c) $2x + 3$ d) $2x^{-6}(5 - 6x)$ e) $2x - 9$ f) $-2x/(x^2 + 1)^2$ g) $-3/(3x - 2)^2$ h) $5x^4$
i) $(3x^2 - 4x - 3)/(3x - 2)^2$ j) $(3 + 4x - 3x^2)/(x^2 + 1)^2$ k) $-(7x^2 + 5)/[x^6(x^2 + 1)^2]$ l) $2(5 - 9x)/[x^6(3x - 2)^2]$
m) $2x(9x^4 - 5x^3 + 1)$ n) $(-3x^8 + 4x^7 + 3x^6 - 7x^2 - 5) / [x^6(x^2 + 1)^2]$ o) $-10x^{-11}$

3.3.4 a) $x^2 + 2x - 3$ b) $3x^2 - 6 + x^{-1}$ c) $x^{-1} + 1 - 2x$ d) $x^2 + 2x - 3 + x^{-1}$ e) $x^2 - 2 - 2x^{-1}$ f) $x^{-1} + 3 - 2x - x^2$

3.3.5   a) $9x^2 + 4x - 3$ b) $1/2\, x^{1/2}(5x + 3)$ c) $4t^3$ d) $s^4[7s^2 - 18s + 35]$ e) $2(t + 1)$ f) $2t + 4t^3$ g) $4t^3 - 2t + 6$
   h) $7s^6 - 18s^5 + 35s^4 - 12s^2 + 24s - 28$

3.3.6   a) $y = -5x - 1$ b) $y = 2$ c) $y = -9x + 16$ d) $y = 1/80\, x + 28/5$ e) $y = 19x + 11$ f) $y = 3x - 2$
   g) $y = 51/49\, x - 144/49$ h) $y = -8/9\, x + 31/9$

3.3.7   a) 96.3 b) 91.3 c) -0.9245^0 d) -5.1^0

Exercise Set 3.4

3.4.1   a) $2^x \ln(2)$ b) $3^x \ln(3)$ c) $3^{2x}\, 2\ln(3)$ d) $2^x[x\ln(2) + 1]$ e) $-\ln(3)/3^x$
   f) $3^x 2^x(\ln(2) + \ln(3))$ g) $7^x \ln(7) - 10^x \ln 10$ h) $3^x \ln(3)/(1 + 3^x)^2$ i) $\ln(5)(5^x - 5^{-x})$

3.4.2   a) $4e^x$ b) $e^x(x + 1)$ c) $(1 - x)/e^x$ d) $3e^{3x}$ e) $-e^{-x}$ f) $5(1 - x)/e^x$ g) $4e^x - e^{-x}$ h) $ex^{e-1} + e^x$
   i) $e^x + e^{-x}$

3.4.3   a) $3/x$ b) $(1/x) + 3$ c) $2/x$ d) $1/x$ e) $-1/x$ f) $-7/x$ g) $1 + \ln(x)$ h) $-2/x$ i) $-1/\{x(\ln(x))^2\}$

3.4.4   a) $1/(x\ln(10))$ b) $1/(x\ln(2))$ c) $1/[x\ln(5)]$ d) $-1/[x\ln(10)(\log(x))^2]$ e) $-1/x\ln(2)$
   f) $\ln(3/5)/[x\ln(5)\ln(3)]$ g) $2\ln x/(x\ln(10))$ h) $(\ln(5) - 1)/x\ln(5)$ i) 0

3.4.5   a) $-8\sin(x)$ b) $2\cos(\theta)$ c) $-5\csc(\theta)\cot(\theta)$ d) $-\sin(x) + \sec^2(x)$ e) $-2\csc(\theta)\cot(\theta)$ f) $-2\sin(\theta) - 2\theta$
   g) $\cos^2(\theta) - \sin^2(\theta)$ h) 0 i) $(t\cos(t) - \sin(t))/t^2$ j) $t[2\sin(t) + t\cos(t)]$ k) $3\cos(x) + \sec(x)\tan(x)$
   l) $\cot(\theta) - \theta\csc^2(\theta)$ m) $-\csc(x)\cot(x) + \cos(x)$ n) $-\csc^2(t)$ o) $2\sin(x)\cos(x)$

3.4.6   a) 1 b) -5 c) $3\ln(3)$ d) 1 e) 0 f) 1

3.4.7   a) $f'(t) = \cos(t)$ b) $f'(t) = 2t\sin(t) + t^2\cos(t)$

	$t_0 = 0$	$t_0 = \pi/2$	$t_0 = \pi$
a)	1	0	-1
b)	0	π	$-\pi^2$

3.4.8   a) $y = 2$; $s = 2$ for all t 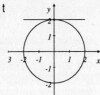   b) $x = -1$; $s(\pi) = 5$

   c) $y = -x + 4$; $s(\pi/6) = 2\sqrt{6}$ 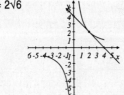   d) $y = \pi x + \pi^2$; $s(\pi) = \text{SQRT}(1 + \pi^2)$

   e) $y = x - 1$; $s(0) = \sqrt{2}$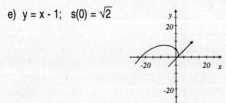

3.4.9   a) $p_t = (\cos(2t), \sin(2t))$ b) HINT: $\Delta[p_t, p_{t+\Delta t}] = 2(t + \Delta) - 2t$ c) $D_t \sin(2t) = 2\cos(2t)$ $D_t \cos(2t) = -2\sin(2t)$

Exercise Set 3.5

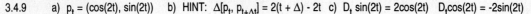

3.5.1   a) $8x$ b) -3 c) $2x + 5$ d) $2x\exp(x^2 + 1)$ e) $2/x$ f) $-2x\exp(-x^2)$ g) $9/2\, x^{-3/2}(x + 1)[3(x^{1/2} - x^{-1/2}) - 2]^2$
   h) $-2x\cos(1 - x^2)$

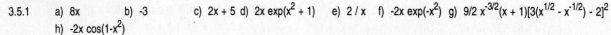

3.5.2   a) $4(2x + 1) + (2x + 1)^{-1/2}$ b) $4(2x + 1) + 1$ c) $2x(x^2 + 2)\exp(x^2 + 1)$ d) $3[\exp(3x - 2) + 1]$ e) $4x(x^2 + 1)^{-2}$
   f) $2[(x + 3)^2 - 1]/(x + 3)$ g) $-[1 + \ln(1 - x)]$ h) $[1 + \exp(e^x + 1 + x)](e^x + 1)$

3.5.3 a) $6(3x + 1)$ b) $8x(2x^2 + 2)$ c) $15x^4(x^5 + 1)^2$ d) $3x^2(x^4 + x^3 + 1)^2(4x + 3)$
e) $1/2(x^2 + 3x - 1)^{-1/2}(2x + 3)$ f) $-(2x + 1)^{-3/2}$ g) $6(x + 3)^5$ h) $2(x + 3)^5(2x - 1)(8x + 3)$
i) $8(2x + 1)^3 + 6x(x^2 - 1)^2$ j) $2(2x - 1)/3x^3$ k) $(3x^2 + 20x + 118)(5x + 1)^{-4}$ l) $(-3x^2 + 20x + 122)(5x - 1)^{-4}$
m) $10x(x^5 - 1)(x^2 - 1)^4(2x^5 - x^3 - 1)$ n) $-2x(x^2 - 1)^{-2}$ o) $(-3x + 1)/[(x + 1)^3(x - 1)^2]$ p) $2(x + 1)(5x^3 + x^2 + 5x - 1)$

3.5.4 a) $y = (-1/\sqrt{3})x + (2/\sqrt{3})$ b) $y = 1.28x - .472$ c) $y = x - 1$ d) $y = \sqrt{2}$ e) $y = 2x + 1$ f) $y = 3(x - 1)$

3.5.5 a) $2x\,e^{2x}$ b) $1/x$ c) $4x\exp(2x^2)$ d) $2/x$ e) $(2 - 3x^2)\exp(2x - x^2)$
f) $(2 - 3x^2)/(2x - x^3)$ g) $2(x - 1)\exp((x - 1)^2)$ h) $2/(x - 1)$ i) $\cos(x)\exp(\sin(x))$ j) $3^{2x}\cdot 2\ln(3)$ k) $1/(x\ln(10))$
l) $\exp_8(x^2 - 4x)(2x - 4)\ln(8)$ m) $2^{\ln x}(1/x)\ln(2)$ n) $5(x^{10} + 1)/[x(x^{10} - 1)\ln(8)]$ o) $9x^2$

3.5.6 a) $\cos(\theta)/\sin(\theta) = \cot(\theta)$ b) $\exp(x^2)[2x^2 + 1]$ c) $\ln(x^2) + 2$ d) $e^{-\theta}(\cos(\theta) - \sin(\theta))$
e) $\cos(\theta)\ln(2\theta) + (\sin(\theta)/\theta)$ f) $e^\theta\cos(e^\theta)$ g) $\cos(\ln(\theta))/\theta$ h) $\exp(e^x)\cdot e^x = \exp(e^x + x)$ i) $1/(x\ln(x))$
j) $6\exp(2e^{3x} + 3x)$ k) $[\cos^2(\theta) - \sin^2(\theta)]/[\sin(\theta)\cos(\theta)] = \cot(\theta) - \tan(\theta)$ l) $\exp(\sin(\theta)\cos(\theta))\cos^2(\theta) - \sin^2(\theta)]$

3.5.7 a) $-2\cos(x)\sin(x)$ b) $-2x\sin(x^2)$ c) $2x\cos(1 + x^2)$ d) $3\cos(x)(\sin(x) + 1)^2$ e) $-2\sec^2(1 - 2x)$
f) $(-\sin(x) - 1)/2(\cos(x) - x)^{1/2}$ g) $\cos(t)\cos(\sin(t))$ h) $-5\sin(t)\cos(\cos(t))\sin^4(\cos(t))$

3.5.8 a) $(3^{2t-7})(2\ln(3))$ b) $(2t\ln 2)\cdot 2^{(-4-t^2)}$ c) $\exp_5(7^t)(7^t\ln(7)\ln(5))$ d) $3/[(3t - 5)\ln(3)]$ e) $-2t/[(9 - t^2)\ln(2)]$
f) $(2t^5 + 3)/[t(t^5 - 1)(\ln(7))^2\log_7(t^2 - t^3)]$ g) $2\ln 3/\ln(10)$ h) $\exp_3[\log_2(4t\text{-}t^{-2})]\ln(3)(4 + 2t^{-3})/[(4t - t^{-2})\ln(2)]$ i) $15t^2/\ln(8)$

3.5.9 a) $1/(x\ln(x))$ b) $2x\cos(x^2)\cos(\sin(x^2))$ c) $-1/[x\ln(1/x)]$ d) $\cos(x)\cdot\cos(\sin(x))\cdot\cos(\sin(\sin(x)))$

3.5.10 a) $x^x(\ln(x) + 1)$ b) $x^{\sin(x)}[\ln(x)\cos(x) + \sin(x)/x]$ c) $\sin(x)^x[\ln(\sin(x)) + x\cot(x)]$
d) $(x^4 - x^{-2})^{3x}[3\ln(x^4 - x^{-2}) + (12x^3 + 6x^{-3})/(x^4 - x^{-2})]$ e) $x^{(x^x)}x^x[(1/x) + (1 + \ln(x))\cdot\ln(x)]$ f) $\ln(x)^{\ln(x)}[1 + \ln(\ln(x))]/x$

3.5.11 $-16\sqrt{\pi}$ cm^2/min   3.5.12 $-25/\pi$ cm^3/min   3.5.13 $-4/5$ cm^2/min 3.5.14 $5/6$ cm/sec
3.5.15 $3125/12 = 260.41\underline{6}$ cm^3/sec 3.5.16 -8% 3.5.17 $20\sqrt{\pi/3}$ cm 3.5.18 $H' = -1/5$ cm/hr

Exercise Set 3.6

3.6.1 a) 0 b) $24x^{-5}$ c) $3/8\,x^{-5/2}$ d) 6 e) $(3/8)x^{-5/2} + (15/8)x^{-7/2} = 3(x + 5)/(8x^{7/2})$ f) $-3/8(1 - x)^{-5/2}$ g) $24x$
h) $0.231x^{0.9}$ i) $-6x^{-4}$ j) $48x(x^2 + 1)(7x^2 + 3)$ k) $-3x(1 - x^2)^{-5/2}$ l) $6(1 + x)^{-4}$

3.6.2 a) $f'(x) = 2x\exp(x^2)$, $f'(0) = 0$, $f''(x) = 2\exp(x^2)(1 + 2x^2)$, $f''(0) = 2$
b) $f'(x) = (2/x)$, $f'(1) = 2$, $f''(x) = (-2/x^2)$, $f''(1) = -2$
c) $f'(x) = e^{-3x}(2 - 3x)$, $f'(\ln(2)) = (1 - \ln(8))/8$ $f''(x) = -3e^{-3x}(2 - 3x)$, $f''(\ln(2)) = -3(2 - \ln(8))/8$
d) $f'(x) = -0.1e^{-0.1x} + 0.005e^{-0.005x}$, $f'(20) \approx -0.009$ $f''(x) = 0.01e^{-0.1x} - 0.000025e^{-0.005x}$, $f''(20) \approx 0.0013$

3.6.3 a) $-\sin(x)$ b) $-9\sin(3x)$ c) $2\sec^2(x)\tan(x)$ d) $-2\sin(x^2) - 4x^2\cos(x^2)$ e) $-4\pi^2\sin(2\pi x - \pi/2)$
f) $2\cos(x) - x\sin(x)$

3.6.4 a) $y' = 2\cos(2t)$, $y'' = -2^2\sin(2t)$, $y''' = -2^3\cos(2t)$, $y^{iv} = 2^4\sin(2t)$ b) $y^{(85)}(t) = 2^{85}\cos(2t)$ NOTE: $85 = 4(21) + 1$

3.6.5 $y' = -4\cos(2\Theta) + 12\sin(2\Theta)$ 3.6.6 The table:

	$f(x_0)$	$f'(x_0)$	$f''(x_0)/2$	$f'''(x_0)/6$
a)	0	3	3	1
b)	1	3	-10	10
c)	-10	-12	-3	0
d)	1	1	1/2	1/6
e)	1	-0.1	0.005	1/6000
f)	9	18	18	12
g)	0	1	0	-1/6
h)	1	0	-1/2	0
i)	-1	0	9/2	0
j)	1	0	-1/2	0
k)	0	-1	0	-1/6
l)	-1	0	2	0

3.6.7 a) $(-\infty,+\infty)$
      b) $D = [3,+\infty)$ for y, y', y''; $D = (3,+\infty)$ for $y^n(x), n \geq 3$
      c) $D = (-\infty,+\infty)$ for y, y', y''; $D = (-\infty,3) \cup (3,+\infty)$ for $y^n(x), n \geq 3$
      d) $D = [-10,10]$ for y; $D = (-10,10)$ for $y^n(x), n \geq 1$
      e) $D = [5,+\infty)$ for y, y'; $D = (5,+\infty)$ for $y^n(x), n \geq 2$
      f) $D = [-512,+\infty)$ for y; $D = [-512,0) \cup (0,+\infty)$ for y'; $D = (-512,0) \cup (0,+\infty)$ for $y^n(x), n \geq 2$

3.6.8 a) i) $x = \pm 2\sqrt{2}/3$ ii) $x = 0$ b) i) $x = (2n+1)/2, n \in I$ ii) $x = n, n \in I$ c) i) $x = 1/2$ ii) $x = 1$
      d) i) none ii) $x = 0$ e) i) $x = n\pi/2, n \in I$ ii) $x = (2n+1)\pi/4, n \in I$ f) i) $x = 1/3\,ln(4)$ ii) $x = 1/3\,ln(16)$

3.6.9 a) $y = [1 + \sqrt{1 + x^3}]/x^2$ $D = [-1,0) \cup (0,+\infty)$, $R = (0,+\infty)$
        $y = [1 - \sqrt{1 + x^3}]/x^2$ $D = [-1,0) \cup (0,+\infty)$, $R = [-1/2,0)$
        $y = 0$ $D = \{0\}$ $R = \{0\}$
      b) $y = 1$ $D = (-\infty,+\infty)$ $R = \{1\}$ $y = -1$ $D = (-\infty,+\infty)$ $R = \{-1\}$
      c) $y = \pm[(x^2 - 1)/(2x)]^{1/4}$ $D = [-1,0) \cup [1,+\infty)$ $R = (-\infty,+\infty)$
      d) $y = \sqrt{(x - 1)/2}$ $D = (1,+\infty)$ $R = (0,+\infty)$ $y = -\sqrt{(x-1)/2}$ $D = (1,+\infty)$ $R = (-\infty,0)$
      e) $y = x\exp(-x^2)$ $D = (-\infty,0) \cup (0,+\infty)$ $R = (-1/\sqrt{2e},0) \cup (0,1/\sqrt{2e})$
      f) $y \pm \{[-1 \pm (1 + 8x^3)^{1/2}]/(4x)\}^{1/4}$ $D = [-1/2,+\infty)$ $R = (-\infty,+\infty)$

3.6.10 a) $y = (-15/13)x + (41/13)$ b) $y = 1/2\,x + 2$ c) $y = (72/11)x + (116/11)$ d) $y = 16x + 10$ e) $y = 1/2$
       f) $y = (-1/9)x + 2019$ g) $y = x$ h) $y = [-1/(10\,ln(10)](x - 10) + 2$

3.6.11 a) $v = \sin(3t) + 3t\cos(3t)$ $a = 6\cos(3t) - 9t\sin(3t)$ b) $v = 5\cos(t)\,e^{\sin(t)}$ $a = 5e^{\sin(t)}[\cos^2(t) - \sin(t)]$ c) $v = -1/(t+1)^2$ $a = 2/(t+1)^3$

3.6.12 a) $y' = y[1/(x-1) + 2x/(x^2-2) + 3x^2/(x^2-3)]$ b) $y' = y[15/(3x-1) + 30x^2/(x^3+1) + 8(2x+6)/(x^2+6x-1)]$
       c) $y' = y[2/(x-1) + 10x/(x^2-2) - 9x^2/(x^3-3)]$ d) $y' = y[10x/(x^2-1) + 6x^2/(x^3-x) + (1/2\,x^{-1/2} - 3x^{3/2})/(x^{1/2} + 6x^{-1/2})]$
       e) $y' = y[10x/(5x^2-1) + 4(-2/3\,x^{-1/3} + 3/2\,x^{1/2})/(x^{2/3} + x^{3/2}) - 10/(5x+4)]$ f) $y' = y[\pi\cot(\pi x) - \tan(x) - 3]$
       g) $y' = y[1/(x\,ln(x)) - 2 + 2\cot(x)]$ h) $y' = y[-3x^2/(2-x^3) + 2/x - 7/(x+5) - 1/(2x-1)]$

CHAPTER 4 ANSWERS

Exercise Set 4.1

4.1.1 a) $f'(2) = 2$, increasing; $f''(2) = 2$, concave up b) $f'(2) = 4$, increasing; $f''(2) = 8$, concave up
      c) $f'(2) = 0$, stationary; $f''(2) = 1$, concave up d) $f'(2) = -2$, decreasing; $f''(2) = 4$, concave up
      e) $f'(2) = -1/4$, decreasing; $f''(2) = 1/4$, concave up f) $f'(2) = 1$, increasing; $f''(2) = 0$, no concavity (straight line)
      g) $f'(2) = -4$, decreasing; $f''(2) = 12$, concave up h) $f'(2) = 2/\sqrt{5}$, increasing; $f''(2) = 1/(5)^{3/2}$, concave up
      i) $f'(2) = -1/4$, decreasing; $f''(2) = 3/8$, concave up j) $f'(2) = -2e^{-2} < 0$, decreasing; $f''(2) = 3e^{-2} > 0$, concave up
      k) $f'(2) = 12e^4 > 0$, increasing; $f''(2) = 54e^4 > 0$, concave up l) $f'(2) = 0$, stationary; $f''(2) = -2e^{-2} < 0$, concave down
      m) $f'(2) = 4ln(2) + 1 > 0$, increasing; $f''(2) = 2ln(2) + (7/2) > 0$, concave up
      n) $f'(2) = -ln(2)$, decreasing; $f''(2) = -1/2$, concave down o) $f'(2) = -1/2$, decreasing; $f''(2) = 1/4$, concave up
      p) $f'(2) = -\sin(2) \approx -.9093$, decreasing; $f''(2) = -\cos(2) \approx .4161$, concave up
      q) $f'(2) = 4e^{\cos(2)}(1 - \sin(2)) \approx 0.2393$, increasing $f''(2) = e^{\cos(2)}[2 - 8\sin(2) - 4\cos(2) + 4\sin^2(2)] \approx -.1995$, concave down
      r) $f'(2) = -2\tan(2) \approx -4.3701$, decreasing; $f''(2) = -2\sec^2(2) \approx -11.5488$, concave down

4.1.2 a) i) 1 ii) none iii) $(1,+\infty)$ iv) nowhere v) none
      b) i) $0, 4/3$ ii) $2/3$ iii) $(-\infty,0) \cup (4/3,+\infty)$ iv) $(-\infty,2/3)$ v) $y = (4/3)x + (89/27)$
      c) i) $-1/2$ ii) none iii) $(-\infty,-1/2)$ iv) $(-\infty,+\infty)$ v) none
      d) i) $-2, 2$ ii) 0 iii) $(-\infty,-2) \cup (2,+\infty)$ iv) $(-\infty,0)$ v) $y = -12x + 3$
      e) i) $4 - \sqrt{17}, 4 + \sqrt{17}$ ii) 4 iii) $(-\infty, 4 - \sqrt{17}) \cup (4 + \sqrt{17},+\infty)$ iv) $(-\infty,4)$ v) $y = -51x + 64$
      f) i) $0, 1$ ii) $0, 1$ iii) $(1,+\infty)$ iv) $(-\infty,0) \cup (1,+\infty)$ v) none
      g) i) $(3-\sqrt{33})/12, (3 + \sqrt{33})/12$ ii) $1/4$ iii) $(-\infty,(3 - \sqrt{33})/12) \cup ((3 + \sqrt{33})/12, +\infty)$ iv) $(-\infty,1/4)$ v) $y = (-11/4)x + (97/16)$
      h) i) $1, 7/3$ ii) $5/3$ iii) $(-\infty,1) \cup (7/3,+\infty)$ iv) $(-\infty,5/3)$ v) $y = -(2/3)x + 77/27$
      i) i) $1/4$ ii) 0 iii) $(0,1/4)$ iv) $(0,+\infty)$ v) none
      j) i) $2(4)^{1/5}$ ii) none iii) $(-\infty,0) \cup (2(4)^{1/5},+\infty)$ iv) nowhere v) none
      k) i) $3/2, 2$ ii) $2, 3$ iii) $(3/2,+\infty)$ iv) $(2,3)$ v) $x = 2$ at $(2,0)$ and $y = 2x - 3$ at $(3,3)$
      l) i) $0, 8$ ii) $0, 1$ iii) $(-\infty,0) \cup (8,+\infty)$ iv) $(-\infty,0) \cup (0,1)$ v) $y = -10/3\,x + 1/3$

m) i) -2, 0, 2   ii) ±2/√3   iii) (-2,0) ∪ (2,+∞)   iv) (-2/√3, 2/√3)   v) y = (64√3/9)x + (192/9) at (-2/√3, 64/9),
y = (-64√3/9)x + (192/9) at (2/√3, 64/9)

n) i) -1/2   ii) 0 iii) (-1/2,+∞)   iv) nowhere   v) none
o) i) 1,2,3   ii) 2 - √3/3, 2 + √3/3 iii) (1,2) ∪ (3,+∞)   iv) (2 - √3/3,2 + √3/3)
v) y = (8√3/9)x - 16√3/9 - 20/3 at   (2 - √3/3, -68/9), and y = (-8√3/9)x + 16√3/9 - 20/3 at (2 + √3/3, -68/9)

4.1.3   a) (-∞,+∞)   b) (1,+∞) c) (-∞,0) ∪ (2,+∞) d) (0,+∞) e) (-∞,*ln*(2))   f) (*ln*(2),+∞)   g) (0,1) h) (e,+∞) i) (0,1/e)
j) (π/2 + 2nπ, 3π/2 + 2nπ), n ε I   k) (1/2 + 2n, 3/2 + 2n), n ε I   l) (n, 1/2 + n), n ε I   3.1.4   GRAPH-HERE

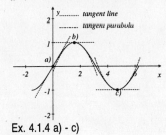

Ex. 4.1.4 a) - c)

Exercise Set 4.2

4.2.1   a) f(x) = -(x - 2)(x - 6) GRAPH-HERE   b) GRAPH-HERE
4.2.2   a) GRAPH-HERE   b) GRAPH-HERE   c) GRAPH-HERE   d) GRAPH-HERE
e) GRAPH-HERE   f) GRAPH-HERE   g) GRAPH-HERE   h) GRAPH-HERE
i) GRAPH-HERE   j) GRAPH-HERE
4.2.3   a) GRAPH-HERE   b) GRAPH-HERE   c) GRAPH-HERE
4.2.4   a) (0,0) is a point of inflection GRAPH-HERE   b) (0,) is a global minimum GRAPH-HERE
c) (2,0) is a point of inflection GRAPH-HERE   d) (-2,0) is a global minimum GRAPH-HERE
e) (-1,-1) is a global minimum (02,0) and (0,0) are points of inflection GRAPH-HERE
f) (5/7,-3 SQRT(95/49)) is a global minimum (0,0) and (2,0) are points of inflection GRAPH-HERE
g) local maximum at x = 0, local minimum at x = 20/19 ≈ 1.05, points of inflection at x = (20 - 15√6)/19 ≈ -0.881, x = 2,
x = (20 + 15√6)/19 ≈ 2.936 GRAPH-HERE
h) local maximum at x = 0, local minimum at x = 40/23 ≈ 1.739, points of inflection at x = (40 - 15√2)/23 ≅ 0.82, x = 2,
x = (40 + 15√2)/23 ≈ 2.66 GRAPH-HERE
4.2.5   a) GRAPH-HERE   b) GRAPH-HERE   c) GRAPH-HERE
4.2.6   a) GRAPH-HERE   b) GRAPH-HERE
4.2.7   GRAPH-HERE
4.2.8 $y = c(e^{-\alpha t} - e^{-\beta t})$ for c > 0, α > 0, β > 0, β > α, t ≥ 0
global maximum at $t_1 = ln(\beta|\alpha)/(\beta - \alpha)$ point of inflection at $t_2 = 2ln(\beta|\alpha)/(\beta - \alpha)$
GRAPH-HERE 3.2.9   GRAPH-HERE
4.2.10 a) GRAPH-HERE   b) GRAPH-HERE   c) GRAPH-HERE   d) GRAPH-HERE   e) GRAPH-HERE   f) GRAPH-HERE
g) logistic equation $y = B/(1 + Ce^{-\lambda Bt})$ GRAPH-HERE
4.2.11 GRAPH-HERE 3.2.12 $y = e^{-ke^{-\lambda x}}$ concave up on (*ln*(k)/λ,t) and concave down on (-∞,*ln*(k)/λ) GRAPH-HERE
4.2.13 a) GRAPH-HERE   b) GRAPH-HERE   c) GRAPH-HERE
4.2.14 a) y = x + sin(x); D = [-π/2,π/2] point of inflection (0,0) GRAPH-HERE
b) y = sin^2(x); D = [-π/2,π/2] points of inflection at (-π/4,1/2),(π/4,1/2) GRAPH-HERE
c) y = sin(x)cos(x); D = [-π/2,π/2] point of inflection at (0,0) maximum at (π/4,1/2) minimum at (-π/4,1/2) GRAPH-HERE
d) y = sin(x) - cos(x) D = [-π/2,π/2] point of inflection at (π/4,o) minimum at (-π/4, - √2) GRAPH-HERE
e) y = sin^2(x) - cos(x) D = [-π/2,π/2] points of inflection (-.936,.055)(.936,.055) minimum at (0,-1) GRAPH-HERE
f) y = sin(x) + cos(x) D = [-π/2,π/2] point of inflection at (-π/4,0) maximums (π/4,√2) GRAPH-HERE
g) y = √cos(x) D = [-π/2,π/2] maximum at (0,1) GRAPH-HERE
h) y = $e^{sin(x)}$ D = [-π/2,π/2] point of inflection at (.67,1.86) GRAPH-HERE

Graphing Solutions to Chapter 4 Section 2

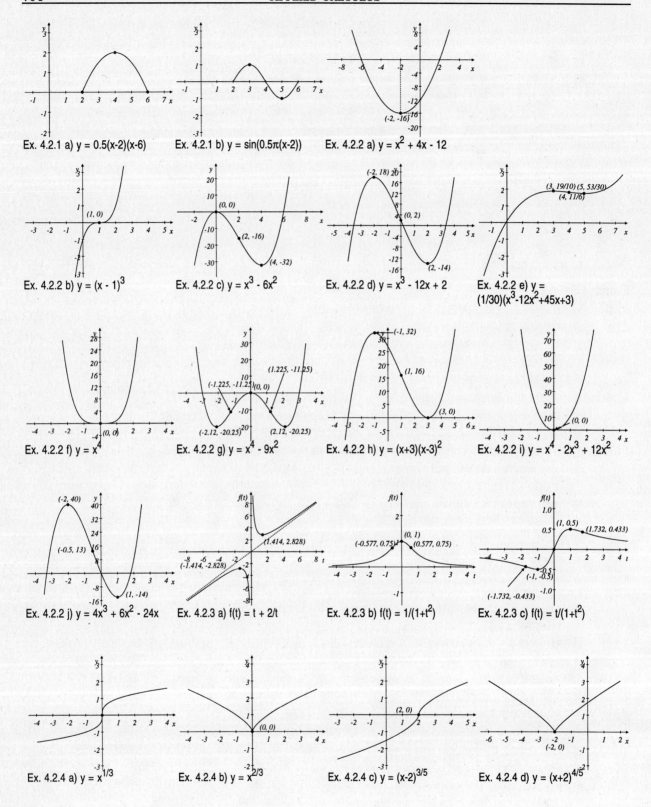

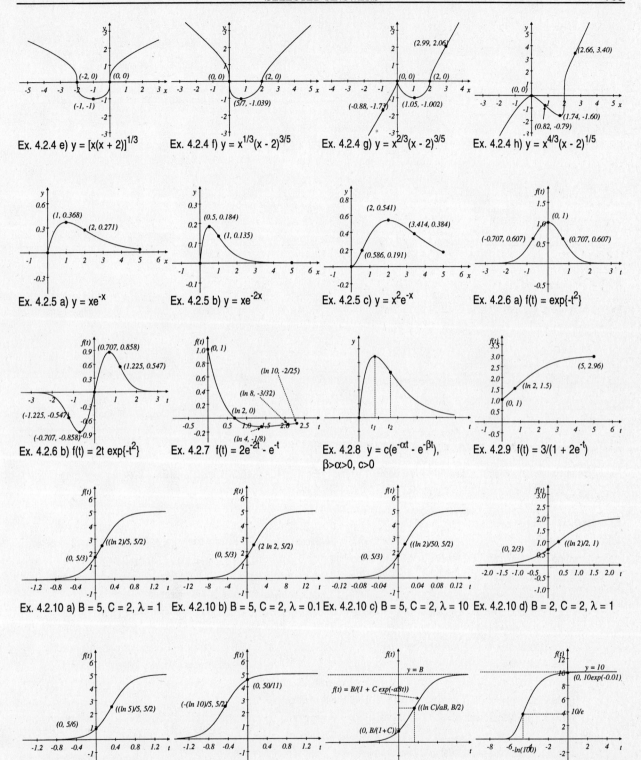

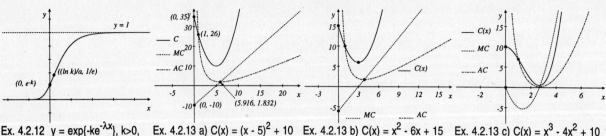

Ex. 4.2.12 $y = \exp\{-ke^{-\lambda x}\}$, $k>0$, $\lambda>0$.   Ex. 4.2.13 a) $C(x) = (x-5)^2 + 10$ Ex. 4.2.13 b) $C(x) = x^2 - 6x + 15$ Ex. 4.2.13 c) $C(x) = x^3 - 4x^2 + 10$

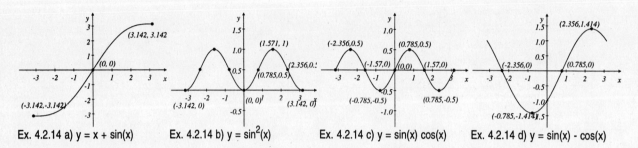

Ex. 4.2.14 a) $y = x + \sin(x)$ Ex. 4.2.14 b) $y = \sin^2(x)$ Ex. 4.2.14 c) $y = \sin(x)\cos(x)$ Ex. 4.2.14 d) $y = \sin(x) - \cos(x)$

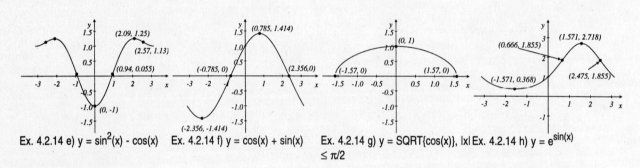

Ex. 4.2.14 e) $y = \sin^2(x) - \cos(x)$ Ex. 4.2.14 f) $y = \cos(x) + \sin(x)$ Ex. 4.2.14 g) $y = \sqrt{\cos(x)}$, $|x| \leq \pi/2$ Ex. 4.2.14 h) $y = e^{\sin(x)}$

Exercise Set 4.3

4.3.1 a) $x = 0$ local min   b) $x = 1/2$ local min   c) $x = -1/\sqrt{3}$ local max; $x = 1/\sqrt{3}$ local min   d) $x = -2$ local max; $x = 2$ local min
e) no critical points; $y' < 0$ always   f) $x = -2, 2$ local min; $x = 0$ local max   g) $x = 0$ local max   h) $x = 0$ local min
i) $x = -3$ local min   j) $x = 0$ local min   k) $x = 1$ local max; $x = -2, 4$ local minimum (end point extrema)
l) $x = -\sqrt{3}$ local max; $x = \sqrt{3}$ local min, $x = 0$ neither   m) $x = 1/2$ local max   n) $x = 0$ local max
o) $x = -1, 1$ local max; $x = 0$ local min   p) $x = 1/2\ln(2)$ local max   q) $x = (\ln 5)/4$ local minimum   r) $x = 2$ local maximum

4.3.2 a) $x = 1/2$ local max   b) $x = 2$ local min   c) $x = -2$ inflection
d) $x = 0$ local max; $x = \sqrt{2}, -\sqrt{2}$ local min; $x = \sqrt{2/3}, -\sqrt{2/3}$ inflection   e) $x = 2/3$ local max; $x = 2$ local min; $x = 4/3$ inflection
f) $x = 2$ inflection   g) $x = 0$ local min; $x = \sqrt{5}, -\sqrt{5}$ inflections
h) $x = 0$ local max; $x = -6/5$ local min; $x = 0, (-6 \pm \sqrt{6})/5$ points of inflection
i) $x = 0$ local max; $x = -7, 7$ local min; $x = -7/\sqrt{3}, 7/\sqrt{3}$ points of inflection
j) $x = 3 - (\sqrt{3}/3)$ local max; $x = 3 + (\sqrt{3}/3)$ local min; $x = 3$ inflection   k) $x = 0$ local max; $x = 2$ local min; $x = 1$ inflection
l) $x = 1/3$ local min, $x = -1/3$ local max; $x = 0, -1/\sqrt{3}, 1/\sqrt{3}$ inflections   m) $x = 1/4^{1/3}$ local min
n) $x = 1/4$ local min; $x = 0$ local max (end point extrema)   o) no local extrema or points of inflection
p) $x = 3/2$ local max; $x = 2, 3$ points of inflection   q) $x = -8/3$ local min; $x = -4$ local max (end point extrema)
r) $x = 2$ local min   s) $x = 0, 1/4$ local min; $x = 3/16$ local max; $x = (15 \pm 3\sqrt{5})/80$ points of inflection
t) $x = 10/3$ local max; $x = 20/3$ inflection   u) $x = 0$ local min; $x = 1/5$ local max; $x = (2 \pm \sqrt{2})/10$ points of inflection
v) $x = \ln(4/3)$ local min; $x = \ln(16/9)$ inflection   w) $x = 0$ local max; $x = \pm\sqrt{50}$ points of inflection
x) $x = 5\sqrt{2}/3$ local max; $x = -5\sqrt{2}/3$ local min, points of inflection at $x = 0$, $x = \pm 5\sqrt{2/3}$

4.3.3 a) at $x = 0$ local max; at $x = 3$ local min   b) at $x = 0$ local max; at $x = 3$ local min   c) at $x = 0, 3$ local max; at $x = 2$ local min
d) at $x = -1, 1$ local max; at $x = -1/2$ local min   e) at $x = 3/2$ local max; at $x = 0, 3$ local min
f) at $x = -4, 4$ local max; at $x = -3/2$ local min   g) at $x = 1$ local max; at $x = -1$ local min

h) at x = 0 local max; at x = *ln*(3) local min i) at x = *ln*(2) local max; at x = 0,5 local min

j) at x = 1 local max; at x = -1 local min   k) at x = 1 local max; at x = 0,*ln*3 local min

l) at x = 0 local max; at x = 5 local min   m) at x = -1, 1/2 local max; at x = -1/2, 1 local min

n) at x = 0, 4 local max; at x = 2 local min o) at x = -3/4, 1/4, 1 local max; at x = -1, -1/4, 3/4 local min

p) at x = π/4 local max; at x = 0,π local min   q) at x = 3π/4 local max; at x = 0, π local min

Exercise Set 4.4

4.4.1   a) M = 7 at x = 5 m = -3 at x = 0   b) M = 49 at x = 4 m = 0 at x = 1/2; local max at x = 0 (end point extrema)

c) M = 5 at x = 0 m = 1 at x = -2, 2   d) M = 3 at x = -1,2 m = -1 at x = -2,1

e) M = 27 at x = 5 m = -8 at x = 0; (2,0) point of inflection

f) M = 5 at x = 1,2.5 m = 0 at x = 0; (2,4) local min

g) M = 6 at x = -1; m = 4 at x = 1; point of inflection at (0,5)

h) M = 31 at x = 4; m = -25 at x = -4; local max at x = $-\sqrt{3}$, local min at x = $\sqrt{3}$

i) M = 11 at x = 1; m = -11 at x = -1

j) M = 16 at x = -2; m = -16 at x = 2; end point extrema at (-3,9) local min (3,-9) local max

k) M = 9 at x = -1 m = 0 at x = 0; (1,1) local max endpoint extrema

l) M = 9 at x = 3 m = 0 at x = 2; (1,1) local max endpoint extrema

m) M = 1 at x = 0 m = 1 - $2^{2/3}$ ≈ -0.5874 at x = 2; (-1,0) local min (endpoint)

n) M = $\sqrt{6}$ at x = 2 m = $\sqrt{2}$ at x = 0; (-1,$\sqrt{3}$) local max (endpoint extrema)

o) M = 2 at x = 2 m = $-2^{2/3}$ at x = -1; local max x = 3/5, local min at x = 0

p) M = $13(10)^{1/3}$ at x = 10 m = $(-3/4)^{1/3}(9/4)$ ≈ -2.044 at x = -3/4 local max x = -8 (endpoint extrema); point of inflection (0,0)

q) M = 2 at x = 2; m = 0 at x = 0, 1; local max at x = 3/7

r) M = $4\sqrt{2}$ = 1.1892 at x = 4; m = -0.6714 at x = 17/7; local max at x = 17/7; point of inflection at (3,0)

s) M = e^2 ≈ 7.389 at x = -1, m = e^{-2} ≈ 0.1353 at x = 1

t) M = 1 at x = 0 m = 1/e ≈ 0.3679 at x = -1, 1

u) M = 0.184 at x = 1/2, m = 0 at x = 0; local max at x = 4 (endpoint extrema)

v) M = 0.0338 at x = 1/2; m = 0 at x = 0; local min at x = 5 (endpoint extrema)

w) M = *ln*5 ≈ 1.6094 at x = 0 m = 0 at x = -2,2

x) M = $(81/4)^{2/3}$ ≈ 7.429 at x = -7/2 m = 0 at x = -8; local max (0,4) endpoint extrema

4.4.2   a) M = 24 at x = 4 m = -56 at x = -4; local max at x = $(1 - \sqrt{19})/3$ ≈ 1.1196 local min at x = $(1 + \sqrt{19})/3$ ≈ 1.7863

b) M = 10 at x = 2 m = -30 at x = -3 c) M = $\sqrt{3}/4$ ≈ 0.433 at t = 1/2 m = $-\sqrt{3}/4$ ≈ -0.433 at t = -1/2

d) M = 1 at s = 1 m = -8 at s = 2; (0,0) local min

e) M ≈ 0.5307 at t = 1/*ln*(2) m = -8 at t = -2; local min(2, 1/2) endpoint extrema f) M = *ln*(2) at t = -1,1 m = 0 at t = 0

g) M = 1/2 at t = π/4 m = -1/2 at t = -π/4; (π/2,0) local min (endpoint) (-π/2, 0) local max (endpoint)

h) M = 1 + $\sqrt{2}$ at (n - 1/4)π, n ε I m = 1 - $\sqrt{2}$ at (n + 1/4)π, n ε I

i) M = 1 + 2π ≈ 7.283 at x = 2π; m = 1 at x = 0; (π/2,π/2) point of inflection

j) M = $\sqrt{2}$ at x = -π/4 + nπ, n ε I   m = -1 at x = π/2 + 2nπ n ε I

4.4.3   a) absolute minimum m = 2 at t = 1; no absolute maximum b) no absolute extrema   c) no absolute extrema

d) absolute minimum m = 0 at x = 0; absolute maximum M = 1/2 at x = 1

e) absolute maximum M = 1/e ≈ 0.3679 at x = e; no absolute minimum

f) absolute maximum M = 0 at x = 0; absolute minimum m = 0 at x = *ln*(3/2)

g) absolute maximum M = *ln*(2) at x = 1; no absolute minimum

h) absolute maximum M = 0.25 at x = *ln*(2), absolute minimum m = 0 at x = 0

Exercise Set 4.5

4.5.1   a) i) $x_E = 10$ ii) $z_E = 8$ iii) $H_0 = 3, H_n = 2$ for n ≥ 1 iv) There is no maximum. As x → 0, H → 5.

b) i) $x_E = 9$ ii) $z_{E1} = (9 - \sqrt{73})/2$ ≈ 0.23 $z_{E2} = (9 + \sqrt{73})/2$ ≈ 8.77 iii) $H_0 = 6, H_n = 18$ for n ≥ 1 iv) $H^* = 20.25$

c) i) $x_{E1} = 5 - 2\sqrt{6}$ ≈ 0.10 $x_{E2} = 5 + 2\sqrt{6}$ ≈ 9.90 ii) z_{E1} ≈ 0.515 z_{E2} ≈ 9.878 iii) $H_0 = 122, H_n = 138$ n≥1 iv) at $x^* = (10 + \sqrt{97})/3, H^*$ ≈ 141.506

4.5.2   b) $x_6 = 5/2$ $x_7 = 23/8$ c) $x_{e1} = 1$ $x_{E2} = 3$ d) at $x^* = 2$ $H^* = 5/2$

4.5.3   b) $x_6 = x_7 = 8$ c) $x_E = 8$ d) i) $z_E = 6$ ii) There is no equilibrium.   e) $H^* = 4$

4.5.5   b) i) $x_3 = 71663616/6103515625 \approx 0.012$ ii) $x_3 = 3456/15625 \approx 0.221$ iii) $x_3 = 6^3 92^4 5^{-14} \approx 2.535$
  c) $x_{E1} = 25/6$ $x_{E2} = (95 + 5\sqrt{73})/12 \approx 11.476$ d) $z_E = 25/3$ when $H = 3$ e) $H^* \approx 3.0416$ at $x^* = 95/12$

4.5.7   a) $x_{n+1} = x_n + 1.2(x_n/2) - 0.2x_n = 1.4x_n$; $x_{n+1} = 1.4x_n - H$ b) ∞ c) Let grow past 2500, then $H^* = 1000$ d) No.

Exercise Set 4.6

4.6.1 $x = \sqrt{10}, y = \sqrt{10}$ 4.6.2 $A = 50/3, B = 25$ 4.6.3 $A = \pm\sqrt{2}$ and $B = 1/\pm\sqrt{2}$

4.6.4   Max area = 6.25   4.6.5   minimum amount of wood is $3 \cdot 7^{2/3}$ m$^2 \approx 10.98$ m^2.

4.6.6   a) $7.5^2 \cdot 20/11 \approx 102.27$ m^2 b) 125 m^2 This is the best possible.

4.6.7 $H = 20\sqrt{15\pi} \approx 137.29$ m and $r = (10\sqrt{15/\pi})^{1/2} \approx 4.67$ m   4.6.8 (-6/5, 3/5)   4.6.9 (2,1) (0,-1)

4.6.10   Max Area 12   4.6.11   max area is $75 - 50\sqrt{2} \approx 4.289$ there is no minimum area.   4.6.12 $t = 40$ $y(40) = 100$

4.6.13 $x = 5/2$ $y = 5/3$ $A = 25/6$ km^2

4.6.14   maximum growth rate $r(18) = 324$; maximum culture size occurs at $t = 36$ 4.6.15   750/year

4.6.16   a) Vol. - S.A. = $w^3 - 7w^2$ b) $W = 14/3, L = 28/3, H = 14/6$ 4.6.17 $U = 150^{1/4} \approx 3.4996$

4.6.18   0.0228 at $t = 200(213)^{1/5}$ 4.6.19   182 trees

4.6.20 $v(2r_0/3) = F(1 + 4r_0^3/27)$ 4.6.21 $27C^4B^4/(256A^3P^3)$

4.6.22   absolute max at $[1/(2|a|)](\pm[4a^4 - c^4]^{1/2}, Bc^2)$; local minimum at $(0, B(c^2 - a^2)^{1/2}$; absolute minimum at $(\pm(a^2 + c^2)^{1/2}, 0)$

4.6.23   a) $G = (B/A)^{1/2}V^2L^{-1} - 1$ Note that as V increases G increases and as L increases G decreases.
  b) V is increasing, hence G increases   c) L is increasing, hence G decreases and $G_{mouse} > G_{dog} > G_{cow} > G_{elephant}$

4.6.24 $10\sqrt{6}$ cm by $20\sqrt{6}$ cm   4.6.25   a) (2000/13, 140)   b) (660.19, 0)   c) (200, 2340/21)

4.6.26 $B(\sqrt{2/3}, 4)$ 4.6.27   a) $D = 60$ b) 80   c) $(85 + 5\sqrt{241})/2$

4.6.28   30 cm by $30\sqrt{3}$ cm   4.6.29 $m = 0.310336$ 2.6.30 $T_c = T/(3\sqrt{2})$ 3.6.31 $\alpha + 3(\beta^2\gamma/4)^{1/3}$

Exercise Set 4.7

4.7.1   a) Yes   b) Yes   c) No; $f(7) \neq f(3) = 0$ d) No, f is not differentiable at $x = 0$ since $f(0)$ does not exist   e) Yes
  f) No since $ln(x - 1)$ does not exist on $[0,1] \subset [0,2]$

4.7.2   a) $c = 3/2$ b) $c = 8/3$ c) $c = 0.615$ d) None. $f(x) = \sqrt{(x-1)^2} = |x - 1|$. Hence $f'(1)$ does not exist   e) $c = 1/3$
  f) $c = 0$ g) $c = ln(2/ln(3)) \approx 0.599$ h) $c = (e^2 - 1)/2$

4.7.3   One root on $((3 + \sqrt{6})/3, +\infty)$. Since $f(2.7) < 0$ and $f(2.8) > 0$ it must be in the interval (2.7, 2.8).

4.7.4   One root on $(0, +\infty)$. since $f(.2) < 0$ and $f(.3) > 0$ it must be in the interval (.2, .3)

4.7.5   At most one.

Exercise Set 4.8

4.8.1   a) $dx = 0.5, df = 0.5, \Delta f = 0.625, |\Delta f - df| = 0.125$ $dx = -1, df = -1, \Delta f = -0.5, |\Delta f - df| = 0.5$
  b) $dx = 0.5, df = 2, \Delta f = 2.125, |\Delta f - df| = 0.125$ $dx = -1, df = -4, \Delta f = -3.5$ $|\Delta f - df| = 0.5$

4.8.2   a) $df = (3x^2 - 2)dx$; 1   b) $df = (3x^2 - 2)dx$; -0.1   c) $df = (3x^2 - 2)dx$; -5
  d) $dg = -x/\sqrt{25 - x^2}\,dx$; -0.15   e) $dh = 3e^{3t}$; 0.75   f) $dM = (-64t + 1/t^2)dt$; 76.7812
  g) $dV = e^{-0.02s}(1 - 0.02s)ds$; 0.5   h) $dA = 10(10 + s)^{-2}ds$; 0.05   i) $df = 3\cos(3\theta)d\theta$; -0.6
  j) $df = [\cos(\theta + \pi\cos(\theta))](1 - \pi\sin(\theta))d\theta$; 0.298032

4.8.3   a) Exact value $y(2) = 63/8 = 7.875$ b) $L_1(x) = 6(x - 1), L_1(2) = 6$; Error = $|7.875 - 6| = 1.875$
  $L_2(x) = (728/27) + (730/27)(x - 3)$; $L_2(2) = -6.07407$; Error = $|7.875 - 0.07407| = 7.94907$

4.8.4   a) $df = 6x\,dx$ $\Delta f = 6x\Delta x + 3(\Delta x)^2$

	df	Δf	|Δf - dx|
$x_0 = 1$ dx = 0.1	.6	0.063	0.03
dx = 1	6	9	3
dx = 10	60	360	300
$x_0 = 10$ dx = 0.1	6	6.03	0.03
dx = 1	60	63	3
dx = 10	600	900	300

  b) $df = -3e^{-3x}dx$ $\Delta f = 3e^{-3x}(e^{-3\Delta x} - 1)$

		df	Δf	$\|\Delta f - dx\|$
$x_0 = 1$	$dx = 0.1$	-0.01494	-0.01290	0.00204
	$dx = 1$	-0.14936	-0.04731	0.10205
	$dx = 10$	-1.49361	-0.04979	1.44382
$x_0 = 10$	$dx = 0.1$	$-2.80728 \cdot 10^{-14}$	$2.42532 \cdot 10^{-14}$	$0.38196 \cdot 10^{-14}$
	$dx = 1$	$-2.80728 \cdot 10^{-13}$	$0.88917 \cdot 10^{-13}$	$1.91811 \cdot 10^{-13}$
	$dx = 10$	$-2.80728 \cdot 10^{-12}$	$-9.35768 \cdot 10^{-14}$	$2.7137 \cdot 10^{-12}$

c) $df = (1/x)dx$ $\Delta f = ln[(x + \Delta x)/x]$

d) $df = (50/(10 + x)^2)dx$ $\Delta f = 50\Delta x/(10 + x + \Delta x)(10 + x)$

		df	Δf	$\|\Delta f - dx\|$			df	Δf	$\|\Delta f - dx\|$
$x_0 = 1$	$dx = 0.1$	0.1	0.09531	0.00469	$x_0 = 1$	$dx = 0.1$	0.04132	0.04095	0.00037
	$dx = 1$	1	0.69315	0.30685		$dx = 1$	0.41322	0.37878	0.03444
	$dx = 10$	10	2.39789	7.60201		$dx = 10$	4.13223	2.16450	1.96773
$x_0 = 1$	$dx = 0.1$	0.01	0.00995	0.00005	$x_0 = 10$	$dx = 0.1$	0.0125	0.01243	0.00006
	$dx = 1$	0.1	0.09531	0.00469		$dx = 1$	0.125	0.11905	0.00595
	$dx = 10$	1	0.69315	0.30685		$dx = 10$	1.25	0.83333	0.41667

4.8.5 $y \approx \cos(3\pi x/2)$

4.8.6   a) $L(x) = 1 + kx$ b) $L(x) = 1 + (k/2)x$ c) $L(x) = 1 + kx$ d) $L(x) = kx$ e) $L(x) = 1 + ln(k)x, k > 0$
f) $L(x) = kx$ g) $L(x) = 1$ h) $L(x) = kx$

4.8.7 $dV = 3x^2 dx$; $dV = 5.4$ mm^3

4.8.8 $dSA = 12x\, dx$; $dSA = 6$ cm^2

4.8.9 $dV = 2\pi rh\, dr/3$; $dV = \pi r^2 h/30$

4.8.10 $dV = 4\pi r^2 dr$; $dV = \pi 10^5$ km^3

4.8.11 $dS = 8\pi r\, dr$; $dS = 40\pi$ km^2

4.8.12 $A = \pi(25 - r^2)$; $dA = -2\pi r\, dr$; $dA = 3\pi$ cm^2

4.8.13 $dV = 2sh\, ds/3$; $dV = 6000$ m^3

4.8.14 $V = 6x^3$; $dV = 18x^2 dx$; $dV = 18(10)^2(2) = 3600$ cm^3

4.8.15   a) 5/3% or 2.92 mm   b) 1% or 1.75 mm

Exercise Set 4.9

4.9.1 $P_0(x) = -8$, $P_1(x) = -8 + 32x$, $P_2(x) = -8 + 32x - 24x^2$
$P_3(x) = -8 + 32x - 24x^2 + 8x^3$, $P_4(x) = -8 + 32x - 24x^2 + 8x^3 - x^4$

4.9.2   a) $P_2(x) = x$ b) $P_2(x) = 2 + 4(x - 1) + 3(x - 1)^2 = 1 - 2x + 3x^2$ c) $P_2(x) = -2 + 4(x + 1) - 3(x + 1)^2 = -1 - 2x - 3x^2$
d) $P_2(x) = 10 + 13(x - 2) + 6(x - 2)^2 = 8 - 11x + 6x^2$
e) $P_2(x) = 1010 + 301(x - 10) + 30(x - 10)^2 = 1000 - 299x + 30x^2$

4.9.3   a) $P_3(x) = 1 - 2x^2$ b) $P_3(x) = 4(x - 1)^2 + 4(x - 1)^3 = 4x - 8x^2 + 4x^3$
c) $P_3(x) = 4(x + 1)^2 - 4(x + 1)^3 = -4x - 8x^2 + 4x^3$ d) $P_3(x) = 9 + 24(x - 2) + 22(x - 2)^2 + 8(x - 2)^3 = -15 + 32x - 26x^2 + 8x^3$
e) $P_3(x) = 9801 + 3960(x - 10) + 598(x - 10)^2 + 40(x - 10)^3 = 9999 + 4000x - 602x^2 + 40x^3$

4.9.4   a) $P_3(x) = -2 + 3x + 2x^2$ b) $P_3(x) = 3 + 7(x - 1) + 2(x - 1)^2$ c) $P_3(x) = 5x$
d) $P_3(x) = 26 + 37(x - 2) + 24(x - 2)^2 + 8(x - 2)^3$ e) $P_3(x) = 1 + 3x + (9/2)x^2 + (9/2)x^3$
f) $P_3(x) = (1/512) + (3/512)(x + ln(8)) + (9/1024)(x + ln(8))^2 + (9/1024)(x + ln 8)^3$ g) $P_3(x) = 1 + x^2$
h) $P_3(x) = 1 - x^2$ i) $P_3(x) = x + x^2 + x^3$ j) $P_3(x) = 1 + (1/2)x - (1/8)x^2 + (1/16)x^3$
k) $P_3(x) = 2 + (1/4)(x - 3) - (1/64)(x - 3)^2 + (1/512)(x - 3)^3$ l) $P_3(x) = 1 - (x - 1) + (x - 1)^2 - (x - 1)^3$
m) $P_3(x) = x - (1/6)x^3$ n) $P_3(x) = -(x - \pi) + (1/6)(x - \pi)^3$

4.9.6   a) n = 10   b) n = 20   c) n = 36

4.9.7   a) $e^x = 1 + x + x^2/2! + x^3/3! + \ldots + x^n/n! + \ldots$ b) $e^x = 1 - x + x^2/2! + x^3/3! + \ldots + (-1)^n x^n/n! + \ldots$
c) $e^{kx} = 1 + kx + k^2 x^2/2! + k^3 x^3/3! + \ldots + k^n x^n/n! + \ldots$ d) $ln(x + 1) = x - x^2/2 + x^3/3 + \ldots + (-1)^{n-1} x^n/n + \ldots$
e) $\sin(x) = x - x^3/3! + x^5/5! + \ldots + (-1)^n x^{2n+1}/(2n + 1)! + \ldots$ f) $\cos(x) = 1 - x^2/2! + x^4/4! + \ldots + (-1)^n x^{2n}/(2n)! + \ldots$
g) $\sin(kx) = kx - k^3 x^3/3! + k^5 x^5/5! + \ldots + (-1)^n (kx)^{2n+1}/(2n + 1)! + \ldots$ h) $\cos(kx) = 1 - k^2 x^2/2! + k^4 x^4/4! + \ldots + (-1)^n (kx)^{2n}/(2n)! +$

4.9.9   a) $P_n(x) = (x - 1) - (1/2)(x - 1)^2 + (1/3)(x - 1)^3 + \ldots + (-1)^{n-1}(x - 1)^n/n$ b) $P_n(x) = 1 - (x - 1) + (x - 1)^2 - (x - 1)^3 + \ldots + (-1)^n(x - 1)^n$

Exercise Set 4.10

4.10.1   a) 1   b) 0   c) $\pm\infty$ d) 1   4.10.2   a) $-\infty$ b) $+\infty$ c) 0   d) -3   e) $\pm\infty$ f) 0

4.10.3   a) 0   b) 5   c) 0   d) -1/2   4.10.4   a) 0   b) 0   c) ∞ d) 0   e) 0   f) ∞

4.10.5   a) 0   b) does not exist ($\lim_{x \to 0+} xe^{1/x} = +\infty$, $\lim_{x \to 0-} xe^{1/x} = 0$)   c) 0   d) 18/e   e) 0   f) 0

4.10.6   a) 0   b) 1   c) 1   d) 1   e) 1   f) 1   g) e^2 h) 1

4.10.7   a) $ln(3/5)$ b) $\pm\infty$ c) $-\infty$ d) ∞ 4.10.8   a) ∞ b) ∞ c) ∞ d) ∞ e) $-\infty$ f) ∞ g) 0   h) ∞

4.10.9   L'Hôpital's rule applies to (e) only, the limit is $+\infty$. The other actual limits are a) $-\infty$ b) ∞ c) 0   d) $\pm\infty$ f) 0   g) ∞
h) does not exist

4.10.10   Let f and g be differentiable in an interval $(b, b + \delta)$, $\delta > 0$ and $g'(x) \neq 0$ on $(b, b + \delta)$
If $\lim_{x \to b+} f(x) = \lim_{x \to b+} g(x) = 0$ (or $\pm\infty$) and if $\lim_{x \to b+} (f'(x)/g'(x)) = L$ then $\lim_{x \to b+} f(x)/g(x) = L$
Let f and g be differentiable in an interval $(b - \delta, b)$, $\delta > 0$ and $g'(x) \neq 0$ on $(b - \delta, b)$
If $\lim_{x \to b-} f(x) = \lim_{x \to b-} g(x) = 0$ (or $\pm\infty$) and if $\lim_{x \to b-} (f'(x)/g'(x)) = L$ then $\lim_{x \to b+} f(x)/g(x) = L$

4.10.11   If a) $f''(x)$ and $g''(x)$ exist in a neighbourhood of b, b) $g''(x) \neq 0$ for $x \neq b$, and c) $\lim_{x \to b} f''(x)/g''(x) = L$
then $\lim_{x \to b} f(x)/g(x) = L$.

Chapter 5 ANSWERS

Exercise Set 5.1

5.1.1   a) $\geq$ b) =   c) $\geq$ d) =   e) $\geq$ f) $\geq$ g) $\geq$ h) $\leq$ i) $\geq$ j) =   k) $\geq$ l) $\leq$ m) =
n) =   o) =

5.1.2   a) $\int_1^3 1 + x\, dx = 6$ b) $2 \cdot 10^3 \leq \int_1^3 10^3 x^2 dx \leq 18 \cdot 10^3$ c) $0 \leq \int_{-2}^2 \sqrt{4 - x^2}\, dx \leq 8$ d) $\int_0^3 2x\, dx = 9$
e) $\int_2^5 3(x - 2) dx = 27/2$ f) $0 \leq \int_0^{2\pi} 1 + \sin(x) \leq 4\pi$

5.1.3   a) 3   b) 6   c) -3

5.1.4   a) $\int_0^6 t^2\, dt$ b) $\int_2^4 (t + 1)^2\, dt$ c) $\int_3^6 t + 7\, dt$ d) $\int_{-2}^3 x^2 - 2x\, dx$ e) $2 \int_3^4 e^x\, dx$ f) $\int_4^5 xe^x dx$

Exercise Set 5.2

5.2.1   25.59   5.2.2   114.543 km/h   5.2.3   1.2   5.2.4   24/7

5.2.5   a) 1/4   c) (2n - 1) units; AVG $\{f, 0, 10^4\}$ = $[1 + (1/3)(3) + (1/5)(5) + ... + (1/199)(199)]/10^4 = \Sigma_{n = 1,100} \{1\} / 10^4 = 100/10^4 = 0.01$

5.2.6   a) 25/8   c) (2n - 1) units; AVG $\{f, 0, 10^4\}$ = $[1(1) + 2(3) + 3(5) + ... + 100(199)]/10^4 = [\Sigma_{n = 1,100}\{n(2n-1)\}]/10^4 = 67.165$

5.2.7   a) $I_1 = [0, 1/8]$, $\Delta t_1 = 1/8$; $I_2 = [1/8, 1/2]$, $\Delta t_2 = 3/8$; $I_3 [1/2, 3/4]$, $t_3 = 1/4$; $I_4 = [3/4, 1]$, $\Delta t_4 = 1/4$
Left endpoints: 0, 1/8, 1/2, 3/4. Right endpoints: 1/8, 1/2, 3/4, 1

b) $I_1 = [0, 1/4]$, $I_2 = [1/4, 1/2]$, $I_3 = [1/2, 3/4]$, $I_4 = [3/4, 1]$, $\Delta t_n = 1/4$, n = 1, 2, 3, 4
Left endpoints: 0, 1/4, 1/2, 3/4. Right endpoints: 1/4, 1/2, 3/4, 1

c) $I_1 = [2,3]$, $I_2 = [3,4]$, $I_3 [4,5]$; $\Delta t_n = 1$, n = 1,2,3,
Left endpoints: 2, 3, 4   Right endpoints: 3, 4, 5

d) $I_1 = [2, 11/4]$, $I_2 = [11/4, 7/2]$, $I_3 = [7/2, 17/4]$, $I_4 = [17/4, 5]$, $\Delta t_n = 3/4$, n = 1, 2, 3, 4
Left endpoints: 2, 11/4, 7/2, 17/4. Right endpoints: 11/4, 7/2, 17/4, 5

e) $I_1 = [-1, -1/2]$, $\Delta t_1 = 1/2$; $I_2 = [-1/2, 1/2]$, $\Delta t_2 = 1$; $I_3 [1/2, 1]$, $t_3 = 1/2$
Left endpoints: -1, -1/2, 1/2   Right endpoints: -1/2, 1/2, 1

f) $I_1 = [-1, -1/3]$, $I_2 = [-1/3, 1/3]$, $I_3 = [1/3, 1]$; $\Delta t_n = 1/3$, n = 1, 2, 3
Left endpoints: -1, -1/3, 1/3   Right endpoints: -1/3, 1/3, 1

g) $I_1 = [0, 1/2]$, $I_2 = [1/2, 1]$, $I_3 = [1, 3/2]$, $I_4 = [3/2, 2]$, $I_5 = [2, 3]$, $\Delta t_1 = \Delta t_2 = \Delta t_3 = \Delta t_4 = 1/2$ and $\Delta t_5 = 1$
Left endpoints: 0, 1/2, 1, 3/2, 2. Right endpoints: 1/2, 1, 3/2, 2, 3

h) $I_1 = [0, 3/5]$, $I_2 = [3/5, 6/5]$, $I_3 = [6/5, 9/5]$, $I_4 = 9/5, 12/5]$, $I_5 = [12/5, 3]$, $\Delta t_n = 3/5$, n = 1, 2, 3, 4, 5
Left endpoints: 0, 3/5, 6/5, 9/5, 12/5. Right endpoints: 3/5, 6/5, 9/5, 12/5, 3

i) $I_i [(i-1)/2, i/2]$, i = 1,...,40, $\Delta t_i = 1/2$, i = 1,...,40 Left endpoints: 0, 1/2, 1, 3/2, ... 39/9   Right endpoints 1/2, 1, 3/2, ... ,20

j) $I_i = [15 + (i - 1)/5, 15 + i/5]$ i = 1, ... , 50 $\Delta t_i = 1/5$,   L. endpoints: 15, 15.2, 15.4, 15.6, ... 24.8 R. endpoints: 15.2, 15.4, ... 25.

5.2.8   a) 4   b) 3   c) 5   d) 101/25   e) 99/25

5.2.9   a) 43/2   b) 37/2   c) 21   d) 19   e) 20 + (6/N)   f) 20 - (6/N)

5.2.10 $(1 + \sqrt{2})\pi \approx 7.5844$

5.2.11   a) 24.5   b) $3601/2520 \approx 1.42896$ c) $e + e^2 + e^3 + e^4 \approx 84.791$
d) $2/3[ln(5/3) + ln(7/3) + ln(3) + ln(11/3) + ln(13/3) + ln(5)] \approx 4.5545$

5.2.12   a) 13/3 ug/l   b) 1,488,000 ug

5.2.13 a) Maximum value occurs when outflow equals inflow: $t = 1$ b) $N = 2$ $S(1) = 3/8$ (Assume $Q(0) = 0$)
         c) $N = 3$ $S(1) = 13/27$; $N = 4$ $S(1) = 17/32$ (Assume $Q(0) = 0$)   d) $N = 10$ $S(1) = 0.615$

5.2.14 a) $T = 2$ b) $N = 2$ $S(2) = 2$ c) $N = 3$ $S(2) = 68/27$; $N = 4$ $S(2) = 11/4$ d) $N = 20$ $S(2) = 3.23$

5.2.15 a) $\int_0^T f(t)dt$; $rpm(T) = rpm(0) + \Sigma_{n=1,N} f(t_n^*) \Delta t_n$ b) $rpm(5) \cong 10e^{-1} + 10e^{-2} + 10e^{-3} + 10e^{-4} + 10e^{-5} \approx 5.78055$

5.2.16 a) $g(T) = g(0) + \Sigma_{n=1,N} g(t_n^*) \Delta t_n$ b) $g(10) = 10$ c) $g(10) \sim 0 + \Sigma_{n=1,20} \sin^2((n-1/2) \cdot (1/2)) = 5$

5.2.17 a) 18   b) 10   c) 20   d) 21   e) 182   f) 19.5   g) 140   h) -6   i) 72.5   j) 8/3   k) 2   l) 99   m) 36
         n) 1/4   0) 34/3

Exercise Set 5.3

5.3.1 a) -5   b) 8   c) 21   d) 3   e) 7   f) $-ln(3)$ g) 1   h) -3

5.3.2 a) $x^3/3 + c$ b) $x^2 - 5x + c$ c) $x^6/2 + 2x + c$ d) $x^3/3 + (1/x) + c$ e) $5x^4/4 - x^6/3 + c$ f) $-x^{-6} - x^7 + c$
        g) $x^3 - (2/15)x^5 + c$ h) $(2/3)x^{3/2} + (4/7)x^{7/4} + c$ i) $4x^{1/2} - x^{-1} + c$ j) $2x^3 + 24x^2 + 42x + c$

5.3.3 a) $(1/3)x^3 - x^2 - 2x + 8$ b) $(2/3)x^{3/2} - (1/2)x^2 + 5/6$ c) $(5/6)x^6 + 2x + 1$ d) $1/3x^3 - 2x^2 + 4x + 16/3$ e) $2x^{1/2} - x + 3$
        f) $-\cos(x) - 1$ g) $\tan(x) - 1$ h) $3e^x - 2$ i) $3ln(x) + 2$ j) $x^2/2 - e^x/e^3 - 3$

5.3.4 a) $(5/2)x^2 + c$ b) $x^3 - x^2 + x + c$ c) $5x^4/4 + c$ d) $-x^{-1} + x^{-2}/2 + 8x + c$ e) $15x^{2/3}/2 + c$
        f) $2x^{1/2} - 2x^{3/2}/3 + c$ g) $4x^3/3 + 2x^2 + x + c$ h) $-x^{-6}/6 + 5x^3/3 - 2x + c$ i) $4t^{3/2}/3 + c$ j) $3t - 4t^{1/2} + c$
        k) $10x^{1.7}/17 - 10x^{0.3}/3 + c$ l) $t^4 - 4t^3 + c$

5.3.5 a) $2e^x + c$ b) $-3e^t + c$ c) $x^2/2 + e^{x+5} + c$ d) $2ln(t) + c$ e) $x^2/2 - ln(x) + c$ f) $2x - 5ln(x) + c$
        g) $2^t/ln(2) + c$ h) $10^x/ln(10) + c$ i) $2^{x/2}/ln(2) + c$

5.3.6 a) $3\sin(\theta) + c$ b) $-\cos(\theta) - 2\sin(\theta) + c$ c) $\tan(\theta) - \sec(\theta) + c$ d) $-\csc(\theta) + c$ e) $\tan(\theta) + c$ f) $-\cos(\theta) - \theta - \cot(\theta) + c$

5.3.7 a) $y = -6x + c$; $y = -6x + 9$ b) $y = -x^2 + 2x + c$; $y = -x^2 + 2x + 2$ c) $y = x^4 - 2x^2 + c$; $y = x^4 - 2x^2 + 4$

5.3.8 a) 6   b) -15   c) 12   d) $(2/3)10^{3/2}$ e) 1/3   f) 14/3   g) $h^2/2 + h^5$ h) 663/4

5.3.9 a) 2   b) 1   c) 0   d) 1   e) 4   f) $\pi^2/2$

5.3.10 b) c) e) and f) can not be evaluated using the FTC.

5.3.11 a) 0   b) 99/2   c) 69/2   d) 20/3   e) $2(e^4 - e^3)$ f) $7(e^4 - e^3)$

5.3.12 a) $e - 1$ b) $1 - (1/e)$ c) $(1/e) - 1$ d) 1   e) $(1/e) - e$ f) -3   g) 1   h) 2   i) $-ln(2)$ j) $1/ln(2)$
         k) $9/(10 ln(10))$ l) $9(10^4)/ln(10)$

5.3.13 $AMC = 228.42$ $AVMC = 3.14$ 4.3.15 $145600/3ln(2) \approx 70018.8$

5.3.16 $A_0(e^{-3} - e^{-7})/4$ 4.3.17 27.296 km/h   4.3.18   125 km

Exercise Set 5.4

5.4.1 a) $F(3) = 4$ GRAPH HERE $F(5) = 16$ GRAPH HERE   4.4.2   a) $F(\pi) = 2$ GRAPH HERE $F(3\pi/4) = (2 + \sqrt{2})/2$ GRAPH HERE

5.4.3 a) $ln(x) = \int_1^x (1/t)dt$ GRAPH HERE
        b) Using a uniform partition with $N = 5$ on [1,2] and left endpoints: 1, 6/5, 7/5, 8/5, 9/5 $A(2) < 0.745 < 3/4$.
           Also using a partition $N = 1$ and right endpoint $A(2) > 1/2$
        c) Using a uniform partition with $N = 2$ and left endpoints $A(2) < 1.5 < 2$
        d) Using a uniform partition with $N = 8$ and right endpoints $A(3) > 1.0198 > 1$

5.4.4 a) $2x^2 + x^3$ b) $5e^t$ c) $4s^2 - 2$ d) $2x(x^2 + 3)$ e) $2x(x^2 + 4)$ f) $[x^2 - 2x + (3/x^2 - 2x)](2x - 2)$
        g) $2[(t+1)^3 - (t+1)^5]$ h) $-2(1 + 2x)$ i) $-2x[2(x^2 - 1) + (x^2 - 1)^4]$ j) $-2x^5 + 3(1 + 3x)^2$
        k) $s(s^2 + 2)/\text{SQRT}(1 + s^2) - s(2 - s^2)/\text{SQRT}(1 - s^2)$ l) $[(s^2 - 2x)^6 - 1/(s^2 - 2s)^6]6(s^2 - 2s)^5(2s - s)$

5.4.5 a) $\sec^2\theta$ b) $(1 + \sin^2(\theta))\cos(\theta)$ c) $2(1 + \theta^2)^{-1/2}$
        d) $(1 - \tan^2(\theta))\sec^2(\theta) - (1 - \sin^2(\theta))\cos(\theta)$ e) $-\sin(\pi - \theta) - \sin(\theta)$ f) $-\sin(\theta)ln(\cos(\theta)) - \cos(\theta)ln(\sin(\theta))$

5.4.6 a) $P_x(\text{ERF}(x)) = (2/\sqrt{\pi})e\{-x^2\}$ b) $y = (2/\sqrt{\pi}) \int_0^{\sqrt{x}} \exp\{-t^2\}dt$; $y' = e^x/\sqrt{\pi}$
        c) $P(X \leq x) = (2/\sqrt{\pi}) \int_0^{e^{\wedge}(-x)} \exp\{-t^2\}dt - (2/\sqrt{\pi})\int_0^{1-x^{\wedge}3} \exp\{-t^2\}dt$ i) $-2/(\sqrt{\pi}e) = -0.4151$ ii) 3.0226

Exercise Set 5.5

5.5.1 a) $(x + 3)^5/5 + c$ b) $(-3/2)(x^2 + 1)^{-2} + c$ c) $(x^2 + x + 2)^3 + c$ d) $(3/2)x^4 + 2x^3 + 6x^2 + c$
        e) $(2x^3 - 1)^6/36 + c$ f) $(2x + 3)^5/10 + (2x + 3)^{-1}/2 + c$ g) $(x^2 + 7x - 2)^{11}/11 + c$ h) $(5x^{1/5} + 1)^5/5 + c$
        i) $(3x^2 + 6)^{21}/126 + c$ j) $2(x^3 + 2x + 7)^{1/2} + c$

5.5.2 a) $e^{2x} + c$ b) $(1/2)e^{2x} + c$ c) $(1/2)e^{2x+1} + c$ d) $(-1/3)e^{-3x} + c$ e) $(1/2)\exp\{x^2\} + c$
        f) $(-1/6)\exp\{-3x^2\} + c$ g) $(-1/6)\exp\{-3x^2 + 1\} + c$ h) $e^{\sin(x)} + c$ i) $(1/6)\exp\{2x^3 + 3x^2 + 6x - 2\} + c$

5.5.3 a) $(2/3)ln(3x + 1) + c$ b) $(1/2)ln(2x + 1) + c$ c) $(1/5)ln(5x^2 + 3) + c$ d) $-ln(x^{-2} + 3) + c$ e) $(1/6)ln(3x^2 - 4) + c$
       f) $(1/3)ln(x^3 + 1) + c$ g) $(1/2)ln(x^2 - 4x + 7) + c$ h) $ln(x^2 - 2x + 1) + c$ i) $ln(x + sinx)) + c$

5.5.4 a) $(x^2 - 3)^5/5 + c$ b) $(-1/4)(x^4 + 5)^{-1} + c$ c) $x^5/5 + x^4 + 2x^3/3 - 2x^2 + x + c$ d) $(x^2 - 1)^4/8 + c$

5.5.5 a) $(1/3)sin(3\theta) + c$ b) $-cos(\theta) - (1/2)sin(2\theta) + c$ c) $tan(\theta) - sec(\theta) + c$ d) $-(1/2)cos(\theta^2 + 3) + c$ e) $-(3/\pi)cos(\pi\theta) + c$
       f) $tan(\theta - \theta^2) + \theta^2 - \theta + c$ g) $(1/4)sin^4(\theta) + c$ h) $(\pi/2)\ tan^2(\theta/\pi) + c$ i) $-cos(\theta) + cos^3(\theta)/3 + c$

5.5.6 a) $2e^t + c$ b) $e^{2x}/2 + c$ c) $exp\{2x^2\}/4 + c$ d) $-100e^{-0.01}x + c$ e) $exp\{x^3 + 6x - 2\}/3 + c$ f) $(5/2)exp\{t^2\} + c$
       g) $-e^{cos(\theta)} + c$ h) $-e^{1/t} + c$

5.5.7 a) $ln(x^2 + 1) + c$ b) $(1/2)ln(t^2 - 2t + 1) + c$ c) $(1/4)ln(2x^2 - 4x + 1) + c$ d) $ln(sin(t) + t^2) + c$ e) $ln(5x^4 + 5x^3 - 2x) + c$
       f) $ln(R\ sin(R)) + c$

5.5.8 a) $[x^4 - 8x - 585]/4$ b) $(2/3)x^3 - 3x$ c) $(1/4)(s^2 + 4)^4 - 324$ d) $(s^2 - 1)^5/5$ e) $(t + 1)^3/3$ f) $(4 + t^3 + 3t)/3$
       g) $((3T^2 + 2)^{3/2} - 2\sqrt{2})/9$ h) $(1/3)[(T^2 - 3T + 6)^3 - 6^3]$

5.5.9 a) $e^t + c$ b) $(1/3)e^{3t+1} + c$ c) $(1/8)exp\{4x^2 - 2\} + c$ d) $(1/3)exp\{x^3\} + c$ e) $(1/2)exp\{(x+3)^2\} + c$ f) $-e^{-x} + c$
       g) $(1/6)exp\{3x^2 - 2\} + c$ h) $2e^{SQRT(x+1)} + c$

5.5.10 a) $3ln(x) + c$ b) $(-1/2)x^{-2} + c$ c) $(1/3)ln(x^3 - 1) + c$ d) $(1/2)ln(x^2 - 2x + 6) + c$ e) $ln(1 + e^x) + c$ f) $ln(x + e^x) + c$
       g) $-ln(-x + x^{-2}) + c$ h) $ln(1 + ln(x)) + c$ i) $(1/2)ln(1 - x^{-2}) + c$ j) $ln(ln(x^2 + 1)) + c$

5.5.11 a) $2^x/ln(2) + c$ b) $2 \cdot 4^{x/2}/ln(2) + c$ c) $3^{x+1}/ln(3) + c$ d) $(1/2ln(3))3^{x^2+1} + c$ e) $2\{e^x\}/ln(2) + c$
       f) $(1/3ln(3))3\{5x^3 + 12x - 2\} + c$ g) $-2^{-x}/ln(2) + c$ h) $-3^{-x}/ln(3) + c$

5.5.12 a) $3/2$ b) 0 c) $7/24$ d) 1 e) 2 f) $ln(11)$ g) $ln(31)/3$ h) $-3ln(2)$ i) 1

Exercise Set 5.6

5.6.1 a) $8/3$ b) $\pm\sqrt{39}/3$ c) $10 + \pm\sqrt{31}$ d) 11 4.6.2 $x = (1/2)(3e^6 - 1)$

5.6.3 a) $e^2 - 3$ b) $9/2$ c) $9/2$ d) $32/3$ e) 6 f) $4\sqrt{2}$

5.6.4 a) $(2,5), (-2,5); 32/3$ b) $(-1,2), (2,5); 9/2$ c) $(\sqrt{2},3), (-\sqrt{2},3); 16\sqrt{2}/3$ d) $(0,0), (1,1), (-1, -1); 2[\ 2/\pi - 1/2] \approx 0.274$
       e) $(-3\pi/4, -1/\sqrt{2}), (\pi/4, \sqrt{2}); 4\sqrt{2}$ f) $(-1,2), (1,2); 4/3$ g) $(0,0), (1,1); 1/3$ h) $(-1, -1), (0, 0), (1, 1); 1/6$

5.6.5 a) $8/3$ b) $5\sqrt{5}/6 \approx 1.8634$ c) $(9\pi^4 - 12\pi^3 + 24\pi^2 + 16)/(12\pi^3) \approx 2.03581$ d) 16 e) 7.52

5.6.6 $8/ln(3)$ 5.6.7 b such that $b^5 - 1 = ln(b)$. Use Newton's Method to show $b \cong 0.370455$

5.6.8 $92/3l; t = (15 + 3\sqrt{41}/4$ minutes   5.6.9   a) $A(0.316754)$ mph   b) $76.0211\ A$ miles

5.6.10 $[ln(11)]^2/[20ln(10)] \approx 0.124857$ cm^2 5.6.11   a) 50 feet   b) 25 cm

5.6.12 a) 2.76% b) $\$39,588$ 5.6.13   a) 0.732561 g/day   b) 477.38 g

Chapter 6

Exercise Set 6.1

6.1.1 a) $2\sqrt{(x + 3)}$ b) $2(x - 6)\sqrt{(x + 3)}/3$ c) $2\sqrt{(x + 3)}(x^2 - 4x + 24)/5$
       d) $(2x + 5)^{3/2}(3x - 5)/15$ e) $5 \cdot 2^{1/5}(1 - 4x)^{6/5}(24x + 5)/1056$ f) $5(3x - 5)^{9/5}(27x + 25)/1134$
       g) $-\sqrt{(1 - 2x)}(3x^2 + 2x + 2)/15$ h) $2(x + 10)\sqrt{(x - 2)}/3$
       i) $(3/10)(2x - 9)(x + 3)^{2/3}$ j) $(1/14)(2x + 1)(3x - 2)^{4/3}$ k) $(3/16)(2x - 15)(2x + 5)^{1/3}$
       l) $(3/1000)(5x + 4)^{5/3}(25x - 12)$

6.1.2 a) $25ln(x^{1/5} - 1)$ b) $5ln(x^{2/5} + 1)/2$ c) $-4(1 - x^3)^{3/2}(3x^3 + 2)/45$ d) same as c)
       e) $-4(1 - x^3)^{3/2}(15x^6 + 12x^3 + 8)/315$ f) same as exer. e)   f)   g) $-2(1/x - 1)^{3/2}/3$
       h) $2(1 - 1/x)(2 + 3/x)(1/x - 1)^{1/2}/15$ i) $ln(e^x - 1)$ j) $6ln(x^{1/6} - 1) + 2\sqrt{x} + 3x^{1/3} + 6x^{1/6}$
       k) $12ln(x^{1/4} - 1) + 6x^{1/4}(x^{1/4} + 2)$ l) $3x^{1/6}(2x^{1/3} - 3x^{1/6} + 6) - 18ln(x^{1/6} + 1)$

6.1.3 a) $Arcsin(x)/2 + x\ SQRT(1 - x^2)/2$ b) $ln(SQRT(x^2 + 1) + x)$ c) $2Arcsin(x/2) + x\ SQRT(4 - x^2)/2$
       d) $ln(SQRT(x^2 + 25) + x)$ e) $(2/3)ln(SQRT(9x^2 + 4) + 3x) + x/2\ SQRT(9x^2 + 4)$ f) $ln(SQRT(9x^2 + 25) + 3x)/3$
       g) $Arctan(3x/2)/6$ h) $ln(SQRT(9x^2 + 4) + 3x)/3$ i) $Arcsin(x/2)$
       j) $-\sqrt{6}ln[(\sqrt{2}x - \sqrt{3})/(\sqrt{2}x + \sqrt{3})]/12$ k) $ln((SQRT(1 - x^2) - 1)/x)$ l) $(x/2)SQRT(x^2 + 4) - 2ln(SQRT(x^2 + 4) + x)$

6.1.5 a) $Arcsin((2x + 1)/3)$ b) $(\sqrt{3}/3)\ ln(SQRT(x^2 - 2x + 3) + x - 1)$
       c) $SQRT(x(x + 2)) - ln(SQRT(x(x + 2)) + x + 1)$ d) $ln(SQRT(x(x - 6)) + x - 3)$

6.1.6 a) $6 - 2\sqrt{3}$ b) $32\sqrt{3} - 160/3$ c) $12/5 - (6/5)3^{2/3}$ d) $(27/448)3^{2/3} + 93/448$
       e) $21/4 - 15/4\ 3^{1/3}$ f) $(-81/128)3^{1/3} - 21/128$ g) $Arcsin(2/3) \approx 0.7297$ h) $3ln(3)$

6.1.7 h) $-25\ Arctan(3/4)/2 + 25\pi/4 - 3/2 \approx 35410/3509$ or 10.09

6.1.8 $sqrt(73) - 3$ 6.1.9 $4/15$

Exercise Set 6.2

6.2.1 a) $e^{3x}(x/3 - 1/9)$ b) $e^{2x+5}(x/2 - 1/4)$ c) $-e^{-6x}(6x + 1)/9$ d) $e^{4x}(x/4 + 11/16)$ e) $e^{kx}(x/k - 1/k^2)$
f) $-e^{-kx}(kx + 1)/k^2$ g) $e^{\wedge}(x^2)(x^2 - 1)/2$ h) $e^{\wedge}(x^3)(x^3 - 1)/3$ i) $-e^{\wedge}(-x^2)(x^2 + 1)$

6.2.2 a) $x^3 \ln(x)/3 - x^3/9$ b) $(x^2/2 + 4x)\ln(x) - x(x + 16)/4$ c) $x^2 \ln(2x) - x^2/2$ d) $(x + 5)\ln(x + 5) - x$
e) $3x^{4/3}\ln(x)/4 - 9x^{4/3}/16$ f) $x\ln(x^{3x}) - 3x^2 \ln(x)/2 - 3x^2/4$ g) $3x^2 \ln(x)/2 - 3x^2/4$ h) $x\ln(x)^2 - 2x\ln(x) + 2x$
i) $x^2 \ln(x)^2/2 - x^2 \ln(x)/2 + x^2/4$ j) $\ln(x)\ln(\ln(x)) - \ln(x)$

6.2.3 a) $3e^4/4 + 1/4$ b) $e^{-3} + 2$ c) $2e^{1/2} - 6e^{\wedge}(-1/2)$ d) $3e^2/4 + 3/4$
e) $-1/4$ f) $29/9 - 16\ln(2)/3$

6.2.4 a) $-e^{-3x}(9x^2 + 6x + 2)/27$ b) $e^{4x-5}(8x^2 - 4x + 1)/32$ c) $e^{\wedge}(x^2)(x^4 - 2x^2 + 2)/2$
d) $-e^{-8x}(32x^2 - 56x + 153)/256$ e) $e^{5x}(125x^3 - 75x^2 + 30x - 6)/625$ f) $e^{5x}(125x^3 - 75x^2 + 30x - 6)/625$
g) $e^{3x}(9x^3 - 9x^2 - 3x + 1)/27$

6.2.5 a) $\cos(3\theta)/9 + \theta\sin(3\theta)/3$ b) $\sin(\pi\theta)/\pi^2 - \theta\cos(\pi\theta)/\pi$ c) $\ln(\cos(3\theta))/9 + \theta\tan(3\theta)/3$
d) $\sin(\theta + 1) - \theta\cos(\theta + 1)$ e) $\cos(\theta^2)/2 + \theta^2\sin(\theta^2)/2$ f) $\sin(\theta)\cos(\theta)/2 + \theta/2$
g) $e^\theta(\sin(\theta)/2 - \cos(\theta)/2)$ h) $\sin(\theta)\cos(\theta)^2/3 + 2\sin(\theta)/3$ i) $e^\theta(\sin(2\theta)/5 - 2\cos(2\theta)/5)$
j) $e^{2\theta}(\sin(\theta)/5 - 2\cos(\theta)/5)$ k) $e^{3\theta}(3\sin(\pi\theta) + \pi\cos(\pi\theta))/(\pi^2 + 9)$ l) $2\theta\cos(\theta) + (\theta^2 - 2)\sin(\theta)$

6.2.6 a) $\theta/2 - \sin(3\theta)\cos(3\theta)/6$ b) $\sin(\theta)^4/4$ c) $-\cos(\theta)(\sin(\theta)^2/3 + 2/3)$
d) $3\theta/8 - \cos(\theta)(\sin(\theta)^3/4 + 3\sin(\theta)/8)$ e) $\tan(\theta) - \theta$ f) $\ln(\tan((\pi + 2\theta)/4))$

6.2.7 a) $2\sin(\sqrt{x+2}) - 2\sqrt{x+2}\cos(\sqrt{x+2})$ b) $e^{\wedge}\sqrt{3-2x}(1 - \sqrt{3-2x})$
c) $2e^{\wedge}\sqrt{2-x}((2-x)^{3/2} + 4\sqrt{2-x} + 3x - 10)$ d) $3e^{(x+1)^{1/3}}((x+1)^{2/3} - 2(x+1)^{1/3} + 2)$

6.2.8 a) $25/8 - 145e^{-24/5}/8$ b) $1000/81 - 34120e^{-36/5}/81$

6.2.9 $-50e^{-73/50}/(73(250000\pi^2 + 1)) + 50/(73(250000\pi^2 + 1)) + 1$

6.2.10 a) $\ln(\sqrt{x^2+1} + x)/2 + x\sqrt{x^2+1}/2$ b) $\ln(\sqrt{4x^2+1} + 2x)/4 + x\sqrt{4x^2+1}/2$
c) $2\ln(\sqrt{x^2+4} + x) + x\sqrt{x^2+4}/2$ d) $25\ln(\sqrt{9x^2+25} + 3x)/6 + x\sqrt{9x^2+25}/2$
e) $2\arcsin(3x/2) + x(4 - 9x^2)^{3/2}/4 + 3x\sqrt{4-9x^2}/2$ f) $x/(25\sqrt{25-9x^2})$

Exercise Set 6.3

6.3.1 a) $\ln((x-1)/(x+1))/2$ b) $\ln((x-1)/2 + (3/2)\ln(x+1)$ c) $5\ln(x-1)/3 - 2\ln(x+2)/3$ d) $5\ln(x-2)/2 - 3\ln(x)/2$
e) $29\ln(x+3)/4 - 14\ln(x+2)/3 + 5\ln(x-1)/12$ f) $(3/5)\ln(x+1) - (2/3)\ln(x-1) + (46/15)\ln(x+4)$
g) $\sqrt{3}\ln((x-\sqrt{3})/(x+\sqrt{3})/4 + 7\ln(x^2 - 3)/4 - \ln(x-1)/2$ h) $2\ln(x-2) - \ln(x-1)$

6.3.2 a) $2\ln(x) + x^2/2$ b) $6\ln(x+2) + x^2/2 - 2x$ c) $7\ln(x-2) + x^3/3 + x^2 + 3x$
d) $-7\ln(x+2) + x^3/3 - x^2 + 4x$ e) $\ln((x-1)/(x+1)) + x$ f) $\ln(x-1) + x^2/2$
g) $7\ln(x+2) + x^2/2 - 3x$ h) $x^{10}/10 + x^9/9 + x^8/8 + x^7/7 + x^6/6 + x^5/5 + x^4/4 + x^3/3 + x^2/2 + x$

6.3.3 a) $\ln(x-2) - 2/(x-2)$ b) $\ln(x) - 1/x^2$ c) $\ln(x-1) - 2/(x-1)$ d) $-2\ln(x+1) + 2\ln(x) + 1/x$
e) $3\ln(x-1) - 1/(x-1) + x^2/2 + 2x$
f) $[\ln(x-1) - \ln(x-2)]/9 - (2x+1)/[3(x+2)(x-1)]$ g) $-\ln(x+3)/16 + \ln(x-1)/16 + (3x-1)/(4(1-x)(x+3))$
h) $2\ln(x-1) - 2\ln(x) + (4x - 7)/(2(x-1)^2)$

6.3.4 a) $\ln(x(x^2 + 1))$ b) $2\ln(x^2 - x + 1) + 6\ln(x - 2)$ c) $\ln((x+1)(x^2 - x + 1))$
d) $3\ln(x) - 3\ln(x^2 + 1)/2$ e) $2\ln(x^2 + x + 1) - 2\ln(x)$ f) $\ln(x)/2 + \ln(x-1) + \ln(x^2 + x + 1)$

Exercise Set 6.4

6.4.1
a) diverges to ∞   b) $e^{-9}/2$ c) diverges to ∞   d) diverges to ∞   e) diverges to ∞   f) $-3/8$
g) diverges to ∞   h) $1/20$ i) $e^{-8}/3$ j) $10/3$ k) diverges to ∞   l) e^{-3}
m) 0 n) 0 o) Does not converge.   p) diverges to ∞   q) 0
r) diverges to ∞   s) $2\ln(2)$ t) 0 u) The integral does not converge.

6.4.2 a) 3 b) diverges to ∞   c) diverges to $-\infty$ d) $-3 \cdot 5^{2/3}/2$ e) diverges to ∞   f) 5 g) $45/2 - 45/(4\ln(3))$
h) $e^{-1/5}$ i) diverges to $-\infty$

6.4.3 a) diverges to ∞   b) $2\sin(\sqrt{\pi}/\sqrt{2})$ c) $1/2$ d) $1/2$ e) Does not converge.
f) diverges to ∞   g) $\lim_{t \to \infty} e^{\sin(t)}$ does not exist.   h) Does not exist.   i) "DOES NOT EXIST"
j) diverges to $-\infty$ k) $1 - e^{\sin(1)}$ l) diverges to ∞   m) -1
n) limit D.N.E   o) The integral does not converge.   p) $8/3$

6.4.4 a) 6 b) not defined for x < 1 c) $3^{4/3}/2+3/2$ d) "DOES NOT EXIST" e) $(x-2)^{-0.75}$ is not defined for x < 2.
f) Does not converge $ln(x^2-1)$ is singular at x = 1. g) integrand not defined for x < 2 h) does not exist
i) does not exist j) $(3/2)[12^{2/3} - 42^{2/3}]$ k) limit at x = 3 does not exist l) limit at x = 3 does not exist.
m) 0 n) does not exist

6.4.5 The solutions are the integral of C(t) over $[0,\infty)$. a) $1/(20k)$ b) $1/(4k^2)$ c) 31/30 d) $14/k^3$
e) 595 f) diverges to ∞ Note C(t) is not defined at t = 0.

Chapter 7

No solutions available at press time. Sorry.

Chapter 8

Section 8.1

8.1.1 a) $y = -e^t + C + t^2$ b) $y = \exp(c + t^2)$ c) $y = \sin(2t)/2 + C$ d) $y = \text{Arcsin}(\exp(c + 5t^2/2))$
e) $y = \exp(e^{3t}/3 + C)$ f) $y = -ln(3(2c + t^2)/2)/3$ g) $y = -\exp(-c - t^2/2) + 1$ h) $y = \pm\sqrt{2}\text{ SQRT}(c - \cos(t))$ i) $y = e^c(t+1)$
j) $y = \exp[\exp\{t^2 ln(t)/2 + C - t^2/4\}]$ k) $y = \pm\sqrt{2}\text{ SQRT}(-t\cos(t) + \sin(t) + C)$ l) $y = \pm(4c + t^4)^{1/4}$

8.1.2 a) $y = ce^{-0.2t}$ b) $y = ce^{3t}$ c) $y = ce^{3t} - 2/3$ d) $y = c\exp(3t^2/2)$
e) $y = \exp(3t^2/2)(\sqrt{6\pi}\text{ ERF}(\sqrt{6}\,t/2)/3 + C)$ See Chapter 5 Section 5.4 Example 4 for the definition of the "error function" ERF(x).
f) $y = ct^5$ g) $y = ct^5 + t/2$ h) $y = ce^{-\cos(t)}$ i) $y = e^{-\cos(t)}(c - t)$
j) $y = ce^{-2t} - 3\cos(3t)/13 + 2\sin(3t)/13$ k) $y = ce^{3t} - (9t^2 + 6t + 2)/27$ l) $y = ce^{-t^2} + (t^2 - 1)/2$

8.1.3 a) $y = e^{7t}$ b) $y = e^{7(t-2)}$ c) $y = (1/3)e^{3(t-5)}$ d) $y = 2/(2\cos(t) - 1)$ e) $y = 2(t+1)$
f) $y = \pm 108^{1/4}(t^3 + 49151)^{1/4}/3$ g) $y = 0.5e^{2(t-5)}$ h) $y = \pm\text{SQRT}(t^2+3)$ i) $y = \text{SQR}(t^2+3/4)$

8.1.4 a) $y = (11/3)e^{3t} - 2/3$ b) $y = (11/3)e^{3(t-1)} - 2/3$ c) $y = 28e^{3t}/9 - t/3 - 1/9$ d) $y = 22e^{3(t-1)}/9 - t/3 - 1/9$
e) $y = t(2 + ln(t))$ f) $y = -e^{-t}/2 + \cos(t)/2 + \sin(t)/2$ g) $y = t(t-1)$ h) $y = t(t+2)$ i) $y = e^t(t-1)/t + 1/t$

8.1.5 The differential equation is $T' = -k(T - T_0)$. Its solution is $T = ce^{-kt} + T_0$; T(t) = 26 at t = $ln(17/11)/2$ = 0.217659

8.1.6 $y(t) = y_0\,\hat{e}^{-kt}$ $k = ln(0.7)$; $y(t) = 20\%y(0)$ at t = $ln(0.2)/ln(0.7)$.

8.1.7 a) The amount of salt is A(t) = concentration * volume. $A'(t) = -25(A(t)/1000)$ A(0) = 10 kilo.
b) $A(t) = 10e^{-0.025t}$. A(15) ≈ 6.87 A(30) ≈ 4.72 A(60) ≈ 2.23 c) A(t) = 2 at t = $ln(5)/0.025$ ≈ 64.38 min

8.1.8 a) $N(t) = K(D - t)^\tau$ b) $3.59\,10^9$ c)

8.1.9 a) $I' = kIS$ Since $I+S = N$ the total population, $S = N - I$ so $I' = kI(N-I)$
b) $S' = -kIS = -k(N-S)S$. General solution is $S(t) = N/(1 - e^{kN(t+c)})$ or $S(t) = N/(1 - Ce^{kNt})$ a Logistic curve.

Exercise Set 8.2

8.2.1
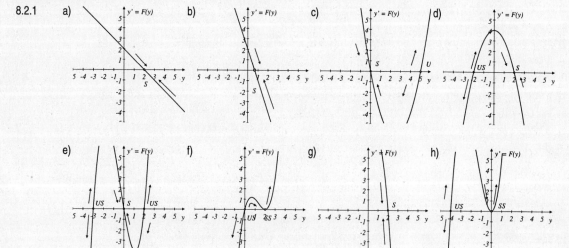

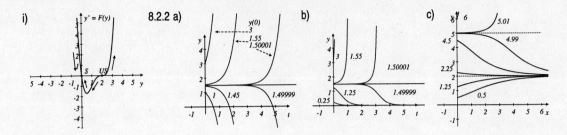

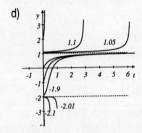

8.2.3 a) $y_E = 1$ $F'(y) = 1/y$; $F'(1) > 0$ means y_E is unstable.

b) $y_E = 4$ $F'(y) = -4/y$; $F'(4) < 0$ and hence y_E is a stable equilibrium.

c) $y_{E1} = \sqrt{2}$ $y_{E2} = -\sqrt{2}$ $F'(y) = 3y^2 - 2$; $F'(0) = -2 < 0$ so y_{E1} is stable $F'(2) = 10 > 0$ so y_{E2} is not stable.

d) $y_{En} = n$ for each interger n ; $F'(y) = \pi\cos(\pi y)$ $F'(n) = \pi(-1)^n$ Thus for even n y_{En} is not stable and for odd n it is stable.

e) $y_E = 1$ $F'(y) = -2(1 - y)$; $F'(1) = 0$ so the stability test does not apply.

f) $y_{En} = n$ for each integer n, $F'(y) = 2\pi \sin(\pi y)\cos(\pi y)$; $F'(n) = 0$ for all n so the test does not apply.

8.2.4 $F'(y) = k$ so zero is a stable equilibrium if $k < 0$.

8.2.5 $F'(y) = (r - 1)y^{r-2}(y - 4)^r + r y^{r-1}(y - 4)^{r-1} = [(r - 1)(y - 4) + r y] y^{r-2}(y - 4)^{r-1}$; $F(4) = 0$ if $r \neq 1$ and the test does not apply. If $r = 1$ $F'(4) > 0$ and 4 is an unstable equilibrium.

8.2.6 The equilibriums are $y_{E1} = 5$, $y_{E2} = 6$ $F(y)$ is only defined for $y > 4$. $F(y) > 0$ for $y \in (4,5) \cup (6,\infty)$ and $F(y) < 0$ for $y \in (5,6)$. Hence $\lim_{t \to \infty} y(t)$ a) is not defined if $y(0) = 2$ a) 5 b) ∞

Exercise Set 8.3

8.3.1 a) $y = c_1 e^t + c_2 e^{-4t}$ b) $y = c_1 e^{2t} + c_2 e^{-t}$ c) $y = c_1 e^t + c_2 e^{-2t}$

d) $y = e^{t/2} (c_1 \cos(\sqrt{7} t/2) + c_2 \sin(\sqrt{7} t/2))$ e) $y = c_1 e^{2t} + c_2 e^{-2t}$ f) $y = c_1 \cos(2t) + c_2 \sin(2t)$

g) $y = e^{t/4} (c_1 \cos(\sqrt{7} t/4) + c_2 \sin(\sqrt{7} t/4))$ h) $y = c_1 e^t + c_2$ i) $y = e^{3t/2} (c_1 e^{\sqrt{21} t/2} + c_2 e^{-\sqrt{21} t/2})$

j) $y = c_1 \cos(\sqrt{2} t) + c_2 \sin(\sqrt{2} t)$ k) $y = c_1 e^{2t} + c_2 e^{-3t}$ l) $y = e^{5t}(c_1 + c_2 t)$

8.3.2 a) $3 e^{3t}/4 + e^{-t}/4$ b) $y = e^t (t + 1)$ c) $y = \cos(\sqrt{5} t) + 2\sqrt{5} \sin(\sqrt{5} t)/5$

d) $y = e^{\sqrt{5}t}(\sqrt{5}/5 + 1/2) + e^{-\sqrt{5}t}(1/2 - \sqrt{5}/5)$ e) $y = 3 e^t - 2 e^{t/2}$ f) $y = e^{(-1+3\sqrt{2}/2)t}(\sqrt{2}/2 + 1/2) + e^{(-1 - 3\sqrt{2}/2)t}(1/2 - \sqrt{2}/2)$

8.3.3 a) $y = ((-\cos(6) + 2)/\sin(6)) \sin(3t) + \cos(3t)$ b) $y = [e^{3t} - e^{-3t}] e^6/(e^{12} - 1)$ c) $y = e^t (2 - t)/2 - t e^{t - 2}/2$

d) $y = -e^{-t} [te^2 + 2 - t]/2$ e) $y = 2 \sin(3t)$ f) $y = 3 \cos(t) - 3 \sin(t)$

g) $y = \cos(5t)/2 + 2 \sin(5t)$ h) $y = \cos(2t) + (2 - \sqrt{3}) \sin(2 t)$ i) $y = (e^{3t} - e^{-3t})e^6/(e^{12} - 1)$

8.3.4 a) $y'' + 3y' + 2y = 0$ b) $y'' - 9y = 0$ c) $y'' - y' + 2y = 0$ d) $y'' - 4y' + 4y = 0$

e) $y'' + 9y = 0$ f) $y'' - 2y' + 10y = 0$ g) $y'' - (a+b)y' + aby = 0$ h) $y'' + 2ky' + k^2 y = 0$

8.3.5 a) $y_1 = y_2 - 100e^{(-DAt/V)}$ b) $y_1(t) = Ce^{(-DAt/V)} + K + [DA / (4V^2 + D^2A^2)] [V^2\sin^2(t) - (2V^2/DA)\sin(t)\cos(t) + 2V^3/(DA)^2]$

c) $y_1(t) = 50(1 - e^{(-20At/V)})$ $y_2 = 50(1 + e^{(-20At/V)})$

8.3.6 a) $u = c_1 + c_2 e^{-3t}$ $v = c_2 + 2c_1 + 2c_2 e^{-3t}$ b) $u = c_1 + c_2 t$ $v = c_1 - c_2/2 + c_2 t$

c) $u = c_1 e^{(-1 + \sqrt{7})t} + c_2 e^{(-1 - \sqrt{7})t}$ $v = c_1(\sqrt{7} + 2)e^{(-1 + \sqrt{7})t} + c_2(2 - \sqrt{7})e^{(-1 - \sqrt{7})t}$

d) $u = c_1 e^{t(\sqrt{13}/2 - 5/2)} + c_2 e^{-t(\sqrt{13}/2 + 5/2)}$ $v = c_1(\sqrt{13}/2 + 1/2)e^{t(\sqrt{13}/2 - 5/2)} - c_2(\sqrt{13}/2 - 1/2)e^{-t(\sqrt{13}/2 + 5/2)}$

e) $u = c_1 e^{(-3 + \sqrt{3})t} + c_2 e^{(-3 - \sqrt{3})t}$ $v = \sqrt{3} c_1 e^{(-3 + \sqrt{3})t} - \sqrt{3} c_2 e^{(-3 - \sqrt{3})t}$

f) $u = c_1 e^{t(\sqrt{133}/2 - 13/2)} + c_2 e^{-t(\sqrt{133}/2 + 13/2)}$ $v = c_1 e^{t(\sqrt{133}/2 - 13/2)}(\sqrt{133}/2 - 11/2) - c_2 e^{-t(\sqrt{133}/2 + 13/2)}(\sqrt{133}/2 + 11/2)$

g) $u = e^t(c_1 \cos(t) + c_2 \sin(t))$ $v = c_2 e^t \cos(t) - c_1 e^t \sin(t)$ h) $u = e^t(c_1 \cos(2t) + c_2 \sin(2t))$ $v = e^t(c_2 \cos(2t) - c_1 \sin(2t))$

8.3.7 a) $y = c_1 \sin(\sqrt{5}t) + c_2 \cos(\sqrt{5}t) + c_3$ b) $y = c_1 e^t + c_2 e^{-2t} + c_3$ c) $y = e^{-t}(c_1 \cos(\sqrt{3}t) + c_2 \sin(\sqrt{3}t)) + c_3 e^{2t}$

Chapter 9

Exercise set 9.1

9.1.1. a) 2 b) 2 c) 9/2 d) 3/2 e) 7/6 f) -7/6 g) 8/3
h) 49/3 i) 1/12 j) -1/12 k) $(e^4 - e - 3)/2$ l) -11/24

9.1.2. a) $\pi^3(3\pi^2 - 5)/15$ b) -255/64 c) $\pi^3(3\pi^2 + 5)/15$ d) 16/3 e) $(1 - \cos(1))/2 \approx 0.2298$
f) $\sin(1)/6 \approx 0.1402$ g) $3/2 - \cos(1) - \sin(1)$ h) $\sin(1) + \cos(1)/2 - 1 \approx 0.1116$

9.1.3. a) 12 b) -15 c) 1 d) 4/3
e) $2e^{-1} + 3e^{-2} - 1 \approx 0.1418$ f) $(1 - 9e^{-2})/2 \approx -0.1090$ g) $3/4 - \ln(2)/2 \approx 0.4034$ h) $2\ln(2) - 3/4 \approx 0.6363$

9.1.4. a) 4 b) 4 c) 8 d) 4 e) 4 f) 4 g) 2 h) 12 i) 12 j) 17
k) 1/12 l) 1/4 m) -10/3 n) 1/4

Exercise Set 9.2 The set notations answers are given first and the corresponding graphs sketched separately afterwards.

9.2.1. For rectangles there is only one set description.

a) $\{(x,y) \mid 0 \leq x \leq 3 \text{ and } 0 \leq y \leq 2\}$ $\int_0^3\int_0^2 F(x,y)\,dy\,dx$; $\int_0^2\int_0^3 F(x,y)\,dx\,dy$; $I = 34.5$
b) $\{(x,y) \mid 0 \leq x \leq 2 \text{ and } 0 \leq y \leq 2\}$ $\int_0^2\int_0^2 F(x,y)\,dy\,dx$; $\int_0^2\int_0^2 F(x,y)\,dx\,dy$; $I = 4$
c) $\{(x,y) \mid -1 \leq x \leq 0 \text{ and } 0 \leq y \leq 4\}$ $\int_{-1}^0\int_0^4 F(x,y)\,dy\,dx$; $\int_0^4\int_{-1}^0 F(x,y)\,dx\,dy$; $I = -9$
d) $\{(x,y) \mid 2 \leq x \leq 3 \text{ and } 1 \leq y \leq 3\}$ $\int_2^3\int_1^3 F(x,y)\,dy\,dx$; $\int_1^3\int_2^3 F(x,y)\,dx\,dy$; $I = 28.5$
e) $\{(x,y) \mid -3 \leq x \leq 2 \text{ and } 1 \leq y \leq 3\}$ $\int_{-3}^2\int_1^3 F(x,y)\,dy\,dx$; $\int_1^3\int_{-3}^2 F(x,y)\,dx\,dy$; $I = -52.5$
f) $\{(x,y) \mid -1 \leq x \leq 2 \text{ and } 1 \leq y \leq 2\}$ $\int_{-1}^2\int_1^2 F(x,y)\,dy\,dx$; $\int_1^2\int_{-1}^2 F(x,y)\,dx\,dy$; $I = -0.75$
g) $\{(x,y) \mid -1 \leq x \leq 2 \text{ and } -1 \leq y \leq 2\}$ $\int_{-1}^2\int_{-1}^2 F(x,y)\,dy\,dx$; $\int_{-1}^2\int_{-1}^2 F(x,y)\,dx\,dy$; $I = 6.75$
h) $\{(x,y) \mid 1 \leq x \leq 8 \text{ and } 2 \leq y \leq 4\}$ $\int_1^8\int_2^4 F(x,y)\,dy\,dx$; $\int_2^4\int_1^8 F(x,y)\,dx\,dy$; $I = 2005.5$

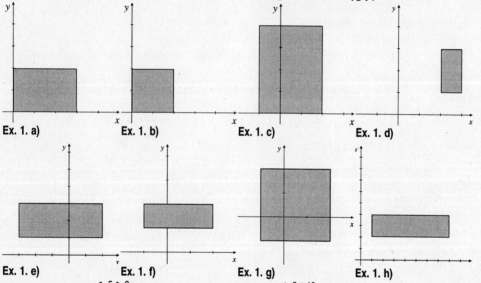

Ex. 1. a)   Ex. 1. b)   Ex. 1. c)   Ex. 1. d)

Ex. 1. e)   Ex. 1. f)   Ex. 1. g)   Ex. 1. h)

9.2.2. a) $I = 0$; $\int_{-5}^5\int_{-2}^2 F(x,y)\,dy\,dx$ b) $I = -252$ $\int_{-3}^3\int_4^{10} F(x,y)\,dy\,dx$ c) $I = 3680$ $\int_{-2}^8\int_8^{12} F(x,y)\,dy\,dx$
d) $I = 3348$ $\int_1^7\int_4^{10} F(x,y)\,dy\,dx$ e) $I = -4120$ $\int_{-8}^2\int_{-1}^3 F(x,y)\,dy\,dx$ f) $I = -3348$ $\int_{-7}^{-1}\int_{-10}^{-4} F(x,y)\,dy\,dx$

9.2.3. a) 1340 b) 156 c) 0   9.2.4. a) 0 b) 0.5 c) $\sqrt{3}$

9.2.5. Since the height is twice the width, if the vertex has coordinates (c,d) the solution must satisfy $\int_1^c\int_2^d x - y\,dy\,dx = -1$. Since, height = d - 2 and width = c - 1, d = 2c. There are three choices for the vertex with c = 0, $(1/2)(1 \pm \sqrt{5})$.

Exercise Set 9.3

9.3.1. a) $\{(x,y) \mid 0 \leq x \leq 2 \text{ and } 3x/2 \leq y \leq 3\}$ or $\{(x,y) \mid 0 \leq x \leq 2y/3 \text{ and } 0 \leq y \leq 3\}$
b) $\{(x,y) \mid 0 \leq x \leq 2 \text{ and } 0 \leq y \leq 3x/2\}$ or $\{(x,y) \mid 2y/3 \leq x \leq 2 \text{ and } 0 \leq y \leq 3\}$
c) $\{(x,y) \mid 2 \leq x \leq 4 \text{ and } (3/2)(x - 2) \leq y \leq 3\}$ or $\{(x,y) \mid 2 \leq x \leq 2 + 2y/3 \text{ and } 0 \leq y \leq 3\}$
d) $\{(x,y) \mid -1 + y/4 \leq x \leq 1 - y/4 \text{ and } 0 \leq y \leq 4\}$ or, splitting the region,
as $\{(x,y) \mid -1 \leq x \leq 0 \text{ and } 0 \leq y \leq 4(x + 1)\}$ plus $\{(x,y) \mid 0 \leq x \leq 1 \text{ and } 0 \leq y \leq -4(x - 1)\}$
e) $\{(x,y) \mid 0 \leq x \leq 2 \text{ and } -3x/2 \leq y \leq 3x/2\}$ or splitting the region as
$\{(x,y) \mid -2y/3 \leq x \leq 2 \text{ and } -3 \leq y \leq 0\}$ plus $\{(x,y) \mid 2y/3 \leq x \leq 2 \text{ and } 0 \leq y \leq 3\}$

f) Each description must be split.

$\{(x,y) \mid 0 \leq x \leq 2 \text{ and } x/4 \leq y \leq 3x/2 \}$ plus $\{(x,y) \mid 2 \leq x \leq 4 \text{ and } x/4 \leq y \leq 3-(x-2) \}$ or

$\{(x,y) \mid 2y/3 \leq x \leq 4y \text{ and } 0 \leq y \leq 1 \}$ plus $\{(x,y) \mid 2y/3 \leq x \leq 2+(y-3) \text{ and } 1 \leq y \leq 3 \}$

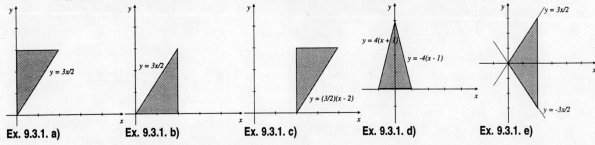

Ex. 9.3.1. a)   Ex. 9.3.1. b)   Ex. 9.3.1. c)   Ex. 9.3.1. d)   Ex. 9.3.1. e)

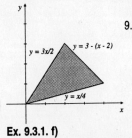

Ex. 9.3.1. f)

9.3.2.
a) $\{(x,y) \mid 0 \leq y \leq 3 \text{ and } -y/3 \leq x \leq 2 \}$
b) Splitting the region: $\{(x,y) \mid 0 \leq x \leq 1 \text{ and } -3x/2 \leq y \leq x \}$
plus $\{(x,y) \mid 1 \leq x \leq 2 \text{ and } -3x/2 \leq y \leq 2x-1 \}$
c) Splitting the region: $\{(x,y) \mid -1 \leq x \leq 0 \text{ and } -x \leq y \leq 2x+3 \}$
plus $\{(x,y) \mid 0 \leq x \leq 2 \text{ and } 3x/2 \leq y \leq 3 \}$
d) Splitting the region: $\{(x,y) \mid -2 \leq x \leq 0 \text{ and } -3x/2 \leq y \leq 3 \}$
plus $\{(x,y) \mid 0 \leq x \leq 2 \text{ and } -2x \leq y \leq 3 \}$
e) Splitting the region: $\{(x,y) \mid -1 \leq y \leq 0 \text{ and } -y \leq x \leq 2y+3 \}$ plus $\{(x,y) \mid 0 \leq y \leq 3 \text{ and } 2y/3 \leq x \leq 3-y/3 \}$
f) Splitting the region: $\{(x,y) \mid -2 \leq x \leq 0 \text{ and } -3x-4 \leq y \leq -x \}$ plus
$\{(x,y) \mid 0 \leq x \leq 2 \text{ and } 7x/2-4 \leq y \leq 3x/2 \}$

9.3.3.
a) $\{(x,y) \mid 0 \leq x \leq 1 \text{ and } x^2 \leq y \leq x \}$ or $\{(x,y) \mid y \leq x \leq \text{SQRT}(y) \text{ and } 0 \leq y \leq 1 \}$
b) $\{(x,y) \mid 0 \leq x \leq 2 \text{ and } x^2 \leq y \leq 2x \}$ or $\{(x,y) \mid y/2 \leq x \leq \text{SQRT}(y) \text{ and } 0 \leq y \leq 4 \}$
c) $\{(x,y) \mid -2 \leq x \leq 1 \text{ and } x \leq y \leq 2-x^2 \}$ or splitting the region as
$\{(x,y) \mid -\text{SQRT}(2-y) \leq x \leq y \text{ and } -2 \leq y \leq 1 \}$ plus $\{(x,y) \mid -\text{SQRT}(2-y) \leq x \leq \text{SQRT}(2-y) \text{ and } 1 \leq y \leq 2 \}$
d) $\{(x,y) \mid -2 \leq x \leq 1 \text{ and } x^2-2 \leq y \leq -x \}$ or splitting the region as
$\{(x,y) \mid -\text{SQRT}(2+y) \leq x \leq -y \text{ and } -1 \leq y \leq 2 \}$ plus $\{(x,y) \mid -\text{SQRT}(2+y) \leq x \leq \text{SQRT}(2+y) \text{ and } -2 \leq y \leq -1 \}$
e) $\{(x,y) \mid -1 \leq x \leq 1 \text{ and } x^2 \leq y \leq 2-x^2 \}$ or splitting the region as
$\{(x,y) \mid -\text{SQRT}(y) \leq x \leq \text{SQRT}(y) \text{ and } 0 \leq y \leq 1 \}$ plus $\{(x,y) \mid -\text{SQRT}(2-y) \leq x \leq \text{SQRT}(2-y) \text{ and } 1 \leq y \leq 2 \}$
f) $\{(x,y) \mid 0 \leq x \leq 2 \text{ and } x^3 \leq y \leq 2x^2 \}$ or $\{(x,y) \mid y^{1/3} \leq x \leq \text{SQRT}(y/2) \text{ and } 0 \leq y \leq 8 \}$
g) $\{(x,y) \mid 0 \leq x \leq 1 \text{ and } x^2 \leq y \leq x^{1/3} \}$ or $\{(x,y) \mid y^3 \leq x \leq \text{SQRT}(y) \text{ and } 0 \leq y \leq 1 \}$
h) $\{(x,y) \mid 0 \leq x \leq 1 \text{ and } x^2 \leq y \leq x^{1/2} \}$ or $\{(x,y) \mid y^2 \leq x \leq y^{1/2} \text{ and } 0 \leq y \leq 1 \}$
i) $\{(x,y) \mid 0 \leq x \leq 4 \text{ and } x/2 \leq y \leq x^{1/2} \}$ or $\{(x,y) \mid y^2 \leq x \leq 2y \text{ and } 0 \leq y \leq 2 \}$
j) $\{(x,y) \mid 0 \leq x \leq 1 \text{ and } x^3 \leq y \leq x \}$ plus $\{(x,y) \mid -1 \leq x \leq 0 \text{ and } x \leq y \leq x^3 \}$ or
$\{(x,y) \mid y^{1/3} \leq x \leq y \text{ and } -1 \leq y \leq 0 \}$ plus $\{(x,y) \mid y \leq x \leq y^{1/3} \text{ and } 0 \leq y \leq 1 \}$

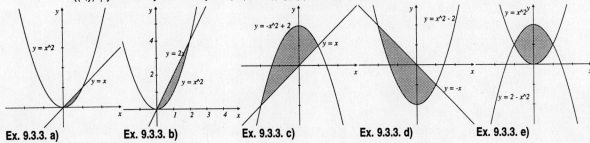

Ex. 9.3.3. a)   Ex. 9.3.3. b)   Ex. 9.3.3. c)   Ex. 9.3.3. d)   Ex. 9.3.3. e)

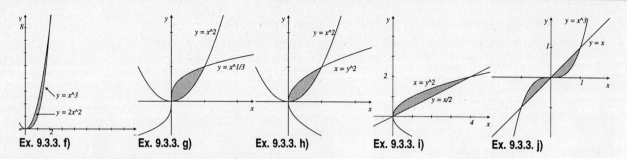

Ex. 9.3.3. f)   Ex. 9.3.3. g)   Ex. 9.3.3. h)   Ex. 9.3.3. i)   Ex. 9.3.3. j)

9.3.4. a) $\{(x,y) \mid -2 \leq x \leq -4/5^{1/2}$ and $-(4 - x^2)^{1/2} \leq y \leq (4 - x^2)^{1/2}\}$ plus
            $\{(x,y) \mid -4/5^{1/2} \leq x \leq 4/5^{1/2}$ and $x/2 \leq y \leq (4 - x^2)^{1/2}\}$
            or
            $\{(x,y) \mid -(4 - y^2)^{1/2} \leq x \leq (4 - y^2)^{1/2}$ and $4/5^{1/2} \leq y \leq 2\}$ plus
            $\{(x,y) \mid -(4 - y^2)^{1/2} \leq x \leq 2y$ and $-4/5^{1/2} \leq y \leq 4/5^{1/2}\}$
    b) $\{(x,y) \mid 0 \leq x \leq 3$ and $-2(1 - x^2/9)^{1/2} \leq y \leq 2(1 - x^2/9)^{1/2}\}$ or
        $\{(x,y) \mid 0 \leq x \leq 3(1 - y^2/4)^{1/2}$ and $-2 \leq y \leq 2\}$
    c) $\{(x,y) \mid 0 \leq x \leq 3/2$ and $-(9 - (x - 3)^2)^{1/2} \leq y \leq (9 - (x - 3)^2)^{1/2}\}$ plus $\{(x,y) \mid 3/2 \leq x \leq 3$ and $-(9 - x^2)^{1/2} \leq y \leq (9 - x^2)^{1/2}\}$
        or $\{(x,y) \mid 3 - (9 - y^2)^{1/2} \leq x \leq (9 - y^2)^{1/2}$ and $-3^{3/2}/2 \leq y \leq 3^{3/2}/2\}$

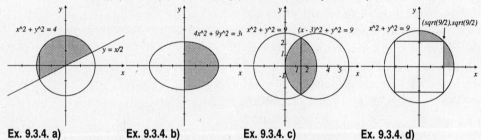

Ex. 9.3.4. a)   Ex. 9.3.4. b)   Ex. 9.3.4. c)   Ex. 9.3.4. d)

9.3.5. a) $\{(x,y) \mid -2 \leq x \leq 3$ and $-2 \leq y \leq 3\}$ for a square both set notations are identical.
    b) $\{(x,y) \mid 0 \leq x \leq 3$ and $0 \leq y \leq 4(-x/3 + 1)\}$ or $\{(x,y) \mid 0 \leq x \leq -3(y/4 - 1)$ and $0 \leq y \leq 4\}$
    c) $\{(x,y) \mid 0 \leq x \leq 2$ and $-3x/2 + 3 \leq y \leq 3\}$ plus $\{(x,y) \mid 2 \leq x \leq 4$ and $3(x - 2)/2 \leq y \leq 3\}$ or
        $\{(x,y) \mid -2y/3 + 2 \leq x \leq 2y/3 + 2$ and $0 \leq y \leq 3\}$
    d) $\{(x,y) \mid -2 \leq x \leq 2$ and $x^2 \leq y \leq 4\}$ or $\{(x,y) \mid -y^{1/2} \leq x \leq y^{1/2}$ and $0 \leq y \leq 4\}$
    e) $\{(x,y) \mid -2 \leq x \leq 2$ and $-(4 - x^2)^{1/2} \leq y \leq (4 - x^2)^{1/2}\}$ or $\{(x,y) \mid -(4 - y^2)^{1/2} \leq x \leq (4 - y^2)^{1/2}$ and $-2 \leq y \leq 2\}$
    f) $\{(x,y) \mid -2 \leq x \leq 2$ and $-(1 - (x/2)^2)^{1/2} \leq y \leq (1 - (x/2)^2)^{1/2}\}$ or $\{(x,y) \mid -2(1 - y^2)^{1/2} \leq x \leq 2(1 - y^2)^{1/2}$ and $-1 \leq y \leq 1\}$

9.3.6. a) $\{(x,y) \mid -2 \leq x \leq 0$ and $0 \leq y \leq -3x/2\}$ plus $\{(x,y) \mid 0 \leq x \leq 1$ and $0 \leq y \leq 2x\}$ or
        $\{(x,y) \mid -2 \leq x \leq -2y/3$ and $0 \leq y \leq 3\}$ plus $\{(x,y) \mid y/2 \leq x \leq 1$ and $0 \leq y \leq 2\}$
    b) $\{(x,y) \mid 0 \leq x \leq 2$ and $0 \leq y \leq 4\}$ plus $\{(x,y) \mid 2 \leq x \leq 4$ and $0 \leq y \leq -x + 4\}$ or
        $\{(x,y) \mid 0 \leq x \leq -y + 4$ and $0 \leq y \leq 2\}$ plus $\{(x,y) \mid 0 \leq x \leq 2$ and $2 \leq y \leq 4\}$
    c) $\{(x,y) \mid -1 \leq x \leq 0$ and $-(1 - x)^{1/2} \leq y \leq (1 - x)^{1/2}\}$ plus $\{(x,y) \mid 0 \leq x \leq 1$ and $-(1 - x)^{1/2} \leq y \leq (1 - x)^{1/2}\}$ or
        $\{(x,y) \mid y^2 - 1 \leq x \leq -y^2 + 1$ and $-1 \leq y \leq 1\}$
    d) $\{(x,y) \mid -1 \leq x \leq 0$ and $-3x \leq y \leq 3\}$ plus $\{(x,y) \mid 0 \leq x \leq 2$ and $-x/2 \leq y \leq 3 - 3x/2\}$ or
        $\{(x,y) \mid -y/3 \leq x \leq 2 - 2y/3$ and $0 \leq y \leq 3\}$ plus $\{(x,y) \mid -2y \leq x \leq 2$ and $-1 \leq y \leq 0\}$
    e) $\{(x,y) \mid 0 \leq x \leq 2y$ and $0 \leq y \leq 1\}$ plus $\{(x,y) \mid 2 \ln(y) \leq x \leq 2y$ and $1 \leq y \leq 3\}$ plus
        $\{(x,y) \mid 2 \ln(y) \leq x \leq 3$ and $3/2 \leq y \leq e^{3/2}\}$
    f) The lower curve is $y = 2x^2 - 1$ and the upper curve is $y = x^2$.
        $\{(x,y) \mid -1 \leq x \leq 1$ and $2x^2 - 1 \leq y \leq x^2\}$ or the sum of three regions:
        $\{(x,y) \mid -((y + 1)/2)^{1/2} \leq x \leq ((y + 1)/2)^{1/2}$ and $-1 \leq y \leq 0\}$ plus $\{(x,y) \mid -((y + 1)/2)^{1/2} \leq x \leq -y^{1/2}$ and $0 \leq y \leq 1\}$ plus
        $\{(x,y) \mid y^{1/2} \leq x \leq ((y + 1)/2)^{1/2}$ and $0 \leq y \leq 1\}$

9.3.7. a) 0   b) -2   c) 0   d) -25.6   e) 0   f) 0
9.3.8. a) 1/3   b) 92/3   c) 0   d) 8.5   e) $e^{3/2} + 2e^{3/2} - 17/8$ f) -4/15

9.3.9.
a) $\int_2^5 \int_1^4 F(x,y)\,dx\,dy$
b) $\int_2^5 \int_1^4 F(x,y)\,dy\,dx$
c) $\int_{1/4}^1 \int_1^{4y} F(x,y)\,dx\,dy$

d) $\int_2^5 \int_1^{y/2} F(x,y)\,dx\,dy$
e) $\int_{-4}^4 \int_{|x|}^4 F(x,y)\,dy\,dx$
f) $\int_2^3 \int_1^5 F(x,y)\,dx\,dy + \int_3^5 \int_{y/3}^{15} F(x,y)\,dx\,dy$

g) $\int_{-4}^{-1} \int_{-x}^4 F(x,y)\,dy\,dx + \int_{-1}^1 \int_1^4 F(x,y)\,dy\,dx + \int_1^4 \int_x^4 F(x,y)\,dy\,dx$

h) $\int_{1/5}^1 \int_{1/y}^5 F(x,y)\,dx\,dy + \int_1^3 \int_1^5 F(x,y)\,dx\,dy + \int_3^{15} \int_{y/3}^5 F(x,y)\,dx\,dy$
i) $\int_{-16}^0 \int_{-x1/2}^4 F(x,y)\,dy\,dx + \int_0^8 \int_{x/2}^4 F(x,y)\,dy\,dx$

j) $\int_{-4}^1 \int_{1-y}^5 F(x,y)\,dx\,dy + \int_1^3 \int_0^5 F(x,y)\,dx\,dy + \int_1^8 \int_{y-3}^5 F(x,y)\,dx\,dy$

k) $\int_{-5}^0 \int_1^{\sqrt{25-x^2}} F(x,y)\,dy\,dx + \int_0^3 \int_1^5 F(x,y)\,dy\,dx + \int_3^{11} \int_{(x-1)/2}^5 F(x,y)\,dy\,dx$

l) $\int_0^3 \int_{\sqrt{9-y^2}}^{\sqrt{9-y}} F(x,y)\,dx\,dy + \int_3^9 \int_0^{\sqrt{9-y}} F(x,y)\,dx\,dy$

9.3.10. $\int_{-1}^8 \int_{(x^2-5x+12)/6}^{x^{1/3}} F(x,y)\,dy\,dx$

Exercise Set 9.4

9.4.1.
a) The area is $\int_{-2}^1 \int_{x^2}^{2-x} 1\,dy\,dx = 9/2$
b) $9/2 = \int_1^4 \int_1^x 1\,dy\,dx$

c) The area is $\int_1^4 \int_{1+(x-1)/3}^x 1\,dy\,dx = 3$
d) $\int_{-\sqrt{2}}^{\sqrt{2}} \int_{x^2-2}^{x^2+2} 1\,dy\,dx = 16\sqrt{2}/3$

e) The area is $\int_0^5 \int_{-5x}^{-x^2} 1\,dy\,dx = 125/6$
f) $\int_{-1}^2 \int_0^{e^{(2x)}} 1\,dy\,dx = (1/2)(e^4 - e^{-2})$

g) The area is $\int_0^1 \int_0^{x(x-1)^2} 1\,dy\,dx = 1/12$

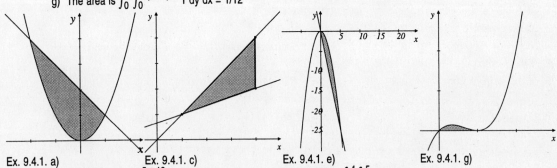

Ex. 9.4.1. a)   Ex. 9.4.1. c)   Ex. 9.4.1. e)   Ex. 9.4.1. g)

9.4.2.
a) The volume is $\int_0^5 \int_0^{10} 0.5x\,dy\,dx = 62.5$
c) The volume is $\int_0^1 \int_0^5 xy\,dy\,dx = 6.25$.

e) The volume is $\int_2^5 \int_3^7 1/(xy)\,dy\,dx = \ln(5/2)\ln(7/3) \approx 0.776$.

g) The region of integration, R, is determined by the condition that $z \geq 0$. Solving $2e^{-xy} - 1 = 0$ gives the upper boundary curve for R, $y = \ln(2)/x$. The lower boundary curve is $y = 0$. The volume is $\int_1^e \int_0^{\ln(2)/x} 2e^{-xy} - 1\,dy\,dx = 1 - \ln(2) \approx 0.3069$.

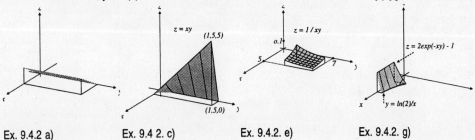

Ex. 9.4.2 a)   Ex. 9.4.2. c)   Ex. 9.4.2. e)   Ex. 9.4.2. g)

Exercise Set 9.5

9.5.1. The mass is the integral of the density over the region. $M = \int_0^{2\pi} \int_0^\pi 1 + 0.5\sin(x-y)\,dy\,dx = 2\pi^2 \approx 19.7392$.

9.5.2. 78 g   9.5.3. 2.35 10^3 bft   9.5.4. 4 µg   9.5.5. a) 37,500 M/l   b) 18,750 M/l

9.5.6.
a) The probability that $0 \leq X \leq 1$ AND $0 \leq Y \leq 2$ is $P = 1 + e^{-8} - e^{-6} - e^{-2} \approx 0.8625$.

b) The probability that x, y satisfy part a) and that $x + y < 1$ is $P = 1 + 2e^{-3} - 3e^{-2} \approx 0.6936$.

9.5.8. a) $(\bar{x},\bar{y}) = (1,3)$ b) $(\bar{x},\bar{y}) = (2,4/3)$ c) $(\bar{x},\bar{y}) = (7/3,7/3)$ d) $(\bar{x},\bar{y}) = (0,3)$

9.5.10. Average = 0. Note that the function F(x,y) is an "odd" function with respect to the y-variable.

9.5.11. 89/6   9.5.12. The average temperature is $T = 17(1 - e^{-20})$

9.5.13. a) 14/3   b) $92/15 \approx 6.1\underline{3}$ c) $7(2e^5 - 37)/\{5(e^5 - 6)\} \approx 2.554$

Chapter 10

Exercise Set 10.1

10.1.1.
a) $F_x(x,y) = 3$ $F_y(x,y) = -4y^3$
b) $F_x(x,y) = y + 1$ $F_y(x,y) = x - 1$

c) $F_x(x,y) = 6x^2$ $F_y(x,y) = -4y^3$
d) $F_x(x,y) = -y/x^2$ $F_y(x,y) = 1/x$

e) $F_x(x,y) = 5 - 2xy^4$ $F_y(x,y) = -4x^2y^3$
f) $F_x(x,y) = y/x^2$ $F_y(x,y) = -1/x$

g) $F_x(x,y) = -2y/(x-y)^2$ $F_y(x,y) = 2x/(x-y)^2$ h) $F_x(x,y) = -2(x+y)^{-3}$ $F_y(x,y) = -2(x+y)^{-3}$

i) $F_x(x,y) = 3 - ye^{xy}$ $F_y(x,y) = -xe^{xy}$ j) $F_x(x,y) = (2x + x^2y)e^{xy}$ $F_y(x,y) = x^3e^{xy}$

k) $F_x(x,y) = 2x\ln(y)$ $F_y(x,y) = x^2/y$ l) $F_x(x,y) = 1/x$ $F_y(x,y) = 1/y$

m) $F_x(x,y) = 4x - 3y$ $F_y(x,y) = -3x + 2/y$ n) $F_x(x,y) = -14x\log(y^2)$ $F_y(x,y) = -14x^2/(\ln(10)y)$

o) $F_x(x,y) = yx^{y-1}$ $F_y(x,y) = x^y \ln(x)$ p) $F_x(x,y) = (1/y)x^{1/y-1}$ $F_y(x,y) = -x^{1/y}\ln(x)/y^2$

q) $F_x(x,y) = -y\sin(x)$ $F_y(x,y) = \cos(x)$ r) $F_x(x,y) = y - y^2\cos(xy^2)$ $F_y(x,y) = x - 2xy\cos(xy^2)$

10.1.2. a) $F_x(1,-2) = 2$ $f_y(1,-2) = -12$ b) $F_x(4,1) = 11/4$ $F_y(4,1) = 12$ c) $F_x(1,3) = 2/9$ $F_y(1,3) = -2/27$

d) $F_x(2,0) = 3$ $F_y(2,0) = 6$ e) $F_x(0.5,\pi) = 0$ $F_y(0.5,\pi) = 0$ f) $F_x(3,1) = \cos(3) \approx -0.99$ $F_y(3,1) = 3\cos(3) \approx -2.97$

g) $F_x(-1,2) = -2$ $F_y(-1,2) = 5$ h) $F_x(2,1) = -3$ $F_y(2,1) = 0$

10.1.3 a) $F_{xx}(x,y) = 0$ $F_{xy}(x,y) = 0$ $F_{yy}(x,y) = 2$

b) $F_{xx}(x,y) = 6xy^2$ $F_{xy}(x,y) = 6x^2y$ $F_{yy}(x,y) = 2x^3$

c) $F_{xx}(x,y) = 9ye^{3x}$ $F_{xy}(x,y) = 3e^{3x}$ $F_{yy}(x,y) = 0$

d) $F_{xx}(x,y) = (2xy^2 + y^3)x^{-4}e^{y/x}$ $F_{xy}(x,y) = -(2x+y)x^{-2}e^{y/x}$ $F_{yy}(x,y) = (2x+y)x^{-2}e^{y/x}$

e) $F_{xx}(x,y) = 3x^{-2}$ $F_{xy}(x,y) = 1$ $F_{yy}(x,y) = 2 - 3y^{-2}$

f) $F_{xx}(x,y) = 9y^2e^{3xy}$ $F_{xy}(x,y) = (9xy + 3)e^{3xy}$ $F_{yy}(x,y) = 9x^2e^{3xy}$

g) $F_{xx}(x,y) = -y\sin(x)$ $F_{xy}(x,y) = -\sin(x)$ $F_{yy}(x,y) = 0$

h) $F_{xx}(x,y) = -y^2\sin(xy)$ $F_{xy}(x,y) = -xy\sin(xy) + \cos(xy)$ $F_{yy}(x,y) = -x^2\sin(xy)$

i) $F_{xx}(x,y) = -\cos(x)$ $F_{xy}(x,y) = 0$ $F_{yy}(x,y) = 2$

j) $F_{xx}(x,y) = e^{3x}[6y\cos(xy) + (9 - y^2)\sin(xy)]$ $F_{xy}(x,y) = e^{3x}[(3x+1)\cos(xy) - xy\sin(xy)]$ $F_{yy}(x,y) = -x^2e^{3x}\sin(xy)$

k) $F_{xx}(x,y) = -y^2e^{xy} - \sin(x) + 4y$ $F_{xy}(x,y) = 4x - e^{xy}(xy + 1)$ $F_{yy}(x,y) = 4 - x^2e^{xy}$

l) $F_{xx}(x,y) = y^2e^{xy} - (x+y)^{-2}$ $F_{xy}(x,y) = e^{xy}(xy + 1) - (x+y)^{-2}$ $F_{yy}(x,y) = x^2e^{xy} - (x+y)^{-2}$

10.1.4. a) $F_x = yz$ $F_y = xz$ $F_z = xy$ $F_{xz} = y$ $F_{zy} = x$

b) $F_x = y + 2z$ $F_y = x + 3z$ $F_z = 2x + 3y$ $F_{xz} = 2$ $F_{zy} = 3$

c) $F_x = 6x - 2y$ $F_y = -2x + 3z^3$ $F_z = 9yz^2$ $F_{xz} = 0$ $F_{zy} = 9y^2$

d) $F_x = 2e^{zy}$ $F_y = 2zxe^{zy}$ $F_z = 2yxe^{zy}$ $F_{xz} = 2ye^{zy}$ $F_{zy} = 2x(1 + zy)e^{zy}$

e) $F_x = y - z$ $F_y = x$ $F_z = -x$ $F_{xz} = -1$ $F_{zy} = 0$

f) $F_x = -y\sin(xy) - z\cos(xz)$ $F_y = -x\sin(xy)$ $F_z = -x\cos(xz)$ $F_{xz} = xz\sin(xz) - \cos(xz)$ $F_{zy} = 0$

g) $F_x = 1/y - z^{2.5}$ $F_y = -x/y^2$ $F_z = -2.5xz^{1.5}$ $F_{xz} = -2.5z^{1.5}$ $F_{zy} = 0$

h) $F_x = y^zx^{y^z-1}$ $F_y = zy^{z-1}\ln(x)x^{y^z}$ $F_z = \ln(x)\ln(y)y^zx^{y^z}$ $F_{xz} = y^zx^{y^z-1}\ln(y)[1 + y^z\ln(x)]$

$F_{zy} = \ln(x)x^{y^z}y^{z-1}[z\ln(y)(y^z\ln(x) + 1) + 1]$ (This has been well factored!)

10.1.5. At (0,0) and (9/2, 3/2).

10.1.12. a) $z_x = 2x = 0$ and $z_y = 2y = 0$ implies $x = 0$ and $y = 0$.

b) $z_x = y = 0$ and $z_y = x = 0$ implies $x = 0$ and $y = 0$.

c) $z_x = 2x = 0$ and $z_y = -2y = 0$ implies $x = 0$ and $y = 0$.

d) $z_x = 9x^2 - 4 = 0$ and $z_y = 2 - 2y = 0$ implies $y = 1$ and $x = \pm 2/3$.

e) $z_x = y^2 - 2xy = y(y - 2x) = 0$ and $z_y = 2xy - x^2 - 3 = 0$ implies $(x_0, y_0) = (1,2)$ or $(-1,-2)$.

f) $z_x = y^2 - 6xy - 9 = 0$ and $z_y = 2xy - 3x^2 = (2y - 3x)x = 0$ implies $(x_0, y_0) = (0, \pm 3)$.

g) $z_x = \sin(\pi y) + 0.5 = 0$ and $z_y = \pi x\cos(\pi y) = 0$ implies $x_0 = 0$ and $y_0 = n + 1/6$ or $y_0 = 2n - 1/6$ for odd integers n.

h) $z_x = 3x^2 - 3y = 0$ and $z_y = -3x + 2 = 0$ implies $x_0 = 2/3$ and $y_0 = 4/9$.

10.1.13 a) 0   b) 1   c) 4   d) For (2/3,1), D = 24. For (-2/3,1), D = -24.

e) For $(1, \pm 2)$, D = 12   f) For $(0, \pm 3)$ D = 36   g) $D = 3\pi^2/4$ h) D = 9

Exercise Set 10.2

10.2.1 a) $\Delta z_x/\Delta x = 6.1$ $\Delta z_y/\Delta y = 1$ b) $z_x = 6$ $z_y = 1$

10.2.2 a) $\Delta z_x/\Delta x = -3$ $\Delta z_y/\Delta y = 2$ $z_x = -3$ $z_y = 2$ b) i) $\Delta z_x/\Delta = 5$ $\Delta z_y/\Delta y = -25/19$ ii) $z_x = 5$ $z_y = -1.25$

10.2.3. At P_1 the tangent plane is $z = 7 - 2(x + 1) + 3(y - 2)$ or $2x - 3y + z = -1$

At P_2 the tangent plane is $z = 7 + 4(x - 2) + 3(y - 1)$ or $4x + 3y - z = 4$

At P_3 the tangent plane is $z = 0 + 0(x - 0) + 3(y - 0)$ or $3y - z = 0$

10.2.4. a) $z = 3 + 4(x - 1) - (y + 2)$ b) $z = 2 + (x - 2) + 2(y - 1)$ c) $z = 3/2 + (1/2)(x - 3) - (3/2)(y - \ln(2))$

d) $z = 6(y - 3)$ e) $z = 16 + 14(x - 2) - 4(y + 2)$ f) $z = 3 - 3(y + 2)$

g) $z = 16 - e^{-4} + (16 + 2e^{-4})(x - 2) - 2e^{-4}(y + 2)$ h) $z = 3x$

Selected Answers

10.2.5 a) $z = 12\ 12(x - 1) + 12(y - 2)$ b) i) 3   ii) 0   iii) -0.087

10.2.7 a) Function $z = 2 + 2x - 3y$ Point-Slope form: $z = 1 + 2(x - 1) - 3(y - 1)$ Intercept form: $x/-1 + y/(2/3) + z/2 = 1$

b) Function $z = 15 - 3x + 5y$ Point-Slope form: $z = 4 - 3(x - 2) + 5(y + 1)$ Intercept form: $x/5 + y/-3 + z/-15 = 1$

c) Function $z = 3 + (1/2)x + (1/4)y$ Point-Slope form: $z = 4 + (1/2)(x - 1) + (1/4)(y - 2)$ Intercept form: $x/-6 + y/-12 + z/3 = 1$

d) Function $z = 29/9 + 28/9\ x - 13/9\ y$ Point-Slope form: $z = 2 + 28/9(x - 1) - 13/9(y - 3)$
Intercept form: $x/(-29/28) + y/(29/13) + z/(29/9) = 1$

e) Function $z = 5 + 3x + 0.5y$ Point-Slope form: $z = 3 + 3(x + 1) + 0.5(y - 2)$ Intercept form: $x/-5/3 + y/-10 + z/5 = 1$

f) Function $z = -4 - 3x + 5y$ Point-Slope form: $z = 3 - 3(x - 1) + 5(y - 2)$ Intercept form: $x/-4/3 + y/4/5 + z/-4 = 1$

g) Function $z = -2 + 2x + y$ Point-Slope form: $z = 3 + 2(x - 1) + (y - 2)$ Intercept form: $x + y/2 + z/-2 = 1$

h) Function $z = 2\pi - 2\pi x + y$ Point-Slope form: $z = 2 - 2\pi(x - 1) + (y - 2)$ Intercept form: $x/1 + y/-2\pi + z/-2\pi = 1$

10.2.9 a) $y = 1$ $z = 4 + 0.2(x - 10)$; $x = 10$ $z = 4 + 3(y - 1)$

b) $y = 2$ $z = 4 + 0.5(x - 4)$; $x = 4$ $z = 2 - (y - 4)$

c) $y = 3$ $z = 6 + 0.75(x - 4)$; $x = 4$ $z = 6 + 2(y - 3)$

Exercise Set 10.3

10.3.1. The approximations are $F(x_1,y_1) \approx F(x_0,y_0) + (x_1 - x_0)F_x(x_0,y_0) + (y_1 - y_0)F_y(x_0,y_0)$.

a) 5 b) -5   c) -1.5   d) -3.5   e) -0.6   f) -2.95   g) $19/96 \approx 0.1979$ h) $53/24 \approx 2.208$ i) 2.1   j) 2.3

10.3.2. a) $P_1(x,y) = 1 + 2(x - 1) - y$ $P_1(1.2, -0.5) = 1.9$ Error = 2.04 - 1.9 = 0.14

b) $P_1(x,y) = 1$ $P_1(1.2,-0.5) = 1$ Error = $\cos(-0.5) - 1 \approx -0.175$

c) $P_1(x,y) = 1 + (x - 1) - y$ $P_1(1.2,-0.5) = 1.7$ Error = $1.2e^{0.5} - 1.7 \approx 0.2785$

d) $P_1(x,y) = -8 + 2(x - 1)$ $P_1(1.2,-0.5) = -7.6$ Error = -7.31 - (-7.6) = 0.29

e) $P_1(x,y) = 2 + 6(x - 1)$ $P_1(1.2,-0.5) = 3.2$ Error = 3.61 - 3.2 = 0.41

f) $P_1(x,y) = 1 + (x - 1)$ $P_1(1.2,-0.5) = 1.2$ Error = 1.3 - 1.2 = 0.1

g) $P_1(x,y) = 3 + 8(x - 1) + y$ $P_1(1.2,-0.5) = 5.1$ Error = $5.76 - e^{-0.6} - 5.1 \approx 0.1112$

h) $P_1(x,y) = y$ $P_1(1.2,-0.5) = -0.5$ Error = -1.5 - (-0.5) = -1.0

i) $P_1(x,y) = 0$ (a constant!) $P_1(1.2,-0.5) = 0$ Error = $-0.5 ln(1.2) - 0 \approx -0.091$

10.3.3 a) 0.015   b) 0.05775   c) 1.2   d) 2.1

10.3.4. a) Using $f(x,y) = x^{1/2} - y^{1/2}$ and the reference point $(x_0,y_0) = (4,25)$, the number $3.9^{0.5} - \sqrt{26}$ is approximately
$f(4,25) + dx \cdot f_x(4,25) + dy \cdot f_y(4,25) = -3 + -0.1 \cdot 0.25 + 1 \cdot 0.1 = -2.925$.   b) 4.3433 c) 0.2

10.3.5. a) The approximate rate of distortion is the differential $dA = 2\pi R \cdot dR - 2\pi r \cdot dr$. When $dR = 0$ (fixed outer radius) and $dr = 10^{-4}$ cm/sec, $dA = -2\pi \cdot 10^{-4} \cdot r$.

b) To maintain a constant cross-sectional area the differential dA must be zero. Thus, $dR = (r/R)dr$.

10.3.6. The differential $dc = c_x(x,t) \cdot dx + c_t(x,t) \cdot dt$ is most easily expressed using the function $c(x,t)$ in the partial derivatives:
$\partial/\partial x\ c(x,t) = -2(x - ut - x_0)/x_0^2 \cdot c(x,t)$,   and $\partial/\partial t\ c(x,t) = [2u(x - ut - x_0)/x_0^2 + 4D - 2D/(\alpha^2 + 4Dt)]c(x,t)$

10.3.8 a) 1.25   b) 1.024   c) 1.455

Exercise Set 10.4

10.4.1 a) The linear expansion of $5\ ln(x - 2y)$ about the point (5,2) is $5 ln(x - 2y) \approx 5(x - 5) - 10(y - 2)$. Thus solving
$xy = 5(x - 5) - 10(y - 2)$ gives the equation $x = -5(2y + 1)/(y - 5)$.

b) The linear expansion of the right side about (2,1) is $5(x^2 - 2xy) \approx 10(x - 2) - 20(y - 1)$. The approximate equation is $x = 2y$.

c) For $(x_0,y_0) = (0,0)$ the approximate integrand is $\exp(-(x^2 + y - 2y^2)) \approx 1 - y$. The approximate integral is thus 1/2.
For $(x_0,y_0) = (1,1)$ the approximate integrand is $\exp(-(x^2 + y - 2y^2)) \approx 1 - 2(x - 1) + 3(y - 1)$. The approximate integral in this case evaluates to 1/2. Which is a bad approximation of the true integral value, approximately 0.935.

10.4.2 a) -0.1869   b) -0.125   10.4.3 $V(b,l) \approx \pi/4\ (l -80)$

10.4.4 a) $D_t x = y/(ma)$ $D_t y = -mgx - cy$ b) $D_t x = y/(ma)$ $D_t y = mg(x - \pi) - cy$

10.4.5 (0,0) is an unstable saddle point ($\lambda_1 = 10, \lambda_2 = -2$); (4,10) is stable equilibrium with $\lambda = \pm 2\sqrt{5}i$

10.4.6 The three equilibrium values are $(H_E, P_E) = (0, 0), (50, 0)$ and $(20, 3/5)$.
For (0,0) the linearized system has an unstable equilibrium with eigenvalues $\lambda = -5$ and $\lambda = 2$.
For (50, 0) the linearized system also has a saddle characteristic with eignevalues $\lambda = -2$ and $\lambda = 15/2$.
For (20, 3/5) the linearized system has complex eigenvalues $\lambda = \{-4 \pm \sqrt{584}\}/10$. As the real part is -0.4, the solutions give a stable spiral about the equilibrium. $\alpha = a + d = -4/5$ and $\beta = ad - bc = 6 > 0$.

10.4.7 The system has two equilibriums, (50, 0) and (20, $ln(5/2)$).

For (50, 0) the eigenvalues are $\lambda = 0$ and $\lambda = 11/2$, hence the equilibrium is unstable.

For (20, $ln(5/2)$) the eigenvalues are complex, $\lambda = -1 \pm \sqrt{1 - 10 ln(5/2)}$ with negative real part and hence the equilibrium is a stable node.

10.4.8 (a) The equilibrium points are (0,0) and (8/3, 3/2).

For (0,0) the eigenvalues are $\lambda = 3$ and $\lambda = 2$. Hence the euqilibrium is unstable.

For (8/3, 3/2) the eigenvalues are complex, $\lambda = -1 \pm \sqrt{7}$ hence the equilibrium is an unstable saddle point as the parameters $\alpha = -2$ and $\beta = -6$.

10.4.9 a) $(X_{E1}, Y_{E2}) = (0,0)$ $(X_{E2}, Y_{E2}) = (2, 3/2)$

b) $(X_{E1}, Y_{E2}) = (2,1)$ $(X_{E2}, Y_{E2}) = (0,0)$ $(X_{E3}, Y_{E3}) = (6,0)$

c) $(X_{E1}, Y_{E2}) = (0,0)$ $(X_{E2}, Y_{E2}) = (0,4)$ $(X_{E3}, Y_{E3}) = (20,0)$ $(X_{E4}, Y_{E4}) = (-12,16)$

d) $(X_{E1}, Y_{E2}) = (5,0)$

e) $(X_{E1}, Y_{E2}) = (0,0)$ $(X_{E2}, Y_{E2}) = (8, 1/2)$

10.4.10 (a) The equilibrium points are $(-5, -\sqrt{10})$ and $(-5, \sqrt{10})$. Both are unstable, the first is a saddle point and the second is a spiral.

(e) The equilibrium points are (0, 3) and (0, -3). Both are unstable, the first is a saddle point and the second is a spiral.

Chapter 11

Exercise Set 11.1

11.1.1. a) i) $\underline{u} = (0,1)$ ii) $\underline{w} = (0,-2)$ iii) $\underline{v} \cdot \underline{r} = 4$ iv) $\theta = \cos^{-1}(4/2\sqrt{5}) \approx 0.46$ v) $\underline{p} = (2,0)$

b) i) $\underline{u} = (-1,0)$ ii) $\underline{w} = (6,0)$ iii) $\underline{v} \cdot \underline{r} = -3$ iv) $\theta = \cos^{-1}(-3/3\sqrt{5}) \approx 2.03$ v) $\underline{p} = (0,2)$

c) i) $\underline{u} = (1/\sqrt{5}, 2/\sqrt{5})$ ii) $\underline{w} = (-2,-4)$ iii) $\underline{v} \cdot \underline{r} = 5$ iv) $\theta = 0$ v) $\underline{p} = (-2/\sqrt{5}, 4/\sqrt{5})$

d) i) $\underline{u} = (-1/\sqrt{2}, -1/\sqrt{2})$ ii) $\underline{w} = (2,2)$ iii) $\underline{v} \cdot \underline{r} = -3$ iv) $\theta = \cos^{-1}(-3/\sqrt{10}) \approx 2.82$ v) $\underline{p} = (2/\sqrt{2}, -2/\sqrt{2})$

e) i) $\underline{u} = (1/\sqrt{26}, 5/\sqrt{26})$ ii) $\underline{w} = (-2,-10)$ iii) $\underline{v} \cdot \underline{r} = 11$ iv) $\theta = \cos^{-1}(11/\sqrt{130}) \approx 0.27$ v) $\underline{p} = (-2/\sqrt{26}, 10/\sqrt{26})$

f) i) $\underline{u} =$ ii) $\underline{w} =$ iii) $\underline{v} \cdot \underline{r} =$ iv) $\theta =$ v) $\underline{p} =$

11.1.2. The point $\underline{p}_1$ is the point $\underline{p}(t_1)$ where $t_1 = 1/||\underline{d}||$

a) i) $\underline{p}(t) = (1,1) + t \cdot (0,-1)$ ii) $\underline{p}_1 = (1,0)$ b) i) $\underline{p}(t) = (1,0) + t \cdot (0,1)$ ii) $\underline{p}_1 = (1,1)$

c) i) $\underline{p}(t) = (1,2) + t \cdot (1,1)$ ii) $\underline{p}_1 = (1 + 1/\sqrt{2}, 2 + 1/\sqrt{2})$ d) i) $\underline{p}(t) = (3,-4) + t \cdot (4,3)$ ii) $\underline{p}_1 = (3.8, -3.4)$

e) i) $\underline{p}(t) = (0,3) + t \cdot (-1,-1)$ ii) $\underline{p}_1 = (-1/\sqrt{2}, 3 - 1/\sqrt{2})$ f) i) $\underline{p}(t) = (2,3) + t \cdot (8,6)$ ii) $\underline{p}_1 = (2.8, 3.6)$

11.1.3. a) $y = x^2 + 1$ b) $y = \pm\sqrt{x} + 1$ or $x = (y - 1)^2$ (parabola opening to the right)   c) $y = (x^2 - 2x - 17)/9$

d) $y = \pm x^{-1/2}$ or $x = y^{-2}$ e) $y = \pm\sqrt{x}/(\pm\sqrt{x} + 1)$ both signs must be positive or both negative to give the two branches of this curve; or $x = (y/(1 - y))^2$ f) $y = (x + 1)^{2/3}$

11.1.4. The x-y equations are: a) $y = x^3 - 9$ b) $y = x^2$, $x > 0$ c) $y = x^{2/3}$, $x > 0$ d) $y = x^6 - 9$, $x \geq 0$

e) $y = 1/x^2$, $x > 0$ f) $y = x^{-2/3}$, $x > 0$

11.1.5. The x-y equations are: a) $y = e^{-2x}$ b) $y = e^{\pm 2\sqrt{x}}$ c) $y = e^{3x/2}$ d) $y = -x$ e) $y = e^x$

f) $y = e^{2x}$ g) $y = -xe^{-x}$ h) $y = ln(x^2)$ i) $y = -ln(x^2)$

11.1.6. The x-y equations are: a) $x^2 + y^2 = 1$ b) $x^2 + y^2 = 1$ c) $x^2 + y^2 = 9$ d) $x^2 + 1/4 \, y^2 = 1$

e) $1/9 \, x^2 + y^2 = 1$ f) $1/9 \, x^2 + 1/4 \, y^2 = 1$ g) $y = \cos(\pi x)$ h) $y = \cos(\pi\sqrt{x})$ i) $y = \cos^2(\pi x)$

11.1.7. $\underline{r}(t) = (4\cos(t), 4\sin(t))$ and $\underline{r}(t) = (4\sin(-t), 4\cos(-t))$ are the simplest parametric curves.

11.1.8. $\underline{r}(t) = (t, 3t^2)$ is the simplest such equation, $\underline{r}(t) = (t^3, 3t^{2/3})$ would be another. There are an infinite choice for such representations.

11.1.9. $\underline{r}(t) = (\sec(\pi t), 3\tan(\pi t))$

Exercise Set 11.2

11.2.1. a) $\underline{r}'(t) = (1, 2t + 3)$ $\underline{r}'(2) = (1,7)$ $SP(2) = 5\sqrt{2}$ b) $\underline{r}'(t) = (-e^{-t}, -2e^{-t})$ $\underline{r}'(2) = (-e^{-2}, -2e^{-2})$ $SP(2) = \sqrt{5} e^{-2}$

c) $\underline{r}'(t) = (\cos(\pi t) - \pi t \sin(\pi t), \sin(\pi t) + \pi t \cos(\pi t))$ $\underline{r}'(2) = (1, 2\pi)$ $SP(2) = \sqrt{1 + 4\pi^2} \approx 6.36$

d) $\underline{r}'(t) = (e^t, 1)$ $\underline{r}'(2) = (e^2, 1)$ $SP(2) = \sqrt{e^4 + 1} \approx 7.46$

e) $\underline{r}'(t) = (2e^{2t}, 4e^{4t})$ $\underline{r}'(2) = (2e^4, 4e^8)$ $SP(2) = 2e^4 \sqrt{1 + 4e^8} \approx 11{,}924.3$

f) $\underline{r}'(t) = (-15\sin(3t), 15\cos(3t))$ $\underline{r}'(2) = (-15\sin(6), 15\cos(6))$ $SP(2) = 15$

11.2.2. a) $\underline{r}'(t) = (1, -\pi \sin(\pi t))$ $SP(t) = \sqrt{1 + \pi^2 \sin^2(\pi t)}$ b) $\underline{r}'(t) = (2t, (t + 1)^{-2})$ $SP(t) = \sqrt{4t^2 + (1 + t)^{-4}}$

c) $\underline{r}'(t) = (3, 2t)$ $SP(t) = \sqrt{9 + 4t^2}$ d) $\underline{r}'(t) = (1/3 \, t^{-2/3}, 1)$ $SP(t) = \sqrt{1 + 1/9 \, t^{-4/3}}$

e) $\underline{r}'(t) = (3e^{3t}, 2e^{2t})$ $SP(t) = SQRT(9e^{6t} + 4e^{4t})$ f) $\underline{r}'(t) = (1/t, 2t)$ $SP(t) = SQRT(1 + 4t^4)/t$

g) $\underline{r}'(t) = (0.5\,t^{-1/2}, 3t^2)$ $SP(t) = SQRT(0.25t^{-1} + 9t^4)$ h) $\underline{r}'(t) = (2, -2e^{-2t})$ $SP(t) = 2\,SQRT(1 + e^{-4t})$

i) $\underline{r}'(t) = (-1/t, ln(t) + 1)$ $SP(t) = SQRT(t^{-2} + ln^2(t) + 2ln(t) + 1)$

11.2.3. If $\underline{r}(t) = (x(t), y(t))$ then $\underline{r}'(t) = (x'(t), y'(t))$ and thus $SP(t) = \sqrt{f(t)}$ where $f(t) = [x'(t)]^2 + [y'(t)]^2$. The critical points on the parametric curve where the speed may have an extremum are the points $\underline{r}(t)$ where t is a critical value at which either (i) $f(t) = 0$, or (ii) $f'(t) = 0$, or (iii) $f'(t)$ does not exist. Notice that $f'(t) = 2[x'\cdot x'' + y'\cdot y'']$

a) $f(t) = 1 + (2t - 3)^2 > 0$ for all t. $f'(t) = 2(2t - 3) = 0$ at $t = 3/2$.
   At the critical point $\underline{r}(3/2) = (3/2, -9/4)$ the trajectory has a minimum speed $SP(2/3) = 1$.

b) $f(t) = 4t^2 + t^{-4} > 0$ for $t \neq 0$. $f(0)$ is not defined. $f'(t) = 8t - 2t^{-5} = 0$ if $t = 0.25^{1/4} \approx 0.707$

c) $f(t) = 1 + 9\sin^2(3t) \geq 1$ for all t. $f'(t) = 54\sin(3t)\cos(3t) = 0$ only if either $\sin(3t) = 0$, i.e. $t = (\pi/3)n$, or $\cos(3t) = 0$, which occurs when $t = \pi/6 + (\pi/3)n$, for n an integer.

d) $f(t) = 4t^2 + 1 \geq 1$ $f'(t) = 8t = 0$ if $t = 0$. At the point $\underline{r}(0) = (-1, 3)$ the trajectory $\underline{r}(t)$ has a minimum speed of $SP(0) = 1$.

e) $f(t) = t^{-2} + 4t^2 > 0$ $f'(t) = -2t^{-3} + 4t = 0$ if $t = 0.25^{1/4}$ At $\underline{r}(0.5^{1/4}) = (-ln(\sqrt{2}), 2)$ the trajectory has a minimum speed $SP(0.5^{1/4}) = 2$

Exercise Set 11.3

11.3.1 $-14/5$ 11.3.2 a) $-1/\sqrt{5}$ b) $1/\sqrt{5}$ c) 1

11.3.3 a) $D_{(2,1)}F(-2,3) = 2/\sqrt{5}$ c) $D_{(-2,1)}F(-2,3) = 42/\sqrt{5}$ c) $D_{(2,1)}F(-2,3) = 38/\sqrt{5}$ $D_{(-2,-1)}F(-2,3) = -38/\sqrt{5}$

e) $D_{(1,1)}F(-2,3) = -9\sqrt{2}\pi/8$ $D_{(-1,1)}F(-2,3) = -9\sqrt{2}\pi/8$

11.3.4 For Ex.3 (a) The maximum directional derivative is $\sqrt{584}$ which is the derivative in the direction $\underline{d} = (1/\sqrt{146})(-5, 11)$. The minimum directional derivative is $-\sqrt{584}$ when the direction is $-\underline{d} = (1/\sqrt{146})(5, -11)$. The directional derivative is zero in the direction perpendicular to $\underline{d}$, $(1/\sqrt{146})(-11, -5)$.

For Ex.3 (c) The maximum derivative is $2\sqrt{2194}$ in the direction $2194^{-1/2}(-13, 45)$. The minimum derivative is then $-2\sqrt{2194}$, in the opposite direction, and the derivative is zero in the normal direction $2194^{-1/2}(-45, -13)$.

For Ex.3 (e) The maximum derivative is $9\pi/4$ in the direction $(-1, 0)$. The minimum is $-9\pi/4$ in the direction $(1, 0)$ while the derivative is zero in the direction $(0, 1)$.

11.3.5 a) $\nabla F = (4x - 4x^3 - 4xy^2, 4y - 4y^3 - 4x^2y)$ b) $z = 3 - 4(x - 1) + 4(y + 1)$

c) i) $\underline{d} = 4(-1, 1)$ ii) $\underline{d} = 4(1, 1)$ iii) $\underline{d} = \pm 4(1, 1)$

Exercise Set 11.4

11.4.1 a) Level curves are $y = 3/2\,x + C$. The steepest curves are $y = -2/3\,x + C$.

c) Level curves are $y = C/x$. Steepest curves are hyperbolas, $x^2 - y^2 = C$

e) Level curves $y = -0.5x^2 + C$; steepest curves $y = ln(x) + C$.

11.4.2 a) Level curves are $y = -ln(x) + C$; steepest curves are $y = 0.5x^2 + C$. No level curve passing through $(0, 0)$.

c) Level curves are $ye^y = Cx$. The steepest curves are $y - ln(1 + y) = 0.5x^2 + C$ While every level curve passes through $(0, 0)$, the only steepest curves through the origin is when $C = 0$.

e) Level curves are circles $x^2 + y^2 = 2C$; steepest curves are lines $y = Cx$. The only level curve through the origin is the degenerate circle, the point when $C = 0$

Chapter 12

Exercise Set 12.1

12.1.1. a) $P_1(x,y) = 3x$ $P_2(x,y) = 3x + y^2$ b) $P_1(x,y) = 0$ $P_2(x,y) = xy$

c) $P_1(x,y) = 7 + 3(x - 1) + 4(y - 2)$ $P_2(x,y) = 7 + 3(x - 1) + 4(y - 2) + (y - 2)^2$

d) $P_1(x,y) = -1 - (x - 1) + (y + 1)$ $P_2(x,y) = -1 - (x - 1) + (y + 1) + (x - 1)(y + 1)$

e) $P_1(x,y) = -4 - (x - 2) + 12(y + 1)$ $P_2(x,y) = -4 - (x - 2) + 12(y + 1) + 6(x - 2)(y + 1) - 12(y + 1)^2$

f) $P_1(x,y) = 1 + 2x$ $P_2(x,y) = 1 + 2x - 4x^2 + x(y - 2)$

g) $P_1(x,y) = (x - 1)$ $P_2(x,y) = (x - 1) - 1/2(x - 1)^2 + (x - 1)(y - 1)$

h) $P_1(x,y) = 1 + x - y$ $P_2(x,y) = 1 + x - y + 1/2\,x^2 - xy + 1/2\,y^2$

i) $P_1(x,y) = 2 + 2ln(2)(x - 1) + (y - 2)$ $P_2(x,y) = 2 + 2ln(2)(x - 1) + (y - 2) + ln(2)^2(x - 1)^2 + ln(2)(x - 1)(y - 2)$

j) $P_1(x,y) = 1$ $P_2(x,y) = 1 + 1/2\,x^2 - x(y - 1)$

k) $P_1(x,y) = 1$ $P_2(x,y) = 1 - 3x^2$ l) $P_1(x,y) = x$ $P_2(x,y) = x + xy$

12.1.2.
a) $P_2(x,y) = 3 + 2(x-1) + 2(y-2) + 3(x-1)^2 - 2(x-1)(y-2) + (y-2)^2$ $P_2(x_1,y_1) = 3.06$ $F(x_1,y_1) = 3.06$
Absolute error $|E| = |P_2(x_1,y_1) - F(x_1,y_1)| = 0$ because F is a quadratic $P_2 = F$ identically.

b) $P_2(x,y) = 1 - \pi^2/8 \,(x-1)^2 - \pi^2/2 \,(x-1)(y-0.5) - \pi^2/2 \,(Y-0.5)^2$ $P_2(x_1,y_1) = 0.9850722$ $F(x_1,y_1) = 0.9842882$
Absolute error $|E| = |P_2(x_1,y_1) - F(x_1,y_1)| = 0.000784$

c) $P_2(x,y) = 1 - x^2 - y^2$ $P_2(x_1,y_1) = 0.95$ $F(x_1,y_1) = 0.9512294$
Absolute error $|E| = |P_2(x_1,y_1) - F(x_1,y_1)| = 0.0012294$

d) $P_2(x,y) = 2(y-1) + (x-2)(y-1) - (y-1)^2$ $P_2(x_1,y_1) = -0.44$ $F(x_1,y_1) = -0.4462871$
Absolute error $|E| = |P_2(x_1,y_1) - F(x_1,y_1)| = 0.0062871$

12.1.3. a) 0   b) $4.8e^{-5}$ c) 0   d) 12   e) $-2^{1/3}/54 \approx 2.331\, 10^{-2}$ f) $1/3750 = 2.66\, 10^{-4}$

12.1.4.
a) $P_3(x,y) = (y-2) + 2(x-1)(y-2) + (x-1)^2(y-2)$

b) $P_3(x,y) = 2 - 2(x-1) + (y-2) + 2(x-1)^2 - (x-1)(y-2) - 2(x-1)^3 + (x-1)^2(y-2)$

c) $P_3(x,y) = 1 + 2(x-1) + (y-2) + (x-1)^2 + (x-1)(y-2) + (1/2)(y-2)^2 + (1/3)(x-1)^2(y-2) + (1/3)(x-1)(y-2)^2 + (1/6)(y-2)^3$

d) $P_3(x,y) = -2\pi(x-1) - (\pi/2)(x-1)(y-2) + (2\pi^3/6)(x-1)^3$

e) $P_3(x,y) = -\pi(x-1) + (\pi^3/6)(x-1)^3 + (\pi^3/6)(x-1)(y-2)^2$

f) $P_3(x,y) = e^2 + 2e^2(x-1) + e^2(y-2) + 2e^2(x-1)^2 + (3e^2/2)(x-1)(y-2) + (e^2/2)(y-2)^2$
     $+ (4e^2/3)(x-1)^3 + (4e^2/3)(x-1)^2(y-2) + (2e^2/3)(x-1)(y-2)^2 + (e^2/6)(y-2)^3$

Exercise Set 12.2

12.2.1.   (1,0) is a minimum point as $D = -2$ and $A = 1$ 12.2.2.   (0,3/2) is a maximum point; $D = -4$, $A = -2$

12.2.3.   (-2,3/2) is a saddle point; $D = 4$ 12.2.4.   (0,3/2) can not be classified; $D = 0$

12.2.5.   critical points $(0, \pi/2 + n\pi)$ for n an integer are all saddle points as $D = 1$

12.2.6. $(\sqrt{3}, -1/\sqrt{3})$ $(-\sqrt{3}, 1/\sqrt{3})$ both are saddle points as $D = 3$

12.2.7. $(2(-4)^{-2/3}, (-4)^{1/3})$ is a saddle point as $D = 12 \cdot 4^{1/3}$.   12.2.8.   (0,0) and (-4,4) are both saddle points with $D = 16$.

12.2.9.   Critical point (0,0) is a saddle point $D = 64$, and (4,4) is a maximum point as $D = -64$.

12.2.10.   The only critical point (2,16/6) is a saddle point as $D = 4/3$.

12.2.11.   (0,0) is a minimum point; $D = -1$, $A = 1$ 12.2.12.   (1,1) is a minimum; $D = -27$, $A = 6$

12.2.13.   (0,0) is a saddle point; $D = 16$, (0,4) and (4,0) are saddle points; $D = 16$, (4/3,4/3) is a maximum point; $D = -48/9$, $A = -8/3$

12.2.14.   12.2.15.   (2,0) is a saddle point; $D = 1/(9 \cdot 2^{4/3})$

12.2.16.   (1/2,1) and (-1/2,-1) are saddle points as $D = 32$. (1/2,-1) is a maximum point as $D = -32$ and $A = -16$.
(-1/2,1) is a minimum point as $D = -32$ and $A = 16$.

12.2.17.   (4,1/2) is a minimum point; $D = -1$, $A = 1/4$ 18.

12.2.19.   (0,0) is a critical point at which $D = 0$, no conclusion may be made. At (-4,-4) $D < 0$ and $A < 0$ so a maximum occurs.

12.2.20.

12.2.21.   (0,0) is a critical point at which $D = 0$, no conclusion may be made. The only other root is at $y_0 = (4^3 \cdot 6)^{1/5} = 2 \cdot 12^{1/5}$ and $x_0 = -y_0^2/4 = -12^{1/5}/2$. At (x_0, y_0) $D < 0$ and $A > 0$ so a minimum occurs.

Exercise Set 12.3

12.3.1.
a)   At $x = y = z = 50/3$ the constrained product xyz has a maximum value $125,000/3$.

b) xy^2z^2 constrained to $x + y + z = 50$ has a maximum value of $4\, 10^5$ at $x = 10$ $z = y = 20$

c) $x^2 + y^2 + z$ constrained to $x + y + z = 50$ has a minimum value or 49.5 at $x = y = 1/2$ $z = 49$.

12.3.6   i) a) Maximum of 25/4 at (5/2, 5/2). Minimum of 0 at (5,0) or (0,5).

i) b) Maximum 2 at $(-\sqrt{2}, -\sqrt{2})$ and at $(\sqrt{2}, \sqrt{2})$ Minimum of -2 at $(\sqrt{2}, -\sqrt{2})$ and at $(-\sqrt{2}, \sqrt{2})$.

i) c) $F(x,y) = 2$ for all points on the constraint curve.

i) d) $\lambda^2 = -4/3$ has no solutions so there are no critical points and hence no extrema.

ii) a) no optimal points   ii) b) maximum at $(2/\sqrt{5}, 4/\sqrt{5})$ is $F = 2\sqrt{5}$. The minimum is $F = -2\sqrt{5}$ which occurs at $(-2/\sqrt{5}, -4/\sqrt{5})$.

12.3.7.
a) $\lambda = 0$: minimum $-1 = F(0,1)$ b) $\lambda = 0$: minimum $0 = F(0,0)$

c)   For $\lambda = 1$: maximum $8 = F(0,2) = F(0,-2)$. For $\lambda = 2$: minimum $4 = F(2,0) = F(-2,0)$ d) $\lambda = -2$: minimum $3 = F(-1,1)$

e)   Correction to problem: should read $g(x,y) = x^2 + y^2 - 2$
For $\lambda = 3/2$: maximum $2 = F(1,1)$. For $\lambda = -3/2$: minimum $-2 = F(-1,-1)$.

f)   For $\lambda = 1/\sqrt{2}$: maximum $= \sqrt{2} = F(1/\sqrt{2}, 1/\sqrt{2})$. For $\lambda = -1/\sqrt{2}$: maximum $= -\sqrt{2} = F(-1/\sqrt{2}, -1/\sqrt{2})$.

g) For $\lambda = -e^{-1}$: maximum $e^{-1} = F(\pm 1, 1)$. Minimum $0 = F(0,0)$ for any λ.

h) For $\lambda = 0$; $x = 3$, $y = 1/6 + N/3$, N an integer. A maximum occurs for N even and a minimum for N odd.

12.3.9 If the compund has length l, width w, and height h = 2, its voulume is $V = 2lw$. The constraint on materials is that the total surface area equals 4,800, assuming the division is along the width: $4l + 6w = 4,800$. Applying the Lagrange method, the optimal volume $V = 480,000$ m^3 occurs when l = 600m and w = 400m.

12.3.13 The Lagrange method with $F(x,y,z) = xyz$ and $G(x,y,z) = 3x + 4y + 3z - 24 = 0$ gives Max $F = F(8/3, 2, 8/3) = 128/3$.

2.1
3 odd
4 a-g
5 acfg
11
13
14

2.2